AF767189

Electromagnetic Processes at High Energies in Oriented Single Crystals

Electromagnetic Processes at High Energies in Oriented Single Crystals

V N Baier
V M Katkov
V M Strakhovenko

Budker Institute of Nuclear Physics, Academy of Sciences of Russia

World Scientific
Singapore • New Jersey • London • Hong Kong

Published by

World Scientific Publishing Co. Pte. Ltd.
5 Toh Tuck Link, Singapore 596224
USA office: 27 Warren Street, Suite 401-402, Hackensack, NJ 07601
UK office: 57 Shelton Street, Covent Garden, London WC2H 9HE

British Library Cataloguing-in-Publication Data
A catalogue record for this book is available from the British Library.

ELECTROMAGNETIC PROCESSES AT HIGH ENERGIES IN ORIENTED SINGLE CRYSTALS

Copyright © 1998 by World Scientific Publishing Co. Pte. Ltd.

All rights reserved. This book, or parts thereof, may not be reproduced in any form or by any means, electronic or mechanical, including photocopying, recording or any information storage and retrieval system now known or to be invented, without written permission from the publisher.

For photocopying of material in this volume, please pay a copying fee through the Copyright Clearance Center, Inc., 222 Rosewood Drive, Danvers, MA 01923, USA. In this case permission to photocopy is not required from the publisher.

ISBN-13 978-981-02-1603-0
ISBN-10 981-02-1603-3

Printed in Singapore

INTRODUCTION

The beauty of phenomena in regular crystal structure always attracts the attention of physicists. Recently, many new phenomena were discovered in the interactions of fast charged particles and photons with crystals. These phenomena are very interesting both for theoretical understanding and practical use. Among them are the channeling of fast charged particles in crystals, radiation of channeled particles, creation of electron–position pairs by photons in a crystalline field, radiation of particles in this field, and connected with these effects, the development of new types of electromagenetic showers in crystals and the deflection of high-energy particles in bent crystals. All of these processes are under intensive investigation by researchers in many countries. There are also many International Conferences devoted to this branch of physics of crystals.

This book is devoted mostly to processes occurring at high energies, previously not considered in monographic literature. It appears that a description of the interactions of charged particles and photons with a crystal by coherent processes is far from complete. There is a region of angles and energies where the processes of quantum electrodynamics in intense external fields manifest themselves in crystals. It is worth mentioning that manifestly quantum effects in the radiation of particles moving in external fields were first observed directly in crystals. One should note that at energies of tens of GeVs and higher a single crystal appears to be the most effective radiation medium — where the energy of charged particles is transformed into radiation — and the most efficient converter for transformation of photons into electron-position pairs. The characteristic lengths in which the radiation and pair processes occur turn out to be one or two orders of magnitude smaller in crystals than in the corresponding amorphous medium. This is the physical basis for the development of a new type of detectors of high–energy electrons and photons.

To evaluate the characteristics of the processes we use the quasiclassical operator approach to quantum electrodynamics in external fields. This permits one to consider processes in any external fields including essentially nonuniform fields as well. This approach appears to be very adequate for consideration processes in crystals. It is shown, that the theory of coherent processes in crystals is the limiting case of the general theory for relatively large angles of incidence of a particle (a photon) with respect to the direction of crystal axes or planes, while for very small incident angles radiation and pair creation occur in the fields of crystal axes or planes. For the convenience of the reader we start with the presentation of the theory of radiation and pair creation in arbitrary external fields which is systematically applied to the consideration of phenomena in crystals.

CONTENTS

Chapter 1

A THEORY OF RADIATION AND PAIR PRODUCTION IN EXTERNAL FIELD

1. CLASSICAL THEORY OF RADIATION

1.1. General Description of Radiation in Classical Electrodynamics

In this section some general results of a theory of radiation of charged particle in classical electrodynamics are given. A detailed description of radiation classical theory can be found for example, in the books by Landau, Lifshifts (1973), Jackson (1962), Panofsky and Phillips (1955), and Baier, Katkov and Fadin (1973).

Let us consider a charged particle travelling inside a confined region in a trajectory $\mathbf{r}_0(t_0)$ with a velocity $\mathbf{v}(t_0)$. An electromagnetic field created by the particle will be considered at distances r large compared to the system dimensions. In this case one can choose the origin of coordinates inside the system (see Fig. 1.1) and make the expansion of all the values over r_0/r:

$$R = |\mathbf{r} - \mathbf{r}_0| \simeq r\left(1 - \frac{\mathbf{n}\cdot\mathbf{r}_0}{r}(t_0)\right) . \tag{1.1}$$

The time in the observation point t is connected with the "particle time" t_0 by the following retardation relation*

*The system of units $c = 1$ is used in this book.

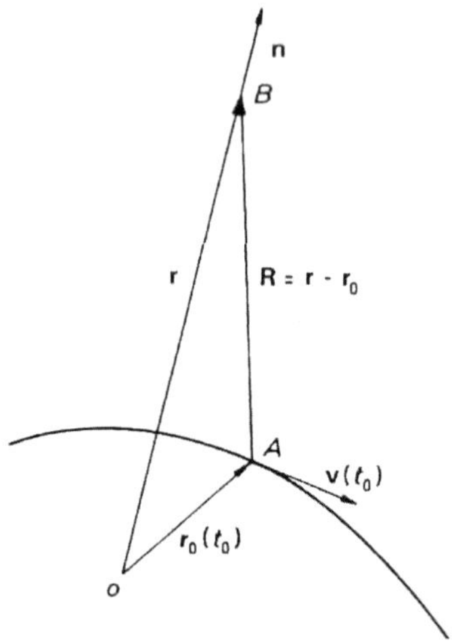

Fig. 1.1. Radiation of a particle moving along the trajectory $\mathbf{r}_0(t)$, which at time t_0 is at point A, is observed at point B with coordinate $\mathbf{r}$ at time t.

$$t_0 = t - R(t_0) \simeq t - r + \mathbf{n} \cdot \mathbf{r}_0(t_0)\,. \tag{1.2}$$

At large distances from the particle, where one should only keep the slowly decreasing terms $\sim 1/r$, the fields have the form

$$\mathbf{E} = \frac{e}{r}\frac{1}{(1-\mathbf{n}\cdot\mathbf{v})^3}(\mathbf{n}\times((\mathbf{n}-\mathbf{v})\times\mathbf{w})); \quad \mathbf{H} = \mathbf{n}\times\mathbf{E}\,, \tag{1.3}$$

where $\mathbf{w} = \dot{\mathbf{v}}$. In this case the values in the right-hand part are taken at a time moment t_0 defined by (1.2). These fields satisfy the following relations

$$\begin{aligned} \mathbf{E}\cdot\mathbf{H} = \mathbf{n}\cdot\mathbf{E} = \mathbf{n}\cdot\mathbf{H} = 0\,, \\ \mathbf{E}\times\mathbf{H} = E^2\,\mathbf{n} = H^2\,\mathbf{n}\,, \end{aligned} \tag{1.4}$$

i.e. $\mathbf{n}$, $\mathbf{E}$, $\mathbf{H}$ form the right-hand triplet of orthogonal vectors that is characteristic for the plane electromagnetic wave propagating at distances which are large compared to the system dimensions and the lengths of radiated waves (in the so-called radiation wave zone). One can consider a small fraction of diverging spherical wave as a plane wave. Since $\mathbf{E} = \mathbf{H}\times\mathbf{n}$ and $\mathbf{H} = \mathrm{rot}\,\mathbf{A}$ it is clear that for the description of the electromagnetic field in the wave zone it is sufficient only to know the vector-potential $\mathbf{A}$ ($\mathbf{r}$,t).

The radiation intensity dI into the element of solid angle $d\Omega$ is defined as an amount of energy passing per unit time through the element of a spherical surface $dS = \mathbf{r}^2 d\Omega$ with a center in the origin of the coordinate. It is expressed via the Poynting's vector $\mathbf{S} = \mathbf{E} \times \mathbf{H}/4\pi = E^2\mathbf{n}/4\pi$:

$$dI = E^2 r^2 d\Omega/4\pi \tag{1.5}$$

If we are interested in the complete radiation for the whole time of the charge motion then we have to integrate the intensity over time. In this case, one should take into account that values in the right-hand part depends on time t_0. Taking into account that (compare with (1.2))

$$dt = (\partial t/\partial t_0)dt_0 = (1 - \mathbf{n}\cdot\mathbf{v})dt_0 \tag{1.6}$$

we get for an energy radiated by a particle per unit time of the "particle's own time"* (at the moment of radiation) t_0:

$$\frac{d\varepsilon(\mathbf{n})}{dt_0} = \frac{e^2}{4\pi}\frac{1}{(1-\mathbf{n}\cdot\mathbf{v})^5}\{2(\mathbf{w}\cdot\mathbf{v})(\mathbf{w}\cdot\mathbf{n})(1-\mathbf{n}\cdot\mathbf{v}) + \mathbf{w}^2(1-\mathbf{n}\cdot\mathbf{v})^2 - (1-v^2)(\mathbf{n}\cdot\mathbf{w})^2\}d\Omega\,. \tag{1.7}$$

The integration of this expression over the angles makes it convenient to carry out in the tensor form together with the use of invariance and with respect to the three-dimensional rotations:

$$\begin{gathered}\int \frac{n_i n_k}{(1-\mathbf{n}\cdot\mathbf{v})^5}d\Omega = \frac{4\pi}{3(1-v^2)^4}[\delta_{ik}(1-v^2) + 6v_i v_k]\,;\\ \int \frac{n_i}{(1-\mathbf{n}\cdot\mathbf{v})^4}d\Omega = \frac{16\pi v_i}{3(1-v^2)^3}\,; \quad \int \frac{d\Omega}{(1-\mathbf{n}\cdot\mathbf{v})^3} = \frac{4\pi}{(1-v^2)^2}\,.\end{gathered} \tag{1.8}$$

By substituting these integrals into (1.7) for the radiation intensity per unit time results in the following expression

$$\frac{d\varepsilon}{dt_0} = \frac{2}{3}\frac{e^2}{(1-v^2)^3}[(\mathbf{v}\cdot\mathbf{w})^2 + w^2(1-v^2)]\,. \tag{1.9}$$

*The physical meaning of an additional factor $(1 - \mathbf{n}\cdot\mathbf{v})$ is the following: the field radiated by the electron in time dt_0 is located between two spherical surfaces with radii r and $r - dt_0$ whose centres, coincide with the positions of electron in the moments of time t_0 and $t_0 + dt_0$. In order to get the value of the energy radiated in the time dt_0 one should calculate the value of the electromagnetic field energy in the spherical layer between the surfaces indicated. The volume element of the spherical layer is $dV = r^2 d\Omega dl$ where $dl = dt_0(1 - \mathbf{n}\cdot\mathbf{v})$, i.e. the factor $(1 - \mathbf{n}\cdot\mathbf{v})$ takes into account the layer deformation caused by the particle motion (i.e. Doppler's effect).

This expression can be written in the convariant form*

$$\frac{d\varepsilon}{dt_0} = -\frac{2e^2}{3}\left(du^\mu(\tau_0)/d\tau_0\right)^2 , \tag{1.10}$$

or

$$\frac{dP^\mu}{d\tau} = -\frac{2e^2}{3}(du^\nu/d\tau)^2 u^\mu ,$$

where τ_0 is the proper time and $u^\mu = dx^\mu/d\tau$ the 4-velocity of the particle.

The expression (1.9) written in terms of the components of electric and magnetic fields has the form:

$$\frac{d\varepsilon}{dt_0} = 2e^4/[3m^2(1-v^2)]\{[\mathbf{E}+\mathbf{v}\times\mathbf{H}]^2 - (\mathbf{E}\cdot\mathbf{v})^2\}. \tag{1.11}$$

or in the convariant form

$$\frac{d\varepsilon}{dt_0} = -2e^4/3m^2(F_{\mu\nu}u^\nu)^2 \tag{1.12}$$

or

$$dP^\mu/d\tau = -2e^4/3m^2(F_{\mu\nu}u^\nu)^2 u^\mu ,$$

where $F^{\mu\nu}$ is field-strength tensor.

Let us consider radiation in some particular cases.

1. The particle velocity and acceleration are parallel. Then from (1.3) it follows that

$$\mathbf{H} = \frac{e}{r}\frac{1}{(1-\mathbf{n}\cdot\mathbf{v})^3}\mathbf{w}\times\mathbf{n}. \tag{1.13}$$

Taking (1.4)–(1.6) into account we have for particle energy radiated per unit time:

$$\frac{d\varepsilon(\mathbf{n})}{dt_0} = \frac{e^2}{4\pi}\frac{w^2\sin^2\vartheta}{(1-v\cos\vartheta)^5}d\Omega , \tag{1.14}$$

where ϑ is an angle between $\mathbf{n}$ and $\mathbf{v}$.

The radiation intensity (as well as an energy radiated per unit time) is symmetric with respect to the direction of motion $\mathbf{v}(\mathbf{w})$ and it vanishes at $\vartheta = 0, \pi$. An energy radiated per unit time

$$\frac{d\varepsilon}{dt_0} = \frac{2e^2}{3}\frac{w^2}{(1-v^2)^3}. \tag{1.15}$$

---

*dP^μ is a radiated 4-momentum.

The motion in the electric field ($\mathbf{H} = 0$) if $\mathbf{v} \parallel \mathbf{E}$ is an example of such a motion. Then

$$\frac{d\varepsilon}{dt_0} = \frac{2}{3}\frac{e^4 E^2}{m^2} \tag{1.16}$$

2. The velocity and acceleration are perpendicular. Then from (1.7) it follows that

$$\frac{d\varepsilon}{dt_0} = \frac{e^2}{4\pi}\frac{w^2}{(1 - v\cos\vartheta)^5}[(1 - v\cos\vartheta)^2 - (1 - v^2)\sin^2\vartheta\cos^2\varphi]d\Omega\,, \tag{1.17}$$

where φ is an azimuthal angle between the vector $\mathbf{n}$ and plane $(\mathbf{v}, \mathbf{w})$. The intensity of radiation as well as the energy radiated per unit time are symmetrical with respect to the plane $(\mathbf{v}, \mathbf{w})$ and vanishes at an angle $\vartheta = \arccos v$. In this case, the energy radiated per unit time is

$$\frac{d\varepsilon}{dt_0} = \frac{2}{3}\frac{e^2 w^2}{(1 - v^2)^2} \tag{1.18}$$

An example of such a motion is the motion in the magnetic field at $\mathbf{E} = 0$. If $\mathbf{v} \perp \mathbf{H}$, then

$$\frac{d\varepsilon}{dt_0} = (2/3)e^4 H^2 v^2/[m^2(1 - v^2)]\,. \tag{1.19}$$

The spectral distribution of radiation intensity is of great interest. Let us have the field expanded into the Fourier integral (see Formula (A.26) in Appendix A). Such an expansion is encountered while considering the radiation accompanying the charged particles collision or the charged particles travelling through external fields. The most interesting is knowing the total amount of energy radiated for the total time of the process. If $d\varepsilon(\mathbf{n}, \omega)$ is an energy radiated into the element of the solid angle $d\Omega$ in the form of waves with frequencies $\omega, \omega + d\omega$ for the total time of a process then an analogue for formula (1.5) has the form*

$$\begin{aligned} d\varepsilon(\mathbf{n}, \omega) &= |\mathbf{E}(\omega)|^2 \frac{d\omega}{4\pi^2} d\Omega r^2 \\ &= |\mathbf{E}(\omega)|^2 (d^3k/4\pi^2\omega^2) r^2\,, \end{aligned} \tag{1.20}$$

*We use here the fact for Fourier transform $f(t) = \frac{1}{2\pi}\int_{-\infty}^{\infty} f_\omega \exp(-i\omega t)d\omega$ we have $\int_{-\infty}^{\infty} f^2(t)dt = \frac{1}{2\pi}\int_{-\infty}^{\infty} |f_\omega|^2 d\omega = \frac{2}{2\pi}\int_0^{\infty} |f_\omega|^2 d\omega$.

where $\mathbf{k} = \mathbf{n}\omega$. And in the case of periodic motion of charged particles, when the radiation field is expanded into the Fourier series, an intensity radiation* into the element of the solid angle $d\Omega$ with frequency $\omega = m\omega_0$ is

$$dI_m = (|\mathbf{E}_m|^2/2\pi)r^2 d\Omega\,. \tag{1.21}$$

By substituting the explicit expression for the Fourier component $\mathbf{E}(\omega)$ (A.26) into (1.20) we have:

$$d\varepsilon(\mathbf{n},\omega) = e^2 \iint [\mathbf{v}(t_1) - \mathbf{n}] \cdot [\mathbf{v}(t_2) - \mathbf{n}] \exp\{i[\omega(t_1 - t_2) - \mathbf{k}[\mathbf{r}_0(t_1) - \mathbf{r}_0(t_2)]]\} dt_1 dt_2 \frac{d^3k}{(2\pi)^2}\,. \tag{1.22}$$

In this case, $\mathbf{r}_0(t_1)$ and $\mathbf{r}_0(t_2)$ are the positions of the particle on the trajectory at the time moments t_1 and t_2 respectively. Integrating the terms containing $\mathbf{v}(t_1)\mathbf{n}$ and $\mathbf{v}(t_2)\mathbf{n}$ in parts we have the spectral distribution of radiated energy

$$\begin{aligned} d\varepsilon(\mathbf{n},\omega) &= e^2 \iint [\mathbf{v}(t_1)\mathbf{v}(t_2) - 1] \exp\{i[\omega(t_1 - t_2) - \mathbf{k}[\mathbf{r}_0(t_1) - \mathbf{r}_0(t_2)]]\} dt_1 dt_2 \frac{d^3k}{(2\pi)^2} \\ &= e^2 \left| \int \frac{mu^\mu}{\varepsilon(t)}(t) \exp[ikx(t)] dt \right|^2 \frac{d^3k}{(2\pi)^2}\,, \end{aligned} \tag{1.23}$$

*Let us take into account that while expanding $f(t)$ into the Fourier series

$$f(t) = \sum_{m=-\infty}^{\infty} f_m \exp(-im\omega_0 t)$$

$$f_m = \frac{1}{T}\int_{-T/2}^{T/2} f(t)\exp(im\omega_0 t)\,,$$

we have

$$\frac{1}{T}\int_{-T/2}^{T/2} f^2(t)dt = \sum_{m=-\infty}^{\infty} |f_m|^2 = 2\sum_{m=1}^{\infty} |f_m|^2$$

where

$$kx(t) = \omega t - \mathbf{k}\mathbf{r}_0(t) .$$

By substituting in (1.12) the explict form of the field expansion in the case of periodic motion we have for the radiation intensity of m-harmonics

$$dI_m = \frac{e^2 2\pi m^2}{T^4} \int_{-T/2}^{T/2} dt_1 \int_{-T/2}^{T/2} dt_2 [\mathbf{v}(t_1) \cdot \mathbf{v}(t_2) - 1] \\ \times \exp\{i[m\omega_0(t_1 - t_2) - \mathbf{k} \cdot [\mathbf{r}_0(t_1) - \mathbf{r}_0(t_2))]]\} d\Omega . \qquad (1.24)$$

When deriving this formula, the terms of the $\mathbf{v}(t_1) \cdot \mathbf{n}$ type were integrated by parts.

We now proceed to the description of the polarization properties of the radiation. Taking into account the formulas (A.29), (A.26), (1.20), we have for the polarization matrix

$$d\varepsilon_{ik}(\mathbf{n}, \omega) = e^2 \int dt_1 \int dt_2 (\mathbf{v}(t_1) - \mathbf{n})_i \cdot (\mathbf{v}(t_2) - \mathbf{n})_k \\ \times \exp[ik(x(t_1) - x(t_2))] \frac{d^3 k}{(2\pi)^2} \qquad (1.25)$$

Instead of the direct use of the quadratic combination of the kind $E_i E_k *$ as in (A.29) in this formula, we have shifted to the direct use of radiation energy (intensity) which naturally includes the quadratic combinations of fields. This is an adequate method namely because the intensity variation is usually used for observing the polarization events. The polarization matrix $d\varepsilon_{ik}(\mathbf{n}, \omega)$ can be presented in the standard form as follows:

$$d\varepsilon_{ik}(\mathbf{n}, \omega) = (1/2) d\varepsilon(\mathbf{n}, \omega) \left(1 + \sum_{n=1}^{3} \xi_n \tau_n \right)_{ik} , \qquad (1.26)$$

where the parameters ξ_n define the radiation polarization. If one introduces some external vector $\mathbf{e}$ and composes the matrix analogue of (A.29), $e_i{}^* e_k = (1/2)(I + \boldsymbol{\eta}\boldsymbol{\tau}), |\boldsymbol{\eta}| = 1$ then $d\varepsilon_{ik}(\mathbf{n}, \omega)$ can be projected onto this matrix

$$d\varepsilon(\mathbf{n}, \omega, \boldsymbol{\eta}) \equiv d\varepsilon_{ik} e_i{}^* e_k = d\varepsilon(\mathbf{n}, \omega) \,\mathrm{Tr}\, [(I + \boldsymbol{\xi}\boldsymbol{\tau}) \cdot (I + \boldsymbol{\eta}\boldsymbol{\tau})]/4 \\ = d\varepsilon(\mathbf{n}, \omega)(1 + \boldsymbol{\eta} \cdot \boldsymbol{\xi})/2 = d\varepsilon(\mathbf{n}, \omega)[1 - |\boldsymbol{\xi}| + |\boldsymbol{\xi}|(1 + \cos 2\varphi)] . \qquad (1.27)$$

This expression can be interpreted as the projecting of radiation with polarization of the parameters ξ_n on the axis corresponding to the completely polarized radiation with parameters η_n. An angle φ introduced in (1.27) is an angle in a plane perpendicular to the vector $\mathbf{n}$ between polarization vectors corresponding to parameters ξ_1, ξ_3 and η_1, η_3. If the radiation is polarized completely ($|\boldsymbol{\xi}| = 1$) then for $\boldsymbol{\eta} = \boldsymbol{\xi}$ we have

$$d\varepsilon(\mathbf{n}, \omega, \boldsymbol{\xi}) = d\varepsilon(\mathbf{n}, \omega) \,. \tag{1.28}$$

In any other case, radiation energy $d\varepsilon(\mathbf{n}, \omega, \boldsymbol{\eta})$ is lower than the total radiation energy $d\varepsilon(\mathbf{n}, \omega)$.

1.2. Features of Radiation in Ultrarelativistic Region*

The radiation of ultrarelativistic particles ($\gamma \gg 1$, $\gamma = \frac{\varepsilon}{m}$) has a remarkable feature: the influence of the longitudinal component of the external force (with respect to the particle motion direction) is negligibly small compared to the role of the perpendicular (transverse) component.

In fact, a force in relativistic mechanics

$$\mathbf{F} = \frac{d\mathbf{P}}{dt} = \frac{m}{\sqrt{1 - v^2}} \left[\dot{\mathbf{v}} + \frac{\mathbf{v}(\mathbf{v} \cdot \dot{\mathbf{v}})}{1 - v^2} \right] . \tag{1.29}$$

Whence for the acceleration longitudinal (with respect to velocity) component we have

$$\dot{v}_{\parallel} = \frac{(\dot{\mathbf{v}} \cdot \mathbf{v})}{v} = \frac{F_{\parallel}}{m\gamma^3} \tag{1.30}$$

and for the transversal one

$$\dot{\mathbf{v}}_{\perp} = \mathbf{F}_{\perp}/(m\gamma) \,. \tag{1.31}$$

From these formulas it follows that the longitudinal force accelerates by γ^2 times less than the transversal one. Substituting these expressions into (1.15) and (1.18) respectively we see that at the same order of magnitude for the longitudinal force and the transverse components the radiation induced by the longitudinal component is negligibly small (of the order of $1/\gamma^2$) compared to the radiation induced by the transverse component of force. Under these conditions the longitudinal force component can be neglected (and,

*In the presentation of Secs. 1.2–1.7 we refer to the book by Baier, Katkov, Fadin (1973).

consequently, longitudinal acceleration) and one could approximately set that radiation intensity is totally determined by the transverse force only (transverse acceleration).* In this approximation $(\mathbf{v}\cdot\dot{\mathbf{v}}) = (1/2)d\mathbf{v}^2/dt = 0$ i.e. $\mathbf{v}^2$ is conserved. For the transversal force component

$$\begin{aligned}\mathbf{F}_\perp^2 &= (d\mathbf{p}/dt)^2 - (1/v^2)(\mathbf{v}\,d\mathbf{p}/dt)^2 = (d\mathbf{p}/dt)^2\\ &\quad - 1/v^2(d\varepsilon/dt)^2 = -1/\gamma^2(dp^\mu/d\tau)^2[1+O(1/\gamma^2)]\,. \end{aligned} \tag{1.32}$$

Whence in the ultrarelativistic limit it follows that

$$(1/m^2)(\mathbf{F}_\perp\gamma)^2 = (\dot{\mathbf{v}}_\perp\gamma^2)^2 = -\frac{1}{m^2}(dp^\mu/d\tau)^2 = \text{inv} \tag{1.33}$$

which is an invariant of Lorentz transformation. This fact turns out to be quite important when considering the radiation of ultrarelativistic particles.

An angular distribution of energy radiated by particle per unit time t_0 follows from the formula (1.7). Taking into account approximations made we have:

$$\frac{d\varepsilon(\mathbf{n})}{dt_0} = \frac{e^2}{4\pi}\frac{1}{(1-\mathbf{n}\cdot\mathbf{v})^5}\left[\dot{\mathbf{v}}_\perp^2(1-\mathbf{n}\cdot\mathbf{v})^2 - \frac{1}{\gamma^2}(\mathbf{n}\cdot\dot{\mathbf{v}}_\perp)^2\right]d\Omega\,. \tag{1.34}$$

In this expression the dominators of both terms contain high powers of a quantity $1-\mathbf{n}\cdot\mathbf{v} = 1-v\cos\vartheta$. This quantity takes its minimum at $\vartheta = 0$ when it is equal to $1-v = (1-v^2)/(1+v) \simeq 1/2\gamma^2$. The denominator's low value leads to a sharp growth of the expression (1.34). Therefore, for separating the main (over $1/\gamma^2$) terms in radiation energy one can perform the following expansion

$$1-\mathbf{n}\cdot\mathbf{v} = 1-v\cos\vartheta \simeq \frac{1}{2}(1/\gamma^2+\vartheta^2) + O(1/\gamma^4)\,. \tag{1.35}$$

Substituting (1.35) into (1.34) leads to the conclusion that angular distribution of ultrarelativistic particle radiation is a narrow, needle-like cone having angle $\vartheta \sim 1/\gamma$ with the axis directed along the particle's instantaneous velocity.

*The case when force $\mathbf{F}$ is directed nearly along the velocity so that $F_\parallel \sim F_\perp\gamma$ requires additional consideration. In this case, the radiation is of the same order as that in the case of longitudinal acceleration.

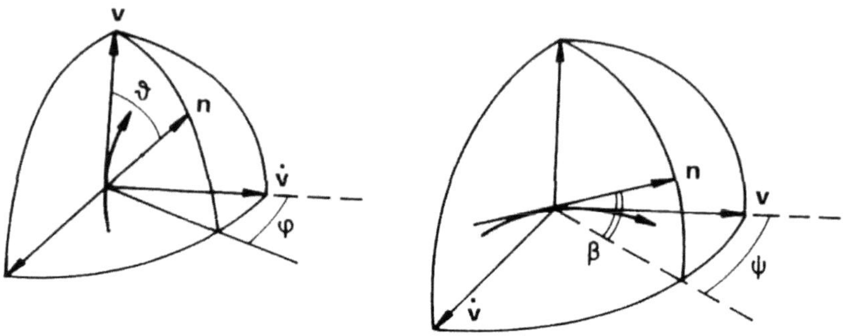

Fig. 1.2. Reference frames for a description of the angular characteristics of the radiation. The angles ψ and β are convenient for the description of the radiation of fast particles.

By this reasoning, in order to get the main order characteristics of radiation, one can always carry out the expansion (1.35) and leave the main terms.

For a detailed consideration of the radiation angular distribution it is convenient to introduce the angle reference system somewhat differently from that taken in Sec. 1.1 (formulas (1.13)–(1.17)). Let us introduce β — an angle between the plane $(\mathbf{v}, \dot{\mathbf{v}})$ and vector $\mathbf{n}$ and ψ — an angle between the projection of vector $\mathbf{n}$ on the plane $(\mathbf{v}, dot\mathbf{v})$ and vector $\mathbf{v}$ (Fig. 1.2). The incoming scalar combinations have the form:

$$\begin{aligned} \mathbf{n} \cdot \mathbf{v} &= v \cos\beta \cos\psi \\ \mathbf{n} \cdot \dot{\mathbf{v}}_\perp &= |\dot{\mathbf{v}}_\perp| \cos\beta \sin\psi \,. \end{aligned} \tag{1.36}$$

Such a reference system of angles is convenient because the main contribution into radiation is given by small ($\sim 1/\gamma$) angles β and ψ corresponding to that by an angle between vector $\mathbf{n}$ and $\mathbf{v}$ of the order of $1/\gamma$. At the same time when angles ϑ and φ are used: $\vartheta \sim 1/\gamma, 0 \leq \varphi \leq 2\pi$.

Substituting (1.36) into (1.34) and taking into account that $\beta, \psi \sim 1/\gamma$ one gets for the angular distribution of radiated energy*

$$\frac{d\varepsilon(\mathbf{n})}{dt} = \frac{2e^2}{\pi} \frac{\dot{v}_\perp^2}{(\mu^2 + \psi^2)^5} [(\mu^2 - \psi^2)^2 + 4\beta^2\psi^2] d\beta d\psi \,, \tag{1.37}$$

*Remember it is considered here that the case when the longitudinal and transversal components of force are of the same order of magnitude.

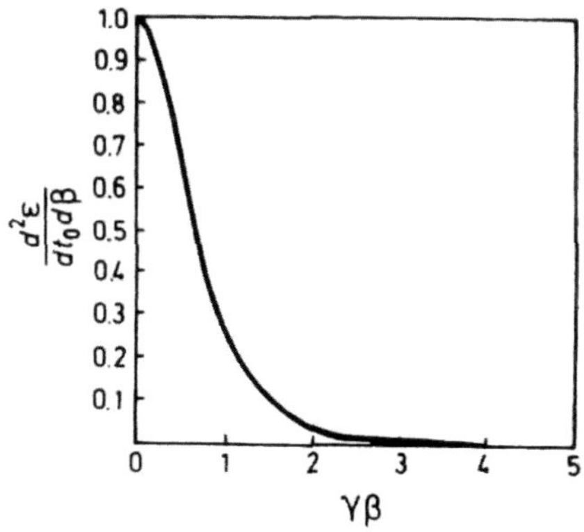

Fig. 1.3. Dependence of the radiation intensity on polar angle (in the $7e^2\dot{v}_\perp^2\gamma^5\,/\,16$ units).

where $\mu^2 = \beta^2 + 1/\gamma^2$ Frequently it is of interest to have the distribution integrated over the azimuthal angle ψ. Carrying out this integration we have* (Fig. 1.3)

$$\frac{d\varepsilon(\beta)}{dt_0} = \frac{e^2}{16}\frac{\dot{v}_\perp^2}{\mu^7}[7\mu^2 + 5\beta^2]d\beta$$

$$= \frac{e^2}{16}\frac{\dot{v}_\perp^2}{\mu^7}\left[12\beta^2 + \frac{7}{\gamma^2}\right]d\beta\,. \tag{1.38}$$

And finally, by integrating angle β we get

$$\frac{d\varepsilon}{dt_0} = \frac{2}{3}e^2\dot{v}_\perp^2\gamma^4 \tag{1.39}$$

This formula coincides with (1.18). The total intensity of the radiation (1.39) is invariant of the Lorentz transformation (see 1.12).

*While integrating angles ψ, β the main contribution is given by $\psi, \beta \sim 1/\gamma$; because of rapid convergency of integrals for the calculation of the main terms of expansion over $1/\gamma$ one can make the integration upper limit tend to infinity. Further calculations are quite elementary if one uses the following formula

$$\int_{-\infty}^{\infty}\frac{x^{2s}dx}{(a^2+x^2)^\nu} = \frac{1}{a^{2\nu-2s-1}}\frac{\Gamma(s+1/2)\Gamma(\nu-s-1/2)}{\Gamma(\nu)}$$

1.3. Polarization of Radiation in Ultrarelativistic Region

It is evident that the radiation energy summed up over all the frequencies can only have the linear polarization that characterises the mean direction of vector **E** in the plane perpendicular to vector **n**. For the description of polarization (see Appendix A.3) let us expand the electric field vector (1.3) over two mutually orthogonal unit vectors:

$$\mathbf{E} = E_1 \mathbf{s} + E_2 \mathbf{n} \times \mathbf{s}\,, \tag{1.40}$$

where $\mathbf{s} = \mathbf{w}/w$, $\mathbf{w} \equiv dot\mathbf{v}$. Such a choice of unit vectors makes it convenient for use with the accepted reference system of angles (see Fig. 1.2). Strictly speaking, vector **E** is not orthogonal to vector **n** though. This inaccuracy is of the order of $1/\gamma$ so, the expansion (1.40) can be used for calculating the main terms (with an accuracy of $\sim 1/\gamma$).

Expanding vector **E** (see (1.3)) over vectors **s**, $\mathbf{n} \times \mathbf{s}$ one can get

$$\begin{aligned} E_1 &= \frac{ew}{r(1-\mathbf{n}\cdot\mathbf{v})^3}\left[(\mathbf{n}\cdot\mathbf{s})^2 - (1-\mathbf{n}\cdot\mathbf{v})\right]; \\ E_2 &= \frac{ew}{r(1-\mathbf{n}\ \mathbf{v})^3}(\mathbf{n}\cdot(\mathbf{v}\times\mathbf{s}))(\mathbf{n}\cdot\mathbf{s})\,. \end{aligned} \tag{1.41}$$

Substituting (1.41) in (A.29) and (1.25) and expressing the incoming combinations in terms of angles β, ψ (1.36) we obtain a general presentation of the polarization matrix for radiation of ultrarelativistic particles:

$$\frac{d\varepsilon_{ik}(\mathbf{n})}{dt_0} = \frac{2e^2w^2 d\psi d\beta}{\pi(\mu^2+\psi^2)^5}\begin{pmatrix} (\mu^2-\psi^2)^2 & 2\psi\beta(\psi^2-\mu^2) \\ 2\psi\beta(\psi^2-\mu^2) & 4\psi^2\beta^2 \end{pmatrix} \tag{1.42}$$

It is natural that

$$\mathrm{Tr}\,(d\varepsilon_{ik}(\mathbf{n})/dt_0) = d\varepsilon(\mathbf{n})/dt_0 \tag{1.43}$$

is a total energy radiated per unit time. The Stokes parameters defined by (A.39) and (1.42) are as follows:

$$\begin{aligned} \xi_1 &= \frac{4\psi\beta(\psi^2-\mu^2)}{(\psi^2-\mu^2)^2 + 4\beta^2\psi^2}\,; \quad \xi_2 = 0\,; \\ \xi_3 &= \frac{(\psi^2-\mu^2)^2 - 4\psi^2\beta^2}{(\psi^2-\mu^2)^2 + 4\psi^2\beta^2} \end{aligned} \tag{1.44}$$

It is seen that there is no circular polarization which was evident from the beginning. From (1.44) it follows that $\xi_1^2 + \xi_3^2 = 1$, i.e. the ultrarelativistic particle radiation is totally polarized linearly at angle φ to an axis s. The angle φ is determined by the relation

$$\mathrm{tg}\,\varphi = 2\beta\psi/(\psi^2 - \mu^2)\,. \tag{1.45}$$

By integration of the polarization density matrix (1.42) over the azimuthal angle ψ we get the polarization properties of the radiation depending on the radiation polar angle β with respect to the plane $(\mathbf{v}, \mathbf{w})$. This integration is performed similarly as in formula (1.38). In this case, nondiagonal terms vanish due to oddness of the integrand

$$\frac{d\varepsilon_{ik}(\beta)}{dt_0} = \frac{e^2}{16}\frac{w^2}{\mu^7}\begin{pmatrix} 7\mu^2 & 0 \\ 0 & 5\beta^2 \end{pmatrix} d\beta\,, \tag{1.46}$$

whence we have the Stokes parameters

$$\xi_1 = \xi_2 = 0\,, \quad \xi_3 = (7 + 2\beta^2\gamma^2)/(7 + 12\beta^2\gamma^2)\,. \tag{1.47}$$

Thus the radiation is completely polarized linearly along the direction of the acceleration $\mathbf{w}$ at $\beta = 0$ and polarized partially along this direction if $\beta \neq 0$ and at $\beta\gamma \gg 1$ the polarization level tends to 1/6 (Fig. 1.4). Finally, by the integration over the polar angle β we get

$$\frac{d\varepsilon_{ik}}{dt_0} = \frac{2}{3}e^2 w^2 \gamma^4 \begin{pmatrix} 7/8 & 0 \\ 0 & 1/8 \end{pmatrix} \tag{1.48}$$

and for the Stokes parameters we have

$$\xi_1 = \xi_2 = 0\,, \quad \xi_3 = 3/4\,. \tag{1.49}$$

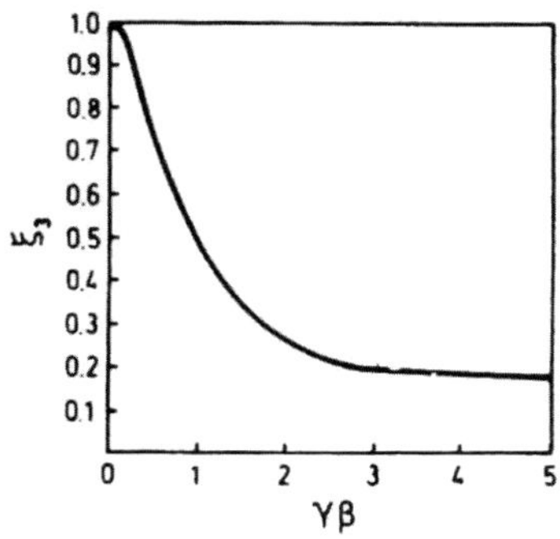

Fig. 1.4. Dependence of the degree of linear polarization of the radiation along the direction of the acceleration on the polar angle.

So, on the whole, the ultrarelativistic particle radiation (independent of the relationship between the radiation direction $\mathbf{n}$ and kinematic characteristics of particles $\mathbf{v}, \mathbf{w}$) is partially polarized along the direction of acceleration $\mathbf{w}$. In this case, the degree of polarization is $\xi_3 = 3/4$.

1.4. Characteristic Cases of Radiation

In Sec. 1.2 it was shown that the ultrarelativistic particle radiates mainly forward into the narrow cone with the axis directed along the particle velocity $\mathbf{v}$ and the cone opening angle $\vartheta \sim 1/\gamma$. Correspondingly, in order to find the radiation spectral expansion of ultrarelativistic particle the relation between the total angle of the particle deflection in external* field and angle $1/\gamma$ becomes essential. The problems of radiation are split into two characteristic cases.

1. The total angle of particle deflection in the field is large compared to $1/\gamma$. Then in the given direction $\mathbf{n}$ the particle radiates from a small fraction of its trajectory where the direction of the particle velocity is changed by the angle $\sim 1/\gamma$. This fraction of the trajectory will be called the radiation formation length (coherence length). In this case, we assume that one can neglect the variation of external field acting on the particle at the radiation length i.e. that

$$|\dot{\mathbf{F}}|\Delta t/F \ll 1\,, \tag{1.50}$$

where $\dot{\mathbf{F}}$ is the external field variation on the trajectory; Δt is the time required to turn at angle $\sim 1/\gamma$. By introducing an instantaneous curvature radius R we have $\Delta t \approx v\Delta t \approx R/\gamma$. Then the criterion (1.50) can be written as

$$n/\gamma \ll 1\,, \tag{1.51}$$

where the dimensionless characteristics of the external field inhomogeneity $n = |\dot{\mathbf{F}}|R/F \sim |\nabla\mathbf{F}|R/F$ is introduced. When this condition is satisfied one can obtain universal expressions valid for any field. All the characteristics of radiation depend on instantaneous values $\mathbf{v}$ and $\dot{\mathbf{v}}$ only, since these values do not vary on the radiation length. Such a situation takes place, for example, in magnetic bremsstrahlung. Therefore we shall often call Case 1 magnetic bremsstrahlung limit.

*The external field below is taken as an electromagnetic field produced by external sources where the radiating charged particle is travelling.

2. The total deflection angle of the particle in an external field is lower or of the order characteristic angle of radiation. Then the whole radiation of the particles passes inside the narrow cone with an opening angle $\sim 1/\gamma$ and it is determined by a noticeable part or nearly the whole trajectory of the particle. Naturally, in this case the characteristics of radiation are more sensitive to the shape of the external field. An important case here is radiation at quasiperiodical motion (by definition, in the frame traveling with an average particle velocity this motion is periodical). This kind of motion is realized in undulators (systems with periodical electric and magnetic fields), for the motion in the field of the electromagnetic wave, for the motion in the single crystal's channels. Another important case is bremsstrahlung in the Coulomb field.

1.5. Qualitative Picture of Radiation in Case 1

The radiation occurs from the small part of the trajectory where the velocity of the particle turns at an angle $\sim 1/\gamma$. The phase difference of waves emitted by the particle in the direction $\mathbf{n}$ at the moments of times t_1 and t_2:

$$\Delta\varphi = \omega\{t_2 - t_1 - \mathbf{n}[\mathbf{r}(t_2) - \mathbf{r}(t_1)]\} \tag{1.52}$$

As soon as the phase difference becomes the order of unity, the radiation from the different points becomes incoherent and breakdown of radiation occurs. Now let us take into account that the radiation occurs from the small part of the trajectory where $\Delta\dot{\mathbf{v}} \approx v/\gamma \approx 1/\gamma$ and let us carry out the expansion of values (included in (1.52)) over the time differences $t_2 - t_1 = \tau$. Then we have

$$\begin{aligned} \mathbf{r}(t_2) &= \mathbf{r}(t_1) + \dot{\mathbf{r}}(t_1)\tau + \ddot{\mathbf{r}}(t_1)\tau^2/2! + \dddot{\mathbf{r}}(t_1)\tau^3/3! + \cdots; \\ \Delta\varphi &\approx \omega\tau[1 - \mathbf{n}\cdot\mathbf{v} - \mathbf{n}\cdot\dot{\mathbf{v}}\tau/2 - \mathbf{n}\cdot\ddot{\mathbf{v}}\tau^2/6] \\ &\approx \omega\tau[1 - \mathbf{n}\cdot\mathbf{v} - \mathbf{n}\cdot\dot{\mathbf{v}}\tau/2 + \dot{\mathbf{v}}^2\tau^2/6]\,. \end{aligned} \tag{1.53}$$

When obtaining the last equality we have used the fact that the force acting on the particle can be considered as transverse i.e. $(\mathbf{v}\ \ \dot{\mathbf{v}}) = 0$. Then with an accuracy up to the terms $\sim 1/\gamma^2$

$$(\mathbf{n}\ \ \ddot{\mathbf{v}}) = -\dot{\mathbf{v}}^2 \tag{1.54}$$

The expression of (1.53) in square brackets has the order of magnitude $1/\gamma^2$. In fact: (a) the value $1 - \mathbf{n}\ \ \mathbf{v} \approx 1 - v \approx 1/2\gamma^2$; (b) as already mentioned, the radiation occurs from the small part of the trajectory where the velocity variation

$$|\Delta \mathbf{v}| = |\dot{\mathbf{v}}|\tau \sim 1/\gamma\,, \quad \tau \sim 1/(|\dot{\mathbf{v}}|\gamma)\,; \tag{1.55}$$

(c) with an accuracy up to the terms of the order $1/\gamma^2$ we have $(\mathbf{v}\dot{\mathbf{v}}) = 0$ but then $(\mathbf{n}\cdot\dot{\mathbf{v}}) \sim |\dot{\mathbf{v}}|/\gamma$ since vector $\mathbf{n}$ is directed nearly along the velocity vector $\mathbf{v}$. Thus, the phase difference

$$\Delta\varphi \sim \omega/|\dot{\mathbf{v}}|\gamma^3\,. \tag{1.56}$$

The condition of radiation breakdown $\Delta\varphi \sim 1$ determines the threshold for frequencies which can be radiated in the given motion (characterized by the velocity $\mathbf{v}$ and acceleration $\dot{\mathbf{v}}$). From (1.56) the boundary frequency

$$\omega \sim \omega_c = |\dot{\mathbf{v}}|\gamma^3\,. \tag{1.57}$$

For higher frequencies the phase difference becomes of the order of unity inside the radiation cone so that there are waves of the given frequency traveling in the opposite phase. The interference of such waves leads to the suppression of the radiation and as a result a decrease in the radiation intensity at frequencies $\omega \gg \omega_c$. The spectral expansions of the radiation field energy include integrals of the kind $\int \exp(i\Delta\varphi)dt$. At high frequencies the integrand oscillates rapidly which results in mutual suppression. Since such kinds of integrals are exponentially small, the radiation intensity drops exponentially at $\omega \gg \omega_c$. For frequencies $\omega < \omega_c$ the phase difference $\Delta\varphi$ is small and the exponential term can be replaced by unity; thus the radiation behavior is determined by the pre-exponential factor. If this factor increases with frequency the maximum radiation intensity distribution over frequencies is in the range when $\omega \sim \omega_c$. Therefore at high energy the particles mainly radiate high harmonics compared to the motion characteristic frequencies which are determined by the value $|\dot{\mathbf{v}}|$ (for example, for circular motion $|\dot{\mathbf{v}}| \approx 1/R \approx \omega_L$, is a frequency of revolution).

The relation (1.57) also results from the following simple consideration. The radiation contributed in the given direction is caused by the fraction of trajectory that is defined by formula (1.55). Whence the characteristic frequencies radiated in the rest frame of the particle are $\omega_r \sim 1/\tau \sim |\dot{\mathbf{v}}|\gamma$. In the laboratory system due to Doppler's effects (particles travel to the observer) the frequency is observed to be γ^2 times higher. In other words, the radiation duration in l-system (see 1.6) is

$$\begin{aligned} \Delta t &= (dt/d\tau)\tau = (1 - \mathbf{n}\cdot\mathbf{v})\tau \sim \tau/\gamma^2\,; \\ \omega &\sim 1/\Delta t \sim |\dot{\mathbf{v}}|\gamma^3 \end{aligned} \tag{1.58}$$

Thus, the picture of radiation in the given direction for arbitrary motion of the ultrarelativistic particle is a short pulse (or short burst) of radiation (the succession of such bursts for periodical motion) of duration Δt (Fig. 1.5) with a set of frequencies up to $\omega \sim |\dot{\mathbf{v}}|\gamma^3$. Since the pulse formation time is short, one is concerned about the instantaneous characteristics of radiation with a frequency ω in the direction $\mathbf{n}$. When the conditions (3.1) and (3.2) are satisfied these characteristics are universal and determined only by the frequency and particle acceleration $\dot{\mathbf{v}}$.

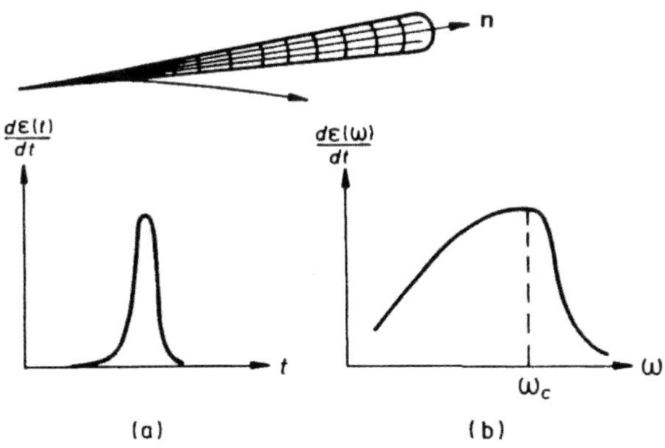

Fig. 1.5. Time (a) and the frequency (b) characteristics of the radiation of ultrarelativistic particle.

1.6. Spectral Distribution and Polarization Properties of Radiation in Magnetic Bremsstrahlung Limit

Here we start from the spectral distribution (1.25). For further consideration of the radiation problem in Case 1, it is convenient to introduce the following variables:

$$\begin{aligned} t_0 &= (t_1 + t_2)/2\,; \quad \tau = t_2 - t_1\,; \\ t_1 &= t_0 - \tau/2\,; \quad t_2 = t_0 + \tau/2\,. \end{aligned} \tag{1.59}$$

As mentioned above the time of radiation in the direction given is much shorter that the time characteristic of particle motion. Having this fact in mind the particle kinematic characteristics can be expanded in a power series in τ:

$$\begin{aligned} \mathbf{v}_1 &= \mathbf{v} - \dot{\mathbf{v}}\tau/2 + \ddot{\mathbf{v}}\tau^2/8 ; \\ \mathbf{v}_2 &= \mathbf{v} + \dot{\mathbf{v}}\tau/2 + \ddot{\mathbf{v}}\tau^2/8 ; \\ \mathbf{r}_2 - \mathbf{r}_1 &= \dot{\mathbf{v}}\tau + \ddot{\mathbf{v}}\tau^3/24 , \end{aligned} \tag{1.60}$$

where $\mathbf{v}_1 \equiv \mathbf{v}(t_1)$ etc. Substituting this expansion in (1.25) we obtain the following expression for radiation intensity of ultrarelativistic particles (energy radiated per unit particle time) with an account of radiation polarization properties

$$\begin{aligned} dI_{ik}(\mathbf{n},\omega) \equiv d\varepsilon_{ik}(\mathbf{n},\omega)/dt_0 = e^2 \int (&\mathbf{v} - \mathbf{n} - \dot{\mathbf{v}}\tau/2 \\ &+ \ddot{\mathbf{v}}\tau^2/8)_i(\mathbf{v} - \mathbf{n} + \dot{\mathbf{v}}\tau/2 + \ddot{\mathbf{v}}\tau^2/8)_k \\ &\times \exp[-i\omega\tau(1 - \mathbf{n}\cdot\mathbf{v} + \dot{\mathbf{v}}^2\tau^2/24)] d\tau d^3k/(2\pi)^2 , \end{aligned} \tag{1.61}$$

where we passed to variables τ, t_0 and at transforming the argument of the exponential we used formula (1.54). Note that although the expansion (1.60) is valid in the narrow interval with respect to the time moment t_0 the intergral over τ in Eq. (1.61) can be extended to an infinite interval as the large region gives the negligibly small contribution because of fast oscillations of the integrand. In order to write down the polarization matrix in the explicit form let us introduce (as in previous sections, see (1.36)) the angles β and ψ (see Fig. 1.2), then (1.61) has the form:

$$\begin{aligned} dI_{ik}(\mathbf{n},\omega) = e^2(d^3k/(2\pi)^2) \int_{-\infty}^{\infty} &L^c_{ik} \exp[-i\omega\tau \\ &\times (\mu^2 + \psi^2 + w^2\tau^2/12)/2] d\tau , \end{aligned} \tag{1.62}$$

where $w = |\mathbf{w}|$, the matrix L^c_{ik} is

$$L^c = \begin{pmatrix} \psi^2 - w^2\tau^2/4 & \beta(\psi + w\tau/2) \\ \beta(\psi - w\tau/2) & \beta^2 \end{pmatrix} \tag{1.63}$$

For the radiation intensity summed over polarization we obtain (compare formulae (1.22) and (1.23))

$$\begin{aligned} dI(\mathbf{n},\omega) = \mathrm{Tr}\,(dI_{ik}) = e^2(d^3k/(2\pi)^2) \int_{-\infty}^{\infty} &[2(1 - \mathbf{n}\cdot\mathbf{v}) - 1/\gamma^2 \\ &- w^2\tau^2/4] \exp[-i\omega\tau(1 - \mathbf{n}\cdot\mathbf{v} + w^2\tau^2/24)] d\tau . \end{aligned} \tag{1.64}$$

Finally, by integrating parts of the term with $1-\mathbf{n}\cdot\mathbf{v}$ analogous to the transition from the formula (1.22) to (1.24) (or similarly by substituting expansions (1.60) in formula (1.23) we have for the radiation intensity*

$$\begin{aligned} dI(\mathbf{n},\omega) &= -e^2(d^3k/(2\pi)^2)\int_{-\infty}^{\infty}(1/\gamma^2+w^2\tau^2/2) \\ &\quad\times\exp\{-i\omega\tau[(1-\mathbf{n}\cdot\mathbf{v})+w^2\tau^2/24]\} \\ &= -e^2(d^3k/(2\pi)^2)\int_{-\infty}^{\infty}(1/\gamma^2+w^2\tau^2/2) \\ &\quad\times\exp[-i\omega\tau(\mu^2+\psi^2+w^2\tau^2/12)/2]d\tau \end{aligned} \tag{1.65}$$

The integral over τ (included in the formulae (1.62)–(1.65)) is taken with the help of the following relations (see Gradstein, Ryzhik (1965), pp. 398, 959):

$$\begin{aligned} &\int_{-\infty}^{\infty}\cos(bx+ax^3)dx = 2/3\sqrt{b/a}K_{1/3}(\sigma)\,; \\ &\int_{-\infty}^{\infty}x\sin(bx+ax^3)dx = (2/3\sqrt{3})(b/a)K_{2/3}(\sigma)\,; \\ &\int_{-\infty}^{\infty}x^2\cos(bx+ax^3)dx = -(b/3a)\int_{-\infty}^{\infty}\cos(bx+ax^3)dx\,, \\ &\sigma = (2/3\sqrt{3})(b^{3/2}/a^{1/2})\,. \end{aligned} \tag{1.66}$$

Here $K_\nu(\sigma)$ is the Bessel function of the imaginary argument (MacDonald's function). Integrating (1.64) and (1.65) over τ we obtain for the radiation intensity summed over polarization:

$$\begin{aligned} dI(\mathbf{n},\omega) &= e^2(d^3k/(2\pi)^2)4\sqrt{2/3}(\sqrt{1-\mathbf{n}\cdot\mathbf{v}}/w)[4\,(1-\mathbf{n}\cdot\mathbf{v}) \\ &\quad -1/\gamma^2]K_{1/3}(\xi) \\ &= e^2(d^3k/\sqrt{3}\pi^2)\frac{\sqrt{\mu^2+\psi^2}}{w}[2(\mu^2+\psi^2) \\ &\quad -1/\gamma^2]K_{1/3}(\xi)\,, \end{aligned} \tag{1.67}$$

where

$$\xi = (4\sqrt{2}/3)\left(\frac{\omega}{w}\right)(1-\mathbf{n}\cdot\mathbf{v})^{3/2} = (2/3)(\omega/w)(\mu^2+\psi^2)^{3/2}\,. \tag{1.68}$$

---

*From the formula (1.65) one can obtain again the estimations (1.51) and (1.58). Indeed the main contribution to the integral is given by the region of τ variation where the integrand does not oscillate, i.e. when $w^2\tau^2 \sim (1-\mathbf{n}\cdot\mathbf{v}) \sim 1/\gamma^2$, whence $\omega \sim \gamma^2/\tau \sim |\dot{\mathbf{v}}|\gamma^3$.

Integrating the polarization matrix (1.62) and (1.63) we get

$$\begin{aligned} dI_{ik} &= e^2(d^3k/(2\pi)^2)4\sqrt{2/3}(\sqrt{1-\mathbf{n}\cdot\mathbf{v}}/w)a_{ik}\,, \\ a_{11} &= [(\mathbf{n}\cdot\mathbf{s})^2 + 2(1-\mathbf{n}\cdot\mathbf{v})]K_{1/3}(\xi)\,, \\ a_{21}^* &= a_{12} = (\mathbf{n}\cdot(\mathbf{n}\times\mathbf{s}))[(\mathbf{n}\cdot\mathbf{s})K_{1/3}(\xi) - i\sqrt{2(1-\mathbf{n}\cdot\mathbf{v})}K_{2/3}(\xi)]\,, \\ a_{22} &= (\mathbf{n}\cdot(\mathbf{v}\times\mathbf{s}))^2K_{1/3}(\xi)\,, \end{aligned} \tag{1.69}$$

where $\mathbf{s} = \mathbf{w}/w$. Taking into account that with the accepted accuracy (see (1.36))

$$\begin{gathered} 1-\mathbf{n}\cdot\mathbf{v} = (\psi^2+\mu^2)/2\,; \quad \mu^2 = \beta^2 + 1/\gamma^2\,; \\ (\mathbf{n}\cdot(\mathbf{v}\times\mathbf{s})) = \beta\,, \quad (\mathbf{n}\cdot\mathbf{s}) = \psi\,, \end{gathered} \tag{1.70}$$

we have the presentation of the polarization matrix in terms of angles ψ, β

$$\begin{aligned} dI_{ik}(\mathbf{n},\omega) &= e^2(d^3k/\sqrt{3}\pi^2)\sqrt{\mu^2+\psi^2}/wb_{ik}\,; \\ b_{11} &= (2\psi^2+\mu^2)K_{1/3}(\xi)\,, \\ b_{21}^* &= b_{12} = \beta[\psi K_{1/3}(\xi) - i\sqrt{\mu^2+\psi^2}K_{2/3}(\xi)]\,, \\ b_{22} &= \beta^2K_{1/3}(\xi)\,. \end{aligned} \tag{1.71}$$

Naturally $dI(\mathbf{n},\omega) = \mathrm{Tr}\,(dI_{ik}(\mathbf{n},\omega))$. The expression (1.67) describes (the differential over all the angles) the spectral distribution of radiation intensity. In other words there are given the instantaneous properties of radiation at the time moment t_0 for which the velocity vector, being of the bench mark of the reference systems of angles, $\mathbf{v}(t_0)$. The argument of $K_y(\xi)$ in formulae (1.67)–(1.71) in the radiation cone i.e. when $1-\mathbf{n}\cdot\mathbf{v} \sim 1/\gamma^2$ has the order of

$$\xi = \omega/w\gamma^3 = \omega/\omega_c\,. \tag{1.72}$$

Let us take into account that the expansion of functions $K_\nu(z)$ at small and large values of arguments (see Gradstein, Ryzhik (1965), pp. 959, 963)

$$\begin{aligned} K_\nu(z) &= \sqrt{\pi/2z}e^{-z}\,, \quad z \gg 1\,; \\ K_\nu(z) &= (\Gamma(\nu)/2)(2/z)^\nu\,, \quad z \ll 1\,. \end{aligned} \tag{1.73}$$

Thus at $\xi \ll 1$ i.e. at low frequencies $\omega \ll \omega_c$ for the fixed angles inside the radiation cone when $\psi^2+\mu^2 \sim 1/\gamma^2[(\psi^2+\mu^2)_{\min} \sim 1/\gamma^2]$ the spectral distribution of radiation intensity has the form

$$dI(\mathbf{n},\omega) = e^2(d\omega d\beta d\psi/2\pi^2) \times (\Gamma(1/3)\omega^{5/3}/3^{1/6}w^{2/3})[2(\psi^2+\mu^2)-1/\gamma^2]\,. \quad (1.74)$$

This expression as a function of radiating frequency grows proportionally to $\omega^{5/3}$ For $\xi \gg 1$ i.e. at high frequencies $\omega \gg \omega_c$ inside the cone of radiation or for large angles of radiation:

$$dI(\mathbf{n},\omega) = e^2(d\omega d\beta d\psi/2\pi^2) \times (\sqrt{\pi}\omega^{3/2}/\sqrt{w}(\mu^2+\psi^2)^{1/4})[2(\psi^2+\mu^2)-1/\gamma^2] \times \exp[-(2\omega/3w)(\mu^2+\psi^2)^{3/2}]\,. \quad (1.75)$$

In this region the intensity decreases exponentially. As mentioned above this is connected with the mutual cancellation of waves radiated in antiphase. For the polarization matrix (1.71) the Stokes parameters defined by the formula (A.39) are the following

$$\begin{aligned} \xi_1 &= 2\psi\beta/[2(\psi^2+\mu^2)-1/\gamma^2]\,; \\ \xi_3 &= (2\psi^2+1/\gamma^2)/[2(\psi^2+\mu^2)-1/\gamma^2]\,; \\ \xi_2 &= 2\beta(\psi^2+\mu^2)^{1/2}K_{2/3}(\xi)/\{[2(\psi^2+\mu^2)-1/\gamma^2]K_{1/3}(\xi)\}\,. \end{aligned} \quad (1.76)$$

Since the circular frequency is defined for monochromatic wave the formula for ξ_2 is valid when the inequality $\psi \geq (\omega_0/\omega)^{1/3}/\gamma$ is satisfied. Thus, generally speaking the radiation in the given direction is partially polarized. At $\beta = 0$ (radiation in the direction $\mathbf{n}$ lying in the plane $(\mathbf{v},\mathbf{w})$, see Fig. 1.2) the parameters $\xi_{1,2} = 0$, $\xi_3 = 1$ i.e. the radiation is completely linearly polarized along the direction of acceleration.

Though, one should have in mind that in the given direction it is observed that the radiation is not emitted at one point but from the part of the trajectory where the particle turns at an angle $\sim 1/\gamma$ (see Fig. 1.5). In order to find the corresponding characteristics which are of special interest while observing the radiation of ultrarelativistic particles, one needs to sum up the radiation over the bending angles i.e. to integrate the expressions (1.67) and (1.71) over angle ψ. Because of the rapid decreasing of functions at large angles ψ, one can extend the integration over ψ to an infinite interval, similarly for the relative time τ. The calculation turns out to be simpler if instead of integration of expressions (1.67) and (1.71) one carries out integration simultaneously over ψ and τ in formulae (1.62) and (1.65).

Let us find out the spectral distribution of radiation intensity (1.65) summed up over polarization

$$dI(\beta,\omega) = \int_\psi dI(\mathbf{n},\omega) = -e^2(\omega^2 d\omega d\beta/(2\pi)^2) \times \int_{-\infty}^{\infty} d\psi \int_{-\infty}^{\infty} d\tau (1/\gamma^2 + w^2\tau^2/2) \times \exp\{-i\omega\tau(\mu^2 + \psi^2 + w^2\tau^2/12)/2\} \quad (1.77)$$

For performing integration let us introduce the variables:

$$w\tau = \delta(x+y), \quad \psi = \delta(x-y)/2, \quad \delta = \left(\frac{2w}{\omega}\right)^{1/3}, \quad \frac{\partial(\tau,\psi)}{\partial(x,y)} = \frac{\delta^2}{w}, \quad (1.78)$$

then

$$dI(\beta,\omega) = -e^2(\omega^2 d\omega d\beta/(2\pi)^2)(\delta^2/w) \times \int_{-\infty}^{\infty} dx \int_{-\infty}^{\infty} dy [1/\gamma^2 + \delta^2(x+y)^2/2] \times \exp\{-i[\mu^2(x+y)/\delta^2 + (x^3+y^3)/3]\}. \quad (1.79)$$

This integal is easy to evaluate with the help of the formulae (1.66)

$$\begin{aligned} dI(\beta,\omega) &= e^2(\omega^2 d\omega d\beta/3\pi^2)(\mu^2/w) \times [\beta^2 K_{1/3}^2(\eta) + \mu^2 K_{2/3}^2(\eta)] \\ &= e^2(\omega^2 d\omega d\beta/3\pi^2)(\mu^2/w) \times [\mu^2(K_{1/3}^2(\eta) + K_{2/3}^2(\eta)) - K_{1/3}^2(\eta)/\gamma^2], \end{aligned} \quad (1.80)$$

where

$$\eta = \omega\mu^3/3w \quad (1.81)$$

Let us consider the properties of the radiation spectral distribution (1.80). Inside the radiation cone the argument of K-functions

$$\eta \sim \omega/(w\gamma^3) = \omega/\omega_c, \quad (1.82)$$

as in (1.72). Then using formulae (1.73) one can obtain asympthotic representation (1.80) at $\eta \ll 1$ and $\eta \gg 1$. In the region of low frequencies ($\eta \ll 1$)

$$dI(\beta, \omega) = e^2(d\omega d\beta/3\pi^2)(9/2)^{2/3}\Gamma^2(2/3)\omega^{2/3}w^{1/3}\,. \tag{1.83}$$

With high frequencies inside the radiation cone ($\omega \gg \omega_c$) or at large radiation angle ($\eta \gg 1$)

$$dI(\beta, \omega) = e^2(d\omega d\beta/2\pi)\omega[\mu + \beta^2/\mu]\exp[-2\omega\mu^3/3w]\,. \tag{1.84}$$

As follows from (1.80)–(1.84) the function $dI(\beta, \omega)/d\beta d\omega$ reaches its maximum at $\eta \sim 1$. Let us consider radiation at various angles β_1 and β_2. From (1.81) it follows that with growing frequency one can get $\eta \sim 1$ earlier as the radiation angle is larger. In other words, the larger an angle β the radiation is more "red" (see Fig. 1.6).

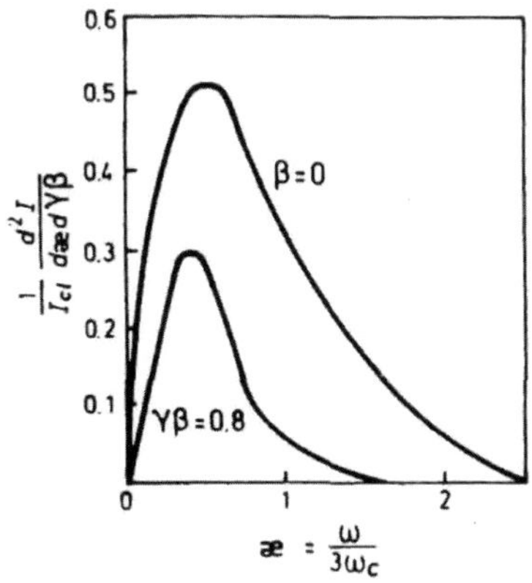

Fig. 1.6. Dependence of the spectrum intensity distribution on a frequency at different polar angles of the radiation. If the angle is larger the radiation is more "red" in its maximum. Value I_{cl} is given by Eq. (1.39).

The polarization matrix integrated over an angle ψ is calculated similarly to that as the expression (1.80). Let us make the replacement of (1.78) in (1.62) and (1.63). Then the matrix L^c (1.63) has the form

$$L^c = \begin{pmatrix} -\delta^2 xy & -\beta\delta y \\ \beta\delta x & \beta^2 \end{pmatrix} \tag{1.85}$$

and polarization matrix:

$$\begin{aligned} dI_{ik}(\beta,\omega) &= e^2(\omega^2 d\omega d\beta/3\pi^2)(\mu^2/w)c_{ik}\,; \\ c_{11} &= \mu^2 K_{2/3}^2(\eta)\,, \\ c_{21} &= c_{12}^* = i\beta\mu K_{1/3}(\eta)K_{2/3}(\eta)\,, \\ c_{22} &= \beta^2 K_{1/3}^2(\eta)\,. \end{aligned} \tag{1.86}$$

It is evident that $dI(\beta,\omega) = \mathrm{Tr}\,(dI_{ik}(\beta,\omega))$. The Stokes parameters for the polarization matrix (1.86) are the following

$$\begin{aligned} &\xi_1 = 0\,, \quad \xi_2 = 2\beta\mu K_{1/3}(\eta)K_{2/3}(\eta)/R\,, \\ &\xi_3 = \left(\mu^2 K_{2/3}^2(\eta) - \beta^2 K_{1/3}^2(\eta)\right)/R \\ &R = \beta^2 K_{1/3}^2(\eta) + \mu^2 K_{2/3}^2(\eta)\,. \end{aligned} \tag{1.87}$$

Radiation is polarized completely since $\xi_2^2 + \xi_3^2 = 1$ (see Appendix A.3). In the general case the polarization is an elliptical and since $\xi_3 > 0$, the larger axis of the ellipse is directed along the direction of acceleration. At $\beta = 0$ (vector $\mathbf{n}$ is in the plane $(\mathbf{v}, \mathbf{w})$) $\xi_1 = \xi_2 = 0$, $\xi_3 = 1$ i.e. the radiation is completely linearly polarized along the direction of acceleration (see (1.76)). The condition $\mu^2 K_{2/3}^2 - \beta^2 K_{1/3}^2 = 0$ cannot be satisfied therefore the polarization cannot tend to the circular one.

An important characteristic of radiation is the spectral distribution of radiation intensity integrated over all the angles of radiation. The relevant calculation is conveniently done by integrating simultaneously over ψ and β the expression (1.67) and (1.71). By changing in (1.67) the variables $u = \psi\gamma, v = \beta\gamma$, we get

$$\begin{aligned} dI(\omega) = \int_\Omega dI(\mathbf{n},\omega) &= e^2(\omega^2 d\omega/\sqrt{3}\pi^2|\mathbf{w}|\gamma^5) \\ &\times \int du \int dv (1+u^2+v^2)^{1/2}[1+2(u^2+v^2)] \\ &\times K_{1/3}(\kappa(1+u^2+v^2)^{3/2})\,, \end{aligned} \tag{1.88}$$

where

$$\kappa = 2\omega/(3w\gamma^3) = 2\omega/3\omega_c\,. \tag{1.89}$$

It is seen that it is reasonable to calculate the integral (1.88) in polar coordinates: $r^2 = u^2 + v^2$, $\varphi = \text{arc tg}\,(v/u)$:

$$dI(\omega) = e^2(\omega^2 d\omega/\sqrt{3}\pi^2 w\gamma^5) \int_0^\infty dr^2/2 \int_0^{2\pi} d\varphi$$

$$\times \sqrt{1+r^2}(1+2r^2)K_{1/3}(\kappa(1+r^2)^{3/2})$$

$$= e^2(2\omega^2 d\omega/3\sqrt{3}\pi w\gamma^5) \int_1^\infty ds(2s^{2/3}-1)K_{1/3}(\kappa s)\,. \qquad (1.90)$$

Using the relation for K-functions (see Gradstein, Ryzhik (1965), p. 970)

$$\frac{d}{ds}[s^{2/3}K_{2/3}(\kappa s)] = -\kappa s^{2/3}K_{1/3}(\kappa s)\,, \qquad (1.91)$$

we get the following expression for the spectral distribution of radiation intensity:

$$dI(\omega) = e^2(\omega d\omega/\sqrt{3}\pi\gamma^2)\left[2K_{2/3}(\kappa) - \int_\kappa^\infty dx K_{1/3}(x)\right] \qquad (1.92)$$

Taking into account that (see Gradstein, Ryzhik (1965), p. 959)

$$2dK_{2/3}(z)/dz + K_{1/3}(z) = -K_{5/3}(z) \qquad (1.93)$$

finally we get

$$dI(\omega) = e^2(\omega d\omega/\sqrt{3}\pi\gamma^2) \int_\kappa^\infty K_{5/3}(\xi)d\xi$$

$$= e^2(3^{3/2}/4\pi)w^2\gamma^4\kappa d\kappa \int_\kappa^\infty K_{5/3}(\xi)d\xi\,. \qquad (1.94)$$

The spectral distribution of radiation intensity behaves differently depending on the relationship between the radiated frequency ω and frequency ω_c. At low frequencies ($\omega \ll \omega_c$)

$$dI(\omega)/d\omega = e^2(3^{1/6}\Gamma(2/3)/\pi)w^{2/3}\omega^{1/3} \qquad (1.95)$$

and high frequencies ($\omega \gg w_c$)

$$dI(\omega)/d\omega = (e^2/2)\sqrt{\omega w/\pi\gamma}\exp[-2\omega/(3w\gamma^3)] \qquad (1.96)$$

so, the qualitative behaviors of spectral distribution (1.74), (1.75), (1.83), (1.84) and (1.95), (1.96) have the same character. Since $dI(\omega)/d\omega$ falls down both at low or high frequencies then its maximum is at $\omega \sim \omega_c$. The main part of radiation is concentrated namely in this region.* The explicit form of combination $(3^{3/2}\kappa/4\pi)\int_\kappa^\infty K_{5/3}(\xi)d\xi$ contained in the formula (1.94) is given in Fig. 1.7.

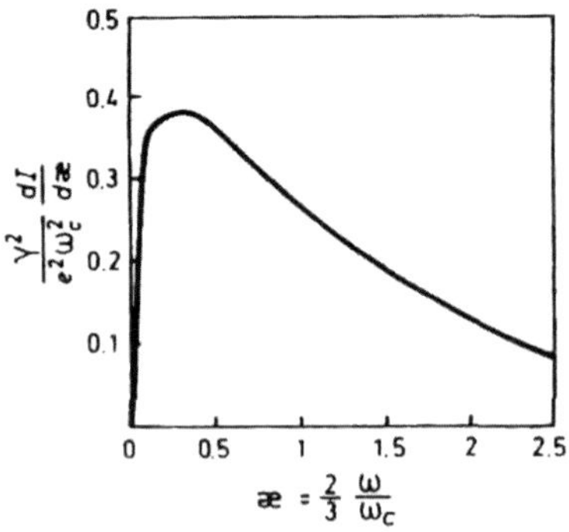

Fig. 1.7. Dependence of the spectral intensity distribution on frequency ($\kappa = 2\omega/3\omega_c$).

If (1.94) is integrated over the frequency by using formulae

$$\int_0^\infty \kappa d\kappa \int_\kappa^\infty f(\xi)d\xi = \int_0^\infty f(\xi)d\xi \int_0^\xi \kappa d\kappa$$
$$= \int_0^\infty f(\xi)(\xi^2/2)d\xi \tag{1.97}$$

and (see Gradstein, Ryzhik (1965), p. 684)

$$\int_0^\infty x^\mu K_\nu(x)dx = 2^{\mu-1}\Gamma((1+\mu+\nu)/2)\Gamma((1+\mu-\nu)/2)\,, \tag{1.98}$$

then we get the formula (1.39).

*Comparing formulae (1.74), (1.83) and (1.95) one can see that for the case of $\omega \ll \omega_c$ the main contribution into radiation is given by the angles $\beta \sim \psi \sim (\omega_c/\omega)^{1/3}(1/\gamma)$ so that the time of the pulse generation (coherence length) is $\Delta t \sim (R/\gamma)(\omega_c/\omega)^{1/3}$ Then condition (1.50) leads to a more rigid limitation (1.54) to the field in homogeneity: $n \ll \gamma(\omega/\omega_c)^{1/3}$. Only by satisfying the last criterion are the formulae (1.83) and (1.95) valid.

The polarization matrix integrated over all angles can be evaluated similarly to (1.90) starting from Eq. (1.77)

$$dI_{ik}(\omega) = e^2(\omega^2 d\omega/\sqrt{3}\pi^2\gamma^5 w)\int_0^\infty dr^2/2\int_0^{2\pi} d\varphi(1+r^2)^{1/2}$$
$$\times K_{1/3}(\kappa(1+r^2)^{3/2})\begin{pmatrix} 1+r^2+r^2\cos^2\varphi & 0 \\ 0 & r^2\sin^2\varphi \end{pmatrix}$$
$$= e^2(2\omega^2 d\omega/3\sqrt{3}\pi w\gamma^5)\int_1^\infty ds K_{1/3}(\kappa s)$$
$$\times\begin{pmatrix} 3s^{2/3}-1 & 0 \\ 0 & s^{2/3}-1 \end{pmatrix} \tag{1.99}$$

The nondiagonal terms vanish due to the oddness of the integrand; it is clear that $dI = \mathrm{Tr}\,[dI_{ik}(\omega)]$. Using (1.91) and (1.93) we get

$$dI_{ik}(\omega) = e^2(\omega d\omega/2\sqrt{3}\pi\gamma^2)$$
$$\times\begin{pmatrix} 3K_{2/3}(\kappa) - \int_\kappa^\infty K_{1/3}(x)dx & 0 \\ 0 & K_{2/3}(\kappa) - \int_\kappa^\infty K_{1/3}(x)dx \end{pmatrix}$$
$$= e^2(\omega d\omega/2\sqrt{3}\pi\gamma^2)$$
$$\times\begin{pmatrix} \int_\kappa^\infty K_{5/3}(x)dx + K_{2/3}(\kappa) & 0 \\ 0 & \int_\kappa^\infty K_{5/3}(x)dx - K_{2/3}(\kappa) \end{pmatrix}. \tag{1.100}$$

In this case the Stokes parameters are:

$$\xi_1 = \xi_2 = 0\,,$$
$$\xi_3 = K_{2/3}(\kappa)\Big/\int_\kappa^\infty K_{5/3}(x)dx\ . \tag{1.101}$$

Thus, after integrating the polar angle the circular polarization disappears so that radiation on the whole is partially linearly polarized along the direction of acceleration. Using the formulae (1.73) we have that at low frequencies ($\omega \ll \omega_c$) the degree of linear polarization is $\xi_3 \to 1/2$, and at high frequencies

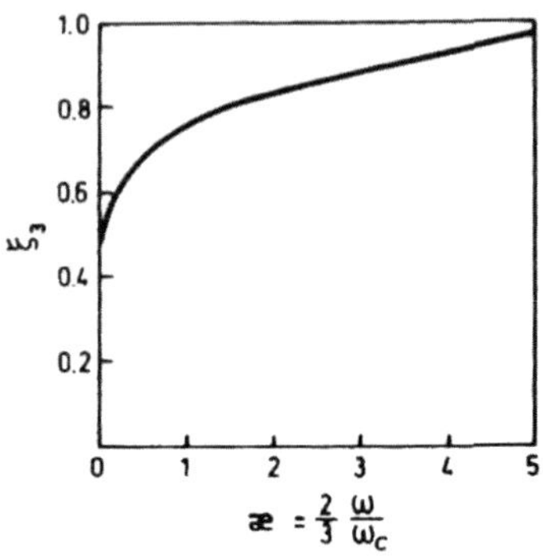

Fig. 1.8. Dependence of the degree of linear polarization of the radiation on frequency.

$(\omega \gg \omega_c)\xi_3 \to 1$. A general picture of polarization as a function of frequency is given in Fig. 1.8. By integrating (1.100) over frequency with the use of formulae (1.57), (1.98) we get formulae (1.48) and (1.49).

1.7. Some Examples of Radiation for Case 1

(a) Magnetic bremsstrahlung ($\mathbf{v} \perp \mathbf{H}$). This is one of the most important kinds of radiation frequently used for applications. As is known in a constant and homogeneous magnetic field, an electron moves along the circle of radius R with frequency ω_L:

$$\begin{aligned} &R = p/eH = mv\gamma/eH\,; \\ &\omega_L = v/R = eH/\varepsilon\,, \quad \mathbf{w} = \frac{e}{\varepsilon}\mathbf{v} \times \mathbf{H}\,, \end{aligned} \tag{1.102}$$

where $\mathbf{w}$ is an acceleration; $\mathbf{H}$ is the magnetic field. In the ultrarelativistic limit we have:

$$w \equiv |\mathbf{w}| = \omega_L = 1/R = eH/\varepsilon + \mathrm{O}\,(1/\gamma^2)\,. \tag{1.103}$$

An angular distribution of radiation intensity is given by formula (1.17), and radiation total intensity is given by formulae (1.18) and (1.19). Since an electron is moving periodically with frequency $\omega_L = 2\pi/\mathrm{T}$, it radiates a set of frequencies multiplied to $\omega_L : \omega = n\omega_L$. One can calculate the intensity of nth harmonics according to (1.24). The total intensity of radiation is a sum of harmonics (the detailed description one can find for example in the book by Landau and Lifshits (1973)). Although, the result obtained in this way can be used directly at low energies ($\gamma \sim 1$) when low harmonics are radiated. The situation changes substantially for the region of high energies ($\gamma \gg 1$). As was shown above in Case 1, in the ultrarelativistic limit very high harmonics are radiated,* $\omega \approx \omega_L \gamma^3$, so that radiation spectrum consists of very large numbers and lines whose frequencies are high compared to the distance between neighboring frequencies ω_L.

In other words, the spectrum of radiation is of quasicontinuous character, consisting of a large number of closely spaced lines. One should also keep in mind that radiation in given directions come from a small section of a circle and it does not depend on the motion along the remaining part of the trajectory. Therefore the fact itself of the motion along the circle becomes insignificant. The universal formulae were obtained in Sec. 3.3 determining radiation in the situation considered.† Thus in the ultrarelativistic limit the magnetic bremsstrahlung including all the spectral and polarization characteristics is described by formulae (1.59)–(1.101) where (1.103) should be placed as an acceleration value. It is evident that these formulae are also valid for motion in an inhomogeneous magnetic field if one can consider it being constant at the length of the pulse formation (Conditions (1.50) and (1.51). In this case the properties of radiation are determined by the magnetic field on the given section of trajectory.‡

(b) Magnetic bremsstrahlung ($\mathbf{v}\cdot\mathbf{H}$) $\neq 0$. In the case when the particle has the velocity component $v_{\parallel} = v \sin\vartheta$ longitudinal with respect to the magnetic field $\mathbf{H}$ it travels in a homogeneous and constant magnetic field along the helix. The radiation is described by the formulae given in Sec. 1.6 where instead of an acceleration value one should put

$$\begin{aligned} \mathbf{w}_{\vartheta} &= (e/\varepsilon)\mathbf{v} \times \mathbf{H} \\ w_{\vartheta} &= (e/\varepsilon) vH \cos\vartheta = (e/\varepsilon) v_{\perp} H = (e/\varepsilon) v H_{\perp} \end{aligned} \tag{1.104}$$

since no assumptions were made of an angle between the velocity and external field. It is necessary only that the condition $\pi/2 - \vartheta \gg 1/\gamma$ is satisfied since if a particle travels at an angle $\sim 1/\gamma$ with respect to the field (nearly along

*This feature of the magnetic bremsstrahlung of high energy particles was first indicated by Arzimovich and Pomeranchuk (1946).

†Note that the trajectory sections were approximated with the circle arcs with instantaneous radius curvature $R = 1/w$.

‡The spectral distribution of the radiation intensity from high energy electrons in magnetic fields has been considered by Ivanenko and Sokolov (1948) and by Schwinger (1949).

the field) the energy of the transverse motion will then be of the order of particle mass and by this reason the particle radiation loses the features intrinsic to ultrarelativistic particle radiation. At helical motion the particle velocity vector $\mathbf{v}$ describes the cone with the vertex angle $\pi - 2\vartheta$, radiation is concentrated around the vicinity of this cone but not in the plane as in the case of rotation along the circle. The presence of longitudinal velocity component on which the external force does not act on leads to a supplementary Doppler's effect connected with the transport motion. So for the stationary observer the frequency is shifted to:

$$\omega_{v_\parallel} = \omega_L/(1 - v_\parallel \sin\vartheta) \approx \omega_L/\cos^2\vartheta\,. \tag{1.105}$$

For revealing the radiation characteristics while in motion along a helical path let us consider two versions. For the first one the perpendicular to the field component of momentum is fixed i.e. the radiation is considered in various inertial systems related by the Lorentz transformation along the magnetic field direction (the field does not vary). Note that the radiation total intensity (1.39) is an invariant of the Lorentz transformation. This is evident from (1.10) and (1.12). While calculating the spectral distribution of radiation intensity (1.94) one should take into account that

$$\begin{aligned} \varepsilon_0 &= \sqrt{\varepsilon^2 - p_\parallel^2} = \varepsilon\sqrt{1 - v_\parallel^2} \simeq \varepsilon\cos\vartheta\,; \\ |w_\vartheta| &= (eH/\varepsilon)\cos^2\vartheta = w_0\cos^2\vartheta\,, \end{aligned} \tag{1.106}$$

where ε_0 and w_0 are the energy and the acceleration in the system where $v_\parallel = 0$. Then for the characteristic frequency we get

$$(\omega_c)_\vartheta = (1/\cos\vartheta)(\omega_c)_0\,, \tag{1.107}$$

i.e. the spectral distribution maximum is shifted into a short-wave region. According to this fact the radiation polarization also changes (see (1.101)). Similarly an analysis for radiation differential characteristics can be done, including polarization where together with energy transformation one should take into account the transformation of angles whose reference system in each individual case is connected with the plane $(\mathbf{v}, \mathbf{w})$.

In the second version the particles with the same energy enter into the magnetic field with a velocity directed at various angles to the field $(\pi/2 - \vartheta)$. In this case the radiation characteristic frequency

$$(\omega_c)_\vartheta = w_\vartheta\gamma^3 = w_0\gamma^3\cos\vartheta = (\omega_c)_0\cos\vartheta\,, \tag{1.108}$$

i.e. the smaller angle between $\mathbf{v}$ and $\mathbf{H}$, the larger the radiation maximum is shifted into the long-wave region. The radiation total intensity also falls down with an increase in the angle ϑ:

$$d\varepsilon_\vartheta/dt = \frac{2}{3}e^2 w_\vartheta^2 \gamma^4 = \frac{2}{3}e^2 w_0^2 \gamma^4 \cos^2\vartheta \,. \tag{1.109}$$

(c) Particle motion at an angle to the electric field (transverse momentum $p_\perp$: $p_\perp^2 \gg m^2$). A particle entered the electric field at an angle to the field travels along the curve turning to the field. In this case an angle between the field and velocity $\pi/2 - \vartheta$ is time dependent:

$$\begin{aligned}\cos\vartheta(t) &= p_\perp/p(t) = p_\perp/(p_\perp^2 + (p_\parallel^0 + eEt)^2)^{1/2} \\ &\approx p_\perp/m\gamma(t)\,,\end{aligned} \tag{1.110}$$

where

$$\gamma(t) = [(p^2(t) + m^2)/m^2]^{1/2} \tag{1.111}$$

Note that the condition $\pi/2 - \vartheta \gg m/\gamma(t)$ is always satisfied if $p_\perp \gg m$ (since the radiation characteristic angle is narrowing proportionally to the energy growth) and namely this is necessary for the use of formulae obtained in Sec. 1.6. Corresponding to that which is mentioned above (see Sec. 1.2) the radiation is determined by the acceleration transverse component $|\mathbf{E}_\perp| = E\cos\vartheta(t)$, so that in the formulae of Sec. 1.6 one should put

$$w(t) = e|\mathbf{E}_\perp|/m\gamma(t) = eE\cos\vartheta(t)/m\gamma(t) = eEp_\perp/m^2\gamma^2(t)\,. \tag{1.112}$$

In this case radiation is characterized by the following features: (1) It is concentrated in a narrow cone around the velocity direction and in addition the open angled cone is permanently narrowing ($\propto 1/\gamma(t)$); (2) the radiation characteristic frequency grows permanently:

$$\omega_c(t) = w(t)\gamma^3(t) = (eEp_\perp/m^2)\gamma(t) = eEp_\perp^2/(m^3\cos\vartheta(t)) \tag{1.113}$$

so that the spectral distribution maximum is shifted into the short-wave region. Formula (1.113) has the same form as (1.107). This is connected with the fact that both of them can be obtained with the help of relativistic transformation of frequency due to motion along the field; (3) the total energy radiated per unit time does not depend on time that is evident if one puts (1.112) into (1.39). This also results from the following considerations. The radiation intensity is invariant with respect to Lorentz transformations (see (1.10)), and for each moment of time one can select the Lorentz transformation along the field in such a way that the longitudinal velocity is equal to zero. Since for such transformations the radiation intensity does not vary, at any time radiation coincides with that of a particle moving transversely to the field with a given momentum.

(d) Radiation of ultrarelativistic particles in crossed fields E⊥H, $\mathbf{E}^2$=$\mathbf{H}^2$. The problem of particle radiation while traveling in the mutually perpendicular electric **E** and magnetic **H** fields if their values are different in magnitude with the relevant Lorentz transformation is reduced to the radiation problem in purely electric or purely magnetic fields ($\mathbf{E}\cdot\mathbf{H}$, $\mathbf{E}^2-\mathbf{H}^2$ are invariant to the Lorentz transformation). These cases were discussed above. Therefore it is also interesting to consider the case of so-called crossed fields

$$\mathbf{E}\cdot\mathbf{H}=0\,,\quad \mathbf{E}^2-\mathbf{H}^2=0\,.$$

For the sake of definiteness consider the magnetic field **H** directed along the z-axis and electric field **E** along the y-axis. Then while traveling in the crossed fields the intergrals of motion are (see Landau, Lifshits (1973), Panofsky, Phillips (1955))

$$p_z=\text{const.}\,;\quad \varepsilon-p_x=\text{const.} \tag{1.114}$$

Substituting the chosen field configuration into (1.11) we get that

$$d\varepsilon/dt_0=(2/3)e^2(eE/m^2)^2(\varepsilon-p_x)^2\,. \tag{1.115}$$

Thus, the radiation total intensity does not vary in time and is naturally Lorentz invariant (1.12). Note that if a particle enters the crossed field with momentum directed along the x-axis (perpendicular to both fields) then at $p_x^2\gg m^2$ we have $\varepsilon-p_x\simeq m/\gamma$, i.e. radiation decreases with the growth of particle energy.

When the conditions $\varepsilon-p_x\gg m/\gamma$ are satisfied the radiation of ultrarelativistic particles in crossed fields is determined by the formulae from Sec. 1.6 where one should put

$$w=(eE/(\gamma^2m^2))(\varepsilon-p_x) \tag{1.116}$$

as the acceleration value (compare (1.33) and (1.115)). The radiation characteristic frequency $\omega_c=w\gamma^3$ grows with an increase of energy like γ so that the intensity spectral distribution maximum is permanently shifted into the short-wave region. The open angle radiation cone is narrowing correspondingly.

1.8. Radiation Qualitative Picture at Quasiperiodical Motion

In the system travelling with the average particle velocity this motion is *periodical* by definition. The results obtained will often be demonstrated by an example of motion in periodical magnetic field (undulator*). As is shown in Secs. 1.4 and 1.5 for Case 1 the radiation is formed on the coherence length l_f (compare (1.50)–(1.51)), while motion in the magnetic field

$$l_f = R/\gamma = m/eH = \frac{1.7\ \ 10^3\mathrm{cm}}{H(Oe)} \tag{1.117}$$

so that l_f is only determined by the particle mass and field value. For wavelength λ where the size of the undulator periodicity component is such that $\lambda \gg l_f$, the radiation is formed inside each of the periodicity components and radiation total intensity is a noncoherent sum of intensities for each component of periodicity (this corresponds to Case 1). In the inverse case $\lambda \ll l_f$ (let us be reminded that the deflection angle for the particle in the field is $\psi \ll 1/\gamma$) radiation is formed on many components of periodicity and has a number of specific features, in particular, the coherence is essential.

Let us discuss the radiation qualitative features in the case of quasiperiodical motion. This consideration is conveniently done in a co-moving frame, in which the mean particle velocity is zero and then to carry out an inverse Lorentz transformation. In the co-moving frame the radiation properties depend strongly on the relation between the kinetic energy and the rest mass of the particle. As is known, in the nonrelativistic limit the radiation is dipole and determined completely by the Fourier components of the particle velocity. In this case, as a rule, one or a few first harmonics, which are multiples of the frequency of the particle motion, are emitted. Transforming back to the laboratory frame we obtain for the frequency of the radiated photon (Doppler effect)

$$\omega \approx \frac{m\omega_0}{1-\mathbf{n}\cdot\mathbf{V}} \approx \frac{2\gamma^2\omega_0 m}{1+\gamma^2\vartheta^2}, \tag{1.118}$$

*Radiation at quasiperiodical motion is considered by the authors (Baier, Katkov, Strakhovenko, 1972, 1981a). Radiation in undulators was considered by Ginzburg (1947), Alferov, Bashmakov, Bessonov (1972, 1975).

where ω_0 is a frequency of the particle's motion in the laboratory system, ϑ is the photon emission angle with respect to the mean velocity of particle $\mathbf{V}, m$ is the number of harmonics, $\mathbf{n} = \mathbf{k}/\omega$ where $\mathbf{k}$ is the photon wave vector.

When the motion in the co-moving frame becomes relativistic the nature of radiation changes. First, in the radiation the higher harmonics turn out to be essential and, second, it becomes necessary to take into account the dependence of particle longitudinal velocity (in the direction of mean velocity) on its transverse motion. Indeed, with an accuracy of terms $\sim U/\varepsilon$ (U is a potential in which the particle moves) $\gamma =$ const. and by definition $\gamma^2(1 - v_\parallel^2 - v_\perp^2) = 1$, so that for the longitudinal $v_\parallel$ and mean V velocities of a particle we have

$$\begin{aligned} v_\parallel &\approx 1 - (1 + v_\perp^2\gamma^2)/2\gamma^2 ; \\ V &\approx 1 - (1 + \bar{v}_\perp^2\gamma^2)/2\gamma^2 \end{aligned} \tag{1.119}$$

Here and below it is assumed that $|\mathbf{v}_\perp| \ll V$. In this case the expression for the frequency of radiation becomes

$$\omega \approx \frac{2\gamma^2\omega_0 m}{1 + \gamma^2\vartheta^2 + \rho/2}, \quad \rho = 2\gamma^2\bar{v}_\perp^2 , \tag{1.120}$$

where $\bar{v}_\perp^2$ is the mean square transverse velocity of the particle.

In the ultrarelativistic limit of the motion in the co-moving frame ($v_\perp\gamma \gg 1$), the main contribution to the radiation is given by high harmonics with $m \gg 1$. Radiation spectrum is quasicontinuous and here the known formulae that describe magnetic bremsstrahlung* (see Sec. 1.6) hold. In this limit the radiation is formed over a short section of the trajectory during the time $\tau \sim 1/|\dot{\mathbf{v}}|\gamma$, and characteristic frequency of the radiation, with allowance for the Doppler effect, is $\omega \sim |\dot{\mathbf{v}}|\gamma^3$ Taking into account that the main contribution to the radiation is made by angles $\vartheta \sim v_\perp$ the estimate for m is

$$m \sim v_\perp^2|\dot{\mathbf{v}}|\gamma^3/\omega_0 \approx (v_\perp\gamma)^3 = (\rho/2)^{3/2} . \tag{1.121}$$

Thus, the parameters ρ entered in the formula (1.120) characterizes the multipolarity of the radiation. At $\rho \ll 1$ the radiation is a dipole character, at $\rho \sim 1$ a noticeable contribution is given by high harmonics and at $\rho \gg 1$ we

*One should remember that these formulae become invalid at sufficiently low frequencies when $m \sim 1$.

have the magnetic bremsstrahlung limit when mainly high harmonics (1.121) are radiated.

1.9. Spectrum and Polarization Characteristics of Radiation at Quasiperiodical Motion* (*Dipole Approximation*)

For determining the radiation characteristics in Case 2 (let us be reminded that the particle deflection angle is $\psi \ll 1/\gamma$) it is convenient to base on the expression for the field in a wave zone (1.3). Let us calcalate the Fourier component:

$$\mathbf{E}_\omega = \int_{-\infty}^{\infty} \mathbf{E} e^{i\omega t} dt \,. \tag{1.122}$$

The expression on the right-hand side is a time retarding function determined by the relation (1.2). Since it is considered Case 2 when the particle trajectory is only slightly deflected from the rectilinear trajectory, one can take the particle velocity to be constant (and equal to average velocity $\mathbf{V}$), so that in (1.2) $\mathbf{r}(t_0) = \mathbf{V}t_0$, only the acceleration is variable $\mathbf{w}(t)$. Then (1.2) can be written in the form

$$t = r + t_0(1 - \mathbf{n} \ \mathbf{V}) \,. \tag{1.123}$$

By changing variables in the integral (1.122)

$$dt = dt_0(1 - \mathbf{n} \ \mathbf{V})$$

we get

$$\mathbf{E}_\omega = \frac{e}{r} \frac{e^{i\omega r}}{(1 - \mathbf{n} \cdot \mathbf{V})^2} \mathbf{n} \times [(\mathbf{n} - \mathbf{V}) \times \mathbf{w}_{\tilde{\omega}}] \,. \tag{1.124}$$

This formula includes the Fourier component of the acceleration

$$\mathbf{w}_{\tilde{\omega}} = \int_{-\infty}^{\infty} w(t_0) e^{i\tilde{\omega} t_0} dt_0 \,, \tag{1.125}$$

where $\tilde{\omega} = \omega(1 - \mathbf{n} \cdot \mathbf{V})$. Substituting (1.124) into the expression for spectral distribution of the radiated energy (compare with (1.20))

$$d\varepsilon_{\mathbf{n}\omega}(e) = \frac{1}{2\pi} |\mathbf{e} \cdot \mathbf{E}_\omega|^2 r^2 d\Omega \frac{d\omega}{2\pi} \,, \tag{1.126}$$

*We follow the paper by Baier, Katkov, Strakhovenko (1972).

where we projected an electric field on the polarization vector $\mathbf{e}$ ($\mathbf{e}\cdot\mathbf{n}=0$, compare (A.27)), we get a general formula for the radiation spectral-angular distribution in Case 2 (in the dipole approximation)

$$d\varepsilon_{\mathbf{n}\omega}(\mathbf{e})=\frac{e^2}{2\pi}\frac{d^3k}{2\pi}\frac{\omega^2}{\tilde{\omega}^4}|(\mathbf{e}^*\cdot\mathbf{V})(\mathbf{n}\cdot\mathbf{w}_{\tilde{\omega}})+(\mathbf{e}^*\cdot\mathbf{w}_{\tilde{\omega}})(1-\mathbf{n}\cdot\mathbf{V})|^2 \quad (1.127)$$

The frequencies are estimated in the region in which the main fraction of the radiation is concentrated. One can obtain from the integral (1.125), in fact the time $t_0 \leq 2\pi/\tilde{\omega}$ should be of the same order of the time T, during which the particle acceleration changes noticeably, i.e. $2\pi/\tilde{\omega}\approx T$, or

$$\omega \simeq \frac{2\pi}{T(1-\mathbf{n}\cdot\mathbf{V})}=\frac{\omega_0}{1-\mathbf{n}\cdot\mathbf{V}}\simeq\frac{2\omega_0\gamma^2}{1+\gamma^2\vartheta^2} \quad (1.128)$$

in accordance with (1.118). Note that the energy dependence of these frequencies ($\omega\propto\gamma^2$) is the same as in Case 1 (see (1.57), $|\dot{\mathbf{v}}|\propto 1/\gamma$), though the coefficient is different. For example, in magnetic undulator for forward radiation in Case 1 $\omega\approx eH_{\perp}\gamma^2/m^2$, for forward radiation in Case 2 $\omega\approx 4\pi\gamma^2/T$, and it does not depend either on particle mass or magnetic field and is determined completely by the undulator period T.

The results obtained above are referred to an arbitrary motion in Case 2. For the quasiperiodical motion (period T) one should take into account that in integral (1.125) $\mathbf{w}(t_0)=\mathbf{w}(t_0+T)=\mathbf{w}(t_0+2T)=\cdots$ Then in (1.125) one can extract the integral over the motion period T

$$\mathbf{w}_{\tilde{\omega}}=\mathbf{w}_{\tilde{\omega}}^T\sum_{n=-\infty}^{\infty}e^{in\varphi_0}=\mathbf{w}_{\tilde{\omega}}^T 2\pi\sum_{k=-\infty}^{\infty}\delta(\varphi_0-2\pi k)\,, \quad (1.129)$$

where

$$\mathbf{w}_{\tilde{\omega}}^T=\int_0^T\mathbf{w}(t)e^{i\tilde{\omega}t}dt\,,\quad \varphi_0=\tilde{\omega}T=\omega T(1-\mathbf{n}\cdot\mathbf{V})\,. \quad (1.130)$$

The transition from (1.125) to (1.129) corresponds to the transition from the integral to the Fourier series.

As was already mentioned the radiation properties depend substantially on the relationship between the particle deflection angle ψ in the period length and the angle $1/\gamma$. In Case 1 when $\psi=|\dot{\mathbf{v}}|T\doteq\Delta v\gg 1/\gamma$, from formula (1.130) it follows that the frequencies mainly radiated are $\omega\sim|\dot{\mathbf{v}}|\gamma/(1-\mathbf{n}\cdot\mathbf{V})$, then $\varphi_0 \gg 1$. Correspondingly, in Case 2, when $\psi = |\dot{\mathbf{v}}|T = \Delta v \leq 1/\gamma$ the frequencies mainly emitted are $\omega \sim 1/T(1 - \mathbf{n}\ \mathbf{V})$, then $\varphi_0 \leq 1$.

In Case 1 in the main frequency region the phase φ_0 is large and then

$$2\pi \sum_k \delta(\varphi_0 - 2\pi k) \to 2\pi \int dk \delta(\varphi_0 - 2\pi k) = 1$$

so that the radiation emitted in each component of the periodic structure is independent and the intensity of the radiation along a given direction can be obtained by adding the intensities at each formation length of the radiation. Note, although, that in Case 1 for frequencies $\omega \leq \gamma^2/T$ the coherence length is comparable with the structure period so that at the emission of these frequencies the coherence should be taken into account. By substituting (1.129) into (1.127) and passing to the radiation intensity* (see (1.61)) we have the general expression that the radiation intensity of the relativistic particle is an infinite undulator

$$\begin{aligned} dI &= e^2 \frac{d^3k}{2\pi T} \frac{\omega^2}{\tilde{\omega}^4} |(\mathbf{e}^*\ \mathbf{V})(\mathbf{n} \cdot \mathbf{w}_{\tilde{\omega}}^T) + (\mathbf{e}^*\ \mathbf{w}_{\tilde{\omega}}^T)(1 - \mathbf{n}\ \mathbf{V})|^2 \Phi\,, \\ \Phi &= \sum_{k=1}^{\infty} \delta(\varphi_0 - 2\pi k)\,. \end{aligned} \tag{1.131}$$

With the fixed vector $\mathbf{n}$ the formula (1.131) is a sum of harmonics.

By summing over polarizations of the radiated photon using the formula (A.42), we find that (ignoring term $\sim 1/\gamma$)

$$\begin{aligned} \langle dI_{\mathbf{n}\omega} \rangle \equiv \sum_{\lambda=1}^{2} dI = e^2 \frac{d^3k}{2\pi T} |\mathbf{w}_{\tilde{\omega}}^T|^2 \frac{\omega^2}{\tilde{\omega}^4} \frac{1}{2\gamma^4} \\ \times \{2y^2 + (1 - 2y)[1 + \cos 2(\beta - \varphi_1) \\ - 2\sin^2(\varphi/2) \sin 2\beta \sin 2\varphi_1]\} \Phi\,, \end{aligned} \tag{1.132}$$

where $y = \gamma^2(1 - \mathbf{n} \cdot \mathbf{V}) = \gamma^2 \frac{\tilde{\omega}}{\omega}$. The Fourier component of the acceleration $\mathbf{w}_{\tilde{\omega}}^T$ is expanded along two mutually perpendicular unit vectors $\mathbf{l}_1$ and $\mathbf{l}_2$ in a plane orthogonal to the vector $\mathbf{V}$:

$$\begin{aligned} \mathbf{w}_{\tilde{\omega}}^T &= |\mathbf{w}_{\tilde{\omega}}^T| e^{i\delta} (\mathbf{l}_1 \cos\beta + \mathbf{l}_2 \sin\beta e^{i\varphi})\,, \\ \mathbf{n} \cdot \mathbf{V} &= V \cos\vartheta\,, \quad \mathbf{n} \cdot \mathbf{l}_1 = \sin\vartheta \cos\varphi_1\,, \quad \mathbf{n} \cdot \mathbf{l}_2 = \sin\vartheta \sin\varphi_1\,. \end{aligned} \tag{1.133}$$

*This transition is most conveniently traced for the undulator of finite length (see (1.147) as follows).

After integration over the azimuthal angle of the photon emission we find that

$$\begin{aligned} \langle dI_{y\omega}\rangle &= \frac{2e^2}{T}\frac{d\omega}{\omega^2}\frac{dy}{y^6}|\mathbf{w}|^2\gamma^6(1-2y+2y^2)F\Phi\,, \\ \Phi &\to \sum_{k=1}^{\infty}\delta\left(\frac{\omega y}{\gamma^2}T-2\pi k\right)\,, \end{aligned} \tag{1.134}$$

where we have introduced the quantity

$$F = |\mathbf{w}_{\tilde{\omega}}^T|^2\tilde{\omega}^2/(4|\mathbf{w}|^2) \tag{1.135}$$

which depends only on the field structure. If the acceleration is variable, then F contains the average over the period: $\overline{|\mathbf{w}|^2} = \int_0^T |\mathbf{w}(t)|^2 dt/T$. Integrating (1.134) over frequency we get the radiation angular distribution

$$\langle dI_y\rangle = e^2\overline{|\mathbf{w}|^2}\gamma^4\frac{dy}{2y^5}(1-2y+2y^2)\frac{1}{\pi^2}\sum_{k=1}^{\infty}F(\varphi_0=2\pi k)/k^2\,. \tag{1.136}$$

Integrating (1.134) over y, we find the spectral distribution of the radiation intensity:

$$\begin{aligned} \langle dI_\omega\rangle &= 4e^2\overline{|\mathbf{w}|^2}\gamma^4\xi d\xi(1/\pi^2) \\ &\times\sum_{k=1}^{\infty}(1-2\xi/k+2\xi^2/k^2)F(\varphi_0=2\pi k)/k^4\,, \end{aligned} \tag{1.137}$$

where $\xi = \omega/\omega_c, \omega_c = 4\pi\gamma^2/T$, $0\le\omega\le k\omega_c$. Finally by integrating (1.136) over y, or (1.137) over ω we get

$$\langle I\rangle = (4/3)e^2\overline{|\mathbf{w}|^2}\gamma^4(1/\pi^2)\sum_{k=1}^{\infty}F(\varphi_0=2\pi k)/k^2\,. \tag{1.138}$$

for the total intensity.

For any structure of the field we find that

$$(1/\pi^2)\sum_{k=1}^{\infty}F(\varphi_0=2\pi k)/k^2 = 1/2\,, \tag{1.139}$$

so that the total intensity of radiation is given by the well-known expression $\langle I\rangle = (2/3)e^2\overline{|\mathbf{w}|^2}\gamma^4$

From the expression (1.131) one can also find the polarization density matrix (compare with (1.61)), which describes the polarization of the radiation:

$$dI_{ij} = \frac{e^2}{2\pi T}\frac{d^3k}{\bar{\omega}^2}|\mathbf{w}_{\bar{\omega}}^T|^2 T_{ij}\Phi\,,$$
$$T_{ij} = s_i s_j^* - \frac{\gamma^2}{y}[u_i s_j^*(\mathbf{us}) + s_i u_j(\mathbf{us}^*)] + \frac{\gamma^4}{y^2}u_i u_j|(\mathbf{us})|^2\,, \tag{1.140}$$

where $\mathbf{s} = \frac{\mathbf{w}_{\bar{\omega}}^T}{|\mathbf{w}_{\bar{\omega}}^T|}$, $\mathbf{u} = \mathbf{n} - \mathbf{V}$ is the 2-dimensional vector in the plane normal to the vector $\mathbf{n}$ (within relativistic accuracy). In the derivation of this formula we took into account that $\mathbf{e}^*\mathbf{V} = -\mathbf{e}^*\mathbf{u}$, and $\mathbf{sn} = \mathbf{su}$ with relativistic accuracy (up to terms $\sim O(\frac{1}{\gamma^2})$). This expression can be written also in the form (see (A.29))

$$dI_{ij} = \frac{d\langle I\rangle}{2}\left(\delta_{ij} + \frac{1}{c}\sum_{n=1}^{3} a_n\tau_{ij}^n\right)\,, \tag{1.141}$$

where $\langle dI\rangle$ is given by Eq. (1.132), τ_{ij}^n are the Pauli matrices and a_n, c are the functions depending on particular choice of axes in the mentioned plane. For the unit vectors $\mathbf{l_1}$ and] $\mathbf{l_2}$ (1.133) one has:

$$a_1 = 2y(1-y)\sin 2\beta\cos\varphi + (2y-1)g\sin 2\varphi_1\,,$$
$$a_2 = 2y(1-y)\sin 2\beta\sin\varphi\,,$$
$$a_3 = 2y(1-y)\cos 2\beta + (2y-1)g\cos 2\varphi_1\,, \tag{1.142}$$
$$c = -2y(1-y) - g\,,$$
$$g = (2y-1)(\cos 2\beta\cos 2\varphi_1 + \sin 2\beta\sin 2\varphi_1\cos\varphi) - 1\,.$$

The Stokes parameters can be found with the help of (A.39)–(A.41):

$$\xi_n = \frac{\sum_{k=1}^{\infty}\delta(\varphi_0 - 2\pi k)|\mathbf{w}_{\bar{\omega}}^T|^2 a_n/k^2}{\sum_{k=1}^{\infty}\delta(\varphi_0 - 2\pi k)|\mathbf{w}_{\bar{\omega}}^T|^2 c/k^2}\,, \tag{1.143}$$

where the parameter φ depends, generally speaking, on the summation index k. In the case, when this dependence is absent, one has $\xi_n = a_n/c$. In the last case, for given $\mathbf{n}$ and ω the radiation is completely polarized (generally speaking, elliptically), i.e. the Stokes parameters satisfy the relationship $\xi_1^2 + \xi_2^2 + \xi_3^2 = 1$ which can be checked by direct calculation. In the case of a planar motion (in the plane $(\mathbf{V}, \mathbf{w})$), one can put without loss of generality $\beta = 0$. Then from (1.143) it follows that $a_2 = 0$, so that for *planar* quasiperiodical motion of ultrarelativistic particles the circular polarization of radiation is absent.

The polarization properties of the radiation are also of interest when the integration is performed over a series of variables. The calculations are carried out similarly to those done in Subsec. 1.6 (compare formulae (1.86) and (1.100)). After integration over the azimuthal angle of the emission and the frequency of radiation we have

$$\xi_1 = \frac{S_1}{2(1-2y+2y^2)}\,, \quad \xi_2 = \frac{2y(1-y)S}{1-2y+2y^2}\,, \quad \xi_3 = \frac{\cos 2\beta}{2(1-2y+2y^2)}\,, \tag{1.144a}$$

where the equality (1.139) is taken into account and

$$S = \frac{2\sin 2\beta}{\pi^2}\sum_{k=1}^{\infty}\frac{F\sin 2\varphi}{k^2}\,, \quad S_1 = \frac{2\sin 2\beta}{\pi^2}\sum_{k=1}^{\infty}\frac{F\cos\varphi}{k^2} \tag{1.145}$$

Finally, after the integration over the radiation polar angle, we find the radiation polarization properties as a whole

$$\xi_1 = \frac{3S_1}{4}\,, \quad \xi_2 = \frac{S}{2}\,, \quad \xi_3 = \frac{3\cos 2\beta}{4}\,. \tag{1.144b}$$

Thus, the undulator radiation in the Case 2 is partially elliptically polarized. For the planar motion ($\beta = 0$) one has a universal result independent of a field structure $\xi_1 = \xi_2 = 0$ and $\xi_3 = 3/4$, whilst for the general case the degree of polarization depends strongly on this structure.

Let us now consider radiation in finite undulators when $\mathbf{w}(t) \neq 0$ within the interval NT. In this case formula (1.129) goes over into

$$\begin{aligned}\mathbf{w}_{\omega} &= \mathbf{w}_{\omega}^{T}\sum_{n=1}^{N} e^{i(n-1)\varphi_0} \\ &= \mathbf{w}_{\omega}^{T}\exp[i(N-1)\varphi_0/2]\sin(N\varphi_0/2)/\sin(\varphi_0/2)\,.\end{aligned} \tag{1.146}$$

Substituting (1.146) into (1.127) and passing to radiation intensity

$$dI_{\mathbf{n}\omega}(\mathbf{e}) = \frac{d\varepsilon_{\mathbf{n}\omega}(\mathbf{e})}{NT} \tag{1.147}$$

we find the general expression for the intensity of relativistic particle radiation in a finite undulator.

This expression is given by the formula (1.131) where Φ should be substituted by

$$\Phi \to \Phi_N = \frac{1}{2\pi N} \frac{\sin^2(N\varphi_0/2)}{\sin^2(\varphi_0/2)} \tag{1.148}$$

It is easy to see that at $N \to \infty$

$$\Phi_N \to \sum_{k=1}^{\infty} \delta(\varphi_0 - 2\pi k) = \Phi \tag{1.149}$$

and we return to the function Φ, defined in (1.131). The distributions (1.132) and (1.134) for finite undulator can also be found with the substitution (1.148). By integrating the angular distribution we obtain the spectral distribution of radiation intensity in the finite undulator

$$\begin{aligned} \langle dI_\omega^N \rangle &= 64\pi^2 e^2 |\mathbf{w}|^2 \gamma^4 \xi d\xi \int_{2\pi\xi}^{\infty} \frac{d\varphi_0}{\varphi_0^4} \\ &\times \left(1 - \frac{4\pi\xi}{\varphi_0} + \frac{8\pi^2\xi^2}{\varphi_0^2}\right) F\Phi_N\,, \end{aligned} \tag{1.150}$$

where the values are defined by (1.137) and (1.148).

Let us give an alternative derivation of formula for the spectral distribution of the radiation in the finite undulator. Substituting the field value (1.124) into the expression for the radiated energy (1.20) and performing the integration over the azimuthal angle with an account of $\mathbf{v} \cdot \mathbf{w} = 0$, we get the general expression for spectral distribution of radiated energy in Case 2 (compare Landau and Lifshits (1973), p. 277).

$$d\varepsilon_\omega = e^2\gamma^2 \frac{d\omega}{4\pi} \int_{1/2}^{\infty} \frac{dy}{y^4}(1 - 2y + 2y^2)|\mathbf{w}_{\tilde{\omega}}|^2\,. \tag{1.151}$$

This expression can be obtained from formula (1.127) if the summation is done over polarizations and integration is performed over the azimuthal angle. If one uses formula (1.151) in the case of the undulator with a finite number of periods (see (1.146)) and considering the intensity (1.147), then we get the general expression for the spectral distribution of intensity

$$\begin{aligned} \langle dI_\omega^N \rangle &= 2\pi e^2\gamma^4 \frac{d\xi}{T^2} \int_{1/2}^{\infty} \frac{dy}{y^4}(1 - 2y + 2y^2) \\ &\times |\mathbf{w}_{\tilde{\omega}}^T|^2 \Phi_N\,, \end{aligned} \tag{1.152}$$

which can also be obtained from formula (1.150) if one substitutes F (1.135) and takes into account that $\varphi_0 = 4\pi y\xi$. At $N \to \infty$ we have the sum of harmonics with $\varphi_0 = 2\pi k, k = 1, 2, \ldots$ see (1.149), then a strict connection between the angle and frequency arises and at a given angle the set of frequencies with a different k is emitted. In the case of finite N such a connection does not already exist and at a given angle the radiation is emitted within some frequency interval dependent on N.

1.10. Examples of Radiation at Quasiperiodical Motion in the Dipole Case

(a) Radiation from flat harmonic motion. Let us consider particles travelling in a periodic magnetic field directed along axis y and varying while shifting along z according to the law

$$H_y = H_0 \sin \nu z\,, \tag{1.153}$$

all the other components of the field are equal to zero. The covariant equations of motion in field (1.153) have the form

$$\begin{aligned} du^x/ds &= -\Omega u^z \sin \nu z\,, \\ du^z/ds &= \Omega u^x \sin \nu z\,, \\ du^y/ds &= du^0/ds = 0\,, \end{aligned} \tag{1.154}$$

where $\Omega = eH_0/m, u^x = dx/ds, u^z = dz/ds, s$ is the proper time. Without loss of generality one can take that a particle is moving in plane (x, z), i.e. $y = 0$. In the case when the mean transverse (along x-axis) velocity is equal to zero, from (1.154) we have the following set of equations

$$\begin{aligned} \frac{dx}{dt} &= -\frac{\Omega}{\nu\gamma}\cos \nu z\,, \\ \frac{d^2 z}{dt^2} &= \frac{\Omega^2}{2\nu\gamma^2}\sin 2\nu z \end{aligned} \tag{1.155}$$

where $\gamma = \varepsilon/m$, ε is the particle's energy, with initial condition $\mathbf{v}(0) = \mathbf{v}_0, \mathbf{r}(0) = 0$. In the case when $\Omega/(\nu\gamma) \ll 1$, the transverse motion is nonrelativistic so that for the consideration of radiation one can use the dipole approximation. Then the second equation (1.155) can be solved with the method of successive approximation. In the first approximation $dz/dt = v_0, z = v_0 t$. Then for transverse motion we have

$$x(t) = -\frac{\Omega}{\gamma \nu^2 v_0} \sin(\nu v_0 t) . \tag{1.156}$$

We are interested in the radiation of ultrarelativistic particles, then $v_0 \simeq 1 - 1/(2\gamma^2) \approx 1$. By substituting the transverse acceleration $w_x(t)$ into formula (1.130), we have for the case under consideration

$$w_{x\tilde{\omega}}^T = -2i\frac{\Omega}{\gamma} e^{i\varphi_0/2} \sin\left(\frac{\varphi_0}{2}\right) \frac{\nu}{\tilde{\omega}^2 - \nu^2} \tag{1.157}$$

Using this result and Φ_N (1.148) we have from (1.152) the intensity spectral distribution in a finite undulator for the case of harmonic motion

$$\langle dI_\omega^h \rangle = \frac{e^4 H_0^2}{\pi m^2} \gamma^2 d\xi \int_{1/2}^{\infty} \frac{dy}{y^4} (1 - 2y + 2y^2) \frac{\sin^2(2\pi N \xi y)}{\pi N[(2\xi y)^2 - 1]^2} . \tag{1.158}$$

At $N \to \infty$, using the known formula

$$\lim_{N \to \infty} \frac{\sin^2(2\pi N \xi y)}{\pi N[(2\xi y)^2 - 1]^2} = \frac{\pi}{4} \delta(2\xi y - 1) \tag{1.159}$$

we have from (1.158) for the spectrum in an infinite undulator

$$\langle dI_\omega^h \rangle = (e^4 H_0^2 \gamma^2 / m^2) \xi d\xi (1 - 2\xi + 2\xi^2) , \quad \xi \leq 1 . \tag{1.160}$$

This expression can be easily obtained from (1.136) if in F (1.135) one substitutes the explicit expression for w_ω^T (1.157). The plots of function $\langle dI_\omega^h \rangle / d\xi$ (1.158) at various N are given in Fig. 1.9. It is seen that already at $N = 2$ the form of radiation spectrum is fairly close to that at $N \to \infty$, where the spectrum is concentrated within the frequency interval $0 \leq \xi \leq 1$.

Let us now pass to the polarization of the radiation. As already shown above in the analysis after Eq. (1.143), there is no circular polarization in the case under consideration. At $\beta = 0$ we have from (1.143)

$$\xi_1 = \frac{(2y-1) g_0 \sin 2\varphi_1}{2y^2 - 2y - g_0}, \quad \xi_3 = \frac{2y(1-y) + (2y-1) g_0 \cos 2\varphi_1}{2y^2 - 2y - g_0}, \tag{1.161}$$
$$g_0 = (2y-1)\cos 2\varphi_1 - 1 ,$$

i.e. radiation is linearly polarized at an angle φ_p with respect to the axis x (see (A.34)), $\tan 2\varphi_p = \xi_1/\xi_3$.

After integration over the azimuthal angle and the frequency of the radiation the polarization density matrix has an especially simple form:

$$\langle dI_{ij}\rangle = \frac{e^2\gamma^2\Omega^2}{16}\frac{dy}{y^5}\left[(2y^2 - 2y + 1)\delta_{ij} + \frac{\tau^3_{ij}}{2}\right],$$

which is universal and has the same form for the finite undulator with any N, so that $\xi_3 = \frac{1}{2(2y^2-2y+1)}$, see Eq. (1.144a) for $\beta = 0$. After the integration over the radiation polar angle, we find for the polarization density matrix

$$\langle dI_{ij}\rangle = \frac{e^2\gamma^2\Omega^2}{8}\left[\frac{4}{3}\delta_{ij} + \tau^3_{ij}\right], \quad \xi_3 = \frac{3}{4},$$

compared with (1.144b), where $S = S_1 = 0$.

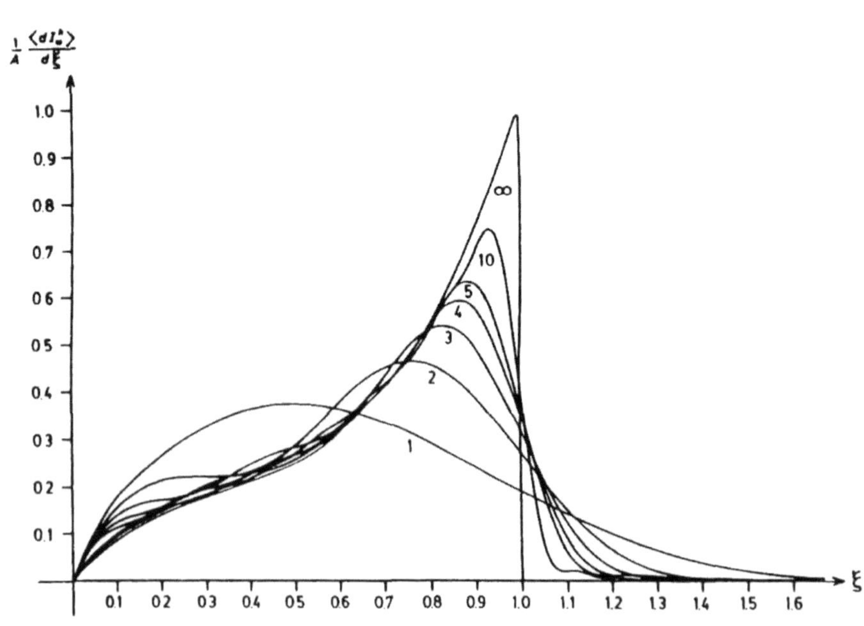

Fig. 1.9. Spectral distribution of the intensity in the undulator with harmonic motion for a different number of motion periods N (numbers near curves), $A = e^4H_0^2\gamma^2/m^2, \omega = \xi\omega_c$.

(b) Radiation at motion along the helix (in a helical undulator). Let us consider a particle travelling in a helical magnetic field

$$\begin{aligned} F^{13} &= -F^{31} = H_y = H_0 \sin \nu z \\ F^{23} &= -F^{32} = H_x = H_0 \cos \nu z \, . \end{aligned} \tag{1.162}$$

The covariant equations of motion have the form (compare with (1.154))

$$\begin{aligned} &\frac{du^0}{ds} = 0\,, \quad \frac{du^x}{ds} = -\Omega u^z \sin \nu z\,, \\ &\frac{du^y}{ds} = -\Omega u^z \cos \nu z\,, \\ &\frac{du^z}{ds} = \Omega(u^x \sin \nu z + u^y \cos \nu z)\,. \end{aligned} \tag{1.163}$$

If the mean transverse velocity (with respect to the z-axis) is equal to zero, then from (1.163) it follows that (compare (1.155))

$$\begin{aligned} u^x/\gamma &= dx/dt = (\Omega/\nu\gamma) \cos \nu z\,; \\ u^y/\gamma &= dy/dt = -(\Omega/\nu\gamma) \sin \nu z\,. \end{aligned} \tag{1.164}$$

Substituting the transverse velocities into equation for u^z we have $du^z/ds = 0, dz/dt = v_0, z = v_0 t$. By using the expression for z obtained and (1.164) we have explicit expressions for transverse velocities (as well as for coordinates and accelerations). By substituting transverse accelerations into formula (1.130) we have

$$\begin{aligned} &w^T_{x\tilde{\omega}} = ig\nu\,, \quad w^T_{y\tilde{\omega}} = -g\tilde{\omega} \\ &g = (2\Omega/\gamma)\exp(i\varphi_0/2)\sin(\varphi_0/2)/(\tilde{\omega}^2 - \nu^2)\,. \end{aligned} \tag{1.165}$$

Using this result and the expression for Φ_N (1.148) we get from (1.152) the spectral distribution of intensity in a finite helical undulator

$$\begin{aligned} \langle dI^{he}_\omega \rangle = &\frac{e^4 H_0^2 \gamma^2}{\pi m^2} d\xi \int_{1/2}^{\infty} \frac{dy}{y^4}(1 - 2y + 2y^2) \\ &\times (1 + (2\xi y)^2) \frac{1}{\pi N} \frac{\sin^2(2\pi N \xi y)}{[1 - (2\xi y)^2]^2} \end{aligned} \tag{1.166}$$

At $N \to \infty$ using the formula (1.159) we have from (1.166)

$$\begin{aligned} \langle dI^{he}_\omega \rangle &= (2e^4 H_0^2 \gamma^2/m^2)\xi d\xi(1 - 2\xi + 2\xi^2) \\ &= 2\langle dI^h_\omega \rangle \end{aligned} \tag{1.167}$$

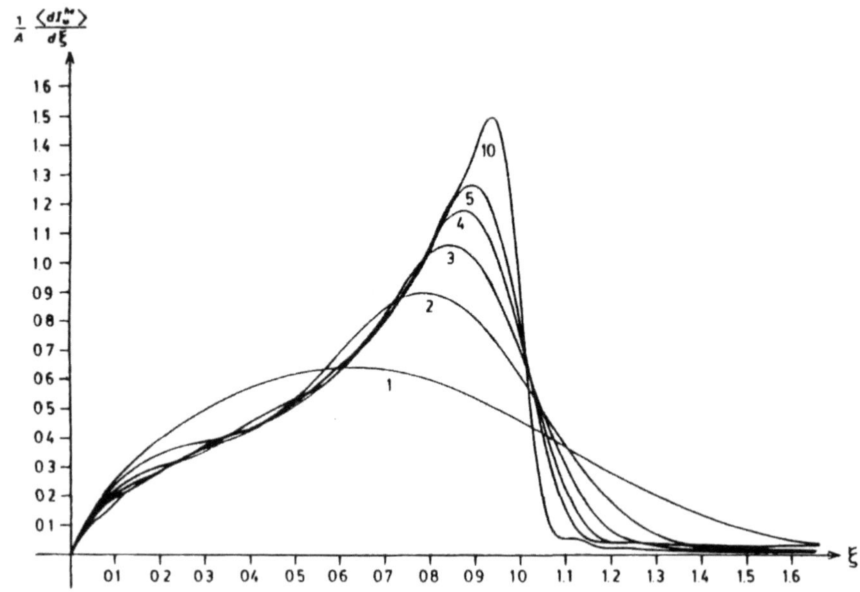

Fig. 1.10. Spectral distribution of the radiation intensity in the undulator with helicoidal magnetic field for a different number of motion periods (numbers near curves), $A = e^4 H_0^2 \gamma^2/m^2, \omega = \xi\omega_c$.

This expression can be easily obtained from (1.136) by substitution of the explicit form of $\mathbf{w}_{\bar{\omega}}^T$ (1.165). The plots of function $\langle dI_\omega^{he}\rangle/d\xi$ for various values of N are given in Fig. 1.10. As in the case of flat harmonic undulator it is seen that even at small N the spectrum form does not differ strongly from the spectrum at $N \to \infty$.

Comparing the Fourier components of acceleration (1.165) with (1.133) we have

$$\varphi = \pi/2, \quad \sin\beta = \bar{\omega}/\sqrt{\nu^2 + \bar{\omega}^2}, \quad \cos\beta = \nu/\sqrt{\nu^2 + \bar{\omega}^2} \tag{1.168}$$

Substituting these angles into (1.142)-(1.145) we find explicit expressions for the degree of polarization. At $N \to \infty$ we have that $\bar{\omega}/\nu = 2y\xi \to 1$ and $\beta \to \pi/4$.

Then it follows from (1.142)

$$\begin{aligned} a_1 &= -(2y-1)\sin 2\varphi_1, \quad a_2 = 2y(1-y), \\ a_3 &= -(2y-1)\cos 2\varphi_1, \quad c = 2y^2 - 2y + 1. \end{aligned} \tag{1.169}$$

Under these conditions we have $S = 1$ (see (1.145)) and after integration over the azimuthal angle of the radiation φ_1 we obtain $a_1 = a_3 = 0$, $a_2 = 2y(1-y)$. This means that after this integration the radiation is partially circularly polarized. In the forward direction ($y = 1/2$) the radiation is completely circularly polarized. Finally, after the integration over the polar angle of the radiation we find $\xi_1 = \xi_3 = 0$, $\xi_2 = 1/2$ (compare with (1.144b)).

(c) Radiation in a structure with sections of homogeneous magnetic field. We consider two more specified case when the periodic structure consists of sections with a homogeneous magnetic field of different directions. The angle of rotation of a particle in a field of given direction is $\psi = |\mathbf{w}|T \ll 1/\gamma$, the z-axis is directed along $\mathbf{v}$, the x-axis along $\mathbf{l}_1$, and the y-axis along $\mathbf{l}_2$.

Case A:

$$\begin{aligned} &H_y = H\,, \quad 0 < z < l/2\,, \quad H_y = -H\,, \quad l/2 < z < l\,; \\ &H_x = H_z = 0\,. \end{aligned} \tag{1.170}$$

In this case, we have

$$F_A(\varphi_0 = 2\pi k) = 4\sin^4(\pi k/2)\,, \quad S = 0\,. \tag{1.171}$$

Substituting F_A in Eqs. (1.134)–(1.138) we obtain the spectrum of the emitted radiation. Figure 1.11(a) shows the graph of the function $2\langle I\rangle^{-1}\langle dI\rangle_A/d\xi$, where $\xi = \omega/w_c$ and $\omega_c = 4\pi\gamma^2/T(l = v^T)$.

The radiation is partially and linearly ($|\xi| = 3/4$) polarized along the direction of acceleration and there is no circular polarization as it was explained above. However, the circular polarization can be obtained in a magnetic field of more complex configuration, discussed below.

Case B:

$$\begin{aligned} &H_y = -H\,, \quad 0 < z < l/8, 5l/8 < z < 3l/4\,; \\ &H_y = H\,, \quad l/8 < z < l/4, l/2 < z < 5l/8\,; \\ &H_x = -H\,, \quad l/4 < z < 3l/8, 7l/8 < z < l\,; \\ &H_x = H\,, \quad 3l/8 < z < l/2, 3l/4 < z < 7l/8\,. \end{aligned} \tag{1.172}$$

In this case, we find that $\beta = \pi/4, \varphi = -\varphi_0/4$ and

$$\begin{aligned} F_B(\varphi_0 = 2\pi k) &= 32\sin^2(\pi k/2)\sin^4(\pi k/8)\,, \\ S &= -(8/\pi^2)[3G - 4G_1] = 0.214\ldots\,. \end{aligned} \tag{1.173}$$

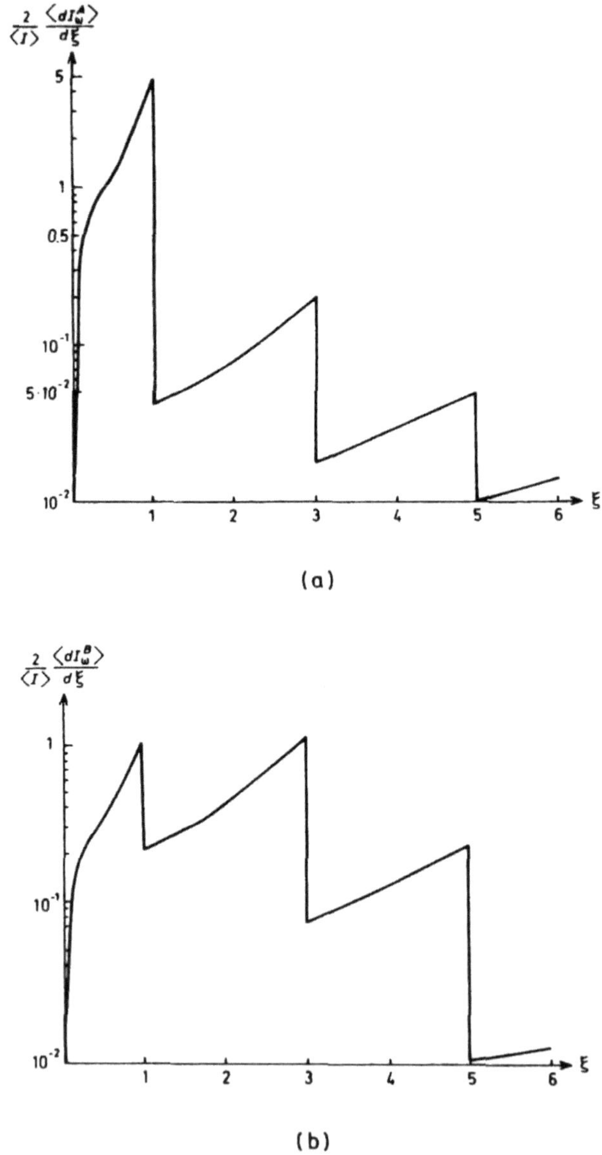

Fig. 1.11. Spectral distribution of the radiation intensity in an infinite undulator which is constructed from pieces with homogeneous magnetic field: *a* is for case A and *b* is for case B.

Here,

$$G = \sum_{k=0}^{\infty} (-1)^k/(2k+1)^2 = 0.916\ldots$$

$$G_1 = \sum_{k=0}^{\infty} (-1)^k \cos((2k+1)\pi/4)/(2k+1)^2 = 0.753\ldots.$$

If we substitute F_B in Eqs. (1.134)–(1.138), we find the spectrum of the emitted radiation, which is plotted in Fig. 1.11(b) (the units are the same as in Case A).

The radiation in this case is partially (elliptically) polarized (see Eq. (1.142)), and after integration over the azimuthal angle of the radiation φ_1 we obtain $S_1 = 0, \xi_1 = \xi_3 = 0$ and ξ_2 is given by (1.144a). This means that after this integration the radiation is partially circularly polarized. The same is true for polarization of the radiation as a whole (1.144b).

Since the motion in both Cases A and B, is not a harmonic one, the higher harmonics ($\xi > 1$) give appreciable contributions in the spectral distribution of radiation seen in Fig. 1.11. In this subsection we have considered radiation in an infinite undulator. The transition to finite undulator can be done according to (1.148).

1.11. Spectral and Angular Distribution of the Radiation in Quasiperiodic Motion* (the General Case)

We start from the general expression for the radiated energy (1.23). We represent the integral contained in (1.23) in the form

$$\int_{-\infty}^{\infty} v^\mu(t) e^{ikx(t)} dt = v^{T\mu} \sum_{m=-\infty}^{\infty} e^{im\varphi_0} = \sum_n 2\pi v^{T\mu} \delta(\varphi_0 - 2\pi n), \qquad (1.174)$$

where

$$\varphi_0 = \omega T(1 - \mathbf{n}\cdot\mathbf{V}), v^{T\mu} = \int_0^T v^\mu(t) e^{ikx(t)} dt, T = \frac{2\pi}{\omega_0}. \qquad (1.175)$$

Here T is the period of the motion, for other notations see (1.118). Substituting (1.174) in (1.23), we obtain for the intensity of the radiation per unit time

*Secs. 1.11 and 1.12 are based on the paper by Baier, Katkov, Strakhovenko (1981a).

(compare with (1.61))

$$dI = e^2 \frac{d^3k}{4\pi T} \sum_{n=1}^{\infty} \delta(\varphi_0 - 2\pi n)\{2(|\mathbf{v}^T|^2 - |v^{T0}|^2)\}\,, \qquad (1.176)$$
$$v^\mu = (v^0, \mathbf{v}) = u^\mu m/\varepsilon\,.$$

Using (1.119), introduce the notation

$$\Delta(t) = \omega_0 \int_0^t (\mathbf{v}_\perp^2(t') - \overline{\mathbf{v}_\perp^2})dt'\,, \qquad (1.177)$$

where $\omega_0 = 2\pi/T$, and ignoring terms of higher order in $1/\gamma^2$ and $v_\perp^2$, we obtain the following expression for the spectral and angular distributions of the radiation:

$$dI = e^2 \frac{d^3k}{(2\pi)^2} \frac{1}{\omega_0\gamma^2} \sum_{n=1}^{\infty} \delta(\varphi_0 - 2\pi n)[-|I_0|^2 + \gamma^2(|\mathbf{I}_\perp|^2 - \mathrm{Re}\, I_0^* I_\parallel)]\,, \quad (1.178)$$

where

$$I_\parallel = \int_0^{2\pi} \mathbf{v}_\perp^2(\psi)e^{if(\psi)}d\psi\,, \quad I_0 = \int_0^{2\pi} e^{if(\psi)}d\psi\,,$$
$$\mathbf{I}_\perp = \int_0^{2\pi} \mathbf{v}_\perp(\psi)e^{if(\psi)}d\psi\,; \quad \psi = \omega_0 t\,, \qquad (1.179)$$
$$f(\psi) = n\psi + \omega(\Delta(\psi)/2\omega_0 - \mathbf{n}\cdot\mathbf{x}_\perp(\psi))\,.$$

Here we have considered the case $\varphi_0 = \omega T(1 - \mathbf{n}\ \ \mathbf{V})$, and the frequency of the emitted radiation is determined by the relation (cf. Eq. (1.120))

$$\omega = \frac{2\gamma^2\omega_0 n}{1 + \gamma^2\vartheta^2 + \rho/2}\,, \qquad (1.180)$$

where ϑ is the angle of emission with respect to $\mathbf{V}$, and $\rho = 2\gamma^2\bar{v}_\perp^2$. The nature of the radiation depends strongly on ρ: for $\rho \ll 1$, we have the dipole radiation, for $\rho \gg 1$ the radiation is of the magnetic bremsstrahlung nature.

For calculations of the polarization characteristics of the radiation the same procedure should be carried out starting from Eq. (1.25).

The integrals in (1.179) have a relatively simple form only for the circular transverse motion, when $|\mathbf{v}_\perp| = \text{const}$. In this case $\Delta(\psi) = 0$,

$$f(\psi) = n\psi - \kappa(\psi - \varphi), \quad I_\parallel = v_\perp^2 I_0,$$
$$|I_0|^2 = (2\pi)^2 J_n^2(\kappa), \quad \kappa = \omega v_\perp \vartheta/\omega_0, \tag{1.181}$$
$$|\mathbf{I}_\perp|^2 = (2\pi)^2 v_\perp^2 [J_{n+1}^2(\kappa) + J_{n-1}^2(\kappa)],$$

where $J_n(k)$ is the Bessel function and φ is the azimuthal angle of emission of the radiation. Substituting these results in (1.178), we obtain the differential distribution of the radiation in the case of the circular motion of the particle (index cr)

$$dI_{cr}^{(i)} = \frac{e^2}{\gamma^2}\frac{d^3k}{\omega_0}\sum_{n=1}^{\infty}\delta(\varphi_0 - 2\pi n)G_{cr}^{(i)},$$
$$G_{cr}^{(0)} = -J_n^2(\kappa) + \frac{\rho}{4}(J_{n-1}^2(\kappa) + J_{n+1}^2(\kappa) - 2J_n^2(\kappa)),$$
$$G_{cr}^{(1)} = -A\sin 2\varphi, G_{cr}^{(3)} = -A\cos 2\varphi, \tag{1.182}$$
$$A = \left(1 + \frac{\rho}{2}\right) J_n^2(\kappa) - \frac{\rho}{2} J_{n+1}(\kappa)J_{n-1}(\kappa),$$
$$G_{cr}^{(2)} = \left(\frac{n\rho}{2\kappa} - \frac{\kappa}{2\xi}\right) J_n(\kappa)(J_{n-1}(\kappa) - J_{n+1}(\kappa)).$$

Here and below, the index 0 is used for the expressions summed over the polarization of the radiation,

$$\xi^{(i)} = \frac{dI^{(i)}}{dI^{(0)}}$$

are the Stokes parameters, and

$$\omega_c = \frac{2\gamma^2\omega_0}{1+\rho/2}, \quad \xi = \frac{\omega}{2\gamma^2\omega_0}$$

For a description of the polarization of radiation one have to define two orthogonal unit vectors in the plane transverse to vector $\mathbf{n}$:

$$\mathbf{e}_{(1)} = \frac{\mathbf{n}\times\mathbf{V}}{|\mathbf{n}\times\mathbf{V}|}, \quad \mathbf{e}_{(2)} = n \times \mathbf{e}_{(1)}.$$

The vector $\mathbf{e}_{(1)}$ is orthogonal to the radiation plane formed by vectors $\mathbf{V}$ and $\mathbf{n}$, the vector $\mathbf{e}_{(2)}$ is lying in this plane. Another possibility is to connect these unit vectors with dynamical vector characteristics, e.g. with vector of acceleration as it was done above (see Subsecs. 1.6 and 1.9). In the problem under consideration there are no constant vectors in the transverse plane. However, one can introduce them using vectors $\mathbf{e}_{(1)}$ and $\mathbf{e}_{(2)}$, e.g. $\mathbf{e}_x = \mathbf{e}_{(2)} \cos\varphi + \mathbf{e}_{(1)} \sin\varphi$ and $\mathbf{e}_y = -\mathbf{e}_{(2)} \sin\varphi + \mathbf{e}_{(1)} \cos\varphi$ so that the transverse velocity is $\frac{\mathbf{v}_\perp}{|\mathbf{v}_\perp|} = \mathbf{e}_x \cos\omega_0 t + \mathbf{e}_y \sin\omega_0 t$. In Eq. (1.182) we used just vectors $\mathbf{e}_x$ and $\mathbf{e}_y$. It is evident that after integration over φ the linear polarization vanishes. If one uses vectors $\mathbf{e}_{(1)}$ and $\mathbf{e}_{(2)}$ then $G_{cr}^{(1)} = 0$ and $G_{cr}^{(3)} = A$ (compare with (A.34)).

In the case of plane motion of the particle, we can, without loss of generality, take $v_x(\psi)$ to be an even function of $\psi(v_y = 0)$; then $f(\psi)$ in the expression (1.179) becomes an odd function of ψ. Thus, with an accuracy to a common phase factor, the quantities I_0, I_x, and $I_\parallel$ are real. This means that for plane quasiperiodic motion the radiation can only be linearly polarized. From the physical point of view, this fact is obvious, since for the motion there is no pseudovector with which circular polarization of the radiation could be associated.

With some allowance for what we have said above, we obtain for plane motion (subscript p)

$$
\begin{aligned}
dI_p^{(i)} &= \frac{e^2}{\omega_0} \frac{d^3k}{(2\pi)^2} \sum_{n=1}^{\infty} \delta(\varphi_0 - 2\pi n) G_p^{(i)} \,, \\
G_p^{(0)} &= -\frac{I_0^2}{\gamma^2} + I_x^2 - I_0 I_\parallel \,, \\
G_p^{(1)} &= 2I_0\vartheta \sin\varphi (I_0\vartheta\cos\varphi - I_x) \,, \quad G_p^{(2)} = 0 \,, \\
G_p^{(3)} &= -(I_0\vartheta\sin\varphi)^2 + (I_0\vartheta\cos\varphi - I_x)^2
\end{aligned}
\tag{1.183}
$$

After integrating the azimuthal emission angle of the radiation $G_p^{(1)} \to 0$.

Let us take into consideration that, in the frame of the operator quasiclassical method, which will be presented in the next section, the same problems could be solved as in classical radiation theory. The formulae with the exact allowance of quantum properties of the radiation do not appear more complicated than as presented above. Therefore, to avoid unnecessary repetition, the presentation of the radiation theory in the quasiperiodic motion will be done in Sec. 5 (Subsecs. 5.2 and 5.3). The results of classical theory can be obtained from the general formulae of these sections putting $\omega/\varepsilon \to 0 (u \to 0)$.

1.12. Angular Distribution of the Radiation in Quasiperiodic Motion (the General Case)

The angular distribution of the energy radiated and the corresponding polarization properties of the radiation may be obtained from the general formula (1.25). In the ultrarelativistic limit, integrating (1.25) over the frequency ω one finds

$$d\varepsilon_{ij} = \frac{4ie^2\gamma^2 d^2\eta}{\pi^2}\int dt_1 \int dt_2 \frac{[\eta - \gamma v_\perp(t_1)]_i[\eta - \gamma v_\perp(t_2)]_j}{(\tau - i0)^3\left[\Omega - 2\eta \mathbf{v}_\perp(t)\gamma\frac{\sin(\omega_0\tau/2)}{(\omega_0\tau/2)}\right]^3}, \quad (1.184)$$

where $\tau = t_2 - t_1, t = (t_1 + t_2)/2, \boldsymbol{\eta} = \gamma\mathbf{u}, \mathbf{u}$ is the 2-dimensional vector introduced in (1.140), $|\mathbf{u}| = \vartheta(\mathbf{n}\bar{\mathbf{v}} = \bar{v}\cos\vartheta)$, $\Omega = 1 + \rho/2 + \eta^2, \rho = 2\gamma^2\overline{v_\perp^2}$, the contour around the point $\tau = 0$ (the sign of the imaginary part) follows from the convergence condition of the integral over ω.

In the case of the helical motion of the radiating particle: $\frac{\mathbf{v}_\perp}{|\mathbf{v}_\perp|} = \mathbf{e}_x \cos\omega_0 t + \mathbf{e}_y \sin\omega_0 t$, we find for the radiation intensity after substitution of mentioned formulae into (1.184), integration over τ (because integrand is an odd function of τ, the integral over τ may be taken as a half-residue in the point $\tau = 0$) and subsequent integration over the azimuthal angle of radiation*

$$dI^{(0)} = \left\langle\frac{d(\varepsilon_{11}+\varepsilon_{22})}{dt}\right\rangle = 4e^2\rho\omega_0^2\gamma^2\eta d\eta\frac{B^{(0)}}{(\Omega^2 - 2\rho\eta^2)^{5/2}},$$
$$B^{(0)} = \tfrac{\Omega^2}{2} + \left(\tfrac{\rho}{2} - 1\right)\eta^2 - \tfrac{5\rho\eta^4}{2(\Omega^2 - 2\rho\eta^2)} \quad (1.185)$$

Here, $\langle\cdots\rangle$ means averaging over the period of the motion, as in the preceding subsections, the index 0 is used for the total intensity of the radiation, and the remaining indices are used for the quantities needed to find the Stokes parameters $\xi^{(i)} = dI^{(i)}/dI^{(0)}$. Note, that after integration over the azimuthal angle $dI^{(1)} = dI^{(3)} = 0$ (compare with discussion after Eq. (1.182)).

---

*For integration of the azimuthal angle it is convenient to use integral (Gradstein, Ryzhik (1962), p. 397):

$$\int_0^\pi \frac{d\varphi}{(a + b\cos\varphi)^n} = \frac{\pi}{(a^2 - b^2)^{n/2}}P_{n-1}\left(\frac{a}{\sqrt{a^2 - b^2}}\right)$$

For $dI^{(2)}$, we have

$$dI^{(2)} = \frac{4}{\pi} e^2 \rho \omega_0^2 \gamma^2 \eta d\eta \int_0^\infty dt$$
$$\times \left[\frac{-(1/2)\sin 2t(\rho\eta^2 \sin^2 t + t^2\Omega^2) + 3\Omega t \eta^2 \sin^2 t}{(t^2\Omega^2 - 2\rho\eta^2 \sin^2 t)^{5/2}} \right. \tag{1.186}$$
$$\left. + \frac{(1+\rho/2)^2 + \eta^2(\rho/2 - 1 - 2\eta^2)}{t^2(\Omega^2 - 2\rho\eta^2)^{5/2}} \right]$$

Note that sign of $\xi^{(2)}$ depends on the direction of the rotation of the particle in a helix (clockwise or counter-clockwise).

Integrating (1.185) and (1.186) over η, we obtain for the total intensity of the radiation and Stokes parameter $\xi^{(2)}$

$$I = \frac{2}{3} e^2 \frac{\rho}{2} \omega_0^2 \gamma^2 , \quad \xi^{(2)} = -\frac{3}{\pi} \int_0^\infty dt \frac{\sin t(t\cos t - \sin t)}{[t^2(1+\frac{\rho}{2}) - \frac{\rho}{2}\sin^2 t]^2} \tag{1.187}$$

For $\rho \ll 1$, the expressions (1.185)–(1.187) go over into the expressions of the dipole approximation, particularly $dI^{(0)}$ (1.185) goes over into (1.136), for intensities integrated over angle we obtain for $\xi^{(2)}$ (1.144) with $S = 1$. In the case $\rho \gg 1$, we obtain the corresponding results for the magnetic bremsstrahlung limit (see Subsecs. 1.2, 1.3). In this case one should take into

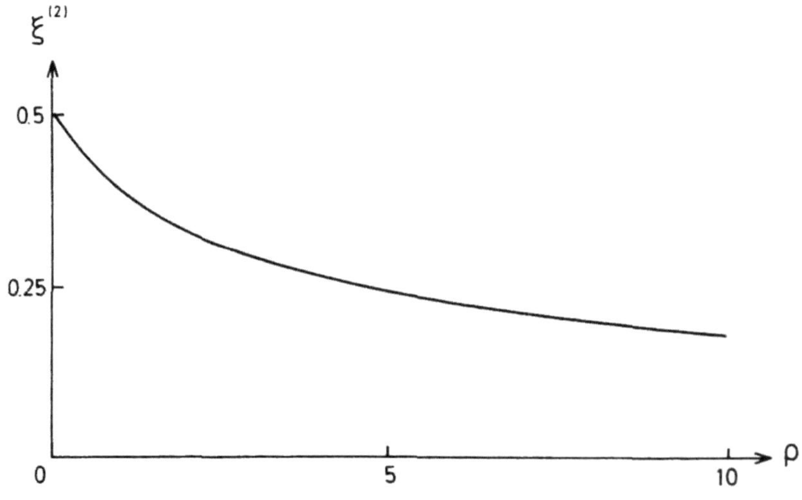

Fig. 1.12. Dependence of degree of the circular polarization on parameter ρ.

account, that $\vartheta_0 \simeq |v_\perp| = \sqrt{\frac{\varrho}{2}}/\gamma \gg 1/\gamma$. Then the radiation is concentrated on the surface of the cone with an angle $\vartheta_0 \simeq \sqrt{\frac{\varrho}{2}}/\gamma$ and its angular width is $(\vartheta - \vartheta_0)^2 \sim \beta^2 \sim 1/\gamma^2$ Taking this into account we go from the total intensity $dI^{(0)}$ (1.185) to the expression (1.37). The degree of circular polarization of the radiation $\xi^{(2)}$ vanishes in the limit $\rho \to \infty$, but the function $\xi^{(2)}(\rho)$ with increasing ρ tends to zero rather slowly. The dependence of $\xi^{(2)}(\rho)$ on ρ given by (1.187) is shown in Fig. 1.12.

In the plane case for sinusoidal transverse motion of the particle (compare with (1.156)) the angular distribution of the radiation intensity is

$$\frac{dI}{d\varphi d\eta} = \frac{2e^2}{\pi}\rho\omega_0^2\gamma^2\eta\int_0^{2\pi}\frac{d\psi}{2\pi}\sin^2\psi\left[\frac{1}{F^3} - \frac{4}{F^4} + \frac{4(1+\eta^2\sin^2\varphi)}{F^5}\right], \tag{1.188}$$
$$F = 1 + \eta^2 + \rho\cos^2\psi - 2\sqrt{\rho}\eta\cos\psi\cos\varphi\,.$$

The integrals over ψ in (1.188) can be calculated by means of the relations

$$\frac{1}{2\pi}\int_0^{2\pi}\frac{\sin^2\psi}{F^k}d\psi = \frac{(-1)^{k-1}}{(k-1)!}\frac{d^{k-1}}{dx^{k-1}}f(x)\bigg|_{x=1},$$
$$f(x) = \frac{1}{\rho}\left\{\left[\frac{h_+ + (h_+^2 - 4\rho\eta^2\cos^2\varphi)^{1/2}}{h_- + (h_+^2 - 4\rho\eta^2\cos^2\varphi)^{1/2}}\right]^{1/2} - 1\right\}, \tag{1.189}$$
$$h_\pm = x + \eta^2 \pm \rho\,.$$

In the dipole case $\rho \ll 1$ and in (1.188) we can set $F \approx 1 + \eta^2$; then the integration of ψ in (1.188) becomes trivial. At the same time, for both plane (1.188) and circular (1.185) motions the main contribution to the radiation is given by angles $\vartheta \sim \gamma^{-1} (\eta \sim 1)$ and arbitrary angles φ. For relativistic transverse motion $\rho \gg 1$, and the main contribution to the radiation in the case of plane motion is given by the region of angles $\vartheta \leq \rho^{1/2}/\gamma$, $|\sin\varphi| \leq \rho^{-1/2}$, i.e. the radiation is concentrated in the plane of the motion.

2. THE QUASICLASSICAL METHOD OF DESCRIBING THE RADIATION OF HIGH-ENERGY PARTICLES

2.1. Types of Quantum Effects in the Radiation Problem

In the quantum theoretical description* of particles in an external electromagetic field, the effects of the interaction of the particles with the radiation

*There are many books devoted to quantum field theory. Here we mention some of them: Berestetskii, Lifshits and Pitaevskii (1982), Akhiezer and Berestetskii (1963), Bogolubov and Shirkov (1959), Bjorken, Drell (1965).

field are usually calculated in the Furry representation without using the perturbation theory with respect to the external field. This is constructed in accordance with the theory of free paticles using the solutions of the corresponding wave equations (Dirac, Klein–Gordon, etc.) in the given field. In particular, the S matrix in spinor electrodynamics has the following form in the Furry representation:

$$S = T\exp\left[-\left(\frac{ie}{\hbar}\right)\int d^4x\bar{\Psi}_F(x)\gamma^{\mu}\Psi_F(x)A_{\mu}(x)\right], \tag{2.1}$$

where $\Psi_F(x)$ is the operator of the electron–positron field in the given external field.* The expansion of the T product as a sum of the normal products is formulated in the same way as for free particles, but the contraction of the fermion operators will now be the Green function in the given field. It should be borne in mind that the solution of wave equations is known only for some specific fields (the Coulomb field, the constant and homogeneous magnetic field, the plane electromagnetic wave field, the constant crossed electrical and magnetic fields). But even for the fields cited, the use of solutions of wave equations for calculating the characteristics of radiation is a sufficiently complex problem from the mathematical point of view. However, in the case of high energy particles ($\gamma \gg 1$) the most essential factor is that all the results may be obtained using the asymptotic approximations of the functions found, and this corresponds to the transition to the quasiclassical approximation.

From the above it follows that a more consistent approach to the problem of radiation of the high-energy particle in an external electromagnetic field will be a formalism which uses a quasiclassical approach from the outset. Such an approach was proposed by two of the authors (Baier, Katkov, (1967(b), (c); 1968). Its basis is that there are two types of quantum effects at the radiation of the high-energy particle in the external field. The first is associated with the quantization of the motion of the particle in this field. The associated noncommutability of the dynamic variables of the particle was investigated in detail in Appendix B.2, where it was shown on the basis of the example of motion in an arbitrary magnetic field that it has the order $\hbar\omega_0/\varepsilon = H/(H_0\gamma^2)$ (in the general case the series over $\hbar\omega_0/\varepsilon$), with the result that with the energy rise the motion becomes increasingly classical.†

---

*We use unit system $c = 1$ in Secs. 2–4.

†The mass operator calculated exactly in α-order in a constant and homogeneous external field gives the complete solution of the radiation problem (see, Schwinger, 1973; Tsai, Yildiz

We shall rewrite the expansion parameter in the form $\hbar\omega_0/\varepsilon = \hbar/(\varepsilon\rho)$, where ρ is the characteristic dimension of the problem; $\hbar/(\varepsilon\rho) = \lambda\!\!\!^{-}_c/(\rho\gamma)$, $\lambda\!\!\!^{-}_c = \hbar/m$ is the Compton wavelength. Taking into account that $\varepsilon\rho \sim \hbar l$ is the angular momentum of the particle, we obtain

$$\frac{\hbar\omega_0}{\varepsilon} = \frac{\hbar}{\varepsilon\rho} = \frac{\lambda\!\!\!^{-}_c}{\rho\gamma} = \frac{1}{l} \tag{2.2}$$

In the magnetic field, where ρ is the radius of curvature, which is usually the macroscopic value, this ratio is very small (at $\varepsilon = 100$ MeV and at a magnetic field strength of $H = 10^4$ Oe, $\hbar\omega_0/\varepsilon \sim 10^{-14}$). At bremsstrahlung on the Coulomb centre, where ρ is the impact parameter, the fundamental contribution to the radiation is given by $\rho > \lambda\!\!\!^{-}_c$, then we obtain $l \gtrsim \gamma$.

The second type of quantum effects is associated with the recoil of the particle when it radiates and is of the order $\hbar\omega/\varepsilon$, where ω is the frequency of the radiated photon. During radiation in the magnetic field, the energy of the radiated photon depends on the value of the parameter χ (see (B.25)). At $\chi \ll 1$, the recoil, and consequently also the value of the quantum effects, is small, where the characteristic radiation frequencies $\omega \sim \omega_0\gamma^3$ (as in the classical theory (1.57)). At $\chi \gtrsim 1$, the energy of the radiated photon $\hbar\omega \sim \varepsilon$, this range is essentially quantum. Thus, at high energies of particles ($\gamma \gg 1$), the quantum effects of this first type are negligibly small compared with the radiation effects, so that they may be disregarded. If the theory contains quantum effects of both types, one of which we neglect from the very beginning, then for this the operator formulation of quantum mechanics is especially convenient. In fact, it is necessary to disregard the noncommutability with each other of the operator dynamic variables of the particle (the value $\sim \hbar\omega_o/\varepsilon$) and take into account only their commutators with the operator $\exp(-i\mathbf{k}\cdot\mathbf{r})$ associated with the field of the radiated photon (the value $\sim \hbar\omega/\varepsilon$).

Since the large orbital momenta are significant, the spin–field interaction terms may be disregarded (for particles with 1/2 spin these are the terms $e\hbar\sigma^{\mu\nu}F_{\mu\nu}/2$ (cf. (B.6)), whose relative contribution to the energy is $s\hbar\omega_0/\varepsilon$ (s is the spin value), i.e. for particles with lower spins the spin effects is of the same order as the quantum effects of the first type (2.2). To sum up, it proves possible to carry out an investigation for particles with any spin, including the

---

1973; Tsai, 1973 in the magnetic field and Baier, Katkov and Strakhovenko, 1974 in the arbitrary external field $F_{\mu\nu}$). From the results obtained, exact over the particle motion, one can find the expressions of quasiclassical approximation in the region of its applicability. Thus, in some cases the quasiclassical method is under control of the more general theory.

calculation of all spin effects, in a single method. The last effects are defined by the structure of the current which coincides in form with the calculation for free particles. In this method no knowledge of the explicit form of the wave equation solutions is required, while the specific calculations from some stage become similar to the classical ones.

2.2. The Probability of Radiation in a Stationary External Field

The matrix element of the photon radiation by the charged particle in the external field in the first order of the perturbation theory with respect to the interaction with the radiation field follows from the power expansion of the S matrix in the form (2.1) and may, for particles with any spin, be represented in the form

$$U_{fi} = \frac{ie}{(2\pi)^{3/2}\sqrt{2\hbar\omega}} \int dt \int d^3r\, F^+_{fs'}(\mathbf{r}) \exp(i\varepsilon_f t/\hbar) \tag{2.3}$$
$$\times\, (e^* J) \exp[i(\omega t - \mathbf{k}\cdot\mathbf{r})] \exp(-i\varepsilon_i t/\hbar) F_{is}(\mathbf{r})\,,$$

where $F_{is}(\mathbf{r})$ is the solution of the wave equation* in a given field with energy ε_i and the spin state s, e^μ is the photon polarization vector; $k^\mu(\omega, \mathbf{k})$ is the photon 4-momentum; J^μ is the current vector.

For the states with large orbital momenta l with which we are concerned the following approximation may be made:

$$\exp(-i\varepsilon_i t/\hbar) F_{is}(\mathbf{r}) = \Psi_s(\mathbf{P}) \exp(-i\mathcal{H}t/\hbar)|i\rangle\,, \tag{2.4}$$

where $\Psi_s(\mathbf{P})$ is the operator form of the wave function of the particle in spin state s in the external field. This form may be obtained from the free wave functions via substitution of the variables by the operators: $\mathbf{p} \to \mathbf{P}, \varepsilon \to \mathcal{H} = \sqrt{\mathbf{P}^2 + m^2}$ The vector of state $|i\rangle$ determines the particle state in the field (with the exception of the spin ones). In formula (2.4) we disregard the interaction terms of the spin-field type. We illustrate the formula (2.4) on the example of the Dirac equation in the external field (B.5). If the interaction terms of the spin-field kind ($e\hbar\sigma^{\mu\nu}F_{\mu\nu}/2$) are discarded in the "squared" Dirac equation (B.6), this equation converts to a Klein–Gordon equation in the same

*It should be noted that in contrast to the case of free particles, where the matrix element of the first order becomes zero since the law of energy-momentum conservation cannot be fulfilled, here part of the recoil momentum takes up the field, in consequence of which $U_{fi} \neq 0$.

field (B.7), so that the formula

$$\Psi(x) = C(\gamma^\mu P_\mu + m)\Phi(x)\,, \tag{2.5}$$

where $\Phi(x)$ is the solution of the Klein–Gordon equation and C is the normalization constant, automatically satisfying the Dirac equation in the field. Comparing this with the formula (A.12), we verify that (2.5) has the same matrix structure as the solution of the free equation where the values p and ε are substituted by operators. Thus, in the coordinate representation $|i\rangle$ is the solution of the Klein–Gordon equation in a given field.

Substituting (2.4) in (2.3) and taking into account that the Schrödinger operators standing between the exponential factors $\exp(\pm i\mathcal{H}t/\hbar)$ convert into explicitly time-dependent Heisenberg operators of the dynamic variables of the particle in the given field, we obtain the following formula for the matrix element (2.3):

$$U_{fi} = \frac{ie}{(2\pi)^{3/2}\sqrt{2\hbar\omega}}\left\langle f|\int dt\exp(i\omega t)M(t)|i\right\rangle\,, \tag{2.6}$$

where

$$M(t) = \frac{1}{\sqrt{\mathcal{H}}}\Psi^+_{s'}(\mathbf{P})\{(e^*J),\exp[-i\mathbf{kr}(t)]\}\quad \Psi_s(\mathbf{P})\frac{1}{\sqrt{\mathcal{H}}}\,; \tag{2.7}$$

$\mathbf{P}(t), J^\mu(t), \mathbf{r}(t)$ are the Heisenberg operators of the particle momentum, current and coordinates respectively, the brackets $\{\,,\,\}$ denote the symmetrized product of the operators (half of the anticommutator). For example, for a particle with spin 0 (s) we have*

$$M_s = \frac{1}{\sqrt{\mathcal{H}}}\{(e^*P),\exp(-i\mathbf{kr})\}\frac{1}{\sqrt{\mathcal{H}}}\,. \tag{2.8}$$

For a particle with spin 1/2 (e)

$$M_e = \sqrt{m/\mathcal{H}}u_{s'}(P)\hat{e}^*\exp(-i\mathbf{kr})u_s(P)\sqrt{m/\mathcal{H}}\,, \tag{2.9}$$

where for the canonical set of γ-matrices† (see (A.8), Appendix A)

$$u_s(P) = \sqrt{\frac{\mathcal{H}+m}{2m}}\begin{pmatrix}\varphi[\zeta(t)]\\ \frac{1}{\mathcal{H}+m}\,\boldsymbol{\sigma}\mathbf{P}\varphi[\zeta(t)]\end{pmatrix}\,, \tag{2.10}$$

---

*See the rule of the diagram technique for the single-photon vertex for scalar particles.

†For particles with spin 0, 1/2 it is appropriate to use the Coulomb gauge of the electromagnetic field.

$\varphi[\zeta(t)]$ is a two-component spinor (see (A.13), (A.14) Appendix A), describing the particle spin states at the moment of time t (in the quasiclassical limit the evolution of vector $\zeta(t)$ in time may be defined by the equation for the classical spin vector (see, e.g. Baier, Katkov and Fadin, 1973).

For a particle with spin 1(v) we have*

$$M_v = \frac{1}{\sqrt{\mathcal{H}}}(e_f^v)_\mu\{((e^*P)g^{\mu v} - e^{*\mu}P^v - e^{*v}P^\mu), \qquad (2.11)$$

$$\exp(-i\mathbf{kr})\}(e_i^v)_v \frac{1}{\sqrt{\mathcal{H}}},$$

where e_i^v, e_f^v are the polarization 4-vectors of the initial and final vector particles. Expressions for particles with higher spins may also be written out in an analogous way.

We are interested in the transition probability with photon radiation summed over the final particle states. Fulfilling this summation† using the condition of the completeness‡ of the vectors of the state $\sum_f |f\rangle\langle f| =$, we obtain for the transition probability with photon radiation:

$$dw = \frac{e^2}{\hbar}\frac{d^3k}{(2\pi)^2\omega}\langle i| \int dt_1 \int dt_2 \qquad (2.12)$$

$$\times \exp[i\omega(t_1 - t_2)]M^+(t_2)M(t_1)|i\rangle$$

Multiplying this formula by the energy of the radiated photon, we obtain the differential formula for the radiated energy:

$$d\varepsilon(\omega) = \hbar\omega dw(\omega). \qquad (2.13)$$

For further calculation, it is necessary to perform a series of actions with the operators appearing in the formula (2.12). We shall take out the operator

*Compare the rule of the diagram technique for the single-photon vertex for vector particles.
†It should be noted that it is possible to fix, if this is required, any of the quantum numbers of the final particle, using the projection operators.
‡In the coordinate representation:

$$\langle f|\cdots|i\rangle = \int d^3x\Phi_f^+(\mathbf{x})\cdots\Phi_i(\mathbf{x}),$$

$$\sum_f \Phi_f(\mathbf{x})\Phi_f^+(\mathbf{x}') = \delta(\mathbf{x}-\mathbf{x}')$$

$\exp[-i\mathbf{kr}(t_1)]$ in $M(t_1)$ on the left, and the operator $\exp[i\mathbf{kr}(t_2)$ in $M^+(t_2)$ on the right, for which we use the relation

$$f(\mathbf{P})\exp(-i\mathbf{kr}) = \exp(-i\mathbf{kr})f(\mathbf{P} - \hbar\mathbf{k}),$$

$$\exp(i\mathbf{kr})f(\mathbf{P}) = f(\mathbf{P} - \hbar\mathbf{k})\exp(i\mathbf{kr}), \tag{2.14}$$

where $f(\mathbf{P})$ is an arbitrary function, which is a consequence of the fact that $\exp(-i\mathbf{kr})$ is the displacement operator in momentum space, which may be verified using the commutation relations $[r_m, p_n] = i\hbar\delta_{mn}$ and expanding $\exp(-i\mathbf{kr})$ and $f(\mathbf{P})$ in a Taylor series. The variation of the function $f(\mathbf{P})$ in (2.14) at commutation with $\exp(-i\mathbf{kr})$ (the "field" of the radiated photon) corresponds to making allowance for the recoil during radiation. After this operation has been performed, in each of the matrix elements only the commutative (with accuracy to the terms $\sim \hbar\omega_0/\varepsilon(1/e)$) operators remain, and the subsequent problem consists in the consideration of the occurring combination $\exp[i\mathbf{kr}(t_2)]\exp[-i\mathbf{kr}(t_1)]$.

The Heisenberg operators $\mathbf{r}(t_2)$ and $\mathbf{r}(t_1)$, taken at various moments of time, do not commute with each other. This noncommutability is essential, so that to find the indicated combination one must not be limited by the expansion over the lower commutators. The calculation of the operator exponential formulae is generally referrred to as "disentanglement". One of the basic propositions of the method described is the disentanglement of the operator formula $\exp[i\mathbf{kr}(t_2)] \cdot \exp[-i\mathbf{kr}(t_1)]$. We shall represent this formula in the form:

$$\exp(i\mathbf{kr}_2)\exp(-i\mathbf{kr}_1) = \exp[i\mathcal{H}\tau/\hbar]\exp(i\mathbf{kr}_1) \times \exp[-i\mathcal{H}\tau/\hbar]\exp(-i\mathbf{kr}_1), \tag{2.15}$$

where $\tau = t_2 - t_1$; here and below the indices 1 and 2 denote the dependence on the corresponding times. This representation is the consequence of the fact that $\exp[i\mathcal{H}\tau/\hbar]$ is the displacement operator in time. Taking account of the relation (2.14), we have:

$$\exp(i\mathbf{kr}_1)\exp(-i\mathcal{H}(\mathbf{P}_1)\tau/\hbar)\exp(-i\mathbf{kr}_1) = \exp[-i\mathcal{H}(\mathbf{P}_1 - \hbar\mathbf{k})\tau/\hbar]. \tag{2.16}$$

Substituting this formula in (2.15) and introducing the notation

$$L_e(\tau) = \exp(-i\omega\tau)\exp(i\mathbf{kr}_2)\exp(-i\mathbf{kr}_1) = \exp(-ikx_2)\exp(ikx_1) \tag{2.17}$$

where $kx_1 = \omega t_1 - \mathbf{k}\mathbf{r}_1$, we obtain

$$L_e(\tau) = \exp[i(\mathcal{H} - \hbar\omega)\tau/\hbar]\exp[-i\mathcal{H}(\mathbf{P} - \hbar\mathbf{k})\tau/\hbar]\,. \tag{2.18}$$

To perform disentanglement in the combination (2.15), it is obviously sufficient to calculate $L_e(\tau)$. In order to obtain an explicit formula for $L_e(\tau)$ it is necessary to differentiate (2.18) over τ:

$$\begin{aligned}\frac{dL^e}{d\tau} &= \frac{i}{\hbar}\exp\left[i\frac{(\mathcal{H} - \hbar\omega)}{\hbar}\tau\right][\mathcal{H} - \hbar\omega - \mathcal{H}(\mathbf{P}_1 - \hbar\mathbf{k})] \\ &\times \exp\left[-i\frac{\mathcal{H}(\mathbf{P} - \hbar\mathbf{k})\tau}{\hbar}\right] = i[\mathcal{H} - \hbar\omega - \mathcal{H}(\mathbf{P}_2 - \hbar\mathbf{k})]L_e(\tau)/\hbar\,,\end{aligned} \tag{2.19}$$

where we once again take advantage of the fact that $\exp[i\mathcal{H}\tau/\hbar]$ is the displacement operator in time. We shall take into account that

$$\begin{aligned}\mathcal{H}(\mathbf{P}_2 - \hbar\mathbf{k}) &= \sqrt{(\mathbf{P}_2 - \hbar\mathbf{k})^2 + m^2} \\ &= \sqrt{(\mathcal{H} - \hbar\omega)^2 + 2\hbar kP_2 - \hbar^2k^2}\,,\end{aligned} \tag{2.20}$$

where

$$kP_2 = \omega\mathcal{H} - \mathbf{k}\mathbf{P}_2\,, \quad k^2 = \omega^2 - \mathbf{k}^2$$

For the real photons $k^2 = 0$, moreover, in accordance with (B.12)

$$kP_2 = \mathcal{H}\omega(1 - \mathbf{n}\mathbf{P}_2/\mathcal{H}) = \mathcal{H}\omega\left[1 - \mathbf{n}\mathbf{V}_2 + O\left(\frac{\hbar\omega_0}{\varepsilon}\right)\right]\,, \tag{2.21}$$

where $\mathbf{n} = \mathbf{k}/\omega$ is the unit vector in the direction of the photon's propagation. The equation (2.19) is analogous in form to the equation for the S matrix (the operators at $L_e(\tau)$ in the right part do not commute with each other at various moments of time). Therefore, the formal solution of Eq. (2.19) has the form

$$L_e(\tau) = T\exp\left[(i/\hbar)\int_0^\tau (\mathcal{H} - \hbar\omega - \sqrt{(\mathbf{P}(t_1 + \tau') - \hbar\mathbf{k})^2 + m^2})d\tau'\right]\,, \tag{2.22}$$

where T is the operator of the chronological product,* taking into account that $L_e(0) = 1$. The representation (2.20) and the solution (2.22) are precise;

*It should be noted that for the free particles the symbol T may be omitted since the momentum operators, taken at various moments of time, commute with each other.

formula (2.21) is correct with accuracy to the terms $\sim \hbar\omega_0/\varepsilon$ (this is the accuracy of the accepted approximation). However, subsequently it is evidently necessary to take account of the characteristics of the radiation of ultrarelativistic particles, i.e. as in Sec. 1, to perform an expansion over the parameter $1/\gamma$. In the classical theory, the particle emits at an angle of $\sim 1/\gamma$, then $(1-\mathbf{nV}) \sim 1/\gamma^2$. In the quantum theory, the mean values of the operators are of the type $\langle 1-\mathbf{nV}\rangle \sim 1/\gamma^2$; in this sense it may be assumed that the operator $1-\mathbf{nV}$ has the smallness $\sim 1/\gamma^2$. Taking this circumstance into account, one may perform the expansion

$$\mathcal{H}(\mathbf{P}-\hbar\mathbf{k}) = (\mathcal{H}-\hbar\omega)[1+\hbar k P_2/(\mathcal{H}-\hbar\omega)^2 - \hbar^2 k^2/[2(\mathcal{H}-\hbar\omega)^2]+\cdots]\,. \tag{2.23}$$

It is evident that this expansion is valid if the mean value $\langle 1-\hbar\omega/\mathcal{H}\rangle \sim 1$, i.e. the photon does not remove all the energy of the initial particle, so that the final particle is ultrarelativistic. Substituting the expansion (2.23) in the solution in the form (2.22), we see that the main terms in the exponent index cancel each other; in this way, for the real photons ($k^2 = 0$)

$$L_e(\tau) = T\exp\left\{-\frac{i}{\hbar}\int_0^\tau \frac{\mathcal{H}}{\mathcal{H}-\hbar\omega}\hbar kv(t_1+\tau')d\tau'\right\}, \tag{2.24}$$

where $kv = \omega(1-\mathbf{nv})$. In the integrand in (2.24) stands the value $kv \sim 1/\gamma^2$, and for calculation with this accuracy the noncommutability of the velocity operator at various moments of time may be neglected and the sign of the T product may be omitted, then

$$L_e(\tau) = \exp\{[-i\mathcal{H}/(\mathcal{H}-\hbar\omega)](kx_2-kx_1)\} \tag{2.25a}$$

In the case where the photons are not real, we have out of (2.23)

$$L_e(\tau) = \exp\{[-i\mathcal{H}/(\mathcal{H}-\hbar\omega)](kx_2-kx_1)+i\hbar k^2\tau/[2(\mathcal{H}-\hbar\omega)]\} \tag{2.26}$$

The derivation of the formula for $L_e(\tau)$ was not based on any particluar characteristics of the external field, only the fact that the ultrarelativitistic particle radiates into a cone with the angle $\sim 1/\gamma$ was used. Therefore, the result obtained is correct for the radiation of high-energy particles in the arbitrary exteral field, while the variation in values during commutation (2.14) corresponds to taking into account the energy and particle momentum variation (recoil) during radiation.

In this way, as a result of performing the desintanglement operation we have from (2.17) and (2.25):

$$\exp(-i\omega\tau)\exp(i\mathbf{k}\mathbf{r}_2)\exp(-i\mathbf{k}\mathbf{r}_1) = \exp(-ikx_2)\exp(ikx_1)$$
$$= \exp\{[-i\mathcal{H}/(\mathcal{H}-\hbar\omega)](kx_2 - kx_1)\}\,. \tag{2.25b}$$

In view of the compensation of the main terms in the argument of the exponential in (2.26), the combination $\exp(i\mathbf{k}\mathbf{r}_2)\exp(-i\mathbf{k}\mathbf{r}_1)$ commutes with accuracy to the terms $\sim 1/\gamma^2$ with all the operators entering in $M_2^+M_1$. So, all the operators in the formula (2.12) prove to be commutative (with accuracy to the terms of the highest order over $\hbar\omega_0/\varepsilon$ and $1/\gamma^2$, and therefore after the disentanglement operation has been performed all those standing in the brackets corresponding to the initial state (the mean ones in the state with high quantum numbers), may be substituted for the corresponding classical values (c-numbers).

Taking account of this formula for the transition probability (2.12) may be written in the form

$$dw = \frac{\alpha}{(2\pi)^2}\frac{d^3k}{\omega}\int dt_1 \int dt_2 R_2^* R_1 \exp\left[-i\frac{\varepsilon(kx_2 - kx_1)}{(\varepsilon-\hbar\omega)}\right], \tag{2.27}$$

where $\alpha = \frac{e^2}{\hbar} = \left(\frac{e^2}{\hbar c}\right)\frac{1}{137}$,

$$R(t) = (1/2\sqrt{\varepsilon'\varepsilon})\Psi_{s'}^+(p'(t))[e^* J(p'(t)) + e^* J(p(t))]\Psi_s(p(t))\,; \tag{2.28}$$

$\varepsilon, \mathbf{p}, \varepsilon' = \sqrt{\mathbf{p}'^2 + m^2}, \mathbf{p}' = \mathbf{p} - \hbar\mathbf{k}$ are no longer operators but c-numbers.

During the disentanglement operation the spin characteristics of the particles contained in the function $R(t)$ are not touched upon. This is linked to the fact that in our approximation the influence on the motion of the particle of the spin interaction with the external field (the terms $\sim \hbar\omega_0/\varepsilon(1/l)$) may be disregarded. The function $R(t)$ describing the spin states has the form of a transition matrix element for free particles taking into account the law of momentum conservation. This makes it possible to commonly study the problems for any spin, so that the formulae (2.27) and (2.28) are applicable for particles with arbitrary spin, in particular for the description of all types of polarization and spin phenomena.

Consequently, all the characteristics of radiation in the external field in (2.28) consist of the following: (1) the argument of the exponent contains the factor $\varepsilon/(\varepsilon - \hbar\omega)$ (taking into account the recoil, which is universal for all external fields); (2) the characteristics of the field appear in (2.28) in the dependence $\mathbf{p} = \mathbf{p}(t), kx = kx(t)$, since the evolution of the momentum and the coordinates in time are taken in the given field.

The formula for the quasiclassical matrix element $R(t)\exp[i\varepsilon kx(t)/(\varepsilon - \hbar\omega)]$ (2.27) is considerably simpler than the one obtained with the immediate integration of the exact solutions of the wave equation in (2.1), and the fact that this formula is time-dependent via $\mathbf{p}(t)$ and $\mathbf{x}(t)$ in a defined sense is analogous to the classical theory (see (1.23)), since it describes the radiation in terms of the trajectory. Therefore, in the given method, as in the classical theory of radiation (see Sec. 1), the interrelation between the total angle of deviation of the particle in the external field and the angle $1/\gamma$ (see Sec. 1.4) proves to be significant. As a result of this we shall separately investigate two characteristic cases: (1) the total angle deviation is large in comparison to $1/\gamma$ (the case of motion in a macroscopic external field, for example in the magnetic field) and (2) the total angle of deviation of the particle in a field $\lesssim 1/\gamma$ (the case of radiation in quasiperiodic motion is an example of this situation).

2.3. Radiation in Nonstationary Conditions

We shall now investigate the radiation process in an arbitrary external field, when $\mathcal{H} = \sqrt{\mathbf{P}^2 + m^2} + e\varphi$, the energy is time-dependent and the state is nonstationary. We shall write the solution of the equation

$$i\hbar\partial_t\Psi = \mathcal{H}\Psi \tag{2.29}$$

in the operator form

$$\Psi(t) = T\exp\left\{(-i/\hbar)\int_0^t \mathcal{H}dt\right\}\Psi(0)\,. \tag{2.30}$$

We shall transform from the operators in the Schrödinger representation to the Heisenberg operators:

$$A(t) = \left\{T\exp\left[(i/\hbar)\int_0^t \mathcal{H}dt\right]\right\}A(0)\left\{T\exp\left[(-i/\hbar\int_0^t \mathcal{H}dt\right]\right\} \tag{2.31}$$

The matrix element of radiation in the arbitrary external electromagnetic field $F_{\mu\nu}$, taking into account of what has been said above, coincides with respect to the form, with the matrix element in the stationary case (2.3),

(2.6) and (2.7), if it is understood that under the operators $\mathbf{p}$ and $\mathbf{r}$ are the corresponding Heisenberg operators in the sense (2.31).

We shall turn to the disentangling operation. The basic formula (2.15) may now be written in the form

$$\exp(i\mathbf{kr}_2)\exp(-i\mathbf{kr}_1)=\left\{T\exp\left[(i/\hbar)\int_0^T \mathcal{H}(\tau)d\tau\right]\right\}$$
$$\times\exp(i\mathbf{kr}_1)\left\{T\exp\left[(-i/\hbar)\int_0^T \mathcal{H}(\tau)d\tau\right]\right\}\exp(-i\mathbf{kr}_1), \tag{2.32}$$

where $\mathcal{H}(0)=\mathcal{H}(\mathbf{P}_1)$. The small times defined by the condition $|\Delta\mathbf{v}|\sim 1/\gamma$, give the main contribution to the radiation matrix element, so that it is possible to perform the expansion over the power of τ of the form

$$\int_0^T \mathcal{H}d\tau=\sum_{n=0}^{\infty}\int_0^T \left.\frac{d^n\mathcal{H}}{d\tau^n}\right|_{\tau=0}\frac{\tau^n}{n!}d\tau=\sum_{n=0}^{\infty}\frac{d^n\mathcal{H}}{d\tau^n}\frac{\tau^{n+1}}{(n+1)!}. \tag{2.33}$$

If the angle between the momentum vector $\mathbf{p}$ and the Lorentz force $\mathbf{F}$, $\vartheta_{pF}\gg 1/\gamma$, then the condition $|\Delta\mathbf{v}|\lesssim 1/\gamma$ leads to $\Delta v\ll 1/\gamma^2$(at $\vartheta_{pF}\sim 1, \Delta v\sim 1/\gamma^3$) and the series (2.33) is a series over the powers $1/\gamma$; with this accuracy for the performance of the disentangling operation, it is possible to be limited by the first term of the expansion in (2.33).

$$\int_0^T \mathcal{H}(\tau)d\tau=\mathcal{H}(0)\tau(1+0(1/\gamma)) \tag{2.34}$$

after which the subsequent consideration (see (2.16)) is identical to the one performed for the stationary case.

As it follows from Sec. 1.4, in the problem of radiation of the ultrarelativistic particle in an external field in the classical theory there are two characteristic cases: I. The total angle of deviation in a field $\vartheta_F\gg 1/\gamma$ is the case, for example, of the radiation in a constant external field. II. A deviation angle $\vartheta_F\lesssim 1/\gamma$. This is the case, for example, of the radiation in the Coulomb field as well as in undulators. This classification also applies to the considered quasiclassical method. In the analysis made above (see (2.22), (2.24) and (2.32)) we treated that $\langle 1-\mathbf{nv}\rangle\sim 1/\gamma^2$ and $\langle\Delta\mathbf{v}\rangle\sim 1/\gamma$. These conditions are fulfilled both in Case I and in Case II. Thus, the basic formula (2.24) is valid also in the cases of nonstationary and inhomogeneous fields, including particularly the case when the length of field inhomogeneity is much shorter than the characteristic formation length of the magnetic bremsstrahlung radiation. The criterion for applicability of quasiclassical approximation in this case is $\hbar\omega_0/\varepsilon \equiv \frac{2\pi\hbar}{T\varepsilon} \ll 1$, i.e. the energy of the external field quanta is much smaller than the electron energy (T is the period of variation of the external field).

In the case of the radiation in quasiperiodic motion the energy of the radiated quanta is $\hbar\omega \sim \hbar\omega_0\gamma^2$ and the classical theory is applicable when $\hbar\omega \ll \varepsilon$. Therefore, the quantum effects in the radiation may be defined by the parameter $\eta = 2\hbar\omega_0\gamma^2/\varepsilon$. At $\eta \ll 1$ we have the classical case. When $\eta \gtrsim 1$ the quantum effects are essential in the radiation process (strictly speaking, it follows from the derivation made above that in quantum region, particularly in criteria (1.118), (1.120), one should change ω by $\omega' = \omega\varepsilon/(\varepsilon - \hbar\omega)$).

Finally, we point out that we shall apply the method under investigation when the particle after radiation is ultrarelativistic (see (2.21) and (2.23)). Below we shall show that this condition is always fulfilled if the fields satisfy the condition

$$H/H_0 \ll 1\,, \quad E/E_0 \ll 1\,. \tag{2.35}$$

In conclusion of the section let us comment on the contributions of higher orders. As a result of disentanglement we obtained T-exponent (2.24) which can be expanded over commutators:

$$T\exp\left[i\int_{t_1}^{t_2} A(t)dt\right] = \exp\left[i\int_{t_1}^{t_2} A(t)dt + \frac{i^2}{2!}\int_{t_1}^{t_2} dt\int_{t_1}^{t_2} dt'[A(t), A(t')] + \cdots\right] \tag{2.36}$$

The expression (2.25a) is the first term of this expansion. In an inhomogeneous field the contribution of the terms containing commutators depends on specific conditions of the problem, although in all the considered cases this contribution is small. For the Coulomb field this problem was analysed in the paper Baier, Katkov, 1976, where the correction of the order $\sim 1/\gamma$ is found to the cross-section of bremsstrahlung of the electron in the Coulomb field.

2.4. The Description of the Radiation of the Spin 1/2 Particle in Quasiclassical Approximation

We present here the explicit expressions for the case of spin 1/2 particle in view of their importance for the applications. Substituting in (2.28) the explicit form of $u(\mathbf{P}(t))$ (2.10) we find:

$$R(t) = \varphi_f^+[A(t) + i\boldsymbol{\sigma}\mathbf{B}(t)]\varphi_i\,,$$

$$A(t) = \frac{\mathbf{e}^*\mathbf{p}(t)}{2\sqrt{\varepsilon\varepsilon'}}\left[\left(\frac{\varepsilon'+m}{\varepsilon+m}\right)^{1/2} + \left(\frac{\varepsilon+m}{\varepsilon'+m}\right)^{1/2}\right],$$

$$\mathbf{B}(t) = \frac{1}{2\sqrt{\varepsilon\varepsilon'}}\left\{\left(\frac{\varepsilon'+m}{\varepsilon+m}\right)^{1/2}\mathbf{e}^*\times\mathbf{p}(t) - \left(\frac{\varepsilon+m}{\varepsilon'+m}\right)^{1/2}\mathbf{e}^*\times(\mathbf{p}(t)-\hbar\mathbf{k})\right\} \tag{2.37}$$

If one uses the equation for classcial spin vector $\boldsymbol{\zeta}(t)$ (see for example, Baier, Katkov, Fadin, 1973) it may easily be verified that $\boldsymbol{\zeta}(t)$ with accuracy to the terms $\sim 1/\gamma$ precesses with the same frequency as the velocity. In Case I the characteristic radiation time has the same order as in classical electrodynamics: $\tau \sim 1/(w\gamma) = 1/(\omega_L\gamma)$. Then

$$\varphi(\boldsymbol{\zeta}(t+\tau/2)) = \varphi(\boldsymbol{\zeta}(t)) + \frac{\tau}{2}\dot{\zeta}_i\frac{\partial\varphi}{\partial\zeta_i} = \varphi(\boldsymbol{\zeta}(t)) + O(1/\gamma) \tag{2.38}$$

since in accordance with what has been said above $|\dot{\boldsymbol{\zeta}}|\tau \simeq |\dot{\mathbf{v}}|\tau \sim 1/\gamma$. In Case II the particle deviation angle $\lesssim 1/\gamma$ and so the estimation (2.38) is valid. Taking into account this result we obtain for combination $R_2^*R_1$, occurring in the expression for the radiation probability (2.27):

$$\begin{aligned} R_2^*R_1 \equiv N_{21}(\boldsymbol{\zeta}_i, \boldsymbol{\zeta}_f, \mathbf{e}) = (1/4)\,\mathrm{Tr}\,[1+\boldsymbol{\zeta}_i\boldsymbol{\sigma}) \\ \times (A_2^* - i\boldsymbol{\sigma}\mathbf{B}_2^*)(1+\boldsymbol{\zeta}_f\boldsymbol{\sigma})(A_1 + i\boldsymbol{\sigma}\mathbf{B}_1)]\,, \end{aligned} \tag{2.39}$$

where $(1+\boldsymbol{\zeta}\boldsymbol{\sigma})/2$ is the two-component polarization density matrix. Substituting (2.39) in (2.37) we can calculate any radiation characteristics, including polarization and spin ones. The intensity of the radiation is $dI = \hbar\omega dw$. After summation over spin states of the final electron and averaging over spin states of the initial electron the radiated energy has the form:

$$d\varepsilon = e^2(d^3k/(2\pi)^2)\int dt_1\int dt_2 N_{21}(\mathbf{e})\exp[ik'(x_1-x_2)]\,, \tag{2.40}$$

where

$$N_{21}(\mathbf{e}) = (1/2)\sum_{s_i,s_f} R_2^* R_1 = A_2^* A_1 + \mathbf{B}_2^*\mathbf{B}_1 = [(\varepsilon+\varepsilon')^2(\mathbf{e}^*\mathbf{v}_1)(\mathbf{e}\mathbf{v}_2)$$

$$- \hbar^2\omega^2(\mathbf{e}^*\mathbf{v}_2)(\mathbf{e}\mathbf{v}_1) + \hbar^2\omega^2(\mathbf{v}_1\mathbf{v}_2 - 1 + 1/\gamma^2)(\mathbf{e}^*\mathbf{e})]/(4\varepsilon'^2)\,. \tag{2.41}$$

Here we use notations: $k'^\mu = k^\mu\varepsilon/(\varepsilon - \hbar\omega), k^\mu$ is the photon 4-momentum $k^\mu = (\omega, \mathbf{k}), \varepsilon' = \varepsilon - \hbar\omega, x_{1,2} \equiv x(t_{1,2}), \mathbf{v}_{1,2} = \mathbf{v}(t_{1,2})$, the terms of the order $O(1/\gamma^2)$ are neglected. Summing over the photon polarization we find

$$\overline{N_{21}} = \sum_{\lambda=1}^{2} N_{21}(\mathbf{e}) = [(\varepsilon^2+\varepsilon'^2)(\mathbf{v}_1\mathbf{v}_2 - 1) + \hbar^2\omega^2/\gamma^2]/(2\varepsilon'^2)$$

$$= -(m^2/\varepsilon\varepsilon')[1 + (\varepsilon^2+\varepsilon'^2)\gamma^2(\mathbf{v}_1 - \mathbf{v}_2)^2/(4\varepsilon\varepsilon')]\,. \tag{2.42}$$

The argument of the exponent in (2.40) can be reduced to the form

$$k'(x_2 - x_1) \equiv A = \frac{\omega\varepsilon}{2\varepsilon'}\int_{t_1}^{t_2}\left[\frac{1}{\gamma^2} + (\mathbf{n} - \mathbf{v}(t))^2\right]dt\,, \tag{2.43}$$

where $\mathbf{n} = \mathbf{k}/\omega$.

The particle velocity can be represented in the form $\mathbf{v}(t) = \mathbf{v}_0 + \Delta\mathbf{v}(t, \mathbf{v}_0)$, where $\mathbf{v}_0$ is the average velocity. Then the integrand in phase A (2.43) may be writeen as

$$1/\gamma^2 + \mathbf{u}^2 - 2\mathbf{u}\Delta\mathbf{v} + (\Delta\mathbf{v})^2 \tag{2.44}$$

here

$$\mathbf{u} = \mathbf{n} - \mathbf{v}_0 \simeq \mathbf{n} - \mathbf{v}_0(\mathbf{n}\mathbf{v}_0) = \mathbf{n}_\perp\,, \tag{2.45}$$

i.e. one can state that vector $\mathbf{u}$ lies in the plane perpendicular to $\mathbf{v}_0$. The main contribution to intensity (2.40) is given by sections of the trajectory where the direction of the vector $\mathbf{n}$ is close to $\mathbf{v}_0$: $|\mathbf{u} - \Delta\mathbf{v}(t, \mathbf{v}_0)| \sim 1/\gamma, |\Delta\mathbf{v}| \sim \sqrt{\rho}/\gamma$, here ρ is defined in (1.120). After integration in (2.40) over $\mathbf{u}\,(d^3k = \omega^2 d\omega d\mathbf{u})$ we obtain the radiated energy, intergrated over the photon emission angle:

$$d\varepsilon = \frac{i\alpha m^2}{2\pi\varepsilon^2}\omega d\omega\int\frac{dt d\tau}{\tau - i0}\left\{1 + \frac{\varepsilon^2+\varepsilon'^2}{4\varepsilon\varepsilon'}\gamma^2\right. \tag{2.46}$$

$$\left.\times\,[\Delta\mathbf{v}(t - \tau/2) - \Delta\mathbf{v}(t + \tau/2)]^2\right\}e^{-iA_1}\,,$$

here

$$A_1 = \frac{\omega\varepsilon\tau}{2\varepsilon'}\left[\frac{1}{\gamma^2} + \frac{1}{\tau}\int_{-\tau/2}^{\tau/2} ds(\Delta\mathbf{v}(t+s))^2 - \left(\frac{1}{\tau}\int_{-\tau/2}^{\tau/2} ds\Delta\mathbf{v}(t+s)\right)^2\right],$$

where we going over to the variables t, τ : $t_1 = t - \tau/2, t_2 = t + \tau/2$ the contour around the point $\tau = 0$ (sign of the imaginary part) follows from the convergenece condition for the integral over $\mathbf{u}$. In the derivation of the expression (2.46) we used integral

$$\int_0^\infty \vartheta d\vartheta \int_0^{2\pi} d\varphi \exp[i(a\vartheta^2 + b\vartheta\cos\varphi)] = \left(\frac{i\pi}{a}\right)\exp(-ib^2/4a). \quad (2.47)$$

3. PAIR PRODUCTION BY A PHOTON IN AN EXTERNAL FIELD

3.1. The Method of Investigation

The method developed in Sec. 2 for the investigation of the radiation process in an external field may easily be generalized for the investigation of other electromagnetic processes. In the lower order of the perturbation theory relating to the interaction with the radiation field, such processes are the production of a pair by a photon and single-photon annihilation of the pair in an external field.

The matrix element of the pair production process has the form (compare (2.6)):

$$U_{fi} = [e/((2\pi)^{3/2}\sqrt{\hbar\omega})]\left\langle q\left|\int \exp(-i\omega t)\cdot M(t)dt\right|\bar{q}\right\rangle, \quad (3.1)$$

where

$$M(t) = \frac{1}{2}\exp(i\mathcal{H}t/\hbar)(1/\sqrt{\mathcal{H}})\Psi_s^+(\mathbf{P},\mathcal{H})[eJ(\mathbf{P},\mathcal{H})\exp(i\mathbf{kr})$$
$$+ \exp(i\mathbf{kr})eJ(\mathbf{P},-\mathcal{H})]\Psi_{\bar{s}}(\mathbf{P},-\mathcal{H})(1/\sqrt{\mathcal{H}})\exp(i\mathcal{H}t/\hbar). \quad (3.2)$$

A particle (with positive frequency $\Psi_s(\mathbf{P}, \mathcal{H})$) and an antiparticle (with negative frequency $\Psi_{\bar{s}}(\mathbf{P}, -\mathcal{H})$) are produced here in the electromagnetic vertex; $|q\rangle, |\bar{q}\rangle$ are state vectors describing the wave packets of the particle and the antiparticle; s, $\bar{s}$ are the spin state indices for particle and antiparticle.

Carrying the operator $\exp(i\mathcal{H}t/\hbar)$ in (3.2) on the right, which corresponds to the transition to Heisenberg operators

$$\exp(i\mathcal{H}t/\hbar)A = \exp(i\mathcal{H}t/\hbar)A\exp(-i\mathcal{H}t/\hbar)\exp(i\mathcal{H}t/\hbar)$$
$$= A(t)\exp(i\mathcal{H}t/\hbar)$$

we obtain:

$$\begin{aligned} M(t) = (1/\sqrt{\mathcal{H}})\Psi_s^+(\mathbf{P}(t),\mathcal{H})(1/2)[eJ(\mathbf{P}(t),\mathcal{H})\exp(i\mathbf{kr}(t)) \\ + \exp(i\mathbf{kr}(t))eJ(\mathbf{P}(t),-\mathcal{H})]\Psi_{\bar{s}}(\mathbf{P}(t),-\mathcal{H})(1/\sqrt{\mathcal{H}}) \\ \times \exp(2i\mathcal{H}t/\hbar)\,. \end{aligned} \tag{3.3}$$

Carrying in formula (3.3) $\exp(i\mathbf{kr}(t))$ on the right using the relation (2.14), we have

$$\begin{aligned} M(t) = (1/\sqrt{\mathcal{H}})\Psi_s^+(P)[eJ(P) + eJ(-P')]\Psi_{\bar{s}}(-P') \\ \times (1/\sqrt{\mathcal{H}'})\exp(i\mathbf{kr}(t))\exp(2i\mathcal{H}t/\hbar)\,, \end{aligned} \tag{3.4}$$

where

$$\mathbf{P}' = \hbar\mathbf{k} - \mathbf{P}, \mathcal{H}' = \mathcal{H}(\hbar\mathbf{k} - \mathbf{P}), P = (\mathcal{H}(\mathbf{P}),\mathbf{P})\,,$$
$$P' = (\mathcal{H}(\hbar\mathbf{k} - \mathbf{P}), \hbar\mathbf{k} - \mathbf{P}) = (\mathcal{H}',\mathbf{P}')\,.$$

We shall concern ourselves with the transition probability, summed over the final states of the pair produced. We shall perform the summation procedure in two stages. First, we shall sum over the final states of the antiparticle, using the condition of completeness $\sum|\bar{q}\rangle\langle\bar{q}| = 1$ (compare (2.12)), then we shall obtain

$$dw = \frac{\alpha}{(2\pi)^2\hbar\omega}\sum_q \left\langle q\right| \int dt_1 \int dt_2 \exp[-i\omega(t_2 - t_1)]M(t_2)M^+(t_1)\left|q\right\rangle \tag{3.5}$$

We shall investigate the combination (compare (2.15)) entering this formula:

$$L_p(\tau) = \exp(-i\omega\tau))\exp(i\mathbf{kr}_2)\exp(2i\mathcal{H}\tau/\hbar)\exp(-i\mathbf{kr}_1) \tag{3.6}$$

which we shall represent in the form

$$\begin{aligned} L_p(\tau) &= \exp(-i\omega\tau)\exp(i\mathcal{H}\tau/\hbar)\exp(i\mathbf{kr}_1)\exp(i\mathcal{H}\tau/\hbar)\exp(-i\mathbf{kr}_1) \\ &= \exp[i(\mathcal{H}/\hbar - \omega)\tau]\exp[i\mathcal{H}(\mathbf{P}_1 - \hbar\mathbf{k})\tau/\hbar]\,, \end{aligned} \tag{3.7}$$

where we used the fact that $\exp(i\mathcal{H}\tau/\hbar)$ is the displacement operator over time while $\exp(i\mathbf{kr})$ is the displacement operator in momentum space. The subsequent consideration is analogous to that performed in Sec. 2. Having differentiated (3.7) over τ, we find

$$\begin{aligned} dL_p(\tau)/d\tau &= i\exp[i(\mathcal{H}/\hbar - \omega)\tau][\mathcal{H} - \hbar\omega + \mathcal{H}(\mathbf{P}_1 - \hbar\mathbf{k})] \\ &\quad \times \exp[i\mathcal{H}(\mathbf{P}_1 - \hbar\mathbf{k})\tau/\hbar] \\ &= i[\mathcal{H} - \hbar\omega + \mathcal{H}(\mathbf{P}(t_1 + \tau) - \hbar\mathbf{k})]L_p(\tau)\,. \end{aligned} \tag{3.8}$$

Integrating this equation and taking into account the initial condition $L_p(0) = 1$, we have

$$L_p(\tau) = T\exp\left\{i\left[(\mathcal{H} - \hbar\omega)\tau + \int_{t_1}^{t_2}\mathcal{H}(\mathbf{P}(t) - \hbar\mathbf{k})dt\right]/\hbar\right\}, \tag{3.9}$$

where T is the chronological product operator.

Subsequently, it must be taken into account (as in Sec. 2) that we shall investigate the case where the produced particle and antiparticle are ultrarelativistic. Below, we shall assert that at $H/H_0 \ll 1$, this is the case of main interest for physics. In this situation the main contribution to the probability of the process is made by the range of velocities of the final particle, for which $1 - \mathbf{nv} \sim 1/\gamma^2$, where $\mathbf{n}$ is the direction of motion of the photon. In physics terms, this means that the produced particle moves at an initial moment of time in the direction of the photon motion, while the photon-particle interaction remains essential, until the particle turns on an angle $\sim 1/\gamma$, so that the picture is very similar to that of magnetic bremsstrahlung, whereas the interaction on the length where particle turns at an angle $\sim 1/\gamma$ corresponds to the radiation from the coherence length. Taking into account that for real photons $k^2 = 0$, we have (see (2.20) and (2.23))

$$\begin{aligned} \mathcal{H}(\mathbf{P} - \hbar\mathbf{k}) &= \sqrt{(\hbar\mathbf{k} - \mathbf{P})^2 + m^2} = \sqrt{(\hbar\omega - \mathcal{H})^2 + 2kP} \\ &= (\hbar\omega - \mathcal{H})[1 + kP/(\hbar\omega - \mathcal{H}) + \cdots]\,. \end{aligned} \tag{3.10}$$

On the basis of the arguments which led us from formula (2.22) to formula (2.25), we have

$$L_p(\tau) = \exp\{i(\mathcal{H}/(\hbar\omega - \mathcal{H}))(kx_2 - kx_1)\}\,. \tag{3.11}$$

Substituting this result in formula (3.5) and converting to the classical means (see (2.27)), we obtain the probability of pair production per all time of the interaction:

$$dw_e = [\alpha d^3p/(2\pi)^2\hbar\omega] \int dt_1 \int dt_2 R_p(t_2) R_p^*(t_1) \times \exp[i(\varepsilon/\varepsilon_f)(kx_2 - kx_1)]\,, \tag{3.12}$$

where

$$R_p(t) = (1/2\sqrt{\varepsilon\varepsilon'})\Psi_s^+(p)(e[J(p) + J(-p')])\Psi_{\bar{s}}(-p')\,;$$
$$\varepsilon_f = \hbar\omega - \varepsilon,\ \varepsilon' = \sqrt{\mathbf{p}'^2 + m^2}, \quad \mathbf{p}' = \hbar\mathbf{k} - \mathbf{p}\,. \tag{3.13}$$

In formula (3.12) summation is performed over the final particle states taking into account that $\sum_q \to d^3p$. It is possible to arrive at formulae (3.12) and (3.13) from formulae (2.27) and (2.28) using the crossing symmetry: $p' \to p$ (substitution of notations for the outgoing particle); $p \to -p', k \to -k, s' \to s, s \to \bar{s}$ (substitution rule); $t_1 \leftrightarrow t_2$ (transition to uniform notation). At the substitution $p \to -p'$, $\Psi_{\bar{s}}(p) \to \Psi_{\bar{s}}(-p')$ (the transition from the incoming particle to the outgoing antiparticle, with such a substitution a wave with the positive frequency $\exp(ipx) \to \exp(-ip'x)$, i.e. into a wave with negative frequency), so that $\Psi_{\bar{s}}(-p')$ describes the outgoing antiparticle.

For particles with 1/2 spin the formula for $R(t)$ (see (2.9)) has the form

$$R_p(t) = m\bar{u}_s(p)\hat{e}u_{\bar{s}}(-p')/\sqrt{\varepsilon\varepsilon'} = m\bar{u}_s(p)\hat{e}v_{\bar{s}}(p')/\sqrt{\varepsilon\varepsilon'} \tag{3.14}$$

and for particles with spin 0 (see (2.8))

$$R_p(t) = \frac{1}{2}(ep - ep')/\sqrt{\varepsilon\varepsilon'} \tag{3.15}$$

Let us discuss some characteristic properties of the pair production process in the electric field with potential $\varphi(\mathbf{r})$. In this case instead of $L_p(\tau)$ (3.6) one has to consider expression

$$L_p^E(\tau) = \exp(i\mathcal{H}^-\tau/\hbar)\exp(i\widetilde{\mathcal{H}}^+(t_1)\tau/\hbar)\,, \tag{3.16}$$

where $\tau = t_2 - t_1, \mathcal{H}^\mp = \mathcal{H}_0 \mp U(\mathbf{r})$ is the Hamiltonian for electron (positron), $U(\mathbf{r}) = e\varphi(\mathbf{r})$ is the potential energy, $\mathcal{H}_0(\mathbf{p}) = \sqrt{\mathbf{p}^2 + m^2}$, here as above we omitted the terms of the spin-field type. Note also, that in radiation problem (see (2.17), (2.18)) the combination

$$L_e^E(\tau) = \exp(i\mathcal{H}^-\tau/\hbar)\exp(-i\mathcal{H}^-(t_1)\tau/\hbar)$$

is originated. In the expression (3.16) there is operator

$$\widetilde{\mathcal{H}}^+(t_1) = \mathcal{H}_0(\mathbf{p}^-(t_1) - \hbar\mathbf{k}) + U(\mathbf{r}(t_1)), \tag{3.17}$$

where the evolution of Heisenberg operators $\mathbf{p}(t)$ and $\mathbf{r}(t)$ in time is determined by the Hamiltonian $\mathcal{H}^-$ and averaging of the probability is carried out over the electron states, so that the final result will contain the classical trajectories of the electron in the given field. If we change the order of summation over final states and start from a summation over the electron states then the evolution of operators in time will be determined by the Hamiltonian $\mathcal{H}^+$ and final results will contain the classical trajectories of positron. The trajectories of the electrons and positrons in the electrical field are very different. The results coincide only after summation over all possible states for the given type of particle trajectories.

Now we pass to the disentanglement of the expression (3.16). Let us differentiate $L_P^E(\tau)$ over τ (compare (3.8), (2.19))

$$\begin{aligned} dL_p^E(\tau)/d\tau &= i\exp(i\mathcal{H}^-\tau/\hbar)[\mathcal{H}^- + \mathcal{H}_0(\mathbf{p}(t_1) - \hbar\mathbf{k}) + U(\mathbf{r}(t_1))] \\ &\quad \times \exp(i\widetilde{\mathcal{H}}^+(t_1)\tau/\hbar) \\ &= i\exp(i\mathcal{H}^-\tau/\hbar)\cdot[\mathcal{H}_0(\mathbf{p}(t_1)) + \mathcal{H}_0(\mathbf{p}(t_1) - \hbar\mathbf{k})] \\ &\quad \times \exp(i\widetilde{\mathcal{H}}^+(t_1)\tau/\hbar) \\ &= i[\mathcal{H}_0(\mathbf{p}(t_1+\tau)) + \mathcal{H}_0(\mathbf{p}(t_1+\tau) - \hbar\mathbf{k})]L_p^E(\tau), \end{aligned} \tag{3.18}$$

where we took into account that operator $\mathcal{H}^-$ does not depend on time and substitute $\mathcal{H}^-(t_1) = \mathcal{H}_0(\mathbf{p}(t_1)) - U(\mathbf{r}(t_1))$ instead of $\mathcal{H}^-$. After this the explicit dependence on $U(\mathbf{r})$ disappears and further calculation coincides completely with the above.

The expression for $R_p(t)$ (see (3.14)) which enters in (3.12) is represented in the form

$$R_p(t) = (m/\sqrt{\varepsilon\varepsilon'})\bar{u}_s(p)\hat{e}v_{\bar{s}}(p') = \varphi_s^+ O\varphi_{\bar{s}} = i\varphi_s^+(A(t) - i\boldsymbol{\sigma}\mathbf{B}(t))\varphi_{\bar{s}}, \tag{3.19}$$

where we passed to the two-component spinors using explicit forms for $u_s(p)$, $v_{\bar{s}}(p')$ (2.10) and

$$\begin{aligned} A &= N\mathbf{e}(\hbar\mathbf{k} \times \mathbf{p}(t)), \\ \mathbf{B} &= N\{\mathbf{e}[(\varepsilon' + m)(\varepsilon + m) - \mathbf{p}'(t)\mathbf{p}(t)] \\ &\quad + (\mathbf{e}\mathbf{p}(t))(\mathbf{p}'(t) - \mathbf{p}(t))\}, \\ N &= [4\varepsilon\varepsilon'(\varepsilon + m)(\varepsilon' + m)]^{-1/2} \end{aligned} \tag{3.20}$$

Here two-component spinors are dependent on time: $\varphi_s = \varphi_s(\zeta(t))$, $\varphi_{\bar{s}} = \varphi_{\bar{s}}(\zeta_+(t))$. If one uses the equation for the classical spin vector, then it is easy to see that with accuracy to terms $\sim 1/\gamma$ the vector $\zeta_\pm(t)$ processes with the same frequency as the velocity (the equation with $g = 2$ should be considered, since the anomalous magnetic moments of the electron is a radiative correction proportional to α, so that it would have to be taken into account only in α^2 order). The characteristic time of pair formation is the time during which the components of the pair are deflected in an external field at angle $\sim 1/\gamma$, i.e. $\Delta v \sim |\mathbf{w}|\tau \sim 1/\gamma$. In transverse magnetic (H) or electric (E) field we obtain

$$\begin{aligned} \tau \sim 1/(|\mathbf{w}|\gamma) = \varepsilon/(eH\gamma) = m/eH = \lambda\!\!\!^{-}_c(H_0/H) \\ (\text{or} \quad \tau = \lambda\!\!\!^{-}_c(E_0/E)), \end{aligned} \tag{3.21}$$

where $\lambda\!\!\!^{-}_c = \hbar/mc$, $H_0 = m^2/e\hbar$, $E_0 = m^2/e\hbar (= m^2c^3/e\hbar)$. Thus, (compare (2.38))

$$\varphi(\zeta(t + \tau/2)) = \varphi(\zeta(t)) + \frac{\tau}{2}\dot{\zeta}\frac{d\varphi}{\partial\zeta} + \cdots \simeq \varphi(\zeta(t)) + O(1/\gamma) \tag{3.22}$$

with allowance of result, we find from (3.12), (3.19)

$$\begin{aligned} R_p(t_2)R_p^*(t_1) &= R_p(t + \tau/2)R_p^*(t - \tau/2) \\ &= \frac{1}{4}\,\mathrm{Tr}\,[(1 + \zeta_-\boldsymbol{\sigma})(A_2 - i\boldsymbol{\sigma}\mathbf{B}_2)(1 + \zeta_+\boldsymbol{\sigma}) \\ &\quad \times (A_1^* + i\boldsymbol{\sigma}\mathbf{B}_2^*)], \end{aligned} \tag{3.23}$$

where $(1 + \boldsymbol{\sigma}\zeta)/2$ is two-component spin density matrix, $A_1 = A(t_1)$ etc. The expression (3.12) with substitution (3.23) may be used for calculation of any characteristic of the process of pair production by a photon including polarization and spin characteristics. After summation over spin states of the produced electron and positron, we find from (3.23)

$$N_p(\mathbf{e}) \equiv \sum_{s,\bar{s}} R_p(t_2)R_p^*(t_1) = 2[A_1^*A_2 + \mathbf{B}_1^*\mathbf{B}_2]. \tag{3.24}$$

Substituting into (3.24) the explicit expressions (3.20) and neglecting the terms of order $O(1/\gamma^2)$, we obtain

$$\begin{aligned} N_p(\mathbf{e}) = [(\varepsilon_f - \varepsilon)^2(\mathbf{e}^*\mathbf{v}_1)(\mathbf{e}\mathbf{v}_2) + \hbar^2\omega^2[(\mathbf{v}_1\mathbf{v}_2 - 1 \\ + 1/\gamma^2)(\mathbf{e}^*\mathbf{e}) - (\mathbf{e}^*\mathbf{v}_2)(\mathbf{e}\mathbf{v}_1)]](1/2\varepsilon_f^2), \end{aligned} \tag{3.25}$$

where $\varepsilon_f = \hbar\omega - \varepsilon$, let us remember that $\mathbf{v}_1 = \mathbf{v}(t_1)$ and $\mathbf{v}_2 = \mathbf{v}(t_2)$ are the velocities of the produced particles. After averaging over the photon polarization, we obtain

$$\begin{aligned} N_p &= \frac{\varepsilon_f^2 + \varepsilon^2}{2\varepsilon_f^2}(\mathbf{v}_1\mathbf{v}_2 - 1) + \frac{\hbar^2\omega^2}{2\varepsilon_f^2\gamma^2} \\ &= (m^2/\varepsilon\varepsilon_f)[1 - (\mathbf{v}_1 - \mathbf{v}_2)^2(\varepsilon_f^2 + \varepsilon^2)\varepsilon/(4\varepsilon_f m^2)]. \end{aligned} \tag{3.26}$$

The argument of the exponent in (3.12) can be written as

$$A \equiv (\varepsilon/\varepsilon_f)(kx_2 - kx_1) = (\varepsilon\omega/2\varepsilon_f)\int_{t_1}^{t_2} dt[1/\gamma^2 + (\mathbf{n} - \mathbf{v}(t))^2], \tag{3.27}$$

where $\mathbf{n} = \mathbf{k}/\omega$. In the phase (3.27) at $\varepsilon_f = \hbar\omega - \varepsilon > 0$ there is compensation in γ^2 times, but at $\varepsilon_f < 0 (\varepsilon > \hbar\omega)$ there is a sudden increase of the phase, which is leading to exponential suppression of the probability. In such a way, in this approach, the energy conservation law is realized.

Substituting (3.23)–(3.26) and (3.27) into formula (3.12) we obtain explicit expressions for probability of production of the pair of particles with spin 1/2 in the external field of the general type in quasiclassical approximation

$$dw_e = \frac{\alpha m^2}{(2\pi)^2\hbar\omega}\frac{d^3p}{\varepsilon\varepsilon_f}\int dt_1\int dt_2\left\{1 - \frac{\varepsilon^2 + \varepsilon_f^2}{4\varepsilon\varepsilon_f}\gamma^2[\mathbf{v}(t_1) - \mathbf{v}(t_2)]^2\right\}e^{iA}, \tag{3.28}$$

where A is given by (3.27).

We represent the velocity as $\mathbf{v}(t) = \mathbf{v}_0 + \Delta\mathbf{v}(t, \mathbf{v}_0)$, where $\mathbf{v}_0$ is averaged velocity and then the integrand in phase A (3.27) can be written in the form

$$\frac{1}{\gamma^2} + \mathbf{u}^2 + 2\mathbf{u}\Delta\mathbf{v} + (\Delta\mathbf{v})^2, \tag{3.29}$$

where $\mathbf{u} \equiv \mathbf{n} - \mathbf{v}_0 = \mathbf{n}(1 - \mathbf{n}\mathbf{v}_0) - \mathbf{v}_{0\perp} \simeq -\mathbf{v}_{0\perp}$ so that we can assume that the vector $\mathbf{u}$ lies in the plane perpendicular to $\mathbf{n}$. The main contribution to the probability (3.28) will be from the portions of the trajectory in which the direction of the vector $\mathbf{v}_0$ is close to that of $\mathbf{n}$:

$$|\mathbf{u}| \sim 1/\gamma, \quad |\Delta\mathbf{v}(t, \mathbf{v}_0)| \sim 1/\gamma .$$

But in that case we can make the substitution

$$\Delta\mathbf{v}(t, \mathbf{v}_0) \to \Delta\mathbf{v}(t, \mathbf{n}) \equiv \Delta\mathbf{v}(t) .$$

Accordingly in the coefficient of the exponential expression in (3.28) we make the substitution

$$\mathbf{v}(t_1) - \mathbf{v}(t_2) \to \Delta\mathbf{v}(t_1) - \Delta\mathbf{v}(t_2)$$

with relativistic accuracy we can rewrite the phase space $d^3p \simeq \varepsilon^2 d\varepsilon d^2u$. After these transformation the integration over $\mathbf{u}$ can be carried out. As a result we obtain the general expression for the probability of pair production in quasi-classical approximation integrated over emission angle

$$dw_e = \frac{i\alpha m^2}{2\pi\hbar\omega^2} d\varepsilon \int \frac{dt d\tau}{\tau + i0} \times \left\{1 - \frac{\varepsilon^2 + \varepsilon_f^2}{4\varepsilon\varepsilon_f}\gamma^2[\Delta\mathbf{v}(t - \tau/2) - \Delta\mathbf{v}(t + \tau/2)]^2\right\} e^{iA_1}, \tag{3.30}$$

where

$$A_1 = \frac{\varepsilon\omega\tau}{2\varepsilon_f}\left[\frac{1}{\gamma^2} + \frac{1}{\tau}\int_{-\tau/2}^{\tau/2} dx(\Delta\mathbf{v}(t + x))^2 - \left(\frac{1}{\tau}\int_{-\tau/2}^{\tau/2} dx\Delta\mathbf{v}(t + x)\right)^2\right] \tag{3.31}$$

we have gone over to the variables t, τ : $t_1 = t - \tau/2$; $t_2 = t + \tau/2$; the contour around the point $\tau = 0$ (sign of the imaginary part) follows from the convergence condition for the integral over $\mathbf{u}$.

The analysis of the contributions of higher orders in the problem of the pair production can be made in the same way as in the radiation problem (see end of Sec. 2.3). In all the investigated cases these contributions are small. Particularly, at substitution $\Delta\mathbf{v}(t, \mathbf{v}_0) \to \Delta\mathbf{v}(t)$ the terms of the same order are neglected as commutators in decomposition of the T-product.

3.2. The Production of Pair of Particles by a Polarized Photon

We shall investigate the production of an electron-positron pair by a polarized photon in the constant field (it is necessary actually, that the field is constant on the pair formation length (3.21)). As vectors, on to which we shall project the photon polarization, we shall select unit vectors which are perpendicular* to the direction $\mathbf{n}$ of the photon motion

$$\mathbf{e}_1 = \frac{(\mathbf{n}\times\mathbf{H}) + \mathbf{E}_\perp}{|(\mathbf{n}\times\mathbf{H}) + \mathbf{E}_\perp|}, \quad \mathbf{e}_2 = (\mathbf{n}\times\mathbf{e}_1), \quad \mathbf{E}_\perp = \mathbf{E} - \mathbf{n}(\mathbf{n}\mathbf{E}). \tag{3.32}$$

Projecting polarization on the vectors $\mathbf{e}_1, \mathbf{e}_2$ (3.32) in (3.25) and substituting the result and (3.27) in (3.12), we obtain the general expression for the probability of the pair production by a polarized photon.

In the constant field limit it is convenient to go over variables (compare (1.59))

$$t = \frac{t_1 + t_2}{2}, \quad \tau = t_2 - t_1; \quad t_1 = t - \tau/2, \quad t_2 = t + \tau/2 \tag{3.33}$$

and $\int dt_1 \int dt_2 = \int dt \int d\tau$. The integration over t one can consider as integration of the probability of pair production per unit time: $dW_e \equiv dw_e/dt$. In the constant field limit one can expand the functions contained in the expression for the probability

$$\begin{aligned} \mathbf{v}(t \pm \tau/2) &= \mathbf{v}(t) \pm \mathbf{w}\tau/2 + \dot{\mathbf{w}}\tau^2/8 + \cdots \\ \mathbf{r}(t \pm \tau/2) &= \mathbf{r}(t) \pm \mathbf{v}\tau/2 + \mathbf{w}\tau^2/8 \pm \dot{\mathbf{w}}\tau^3/48 + \cdots \end{aligned} \tag{3.34}$$

Let us take into account, that produced particles are ultrarelativistic, so that (compare (1.54))

$$\mathbf{v}\cdot\mathbf{w} = O(1/\gamma^2), \quad \mathbf{n}\cdot\dot{\mathbf{w}} = -\mathbf{w}^2 + O(1/\gamma). \tag{3.35}$$

---

*These vectors are time-dependent. However, for fields satisfying the inequalities (1.50) and (1.51) with accepted precision they may be regarded as permanent during the time of process.

Then

$$\mathbf{v}_1 \cdot \mathbf{v}_2 = 1 - 1/\gamma^2 + \mathbf{w}^2\tau^2/2$$
$$kx_1 - kx_2 = \omega\tau - \mathbf{k}\cdot(\mathbf{r}_2 - \mathbf{r}_1) = \omega\tau(1 - \mathbf{n}\cdot\mathbf{v} + \mathbf{w}^2\tau^2/24)\,. \tag{3.36}$$

Projecting the polarization on the vectors $\mathbf{e}_1$ and $\mathbf{e}_2$, (3.32) in (3.25), substituting found expansions in (3.25) and (3.27) and then putting the results into (3.12) and converting to angles ψ and β (1.36) we obtain for pair production probability per unit time

$$\begin{aligned} dW_e \equiv \frac{dw_e}{dt} = -\frac{\alpha}{(2\pi)^2}\frac{d^3p}{\hbar\omega}\frac{2\varepsilon}{\varepsilon_f}\int_{-\infty}^{\infty} d\tau\{|\mathbf{e}\mathbf{e}_1|^2 \\ \times[\psi^2 + (\varepsilon^2+\varepsilon_f^2)w^2\tau^2/(8\varepsilon\varepsilon_f)] \\ + |\mathbf{e}\mathbf{e}_2|^2[\beta^2 + \hbar^2\omega^2 w^2\tau^2/(8\varepsilon\varepsilon_f)] \\ + (\mathbf{e}^*\mathbf{e}_1)(\mathbf{e}\mathbf{e}_2)\beta[\psi - (\varepsilon^2+\varepsilon_f^2)w\tau/(4\varepsilon\varepsilon_f)] \\ + (\mathbf{e}^*\mathbf{e}_2)(\mathbf{e}\mathbf{e}_1)\beta[\psi + (\varepsilon^2+\varepsilon_f^2)w\tau/(4\varepsilon\varepsilon_f)]\} \\ \times \exp[i\varepsilon\omega\tau(1-\mathbf{n}\cdot\mathbf{v}+\mathbf{w}^2\tau^2/24)/\varepsilon_f]\,, \end{aligned} \tag{3.37}$$

where the angles β, ψ, characterizing the electron motion, are introduced in the same way as for the radiation case $\lambda, w = |\mathbf{w}|$. The form of notation in (3.37) is equivalent to the notation in the form of the polarization matrix, where the coefficients at $(\mathbf{e}^*\mathbf{e}_i)(\mathbf{e}\mathbf{e}_k)$ are the essence of the component (ik) of the polarization matrix. Averaging over the photon polarizations, we have

$$\begin{aligned} dW_e \equiv dw_e/dt = \frac{\alpha}{(2\pi)^2}\frac{d^3p}{\hbar\omega}\frac{\varepsilon}{\varepsilon_f}\int_{-\infty}^{\infty} d\tau[1/\gamma^2 \\ - (\varepsilon^2+\varepsilon_f^2)w^2\tau^2/(4\varepsilon\varepsilon_f)] \\ \times \exp[i\omega\varepsilon\tau(1-\mathbf{n}\cdot\mathbf{v}+\mathbf{w}^2\tau^2/24)/\varepsilon_f]\,. \end{aligned} \tag{3.38}$$

The main interest presents the probability of pair production integrated over the azimuthal angle (see discussion before formula (1.77)). In this case it also appears to be convenient to do integration over τ and ψ simultaneously extending it to infinite intervals due to rapid diminshing of the integrand. Taking into account, that $1 - \mathbf{n}\cdot\mathbf{v} = (1/\gamma^2 + \beta^2 + \psi^2)/2$ and going over to the variables (see (1.78))

$$\psi = \delta(u-v)/2\,,\ w\tau = \delta(u+v)\,,\ \delta = [2\varepsilon_f w/(\varepsilon\omega)]^{1/3}\,, \tag{3.39}$$

we obtain from (3.38)

$$dW_e = \frac{\alpha}{(2\pi)^2}\frac{p^2dpd\beta}{\hbar\omega}\frac{\varepsilon}{\varepsilon_f}\frac{\delta^2}{w}\int_{-\infty}^{\infty}du\int_{-\infty}^{\infty}dv\left[1/\gamma^2 - \frac{(\varepsilon^2+\varepsilon_f^2)\delta^2(u+v)^2}{4\varepsilon\varepsilon_f}\right]\exp[i[\mu^2(u+v)/\delta^2 + (u^3+v^3)/3]]\,. \tag{3.40}$$

Using the integrals (1.66) we obtain

$$dW_e = \frac{2\alpha m^2}{3\pi^2\hbar\omega\kappa}F(x,y)\cosh^3 x dxdy \tag{3.41}$$

here

$$F(x,y) = K_{1/3}^2(\eta) + \cosh^2 x(2\cosh^2 y - 1) \times [K_{1/3}^2(\eta) + K_{2/3}^2(\eta)]\,. \tag{3.42}$$

In these expressions we go over to the variables

$$\begin{aligned}
&\kappa = H\hbar k_\perp/(H_0 m) = e\hbar\sqrt{|F_{\mu\nu}k^\nu|^2/m^3}\,,\\
&x = \gamma\beta\,, \quad \cosh x = \gamma\mu\,, \quad \cosh^2 y = \hbar^2\omega^2/4\varepsilon\varepsilon_f\,,\\
&\eta = 4\cosh^2 y\cosh^3 x/(3\kappa)\,, \quad -\infty \le y \le \infty\,, \quad -\infty \le x \le \infty\,.
\end{aligned} \tag{3.43}$$

Making the same calculations starting from Eq. (3.37) we find for the probability of pair production by a polarized photon in an external field per unit time:

$$\begin{aligned}
dW_e(\mathbf{e}) = {} & (4/3\pi^2)(\alpha m^2/\hbar\omega\kappa)dxdy\cosh^3 x\{|\mathbf{e}\mathbf{e}_1|^2\cosh^2 x[\cosh^2 y K_{1/3}^2(\eta)\\
& + \sinh^2 y K_{2/3}^2(\eta)] + |\mathbf{e}\mathbf{e}_2|^2[\cosh^2 x[\sinh^2 y K_{1/3}^2(\eta)\\
& + \cosh^2 y K_{2/3}^2(\eta)] + K_{1/3}^2(\eta)] - i[(\mathbf{e}^*\mathbf{e}_1)(\mathbf{e}\mathbf{e}_2) - (\mathbf{e}^*\mathbf{e}_2)(\mathbf{e}\mathbf{e}_1)]\\
& \times \sinh x\cosh x(2\cosh^2 y - 1)K_{1/3}(\eta)K_{2/3}(\eta)\}\,.
\end{aligned} \tag{3.44}$$

It is convenient to represent formula (3.44) via the Stokes parameters of the photon $\boldsymbol{\xi}$ (compare (1.27)):

$$dW_e(\boldsymbol{\xi}) = dW_e(1 + \boldsymbol{\xi}\boldsymbol{\lambda})\,, \tag{3.45}$$

where

$$\lambda_1 = 0, \lambda_2 = [2\sinh x \cosh x(2\cosh^2 y - 1)K_{1/3}(\eta)K_{2/3}(\eta)]/F(x,y)$$
$$\lambda_3 = [\sinh^2 x K_{1/3}^2(\eta) - \cosh^2 x K_{2/3}^2(\eta)]/F(x,y)\,. \tag{3.46}$$

For calculation of the spectrum of the final particles it is convenient to return to the initial formulae (3.37), (3.38) and calculate the integral over τ using the formulae (1.66). From (3.38) we obtain

$$dW_e = \frac{\alpha}{\sqrt{3}\pi^2}\frac{d^3p}{\hbar\omega}\frac{\varepsilon}{\varepsilon_f}\frac{\sqrt{2(1-\mathbf{nv})}}{w\gamma^2}\left[1+2\frac{\varepsilon^2+\varepsilon_f^2}{\varepsilon\varepsilon_f}\gamma^2(1-\mathbf{nv})\right]K_{1/3}(\zeta)\,, \tag{3.47}$$

where

$$\zeta = \frac{2}{3}\frac{\hbar^2\omega^2}{\varepsilon\varepsilon_f\kappa}[2\gamma^2(1-\mathbf{nv})]^{3/2} = \frac{8\cosh^2 y}{3\kappa}[2\gamma^2(1-\mathbf{nv})]^{3/2} \tag{3.48}$$

For the angular integration we shall use angles ϑ and $\varphi(\mathbf{nv} = v\cos\vartheta)$. The integration over the azimuthal angle φ is trivial, since the integrand does not depend on φ. For integration over ϑ we change to the variable

$$z(\vartheta) = [2\gamma^2(1-\mathbf{nv})]^{3/2} \tag{3.49}$$

The lower limit of integration over z is $z(0) = [2\gamma^2(1-v)]^{3/2} = 1 + O(1/\gamma^2)$, while the upper limit is $z(\pi) = 8\gamma^3(1+O(1/\gamma^2))$, in view of the fact that at large z the integrand falls exponentially, the upper limit $z(\pi)$ may be substituted for infinity. Then with accuracy to the terms $O(1/\gamma^2)$ we obtain the following expression for the spectral distribution over energy of one of the final particles

$$dW_e = \frac{\alpha m^2}{\sqrt{3}\pi}\frac{d\varepsilon}{\hbar^2\omega^2}\left[\frac{\varepsilon^2+\varepsilon_f^2}{\varepsilon\varepsilon_f}K_{2/3}(\xi) + \int_\xi^\infty dy K_{1/3}(y)\right]$$
$$= \frac{\alpha}{\sqrt{3}\pi}\frac{m^2 dy}{\hbar\omega\cosh^2 y}\left[4\cosh^2 y K_{2/3}(\xi) - \int_\xi^\infty dy K_{5/3}(y)\right]\,, \tag{3.50}$$

where

$$\xi = \frac{2}{3}\frac{\hbar^2\omega^2}{\varepsilon\varepsilon_f\kappa} = \frac{8\cosh^2 y}{3\kappa}\,, \quad 0 \le y < \infty\,. \tag{3.51}$$

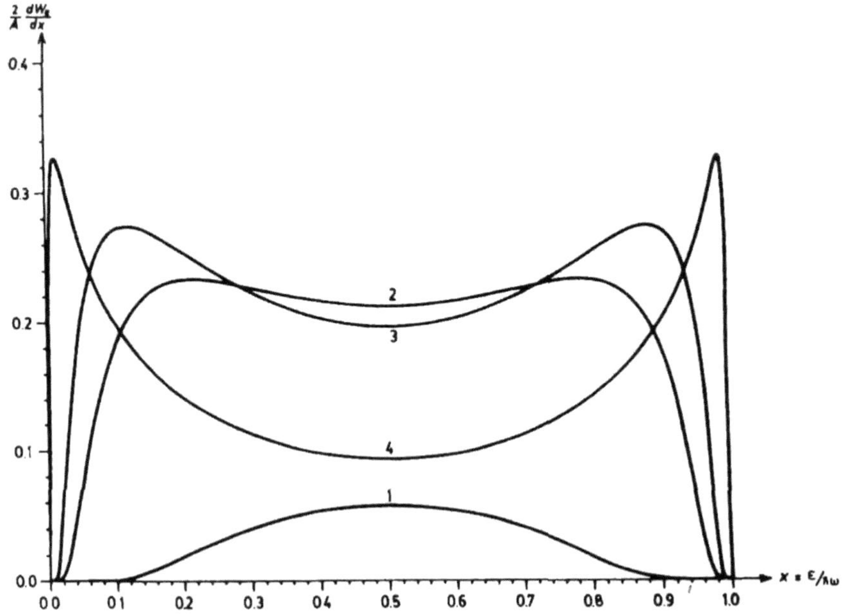

Fig. 3.1. Spectrum of one of the particles of the pair for different values of the parameters $\kappa(1:\kappa=1; 2:\kappa=5; 3:\kappa=10; 4:\kappa=100)$, $A=\alpha mH/H_0$.

In derivation of the second formula (3.50) we have used the recurrent relation (1.93). The form of the spectrum is presented in Fig. 3.1. The spectrum is symmetric with respect to point $\varepsilon/\hbar\omega \equiv x = 1/2$. It is seen that at small κ the spectral distribution is concentrated near $\varepsilon/\hbar\omega \equiv x \simeq 1/2$. If κ gets larger the distribution becomes wider and a plateau arises at all x except the edges $(x \ll 1, 1-x \ll 1)$. If κ gets larger $(\kappa \gg 1)$ a hollow arises in the centre of the plateau a depth of which is increasing with κ, so that at $\kappa \gg 1$ the spectral distribution has peaks near the edges. At $\kappa \gg 1$ the main contribution in expression (3.50) gives the term without the integral. The analysis of this term shows that these peaks near the edges are situated at x or $(1-x) \simeq 1.6/\kappa$, besides the height of distribution in the maximum does not depend on κ : $2A^{-1}dW_e/dx \simeq 0.33 (A = \alpha mH/H_0)$, and the height of the distribution in the minimum at $x = 1/2$ is $2A^{-1}dW_e/dx \simeq 0.41\kappa^{-1/3}$ and it decreases when κ increases. The contribution of the peaks into the integral probability of the pair production accounts for $\kappa^{-2/3}$ of the total probability.

Similar calculations for polarized photon give

$$dW_e(\mathbf{e}) = \frac{\alpha m^2 dy}{\sqrt{3}\pi\hbar\omega\cosh^2 y}\left\{|\mathbf{ee}_1|^2\left[(4\cosh^2 y - 1)K_{2/3}(\xi) - \int_\xi^\infty K_{5/3}(y)dy\right] + |\mathbf{ee}_2|^2\left[(4\cosh^2 y + 1)\right.\right.$$
$$\left.\left.\times K_{2/3}(\xi) - \int_\xi^\infty K_{5/3}(y)dy\right]\right\}$$
$$= \frac{\alpha m^2}{\sqrt{3}\pi}\frac{d\varepsilon}{\hbar^2\omega^2}\left\{|\mathbf{ee}_1|^2\left[\left(\frac{\hbar^2\omega^2}{\varepsilon\varepsilon_f} - 3\right)K_{2/3}(\xi) + \int_\xi^\infty K_{1/3}(y)dy\right]\right.$$
$$\left.+|\mathbf{ee}_2|^2\left[\left(\frac{\hbar^2\omega^2}{\varepsilon\varepsilon_f} - 1\right)K_{2/3}(\xi) + \int_\xi^\infty K_{1/3}(y)dy\right]\right\}. \qquad (3.52)$$

It should be observed that after integration over the polar angle of the emission of the final particle the dependence on circular polarization of the photon is vanished.

At $\kappa \ll 1$, $\xi \gg 1$ and one can use the asymptotic formula (1.73). Then from (3.50) we obtain asymptotic representation of the pair spectrum

$$\frac{dW_e}{dx} = \frac{A}{2}\sqrt{\frac{3\xi}{2\pi}}e^{-\xi}(1 - x + x^2), \qquad (3.53a)$$

where $x = \varepsilon/\hbar\omega, \xi = \frac{2}{3}\frac{1}{\kappa x(1-x)}, A = \frac{\alpha m H}{H_0}$. Even though this expression is derived at $\kappa \ll 1$, it gives the pair spectrum at $\kappa = 1$ with an accuracy better than 3%.

The same asymptotic representation for the polarized photon (3.52) has the form

$$\frac{dW_e}{dx} = \frac{A}{2}\sqrt{\frac{3\xi}{2\pi}}e^{-\xi}\{|\mathbf{ee}_1|^2[x^2 + (1-x)^2] + |\mathbf{ee}_2|^2\} \qquad (3.53b)$$

At $\kappa \gg 1$ and in the interval of $x : x \gg 1/\kappa$ or $(1 - x) \gg 1/\kappa$ we have $\xi \ll 1$ and one can use an asymptotic expansion (1.73). The main contribution to probability (3.50) gives the term without integral. As a result we obtain

$$\frac{dW_e}{dx} = \frac{A 3^{1/6}\Gamma(2/3)}{2\pi\kappa^{1/3}}\frac{1}{[x(1-x)]^{1/3}}[x^2 + (1-x)^2], \qquad (3.53c)$$

where we use the same notation as in (3.53a). The terms $\sim \kappa^{-2/3}$ are dropped. The distribution (3.53c) describes the spectrum in all the regions beyond the peaks near the edges.

To fulfill the integration over the spectrum, it is necessary in the integrals over y containing $\int_\xi^\infty K_{5/3}(x)dx$, to perform integration by parts and to use the recursion formula (Gradstein and Ryzhik, 1965, p. 959)

$$K_{5/3}(z) = K_{1/3}(z) + 4K_{2/3}(z)/(3z)\,.$$

As a result, we find total probability of electron-positron pair production by a polarized photon in an external field

$$W_e(\mathbf{e}) = (\mathbf{e}\mathbf{e}_1)^2 W_{e1} + (\mathbf{e}\mathbf{e}_2)^2 W_{e2}\,, \tag{3.54}$$

where

$$\begin{aligned} W_{e1} &= \frac{\alpha m^2}{\sqrt{3}\pi\hbar\omega}\int_0^\infty [(3 - 5\tanh^2 y/3)K_{2/3}(\xi) - 16\sinh^2 y K_{1/3}(\xi)/3\kappa)]dy \\ &= \frac{2\alpha m^2}{3\sqrt{3}\pi\hbar\omega}\int_0^1 \frac{3+v^2}{1-v^2}K_{2/3}(\xi)dv\,, \\ W_{e2} &= \frac{\alpha m^2}{\sqrt{3}\pi\hbar\omega}\int_0^\infty [(5 - 11\tanh^2 y/3)K_{2/3}(\xi) - 16\sinh^2 y K_{1/3}(\xi)/(3\kappa)]dy \\ &= \frac{4\alpha m^2}{3\sqrt{3}\pi\hbar\omega}\int_0^1 \frac{3-v^2}{1-v^2}K_{2/3}(\xi)dv\,. \end{aligned} \tag{3.55}$$

At derivation of the last equalities we change the variable to $\cosh^2 y = 1/(1-v^2)$, i.e. $\xi = 8/[3\kappa(1-v^2)]$ and using the relation: $-\xi K_{1/3}(\xi) = \xi(d/d\xi)K_{2/3}(\xi) + 2/3K_{2/3}(\xi)$ and do an integration by parts. Finally, averaging over the polarizations, we have

$$\begin{aligned} W_e = (W_{e1}+W_{e2})/2 &= \frac{4\alpha m^2}{\sqrt{3}\pi\hbar\omega}\int_0^\infty \left[\left(1 - \frac{2\tanh^2}{3}y\right)K_{2/3}(\xi)\right. \\ &\left. - 4\sinh^2 y K_{1/3}(\xi)/(3\kappa)\right]dy = \frac{\alpha m^2}{3\sqrt{3}\pi\hbar\omega}\int_0^1 \frac{9-v^2}{1-v^2}K_{2/3}(\xi)dv \\ &= \frac{\alpha m^2}{6\sqrt{3}\pi\hbar\omega}\int_1^\infty \frac{du(8u+1)}{u^{3/2}\sqrt{u-1}}K_{2/3}(\xi)\,. \end{aligned} \tag{3.56}$$

At derivation of the last equality we change once more the variable: $u = 1/(1-v^2), \xi = 8u/3\kappa$. The last integral may be expressed in terms of the generalized hypergeometric functions ${}_2F_3(a,b;c,d,e;g^2/4), g = 8/3\kappa$, and so may be used to obtain the representation of W_e as the series over inverse powers of κ. The explicit formulae in terms of ${}_2F_3$ is rather cumbersome and we do not present it here.

We consider the dependence of the total probability on values of the parameter κ. At $\kappa \ll 1$ and at any $y, \xi = 8\cosh^2 y/3\kappa \gg 1$ therefore in all the range of variations of y one may use the asymptotic expansion $K_\nu(\xi)$ at $\xi \gg 1$ (1.73). Then the integrals in (3.55) and (3.50) may easily be calculated, if it is taken into account that the values of $y \ll 1$ make the essential contribution in them, i.e. it is possible to expand the incoming hyperbolic functions. As a result, we find the series over the powers of κ the main terms of which for polarized photons are (see (3.54))

$$\begin{aligned} W_{e2} = 2W_{e1} &= (\sqrt{3}/4\sqrt{2})(\alpha m^2\kappa/\hbar\omega)\exp(-8/3\kappa) \\ &= (\sqrt{3}/4\sqrt{2})A\exp(-8/3\kappa)\,, \quad A = \alpha m H/H_0 \end{aligned} \tag{3.57}$$

and for the averaged over the polarizations probability

$$\begin{aligned} W_e &= [3\sqrt{3}\alpha m^2\kappa/(16\sqrt{2}\hbar\omega)]\exp(-8/3\kappa) \\ &\quad\times(1-(11/64)\kappa + (7585/73728)\kappa^2 + \cdots)\,. \end{aligned} \tag{3.58}$$

Although the expression (3.58) was obtained under the assumption that $\kappa \ll 1$, comparing the results of numerical calculations of (3.58) and exact formula (3.50) one can be convinced that Eq. (3.58) represents the probability with an accuracy of better 15% up to $\kappa = 2$.

In the range $\kappa \ll 1$, the probability of the process is exponentially small. This situation is characteristic for all processes with a finite discontinuity of the square of the four momentum of the system (invariant mass of the system). For the pair production process, the discontinuity of the four momentum of the system is $\Delta_p^2 = (p+p')^2 - k^2 = (p+p')^2, \Delta^2_{p_{\min}} = 4m^2$

At $\kappa \gg 1$ and $y \sim 1$, we have $dW_e/dy \sim \kappa^{2/3}$ and at $\cosh^2 y \sim \kappa$, $dW_e/dy \sim 1$, so that the main contribution to the value W_e is made by the range $y \sim 1$ (an analogous situation also occurs in magnetic bremsstrahlung). Using the asymptotic expansion of the K_ν functions (1.73) and performing the corresponding integrations, we obtain the probability of pair production by a polarized photon:

$$W_{e2} = (3/2)W_{e1} = C_0 \alpha m^2 \kappa^{2/3}/\hbar\omega = C_0 A \kappa^{-1/3}$$

$$A = \frac{\alpha m H}{H_0}, \quad C_0 = \frac{3\Gamma(5/6)(2/3)^{1/3}}{7\Gamma(7/6)} = 0.45553. \qquad (3.59)$$

The functions W_{e1} and W_{e2} are given in Fig. 3.2(a). For the probability averaged over the polarizations* one has

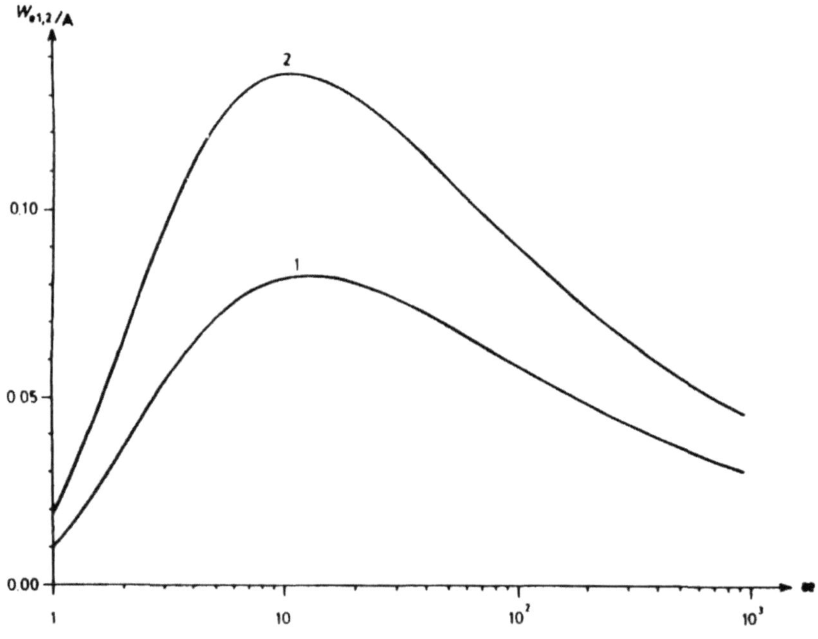

(a)

Fig. 3.2. Probability of the pair creation by (a) polarized and (b) unpolarized photons, $A = \alpha m H/H_0$.

*Using formula (3.56), particularly expressed in terms of ${}_2F_3$ functions, one can obtain the first terms of the series:

$$W_e = A\kappa^{-1/3}[C - C_1\kappa^{-2/3} - C_2\kappa^{-4/3} + C_3\kappa^{-5/3} + \cdots]$$

$$C_1 = 2/3, \quad C_2 = \frac{\Gamma(7/6)(3/2)^{1/3}}{\Gamma(5/6)} = 0.94081, \quad C_3 = \frac{16\sqrt{3}}{5}$$

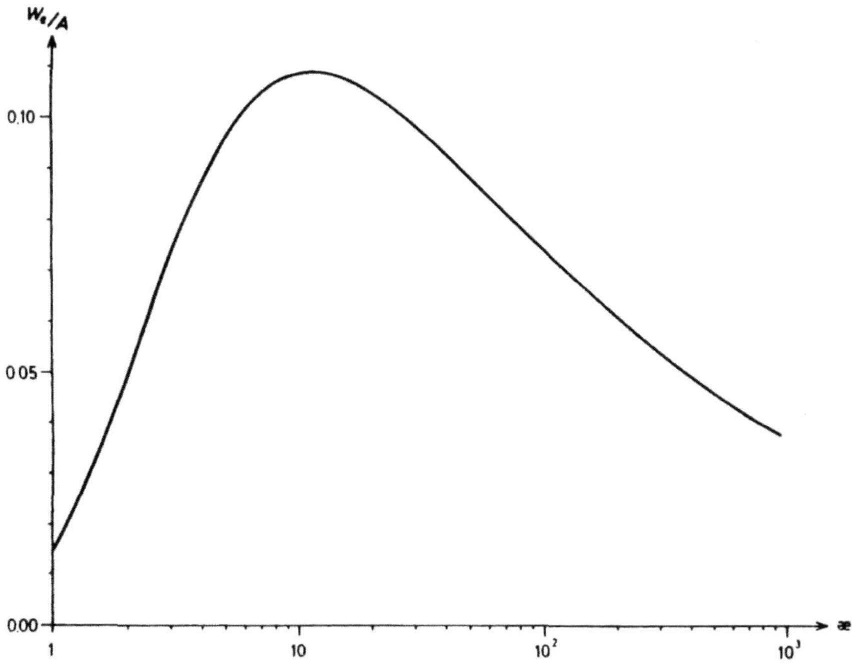

Fig. 3.2 (b) *Continued*

$$W_e = C\alpha m^2\kappa^{2/3}/\hbar\omega = CA\kappa^{-1/3}$$
$$C = \frac{5\Gamma(5/6)(2/3)^{1/3}}{14\Gamma(7/6)} = 0.37961\,, \kappa \gg 1\,. \tag{3.60}$$

This probability may be obtained by also performing the integration of Eq. (3.53c) in the interval $0 \leq x \leq 1$.

The function W_e/A is given in Fig. 3.2.(b). This function reaches a maximum of 0.11 at $\kappa \simeq 11.7$. Thus, in the given field (for definiteness, magnetic field) the probabilities W_{e1}, W_{e2}, W_e have maximum values (somewhat shifted, relative each other) within interval $\kappa \simeq 10 \div 12$. At $\kappa \ll 1$ up to $\kappa = 1$, one can use expansions (3.57), (3.58). At $\kappa \gg 1$ one can use the formulae (3.59), (3.60), with allowance, however, that at $\kappa = 100$ these formulae present the probabilities within an accuracy order of 10%. So, in the rather wide interval $1 \le \kappa \le 100$ one should use the general formulae* (3.55),(3.56).

3.3. Production of The Pair with Polarized Created Particles

We consider now production of the pair when the spin state of the electron is fixed and a summation is carried out over final spin states of the positron. Then we find using (3.23)

$$\sum_{\zeta_+} R_p(t_2)R_p^*(t_1) = A_1^* A_2 + \mathbf{B}_1^* \mathbf{B}_2$$

$$+ i[\zeta_-(A_2\mathbf{B}_1^* - A_1^*\mathbf{B}_2) + (\zeta_-(\mathbf{B}_2 \times \mathbf{B}_1^*))] , \quad (3.61)$$

where the first two terms are one half of the probability of unpolarized pair creation (see (3.24)) and additional terms are dependent on an electron spin. Proceeding from the general formula (3.23) one can easily see that the polarization of the final electron is (see(A.20), Appendix A)

$$\zeta_f = \frac{i(A_2\mathbf{B}_1^* - A_1^*\mathbf{B}_2) + (\mathbf{B}_2 \times \mathbf{B}_1^*)}{A_1^* A_2 + \mathbf{B}_1^* \mathbf{B}_2} . \quad (3.62)$$

As it was mentioned above, the coefficients at $(\mathbf{e}^*\mathbf{e}_i)(\mathbf{e}\mathbf{e}_k)$ in formula (3.37) are the elements of the polarization matrix l_{ik}. In accordance with formula (3.61) we present the total polarization matrix in the form

$$l_{ik}^p = l_{ik} + l_{ik}^\zeta , \quad (3.63)$$

where l_{ik} are given by Eq. (3.37) and terms l_{ik}^ζ follow from formula (3.61), if one substitutes into it the explicit representations of A and $\mathbf{B}$ (3.20). As a result of the substitution we find

$$l^\zeta \equiv l_{11}^\zeta + l_{22}^\zeta = i\omega\tau\hbar\omega[(1 - \varepsilon/\varepsilon_f)\beta(\zeta\mathbf{v}) + (\zeta\mathbf{v}\mathbf{s})/\gamma]/2\varepsilon ;$$

$$l_-^\zeta \equiv l_{11}^\zeta - l_{22}^\zeta = -i\omega\tau m\hbar\omega(\zeta\mathbf{v}\mathbf{s})/(2\varepsilon\varepsilon_f) ;$$

---

*Under the circumstances, the interpolation formula is rather useful, for the probability averaged over the photon polarization, we have (see Baier, Katkov and Strakhovenko, 1991b)

$$W_e = \left(\frac{3}{8}\right)^{3/2} \frac{A}{2} e^{-8/3\kappa} [(1 + 0.094\kappa)^{-1/3} + (1 + 0.94\kappa)^{-1/3}]$$

whose accuracy is better than 3%.

$$
\begin{aligned}
l_{12}^{\zeta} + l_{21}^{\zeta} &= iw\tau m\hbar\omega(\zeta\mathbf{s})/(2\varepsilon\varepsilon_f)\,; \\
l_{12}^{\zeta} - l_{21}^{\zeta} &= i\hbar\omega\{[(1-\varepsilon/\varepsilon_f)w^2\tau^2/4 + 1/\gamma^2](\zeta\mathbf{v}) \\
&\quad - \psi(\zeta\mathbf{s})/\gamma - \beta(\zeta\mathbf{vs})/\gamma\}/\varepsilon\,,
\end{aligned} \tag{3.64}
$$

where $(\zeta\mathbf{vs}) = (\zeta\cdot(\mathbf{v}\times\mathbf{s}), \mathbf{s} = \mathbf{w}/w$.

Substituting the terms with l_{ik}^{ζ} into the expression for the probability written in the form (3.37), picking out the terms at ζ and substituting the result and (3.37) into (3.62) we find the polarization of the final electron for the given polarization of the incoming photon $\mathbf{e}$. The same procedure is valid also in the case, when the integration over variables is carried out.

One can use the same method of integration over final electron states as used in the previous subsection. The integration of (3.64) over the angle ψ is convenient to do simultaneously with integration over τ (see (3.39), (3.40)). The resulting matrix elements are written in the form: $g_{ik}^{p} = g_{ik} + g_{ik}^{\zeta}$. The elements g_{ik} are given by Eq. (3.44), depending on spin additional parts are:

$$
\begin{aligned}
g^{\zeta} \equiv g_{11}^{\zeta} + g_{22}^{\zeta} &\equiv \mathbf{a}_1\zeta = -\hbar\omega\mu[(1-\varepsilon/\varepsilon_f)\beta(\zeta\mathbf{v}) + (\zeta\mathbf{vs})/\gamma] \\
&\quad\times K_{2/3}(\eta)K_{1/3}(\eta)/\varepsilon\,; \\
g_{-}^{\zeta} \equiv g_{11}^{\zeta} - g_{22}^{\zeta} &= m\hbar\omega\mu(\zeta\mathbf{vs})K_{2/3}(\eta)K_{1/3}(\eta)/(\varepsilon\varepsilon_f)\,; \\
g_{12}^{\zeta} + g_{21}^{\zeta} &= -m\hbar\omega\mu(\zeta\mathbf{s})K_{2/3}(\eta)K_{1/3}(\eta)/(\varepsilon\varepsilon_f)\,; \\
g_{12}^{\zeta} - g_{21}^{\zeta} \equiv i\mathbf{a}_2\zeta &= i\hbar\omega\{[(1-\varepsilon/\varepsilon_f)\mu^2(K_{2/3}^2(\eta) + K_{1/3}^2(\eta)/2 \\
&\quad - K_{1/3}^2(\eta)/\gamma^2](\zeta\mathbf{v}) + \beta(\zeta\mathbf{vs})K_{1/3}^2(\eta)/\gamma\}/\varepsilon\,,
\end{aligned} \tag{3.65}
$$

where η is defined in (3.43) and μ is defined in (1.37).

Using (3.65) and (3.44) we can find the polarization of the final electron, as it was mentioned above. We consider this problem for the circular polarization of the incoming photon: $\mathbf{e}_{\mp} = (\mathbf{e}_1 \mp i\mathbf{e}_2)/\sqrt{2}$. Then the elements of the polarization matrix enter probability in the following combinations $(1/2)(g_{11} + g_{22} + i\lambda(g_{12} - g_{21}))$, here λ is the photon helicity ($\lambda = \mp 1$). Substituting these combinations into (3.62) we find,

$$
\zeta^{(\pm)} = (\mathbf{a}_1 - \lambda\mathbf{a}_2)/a_{(\pm)}\,, \tag{3.66}
$$

where $\mathbf{a}_1$ and $\mathbf{a}_2$ are defined in (3.65),

$$a_{(\pm)} = (\hbar\omega/2\varepsilon)\left\{\frac{(\varepsilon^2+\varepsilon_f^2)}{2\varepsilon\varepsilon_f}\left[\frac{\mu^2}{2}(K_{1/3}^2(\eta)+K_{2/3}^2(\eta))\right.\right.$$

$$\left.\left.\pm\beta\mu K_{1/3}(\eta)K_{2/3}(\eta)\right]+K_{1/3}^2(\eta)/\gamma^2\right\}, \qquad (3.67)$$

here index in $\boldsymbol{\zeta}^{(\mp)}$ refers to photon polarization $\mathbf{e}_\mp$, expression for $a_{(\mp)}$ follows from (3.44), all the K_ν functions depend on η. Now we consider a situation when the final electron takes away nearly all the energy of the photon $\varepsilon \approx \hbar\omega, \varepsilon_f \ll \varepsilon$. Then it's following from (3.66) that

$$\boldsymbol{\zeta}^{(\pm)} = \lambda\mathbf{v} \qquad (3.68)$$

and one has to take into account that within accepted accuracy $|\mathbf{v}| \approx 1$. Thus, when the pair is produced by circularly polarized photon "the transfer" of the helicity occurs if the electron takes away nearly all the energy of the photon.

The polarization density matrix integrated over all the angles has the form

$$dW_{e+} \equiv dW_{e11}+dW_{e22} = \frac{A_0}{\hbar\omega}\frac{dy}{2\cosh^2 y}\left\{4\cosh^2 y K_{2/3}(\xi)\right.$$

$$\left.-\int_\xi^\infty K_{5/3}(y)dy+\hbar\omega(\boldsymbol{\zeta}\mathbf{v}\mathbf{s})K_{1/3}(\xi)/\varepsilon\right\};$$

$$dW_{e-} \equiv dW_{e11}-dW_{e22} = -\frac{A_0}{\hbar\omega}\frac{dy}{2\cosh^2 y}$$

$$\times[K_{2/3}(\xi)+\hbar\omega(\boldsymbol{\zeta}\mathbf{v}\mathbf{s})K_{1/3}(\xi)/\varepsilon_f]; \qquad (3.69)$$

$$dW_{e12}+dW_{e21} = \frac{A_0}{\varepsilon_f}\frac{dy}{2\cosh^2 y}(\boldsymbol{\zeta}\mathbf{s})K_{1/3}(\xi);$$

$$dW_{e12}-dW_{e21} = -\frac{iA_0}{\varepsilon}\frac{dy}{2\cosh^2 y}\left[\hbar\omega K_{2/3}(\xi)/\varepsilon_f-\int_\xi^\infty K_{5/3}(y)dy\right](\boldsymbol{\zeta}\mathbf{v}),$$

where $A_0 = \alpha m^2/(\sqrt{3}\pi\hbar), \xi$ is given by Eq. (3.51).

We can find the polarization of the final electron using Eqs. (3.69) and (3.44). For this we should rewrite, depending on the spin terms in (3.69), to the form (3.52) (the result contains all the 4 combinations $(\mathbf{e}^*\mathbf{e}_i)(\mathbf{e}\mathbf{e}_k)$), then we pick out a term at $\boldsymbol{\zeta}$ and divide by expression (3.52). Particularly, for the circular polarization of the incoming photon we find

$$\zeta^{(\pm)} = \zeta_1(x)\lambda\mathbf{v} + \zeta_2(\mathbf{v}\times\mathbf{s}),$$
$$\zeta_1(x) = \frac{F_1(x)}{F(x)}, \quad \zeta_2(x) = \frac{F_2(x)}{F(x)}, \tag{3.70}$$

where

$$F_1(x) = \frac{1}{x}\left[\left(\frac{x}{1-x} - 1\right)K_{2/3}(\xi) + \int_\xi^\infty K_{1/3}(z)dz\right],$$
$$F_2(x) = \frac{1}{x}K_{1/3}(\xi), \tag{3.71}$$
$$F(x) = \left(\frac{x}{1-x} + \frac{1-x}{x}\right)K_{2/3}(\xi) + \int_\xi^\infty K_{1/3}(z)dz;$$
$$x = \frac{\varepsilon}{\hbar\omega},$$

here ξ is given by Eq. (3.51), we used relation (1.93). In the region $x \ll 1$ always exist at intervals where $\xi \gg 1$. Applying asymptotic formulae for $K_\nu(\xi)$ function (the two first terms should be used) one can find that $\zeta_1(x \ll 1)$

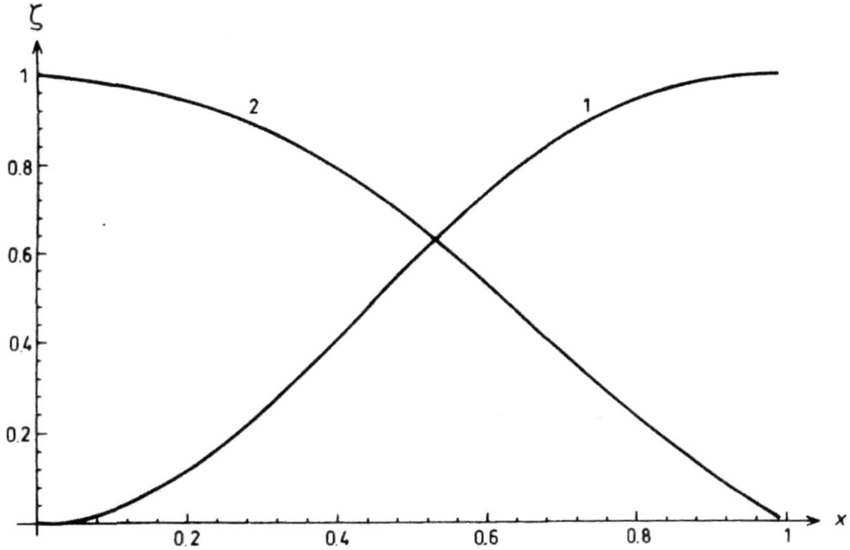

Fig. 3.3. Polarization of created electrons as a function of their relative energy $x = \varepsilon/\hbar\omega$ for $\kappa = 1$. Curve 1 is for longitudinal polarization ζ_1 and curve 2 is for the projection of polarization vector onto $\mathbf{v}\times\mathbf{s}$.

$\propto x(1-\kappa)$ and $\zeta_1'(x) > 0$ for $\kappa < 1$ and $\zeta_1'(x) < 0$ for $\kappa > 1$. When $x \to 1 (1 - x \ll 1)$ from Eqs. (3.70) and (3.71) follow that $\zeta_1(x \gg 1) \to 1$. This fact, as it was said above, is due to "helicity transfer" The functions $\zeta_1(x)$ and $\zeta_2(x)$ at $\kappa = 1$ are shown in Fig. 3.3. One can see that $\zeta_2(x \ll 1) \to 1$. A behavior of the function $\zeta_1(x)$ as function both x and κ is demonstrated by $3D$ plot in Fig. 3.4, where one can observe a formation of deepening at low x with κ increase and rather uniform behavior at $x \to 1$. When $\kappa \gg 1, \zeta_1(x)$ is decreasing rapidly from $\zeta_1(0) = 0$ (at $\kappa = 500$ to $\zeta_1(x) = -0.8$ at $x \simeq 0.1$) and then increasing smoothly to $\zeta_1(1) = 1$; and $\zeta_2(x)$ is decreasing rapidly from $\zeta_2(0) = 1$ (at $\kappa = 500$ to $\zeta_2(x) = 0.4$ at $x = 0.08$) and then decreasing smoothly to $\zeta_2(1) = 0$.

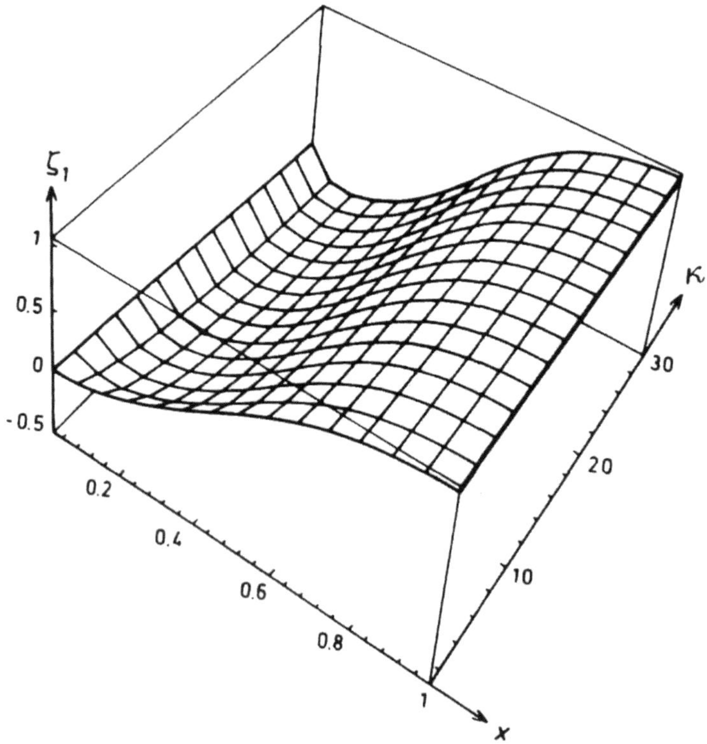

Fig. 3.4. The degree of the longitudinal polarization ζ_1 as a function of the parameters x and κ.

After integration of (3.69) over y (over spectrum) we obtain polarization density matrix for total probabilities. Let us note that at $\kappa \gg 1$ in expression for $W_{e12} - W_{e21}$ the main term ($\propto \kappa^{2/3}$) vanishes.

4. RADIATION IN MAGNETIC BREMSSTRAHLUNG LIMIT

4.1. Differential Probability and Intensity for Electron Radiation

In this section the radiation from ultrarelativistic electrons in Case 1 will be investigated under the same assumptions (the fulfillment of the inequalities (1.50) and (1.51)) with which the corresponding classical problem was considered (field constancy on the coherence length).

In the quantum theory in Case 1, as in the classical theory, radiation takes place from a small part of the trajectory on which the particle turns to an angle of $\sim 1/\gamma$. Therefore once again we shall introduce the variables (1.59):

$$t = (t_1 + t_2)/2, \quad \tau = t_2 - t_1; \quad t_1 = t - \tau/2, \quad t_2 = t + \tau/2 . \tag{4.1}$$

For these variables we have $\int dt_1 \int dt_2 = \int dt \int d\tau$. The integration over t may be regarded as the integration of the radiation characteristics per unit time. Then for the probability of radiation per unit time (2.27) we have

$$dW \equiv dw/dt = \frac{\alpha}{(2\pi)^2} \frac{d^3k}{\hbar\omega} \int d\tau R^*(t+\tau/2) R(t-\tau/2) \times \exp\{-i(\varepsilon/\varepsilon_f)[kx(t+\tau/2) - kx(t-\tau/2)]\} , \tag{4.2}$$

where $\varepsilon_f = \varepsilon - \hbar\omega$. We shall proceed to the calculation of this general formula for the particles with definite spin.

For particles with spin 1/2 in accordance with (2.9) we have (it should be remembered that the Coulomb gauge is used)

$$R_e = \sqrt{\frac{m}{\varepsilon_f}} \bar{u}_{s'}(\mathbf{P}')\mathbf{e}^*\boldsymbol{\alpha} u_s(\mathbf{P}) \sqrt{\frac{m}{\varepsilon}} = \varphi_{s'}^+ Q \varphi_s = \varphi_{s'}^+ (A + i\boldsymbol{\sigma}\mathbf{B})\varphi_s , \tag{4.3}$$

where we have proceeded to the two-component spinors. Then we use explicit representative of $u(\mathbf{P})$ (see (2.10) and (2.37)) and make in (2.37) corresponding expansions. The result can be written as

$$A = (1/2)\left(1 + \frac{\varepsilon}{\varepsilon - \hbar\omega}\right) \mathbf{e}\mathbf{v} \simeq (1/2)\left(1 + \frac{\varepsilon}{\varepsilon - \hbar\omega}\right) \mathbf{e}\boldsymbol{\vartheta} ,$$

$$\mathbf{B} = \frac{\hbar\omega}{2(\varepsilon - \hbar\omega)} \mathbf{e} \times \mathbf{b}, \mathbf{b} = \mathbf{n} - \mathbf{v} + \frac{\mathbf{n}}{\gamma} \simeq -\boldsymbol{\vartheta} + \frac{\mathbf{n}}{\gamma} , \tag{4.4}$$

where the angle $\boldsymbol{\vartheta} = (1/v)(\mathbf{v} - \mathbf{n}(\mathbf{nv})) \approx \mathbf{v}_\perp; \mathbf{v}_\perp$ is the component of velocity transversal to the vector $\mathbf{n}$.

If we take the equation of motion of the spin* (see e.g. Baier, Katkov and Fadin, 1973)

$$\frac{d\boldsymbol{\zeta}}{dt} = \frac{e}{\varepsilon}\boldsymbol{\zeta} \times \mathbf{H}_E, \mathbf{H}_E = \mathbf{H} + \frac{1}{1+1/\gamma}\mathbf{E} \times \mathbf{v}\,, \tag{4.5}$$

and compare it with the equation of motion of the particle $d\mathbf{v}/dt = (e/\varepsilon)[\mathbf{E} - \mathbf{v}(\mathbf{vE}) + (\mathbf{v} \times \mathbf{H})]$, we can see that the vector $\boldsymbol{\zeta}(t)$ with accuracy to the terms $\sim 1/\gamma$ precesses with the same frequency as the velocity. Taking into consideration that the characteristic radiation time is $\tau \approx 1/w\gamma = 1/\omega_L\gamma$ we obtain the estimation (2.38) taking into account (4.2),

$$\begin{aligned} R_e^*(t+\tau/2)R_e(t-\tau/2) \equiv R_2^* R_1 &= (1/4)\mathrm{Tr}[(1+\boldsymbol{\zeta}_i\boldsymbol{\sigma}) \\ &\times (A_2^* - i\boldsymbol{\sigma}\mathbf{B}_2^*)(1+\boldsymbol{\zeta}_f\boldsymbol{\sigma})(A_1 + i\boldsymbol{\sigma}\mathbf{B}_1)]\,, \end{aligned} \tag{4.6}$$

where $(1+\boldsymbol{\sigma}\boldsymbol{\zeta})/2$ is the two-component polarization density matrix. This expression may be used for calculation of any radiation characteristics, including polarization and spin characteristics.

We shall begin by considering the radiation intensity and probability, summed over the photon and final electron polarizations and averaged over the polarization of the initial electron. Then

$$(1/2)\sum_{s_i, s_f} R_{e2}^* R_{e1} = A_2^* A_1 + \mathbf{B}_2^* \mathbf{B}_1\,, \tag{4.7}$$

and after summation over the photon polarization λ

$$\begin{aligned} &\sum_\lambda A_2^* A_1 = (1/4)(1+\varepsilon/\varepsilon_f)^2(\mathbf{v}_2\mathbf{v}_1 - 1)\,, \\ &\sum_\lambda \mathbf{B}_2^* \mathbf{B}_1 = (1/4)(\hbar\omega/\varepsilon_f)^2(\mathbf{v}_2\mathbf{v}_1 - 1 + 2/\gamma^2)\,. \end{aligned} \tag{4.8}$$

Here the terms of the higher order over $1/\gamma$ have been neglected and use has been made of the fact that terms of the type $\mathbf{nv}_1$ and $\mathbf{nv}_2$ may be represented

---

*We want to stress that spin motion equation with $g = 2$ should be used and not the equation taking into account the anomalous magnetic moment, since for electrons the latter is a higher radiative correction and it must be taken into consideration only if the problem is solved in α^2 order.

in the form

$$\mathbf{n}\mathbf{v}_2 \exp(-i(\varepsilon/\varepsilon_f)kx_2) = \left(\frac{i\varepsilon_f d}{\omega\varepsilon dt_2} + 1\right) \exp(-i(\varepsilon/\varepsilon_f)kx_2)\,, \tag{4.9}$$

and the term with derivatives in the right part (4.9) vanishes at integration over time within infinite limits. Performing the expansion of the terms entering (4.2) we obtain (compare (1.60))

$$\begin{aligned} \mathbf{v}(t \pm \tau/2) &= \mathbf{v}(t) \pm \mathbf{w}\tau/2 + \dot{\mathbf{w}}\tau^2/8 + \dots\,, \\ \mathbf{r}(t \pm \tau/2) &= \mathbf{r}(t) \pm \mathbf{v}\tau/2 + \mathbf{w}\tau^2/8 \pm \dot{\mathbf{w}}\tau^3/48 + \dots\,, \end{aligned} \tag{4.10}$$

taking into consideration that $\mathbf{v}\dot{\mathbf{v}} = O(\gamma^{-2})$ and $\mathbf{n}\dot{\mathbf{w}} = -\dot{\mathbf{v}}^2(1 + O(1/\gamma))$ (compare (3.5)), we obtain

$$\begin{aligned} \mathbf{v}_2\mathbf{v}_1 &= 1 - 1/\gamma^2 - w^2\tau^2/2\,, \\ kx_2 - kx_1 &= \omega\tau - \mathbf{k}\mathbf{r}_2 + \mathbf{k}\mathbf{r}_1 = \omega\tau\left(1 - \mathbf{n}\mathbf{v} + \frac{w^2\tau^2}{24}\right) \end{aligned} \tag{4.11}$$

Substituting (4.7), (4.8) and (4.11) into (4.2) we have

$$\begin{aligned} dW_e = &-\frac{\alpha}{(2\pi)^2}\frac{d^3k}{\hbar\omega}\int_{-\infty}^{\infty} d\tau[(1+u)/\gamma^2 + (1 + u + u^2/2)w^2\tau^2/2] \\ &\times \exp\{-i(u\varepsilon\tau/\hbar)(1 - \mathbf{n}\mathbf{v} + w^2\tau^2/24)\}\,, \end{aligned} \tag{4.12}$$

where $u = \hbar\omega/\varepsilon_f = \hbar\omega/(\varepsilon - \hbar\omega)$. At $\hbar \to 0$ this formula converts to $dI(\mathbf{n}, \omega)/\hbar\omega$ (1.65). Taking the integral over τ using Eq. (1.66) we find

$$\begin{aligned} dW_e(\mathbf{k}) = &\frac{\alpha}{\pi^2}\frac{d^3k}{\hbar\omega}\left(\frac{2}{3}\right)^{1/2}\frac{(1 - \mathbf{n}\mathbf{v})^{1/2}}{w}\left\{-\frac{(1+u)}{\gamma^2}\right. \\ &\left. + 2(1 - \mathbf{n}\mathbf{v})[1 + (1+u)^2]\right\}K_{1/3}(\zeta)\,, \end{aligned} \tag{4.13}$$

where

$$\begin{aligned} \zeta &= \frac{4\sqrt{2}}{3}\frac{u\varepsilon}{w\hbar}(1 - \mathbf{n}\mathbf{v})^{3/2} = \left(\frac{2}{3}\right)\left(\frac{u}{\chi}\right)[2\gamma^2(1 - \mathbf{n}\mathbf{v})]^{3/2} \\ &= \left(\frac{2}{3}\right)\left(\frac{u}{\chi}\right)[\gamma^2(\psi^2 + \mu^2)]^{3/2}\,. \end{aligned} \tag{4.14}$$

At $\hbar \to 0$ we have $dI(\mathbf{n},\omega)/\hbar\omega$ in (1.67) and (1.68). About parameter χ see Appendix B, Eqs. (B.26), (B.28).

As it was mentioned (see discussion after Eq. (3.13)) the probabilities of the radiation of the electron and the pair production by the photon in an external field are connected by the crossing symmetry (the substitution law). This law is exact for differential probabilities. However, since in these processes all the final particles are emitted the angle $\sim 1/\gamma$ in the forward direction, the crossing symmetry relations are fulfilled within relativistic accuracy also for probabilities integrated over some of the variables. Particularly, from formulae (4.12) and (4.13) after the substitutions $1+u \to -\varepsilon/\varepsilon_f$, $u^2 \to (\hbar\omega/\varepsilon_f)^2$, $d^3k \to d^3p$ we have (3.34) and (3.43).

As in classical electrodynamics, the probability integrated over the azimuthal angle is of interest. We introduce the same variables as in Sec. 1 (see formula (1.36), Fig. 1.2). As in classical theory, it proves appropriate to perform integration over ψ and τ simultaneously and we spread out the first one over an infinite interval, since the range $\psi \sim 1/\gamma$ gives the main contribution. We introduce the variables x and y (see (1.78)):

$$\psi = \delta(x-y)/2\,, \quad w\tau = \delta(x+y)\,, \quad \delta = (2\hbar w/u\varepsilon)^{1/3}\,. \tag{4.15}$$

In these variables (4.12) takes the form

$$\begin{aligned} dW_e(\beta,\omega) = &-\frac{\alpha}{(2\pi)^2}\frac{\delta^2}{w}\omega d\omega d\beta \int_{-\infty}^{\infty} dx \int_{-\infty}^{\infty} dy[(1+u)/\gamma^2 \\ &+ \delta^2(1+u+u^2/2)(x+y)^2/2]\exp\{-i[\mu^2(x+y)/\delta^2 \\ &+ (x^3+y^3)/3]\} \end{aligned} \tag{4.16}$$

Using the integrals (1.66), we obtain

$$\begin{aligned} dW_e = &\frac{\alpha}{3\pi^2}\frac{\varepsilon^2}{\hbar^2 w}\frac{u\mu^2}{(1+u)^3}\{\mu^2(1+u+u^2/2)[K_{1/3}^2(\eta) \\ &+ K_{2/3}^2(\eta)] - (1+u)K_{1/3}^2(\eta)/\gamma^2\}dud\beta\,, \end{aligned} \tag{4.17}$$

where

$$\eta = \frac{u\varepsilon\mu^3}{3\hbar w} = \frac{u(\gamma\mu)^3}{3\chi}\,, \quad \mu^2 = \beta^2 + 1/\gamma^2\,. \tag{4.18}$$

From formula (4.17) we can obtain formula (3.41) with the help in the above-mentioned substitutions and taking into account that $\kappa = \omega\chi/\varepsilon$. The expressions for the radiation intensity (the energy radiated by particle per unit time) follow from the formulae obtained above (see (2.13)):

$$dI \equiv \frac{d\varepsilon}{dt} = \hbar\omega\,\frac{dw}{dt} = \hbar\omega\, dW = \varepsilon u\, dW/(1+u)\,. \tag{4.19}$$

From the formula (4.13) we have the following formula for the differential intensity distribution of electron radiation in external field

$$dI_e\,(\mathbf{k}) = \frac{e^2}{\pi^2}(2/3)^{1/2}\frac{(1-\mathbf{nv})}{w}^{1/2} d^3k$$
$$\times\,[2(1-\mathbf{nv})[1+(1+u)^2]-(1+u)/\gamma^2]K_{1/3}(\zeta)\,, \tag{4.20}$$

while from the formula (4.17) we have the distribution of the radiation intensity over the polar angle and energy of the radiated photon

$$dI_e(\beta,\omega) = \frac{e^2}{3\pi^2}\frac{\varepsilon^3}{\hbar^3 w}\frac{u^2}{(1+u)^4}\mu^2\{\mu^2(1+u+u^2/2)$$
$$\times\,[K_{1/3}^2(\eta)+K_{2/3}^2(\eta)]-(1+u)K_{1/3}^2/\gamma^2\}du d\beta\,. \tag{4.21}$$

At $\hbar\to 0$ these formulae convert to the corresponding classical distributions (1.67) and (1.68).

To obtain spectral probability distribution it is appropriate to use the formula (4.13) and the angles of radiation ϑ and φ (see Fig. 1.2). The integration over azimuthal angle φ is trivial since the integrand does not depend on φ. For integration over ϑ we change to the variable

$$z = [2\gamma^2(1-\mathbf{nv})]^{3/2} \tag{4.22}$$

The lower limit of the integration over z is $z(0) = [2\gamma^2(1-v)]^{3/2} = 1+O(1/\gamma^2)$, while the upper limit is $z(\pi) = 8\gamma^3(1+O(1/\gamma^2))$, since at large z the integrand falls exponentially, the upper limit may be substituted for infinity. Thus, up to terms $\sim 1/\gamma^2$ we obtain

$$dW_e(u) = \frac{2\alpha m^2}{3\sqrt{3}\pi\hbar\chi\varepsilon}\frac{udu}{(1+u)^3}\int_1^\infty dz\{z^{2/3}[1$$
$$+(1+u)^2]-(1+u)\}K_{1/3}\left(\frac{2u}{3\chi}z\right) \tag{4.23}$$

Using the formula (1.91), we convert this formula to the form

$$dW_e(u) = \frac{\alpha m^2}{\sqrt{3}\pi\hbar\varepsilon}\frac{du}{(1+u)^3}\left\{[1+(1+u)^2]K_{2/3}(2u/3\chi) - (1+u)\int_{2u/3\chi}^{\infty} K_{1/3}(y)dy\right\} \tag{4.24}$$

We shall get another form of the spectral probability distribution from (4.24), if we use the recursion formula (1.93)

$$dW_e(u) = \frac{\alpha m^2}{\sqrt{3}\pi\hbar\varepsilon}\frac{du}{(1+u)^2}\left\{\frac{u^2}{1+u}K_{2/3}\left(\frac{2u}{3\chi}\right) + \int_{2u/3\chi}^{\infty} K_{5/3}(y)dy\right\}. \tag{4.25}$$

Using the above-mentioned substitutions we get from (4.25) the probability (3.50) while $udu/(1+u)^3 \to dy/(2\cosh^2 y)$. To obtain the corresponding formulae for spectral distribution of the radiation intensity we use the formulae (4.19)

$$dI_e(u) = \frac{e^2m^2}{\pi\sqrt{3}\hbar^2}\frac{udu}{(1+u)^4}\left\{[1+(1+u)^2]K_{2/3}(2u/3\chi) - (1+u)\int_{2u/3\chi}^{\infty} K_{1/3}(y)dy\right\}, \tag{4.26}$$

$$\begin{aligned} dI_e(u) &= \frac{e^2m^2}{\pi\sqrt{3}\hbar^2}\frac{udu}{(1+u)^3}\left\{\frac{u^2}{(1+u)}K_{2/3}(2u/3\chi) + \int_{2u/3\chi}^{\infty} K_{5/3}(y)dy\right\} \\ &= \frac{e^2m^2}{\pi\sqrt{3}\hbar^2}\frac{udu}{(1+u)^3}\left\{\left(1+\frac{u^2}{1+u}\right)K_{2/3}(2u/3\chi) + (2/3)\int_{2u/3\chi}^{\infty} K_{2/3}(y)dy\right\} \end{aligned} \tag{4.27}$$

At $\hbar \to 0$ these formulae convert to (1.92) and (1.94).

Before proceeding to the integral characteristics of the radiation process, we perform analysis of the results obtained. The properties of the radiation depend strongly on the relation between the values $u = \hbar\omega/(\varepsilon - \hbar\omega)$ and χ. At $u \ll \chi$ we have (see (1.73))

$$dI_e(u) = \frac{e^2}{\pi} \frac{3^{1/6}\Gamma(2/3)}{2\hbar^2} \frac{\chi^{2/3} m^2 u^{1/3} du}{(1+u)^4} [1 + (1+u)^2], \tag{4.28}$$

and at $u \gg \chi$ in accordance with (1.73) we obtain

$$dI_e(u) = \frac{e^2 m^2}{2\hbar^2} \left(\frac{\chi}{\pi}\right)^{1/2} \frac{u^{1/2} du}{(1+u)^4} (1 + u + u^2) \exp(-2u/3\mathcal{X}). \tag{4.29}$$

In the range $\chi \ll 1$ (or, see (B.28), $\hbar\omega_L\gamma^3 \ll \varepsilon$) in all the essential regions $u \ll 1$ (or $\hbar\omega \ll \varepsilon$) and $u/\chi = \omega/\omega_c$ (see (1.89)). The transition to this range coincides with transition $\hbar \to 0$ ($u \to \hbar\omega/\varepsilon$, $1 + u \to 1$). Then the formulae (4.28) and (4.29) convert to (1.95) and (1.96). Consequently, at $\chi \ll 1$ the radiation picture is the same as in classical electrodynamics (with small quantum corrections) and in essential region ($u \sim \chi$) the photon energy is much smaller than the energy of the radiating particle. As the energy increases, the value of $\hbar\omega_L\gamma^3$, increasing as γ^2, reaches the energy value ε (this is range $\chi = \hbar\omega_L\gamma^3/\varepsilon \geq 1$), and then the qualitative picture of the radiation becomes completely different from the classical one.

At $\chi \sim 1$ in the essential region $u \sim \chi \sim 1$, $\hbar\omega \sim \varepsilon$. A somewhat different situation arises at $\chi \gg 1$, as it can be seen from (4.28), at $u \gg 1$ $dI_e \sim du/u^{5/3}$ and at the integral over u the main contribution gives the range $u \sim 1$. Thus,

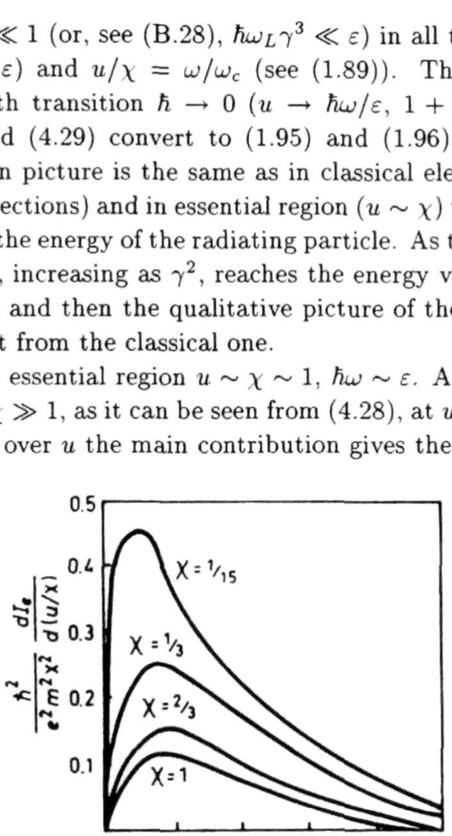

Fig. 4.1. Dependence of the spectral intensity distribution on frequency (4.26) for the indicated values of parameter χ.

at any $\chi \geq 1$ the energy of radiated photons $\hbar\omega \sim \varepsilon$ (but also $\varepsilon - \hbar\omega \sim \varepsilon$), i.e. the energy of the final electron is also of the order of the energy of the initial one.* In this situation the particle loses energy by quanta with the energy of the order ε, but not in such a way that the major part is lost by a single interaction. It is evident that range $\chi \geq 1$ is a quantum one. The picture of dependence of $dI_e(u)$ on u at various χ is given in Fig. 4.1.

We proceed now to the angular distribution of the radiation (4.21). At $\eta \ll 1$ we have (compare (1.73))

$$dI_e(\beta,\omega) = e^2 \frac{3^{1/3}\Gamma^2(2/3)}{2^{2/3}\pi^2}\left(\frac{\varepsilon}{\hbar}\right)^{5/3}\frac{u^{2/3}}{(1+u)^4} d\beta du w^{1/3}(1+u+u^2/2), \quad (4.30)$$

and at $\eta \gg 1$

$$dI_e(\beta,u) = \left(\frac{e^2}{2\pi}\right)\left(\frac{\varepsilon}{\hbar}\right)^2 \frac{udud\beta}{(1+u)^4}\frac{1}{\mu}[2\mu^2(1+u+u^2/2) - (1+u)/\gamma^2]\exp[-2u(\gamma\mu)^3/3\chi]\,. \quad (4.31)$$

The region $\eta \sim 1$ gives the main contribution. At $\chi \ll 1(\chi/u \sim \omega/\omega_c)$ we have $\mu\gamma \sim 1(\beta \sim 1/\gamma)$. This situation coincides with the classical one and the formulae (4.30) and (4.31) convert to (1.83) and (1.84). This angular

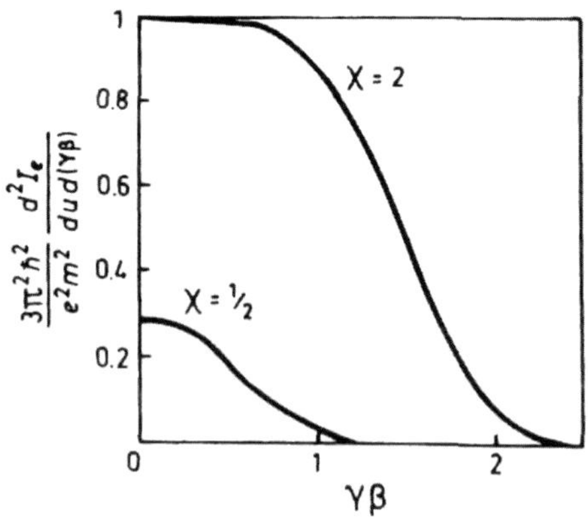

Fig. 4.2. Dependence of the angular distribution on the polar angle (4.21) for different χ.

*At $\chi \gg 1$, if $u \sim \chi \gg 1$ (at larger u the radiation is exponentially suppressed), then $\hbar\omega \gg \varepsilon - \hbar\omega = \varepsilon_f$, i.e. $\varepsilon/\varepsilon_f \sim \gamma H/H_0$ and $\varepsilon_f \sim H_0 m/H$. When $H/H_0 \ll 1$, the final particle is ultrarelativistic and results obtained are valid for the entire spectrum. If this condition is not fulfilled, the end of the spectrum may not be described by the obtained formulae and, consequently, the method applied here ceases to be closed.

distribution is preserved at $\chi \sim 1$. But at $\chi \gg 1$ in the essential range $u \sim 1$, i.e. $\mu\gamma \sim \chi^{1/3}$. This means that the angular distribution expands as $\beta \sim \chi^{1/3}/\gamma$. The same form, as this follows from (4.20), has the angular distribution in the azimuthal angle ψ: at $\chi \gg 1, \psi \sim \chi^{1/3}/\gamma$ (Fig. 4.2).

It follows from (4.29) and (4.31) that at $u \gg \chi$ and $\eta \gg 1$ that the intensity (and the probability) of the radiation is exponentially small. Such a situation is typical for all the processes with finite discontinuity of the square of the four-momentum of the system (invariant mass of the system). From this point of view we compare the radiation process with pair production. For the pair production process, the discontinuity of the square of the four-momentum of the system is $\Delta_p^2 = (p+p')^2 - k^2 = (p+p')^2, (\Delta_p^2)_{\rm min} = 4m^2$ whereas for the radiation process $\Delta_r^2 = (p'+k)^2 - p^2 = 2kp' = 2\varepsilon'\omega(1-\mathbf{nv}) \approx m\omega(1+\gamma^2\vartheta^2)/\gamma'$ So, at given frequency $(\Delta_r^2)_{\rm min} = m\omega/\gamma'$ At $\chi \ll 1$ the characteristic energy of the radiated photons is $\hbar\omega \sim \hbar\omega_L\gamma^3 \ll \varepsilon$, then $\gamma' \sim \gamma$ and $(\Delta_r^2)_{\rm min} \ll (\Delta_p^2)_{\rm min}$. Since the discontinuity of the invariant mass is due to the momentum transfer to the field, it follows from the above consideration that when the momentum transfer to the field (or the discontinuity of the invariant mass) becomes considerable, the probability of the processes at $\kappa \ll 1 (\chi \ll 1)$ is exponentially suppressed (particularly, this relates to the radiation of high-frequency photons).

4.2. The Integral Characteristics of the Electron Radiation Process (1)

We transform the expression (4.25) to a form suitable for integration over u. For this we use the obvious formula

$$\int_0^\infty f(x)dx \int_{x/\xi}^\infty \varphi(y)dy = \int_0^\infty \varphi(y)dy \int_0^{\xi y} f(x)dx \tag{4.32}$$

then

$$\int_0^\infty \frac{du}{(1+u)^2} \int_{u/\xi}^\infty K_{5/3}(y)dy = \int_0^\infty \frac{\xi y}{1+\xi y} K_{5/3}(y)dy\,, \tag{4.33}$$

where $\xi = 3\chi/2$. Using the expression (see, Gradshteyn and Ryzhik, 1965, p. 970)

$$z\frac{dK_{2/3}}{dz} - \frac{2}{3}K_{2/3}(z) = -zK_{5/3}(z)\,, \tag{4.34}$$

we transform the integral (4.33) to the form

$$\int_0^\infty [\xi y/(1+\xi y)]K_{5/3}(y)dy = \int_0^\infty [\xi(5+2\xi y)/3(1+\xi y)^2]K_{2/3}(y)dy$$
$$= \int_0^\infty [(5+2u)/3(1+u)^2]K_{2/3}(u/\xi)du\,. \tag{4.35}$$

Substituting into (4.25) we have the final expression for the total probability of the radiation of the electron

$$W_e = \frac{\alpha m^2}{3\sqrt{3}\pi\hbar\varepsilon}\int_0^\infty \frac{5u^2+7u+5}{(1+u)^3}K_{2/3}\left(\frac{2u}{3\chi}\right)du\,. \tag{4.36}$$

To calculate this integral it is convenient to use representation (see, Gradshteyn and Ryzhik, 1965, p. 657)

$$\frac{1}{(1+u)^m} = \frac{1}{2\pi i}\int_{\lambda-i\infty}^{\lambda+i\infty}\frac{\Gamma(-s)\Gamma(m+s)}{\Gamma(m)}u^s ds\,, \tag{4.37}$$

where $1-m<\lambda<0$, after which the integrals may easily be taken using formula (1.98)

$$W_e = \frac{\alpha m^2}{3\sqrt{3}\pi\hbar\varepsilon 2\pi i}\int_{\lambda-i\infty}^{\lambda+i\infty}\frac{\Gamma(-s)\Gamma(3+s)}{\Gamma(3)}ds\int_0^\infty (5u^2+7u+5)K_{2/3}\left(\frac{2u}{3\chi}\right)$$
$$\times u^s du = \frac{\alpha}{3\sqrt{3}\pi}\frac{m^2}{\hbar\varepsilon}\frac{1}{2\pi i}\int_{\lambda-i\infty}^{\lambda+i\infty}ds\frac{\Gamma(-s)\Gamma(3+s)}{\Gamma(3)}\frac{(3\chi)^{s+1}}{4}$$
$$\times\left\{5\Gamma\left(\frac{s}{2}+\frac{5}{6}\right)\Gamma\left(\frac{s}{2}+\frac{1}{6}\right)+7(3\chi)\Gamma\left(\frac{s+1}{2}+\frac{5}{6}\right)\Gamma\left(\frac{s+1}{2}+\frac{1}{6}\right)\right.$$
$$\left.+5(3\chi)^2\Gamma\left(\frac{s+2}{2}+\frac{5}{6}\right)\Gamma\left(\frac{s+2}{2}+\frac{1}{6}\right)\right\} = \frac{\alpha m^2\chi}{8\sqrt{3}\pi\hbar\varepsilon}\frac{1}{2\pi i}\int_{\lambda-i\infty}^{\lambda+i\infty}(3\chi)^s$$
$$\times(3s^2+3s+10)\Gamma(-s)\Gamma(s+1)\Gamma\left(\frac{s}{2}+\frac{1}{6}\right)\Gamma\left(\frac{s}{2}+\frac{5}{6}\right)ds\,. \tag{4.38}$$

In obtaining the latter expression, the variables were substituted in the second and third terms in the braces: $s+1\to s$, $s+2\to s$.

At $\chi \ll 1$, closing the integration contour on the right, we have the asymptotic series in powers of χ:

$$\begin{aligned} W_e &= \frac{\alpha m^2 \chi}{8\sqrt{3}\pi\hbar\varepsilon} \sum_{n=0}^{\infty} (-3\chi)^n (3n^2 + 3n + 10) \Gamma\left(\frac{n}{2} + \frac{1}{6}\right) \Gamma\left(\frac{n}{2} + \frac{5}{6}\right) \\ &= \frac{5\alpha}{2\sqrt{3}} \frac{m^2\chi}{\hbar\varepsilon} \left(1 - \frac{8\chi}{5\sqrt{3}} + \frac{7\chi^2}{2} + \cdots\right) \end{aligned} \tag{4.39}$$

At $\chi \gg 1$ the integration contour should be closed on the left, then the poles of the Γ functions will lie at $s < 0$, and we obtain a series over the inverse powers of χ

$$\begin{aligned} W_e &= \frac{\alpha\pi m^2 \chi}{2\sqrt{3}\hbar\varepsilon} \sum_{n=0}^{\infty} \left\{ \frac{4(18n^2 - 3n + 14)}{9n!\Gamma(n + 1/3)} (3\chi)^{-2n-1/3} \right. \\ &\quad + \frac{4(18n^2 + 21n + 20)}{9n!\Gamma(n - 5/3)} (3\chi)^{-2n-5/3} \\ &\quad \left. - \frac{(3n^2 + 3n + 10)(3\chi)^{-n-1}}{[2 + (-1)^n]\Gamma(n/2 + 2/3)\Gamma(n/2 + 4/3)} \right\} \\ &= \frac{14\Gamma(2/3)\alpha m^2}{27\hbar\varepsilon} (3\chi)^{2/3} \left[1 - \frac{45}{28\Gamma(2/3)} (3\chi)^{-2/3} \right. \\ &\quad \left. + \frac{15\Gamma(1/3)}{7\Gamma(2/3)} (3\chi)^{-4/3} - \frac{216\sqrt{3}}{35\Gamma(2/3)} (3\chi)^{-5/3} + \cdots \right] \end{aligned} \tag{4.40}$$

In addition to the integral representation for the total probability of the radiation process per unit time and its asymptotic expansions cited above, the closed formulae for W_e may be obtained. We shall represent (4.36) in the form

$$\begin{aligned} W_e &= \frac{\alpha m^2 \xi}{3\sqrt{3}\pi\hbar\varepsilon} \int_0^\infty \left[\frac{3\xi^2 x^2}{(1 + \xi x)^3} - \frac{3\xi x}{(1 + \xi x)^2} + \frac{5}{1 + \xi x} \right] K_{2/3}(x) dx \\ &= \frac{\alpha m^2 \xi}{3\sqrt{3}\pi\hbar\varepsilon} \left[\frac{3\xi^2}{2} \frac{d^2}{d\xi^2} + 3\xi \frac{d}{d\xi} + 5 \right] F(\xi), \end{aligned} \tag{4.41}$$

where

$$F(\xi) = \int_0^\infty \frac{K_{2/3}(x)}{1 + \xi x} dx. \tag{4.42}$$

Taking into account that (see Gradshteyn and Ryzhik, 1965, p. 677)

$$\int_0^\infty [K_\nu(x)/(1+\xi x)]dx = [\pi^2/(2\xi\sin^2\pi\nu)]\Phi_\nu(i/\xi), \tag{4.43}$$

where

$$\Phi_\nu(z) = \exp(-i\pi\nu/2)(J_\nu(z) - \mathbf{J}_\nu(z)) \\ + \exp(i\pi\nu/2)(J_{-\nu}(z) - \mathbf{J}_{-\nu}(z)), \quad \Phi_\nu = \Phi_{-\nu}, \tag{4.44}$$

here $J_\nu(z)$ is the Bessel function and $\mathbf{J}_\nu(z)$ is the Anger function (see Gradshteyn and Ryzhik, 1965, p. 988), we obtain for $F(\xi)$ (see (4.42))

$$F(\xi) = (2\pi^2/3\xi)\Phi_{2/3}(i/\xi). \tag{4.45}$$

Converting to the variable $z = i/\xi$ and using the fact that

$$d/d\xi = iz^2\frac{d}{dz}, \quad d^2/d\xi^2 = -z^4\frac{d^2}{dz^2} - 2z^3\frac{d}{dz},$$

$$d^3/d\xi^3 = -6iz^4\frac{d}{dz} + 6iz^5\frac{d^2}{dz^2} - iz^6 d^3/dz^3; \tag{4.46}$$

it is possible to rewrite the formula (4.41) in the form

$$W_e = \frac{2\alpha\pi m^2}{9\sqrt{3}\hbar\varepsilon z}\left(\frac{3}{2}z^2\frac{d^2}{dz^2} + 5\right)(z\Phi_{2/3}(z)) \\ = \frac{\alpha\pi m^2}{9\sqrt{3}\hbar\varepsilon}(3z^2\Phi''_{2/3}(z) + 6z\Phi'_{2/3}(z) + 10\Phi_{2/3}(z)), \tag{4.47}$$

where $\Phi'_\nu = d\Phi_\nu/dz$ etc. Using the recursion formula for the functions Φ_ν we have

$$\Phi'_{2/3} = -\Phi_{1/3} - 2\Phi_{2/3}/(3z) - \sqrt{3}/(2\pi z),$$

$$\Phi''_{2/3} = \left(\frac{10}{9z^2} - 1\right)\Phi_{2/3} + \frac{i\Phi_{1/3}}{z} + \frac{3i}{2\pi z} + \frac{5\sqrt{3}}{6\pi z^2} \tag{4.48}$$

Substituting these relations in (4.47) and converting to the variable $z = 2i/3\chi$, we obtain a formula for the total probability of radiation from an electron in an external field per unit time

$$W_e = \frac{\alpha\pi m^2}{9\sqrt{3}\hbar\varepsilon}\left[\frac{4}{3}\left(7 + \frac{1}{\chi^2}\right)\Phi_{2/3}\left(\frac{2i}{3\chi}\right) + \frac{2}{\chi}\Phi_{1/3}\left(\frac{2i}{3\chi}\right) - \frac{3}{\pi\chi} - \frac{\sqrt{3}}{2\pi}\right]. \tag{4.49}$$

We proceed now to the consideration of the total radiation intensity. Starting from the formula (4.27) and repeating the arguments (see (4.32) – (4.35)), which lead us to the formula (4.36), we obtain for the total radiation intensity

$$I_e = \frac{e^2m^2}{3\sqrt{3}\pi\hbar^2}\int_0^\infty \frac{u(4u^2+5u+4)}{(1+u)^4}K_{2/3}\left(\frac{2u}{3\chi}\right)du\,. \tag{4.50}$$

Using the formulae (4.37) and (1.98) and performing the calculation as in (4.38), we find the following formula for the total radiation intensity:

$$I_e = \frac{e^2m^2\chi^2\sqrt{3}}{8\pi\hbar^2 2\pi i}\int_{\lambda-i\infty}^{\lambda+i\infty}(3\chi)^s(s^2+2s+8)\Gamma(-s)$$
$$\times\,\Gamma(s+2)\Gamma(s/2+2/3)\Gamma(s/2+4/3)ds\,. \tag{4.51}$$

At $\chi \ll 1$, closing the integration contour on the right, we obtain the asymptotic series over the powers of χ

$$I_e = \frac{e^2m^2\chi^2\sqrt{3}}{8\pi\hbar^2}\sum_{k=0}^{\infty}(-1)^k(k+1)(k^2+2k+8)\Gamma(k/2+2/3)\Gamma(k/2+4/3)(3\chi)^k$$
$$= \frac{2e^2m^2\chi^2}{3\hbar^2}\left[1-\frac{55\sqrt{3}}{16}\chi+48\chi^2-\cdots\right]. \tag{4.52}$$

The first term in this expression does not contain the Planck constant and represents the classical radiation intensity (1.39) (see (B. 28)) which does not naturally depend on the particle spin; the second term is the first quantum correction,* which similarly does not depend on the particle spin. This dependence appears from terms χ^2 (terms $\propto \hbar^2$).

*The first quantum correction was found by Sokolov, Klepikov and Ternov, 1952 and somewhat later by Schwinger, 1954. The complete solution of the quantum problem of the magnetic bremsstrahlung was first obtained (using exact wave functions) by Klepikov, 1954. The radiation intensity and probability in the field plane electromagnetic wave or constant electric and magnetic fields perpendicular to each other were calculated using exact wave functions by Nikishov and Ritus, 1964a,b. In the quasiclassical limit both these results coincide (see Appendix B.3, (B. 27)). The complete solution of the radiation problem in a constant external field $F_{\mu\nu}$ of an arbitrary intensity (applicable also for $H \geq H_0$) was obtained at the calculation of the mass operator of a particle in the magnetic field (Schwinger, 1973; Tsai and Yildiz, 1973) and in arbitrary constant external field $F_{\mu\nu}$ (Baier, Katkov and Strakhovenko, 1974). In the case when invariant $\chi \gg f, g$ (see (B. 26), Appendix B) the complete formulae convert into formulae obtained in this section in quasiclassical approximation.

At $\chi \gg 1$, the integration contour should be closed on the left, since the poles of the Γ functions lie at $s < 0$ and the series over the inverse powers of χ are obtained

$$I_e = \frac{\sqrt{3}\pi e^2 m^2 \chi^2}{2\hbar^2} \sum_{n=0}^{\infty} \left\{ \frac{8(6n+1)(9n^2+3n+16)}{81 n! \Gamma(n+1/3)} (3\chi)^{-2n-4/3} \right.$$

$$+ \frac{8(6n+5)(9n^2+15n+22)}{81 n! \Gamma(n+5/3)} (3\chi)^{-2n-8/3}$$

$$\left. - \frac{n(n^2+7)(3\chi)^{-n-1}}{[2-(-1)^n]\Gamma(n/2+1/6)\Gamma(n/2+5/6)} \right\} = \frac{32\Gamma(2/3) e^2 m^2 (3\chi)^{2/3}}{3^5 \hbar^2}$$

$$\times \left\{ 1 - \frac{81}{16\Gamma(2/3)} (3\chi)^{-2/3} + \frac{165\Gamma(1/3)}{16\Gamma(2/3)} (3\chi)^{-4/3} - \frac{11 \cdot 3^6}{80\sqrt{3}\Gamma(2/3)} \right.$$

$$\left. \times (3\chi)^{-5/3} + \frac{147}{4} (3\chi)^{-2} - \cdots \right\} = 0.178 \frac{e^2 m^2}{\hbar^2} (3\chi)^{2/3}$$

$$\times [1 - 3.74(3\chi)^{-2/3} + 20.4(3\chi)^{-4/3} - 42.7(3\chi)^{-5/3}$$

$$+ 36.8(3\chi)^{-2} - \cdots] . \tag{4.53}$$

In addition to the formulae found above, it is possible, as for the probability of the radiation, to obtain closed formulae for the radiation intensity. Analogous to (4.41) we represent (4.50) in the form

$$I_e = -\frac{e^2 m^2 \xi^2}{3\sqrt{3}\pi\hbar^2} \left[\frac{\xi^2}{2} \frac{d^3}{d\xi^3} + \frac{3\xi}{2} \frac{d^2}{d\xi^2} + 4 \frac{d}{d\xi} \right] F(\xi) , \tag{4.54}$$

where $F(\xi)$ is given by (4.42). Converting to the variable $z = i/\xi$ and using the formulae (4.46) we have

$$I_e = \frac{2e^2 m^2}{9\sqrt{3}\pi\hbar^2} \left[\frac{z^3}{2} \Phi_{2/3}''' + 3z^2 \Phi_{2/3}'' + 7z\Phi_{2/3}' + 4\Phi_{2/3} \right] . \tag{4.55}$$

Using the formulae (4.48), and also the fact that

$$\Phi_{2/3}''' = \left(\frac{5}{3z} - \frac{80}{27z^3} \right) \Phi_{2/3} + i \left(1 - \frac{22}{9z^2} \right) \Phi_{1/3}$$

$$+ \left(1 - \frac{40}{9z^2} \right) \frac{\sqrt{3}}{2\pi z} - \frac{3i}{\pi z^2} ,$$

we obtain the final formula for the intensity of the magnetic bremsstrahlung of the electron ($z = 2i/3\chi$)

$$I_e = \frac{2e^2m^2}{3^4\hbar^2}\left\{\left[\left(16+\frac{13}{\chi^2}\right)\Phi_{2/3}\left(\frac{2i}{3\chi}\right)+\left(47+\frac{2}{\chi^2}\right)\frac{1}{\chi}\right.\right.$$
$$\left.\left.\times\Phi_{1/3}\left(\frac{2i}{3\chi}\right)\right]\frac{2\pi}{3\sqrt{3}}-19-\frac{6\sqrt{3}}{\chi}-\frac{1}{\chi^2}\right\} \qquad (4.56)$$

Another representation of the integral probability and intensity of radiation will be given in Sec. 4.5.

To find the asymptotic expansions of I_e at $\chi \ll 1$ and $\chi \gg 1$, it is convenient to use integral representation (4.51), while for the analysis of the behavior of I_e near $\chi \sim 1$, the integral representation (4.50) or the closed formula (4.56) should be used, but the most convenient is to use integral representation obtained in Sec. 4.5. The ratio $I_e/I_{cl}(I_{cl} = 2e^2m^2\chi^2/3\hbar)$ is always smaller than 1 (see Fig. 4.3). It should be observed that the series (4.52) is an asymptotic one, the numerical values of the coefficients are increasing rapidly and at $\chi = 0.1$ one cannot use it. As one can see from Fig. 4.3, at $\chi = 0.1$ the quantum effects are essential and the intensity of radiation is 1.5 times less than the classical value of the intensity. Actually, the series (4.52) can be used if the corrections (to the classical value) are not larger than 10–15%. Thus, the quantum effects in radiation are "turned on" rather early, when $\chi \ll 1$, while the region $\chi \sim 1$ is essentially a quantum one. The series (4.53) is a convergent one, but at $\chi = 10$ it gives the function I_e with rather poor accuracy. So, in very wide intervals (at least, from $\chi = 10^{-2}$ to $\chi = 10^2$) one has to use formulae (4.50) and (4.56) or (4.99). Under the circumstances, the interpolation formulae are rather useful (see Baier, Katkov and Strakhovenko, 1991b)*

*In numerical calculations the following approximate expressions are useful

$$K_{2/3}(z) = f(z)[(1+a_1/z)^{1/6} + (1+a_2/z)^{1/6}],$$

$$K_{1/3}(z) = f(z)[(1+b_1/z)^{-1/6} + (1+b_2/z)^{-1/6}],$$

$$\int_z^\infty K_{1/3}(y)dy = 2f(z)[1+c_1/z+c_2/z^2]^{-1/4}, \quad f(z) = (\pi/2z)^{1/2}e^{-z}/2,$$

where $a_1 = 0.1265$, $a_2 = 1.040$, $b_1 = 0.054$, $b_2 = 0.78$, $c_1 = 0.245$, $c_2 = 2.53$. The accuracy of these formulae: for $K_{2/3}(z)$ it is better than 0.2%, for $K_{1/3}(z)$ it is better than 1% and for the last integral it is better than 2%.

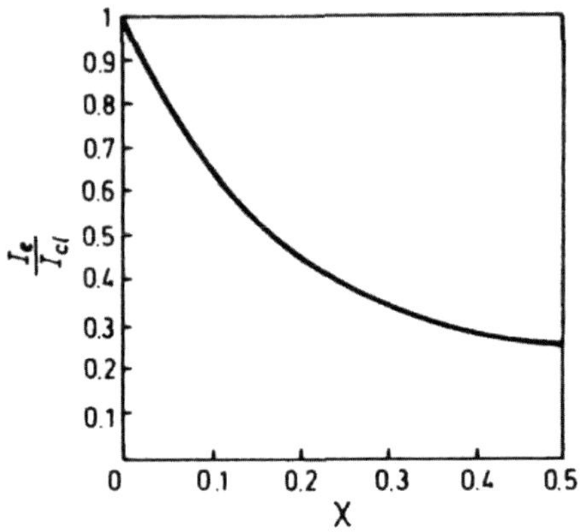

(a)

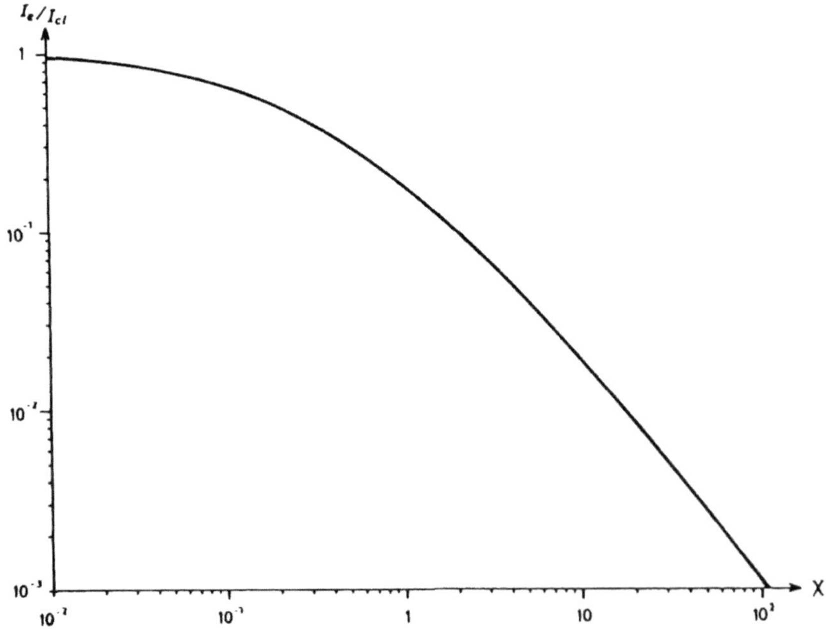

(b)

Fig. 4.3(a), (b). Ratio I_e/I_{cl} as a function of the parameter χ.

$$I_e(\chi) = \frac{2}{3\hbar^2} \frac{e^2 m^2 \chi^2}{[1 + 4.8(1+\chi)\ln(1+1.7\chi) + 2.44\chi^2]^{2/3}}$$

$$W_e(\chi) = \frac{1}{\varepsilon\hbar^2} \frac{1.45 e^2 m^2 \chi}{[1 + 7.2\chi + \chi^2]^{1/6}} \tag{4.57}$$

which present the total intensity and probability of the radiation of the electron with accuracy better than 2% for arbitrary χ.

4.3. Polarization Characteristics of Magnetic Bremsstrahlung

In Sec. 1 we considered the polarization characteristics of the radiation of high-energy particles in the classical theory. Here we investigate polarization effects in the quantum theory.* First we shall obtain formula for polarization density matrix which are summed up over the spins of final electrons and averaged over the spins of the initial ones. Using the formulae (4.2), (4.4) and (4.6) and taking into account that $dI = \hbar\omega dW$, we have

$$dI = e^2 \frac{d^3k}{(2\pi)^2} \int_{-\infty}^{\infty} d\tau N(\tau) \exp\left[-i\frac{\varepsilon}{\varepsilon_f}(kx_2 - kx_1)\right], \tag{4.58}$$

where

$$N(\tau) = \frac{1}{4\varepsilon'^2}\{(\varepsilon+\varepsilon')^2(\mathbf{e}^*\mathbf{v}_1)(\mathbf{e}\mathbf{v}_2)$$
$$+ (\varepsilon-\varepsilon')^2[-(\mathbf{e}^*\mathbf{v}_2)(\mathbf{e}\mathbf{v}_1) + (\mathbf{e}^*\mathbf{e})(\mathbf{v}_2\mathbf{v}_1 - 1 + 1/\gamma^2)]\} \tag{4.59}$$

In these formulae we discard terms $\sim 1/\gamma^3$; $\varepsilon_f = \varepsilon - \hbar\omega$, $\varepsilon' = (\mathbf{p}'^2 + m^2)^{1/2}$, $\mathbf{p}' = \mathbf{p} - \hbar\mathbf{k}$; the indexes 1 and 2 mean the time dependence on $t_1 = t - \tau/2, t_2 = t + \tau/2$.

Proceeding from (4.58), we may present the polarization matrix in the form

$$dI_{ik} = e^2 \frac{d^3k}{(2\pi)^2} \int d\tau N_{ik}(\tau) \exp\left[-i\frac{\varepsilon}{\varepsilon_f}(kx_2 - kx_1)\right], \tag{4.60}$$

where

$$N_{ik} = \frac{1}{4}[(2+u)^2(v_1 - n)_i(v_2 - n)_k - u^2(v_2 - n)_i(v_1 - n)_k$$
$$+ u^2(\delta_{ik} - n_i n_k)(\mathbf{v}_1\mathbf{v}_2 - 1 + 1/\gamma^2)]. \tag{4.61}$$

*A series of results for magnetic bremsstrahlung, taking into account the polarization of the photons and electron spin states, is given in the book by Socolow and Ternov, 1983, in which there is a detailed bibliography.

The tensor dI_{ik} is transverse ($n_i dI_{ik} = n_k dI_{ik} = 0$) since the expressions $(\mathbf{nv}_1 - 1)$, $(\mathbf{nv}_2 - 1)$, occurring at the contraction in the first two terms of $N_{ik'}$, in accordance with the formula (4.9), are proportional to the total derivative and, consequently, vanish when integration over ψ and τ is performed, the third term in N_{ik} is transverse explicitly.

Performing the expansions over the powers of τ (see (4.10) and (4.11)) and introducing the angles β and ψ (see (1.36), Fig. 1.2) we find

$$\begin{aligned} N_{11} &= (1+u)\left[\psi^2 - \left(1 + \frac{u^2}{2(1+u)}\right)\frac{w^2\tau^2}{4}\right], \\ N_{\substack{12\\(21)}} &= (1+u)\left[\beta\psi \pm \beta\left(1 + \frac{u^2}{2(1+u)}\right)\frac{w\tau}{2}\right], \\ N_{22} &= (1+u)\left[\beta^2 - \frac{u^2}{8(1+u)}w^2\tau^2\right] \end{aligned} \tag{4.62}$$

Using the substitutions mentioned after Eq. (4.14) from (4.60) and (4.62) one obtains formula (3.37). If $u = \hbar\omega/(\varepsilon - \hbar\omega) \to 0$ (the transition to the classical limit), then N converts to the formula (1.63).

Integrating over relative time τ using formula (1.66), we have (compare (4.20))

$$dI_{ik} = \frac{e^2}{\pi^2}\sqrt{\frac{2}{3}}\frac{\sqrt{1-\mathbf{nv}}}{w}F_{ik}d^3k\,, \tag{4.63}$$

where

$$\begin{aligned} F_{11} &= \{\psi^2 + [2 + u^2/(1+u)](1 - \mathbf{nv})\}K_{1/3}(\zeta)\,, \\ F_{22} &= \left[\beta^2 + \frac{u^2}{1+u}(1 - \mathbf{nv})\right]K_{1/3}(\zeta)\,, \\ F_{12} &= F_{21}^* = \beta\left\{\psi K_{1/3}(\zeta) - i\left(1 + \frac{u^2}{2(1+u)}\right)\right. \\ &\quad \left. \times\sqrt{2(1-\mathbf{nv})}K_{2/3}(\zeta)\right\}, \end{aligned} \tag{4.64}$$

here $1 - \mathbf{nv} = (\psi^2 + \beta^2)/2$, ζ is given by the formula (4.14), at $u \to 0$, (4.64) converts to (1.71).

As has already been pointed out, to obtain the observable characteristics (the radiation from the coherence length), it is necessary to integrate over the azimuthal angle ψ, which, as before, it is convenient to perform simultaneously with integration over τ. Converting to the variables x and y (see (4.15)) and performing the integration as in (4.16), we obtain

$$dI_{ik} = \frac{e^2}{3\pi^2}\frac{\varepsilon^3}{w\hbar^3}\frac{u^2 du d\beta}{(1+u)^3}\mu^2 G_{ik}\,, \tag{4.65}$$

where

$$\begin{aligned} G_{11} &= \mu^2 K_{2/3}^2(\eta) + \frac{u^2\mu^2}{4(1+u)}[K_{1/3}^2(\eta) + K_{2/3}^2(\eta)]\,, \\ G_{22} &= \beta^2 K_{1/3}^2(\eta) + \frac{u^2\mu^2}{4(1+u)}[K_{1/3}^2(\eta) + K_{2/3}^2(\eta)]\,, \\ G_{12} &= G_{21}^* = -i\left(1 + \frac{u^2}{2(1+u)}\right)\beta\mu K_{1/3}(\eta)K_{2/3}(\eta)\,. \end{aligned} \tag{4.66}$$

Here η is given by the formula (4.18). The Stokes parameters, defined in accordance with formula (A.39), for the polarization matrix (4.65) and (4.66) have the form

$$\begin{aligned} &\xi_1 = 0\,, \quad \xi_2 = 2G^{-1}\left[1 + \frac{u^2}{2(1+u)}\right]\beta\mu K_{1/3}(\eta)K_{2/3}(\eta)\,, \\ &\xi_3 = G^{-1}[\mu^2 K_{2/3}^2(\eta) - \beta^2 K_{1/3}^2(\eta)]\,, \end{aligned} \tag{4.67}$$

where

$$G = G_{11} + G_{22} = \mu^2\left(1 + \frac{u^2}{2(1+u)}\right)[K_{1/3}^2(\eta) + K_{2/3}^2(\eta)] - K_{1/3}^2(\eta)/\gamma^2\,. \tag{4.68}$$

At $u \to 0$ the formulae (4.65) and (4.67) convert respectively to (1.86) and (1.87). From (4.65) and (4.66), using substitutions mentioned after Eq. (4.14), we obtain (3.44).

At $\chi \ll 1$ (see analysis after Eq. (4.29)) all the formulae convert to the corresponding classical formulae (with small quantum corrections representing series over χ, the calculation for which is performed in the same way as the corresponding calculation for total intensity and probability in Secs. 4.1 and 4.2).

At $\chi \geq 1$ the polarization characteristics of radiation are different from classical ones. Primarily, the radiation is not completely polarized ($\xi^2 < 1$). In a certain sense, this is linked to the fact that we consider radiation of unpolarized electrons, while at $\chi \geq 1$ the spin properties of particles become essential ones (at $\chi \ll 1$ the spin terms are of the order $u^2 \ll 1$). At $\chi \geq 1$ the main contribution gives the region $u \sim 1$ and the radiated photons are partially elliptically polarized, while at $\beta = 0$ the circular part of the polarization vanishes. On the whole, at $u \sim 1$ the distinction of the polarization characteristics from classical ones is basically quantitative (see the discussion after formula (1.87)).

Another situation arises at $\chi \gg 1$ and $u \gg 1$. In this region, where the intensity is of order $u^{-2/3}$ in comparison with the main contribution and the photons take away most of the particle energy $\hbar\omega \gg \varepsilon - \hbar\omega$, new qualitative polarization characteristics appear.* In particular, the polarization is only circular

$$\xi_1 = \xi_3 = 0, \quad \xi_2 = \frac{\beta\gamma}{\sqrt{1+(\beta\gamma)^2}} \frac{2K_{1/3}(\eta)K_{2/3}(\eta)}{K_{1/3}^2(\eta) + K_{2/3}^2(\eta)} \tag{4.69}$$

Finally, integrating over angle β, which is convenient to do in accordance with formula (4.63) in conjuction with integration over ψ, we obtain

$$dI_{12} = dI_{21} = 0\,,$$

$$dI_{11} = \frac{e^2 m^2 u du}{2\sqrt{3}\pi\hbar^2(1+u)^3}\left\{\int_{2u/3\chi}^{\infty} K_{5/3}(y)dy + \left[1 + \frac{u^2}{1+u}\right] K_{2/3}\left(\frac{2u}{3\chi}\right)\right\},$$

$$dI_{22} = \frac{e^2 m^2 u du}{2\sqrt{3}\pi\hbar^2(1+u)^3}\left\{\int_{2u/3\chi}^{\infty} K_{5/3}(y)dy + \left[-1 + \frac{u^2}{1+u}\right] K_{2/3}\left(\frac{2u}{3\chi}\right)\right\} \tag{4.70}$$

At $u \ll 1$ these formulae convert to (1.100). From (4.69) using the substitutions $udu/(1+u)^3 \to dy/(2\cosh^2 y)$ mentioned after (4.14), we obtain (3.52).

Integrating over u, we have

$$I \equiv I_{11} + I_{22} = I_e\,,$$

*This region is discussed in greater detail when the radiation of polarized electrons is considered.

$$I_- \equiv I_{11} - I_{22} = \frac{2e^2m^2\pi}{3\sqrt{3}\hbar^2}\left[\frac{2}{9}\left(1+\frac{1}{\chi^2}\right)\Phi_{2/3}\left(\frac{2i}{3\chi}\right)\right.$$

$$\left. + \frac{1}{\chi}\Phi_{1/3}\left(\frac{2i}{3\chi}\right) - \frac{7\sqrt{3}}{12\pi} - \frac{1}{2\pi\chi}\right], \tag{4.71}$$

where I_e is given by the formula (4.56). As in the classical theory, circular polarization disappears after integration over the angles. The degree of linear polarization depends on u and χ in (4.70) and on χ in (4.71). At $\chi \ll 1$, these formulae convert into the corresponding classical ones. At $\chi \gg 1$, taking into account that the region $u \sim 1$ gives the main contribution and using the expansion of K-functions in (4.70) for small arguments (1.79), we have

$$I_- \approx \frac{e^2m^2}{2\sqrt{3}\pi\hbar^2}\Gamma(2/3)(3\chi)^{2/3}\int_0^\infty \frac{u^{1/3}du}{(1+u)^3}$$

$$= \frac{e^2m^2}{27\hbar^2}\Gamma(2/3)(3\chi)^{2/3} \tag{4.72}$$

The same result may be obtained if the asymptotic expansions of the Φ functions are applied in (4.71) at $\chi \gg 1$. Using the formula (4.53) we have at $\chi \gg 1$

$$\xi_1 = \xi_2 = 0\,, \quad \xi_3 = \frac{I_-}{I_e} = \frac{9}{32}\,. \tag{4.73}$$

4.4. The Properties of Polarized Electron Radiation

We shall consider the radiation in the case where the initial electrons are polarized and we sum over the spin states of the final electrons.

Using formula (4.6), where $\zeta = \zeta_i$ and we sum over ζ_f, we have

$$\sum_{\zeta_f} R_2^* R_1 = A_2^* A_1 + \mathbf{B}_2^*\mathbf{B}_1 + i[A_2^*(\boldsymbol{\zeta}\mathbf{B}_1) - A_1(\boldsymbol{\zeta}\mathbf{B}_2^*) + \boldsymbol{\zeta}\cdot(\mathbf{B}_2^*\times\mathbf{B}_1)]\,, \tag{4.74}$$

where the first term describes the radiation of the unpolarized electrons and the second is an addition dependent on spin.

We shall represent the tensor N_{ik}^p in formula (4.60) in the form

$$N_{ik}^p = N_{ik} + N_{ik}^\zeta\,, \tag{4.75}$$

where N_{ik}^p is the polarization matrix for the electron with spin $\boldsymbol{\zeta}$ and N_{ik} is given by the formula (4.62) (the term describing radiation of the unpolarized electrons), while N_{ik}^{ζ} are spin-dependent additions, for which we have from (4.74)

$$\begin{aligned}
N^{\zeta} &\equiv N_{11}^{\zeta} + N_{22}^{\zeta} = iw\tau u[(2+u)\beta(\boldsymbol{\zeta}\mathbf{v}) - (\boldsymbol{\zeta}\mathbf{vs})/\gamma]/2(1+u)\,,\\
N_{-}^{\zeta} &\equiv N_{11}^{\zeta} - N_{22}^{\zeta} = -iw\tau u(\boldsymbol{\zeta}\mathbf{vs})/2\gamma\,,\\
N_{12}^{\zeta} &+ N_{21}^{\zeta} = iw\tau u(\boldsymbol{\zeta}\mathbf{s})/2\gamma\,, \qquad (4.76)\\
N_{12}^{\zeta} &- N_{21}^{\zeta} = i\{[u(2+u)w^2\tau^2/4(1+u) + u/(\gamma^2(1+u))](\boldsymbol{\zeta}\mathbf{v})\\
&\qquad + u\psi(\boldsymbol{\zeta}\mathbf{s})/(\gamma(1+u)) + u\beta(\boldsymbol{\zeta}\mathbf{vs})/(\gamma(1+u))\}\,,
\end{aligned}$$

where $(\boldsymbol{\zeta}\mathbf{vs}) = (\boldsymbol{\zeta}\cdot(\mathbf{v}\times\mathbf{s}))$, $\mathbf{s} = \mathbf{w}/w$. Using the substitutions mentioned after Eq. (4.14) we have from (4.76) the formula (3.64).

Integrating over relative time τ, we arrive at the formula (4.63) with $F_{ik}^{p} = F_{ik} + F_{ik}^{\zeta}$, where spin-dependent additions to F_{ik} (4.64) have the form

$$\begin{aligned}
F^{\zeta} &\equiv F_{11}^{\zeta} + F_{22}^{\zeta} = \frac{u\sqrt{2(1-\mathbf{nv})}}{1+u}[(2+u)\beta(\boldsymbol{\zeta}\mathbf{v}) - (\boldsymbol{\zeta}\mathbf{vs})/\gamma]K_{2/3}(\zeta)\\
F_{-}^{\zeta} &\equiv F_{11} - F_{22}^{\zeta} = -\frac{u\sqrt{2(1-\mathbf{nv})}}{\gamma}(\boldsymbol{\zeta}\mathbf{s})K_{2/3}(\zeta)\\
F_{12}^{\zeta} &+ F_{21}^{\zeta} = \sqrt{2(1-\mathbf{nv})}u(\boldsymbol{\zeta}\mathbf{s})K_{2/3}(\zeta)/\gamma\,, \qquad (4.77)\\
F_{12}^{\zeta} &- F_{21}^{\zeta} = \frac{i}{1+u}\left\{\left[\frac{u}{\gamma^2} - 2u(2+u)(1-\mathbf{nv})\right](\boldsymbol{\zeta}\mathbf{v})\right.\\
&\qquad \left. + [u\psi(\boldsymbol{\zeta}\mathbf{s}) + u\beta(\boldsymbol{\zeta}\mathbf{vs})]/\gamma\right\}K_{1/3}(\zeta)\,.
\end{aligned}$$

Integrating over the angle ψ, which is convenient to do in formula (4.76) and simultaneously integrating over τ and ψ, we obtain for $G_{ik}^{p} = G_{ik} + G_{ik}^{\zeta}$ and the spin-dependent addition to G_{ik} (4.66) in formula (4.65) has the form

$$\begin{aligned}
G^{\zeta} &\equiv G_{11}^{\zeta} + G_{22}^{\zeta} = u\mu[(2+u)\beta(\boldsymbol{\zeta}\mathbf{v}) - (\boldsymbol{\zeta}\mathbf{vs})/\gamma]\frac{K_{1/3}(\eta)K_{2/3}(\eta)}{(1+u)}\,,\\
G_{-}^{\zeta} &\equiv G_{11}^{\zeta} - G_{22}^{\zeta} = -u\mu(\boldsymbol{\zeta}\mathbf{vs})K_{1/3}(\eta)K_{2/3}(\eta)/\gamma\,,\\
G_{12}^{\zeta} &+ G_{21}^{\zeta} = u\mu(\boldsymbol{\zeta}\mathbf{s})K_{1/3}(\eta)K_{2/3}(\eta)/\gamma\,, \qquad (4.78)\\
G_{12}^{\zeta} &- G_{21}^{\zeta} = \frac{-iu}{1+u}\{[(1+u/2)\mu^2[K_{1/3}^2(\eta)K_{2/3}^2(\eta)]\\
&\qquad - K_{1/3}^2(\eta)/\gamma^2](\boldsymbol{\zeta}\mathbf{v}) - \beta(\boldsymbol{\zeta}\mathbf{vs})K_{1/3}^2(\eta)/\gamma\}\,,
\end{aligned}$$

and the Stokes parameters are defined by the relations (compare with (A.39)–(A.41))

$$\xi_1 = \frac{G_{12}^p + G_{21}^p}{G^p}, \quad \xi_2 = i\frac{G_{12}^p - G_{21}^p}{G^p}, \quad \xi_3 = \frac{G_{11}^p - G_{22}^p}{G^p}, \tag{4.79}$$

where $G^p = \mathrm{Tr}\, G_{ik}^p = G_{11}^p + G_{22}^p = G + G^\zeta$ Using the substitutions mentioned after Eq. (4.14) we have formula (3.65) from (4.78).

In the case $\chi \ll 1$ and $u \sim \chi \ll 1$, all the terms depending on the polarization vector $\boldsymbol{\zeta}$ (and having the order $u \sim \chi \ll 1$) may be omitted (naturally, the spin-dependent terms contain additional power $\hbar$). In the case $\chi \gg 1$, the main contribution to radiation gives the region $u \sim 1$ and terms with u become essential. In particular, at $\beta = 0$ the circular part of the radiation polarization does not vanish and becomes proportional to the value of the electron spin projection on the direction of motion $(\boldsymbol{\zeta}\mathbf{v})$.

In the region $1 \ll u \sim \chi$, which, as has been pointed out, contributes to the radiation intensity of the order $\chi^{-2/3}$ in comparison with the main region $u \sim 1$, we have from (4.79)

$$\begin{aligned}
\xi_1 &= 2(\boldsymbol{\zeta}\mathbf{s})K_{1/3}(\eta)K_{2/3}(\eta)/(\mu\gamma\widetilde{G}), \\
\xi_2 &= [(\boldsymbol{\zeta}\mathbf{v})[K_{1/3}^2(\eta) + K_{2/3}^2(\eta)] + 2\beta K_{1/3}(\eta)K_{2/3}(\eta)/\mu]\widetilde{G}^{-1}, \\
\xi_3 &= -2(\boldsymbol{\zeta}\mathbf{v}\mathbf{s})K_{1/3}(\eta)K_{2/3}(\eta)/(\mu\gamma\widetilde{G}),
\end{aligned} \tag{4.80}$$

where

$$\widetilde{G} = 2G^p(u \gg 1)/(u\mu^2) = K_{1/3}^2(\eta) + K_{2/3}^2(\eta) + 2\beta K_{1/3}(\eta)K_{2/3}(\eta)(\boldsymbol{\zeta}\mathbf{v})/\mu.$$

It is clear, that in the case under consideration the spin characteristics to a significant degree determine the polarization of the radiation. For example, in the case where the spin vector is parallel to the velocity $\mathbf{v}$ and $\boldsymbol{\zeta}^2 = 1$ (with relativistic accuracy $|\mathbf{v}| = 1$) we have

$$\xi_1 = \xi_3 = 0, \quad \xi_2 = (\boldsymbol{\zeta}\mathbf{v}) = \pm 1, \tag{4.81}$$

so that the radiation is completely circularly polarized.*

*At radiation of hard photons (near the end of the spectrum) by longitudinally polarized relativistic electrons in collisions of particles also takes place "transfer of helicity" (Baier *et al.*, 1981, p. 372).

If $\zeta\mathbf{v} = 0$, the radiation is partially elliptically polarized while the degree of linear polarization $(\xi_1^2 + \xi_3^2)^{1/2}$ and the degree of circular polarization ξ_2 are as follows

$$
\begin{aligned}
(\xi_1^2 + \xi_3^2)^{1/2} &= \frac{2K_{1/3}(\eta)K_{2/3}(\eta)}{\mu\gamma[K_{1/3}^2(\eta) + K_{2/3}^2(\eta)]}, \\
\tan 2\varphi &= -(\zeta\mathbf{s})/(\zeta\mathbf{v}\mathbf{s}); \\
\xi_2 &= (2\beta/\mu)K_{1/3}(\eta)K_{2/3}(\eta)/[K_{1/3}^2(\eta) + K_{2/3}^2(\eta)]; \\
\xi \equiv (\xi^2)^{1/2} &= 2K_{1/3}(\eta)K_{2/3}(\eta)/[K_{1/3}^2(\eta) + K_{2/3}^2(\eta)],
\end{aligned}
\tag{4.82}
$$

where the angle φ characterizes the direction of linear polarization (see (A. 34)) and is measured from the direction of the acceleration in the plane perpendicular to $\mathbf{v}$, ξ is the degree of polarization.

The formula for the polarization density matrix integrated over all the angles of radiation, has the form

$$
\begin{aligned}
dI &= \frac{e^2m^2udu}{\pi\sqrt{3}\hbar^2(1+u)^3}\left[\int_\sigma^\infty K_{5/3}(y)dy + \frac{u}{(1+u)}\right. \\
&\quad \left.\times\left(uK_{2/3}(\sigma) - (\zeta\mathbf{v}\mathbf{s})K_{1/3}(\sigma)\right)\right], \\
dI_- \equiv dI_{11} - dI_{22} &= \frac{e^2m^2udu}{\hbar^2\pi\sqrt{3}(1+u)^3}[K_{2/3}(\sigma) \\
&\quad - u(\zeta\mathbf{v}\mathbf{s})K_{1/3}(\sigma)], \\
dI_{12} + dI_{21} &= \frac{e^2m^2u^2du}{\hbar^2\pi\sqrt{3}(1+u)^3}(\zeta\mathbf{s})K_{1/3}(\sigma), \\
dI_{12} - dI_{21} &= -i\frac{e^2m^2u^2du}{\pi\sqrt{3}\hbar^2(1+u)^4}\left[\int_\sigma^\infty K_{5/3}(y)dy + uK_{2/3}(\sigma)\right](\zeta\mathbf{v}),
\end{aligned}
\tag{4.83}
$$

where $\sigma = 2u/3\chi$. The formulae (4.83) by means of substitutions are connected with formulae (3.69).

Finally, integrated over the frequency expression for the polarization density matrix (the total intensities) has the form

$$
\begin{aligned}
I &= \frac{e^2m^2}{3\sqrt{3}\pi\hbar^2}\int_0^\infty \frac{udu}{(1+u)^4}[(4u^2+5u+4)K_{2/3}(\sigma) \\
&\quad - 3u(\zeta\mathbf{v}\mathbf{s})K_{1/3}(\sigma)],
\end{aligned}
$$

$$I_- = \frac{e^2m^2}{\sqrt{3}\pi\hbar^2}\int_0^\infty \frac{udu}{(1+u)^3}[K_{2/3}(\sigma) - u(\zeta\mathbf{vs})K_{1/3}(\sigma)]\,,$$

$$I_{12}+I_{21} = \frac{e^2m^2}{\sqrt{3}\pi\hbar^2}(\zeta\mathbf{s})\int_0^\infty \frac{u^2du}{(1+u)^3}K_{1/3}(\sigma)\,, \tag{4.84}$$

$$I_{12}-I_{21} = -i\frac{11e^2m^2}{9\sqrt{3}\pi\hbar^2}(\zeta\mathbf{v})\int_0^\infty \frac{u^2du}{(1+u)^3}K_{2/3}(\sigma)\,.$$

The circular polarization at the radiation of the longitudinally polarized electrons $\xi_2 = i(I_{12} - I_{21}/I$ is given in Fig. 4.4.

At $\chi \ll 1$ the main contributor to radiation is the region $u \sim \chi$ and since all the spin-dependent terms contain additional powers of u, this means that in the main region they contain the additional power χ. This statement holds also for formulae integrated over the frequencies u.

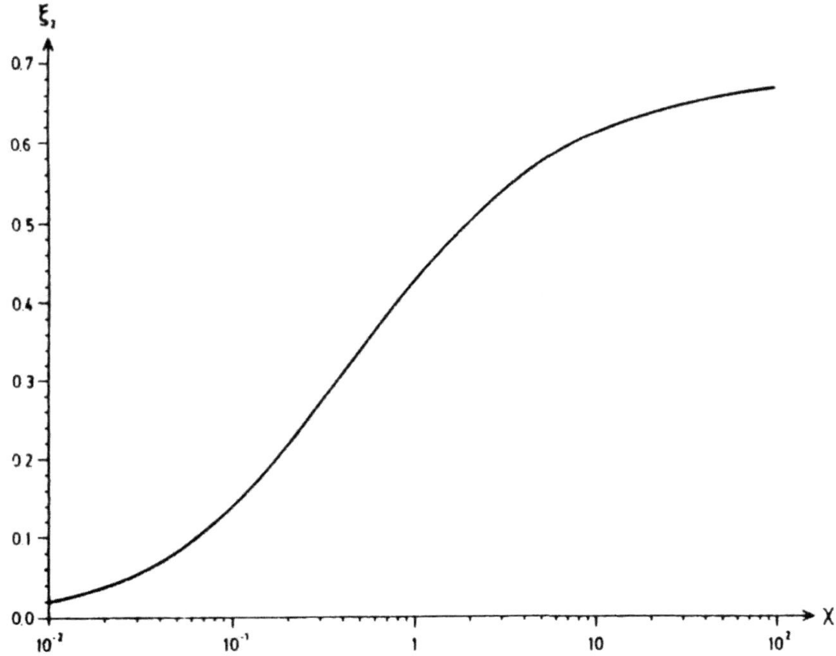

Fig. 4.4. Circular polarization of the radiation of longitudinally polarized electrons.

At $\chi \gg 1$ the main contribution to the radiation is given by the region $u \sim 1$. Expanding in (4.83) the K_ν-function in the range of the small values of the argument (1.73), we obtain

$$
\begin{aligned}
dI &= \frac{e^2m^2}{2\sqrt{3}\pi\hbar^2}\frac{u^{1/3}du}{(1+u)^4}[1+(1+u)^2]\Gamma(2/3)(3\chi)^{2/3}\,, \\
dI_- &= \frac{e^2m^2}{2\sqrt{3}\pi\hbar^2}\frac{u^{1/3}du}{(1+u)^3}\Gamma(2/3)(3\chi)^{2/3}\,, \\
& dI_{12}+dI_{21}=0\,, \\
dI_{12}-dI_{21} &= -i\frac{e^2m^2}{2\sqrt{3}\pi\hbar^2}\frac{u^{4/3}du}{(1+u)^4}(2+u)(\boldsymbol{\zeta}\mathbf{v})\Gamma(2/3)(3\chi)^{2/3}
\end{aligned}
\tag{4.85}
$$

Consequently, at $\chi \gg 1$ the Stokes parameters are

$$
\begin{aligned}
\xi_1 &= 0\,, \quad \xi_2=(\boldsymbol{\zeta}\mathbf{v})u(2+u)/(2+2u+u^2)\,, \\
\xi_3 &= \frac{1+u}{(2+2u+u^2)}
\end{aligned}
\tag{4.86}
$$

Having integrated (4.84) over u, we find the main terms of the expansion of the values over inverse powers of χ (see also Sec. 4.5):

$$
\begin{aligned}
I &= \frac{e^2m^2}{\hbar^2}\left(\frac{2}{3}\right)^5\Gamma(2/3)(3\chi)^{2/3}\,, \\
I_- &= \frac{e^2m^2}{\hbar^2}\frac{\Gamma(2/3)}{3^3}(3\chi)^{2/3}\,, \\
I_{12}-I_{21} &= -i\frac{e^2m^2}{\hbar^2}\frac{22\Gamma(2/3)}{3^5}(3\chi)^{2/3}(\boldsymbol{\zeta}\mathbf{v})\,,
\end{aligned}
\tag{4.87}
$$

so that Stokes parameters

$$
\xi_1=0\,, \quad \xi_2=11(\boldsymbol{\zeta}\mathbf{v})/16\,, \quad \xi_3=9/32\,. \tag{4.88}
$$

At $\boldsymbol{\zeta}\mathbf{v}=0$ we arrive to (4.73), at $\boldsymbol{\zeta}\mathbf{v}=\pm 1$ the degree of the polarization of the radiation is $\xi=\sqrt{\xi^2}=\sqrt{565}/32 \cong 3/4$.

Now, we discuss the results obtained above from the point of view of the invariance under discrete P and T transforms. The properties of the constituent combinations are given in Table 4.1. These properties are evident and perhaps only the Stokes parameters ξ_i require special comment.

Table 4.1. Properties of some quantities under P- and T- transformations.

	$\zeta\mathbf{v}$	$\zeta\mathbf{s}$	$\zeta(\mathbf{v}\times\mathbf{s})$	$\mathbf{ns}$	$\mathbf{n}(\mathbf{v}\times\mathbf{s})$	ξ_1	ξ_2	ξ_3
P	$-$	$-$	$+$	$+$	$-$	$-$	$-$	$+$
T	$+$	$-$	$+$	$-$	$+$	$-$	$+$	$+$

We have introduced mobile polarization axes $\mathbf{s} = \mathbf{w}/w$ and $\mathbf{n} \times \mathbf{s}$ (see (1.40)), while the combination $(\mathbf{es})(\mathbf{e}(\mathbf{n} \times \mathbf{s}))$ change the sign under P and T transforms. Therefore the Stokes parameter ξ_1 linked to the components of the polarization matrix $\rho_{12} + \rho_{21}$, has the properties indicated in the table. The Stokes parameter ξ_i linked to components $\rho_{11} - \rho_{22}$ (or to combinations $(\mathbf{es})(\mathbf{es})$ and $(\mathbf{e}(\mathbf{n} \times \mathbf{s}))$ $(\mathbf{e}(\mathbf{n} \times \mathbf{s})))$, obviously does not vary with P and T transforms. The Stokes parameter ξ_2 is the mean helicity value and this determines its characteristics.

With accepted accuracy $\mathbf{ns} = \psi$ and $\mathbf{n}(\mathbf{v} \times \mathbf{s}) = \beta$ and it may also be easily verified that formulae (4.62) – (4.88) obtained above include combination of values in agreement with Table 4.1. We shall discuss formulae (4.83) as an example. The intensities dI and $dI_-(\xi_3)$ must be P and T even and because of this the spin in them enters in the form P and T even combination $(\zeta\mathbf{vs})$; the combination $dI_{12} + dI_{21}(\xi_1)$ must be P and T odd (the only possible combination is $\zeta\mathbf{s}$); while the combination $dI_{12} - dI_{21}(\xi_2)$ is P odd and T even (the combination $\zeta\mathbf{v}$).

4.5. The Integral Characteristics of the Electron Radiation Process (2)

In the above we have obtained two types of expressions for the total probabilities or intensities of the magnetic bremsstrahlung:

1. Integral representation (4.36), (4.50), (4.84) containing McDonald's (K_ν) functions. For these functions one can use integral representation with rapidly decreasing integrand, so that we have two-fold integral with good convergence.

2. Formulae with Φ_ν functions which can be expressed in terms of the difference of the Bessel and the Anger functions ((4.49), (4.56), (4.71)). This difference may be expressed as a one-fold integral but with oscillating integrand (Bateman

and Erdelyi, 1953, p. 35):

$$\Phi_\nu(i/\xi) = \frac{\sin \pi\nu}{\pi} \int_0^\infty \exp(-i \sinh t/\xi)[\exp(\nu t + i\pi\nu/2) - \exp(-(\nu t + i\pi\nu/2))]dt\,. \qquad (4.89)$$

Since formulae for the integral characteristics of the magnetic bremsstrahlung have very vast fields of applications some of which will be discussed in this book, it is very important to obtain for them the simplest representations. One of the possibilities is the use of the approximate expressions for McDonald's functions (see footnote on page 107). However it is desirable, of course, to obtain an exact simple expression for the integral characteristics of the radiation. Such formulae will be discussed in this section. They are one-fold integrals without oscillating functions and with rapidly decreasing integrands (Baier, Katkov and Strakhovenko, 1993).

For our purpose we need integral representation of McDonald's functions (1.66) and the following formula

$$\int_{-\infty}^\infty \frac{dx}{x+i0} \exp[ib(x+x^3/3)]dx = -\frac{2i}{\sqrt{3}} \int_{2b/3}^\infty K_{1/3}(z)dz\,. \qquad (4.90)$$

To find expressions for the total intensity and probability we can proceed from expressions for spectral distributions (4.23) – (4.27). But more compact expressions one can obtain from representations (4.36) – (4.50). We consider first the total probability of the radiation. Substituting representation (1.66) into formula (4.36) we obtain

$$W_e = \frac{e^2 m^2}{6\pi\varepsilon i\hbar^2} \int_0^\infty du \frac{(5+7u+5u^2)}{(1+u)^3} \int_{-\infty}^\infty d\tau\tau \exp[iu(\tau+\tau^3/3)/\chi]\,. \qquad (4.91)$$

Now we change variables:

$$v = u(1+\tau^2/3)\,, \quad \tau = \tau \qquad (4.92)$$

In these variables the expression for the total probability has the form

$$W_e = \frac{e^2 m^2}{2\pi\varepsilon i\hbar^2} \int_0^\infty dv \int_{-\infty}^\infty d\tau\tau \frac{(a+b\tau^2+c\tau^4)}{(3(1+v)+\tau^2)^3} \exp(i\tau v/\chi) \qquad (4.93)$$

here

$$a = 9(5v^2+7v+5)\,, \quad b = 3(10+7v)\,, \quad c = 5\,.$$

The integration over τ in (4.93) may be carried out using the residue method and the result is

$$W_e = \frac{e^2m^2}{2^4\varepsilon\hbar^2}\int_0^\infty dv\frac{\exp(-f)}{(1+v)^2}[40(1+v)^2 - 3fv(4+3v) + 3f^2v^2], \qquad (4.94)$$

where $f = v(3(1+v))^{1/2}/\chi$. One can see, that the integrand contains the elementary functions only, the external field (χ) enters in a very simple form and the integrand is decreasing very rapidly with v increasing. After integration by parts we obtain another representation of the total probability of the radiation

$$W_e = \frac{e^2m^2}{\varepsilon\hbar^2}\int_0^\infty dv\exp(-f)\left[4 - \left(\frac{3v}{2\chi}\right)^2(1+2v)\right] \qquad (4.95)$$

From the formulae (4.94) and (4.95) the expansions for $\chi \ll 1$ and $\chi \gg 1$ may be obtained in a direct way. The systematic procedure of the expansion was presented in Sec. 4.2. Here we discuss some alternative approaches. At $\chi \ll 1$ the main contribution gives the region $v \sim \chi \ll 1$. Expanding integrand over powers of v ($v3^{1/2}/\chi$ is retained in the exponent) and carrying out elementary integration over v (gamma functions) we find the asymptotic series (4.39). At $\chi \gg 1$ the main contribution gives the region $v \sim \chi^{2/3} \gg 1$. So here we have to expand in inverse powers of v. The integrals obtained are reduced also to gamma functions and we obtain the first terms of the series (4.40). However in the expansion of the denominator of the integrand in (4.94) in powers of $1/v$ the powers of v are accumulated in the denominator and starting from some term the expansion ceases to be valid (formally the integral is divergent at $v = 0$). We discuss how to overcome this difficulty for the integral, where it appears in the main term

$$I_1 = \int_0^\infty \frac{dvv^2}{(1+v)^3}\exp(-f). \qquad (4.96)$$

Changing in this integral variable $1 + v = z^{2/3}$, we transform the integral to the form

$$I_1 = (2/3)\int_1^\infty dz\exp(-\sqrt{3}z/\chi)[z^{-1} - 2z^{-5/3} + z^{-7/3}](1 + \sqrt{3}z^{1/3}/\chi) \quad (4.97)$$

where the higher order terms in $1/\chi$ are omitted. All the integrals in (4.97) are expressed in terms of incomplete gamma functions (see Gradshteyn and

Ryzhik, 1965, p. 940). Using the expansions of these functions over inverse powers of χ we find

$$I_1 = (2/3)[\ln(\chi/\sqrt{3}) - C - 9/4 + 4\Gamma(1/3)3^{1/3}\chi^{-2/3}] + O(1/\chi) \qquad (4.98)$$

here C is the Euler constant.

So, in a such way one can obtain all the known series.

The scheme presented may be used for all the expressions for integral characteristics of the radiation. Proceeding from (4.50) and deriving the same way as above we obtain for the total intensity of the radiation

$$I_e = \frac{e^2m^2}{2^5\hbar^2}\int_0^\infty dv \frac{\exp(-f)}{(1+v)^3} vf[3 + 29(1+v)^2 - 3fv(2+v) + f^2v^2], \qquad (4.99)$$

where f is defined in (4.94). For $\chi \ll 1$ and $\chi \gg 1$ from formula (4.99) one can obtain known expansions (4.52) and (4.53). In Fig. 4.5 the radiation intensity

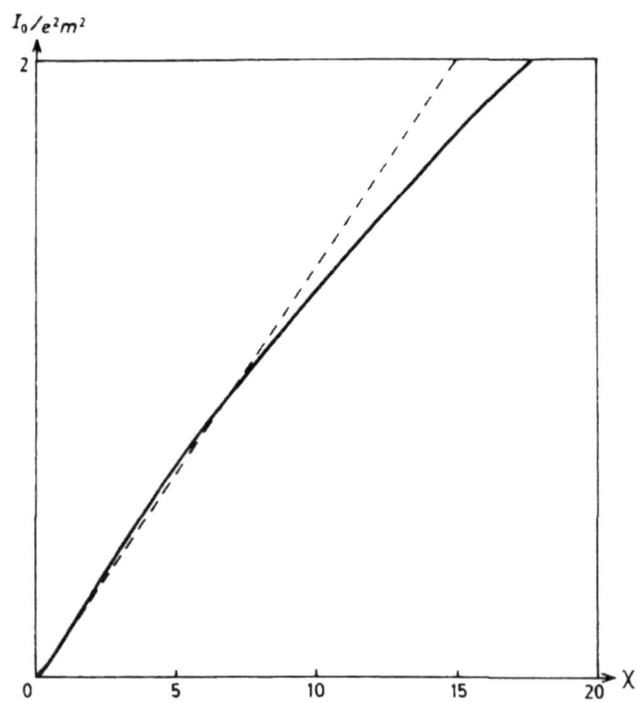

Fig. 4.5. Total intensity of the magnetic bremsstrahlung in quantum region. The dashed straight line is the approximation given in the text.

I_0/e^2m^2 is shown in the interval $0 \le \chi \le 20$. The dashed straight line is the linear approximation $I_e \cong 2e^2m^2\chi/15$ used in the paper by Baier, Katkov and Strakhovenko, 1986d, in the interval $1 \le \chi \le 15$.

Now we consider the polarization of the radiation unpolarized electrons (see (4.71) - (4.73)). Proceeding in accordance with the above presented procedure we obtain for $I_- = I_{11} - I_{22}$

$$\begin{aligned} I_- &= \frac{e^2m^2}{\sqrt{3}\pi\hbar^2}\int^{\infty}\frac{udu}{(1+u)^3}K_{2/3}(2u/3\chi) \\ &= 3\frac{e^2m^2}{2^4\hbar^2}\int_0^{\infty}dv\frac{\exp(-f)}{(1+v)^2}fv[4+3v-fv]\,, \end{aligned} \tag{4.100}$$

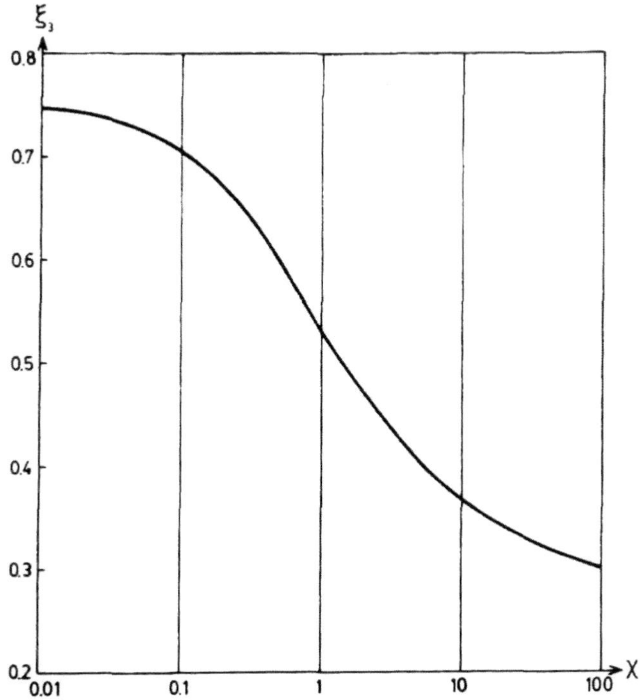

Fig. 4.6. Degree of linear polarization of integral magnetic bremsstrahlung (the Stokes parameter ξ_3) directed along the acceleration.

where the function f is defined in Eq. (4.94). The Stokes parameter $\xi_3 = I_-/I_e$ characterizing polarization of the radiation as a whole (along the direction of the acceleration) which is found using Eqs. (4.99), (4.100) is presented in Fig. 4.6. For small and large χ from formulae given here one can find the asymptotic expansions

$$\begin{aligned} \xi_3 &= \frac{3}{4}(1 - 5\sqrt{3}\chi/16 + \cdots) \quad \chi \ll 1\,, \\ \xi_3 &= \frac{9}{32}\left[1 + \frac{81}{16\Gamma(2/3)(3\chi)^{2/3}}\right] \quad \chi \gg 1\,. \end{aligned} \tag{4.101}$$

Finally we consider polarization of the radiation of the polarized electrons. Formulae (4.84) can be written in the form

$$\begin{aligned} I &= I_{11} + I_{22} = I_e - (\boldsymbol{\zeta}\mathbf{v}\mathbf{s})J_0 \\ I_-^p &= I_{11} - I_{22} = I_- - (\boldsymbol{\zeta}\mathbf{v}\mathbf{s})J_1 \\ I_1 &= I_{12} + I_{21} = (\boldsymbol{\zeta}\mathbf{s})J_1 \\ I_2 &= I_{12} - I_{21} = (\boldsymbol{\zeta}\mathbf{v})J_2\,, \end{aligned} \tag{4.102}$$

where the intensities I_e and I_- are given in Eqs. (4.99) and (4.100). The integrals in (4.103) are calculated according to the above given procedure

$$\begin{aligned} J_0 &= \frac{e^2m^2}{\sqrt{3}\pi\hbar^2}R_0\,, \quad R_0 = \int_0^\infty \frac{u^2du}{(1+u)^4}K_{1/3}(2u/3\chi) \\ &= \frac{\pi}{2^5}\int_0^\infty dv\frac{\exp(-f)}{(1+v)^{7/2}}v^2[3(6+v)(1+f) + 6f^2 - vf^3] \end{aligned} \tag{4.103}$$

$$\begin{aligned} J_1 &= \frac{e^2m^2}{\sqrt{3}\pi\hbar^2}R_1\,, \quad R_1 = \int_0^\infty \frac{u^2du}{(1+u)^3}K_{1/3}(2u/3\chi) \\ &= \frac{3\pi}{2^4}\int_0^\infty dv\frac{\exp(-f)}{(1+v)^{5/2}}v^2[3 + 3f + f^2] \end{aligned} \tag{4.104}$$

$$\begin{aligned} J_2 &= -i\frac{11e^2m^2}{9\sqrt{3}\pi\hbar^2}R_2\,, \quad R_2 = \int_0^\infty \frac{u^2du}{(1+u)^3}K_{2/3}(2u/3\chi) \\ &= \frac{3\sqrt{3}\pi}{2^4}\int_0^\infty dv\frac{\exp(-f)}{(1+v)^2}v^2 f(1+f)\,, \end{aligned} \tag{4.105}$$

where the function f is defined in (4.94).

These expressions have the following asymptotic expansions at $\chi \ll 1$

$$R_0 = \sqrt{3}\pi\chi^3 \left[1 - \frac{35\sqrt{3}}{4}\chi + 200\chi^2 - \cdots\right],$$

$$R_1 = \sqrt{3}\pi\chi^3 \left[1 - \frac{105\sqrt{3}}{16}\chi + 120\chi^2 - \cdots\right], \tag{4.106}$$

$$R_2 = \frac{15\pi}{8}\chi^3 \left[1 - \frac{32\sqrt{3}}{5}\chi + \frac{231}{2}\chi^2 - \cdots\right],$$

and at $\chi \gg 1$

$$R_0 = \frac{5\sqrt{3}}{27}\pi\Gamma(1/3)(3\chi)^{1/3} \left[1 - \frac{42\Gamma(2/3)}{5\Gamma(1/3)}(3\chi)^{-2/3} + \frac{729}{20\Gamma(1/3)}(3\chi)^{-4/3} - \cdots\right],$$

$$R_1 = \frac{5\sqrt{3}}{27}\pi\Gamma(1/3)(3\chi)^{1/3} \left[1 - \frac{27\sqrt{3}}{5\Gamma(1/3)}(3\chi)^{-1/3} + \frac{42\Gamma(2/3)}{5\Gamma(1/3)}(3\chi)^{-2/3} - \cdots\right],$$

$$R_2 = \frac{2\sqrt{3}}{27}\pi\Gamma(2/3)(3\chi)^{2/3} \left[1 - \frac{27}{4\Gamma(2/3)}(3\chi)^{-2/3} + \frac{15\Gamma(1/3)}{\Gamma(2/3)}(3\chi)^{-4/3} - \cdots\right] \tag{4.107}$$

Functions J_0, J_1 and iJ_2 for intermediate values of the parameter χ are given in Fig. 4.7. Let us now discuss obtained results. For $\chi \gg 1$ all the functions are the quantum corrections (spin effects), so all the expansions begin from the term $\propto \chi^3$ For $\chi \gg 1$ the functions J_0 and J_1 are increasing as $\chi^{1/3}$, while the function J_2 is increasing as $\chi^{2/3}$ just as the total intensity I_e. From this it follows that the circular polarization of the radiation ($\xi_2 = iJ_2/I$) of the longitudinally polarized electrons tend to the fixed limit (compare with (4.88))

$$\xi_2 = \frac{11}{16}\left[1 - \frac{27}{16\Gamma(2/3)}(3\chi)^{-2/3} + \frac{3(800\pi - 729\sqrt{3})}{256\Gamma^2(2/3)}(3\chi)^{-4/3} - \cdots\right] \tag{4.108}$$

This is the consequence of the helicity transfer from the electron to photon (see Sec. 4.4).

Depending on the electron spin addition to the total intensity (J_0) and to the degree of the linear polarization, (J_1) vanishes asymptotically comparing with the main terms at $\chi \gg 1$. One can see from Fig. 4.7 that the addition to the total intensity is always rather small (compare with Fig. 4.5) and does not exceed 15% of I_e. But depending on spin the addition (J_1) to the intensity I_- defining the linear polarization of the radiation has nearly the same magnitude as I_- at $1 \leq \chi \leq 20$. Let us note also that at $\chi \gg 1$ in the series for R_1 the three first terms are the decomposition over the powers $(3\chi)^{-1/3}$ As a result this series is converging very slowly.

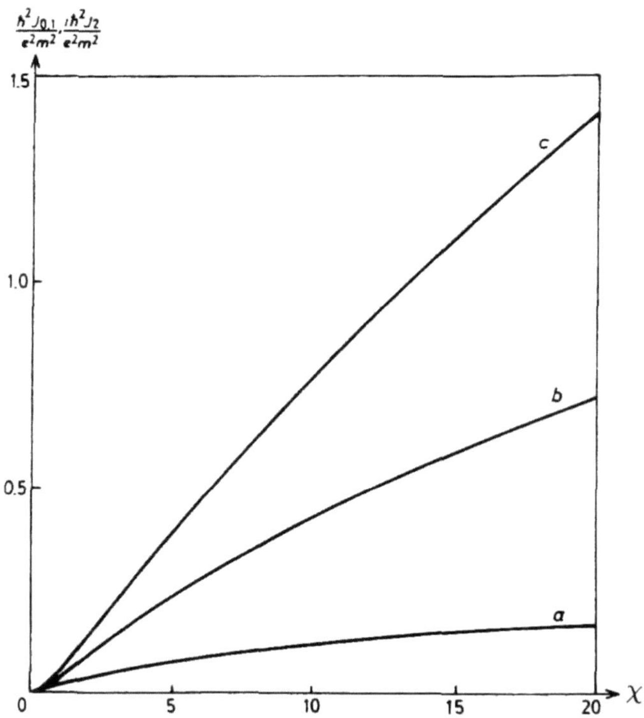

Fig. 4.7. Contributions to the density matrix in terms of integral intensities (Eq. (4.103)) of terms depending on the polarization of initial electrons: $a(J_0)$; $b(J_1)$; $c(iJ_2)$.

5. RADIATION OF PARTICLES MOVING QUASIPERIODICALLY AND IN SUPERPOSITION OF A PLANE WAVE AND A CONSTANT FIELD

5.1. General Formulae

In Secs. 1.8 – 1.12 the radiation at quasiperiodic motion has been considered in the framework of classical electrodynamics. The results obtained may be used for the description of the radiation in undulators or in a plane electromagnetic wave and their combination, in crystals, etc. Here we consider this problem in quantum theory. In the framework of the quasiclassical method (see Sec. 2) the general picture presented in Sec. 1.8 is preserved and distinctions which appear we discuss below. We proceed from the general expression for the radiation intensity (2.40)–(2.43) which substitute for the standard formula of the classical electrodynamics (1.23). The integral entering into (2.40) is represented in the form* (cf. (1.174–1.179))

$$\int_{-\infty}^{\infty} v^\mu(t) e^{ik'x(t)} dt = v_T^\mu \sum_{m=-\infty}^{\infty} e^{im\varphi_0'} = \sum_{m=1}^{\infty} 2\pi v_T^\mu \delta(\varphi_0' - 2\pi m)\,, \tag{5.1}$$

where

$$\varphi_0' = \omega' T(1 - \mathbf{nv}), \quad T = \frac{2\pi}{\omega_0}, \quad \omega' = \frac{\omega\varepsilon}{\varepsilon - \omega}, \quad v_T^\mu = \int_0^T v^\mu(t) e^{ik'x(t)} dt\,, \tag{5.2}$$

here T is the period of the motion. Substituting (5.1) into (2.40), we obtain for the intensity of the radiation per unit time

$$dI = e^2 \frac{d^3k}{4\pi T} \sum_{m=1}^{\infty} \delta(\varphi_0' - 2\pi m)\{[1 + (1+u)^2](|\mathbf{v}_T|^2 - |v_T^0|^2) + (u^2/\gamma^2)|v_T^0|^2\}\,; \tag{5.3}$$

here

$$v^\mu = (v^0, \mathbf{v})\,, \quad v^0 = 1\,, \quad u = \frac{\omega}{\varepsilon'} = \frac{\omega}{\varepsilon - \omega}$$

Using Eq. (1.119) and introducing the notation

$$\Delta(t) = \omega_0 \int_0^t (v_\perp^2(t') - \bar{v}_\perp^2) dt'\,, \tag{5.4}$$

where $\omega_0 = 2\pi/T$, and ignoring terms of higher order in $1/\gamma^2$ and $v_\perp^2$, we obtain the following expression for the spectral and angular distribution of the radiation

$$dI = \frac{e^2}{(2\pi)^2} \frac{d^3k}{\omega_0 \gamma^2} \sum_{m=1}^{\infty} \delta(\varphi_0' - 2\pi m)(1+u)\left[-|I_0|^2 + \gamma^2 \left(1 + \frac{u^2}{2(1+u)}\right) \times [|\mathbf{I}_\perp|^2 - Re I_0^* I_\parallel]\right], \tag{5.5}$$

*Here and below we set $\hbar = c = 1$.

where

$$I_0 = \int_0^{2\pi} e^{i\mathcal{F}(\psi)} d\psi\,, \quad \mathbf{I}_\perp = \int_0^{2\pi} \mathbf{v}_\perp(\psi) e^{i\mathcal{F}(\psi)} d\psi\,,$$

$$I_\| = \int_0^{2\pi} v_\perp^2(\psi) e^{i\mathcal{F}(\psi)} d\psi\,, \quad \psi = \omega_0 t\,, \tag{5.6}$$

$$\mathcal{F}(\psi) = m\psi + \omega'(\Delta(\psi)/2\omega_0 - \mathbf{n}\mathbf{x}_\perp(\psi))\,.$$

Here, we have used the circumstances that $\varphi_0 = \omega' T(1-\mathbf{n}\mathbf{v})$, and the frequency of the emitted photons is determined by the relation (cf. Eq. (1.120))

$$\omega' = \omega\frac{\varepsilon}{\varepsilon-\omega} = \frac{2\gamma^2\omega_0 m}{1+\gamma^2\vartheta^2+\rho/2}\,, \tag{5.7}$$

where ϑ is the angle of emission of the photon with respect to $\mathbf{V}$, and $\rho = 2\gamma^2\bar{v}_\perp^2$. The nature of radiation depends strongly on ρ for $\rho \ll 1$, we have the dipole radiation, for $\rho \gg 1$ the radiation is of magnetic bremsstrahlung nature.

5.2. Dipole Radiation at Quasiperiodic Motion*

As in classical electrodynamics, we will consider case II (see Sec. 1.8), where the length of an inhomogeneity (the structure period) is much shorter than the coherence length in magnetic bremsstrahlung limit: $\omega_0 = 2\pi/T \gg Hm/H_0$ (cf. Eq. (1.117)). The intensity of the radiation in the dipole approximation is given by Eqs. (2.40) and (2.41). Passing on in this formulae to the Fourier components of the acceleration (1.125) (cf. derivation of Eq. (1.124)), we have (cf. Eq. (1.131)):

$$dI = e^2\frac{d^3k}{2\pi T}\frac{\varepsilon'}{\varepsilon}\frac{\omega^2}{\tilde{\omega}^4}\Big\{|(\mathbf{e}^*\mathbf{w}_q)(1-\mathbf{n}\mathbf{V}) + (\mathbf{e}^*\mathbf{V})(\mathbf{n}\mathbf{w}_q)|^2 + \frac{\omega^2}{4\varepsilon\varepsilon'}|\mathbf{w}_q|^2(1-\mathbf{n}\mathbf{V})^2(\mathbf{e}^*\mathbf{e})\Big\}\Phi'\,,$$

$$\varepsilon' = \varepsilon - \omega\,, \quad \mathbf{w}_q = \mathbf{w}^T_{\tilde{\omega}'}\,, \quad \Phi' = \sum_{k=1}^{\infty}\delta(\varphi_0' - 2\pi k)\,, \tag{5.8}$$

$$\varphi_0' = \omega'(1-\mathbf{n}\mathbf{V})T = \tilde{\omega}' T\,.$$

For description of polarization of the radiation we use the same approach as in classical electordynamics (see Eqs. (1.140)–(1.143)). For spinor particles under

*See, Baier, Katkov and Stakhovenko, 1972.

consideration the Stokes parameters are given by Eq. (1.143), where one has to substitute functions a_n, c by

$$a_1^q = a_1, \quad a_3^q = a_3, \quad a_2^q = a_2\left(1 + \frac{\omega^2}{2\varepsilon\varepsilon'}\right), \quad c^q = c + \frac{\omega^2 y^2}{\varepsilon\varepsilon'} \tag{5.9}$$

After summation over the photon polarization and integration over the azimuthal angle of emission of the photon (factor 2π) we find that

$$dI = \frac{e^2 m^2}{4\pi^2} \sum_{k=1}^{\infty} \frac{\eta^2 \chi^2 du dy F}{y^6 u^2 (1+u)^3} (y^2 f(u) + 1/2 - y) \delta(uy - \eta k/2)\,. \tag{5.10}$$

Here we introduced the following notations:

$$u = \frac{\omega}{\varepsilon'}, \quad \eta = \frac{\omega_c}{\varepsilon} = \frac{4\pi\gamma}{Tm}, \quad f(u) = 1 + \frac{u^2}{2(1+u)}, \quad \chi^2 = \frac{\mathbf{w}^2 \gamma^4}{m^2}$$

The function F used here and below can be obtained from earlier sections (see (1.135), (1.171) and (1.173)) making in (1.135) the substitutions: $\tilde{\omega} \to \tilde{\omega}', \mathbf{w}_{\tilde{\omega}}^T \to \mathbf{w}_{\tilde{\omega}'}^T$.

After integration (5.10) over the polar angle of the photon emission, we obtain spectral distribution of the radiation:

$$dI = \frac{4e^2 m^2}{\pi^2} \sum_{k=1}^{\infty} \frac{\chi^2 u du F}{\eta^2 k^4 (1+u)^3} (f(u) + 2(u/\eta k)^2 - 2u/\eta k) \vartheta(\eta k - u)\,. \tag{5.11}$$

The classical expressions for the intensity can be obtained from (5.10), (5.11) by the substitutions $(1+u) \to 1, f \to 1$ (cf. (1.134), (1.137)). Let us rewrite the last formula in terms of $x = \omega/\varepsilon$. Then we obtain from (5.11)

$$\frac{dI}{dx} = \frac{4e^2 m^2 \chi^2}{\pi^2 \eta^2} \sum_{k=1}^{\infty} \frac{F}{k^4} \left[1 + \frac{x^2}{2(1-x)} + \frac{2x^2}{(1-x)^2 \eta^2 k^2} - \frac{2x}{(1-x)\eta k}\right] \vartheta\left(\eta k - \frac{x}{1-x}\right) \tag{5.12}$$

The term with $x^2/2(1-x)$ in brackets is growing at $x \to 1$ and for $\eta \gg 1$ it gives a peak in the spectral distribution of the intensity for a value $x = x_1 = \eta(1+\eta)^{-1} \approx 1 - 1/\eta$ (which is defined by the ϑ function) or $\omega \approx (1 - 1/\eta)\varepsilon$. In the vicinity of the peak the photon takes away nearly all the energy of the radiating particle. The first harmonics of the radiation covers nearly all the spectrum, while higher harmonics appear in the narrow interval near the end of the spectrum with typical width $\sim 1/\eta k$ and in the rest of the spectrum they give only a small correction to the contribution of the first harmonics. The width of the discussed peak is $\Delta\omega \sim \varepsilon/\eta$ and the height of the peak is

$$\left.\frac{dI}{dx}\right|_{\max} \simeq \frac{2e^2m^2\chi^2}{\pi^2\eta} \tag{5.13}$$

From formulae (5.13) it follows that the value $dI/d\omega|_{\max}$ does not depened on energy since $\chi \propto \varepsilon$ and $\eta \propto \varepsilon$. The results obtained are illustrated in Fig. 5.1, where the spectral distribtuion of the intensity for quantum undulator is presented for $\eta = 10$ and $\eta = 50$ for harmonic motion (1.156), $F = \delta_{k1}/2$, and cases A (1.170) and B (1.172). For the last cases, the functions F are given by Eqs. (1.171) and (1.173). Using (5.9), one can see that in the vicinity of the peak the radiation has substantial circular polarization for case B.

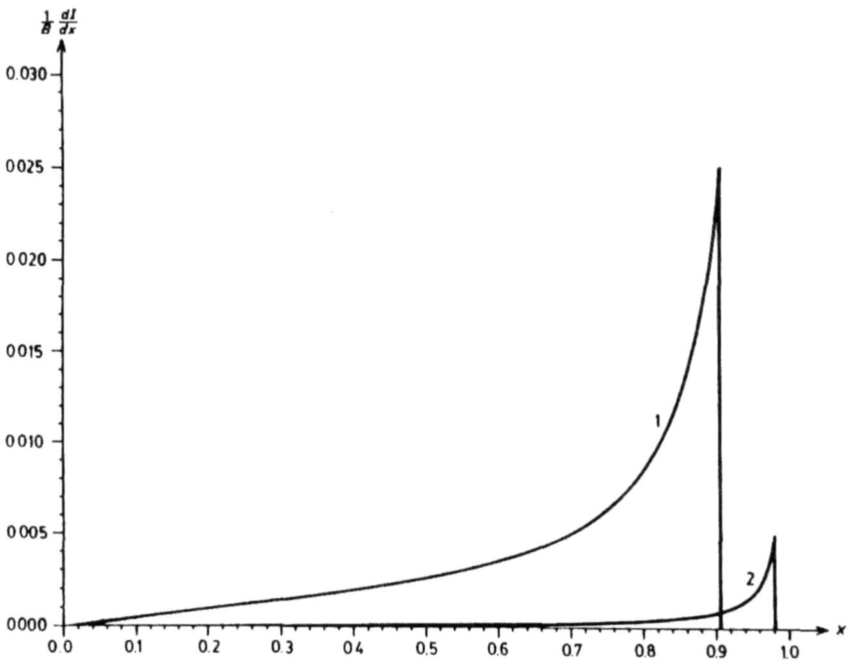

(a)

Fig. 5.1. Spectral distribution of intensity; *a*: in quantum undulator with harmonic motion; *b*: in quantum undulator constructed from pieces with homogeneous magnetic filed, case A; *c*: the same for case B. The curve 1 if for $\eta = 10$ and curve 2 is for $\eta = 50$, $B = 4e^2m^2\chi^2/\pi^2$

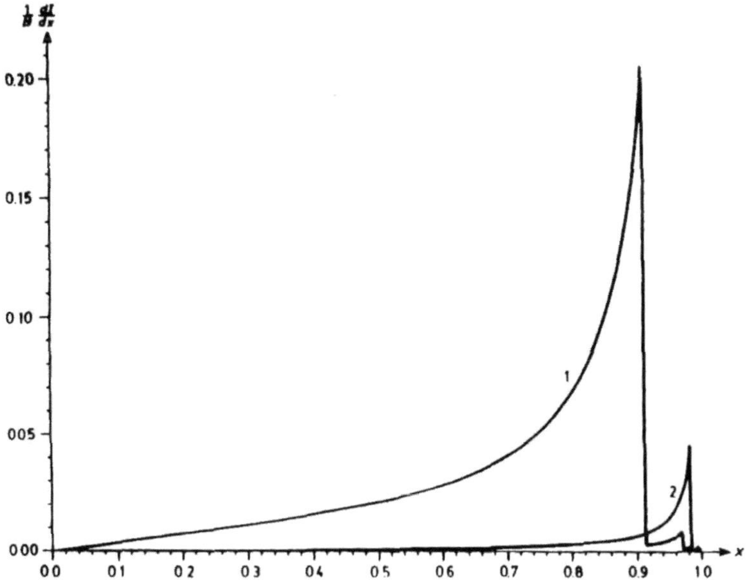

Fig. 5.1 (b) *Continued*

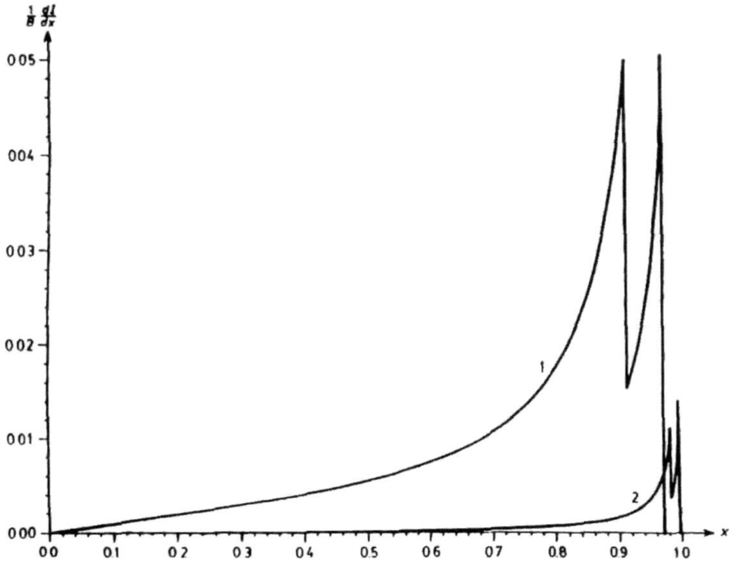

Fig. 5.1 (c) *Continued*

To obtain the total intensity of the radiation it's convenient to integrate first (5.10) over u. As a result we find angular distribtuion of the radiation (cf. (1.136)):

$$dI = \frac{4e^2m^2\chi^2}{\pi^2}\sum_{k=1}^{\infty}\frac{Fzdz}{k^2(1+\eta kz)^3}(f(\eta kz)+2z^2-2z)\,, \tag{5.14}$$

where $z = 1/2y(0 \le z \le 1)$.

All the formulae obtained above describe the radiation in the infinite undulator. The transition to the finite undulator can be done in the same manner as in Sec. 1.9 (see, particularly, Eqs. (1.148) and (1.149)).

We calculate now the total intensity of the radiation in the asymptotic cases $\eta \ll 1$ and $\eta \gg 1$. If $\eta \ll 1$, we obviously obtain the classical expression for the intensity (Sec. 1.10). If $\eta \gg 1$ we obtain with power accuracy in Case A (1.170) and in Case B (1.172)

$$\begin{aligned}
I_A &= I_A^{(0)}(\ln\eta - 5/6 + 96L_A/\pi^4)\,,\\
L_A &= \sum_{k=0}^{\infty}\frac{\ln(2k+1)}{(2k+1)^4} = 0.018\,,\\
I_B &= I_B^{(0)}(\ln\eta - 5/6 + 348L_B/\pi^4), \quad L_B = L_4 = 0.096\,,\\
\mathcal{L}_m &= \sum_{k=0}^{\infty}\frac{[3-4\cos\pi(k/2+1/4)]\ln(2k+1)}{(2k+1)^m}\,,
\end{aligned} \tag{5.15}$$

where the functions $I_A^{(0)}$ and $I_B^{(0)}$ are

$$\begin{aligned}
I_A^{(0)} &= \frac{8e^2m^2\chi^2}{\pi^2\eta^2}\sum_{k=0}^{\infty}\frac{1}{(2k+1)^4} = \frac{8e^2m^2\chi^2}{\pi^2\eta^2}\frac{\pi^4}{96}\\
&= \frac{1}{3}\left(\frac{\pi e m\chi}{2\eta}\right)^2 = e^2\frac{(eHT)^2}{192}\,,
\end{aligned} \tag{5.16}$$

$$I_B^{(0)} = \frac{8e^2m^2\chi^2}{\pi^2\eta^2}\Lambda_4 = \frac{I_A^{(0)}}{4}, \quad \Lambda_m = \sum_{k=0}^{\infty}\frac{3-4\cos\pi(k/2+1/4)}{(2k+1)^m}\,.$$

It is evident from the above expressions that the functions $I_A^{(0)}$ and $I_B^{(0)}$ (they are the intensities of the radiation of spin 0 particles) are independent of the mass and energy of the radiating particle. In the contribution of these functions is dominated in the range of angles $\vartheta \sim \sqrt{\eta}/\gamma$ which are large compared with $1/\gamma$. The whole integration domain over u contributes to logarithmic terms in (5.15), if one integrates over ω, these terms contribute mainly a peak region.

In the range of high values of η, the radiation is concentrated more strongly in the first harmonics than in the classcial case.

If the energy of the photon is comparable with the energy of the radiating particle, it would be interesting to determine the probability of the photon emission. It is necessary to multiply the intensity of the radiation by $1/\omega = (1+u)/u\varepsilon$ to obtain it. Then the probability of the radiation of the photon with energy ω is

$$dW = \frac{4e^2m^2\chi^2}{\pi^2\eta^2\varepsilon}\sum_{k=1}^{\infty}\frac{F du}{k^4(1+u)^2}(f(u) + 2(u/\eta k)^2 - 2u/\eta k)\vartheta(k\eta - u) \quad (5.17)$$

and the angular distribution of the probability is

$$dW = \frac{4e^2m^2\chi^2}{\pi^2\eta^2\varepsilon}\sum_{k=1}^{\infty}\frac{F dz}{k^3(1+\eta kz)^2}(f(\eta kz) + 2z^2 - 2z)$$

$$z = 1/2y = [2\gamma^2(1 - \mathbf{nV})]^{-1}\,. \quad (5.18)$$

If $\eta \ll 1$, we find that the total probability of the radiation is

$$W_A = \frac{32}{3}\frac{e^2m^2\chi^2}{\pi^2\eta\varepsilon}\sum_{k=0}^{\infty}\frac{1}{(2k+1)^3} = \frac{28}{3}\zeta(3)\frac{e^2m^2\chi^2}{\pi^2\eta\varepsilon}\,.$$

$$W_B = \frac{32}{3}\frac{e^2m^2\chi^2}{\pi^2\eta\varepsilon}\Lambda_3\,, \quad \Lambda_3 \simeq 0.445\,, \quad (5.19)$$

where $\zeta(n)$ is the Rieman zeta function. If $\eta \gg 1$, the total probability of the radiation is

$$W_{A,B} = W_{A,B}^{(0)}(\ln\eta + 1/2 + \delta_{A,B})/2\,,$$

$$\delta_A = 96L_A/\pi^4 \simeq 0.0174, \quad \delta_B = 384L_B/\pi^4 \simeq 0.379\,, \quad (5.20)$$

where

$$W_A^{(0)} = \frac{16e^2m^2\chi^2}{\pi^2\eta^2\varepsilon}\sum_{k=1}^{\infty}\frac{1}{(2k+1)^4} = \frac{2I_A^{(0)}}{\varepsilon}$$

$$W_B^{(0)} = \frac{16e^2m^2\chi^2}{\pi^2\eta^2\varepsilon}\Lambda_4 = \frac{2I_B^{(0)}}{\varepsilon} \quad (5.21)$$

If we substitute $\delta = 0$ in Eq. (5.20), we find that the expression obtained coincides, to within numerical factor, with the probability of Compton scattering in the case when incident photon is soft. This is due to the fact that an undulator can be represented by a superposition of standing waves of definite frequencies and these, in turn, can be regarded as a sum of plane waves "traveling" in opposite directions, and the dominant contribution being made by that wave whose wave vector is opposite to the electron velocity. The softness of the quanta of this wave follows from the condition $\hbar\omega \ll \varepsilon$ (this is the criterion of the quasiclassical case). The peak near the end of the spectrum discussed above also has an analog in the Compton effect of the soft photon on the high-energy electron moving in the opposite direction.

5.3. Radiation at Quasiperiodic Motion in Nondipole Domain (General Case)

In this section we will analyze the general case of the radiation.

The integrals in (5.6) have a relatively simple form only for circular transverse motion, when $|\mathbf{v}_\perp| = \text{const}$. In this case $\dot{\Delta}(\psi) = 0$,

$$\begin{aligned} &\mathcal{F}(\psi) = m\psi - \kappa\sin(\psi - \varphi)\,, \quad I_{\parallel} = v_\perp^2 I_0\,, \\ &|I_0|^2 = (2\pi)^2 J_m^2(\kappa)\,, \quad \kappa = \omega' v_\perp \vartheta/\omega_0\,, \\ &|\mathbf{I}_\perp|^2 = 2\pi^2 v_\perp^2 [J_{m+1}^2(\kappa) + J_{m-1}^2(\kappa)]\,, \end{aligned} \tag{5.22}$$

where $J_n(\kappa)$ is the Bessel function and φ is the azimuthal angle of emission of the radiation. Substituting these results in (1.178), we obtain the differential distribution of the radiation in the case of the circular motion of the particle (index cr)

$$\begin{aligned} dI_{cr}^{(i)} &= \frac{e^2}{\gamma^2}\frac{d^3k}{\omega_0}\frac{\varepsilon}{\varepsilon'}\sum_{n=1}^{\infty}\delta(\varphi_0' - 2\pi n)G_{cr}^{(i)}\,, \\ G_{cr}^{(0)} &= -J_n^2(\kappa) + \frac{\rho}{4}f(u)(J_{n-1}^2(\kappa) + J_{n+1}^2(\kappa) - 2J_n^2(\kappa))\,, \\ G_{cr}^{(1)} &= -A\sin 2\varphi, \quad G_{cr}^{(3)} = -A\cos 2\varphi\,, \\ A &= \left(1 + \frac{\rho}{2}\right)J_n^2(\kappa) - \frac{\rho}{2}J_{n+1}(\kappa)J_{n-1}(\kappa)\,, \\ G_{cr}^{(2)} &= f(u)\left(\frac{n\rho}{2\kappa} - \frac{\kappa}{2\xi}\right)J_n(\kappa)(J_{n-1}(\kappa) - J_{n+1}(\kappa))\,. \end{aligned} \tag{5.23}$$

Here and below, the index 0 is used for the expressions summed over the polarization of the radiation,

$$\xi^{(i)} = \frac{dI^{(i)}}{dI^{(0)}}$$

are the Stokes parameters, and

$$f(u) = 1 + \frac{u^2}{2(1+u)}, \quad \omega_c = \frac{2\gamma^2\omega_0}{1+\rho/2}, \quad \xi = \frac{\omega'}{2\gamma^2\omega_0}$$

For description of the polarization of the radiation we use the same vectors as in the classical theory, see discussion after Eq. (1.182).

In the case of the plane motion of the particle, we can, without loss of generality, take $v_x(\psi)$ to be an even function of ψ ($v_y = 0$). Then $\mathcal{F}(\psi)$ in the formula (5.6) becomes (apart from a constant term) an odd function of ψ. Thus, apart from a common phase factor, the functions I_0, I_x and $I_{\parallel}$ are real. This means that for plane quasiperiodic motion the radiation can be only linearly polarized. From the physical point of view, this fact is obvious, since for such motion there is no pseudovector with which circular polarization of the radiation could be associated.

With allowance for what we have said above, we obtain for plane motion (subscript p)

$$\begin{aligned}
dI_p^{(i)} &= \frac{e^2}{\omega_0}\frac{d^3k}{(2\pi)^2}\frac{\varepsilon}{\varepsilon'}\sum_{n=1}^{\infty}\delta(\varphi_0' - 2\pi n)G_p^{(i)}, \\
G_p^{(0)} &= -\frac{I_0^2}{\gamma^2} + f(u)(I_x^2 - I_0 I_{\parallel}), \\
G_p^{(1)} &= 2I_0\vartheta\sin\varphi(I_0\vartheta\cos\varphi - I_x), \quad G_p^{(2)} = 0, \\
G_p^{(3)} &= -(I_0\vartheta\sin\varphi)^2 + (I_0\vartheta\cos\varphi - I_x)^2.
\end{aligned} \tag{5.24}$$

After integration over the azimuthal emission angle of the photon $G_p^{(1)} = 0$. If $\omega \ll \varepsilon (u \ll 1)$, the formulae obtained above (5.22) – (5.24) turns into classical expressions (see (1.181) – (1.183)).

Assuming that the plane motion of the particle is determined by some potential $U(x)$ and using the quasiclassical conditions of quantization of the energy in this one-dimensional well, we can represent $\bar{v}_x^2$ in the form

$$\bar{v}_x^2 = \frac{1}{T\varepsilon}\oint p_x dx \approx \frac{\omega_0 n_x}{\varepsilon} = -2\frac{\partial\varepsilon_\perp(\varepsilon, n_x)}{\partial\varepsilon}, \tag{5.25}$$

where n_x is the number of the level, and $\varepsilon_\perp$ is the energy of the transverse motion.

In the special case of motion in an oscillator potential, the spectral density of the radiation can be written in the form

$$dI_p = e^2 \frac{\omega d\omega d\varphi}{(2\pi)^3\gamma^2} \sum_{n=1}^{\infty} \vartheta(n\omega_c - \omega')[-A_0^2 + \rho f(u)(A_1^2 - A_0 A_2)]\,; \tag{5.26}$$

here

$$A_m = \int_0^{2\pi} e^{i\mathcal{F}(\psi)} \cos^m \psi d\psi\,,$$

$$\mathcal{F}(\psi) = n\psi - \alpha_0 \sin\psi + \beta_0 \sin 2\psi\,,$$

$$\alpha_0 = \frac{\omega'}{\omega_0}\left(\frac{2\varepsilon_\perp}{\varepsilon}\right)^{1/2} \vartheta \cos\varphi\,, \quad \beta_0 = \frac{\omega'\varepsilon_\perp}{4\omega_0\varepsilon} \tag{5.27}$$

In this case, $\rho = 2\gamma\varepsilon_\perp/m$.

The formulae (5.23) and (5.26) are identical to the expressions for the spectral density of the radiation in the field of a monochromatic plane wave of circular and linear polarization, respectively. This agreement is not fortuitous and is due to the circumstance that in the comoving frame of the electron the field in which the electron moves is, with relativistic accuracy, the field of a plane wave. The spectral distribution of the radiation in the field of a plane wave has been investigated in detail.* The theory of the radiative effects in the field of a plane wave is presented in Appendix C.

As one further example of the use of the general expressions (5.5) and (5.6), we consider the case when the transverse motion takes place in an ellipse, but the time dependence of the coordinate is not harmonic

$$x = a(\cos\xi - \delta)\,, \quad y = a(1-\delta^2)^{1/2} \sin\xi\,,$$

$$\psi = \omega_0 t = \xi - \delta \sin\xi\,. \tag{5.28}$$

Such a motion can be realized if the potential $U(\mathbf{x}_\perp)$ has a Coulomb form. In

*The processes in the field of a plane electromagnetic wave has been considered by Nikishov and Ritus (1964a,b), see also Nikishov (1979) and Ritus (1979), with the use of the solution of the Dirac equation in the field of a plane wave. The authors jointly with A.I. Milstein have analyzed the radiative effects in the field of plane electromagnetic wave based on operator approach (see Baier *et al.*, 1975; Baier, Milstein and Strakhovenko, 1975).

this case

$$\mathcal{F}(\psi) = n\xi - \mu\sin(\xi + \xi_0),$$

$$\cot\xi_0 = \frac{n\delta - \delta\omega'\omega_0 a^2 + \omega' a(1-\delta^2)^{1/2}\vartheta\sin\varphi}{a\omega'\vartheta\cos\varphi},$$

$$\mu^2 = [\delta(n - \omega'\omega_0 a^2) + a\omega'\vartheta(1-\delta^2)^{1/2}\sin\varphi]^2 + (a\omega'\vartheta\cos\varphi)^2, \quad \rho = 2\gamma^2 a^2\omega_0^2. \tag{5.29}$$

Substituting (5.28) and (5.29) in Eqs. (5.5) and (5.6) and integrating over the polar angle ϑ, we obtain the spectral distribution of the radiation intensity

$$dI = \frac{e^2\omega d\omega d\varphi}{2\pi\gamma^2}\sum_{n=1}^{\infty}\vartheta(n\omega_c - \omega')\left\{\frac{\rho f(u)}{2}\left[\left(\frac{n^2}{\mu^2} - 1\right)J_n^2(\mu) + J_n'^2(\mu)\right] - \left(1 - \frac{\delta n}{\mu}\cos\xi_0\right)^2 J_n^2(\mu) - \delta^2\sin^2\xi_0 J_n'^2(\mu)\right\} \tag{5.30}$$

For $\delta = 0$, the expression (5.30) goes over into the spectral distribution for circular transverse motion (cf. (5.23)), and in the limit $\rho \to 0$ we have the dipole case (cf. (5.12)).

Now we consider one more method of calculation of the spectral distribution of the radiation intensity. We can integrate over the photon emission angles in the formula (2.40) (using (2.42) and (2.43)) with help of Eq. (2.47). Going over also to the variables: $t = \omega_0(t_2 - t_1)/2, \tau = \omega_0(t_2 + t_1)/2$, we find for the spectral distribution

$$dI = \frac{ie^2m^2udu}{(2\pi)^2(1+u)^3}\int_{-\infty}^{\infty}\frac{dt}{t}\int_0^{2\pi}d\tau\left[\gamma^2(1 - \mathbf{v}_1\mathbf{v}_2)f(u) - \frac{u^2}{2(1+u)}\right] \times\exp\left\{-\frac{iu\varepsilon}{2\omega_0\gamma^2}\left[2t(1+\rho/2) + (\Delta_2 - \Delta_1)\gamma^2 - \frac{1}{2t}\omega_0^2\gamma^2(\mathbf{x}_2 - \mathbf{x}_1)_\perp^2\right]\right\} \tag{5.31}$$

The subscripts 1 and 2 denote the dependence on the variables $t_1 = (\tau - t)/\omega_0, t_2 = (\tau + t)/\omega_0$, other notations see in (5.3)–(5.7).

Note an important circumstance. In order to define the integral over t in (5.31) correctly, it is sufficient to assume that the contour of integration with respect to t (in complex domain) is displaced below the real axis. This operation corresponds to the substructional procedure in the finding of the mass operator in an external field (see Baier *et al.*, 1975) and ensures, in particular, disappearance of the radiation intensity when the field is switched off.

We consider the case of elliptic transverse motion of the particle

$$v_x(t) = a\cos\omega_0 t\,, \quad v_y = b\sin\omega_0 t\,.$$

We reduce the expression in (5.31) to the form

$$\begin{aligned} \gamma^2(1-\mathbf{v}_1\mathbf{v}_2) &= 1+\rho\sin^2 t(1-\alpha\cos 2\tau)\,,\\ \gamma^2(\Delta_2-\Delta_1) &= \alpha\rho\sin 2t\cos 2\tau/2\,,\\ \gamma^2\omega_0^2(\mathbf{x}_2-\mathbf{x}_1)_\perp^2 &= 2\rho\sin^2 t(1+\alpha\cos 2\tau)\,,\\ \rho = \gamma^2(a^2+b^2)\,, &\quad \alpha = (a^2-b^2)/(a^2+b^2)\,. \end{aligned} \tag{5.32}$$

After this, the integration over τ can be readily performed*

$$\begin{aligned} dI = \frac{ie^2m^2}{2\pi}\frac{u\,du}{(1+u)^3}\int_{-\infty}^{\infty}\frac{dt}{t}\Big\{ & J_0(\alpha z_0) + f(u)\rho\sin^2 t\\ & \times\left[J_0(\alpha z_0) - i\alpha J_1(\alpha z_0)\right] + \frac{iu(2+u)}{2(1+u)}\rho\beta(\boldsymbol{\zeta}\mathbf{v})\sin t\\ & \times\left(\frac{\sin t}{t} - \cos t\right)J_0(\alpha z_0)\Big\}e^{-iz_1}\,,\\ z_0 = x_0 - x_1\,, \quad z_1 = 2\lambda t - x_0\,, &\quad x_0 = \rho\xi\sin^2 t/t\,,\\ x_1 = \rho\xi\sin 2t/2\,, \quad \xi = u\varepsilon/(2\gamma^2\omega_0)\,, &\quad \beta = 2ab/(a^2+b^2)\,,\\ \lambda = \xi(1+\rho/2)\,, \quad f(u) &= 1 + \frac{u^2}{2(1+u)} \end{aligned} \tag{5.33}$$

The formula (5.33) corresponds to the imaginary part of the expression for mass operator of an electron in the field of a plane wave obtained in Baier *et al.*, 1975 (see Appendix C).

Note that, using the expressions (2.40) and (2.41) and the calculation procedure adopted above, we can obtain all the polarization properties of the

*For greater generality, we have included in the expression (5.33) the spin term; $\boldsymbol{\zeta}$ is the spin vector in the electron rest frame, as it is accepted in the book.

radiation. For example, in the case of helical motion of the particle ($\alpha = 0$), we have

$$\begin{aligned} dI_{cr}^{(i)} &= \frac{e^2m^2}{2\pi}\frac{udu}{(1+u)^3}\int_{-\infty}^{\infty}\frac{dt}{t}A_{cr}^{(i)}(t)\exp[-2i\lambda t + ix_0(t)]\,, \\ A_{cr}^{(0)} &= i[1+\rho f(u)\sin^2 t]\,, \\ A_{cr}^{(2)} &= f(u)\rho\sin t(\cos t - \sin t/t)\,, \\ \xi_{cr}^{(i)} &= \frac{dI_{cr}^{(i)}}{dI_{cr}^{(0)}} \end{aligned} \tag{5.34}$$

In the case of plane motion ($\alpha = 1$), we obtain the following expressions for the intensity and polarization of the radiation

$$\begin{aligned} dI_p = \frac{ie^2m^2}{2\pi}\frac{udu}{(1+u)^3}\int_{-\infty}^{\infty}\frac{dt}{t}[J_0(z_0) \\ + f(u)\rho\sin^2 t(J_0(z_0) - iJ_1(z_0)]e^{-iz_1}\,. \end{aligned} \tag{5.35}$$

The degree of linear polarization $\xi^{(3)}$ is given by

$$\begin{aligned} \xi^{(3)} &= \frac{dI^{(3)}}{dI_p}\,, \\ dI^{(3)} &= \frac{e^2m^2}{4\pi}\rho\frac{udu}{(1+u)^3}\int_{-\infty}^{\infty}\frac{dt}{t-i0}e^{-iz_1} \\ &\times\left\{J_1(z_0)\left[1-\frac{\sin 2t}{t}+\left(\frac{\sin t}{t}\right)^2\right]\right. \\ &\left.-iJ_0(z_0)\left[\cos 2t-\frac{\sin 2t}{t}+\left(\frac{\sin t}{t}\right)^2\right]\right\} \end{aligned} \tag{5.36}$$

Let us analyze this expression for the case of plane motion. For simplicity, we shall assume that soft photons are emitted, i.e. $u \ll 1$ (this is valid if $2\omega_0\gamma^2(1+\rho)^{1/2}/\varepsilon \ll 1$). In this case the formula (5.35) becomes

$$dI_p = \frac{2ie^2}{\pi}\omega_0^2\gamma^2\xi d\xi\int_{-\infty}^{\infty}\frac{dt}{t}\{J_0(z_0)+\rho\sin^2 t[J_0(z_0)-iJ_1(z_0)]\}e^{-iz_1}\,, \tag{5.37}$$

here

$$\xi = \frac{\omega}{\omega_c} = \frac{\omega}{2\gamma^2\omega_0}\,, \qquad z_0 = \rho\xi\left(\frac{\sin^2 t}{t} - \frac{\sin 2t}{2}\right)\,,$$

$$z_1 = 2\xi t\left[1 + \frac{\rho}{2}\left(1 - \frac{\sin^2 t}{t^2}\right)\right]\,, \qquad \rho = \gamma^2 a^2\,. \tag{5.38}$$

In general, the integral in (5.37) is a discontinuous function of the parameter $\lambda = \xi(1 + \rho/2)$ at the point where $\lambda = n$ with n an integer. To separate the discontinuity we use the relation

$$\int_{-\infty}^{\infty} \frac{dt}{t} F(t) = \int_{-\infty}^{\infty} \frac{dt}{t}[F(t) - F_a(t)] + \int_{-\infty}^{\infty} \frac{dt}{t} F_a(t)\,. \tag{5.39}$$

From (5.37), we have

$$\begin{aligned} F_a(t) \equiv F(t)_{t\to\infty} &= \{J_0(y_0) + \rho\sin^2 t[J_0(y_0) + iJ_1(y_0)]\}e^{-2i\lambda t}\,, \\ y_0 &= \xi\rho\sin 2t/2\,. \end{aligned} \tag{5.40}$$

Using the Fourier expansion of the Bessel functions,

$$\begin{aligned} J_0(a\sin 2t) &= J_0^2(a/2) + 2\sum_{n=1}^{\infty} J_n^2(a/2)\cos 4nt\,, \\ J_1(a\sin 2t) &= 2\sum_{n=0}^{\infty} J_n(a/2)J_{n+1}(a/2)\sin 2(2n+1)t\,, \end{aligned} \tag{5.41}$$

we can explicitly separate the discontinuities at the points $\lambda = n$. The remaining integral is a continuous function of λ, and in the limit $t \to \infty$ the integrand behaves as $1/t^2$, which is convenient for numerical calculations. The separated part of the spectral distribution of the radiation intensity has the form

$$\begin{aligned} \frac{dI_{p(a)}}{d\xi} &= 4e^2\gamma^2\omega_0^2 \sum_{n=0}^{\infty}\{\vartheta(2(n+1) - \lambda)[2(n+1) - \lambda]J_{n+1}^2(y) \\ &\quad + \vartheta(2n + 1 - \lambda)y[J_n(y) - J_{n+1}(y)]^2\}\,, \\ y &= \rho\xi/4\,. \end{aligned} \tag{5.42}$$

It can be seen from the expression (5.42) that discontinuities of the integral exist only for odd n, i.e. for $\lambda = 2m+1$. In the language of the radiation in the field of a plane wave, this circumstance is due to the fact that if the angle of the emitted photon is $\vartheta = 0$ only an odd number of photons can be absorbed from linearly polarized wave, this, in turn, being due to the conservation of the projection of the angular momentum onto the momentum.

In the case of helical motion, the integral in (5.34) is a discontinuous function at $\lambda = 1$. Proceeding as in the case of plane motion, we separate explicitly the terms containing this discontinuity. So, we have

$$dI_{cr}^{(i)} = \frac{e^2m^2}{2}\frac{udu}{(1+u)^3}\left\{\frac{1}{\pi}\int_{-\infty}^{\infty}\frac{dt}{t}A_{cr}^{(i)}e^{-2i\lambda t}[e^{ix_0}-1-ix_0(t)]+F_a^{(i)}\right\};$$

$$A_{cr}^{(0)} = i(1+f\rho\sin^2 t)\,, A_{cr}^{(2)} = f\rho\sin t\left(\cos t - \frac{\sin t}{t}\right),$$

$$F_a^{(0)} = \frac{\rho}{2}\left\{[f-2\xi(1+\rho f)(1-\lambda)]\vartheta(1-\lambda)+\frac{f\rho\xi}{2}(2-\lambda)\vartheta(2-\lambda)\right\}, \quad (5.43)$$

$$F_a^{(2)} = \frac{f\rho}{2}\left\{\left[1-2\left(1-\frac{1}{2}\rho\xi(1-2\lambda)\right)(1-\lambda)\right]\vartheta(1-\lambda)\right.$$
$$\left.-\frac{\rho\xi}{2}(1-\lambda)(2-\lambda)\vartheta(2-\lambda)\right\}$$

For soft photons ($\omega \ll \varepsilon$) and in dipole approximation ($\rho \ll 1$), we can ignore the integral in (5.43), and in the expressions for F_a we retain only the leading terms in ρ; then

$$F_a^{(0)} \simeq \frac{\rho}{2}[1-2\xi(1-\xi)]\vartheta(1-\xi)\,,$$
$$F_a^{(2)} \simeq \frac{1}{2}\rho(2\xi-1)\vartheta(1-\xi)\,. \quad (5.44)$$

As can be seen from (5.44), $\xi^{(2)} = 1$ at $\xi = 1$ and $\xi^{(2)} = -1$ at $\xi = 0$. This is due to the helicity conservation law (from the point of view of the Compton effect), as can be seen by noting that in the electron rest frame a circularly polarized photon is scattered backward or almost forward.

To conclude this section, we give the asymptotic expansions for the spectral distribution of the radiation intensity (5.33) in the case when $\rho \gg 1$. Using integral representation of the Bessel functions, and also the method of stationary phase for $\rho\xi \gg 1$, we obtain for unpolarized particles

$$dI = \frac{e^2m^2udu}{\pi\sqrt{3}(1+u)^3}\int_0^{2\pi}\frac{d\psi}{2\pi}\left\{\int_\kappa^\infty dyK_{5/3}(y) + 2[f(u)-1]K_{2/3}(\kappa)\right.$$
$$-\frac{4}{3\rho(1-\alpha\cos\psi)\kappa}\left[\left(1-\frac{8}{15}g(\alpha)\right)f(u)\frac{d}{d\kappa}(\kappa K_{1/3}(\kappa))\right.$$
$$\left.\left.+\frac{2f(u)-1}{10}g(\alpha)\frac{d}{d\kappa}(\kappa^2K_{2/3}(\kappa))\right]\right\},$$
$$g(\alpha) = \frac{1-2\alpha\cos\psi}{1-\alpha\cos\psi}, \quad \kappa^2 = \frac{32\xi^2}{9\rho(1-\alpha\cos\psi)},$$
$$f(u) = 1 + \frac{u^2}{2(1+u)}, \tag{5.45}$$

where α is determined in Eq. (5.32) and ξ in (5.33). The first two terms in the curly bracket coincide with the expression for the intensity of the magnetic bremsstrahlung of an ultrarelativistic electron (see(4.27)) and the

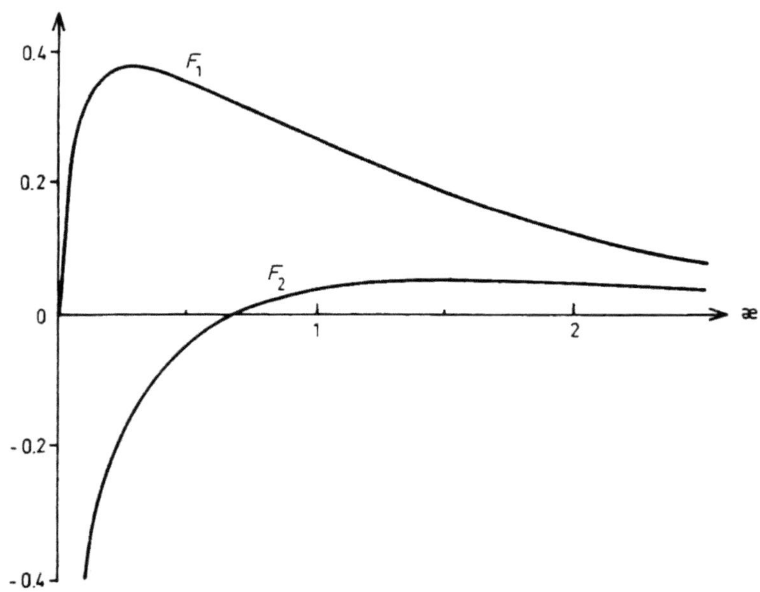

Fig. 5.2. Dependence of the functions F_1 and F_2 (Eq. (5.46)) on κ.

remainder are corrections $\propto 1/\rho$, which control the transition to the magnetic bremsstrahlung limit. For $\alpha = 0, u \ll 1$, we can write (5.45) in the form

$$\frac{\gamma^2}{e^2\omega_c^2}\frac{dI}{d\kappa} = F_1(\kappa) + \frac{1}{\rho}F_2(\kappa)\,. \qquad (5.46)$$

The functions F_1 and F_2 are shown in Fig. 5.2.

5.4. Integral Characteristics of the Radiation at a Quasiperiodic Motion

In classical electrodynamics the total intensity of the radiation depends on local characteristics and may be presented in the form

$$I(\mathbf{x}) = (2/3)e^2m^2\chi^2(\mathbf{x})\,, \qquad (5.47)$$

where $\chi(\mathbf{x}) = e|\mathbf{F}_\perp(\mathbf{x})|\varepsilon/m^3$, here $\mathbf{F}_\perp(\mathbf{x})$ is the local value of the external field at a point $\mathbf{x}, \mathbf{F}_\perp = \mathbf{E} - \mathbf{v}(\mathbf{v}\mathbf{E}) + \mathbf{v}\times\mathbf{H}$, $\mathbf{v}$ is the particle velocity, $\mathbf{E}$ and $\mathbf{H}$ are an electric and a magnetic field. Due to this fact the total intensity of radiation is obtained by integration over $\mathbf{x}$ of the local intensity $I(\mathbf{x})$ irrespective of the set of trajectories. The distribution function of the particle in phase space should be used as the weight in this integration.

Generally speaking, in the quantum region the simple formula of the type (5.47) for the total intensity does not exist. But in the magnetic bremsstrahlung limit ($\rho \gg 1$) the description of the radiation is once again becoming local. This means that the strongest effects of nonlocality in the total intensity of radiation (or in the total energy loss) become apparent just inside of the intermediate interval of ρ. In many situations, within this interval one can put $\chi \ll 1$. For analysis of the problem we will use the general quasiclassical formula for energy losses (see Eq. (2.46)).
Then the energy losses E are

$$E = \frac{ie^2m^2}{2\pi}\int_0^\infty \frac{udu}{(1+u)^3}\int dt\int_{-\infty}^{\infty}\frac{d\tau}{\tau - i0}[1 + \gamma^2 f(u)(\Delta\mathbf{v}_1 - \Delta\mathbf{v}_2)^2/2]e^{-iu\phi}\,, \qquad (5.48)$$

where we take into account that $f(u) = (\varepsilon^2 + \varepsilon'^2)/2\varepsilon\varepsilon', A_1 = u\phi,\ \omega = \varepsilon u/(1+u)$, as above (see (2.46)) $\mathbf{v}_{1,2} = \mathbf{v}(t_{1,2}),\ t_{2,1} = t \pm \tau/2$.

The integral over u in (5.48) can be evaluated in the general form. As a result we find the formula for the energy losses for any type of charged particle motion

$$E = \frac{ie^2m^2}{2\pi}\int dt \int_{-\infty}^{\infty}\frac{d\tau}{\tau - i0}\left[\left(1+\frac{\gamma^2}{2}(\Delta\mathbf{v}_1 - \Delta\mathbf{v}_2)^2\right)R(\phi) + \frac{\gamma^2}{4}(\Delta\mathbf{v}_1 - \Delta\mathbf{v}_2)^2 S(\phi)\right], \tag{5.49}$$

where

$$R(\phi) = i\phi(1+i\phi/2)\exp(i\phi)^{\mathrm{Ei}}(-i\phi) + (1+i\phi)/2;$$

$$S(\phi) = -(1+3i\phi+3(i\phi)^2/2+(i\phi)^3/6)\exp(i\phi)^{\mathrm{Ei}}(-i\phi) - (11+8i\phi+(i\phi)^2)/6.$$

${}^{\mathrm{Ei}}(z)$ is the integral exponential function defined as in Gradshteyn and Ryzhik (1965).

Expanding the integrand in Eq. (5.48) into a series of powers of u (the main contribution is given by the region $u \sim \chi \ll 1$) and retaining the terms which give the contribution up to χ^4, we obtain after integration over u

$$E = E_0 + E_1 + E_2 = \frac{e^2m^2}{2\pi}\int dt \int \frac{d\tau}{\tau - i0}\{[1+\gamma^2(\mathbf{v}_1 - \mathbf{v}_2)^2/2] \times [-i/\phi^2 + 6/\phi^3] + (3i/\phi^4)[12 + (13/2)\gamma^2(\mathbf{v}_1 - \mathbf{v}_2)]\}. \tag{5.50}$$

This decomposition may be obtained also from the general formula (5.49) using expansion of the function ${}^{\mathrm{Ei}}(z)$ at $z \gg 1$.

In this expression the term $1/\phi^2$ gives the classical formula for the energy losses. Because the function $\phi(\tau)$ is odd the integral over τ may be taken as half-residue in the point $\tau = 0$. As a result we obtain the known local expression

$$I_{cl} \equiv dE_0/dt = (2/3)e^2\dot{\mathbf{v}}^2\gamma^4 = (2/3)e^2m^2\chi^2 \tag{5.51}$$

Precisely in the same way one can calculate the term with $1/\phi^4$ which gives the second quantum correction ($\propto \chi^4$). It is also universal and local

$$I_2 \equiv dE_2/dt = 32e^2m^2[(\dot{\mathbf{v}}\gamma^2/m)^4 + (\gamma^6/160m^4)(8(d^2\mathbf{v}/dt^2)^2 - 41(d\mathbf{v}/dt \cdot d^3\mathbf{v}/dt^3))]. \tag{5.52}$$

But the term containing $1/\phi^3$, is, generally speaking, nonlocal. For its calculation one has to substitute into (5.50) the given trajectory (the explicit expression for $\mathbf{v}(t)$). We consider here the rather general situation of quasi-classical motion following an elliptic trajectory discussed in Sec. 5.3. For this case one has for the first quantum correction, in terms of mean radiation intensity, $\langle I_1\rangle \equiv (1/T)\int_0^T dt, \ldots$, where the explicit representation of the integrand is given in Eq. (5.50):

$$\langle I_1\rangle = -(55e^2m^2/8(3)^{1/2}\lambda)\langle\chi^3\rangle k_1(\rho)\,. \tag{5.53}$$

Here ω_0 is the motion frequency, $T = 2\pi/\omega_0$ is the period of the motion

$$k_1(\rho) = -\frac{B}{\rho^{3/2}}\int_0^\infty \frac{d\tau}{\tau^4}\left\{\frac{[(1+\rho\sin^2\tau)(2g_1^2+\alpha^2g_2^2)+3\rho\alpha^2g_1g_2\sin^2\tau]}{(g_1^2-\alpha^2g_2^2)^{5/2}}\right.$$
$$\left. -\,2-2\rho\tau^2\right\}, \tag{5.54}$$

where the components of the transverse velocity of the particle are (see (5.32)) $v_x = a\cos\omega_0 t, v_y = b\sin\omega_0 t$;

$$\rho = \gamma^2(a^2+b^2)\,,\quad \alpha = (a^2-b^2)/(a^2+b^2)\,,$$
$$g_1 = 1+\rho(1-\sin^2\tau/\tau^2)/2\,,\quad g_2 = \frac{\rho}{2}\left[\frac{\sin 2\tau}{2\tau} - \frac{\sin^2\tau}{\tau^2}\right],$$
$$B = (48(6)^{1/2}/55\pi)\lambda\,,\quad \lambda = \langle v_\perp^2\rangle^{3/2}/\langle|v_\perp|^3\rangle\,,$$

$\langle\ldots\rangle$ means averaging over time, two last terms in brace in Eq. (5.54) are subtraction terms. Let us stress, that Eqs. (5.51) – (5.53) are obtained as expansions in terms of χ, but are exact functions of the parameter ρ characterizing multipolarity of the radiation.

In Eq. (5.53) at $\alpha = 0$ we have circular motion and at $\alpha = 1$ we have planar motion. For these cases the function $k_1(\rho)$ (Eq. (5.54)) is presented in Fig. 5.3. We use a normalization of this function such that $k_1(\rho) \Rightarrow 1$ when $\rho \gg 1$, what corresponds to magnetic bremsstrahlung limit. Curve 1 is for circular motion and curve 2 is for planar motion. It is seen that deviation from the magnetic bremsstrahlung limit is larger for circular motion, but the interval between curves 1 and 2 is rather narrow and intermediate values of α is lying inside this interval. For circular motion the asymptotic expansions of the function $k_1(\rho)$ are

$$k_1(\rho) = \begin{cases} (112(6)^{1/2}/275\rho^{1/2}) & \rho \ll 1 \\ 1+(114/275\rho) & \rho \gg 1 \end{cases} \tag{5.55}$$

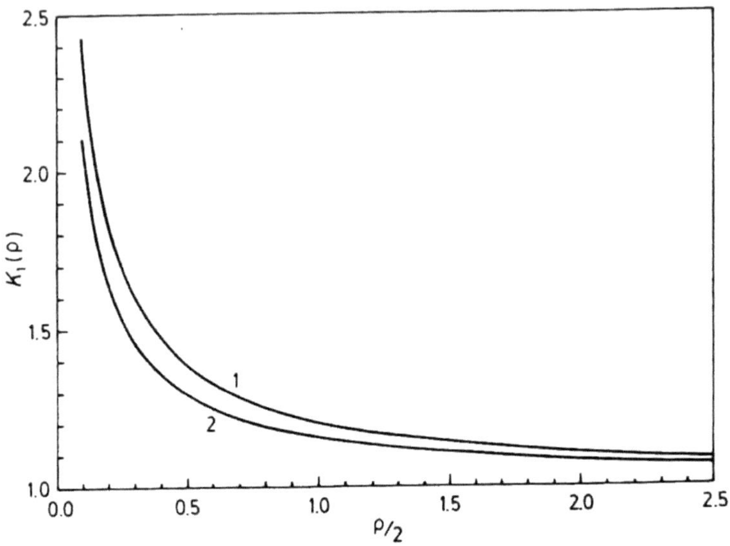

Fig. 5.3. Function $k_1(\rho)$ (Eq. (5.54)) for circular (1) and planar (2) motion.

Let us note, that the asymptotic expansion at $\rho \gg 1$ is valid actually up to $\rho \approx 1$ (at $\rho = 1$ the accuracy is better than 2%).

Let us note also, that for arbitrary ρ the function $\kappa_1(\rho)$ may be presented by simple interpolation formulae $k_1(\rho) = [1 + 1/\rho]^{1/2}$ for the circular motion (the accuracy is better than 2%) and $k_1(\rho) = [1 + \ln 2/\rho]^{1/2}$ for planar motion (the accuracy is better than 1%). This may be useful for numerical calculations.

Finally, we present the explicit form of the mean intensity at the elliptic quasiclassical motion as a series in powers of χ which is valid for arbitrary ρ

$$\langle I \rangle = \frac{2}{3} e^2 m^2 \left[\langle \chi^2 \rangle - \frac{55(3)^{1/2}}{16} \langle \chi^3 \rangle k_1(\rho) + 48 \langle \chi^4 \rangle k_2(\rho) \right] , \qquad (5.56)$$

where the function $k_1(\rho)$ is given by Eqs. (5.54) and (5.55), and

$$k_2(\rho) = 1 + (49/40\rho(2 + \alpha^2)) \qquad (5.57)$$

follows from Eq. (5.52) with allowance for Eq. (5.54).

Now, we will use a different approach. We will use results of Sec. 5.3 (formula (5.45)). Let us take into account that the classical radiation intensity is proportional to the square of acceleration (quantum corrections contain higher powers of acceleration). So, on the strongly stretched trajectories the particle radiates mostly in the region where acceleration is maximal, i.e. on the small part of the trajectory. In other words, the radiation is quasilocal. The least favorable situation is in this sense for circular trajectories, as we have already seen at the discussion of the function $k_1(\rho)$ (see Fig. 5.3). Therefore we consider the radiation intensity with $1/\rho$ corrections just for circular motion (see Eq. 5.45)):

$$dI = dI_m - dI_c/\rho\,, \tag{5.58}$$

where

$$dI_c = \frac{4e^2m^2udu}{15\sqrt{3}\pi\kappa(1+u)^3}\left[\frac{7}{3}f(u)\frac{d}{d\kappa}(\kappa K_{1/3}(\kappa)) + (f(u)-1/2)\frac{d}{d\kappa}(\kappa^2K_{2/3}(\kappa))\right], \tag{5.59}$$

here dI_m is the radiation intensity in the magnetic bremsstrahlung limit, in our case $\chi = \sqrt{(\rho/2)}\omega_0\varepsilon/m^2, \omega_0$ is the motion frequency in l-system. Let us stress that this expression is valid for any value of the parameter χ, but contains only the first term of the expansion over $1/\rho$. Taking in (5.58) the integral over u we obtain the correction to the total intensity

$$I_c = \frac{2e^2m^2}{15\sqrt{3}\pi}\int_0^\infty [7\chi u(3+2u+u^2)K_{1/3}(2u/3\chi) + u^2(3+u+2u^2)K_{2/3}(2u/3\chi)]\frac{du}{(1+u)^5}\,, \tag{5.60}$$

where integration by parts was carried out. The integral in this expression is of the type encountered in the theory of the magnetic bremsstrahlung. So, using the standard methods (see Sec. 4.2) one can obtain the asymptotic expansion of I_c at small and large χ. For $\chi \ll 1$ one obtains from Eq. (5.60)

$$I_c = e^2m^2\chi^3[(19\sqrt{3}/20) - (98/5)\chi + \cdots]\,. \tag{5.61}$$

Using known asymptotic expansion of the intensity of the magnetic bremsstrahlung at $\chi \ll 1$ (see (4.52)) we find with allowance for (5.58) and (5.61) the expression for total intensity at $\chi \ll 1$, which agrees, naturally, with (5.56),

(5.57) and (5.55):

$$I = \frac{2}{3}e^2m^2\chi^2\left[1 - \frac{55\sqrt{3}}{16}\chi\left(1 + \frac{114}{275\rho}\right) + 48\chi^2\left(1 + \frac{49}{80\rho}\right) + \cdots\right]. \quad (5.62)$$

As is known, the series (5.62) is an asymptotic series, but the corrections of the order $1/\rho$, which control the transition to the magnetic bremsstrahlung limit, have coefficients smaller than 1.

5.5. Radiation in the Superposition of a Plane Wave and a Constant Field

The radiation theory above has been developed for a constant external field (Sec. 4) and a case of quasiperiodic motion, which is connected, in particular, with motion in an electromagnetic plane wave (see this section). The radiation process in an electromagnetic field of more complex configuration is of evident interest. One such field is a superposition of a plane wave and a constant field. This problem can be considered based on the quasiclassical theory derived in Sec 2.

We add to Eq. (2.46) a spin term, depending on the spin of the initial electron $\boldsymbol{\zeta}$ (spin effects were considered for both separate types of fields) and rewrite this formula in slightly different form:

$$\frac{dw_\gamma}{d\omega} = \frac{ie^2}{8\pi\gamma^2}\int\int\frac{dt d\tau}{\tau - i0}\left[4 + \left(\frac{\varepsilon}{\varepsilon'} + \frac{\varepsilon'}{\varepsilon}\right)(\boldsymbol{\Delta}_1 - \boldsymbol{\Delta}_2)^2 + \frac{i\omega}{\varepsilon}\boldsymbol{\zeta}\mathbf{N}\right]$$
$$\times \exp\left\{-\frac{i\tau}{l_\omega}\left(1 + \frac{1}{\tau}\int_{t_1}^{t_2}\boldsymbol{\Delta}^2(t')dt'\right)\right\},$$
$$\boldsymbol{\Delta}(t') = \frac{1}{m}(\mathbf{p}(t') - \boldsymbol{\pi}), \quad \boldsymbol{\pi} = \frac{1}{\tau}\int_{t_1}^{t_2}\mathbf{p}(t)dt, \quad l_\omega = \frac{2\varepsilon\varepsilon'}{m^2\omega},$$
$$\mathbf{N} = 2\left[\left(2 + \frac{\omega}{\varepsilon'}\right)(\boldsymbol{\Delta}_2 \times \boldsymbol{\Delta}_1) + ((\boldsymbol{\Delta}_1 - \boldsymbol{\Delta}_2) \times \mathbf{v}(t))\right], \quad (5.63)$$

where $t_{2,1} = t \pm \tau/2$ and $\mathbf{p}(t) = \varepsilon\mathbf{v}(t)$ is the momentum of the particle. Obviously, the quantity $\boldsymbol{\Delta}(t)$ is unaffected by the substitution $\mathbf{p}(t) \to \mathbf{p}(t) + \mathbf{p}_0$, where $\mathbf{p}_0$ is time-independent momentum. To pursue the analysis we need explicit expressions for the momentum $\mathbf{p}(t)$ and for the vector $\boldsymbol{\Delta}(t)$ as functions of the time in a field under consideration. We will carry out calculations in the frame of reference in which the monochromatic plane wave, with a wave vector $q = q(q_0, \mathbf{q})$ is propagating in the direction $\mathbf{n} = \mathbf{q}/q_0$, opposite the electron momentum. One can always find a relativistic frame of reference ($\gamma = \varepsilon/m \gg 1$) in which the condition $q_0 \ll \varepsilon$ holds. This is a necessary condition if we wish to treat the plane wave as classical. Solving the equation of motion of the particle in the field under consideration, we find, to within relativistic accuracy,

$$\begin{aligned}
&\mathbf{p}_\perp(t)/m = \boldsymbol{\Omega} t + \boldsymbol{\xi}(t), \quad \boldsymbol{\xi}(t) = \boldsymbol{\xi}_2 \sin(\nu t + \varphi_0) + \boldsymbol{\xi}_1 \cos(\nu t + \varphi_0), \\
&\boldsymbol{\Omega} = \frac{e}{m}\mathbf{F}_\perp, \quad \mathbf{F}_\perp = \mathbf{E} - \mathbf{n}(\mathbf{n}\mathbf{E}) + \mathbf{H} \times \mathbf{n}, \\
&\nu = 2q_0, \quad \boldsymbol{\xi}\mathbf{n} = 0.
\end{aligned} \tag{5.64}$$

Here $\mathbf{E}$ and $\mathbf{H}$ are the electric and magnetic fields, both independent of the time, and orthogonal vectors $\boldsymbol{\xi}_1$ and $\boldsymbol{\xi}_2$ characterize the intensity $\xi_0^2 = (\xi_1^2 + \xi_2^2)/2$ and the polarization of the wave. The corresponding Stokes parameters are

$$\lambda_3 = \frac{\xi_1^2 - \xi_2^2}{\xi_1^2 + \xi_2^2}, \quad \lambda_2 = \frac{(\boldsymbol{\xi}_1 \times \boldsymbol{\xi}_2)\mathbf{n}}{\xi_0^2} \tag{5.65}$$

Evaluating the combinations which appear in the expression for the spectral density of the probability (5.63), we find

$$\begin{aligned}
&\boldsymbol{\Delta}_2 - \boldsymbol{\Delta}_1 = \boldsymbol{\Omega}\tau + 2\sin\frac{\nu\tau}{2}\boldsymbol{\eta}(t) \\
&\qquad \boldsymbol{\eta}(t) = \boldsymbol{\xi}_2 \cos(\nu t + \varphi_0) - \boldsymbol{\xi}_1 \sin(\nu t + \varphi_0), \\
&\int_{t_1}^{t_2} \boldsymbol{\Delta}^2(t')dt' = \Omega^2\frac{\tau^3}{12} + \frac{4}{\nu^2}\boldsymbol{\Omega}\boldsymbol{\eta}(t)\left(\sin\frac{\nu\tau}{2} - \frac{\nu\tau}{2}\cos\frac{\nu\tau}{2}\right) \\
&\qquad + \xi_0^2\left[\tau + \frac{2}{\nu^2\tau}(\cos\nu\tau - 1) + \frac{\lambda_3}{\nu}\cos 2(\nu t + \varphi_0)\right. \\
&\qquad \left.\times\left(\sin\nu\tau + \frac{2}{\nu\tau}(\cos\nu\tau - 1)\right)\right], \\
&(\boldsymbol{\Delta}_2 - \boldsymbol{\Delta}_1)^2 = \Omega^2\tau^2 + 4\boldsymbol{\Omega}\boldsymbol{\eta}(t)\tau\sin\frac{\nu\tau}{2} + 4\xi_0^2(1 - \lambda_3\cos 2(\nu t + \varphi_0)) \\
&\qquad \times\sin^2\frac{\nu\tau}{2}, \\
&(\boldsymbol{\Delta}_2 \times \boldsymbol{\Delta}_1) = \left(\cos\frac{\nu\tau}{2} - \frac{2}{\nu\tau}\sin\frac{\nu\tau}{2}\right)\left\{\tau(\boldsymbol{\Omega} \times \boldsymbol{\xi}(t)) - 2\lambda_2\mathbf{n}\xi_0^2\sin\frac{\nu\tau}{2}\right\}.
\end{aligned} \tag{5.66}$$

We make the replacements

$$\nu t + \varphi_0 = \varphi\,, \quad \tau \to l_0\tau\,, \quad l_0 = \frac{2\varepsilon}{m^2} = 2\gamma \lambda_c$$

and take into account that

$$\Omega l_0 \approx \frac{2e\mathbf{F}_\perp\varepsilon}{m^3} = 2\chi\,, \quad \nu l_0 = \frac{4q_0\varepsilon}{m^2} \approx \frac{2qp}{m^2} \equiv s\,. \tag{5.67}$$

As a result, we find the following expression for the spectral distribution of the probability per unit time:

$$\frac{d^2 w_\gamma(t)}{dtdx} = \frac{dW_{\gamma(t)}}{dx} = \frac{ie^2m^2}{2\pi\varepsilon}\int_{-\infty}^{\infty}\frac{d\tau}{\tau - i0}\left[1 + \left(\frac{\varepsilon'}{\varepsilon} + \frac{\varepsilon}{\varepsilon'}\right)m_0 + ix\boldsymbol{\zeta}\mathbf{m}\right]$$

$$\times \exp[-iu\tau\Phi(\varphi,\tau)]\,, \quad u = \frac{\omega}{\varepsilon'}\,, \quad x = \frac{\omega}{\varepsilon}\,,$$

$$m_0^2 = \chi^2\tau^2 + 2\chi\eta(\varphi)\tau\sin\frac{s\tau}{2} + \xi_0^2(1 - \lambda_3\cos 2\varphi)\sin^2\frac{s\tau}{2}\,,$$

$$\mathbf{m} = 2\left(1 + \frac{u}{2}\right)\left\{(\boldsymbol{\chi}\times\boldsymbol{\xi}(\varphi))\left(\tau\cos\frac{s\tau}{2} - \frac{2}{s}\sin\frac{s\tau}{2}\right)\right. \tag{5.68}$$

$$\left. + \frac{1}{2}\lambda_2\xi_0^2\mathbf{v}\left(\sin s\tau + \frac{2}{s\tau}(\cos s\tau - 1)\right)\right\} - (\boldsymbol{\chi}\times\mathbf{v})\tau$$

$$- (\boldsymbol{\eta}(\varphi)\times\mathbf{v})\sin\frac{s\tau}{2}\,,$$

$$\Phi(\varphi,\tau) = \frac{1}{3}\chi^2\tau^2 + \frac{8}{s^2\tau}\chi\eta(\varphi)\left(\sin\frac{s\tau}{2} - \frac{s\tau}{2}\cos\frac{s\tau}{2}\right)$$

$$+ \xi_0^2\left[1 + \frac{2}{s^2\tau^2}(\cos s\tau - 1)\right.$$

$$\left. + \frac{\lambda_3}{s\tau}\cos 2\varphi\left(\sin s\tau + \frac{2}{s\tau}(\cos s\tau - 1)\right)\right] + 1\,.$$

The functions $\mathbf{v}(t), \boldsymbol{\chi}(t)$ and $\boldsymbol{\zeta}(t)$ in this expression are determined by the classical trajectory of the particle. If the wave period is smaller than or of the order of the formation time of the process (which is determined by the characteristic values of the variable τ in the integral (5.68)), this expression must be averaged over the phase of the wave, φ.

The comparison of the spectrum (5.68) at $\chi = 0$ with a general case (see Eq. (5.33)) of quasi-periodic motion shows that the latter coincides with quasi-classical accuracy (the terms ω_0/ε are neglected, ω_0 is the frequency of motion) with that in a field of the corresponding wave. The vector potential of the wave is $e\mathbf{A}(\kappa x) = \langle \mathbf{p} \rangle - \mathbf{p}(\kappa x), A_0 = 0, e$ is the particle charge, where $\mathbf{p}(\kappa x)$ means the solution of the motion equation for momentum in which the argument (time) is substituted by $\kappa x/\omega_0$; κ is 4-momentum of the wave.

To write expression (5.68) in covariant form, we introduce 4-vectors which characterize the wave:

$$\begin{aligned} a^\mu(\varphi) &= a_1^\mu \cos\varphi + a_2^\mu \sin\varphi\,, \quad b^\mu(\varphi) = \frac{da^\mu(\varphi)}{d\varphi}\,, \quad \varphi = (qx)\,, \\ q^2 &= qa_1 = qa_2 = a_1 a_2 = 0\,, \quad f^{\mu\nu} = q^\mu b^\nu - q^\nu b^\mu\,, \\ \xi_{1,2}^2 &= -e^2 a_{1,2}^2/m^2 \end{aligned} \tag{5.69}$$

Here $a^\mu(\varphi)$ is the vector potential, and $f^{\mu\nu}$ is the electromagnetic field tensor of the wave. Within relativistic accuracy, the combinations which appear in (5.68) are

$$\begin{aligned} \chi^2 &\approx -\frac{e^2}{m^6}(F^{\mu\nu}p_\nu)^2\,, \quad \chi\eta \approx -\frac{e^2}{m^4(qp)} F^{\mu\nu} p_\nu f_{\mu\rho} p^\rho \\ (\zeta \mathbf{v} \mathbf{s}) &\equiv \zeta(\mathbf{v} \times \mathbf{s})\,, \\ (\zeta\chi\mathbf{v}) &\approx \frac{e}{m^3} s^\mu F^*_{\mu\nu} p^\nu\,, \quad (\zeta\eta\mathbf{v}) \approx \frac{e}{m(qp)} s^\mu f^*_{\mu\nu} p^\nu \\ (\zeta\chi\xi) &\approx \frac{e^2}{m^3} s^\mu F^*_{\mu\nu} a^\nu\,, \\ -\lambda_2 \xi_0^2(\zeta\mathbf{n}) &= \frac{e^2}{m(qp)} \varepsilon_{\mu\nu\rho\sigma} s^\mu a_1^\nu a_2^\rho q^\sigma\,, \end{aligned} \tag{5.70}$$

where s^μ is the spin 4-vector, $F_{\mu\nu}$ is the tensor of the static electromagnetic field, and asterisk (*) denotes a dual tensor.

Integrating (5.68) over $x = u/(1+u)$, we find the total probability of the radiation per unit time, W_γ. This probability is related in known way with radiative correction to the electron mass: $\operatorname{Im}\Delta m = (-\varepsilon/2m)W_\gamma$. The real part of this correction can be restored by a dispersion relation. So we find the radiative correction to the mass of the particle

$$\begin{aligned} \frac{\Delta m}{m} = \frac{\alpha}{2\pi} \int_0^1 dx \int_0^\infty \frac{d\tau}{\tau} \bigg\{ \left[1 + \left(\frac{\varepsilon'}{\varepsilon} + \frac{\varepsilon}{\varepsilon'} \right) m_0 + ix(\zeta\mathbf{m}) \right] \\ \times \exp[-iu\tau\Phi(\tau)] - \exp(-iu\tau) \bigg\} \end{aligned} \tag{5.71}$$

In our approximation, part of this expression which depends on the electron spin, $\Delta m_\zeta(t) = \zeta(t)\mathbf{M}(t)$, determines the radiation corrections to the equation of motion of spin

$$\frac{d\zeta_{\rm rad}}{dt} = 2(\mathbf{M} \times \zeta)/\gamma\,. \tag{5.72}$$

If the plane wave is circularly polarized ($\lambda_2 = 0, \lambda_3 = \pm 1$), the expressions for $m_\mu(\varphi, \tau)(m_\mu = (m_0, \mathbf{m}))$ and $\Phi(\varphi, \tau)$ become

$$\begin{aligned}
m_0^{cr} &= \chi^2\tau^2 + 2\chi\eta\tau \sin\frac{s\tau}{2} + \xi_0^2 \sin^2\frac{s\tau}{2}\,,\\
\zeta\mathbf{m}^{cr} &= 2\left(1 + \frac{u}{2}\right)\lambda_2(\zeta\mathbf{v})\left[(\chi\eta)\left(\tau\cos\frac{s\tau}{2} - \frac{2}{s}\sin\frac{s\tau}{2}\right)\right.\\
&\left. + \frac{\xi_0^2}{2}\left(\sin s\tau + \frac{2}{s\tau}(\cos s\tau - 1)\right)\right] - (\zeta\chi\mathbf{v})\tau - (\zeta\eta\mathbf{v})\sin\frac{s\tau}{2}\,,\\
\Phi^{cr} &= \frac{1}{3}\chi^2\tau^2 + \frac{8}{s^2\tau}(\chi\eta)\left(\sin\frac{s\tau}{2} - \frac{s\tau}{2}\cos\frac{s\tau}{2}\right)\\
&+ \xi_0^2\left(1 + \frac{2}{s^2\tau^2}(\cos s\tau - 1)\right) + 1\,.
\end{aligned} \tag{5.73}$$

If we set the wave intensity equal to zero ($\xi_0^2 = 0$), Eqs. (5.68) and (5.71) become expressions for the radiation probability (4.36) and the correction to the mass in a constant external field in quasiclassical approximation (see Baier, Katkov and Fadin, 1973). In the case $\chi = 0$, these expressions are the same as the corresponding formulae in the field of a monochromatic plane wave (see Ritus 1979, Appendix C).

The parameter ξ_0 characterizes the change induced by the wave field in the momentum of the particle in comparison with the mass of the particle over a time of the order of the wave oscillation period. At the large value $\xi_0 \gg 1$, the radiation is formed in a time much shorter than this period ($\tau \ll 1/s$). In this case, we can carry out an expansion in the quantity $s\tau$ in Eqs. (5.68), (5.71) and (5.73). We find as a result that the process is determined by the local value of the field strength with the corresponding value of the parameter χ_1:

$$\chi_1(t) = \chi(t) + \eta(\varphi)\frac{s}{2}\,, \quad \chi_1^2 = -\frac{e^2}{m^6}[(F^{\mu\nu} + f^{\mu\nu})p_\nu]^2 \tag{5.74}$$

The radiation is of the same nature in the case $\xi_0 \leq 1$, but at large values of the parameter $\mu = \chi/s(\mu \gg 1)$. Under the condition $\mu \ll 1(\xi_0^2 \leq 1)$, the momentum transferred by the static field over the wave oscillation period is much smaller than the mass of the particle ($\tau \sim 1/s, \chi\tau \ll 1$), and corresponding expansions must be carried out in $\chi\tau$. Retaining the leading terms of this expansion, we find the corrections to the probability for the process in a plane wave which result from an influence of the constant field.

5.6. Compton Effect in a Static Field

In the case $\xi_0 \ll 1$, we can expand the exponential factor in formula (5.68) in the parameter $\xi(\eta)$ and retain terms $\propto \xi_0^2$ only. The zeroth terms in the expansion in ξ_0 in Eq. (5.68) then give us the probability for radiation in a constant field $F_{\mu\nu}$, and the corrections $\propto \xi_0^2$ describe Compton scattering in this field. Let us examine this Compton scattering in more detail.

Making use of the relationship between the electric and magnetic fields of the wave, on the one hand, and the wave energy, on the other, we express the parameter ξ_0^2 in terms of the photon density n_{ph} and frequency q_0:

$$\frac{E^2+H^2}{8\pi} = \frac{1}{4\pi}\left(\frac{m\xi_0}{e}\right)^2 = q_0 n_{ph}\,, \quad \xi_0^2 = \frac{4\pi\alpha n_{ph}}{m^2 q_0} \approx \frac{16\pi n_{ph}\varepsilon}{m^4 s}\,. \tag{5.75}$$

We also note that the term proportional to ξ_0^2 in the radiation probability in (5.68),

$$dW_\gamma = dW_\gamma^{ex} + \xi_0^2 dW_\gamma^\xi$$

is related to the cross section for the Compton effect by

$$\xi_0^2 dW_\gamma^\xi = 2n_{ph} d\sigma_c\,, \quad d\sigma_c = \frac{8\pi\alpha\varepsilon}{m^4 s} dW_\gamma^\xi\,. \tag{5.76}$$

As a result we find the following expression for the cross-section for Compton scattering in a constant field:

$$\begin{aligned}\frac{d\sigma_c}{dx} = \frac{4\alpha^2}{m^2 si}\int_{-\infty}^{\infty}\frac{d\tau}{\tau - i0}e^{-i\rho(\tau)}\bigg\{&(1+\beta\chi^2\tau^2)F + \frac{4i}{s^3}u\beta\chi^2\Lambda_1 z g_2(z)\\ &+\frac{\beta}{2}g_3(z) + 2\left(1+\frac{u}{2}\right)\lambda_2(\boldsymbol{\zeta}\mathbf{v})\\ &\times x\left[\frac{4u}{s^3}\chi^2 g_1(z) + \frac{i}{z}g_2(z)\right] - x(\boldsymbol{\zeta}\boldsymbol{\chi}\mathbf{v})\left[i\frac{z}{s}F - \frac{2u}{s^2}\Lambda_1 g_2(z)\right]\\ &+\frac{2u}{s^2}x(\boldsymbol{\zeta}\boldsymbol{\chi})\lambda_3 g_2(z)\sin 2\varphi_1\bigg\}\,,\end{aligned} \tag{5.77}$$

where

$$\rho(\tau) = u\tau\left(1 + \chi^2\frac{\tau^2}{3}\right), \quad \beta = \frac{\varepsilon}{\varepsilon'} + \frac{\varepsilon'}{\varepsilon}, \quad z = s\tau,$$

$$F = \frac{8u^2}{s^4}\chi^2\Lambda_1 g_1(z) + \frac{2iu}{sz}g_3(z) + iu\frac{z}{s}, \quad \Lambda_1 = 1 + \lambda_3\cos 2\varphi_1,$$

$$g_1(z) = 1 + z^2/4 + \left(\frac{z^2}{4} - 1\right)\cos z - z\sin z, \tag{5.78}$$

$$g_2(z) = 1 - \cos z - \frac{z\sin z}{2}, \quad g_3(z) = \cos z - 1$$

and φ_1 is the angle between χ and ξ_1. Setting in (5.77) and (5.78) $\lambda_{2,3} = 0$, we find the cross-section for the Compton scattering for the case of unpolarized initial photons. The integrand in (5.77) has no singularities as $\tau \to 0$, so the integral has the usual meaning. In deriving the asymptotic behavior in the case of a weak field, however, it is convenient to use a contour representation of (5.77).

The overall effect of the external field on the Compton scattering is determined by the value of the parameter $\mu = \chi/s$. For $\mu \ll 1$ the external field is unable to cause any substantial change in the motion of the charged particles over the timescale of the process, $T \sim 1/q_0$:

$$\frac{\Delta p_\perp}{m} \sim \frac{eF_\perp T}{m} \sim \frac{eF_\perp}{mq_0} \sim \frac{\chi}{s} \equiv \mu \ll 1. \tag{5.79}$$

Even if the condition $\mu \ll 1$ holds, near the maximum of the spectral distribution, at $u \approx s$, the effect of the external field becomes important when $\delta = 1 - u/s$ is sufficiently small. The reason is that the value $u = s$ corresponds to the Compton back scattering, and a photon with frequency ω near the extreme value $\varepsilon s/(1+s)$ is formed over a time

$$\frac{T}{\delta} \sim \frac{1}{q_0\delta}, \quad \left(\tau \sim \frac{1}{s\delta}\right).$$

The field-dependent term in the phase $\rho(\tau)$ in Eq. (5.78) can then be ignored only if

$$u\chi^2\tau^3 \sim \frac{\chi^2}{s^2\delta^3} = \frac{\mu^2}{\delta^3} \ll 1. \tag{5.80}$$

The parameter $\mu\delta^{-3/2}$ thus serves as a measure of the effect of the external field on the spectrum of the Compton scattering near the limiting frequency. If the parameter μ is small ($\mu \ll 1$), but δ is also small, so that we have $\delta \leq \mu^{2/3}$, the leading term in (5.77) simplifies substantially:

$$\frac{d\sigma_c}{dx} = \frac{\alpha^2}{im^2 s}\int_{-\infty}^{\infty}\frac{d\tau}{\tau - i0}[\beta - x(2+u)\lambda_2(\zeta \mathbf{v})]\exp\{i[z-\rho(\tau)]\}\,. \qquad (5.81)$$

Using known integrals (see Gradshteyn and Ryzhik, 1965) we find (cf. (4.90))

$$\frac{1}{2\pi i}\int_{-\infty}^{\infty}\frac{d\tau}{\tau - i0}\exp\{i[z-\rho(\tau)]\}$$

$$= \begin{cases} (1/\sqrt{3}\pi)\int_a^\infty K_{1/3}(y)dy\,, & u \geq s\,, \\ 1-(1/3)\int_a^\infty (J_{1/3}(y) + J_{-1/3}(y))dy\,, & u \leq s\,, \end{cases}$$

$$a = \frac{2u}{3\chi}\left|\frac{s}{u} - 1\right|^{3/2} \approx \frac{2s}{3\chi}\left|1 - \frac{u}{s}\right|^{3/2} \qquad (5.82)$$

We see from the last expression that the field "smears out" the frequency boundary in the region $u > s$, replacing it by an exponential decay at $a \gg 1$. At $u = s$, the height of the spectral curve is a third of the maximum value in the absence of a field, and it is independent of the strength of the external field. These results are illustrated in Fig. 5.4, where the spectral distribution of the intensity xdW_γ^ξ/dx is plotted for $s = 4$, the fine curves are in absence of an external field, the heavy solid curves are for $\chi = 0.04$ and $\chi = 0.3$ in a and b, respectively, their ratio is given by a solid curve. In the region $u < s$, with $a \gg 1$, the field has only a slight effect on the Compton scattering. This conclusion agrees with the qualitative analysis above.

Since

$$\int_0^\infty dy\int_{2y^{3/2}/3\mu}^\infty dz\left[\frac{\sqrt{3}}{\pi}K^{1/3}(z) - J_{1/3}(z) - J_{-1/3}(z)\right] = 0\,,$$

the external field basically causes only a redistribution of the spectral probability for the process, leaving its overall value unchanged in this frequency region. Correspondingly, in evaluating the corrections to the total cross-section

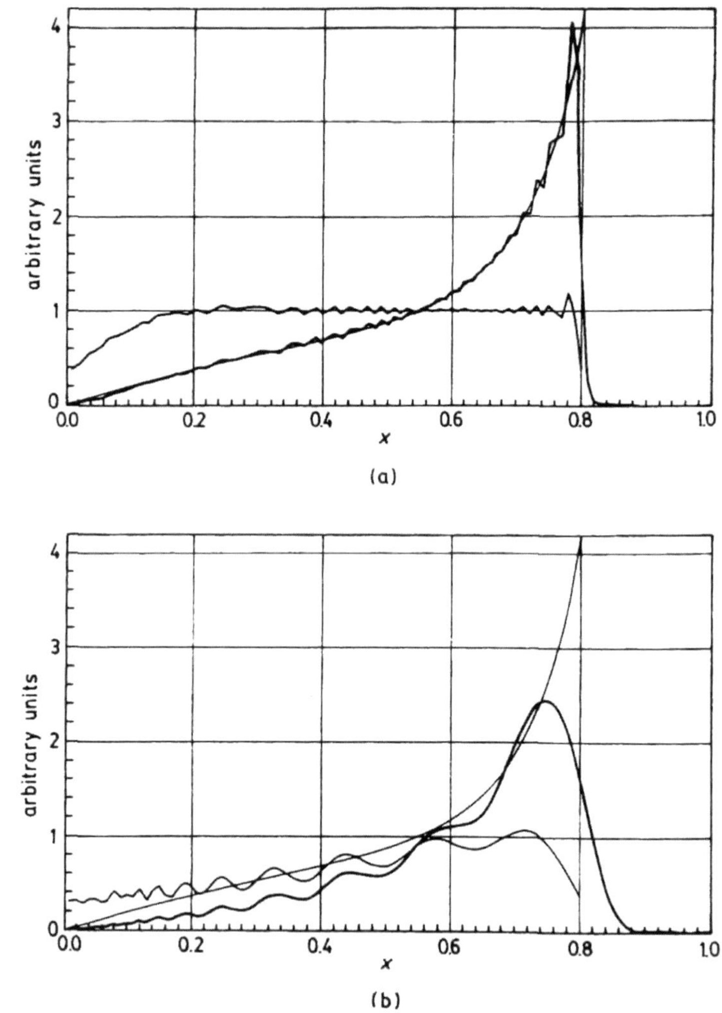

Fig. 5.4. Intensity spectra xdW_{γ}^{ξ}/dx for $s = 4$. In the absence of an external field ($\chi = 0$, fine curves); in the presence of a field (heavy solid), and their ratio (solid at $\chi = 0.04$ in (a); and at $\chi = 0.3$ in (b)).

we can carry out an expansion in $\chi\tau$ in (5.77) at all values of x. We retain the first terms of the expansion in μ and $\chi\tau$ and we evaluate the integral over τ by means of the theory of residues. It is then an elementary matter to evaluate the integral over x. As a result we find the following expression for the total cross-section for the Compton effect with field corrections:

$$\sigma_c = \frac{\pi r_e^2}{s}\left\{2\left(1-\frac{4}{s}-\frac{8}{s^2}\right)L+1+\frac{16}{s}-\frac{1}{z^2}+\lambda_2(\boldsymbol{\zeta}\mathbf{v})\left[2\left(1+\frac{2}{s}\right)L\right.\right.$$
$$-5+\frac{2}{z}-\frac{1}{z^2}\Bigg]+4\mu^2\left[\frac{16}{s^2}(L-s)-2+\frac{59}{3z}-\frac{59}{3z^2}+\frac{13}{z^3}-\frac{1}{z^4}\right.$$
$$-\frac{2}{z^5}+2\lambda_3\cos(2\varphi_1)\left(\frac{8}{s^2}(L-s)-2+\frac{11}{z}-\frac{11}{z^2}+\frac{9}{z^3}-\frac{3}{z^4}\right)$$
$$\left.+\lambda_2(\boldsymbol{\zeta}\mathbf{v})\left(\frac{4}{s}(L-s)+4-\frac{7}{3z}-\frac{19}{3z^2}+\frac{11}{3z^3}+\frac{3}{z^4}-\frac{2}{z^5}\right)\right]$$
$$+4(\boldsymbol{\zeta}\boldsymbol{\chi}\mathbf{v})\left[\frac{1}{z^2}-y\frac{1}{z^3}\right]-8\lambda_3[(\boldsymbol{\zeta}\boldsymbol{\chi}\mathbf{v})\cos(2\varphi_1)+(\boldsymbol{\zeta}\boldsymbol{\chi})\sin(2\varphi_1)]$$
$$\left.\times\left(\frac{1}{s^2}(L-s)+\frac{1}{z}-\frac{1}{z^2}+\frac{1}{2z^3}\right)\right\} \tag{5.83}$$

Here $r_e = \frac{\alpha}{m}$, $z = 1+s$, $L = \ln(1+s)$, λ_3 and λ_2 are the degree of linear polarization and the degree of circular polarization, respectively, of the initial photon, and $\boldsymbol{\zeta}$ is the vector of the polarization of the initial electron. With $\chi = 0$, formula (5.83) becomes the cross-section for the polarized Compton effect (see, for example, Ritus, 1979; Appendix C). The correction to the Klein–Nishina formula for unpolarized photons were found by Zhukovskii and Nikitina, 1973. They agree with Eq. (5.83) if we set $\lambda_2 = \lambda_3 = 0$ in the latter.

In the other limiting case, in which the photon formation length in the constant field

$$l_c = \frac{m}{eF_\perp}(1+\chi/u)^{1/3},$$

is much shorter than the wavelength $\lambda = 1/q_0$, the field of the wave can be assumed to remain constant over this distance. This situation corresponds to an expansion in powers $s\tau$ in Eqs. (5.68) and (5.73):

$$s\tau \sim \frac{l_c}{\lambda} = \frac{mq_0}{eF_\perp}\left(1+\frac{\chi}{u}\right)^{1/3} = \frac{s}{\chi}\left(1+\frac{\chi}{u}\right)^{1/3} \ll 1. \tag{5.84}$$

The probability for the process is determined by the expression for the magnetic bremsstrahlung in the resultant field (5.74) (see Eqs. (4.24), (4.70) and (4.84)).

For unpolarized electrons this probability is

$$\frac{dW_\gamma}{dx} = \frac{\alpha m^2}{\sqrt{3}\pi\varepsilon}\left(\beta K_{2/3}(z_1) - \int_{z_1}^{\infty} K_{1/3}(y)dy\right),$$

$$\beta = \frac{\varepsilon'}{\varepsilon} + \frac{\varepsilon}{\varepsilon'}, \quad z_1 = \frac{2u}{3\chi_1}, \quad \chi_1 = \chi + \eta(\varphi)\frac{s}{2}. \tag{5.85}$$

The condition of the applicability of formula (5.85), in accordance with (5.84), is that the value of the parameter μ_1 be large

$$\mu_1 = \frac{\chi_1}{s}(\mu + \xi_0) \gg 1.$$

If the wave intensity is sufficiently small ($\xi_0 \ll \mu$) in this case, we can carry out an expansion in the wave field in (5.85), and we can switch to the cross-section for the Compton effect in accordance with (5.75) and (5.76). As a result we find, for example, the following expression for the unpolarized case:

$$\frac{d\sigma_c}{dx} = \frac{r_e^2}{2\sqrt{3}\mu^2 s}\left\{\left[(\beta - 1)z^2 + \frac{4}{9}\beta\right] K_{2/3}(z) + \frac{2}{3} z K_{1/3}(z)\right\}$$

$$z = 2u/3\chi \tag{5.86}$$

In deriving (5.86) we used recurrence relations for the Bessel functions. With the help of these functions, we can write the total cross-section of the Compton effect as

$$\sigma_c = \frac{r_e^2}{6\sqrt{3}\mu\chi}\int_0^{\infty}\left(\frac{4}{9} + z^2\right)\frac{5u^2 + 7u + 5}{(1+u)^3} K_{2/3}(z)du. \tag{5.87}$$

In the case $\chi \ll 1$, values $u \sim \chi \ll 1$ contribute to the integral in (5.87), and the total cross-section in this case is

$$\sigma_c = \frac{5\pi}{4\sqrt{3}}\frac{r_e^2}{\mu}, \quad \chi \ll 1, \quad \mu \gg 1. \tag{5.88}$$

Under the conditions $\chi \gg 1$, values $u \sim 1$ contribute, and we can use the asymptotic expressions for the Bessel functions at small values of their arguments ($z \ll 1$). In this case we find

$$\sigma_c = \frac{28}{81}\frac{\Gamma(2/3)}{3^{1/3}}\left(\frac{s}{\chi^{2/3}}\right)^2 \frac{\pi r_e^2}{s}, \quad \chi \gg 1. \tag{5.89}$$

It can be seen from (5.89) that in the case $\chi \gg 1$, $s \gg 1$ the expansion parameter is the quantity $s/\chi^{2/3}$, in agreement with (5.84).

5.7. Photoproduction of an Electron-Positron Pair

The spectral distribution of the probability for pair production in a given field can be found from the spectral distribution of the probability of the radiation in this field through the substitutions: $\varepsilon \to -\varepsilon, \omega \to -\omega, z \to -z, \omega d^2\omega \to -\varepsilon^2 d\varepsilon$ (see discussion after Eq. (4.14)). In our case we have

$$\begin{gathered} u \to -(1-x)^{-1}, \quad \mu = \frac{\chi}{s} \to \frac{\kappa}{4\Lambda}, \\ \kappa^2 = -\frac{e^2}{m^6}(F^{\mu\nu}k_\nu)^2, \\ \frac{u}{s} \to (4x(1-x)\Lambda)^{-1} = \cosh^2 y/\Lambda, \\ x = \frac{\varepsilon}{\omega} = \frac{1+\tanh y}{2}, \quad \Lambda = \frac{qk}{2m^2} \end{gathered} \tag{5.90}$$

According to the analysis above, in the case $\kappa \ll \Lambda$ the field has a strong effect on pair production by two photons only near the threshold, with $\cosh^2 y \approx \Lambda$. In this case the cross-section for the two-photon electron-positron production is

$$\begin{aligned} \frac{d\sigma_{\gamma\gamma}}{dy} &= \frac{\pi r_e^2}{\Lambda}\left[1 - \frac{1}{2\cosh^2 y} + \frac{1}{\Lambda}\left(1 - \frac{\cosh^2 y}{\Lambda}\right)\right]\Big[\vartheta(\Lambda - \cosh^2 y) \\ &\times \left(1 - \frac{1}{3}\int_b^\infty dx(J_{1/3}(x) + J_{-1/3}(x))\right) + \frac{\vartheta(\cosh^2 y - \Lambda)}{\sqrt{3}\pi} \\ &\times \int_b^\infty K_{1/3}(x)dx\Big], \\ b &= \frac{8\cosh^2 y}{3\kappa}\left|\frac{\Lambda}{\cosh^2 y} - 1\right|^{3/2}, \end{aligned} \tag{5.91}$$

where $\vartheta(x)$ is the Heaviside function: $\vartheta(x) = 1, x > 0; \vartheta(x) = 0, x < 0$.

Let us consider the case $\Lambda < 1$. In other words, we consider the situation outside the kinematic region of pair production in the absence of a field, with an exponential suppression of the one-photon process in a constant field ($\kappa \ll 1$). Using Eqs. (5.81), (5.82) and (5.91), we find the following expression for the cross-section of the two-photon process:

$$\begin{gathered} \frac{d\sigma_{\gamma\gamma}}{dy} \approx \frac{r_e^2}{2\sqrt{3}}\int_{b_1}^\infty K_{1/3}(z)dz, \\ b_1 = \frac{8}{3\kappa}(\Delta + y^2)^{3/2}, \quad \Delta = 1 - \Lambda \ll 1, \quad y \ll 1. \end{gathered} \tag{5.92}$$

Integrating the cross-section (5.92), using the integration by parts in it, we find the total cross-section under the condition $\Lambda < 1$:

$$\sigma_{\gamma\gamma} = r_e^2 \left(\frac{\Delta}{3}\right)^{1/2} \int_{z_0}^{\infty} \left[\left(\frac{z}{z_0}\right)^{2/3} - 1\right]^{1/2} K_{1/3}(z)dz\,, \qquad (5.93)$$

$$z_0 = 8\Delta^{3/2}/3\kappa\,.$$

In the limiting cases $z_0 \gg 1$ and $z_0 \ll 1$ we have, respectively:

$$\begin{aligned} \sigma_{\gamma\gamma} &\approx \frac{\pi r_e^2}{6}\frac{\Delta^{1/2}}{z_0^{5/6}}e^{-z_0}\,, \quad z_0 \gg 1\,, \\ \sigma_{\gamma\gamma} &\approx \frac{r_e^2}{4}(6\kappa)^{1/3}\sqrt{\frac{\pi}{3}}\Gamma\left(\frac{5}{6}\right)\,, \quad z_0 \ll 1\,. \end{aligned} \qquad (5.94)$$

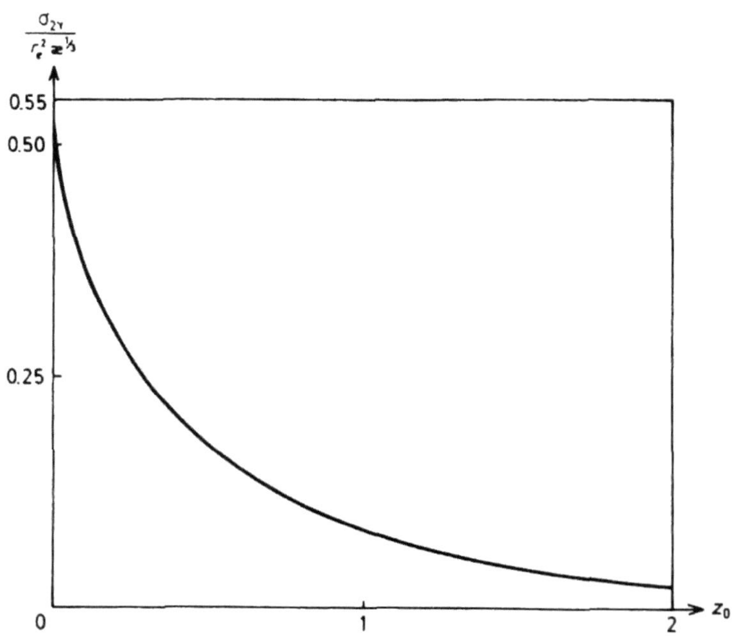

Fig. 5.5. Function $\sigma_{\gamma\gamma}/\kappa^{1/3}$ (Eq. (5.93)).

It follows from (5.93) and (5.94) that near the threshold ($\Delta \ll 1$) the function $\sigma_{\gamma\gamma}/\kappa^{1/3}$ has a scaling form which depends on z_0 only, i.e. on only a certain combination of Δ and κ. This function is plotted in Fig. 5.5. If the channel for two-photon process is open ($\Lambda - 1 \geq 1$), the corrections to the total cross-section are small (they contain a small factor $(\kappa/\Lambda)^2$).

In the other limit $\kappa \gg \Lambda$, the differential cross-section for two-photon pair production can be found from Eq. (5.86) with the help of the substitutions (5.90). In this case the total cross-section can be written

$$\sigma_{\gamma\gamma} = \frac{2r_e^2\Lambda}{3^{3/2}\kappa^2}\int_0^1 \frac{9-v^2}{1-v^2}\left(\frac{4}{9}+\lambda^2\right)K_{2/3}(\lambda)dv\,,$$
$$\lambda = (3\kappa(1-v^2)/8)^{-1} \tag{5.95}$$

For $\kappa \gg 1$, we can use the asymptotic expression for $K_{2/3}(\lambda)$ at small values of its argument (see Eq. (1.73)) in the integral over v. As a result we find

$$\sigma_{\gamma\gamma} \approx \frac{20}{63}\left(\frac{2}{3}\right)^{1/3}\frac{\Gamma(5/6)}{\Gamma(7/6)}\frac{\Lambda}{\kappa^{4/3}}\pi r_e^2\,. \tag{5.96}$$

6. HIGHER-ORDER EFFECTS IN EXTERNAL FIELDS

6.1. Creation of Neutrino Pairs by Electrons

According to the standard theory of electroweak interaction (see, e.g., Okun, 1982) this process is described by two diagrams (Fig. 6.1) with charged and neutral currents. At low energies they are reduced to the diagram in Fig. 6.2 describing the Fermi's point-like interaction with the dimensional Fermi constant $G/\sqrt{2} = g^2/8m_W^2$. Then the process formation is mainly the same as at emission of a massive photon with four-momentum $k = k_1 + k_2$(k_1 and k_2 are the momenta of neutrino and antineutrino). Using the expression for the phase φ of exponential function $\exp(-i\varphi)$ in Eq. (2.26) and expanding it in powers of $w\tau$ (magnetic bremsstrahlung limit) we have

$$\varphi(\tau) = \frac{\omega\varepsilon\tau}{2(\varepsilon-\omega)}\left[\frac{1}{\gamma^2}+\vartheta^2+\frac{k^2(\varepsilon-\omega)}{\omega^2\varepsilon}+\frac{w^2\tau^2}{12}\right]$$
$$= \frac{u\varepsilon\tau}{2\gamma^2}\left[1+\vartheta^2\gamma^2+\frac{k^2}{m^2}\frac{1+u}{u^2}+\frac{1}{12}w^2\tau^2\gamma^2\right], \tag{6.1}$$
$$u = \frac{\omega}{\varepsilon-\omega}\,.$$

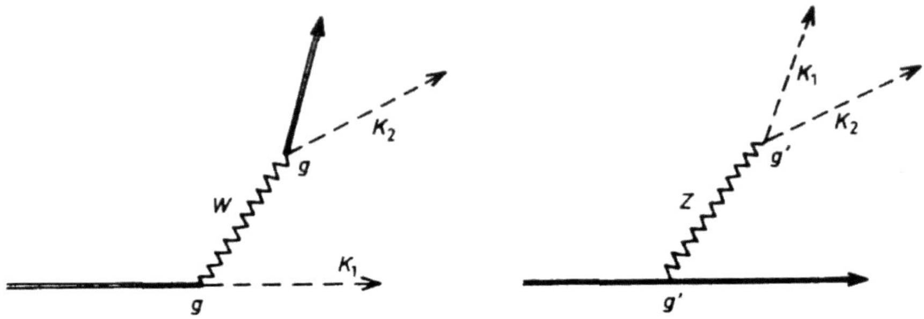

Fig. 6.1. Diagrams for the neutrino pair production through intermediate bosons.

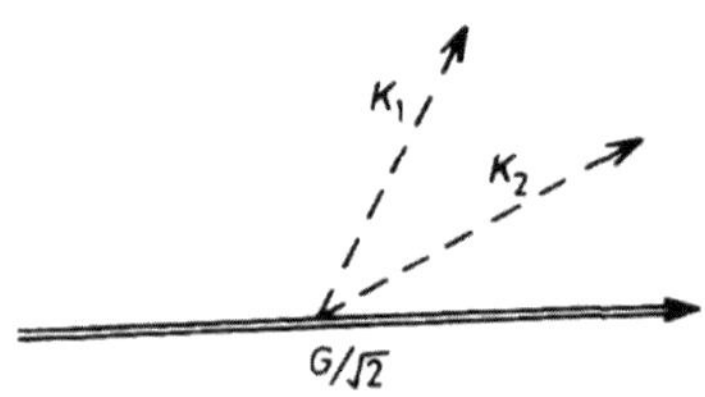

Fig. 6.2. Contact diagram for the neutrino pair production.

The basic characteristics of the process as the formation time τ, the mean emission angle ϑ, and the invariant mass of the pair $(k^2)^{1/2}$ are determined by the condition $\varphi(\tau) \sim 1$. In particular, Eq. (6.1) implies

$$
\begin{aligned}
k^2 \sim k_{ef}^2 &\equiv m^2 \frac{u^2}{1+u}\left(1+\left(\frac{\chi}{u}\right)^{2/3}\right) \\
k_\perp^2 &\simeq \omega^2\vartheta^2 \sim q_{ef}^2 \\
q_{ef}^2 &\equiv \frac{\omega^2}{\gamma^2}\left(1+\left(\frac{(\varepsilon-\omega)\chi}{\omega}\right)^{2/3}\right) \\
&= m^2 \frac{u^2}{(1+u)^2}\left(1+\left(\frac{\chi}{u}\right)^{2/3}\right).
\end{aligned}
\tag{6.2}
$$

The point-like description of the process by Fig. 6.2 is valid if $k^2 \ll m_W^2$. This condition with regard for Eq. (6.2) leads to the constraint

$$m^2\chi \ll m_W^2 . \tag{6.3}$$

Starting from the known probability of photon emission (4.24) and taking into account that the constant G is contained in dimensionless combination $G^2k^4(u)$, where $k^2(u)$ is the squared invariant mass of neutrino and antineutrino, we have the following estimate for the probability of neutrino pair creation

$$\frac{dW_\nu}{du} \sim \frac{dW_\gamma}{du}\frac{G^2k^4(u)}{\alpha} \tag{6.4}$$

In the classical limit ($\chi \ll 1$) we have $W_\gamma \sim \alpha m^2\chi/\varepsilon$ and $k^2 \sim m^2\chi^2$, then the estimate for the total probability of neutrino pair emission by an electron reads

$$\begin{aligned} W_\nu &\sim \frac{G^2m^6\chi^5}{\varepsilon} \\ \omega &\sim \varepsilon\chi \\ \chi &\ll 1 . \end{aligned} \tag{6.5}$$

In the ultra-quantum case ($\chi \gg 1$) the asymptotic expression (4.28) for the spectral probability dW_γ/du can be used in the photon frequency region up to $u \sim \chi$, then, taking into account Eqs. (6.2), (6.4), we obtain

$$\begin{aligned} W_\nu &\sim \frac{W_\gamma}{\alpha}G^2m^4\chi^{4/3}\int_1^\chi \frac{du}{u} \\ &\sim \frac{G^2m^6\chi^2}{\varepsilon}\ln\chi , \quad \chi \gg 1 . \end{aligned} \tag{6.6}$$

Going over to the direct calculation of the probability of neutrino pair creation within quasiclassical operator method, we represent the amplitude of the process described by Fig. 6.2 in the form

$$\begin{aligned} M = &\frac{G}{\sqrt{2}}\int dt J_\mu(t)N^\mu \\ &\times \exp\left\{i\frac{\varepsilon}{\varepsilon-\omega}\left(kx(t) - \frac{k^2t}{2\varepsilon}\right)\right\} ; \end{aligned} \tag{6.7}$$

here

$$\begin{aligned} J_\mu(t) &= \bar{u}_{s'}(\varepsilon(\mathbf{p}'(t)), \mathbf{p}'(t)) \\ &\quad \times \gamma_\mu(g_V + g_A\gamma_5)u_s(\varepsilon\mathbf{p}(t)) , \quad \mathbf{p}' = \mathbf{p} - \mathbf{k} \\ N_\mu &= \bar{u}(k_1)\gamma_\mu(1+\gamma_5)u(k_2) , \end{aligned} \tag{6.8}$$

where we have used the explicit structure of the four-fermion interaction (see Okun, 1982).

The probability of the process is expressed by means of the amplitude in the usual way

$$dw = \frac{1}{(2\pi)^6}\frac{d^3k_1}{2\omega_1}\frac{d^3k_2}{2\omega_2}|M|^2 \tag{6.9}$$

Carrying out the summation over particle polarizations and integration over the relative momentum of the created pair (it is convenient to accomplish this integration in invariant form), we get for massless neutrinos

$$\int \frac{d^3k_1}{2\omega_1}\frac{d^3k_2}{2\omega_2}\sum N_\mu N_\nu^* = \frac{4\pi}{3}(k_\mu k_\nu - k^2 g_{\mu\nu})d^4k \equiv \frac{4\pi}{3}N_{\mu\nu}d^4k\,. \tag{6.10}$$

Expanding the expression arising after convolution of the tensor $N_{\mu\nu}$ with the tensor $J_\mu^*(t_2)J_\nu(t_1)$ in powers of $w\tau$ ($\tau = t_2 - t_1$) and operating later on as in Sec. 4.1 (similar to Eq. (1.22) the integration by parts of the terms like $\mathbf{nv}(t_1)$ and $\mathbf{nv}(t_2)$ is carried out), we obtain with relativistic accuracy

$$\mathcal{L} = J_\mu^*(t_2)J_\nu(t_1)N^{\mu\nu} = \frac{(1+u)}{\gamma^2}k^2\left[3g_A^2 - (g_A^2+g_V^2)\left(1+\frac{k^2}{2m^2}\right)\right] - \frac{(1+u)w^2\tau^2}{2}\left[(g_A^2+g_V^2)\left(1+\frac{u^2}{2(1+u)}\right)k^2 + m^2 g_A^2\frac{u^2}{1+u}\right]. \tag{6.11}$$

The dependence of the phase $\varphi(\tau)$ is given by Eq. (6.1). Substituting Eqs. (6.11) and (6.1) into Eq. (6.9), integrating over ϑ and τ as in Sec. 4.1 and going over to the probability per unit time, we have

$$dW_{\nu\bar{\nu}} = \frac{G^2m^2dk^2}{48\sqrt{3}\pi^4\varepsilon}\cdot\frac{du}{(1+u)^3}\left[F_{2/3}\cdot(1+\delta)K_{2/3}(\eta) + F_{1/3}\cdot\int_\eta^\infty K_{1/3}(x)dx\right];\quad \eta = \frac{2u}{3\chi}(1+\delta)^{3/2}, \tag{6.12}$$

$$\delta = \frac{k^2}{m^2}\frac{1+u}{u^2},$$

where

$$F_{2/3} = g^2(2 + 2u + u^2) + 2g_A^2 m^2 u^2$$

$$F_{1/3} = \left[3g_A^2 - g^2\left(1 + \frac{k^2}{2m^2}\right)\right](1 + u) \cdot k^2 \tag{6.13}$$

$$g^2 \equiv g_A^2 + g_V^2$$

At $\chi \ll 1$ the contribution is given by $u \sim \chi$, then $k^2 \simeq \delta m^2 u^2$ and the probability gets the form

$$dW_{\nu\bar{\nu}} \simeq \frac{G^2 m^6 u^4 du d\delta}{48\sqrt{3}\pi^4 \varepsilon}\left[2(1+\delta)(\delta g^2 + g_A^2)K_{2/3}(\eta) + \delta(3g_A^2 - g^2)\int_\eta^\infty K_{1/3}(x)dx\right] \tag{6.14}$$

The integration over u in (6.14) is carried out by means of Eq. (1.98) first integrating the term containing $K_{1/3}$ by parts. After that the integration over δ becomes trivial and we have

$$W_{\nu\bar{\nu}} = \frac{7\sqrt{3}G^2 m^6 \chi^5}{16(6\pi)^3 \varepsilon}(7g^2 + 54g_A^2)\,. \tag{6.15}$$

In the case $\chi \gg 1$ the contribution is given by $k^2 \gg m^2 (k^2 \sim m^2\chi^{2/3}u^{1/3})$ and up to terms $\sim m^2/k^2$ we obtain

$$dW_{\nu\bar{\nu}} \simeq \frac{G^2 m^6 u^4 du \delta d\delta g^2}{48\sqrt{3}\pi^4 \varepsilon (1+u)^5}\left[(1+\delta)(2 + 2u + u^2)K_{2/3}(\eta) - \frac{\delta}{2}\int_\eta^\infty K_{1/3}(x)dx\right] \tag{6.16}$$

The constants g_A^2 and g_V^2 are contained in Eq. (6.16) with equal weights. It is connected to the conservation of the axial current at $k^2 \gg m^2$ up to terms $\sim m^2/k^2$ giving the same contribution to the process as the vector current. With logarithmic accuracy, large values of variables $u \gg 1$ and $\delta \gg 1$ contribute, so that Eq. (6.16) is essentially simplified. The calculation of the probability with power accuracy is much more cumbersome. This calculation can be carried out as in Sec. 4.2 if we represent $(1 + u)^{-5}$ in the form of Eq. (4.37). As a result we obtain for the probability of the process at $\chi \gg 1$ the following expression

$$W_{\nu\bar{\nu}} = \frac{G^2 g^2 m^6 \chi^2}{(6\pi)^3 \varepsilon} \cdot \left(\ln \chi - C - \frac{1}{2}\ln 3 - \frac{5}{6}\right) \tag{6.17}$$

The creation of the neutrino pair by an electron was considered in the paper by Baier and Katkov, 1966, in a constant magnetic field and in the paper by Ritus, 1969, in a plane wave field. Only charged currents when $g_A = g_V = 1$ were taken into account in the indicated papers. Expressions for the process probability obtained in these papers are consistent with Eqs. (6.15), (6.17) in this case. The probability in the form of (6.15), (6.17) was obtained by Merenkov, 1985.

Let us compare the obtained results for electron $\nu_e + \bar{\nu}_e$ neutrinos creation to the probability of photon emission. Taking into account that according to the standard theory of electroweak interaction

$$\begin{aligned} g_A^{\nu_e\bar{\nu}_e} &= \frac{1}{2}, \\ g_V^{\nu_e\bar{\nu}_e} &= \frac{1}{2} + 2\sin^2\vartheta_W, \\ G^2 &= \frac{\pi^2\alpha^2}{2m_W^2 \sin^4\vartheta_W}, \end{aligned} \tag{6.18}$$

where ϑ_W is Weinberg mixing angle ($\sin^2\vartheta_W \simeq 0.23$), α is the fine structure constant, we have

$$\begin{aligned} \frac{W_\nu}{W_\gamma} &\simeq \frac{7\alpha[7(1+2\sin^2\vartheta_W)\sin^2\vartheta_W + 17/2]}{45\cdot 2^6\pi} \\ &\times \left(\frac{m\chi}{m_W \sin\vartheta_W}\right) \sim 10^{-24}\chi^4, \quad \chi \ll 1; \\ \frac{W_\nu}{W_\gamma} &\simeq \frac{\alpha[(1+2\sin^2\vartheta_W)\sin^2\vartheta_W + \frac{1}{4}]\chi^{4/3}\ln\chi}{112\cdot 3^{2/3}\Gamma(\frac{2}{3})\pi} \\ &\times \left(\frac{m}{m_W \sin\vartheta_W}\right)^4 \sim 10^{-25}\chi^{4/3}\ln\chi, \quad \chi \gg 1. \end{aligned} \tag{6.19}$$

It should be noted that the expansions (6.1) and (6.11) become invalid if the field strength varies essentially during the process formation time. In this case the quantity $(\mathbf{v}(t_2) - \mathbf{v}(t_1))^2$ should be substituted into Eqs. (6.1), (6.11) instead of terms $w^2\tau^2$ and in other respects the calculation can be carried out as in Sec. 5.

6.2. Poles of Photon Propagators. Real and Quasireal Photons in External Fields

The basic processes taking place in the lowest approximation in the electromagnetic interaction were considered in previous sections. The analysis of photon emission and pair creation in external fields was greatly simplified by the use of quasiclassical operator method. It is naturally of interest to analyze the higher orders in the electromagnetic interaction (with exact account taken of the field), which makes it possible to judge on the general structure of quantum electrodynamics in external fields. The specific type of higher-order processes is described by the diagram in Fig. 6.3. However, considering this diagram, we encounter the divergence of the photon propagator at $k^2 = 0$. In an external field the value $k^2 = 0$ (kinematically forbidden for free particles) can be achieved owing to the additional momentum transfer by the field. We assume the homogeneity of the field which really means that it does not vary during the process formation time.

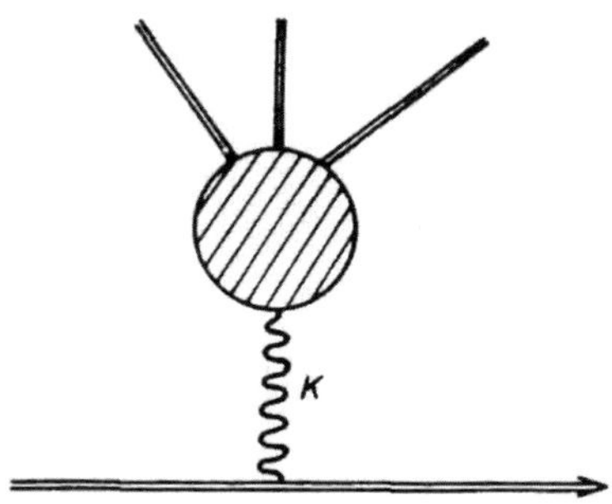

Fig. 6.3. Block diagram for the amplitude of a process with a virtual photon.

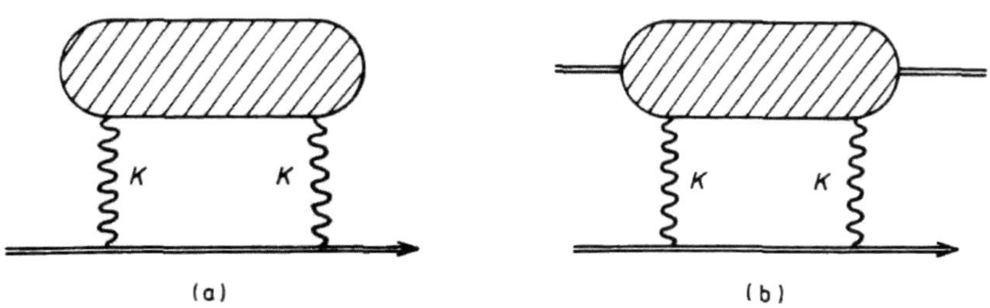

Fig. 6.4. Squared diagrams for a process with a virtual photon.

Then in the amplitude of Fig. 6.4, with imaginary part connected with the probability of the process described by Fig. 6.3, the integral has the form

$$\int \frac{dk^2}{k^4} f(k^2) \,. \tag{6.20}$$

This integral is not defined mathematically. Taking into account the Feynman circuiting rules, we have

$$\begin{aligned} \frac{1}{k^4} \to \frac{1}{k^2+i\varepsilon} \cdot \frac{1}{k^2-i\varepsilon} &= \left[\frac{\mathcal{P}}{k^2} - i\pi\delta(k^2)\right] \times \left[\frac{\mathcal{P}}{k^2} + i\pi\delta(k^2)\right] \\ &= \frac{\mathcal{P}}{k^2} \cdot \frac{\mathcal{P}}{k^2} + \pi^2\delta^2(k^2) \,. \end{aligned} \tag{6.21}$$

The terms in the right-hand side of this formula have different physical meanings. The term $\pi^2\delta(k^2)$ leads to a value of $dw_r^{(2)}/dt \equiv W_r^{(2)}$ which diverges linearly in time. The reason is that a real photon radiated in an external field is unstable and must form a pair if the field is sufficiently extended in space. The electroproduction going through a real photon can be described by the perturbation theory only if $W_r^{(2)} \ll W_\gamma^{(1)}$, where $W_\gamma^{(1)}$ is the probability of photon emission in the first order in α. On the other hand, the particle's path in the field should be much larger than the characteristic length of the process formation

$$l_f \ll t \ll W_\gamma^{(1)}/(dW_r^{(2)}/dt) \,. \tag{6.22}$$

In the case $W_r^{(2)} \geq W_\gamma^{(1)}$ the problem should be solved within the frame of cascade theory of radiation and pair creation (see for e.g. Baier, Katkov, and Strakhovenko (1987b)) using the probabilities of elementary processes obtained in Secs. 3–5.

The first term in the right-hand side of the formula (6.21) corresponds to the contribution of a virtual intermediate photon. We shall calculate it in accordance with the rule

$$\begin{aligned} \int dk^2 \frac{\mathcal{P}}{k^2}\frac{\mathcal{P}}{k^2} f(k^2) &= \lim_{a^2\to 0} \int dk^2 \frac{\mathcal{P}}{k^2+a^2} \cdot \frac{\mathcal{P}}{k^2} f(k^2) \\ &= \int dk^2 \frac{\mathcal{P}}{k^2} f'(k^2) \\ &= \int dk^2 \frac{\mathcal{P}}{k^4} [f(k^2) - f(0)] \,, \end{aligned} \tag{6.23}$$

which is based on the fact that for real fields there are always inhomogeneities "smearing out" the δ-function, and this is equivalent to the rule (6.23).

Let us now briefly discuss the equivalent photon method. In the presence of an external field this method is applicable if the field influence on the intermediate photon formation can be neglected up to logarithmic accuracy. It takes place for $k_\perp^2 \gg q_{ef}^2$, where the quantity q_{ef}^2 is defined in Eq. (6.2). According to this method the cross section of any desired electroprocess can be connected with the cross section of the corresponding photon process:

$$d\sigma = n(k_\perp^2, \omega) dk_\perp^2 d\omega d\sigma_\gamma(\omega) \,, \tag{6.24}$$

where

$$n(k_\perp^2, \omega) = \frac{\alpha}{\pi} \cdot \frac{\varepsilon - \omega}{\varepsilon\omega} \cdot \frac{\left[k_\perp^2 + \frac{\omega^2}{2\varepsilon(\varepsilon-\omega)}\left(k_\perp^2 + \frac{m^2\omega^2}{\varepsilon^2}\right)\right]}{\left(k_\perp^2 + \frac{m^2\omega^2}{\varepsilon^2}\right)^2} \tag{6.25}$$

Equation (6.24) is valid for $k_\perp^2 \le \Delta^2$ The boundary value Δ^2 is determined by the upper block of Fig. 6.3 provided that

$$d\sigma_\gamma(\omega, k_\perp) \simeq d\sigma_\gamma(\omega, 0) \,. \tag{6.26}$$

Integrating Eq. (6.25) over transverse momentum $k_\perp$ we find

$$n(\omega) = \int_{q_{ef}^2}^{\Delta^2} dk_\perp^2 n(k_\perp^2, \omega) \,, \tag{6.27}$$

where $\Delta^2 \gg q_{ef}^2$. As a result we obtain the cross section of the electroprocess with logarithmic accuracy

$$d\sigma = n(\omega) d\omega d\sigma_\gamma(\omega) = \frac{2\alpha}{\pi} \; \frac{\varepsilon - \omega}{\varepsilon} \left(1 + \frac{\omega^2}{2\varepsilon(\varepsilon - \omega)}\right)$$
$$\times \ln \frac{\varepsilon\Delta}{m\omega\left[1 + \left(\frac{\chi}{u}\right)^{1/3}\right]} \cdot d\sigma_\gamma(\omega) \cdot \frac{d\omega}{\omega} \tag{6.28}$$

Note that one should omit the term $\omega^2/2\varepsilon(\varepsilon - \omega)$ in Eq. (6.28) to obtain the equivalent photon spectrum generated by the spin 0 particle. Both the boundary momentum transfer Δ and the characteristic frequency ω may be dependent on the strength of an external field. The equivalent photon spectrum is simplified if small frequencies $\omega \ll \varepsilon$ contribute to the considered process

$$n(\omega) d\omega = \frac{2\alpha}{\pi} \cdot \frac{d\omega}{\omega} \cdot \ln \frac{\varepsilon}{m\omega\left[1 + \left(\frac{\chi}{u}\right)^{1/3}\right]} \,. \tag{6.29}$$

This formula was obtained by Baier and Katkov (1972).

6.3. Electroproduction of Charged Particles

As an important example of a higher-order process, we consider in detail, within the framework of a quasiclassical operator method stated in Secs. 2 and 3, the process of pair production by a particle moving in an external field. This process is represented by a diagram (see Fig. 6.5) on which the double lines represent the particles in the field, and the numbers denote the indices of the coordinates at the corresponding verticles. If we neglect interference effects in the case of identical particles, the probability of the process directly connected to the imaginary part of the amplitude on Fig. 6.6 can be represented in the form

$$w = \frac{e^4}{(2\pi)^8} \cdot \int \frac{d^4k_1}{k_1^2}\frac{d^4k_2}{k_2^2} \sum_F \langle p| \int dt_3 dt_4 I^*_{e\mu}(k_2, t_4) I_{e\nu}(k_1, t_3)|p\rangle$$
$$\times \langle p^-| \int dt_1 dt_2 I_p^{\mu *}(k_2, t_2) I_p^{\nu}(k_1, t_1)|p^-\rangle\,, \tag{6.30}$$

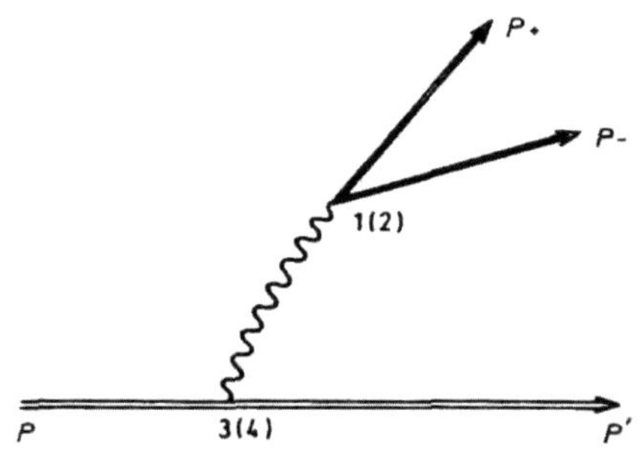

Fig. 6.5. Diagram for the electroproduction of a charged particle pair.

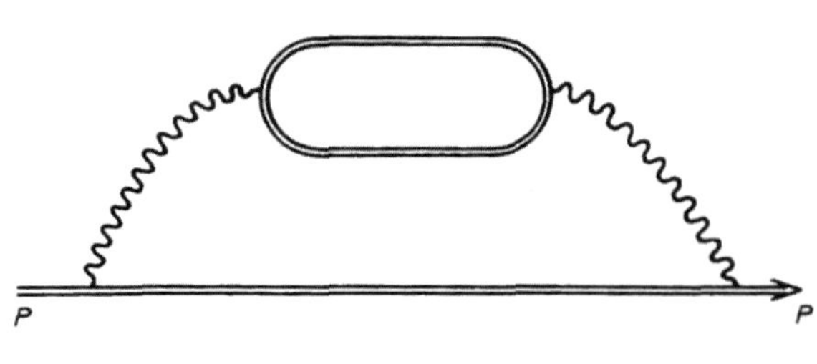

Fig. 6.6. Higher order diagram contributing to the mass operator.

where $\sum_F$ denotes summation over the final states (summation over the momenta of the radiating particle and one of the produced particles has already been carried out):

$$I_e^\mu(t) = \sqrt{\frac{m}{\mathcal{H}}}\psi_{s'}^+(\mathbf{P},\mathcal{H})\{j^\mu(\mathbf{P},\mathcal{H}), e^{ikx}\}\psi_s(\mathbf{P},\mathcal{H})\sqrt{\frac{m}{\mathcal{H}}},$$

$$I_p^\mu(t) = \frac{1}{2}\sqrt{\frac{\mu}{\mathcal{H}}}\psi_{s_-}^+(\mathbf{P},\mathcal{H})\cdot[j^\mu(\mathbf{P},\mathcal{H})e^{-ikx}$$

$$+ e^{-ikx}j^\mu(\mathbf{P},-\mathcal{H})]\psi_{s_+}(\mathbf{P},-\mathcal{H})\sqrt{\frac{\mu}{\mathcal{H}}}\cdot e^{2i\mathcal{H}t}. \tag{6.31}$$

We use here the notation of Secs. 2 and 3.

The tensors of the radiating and produced particles, which are contained in the expression for the probability, will be separated into two parts:

$$\begin{aligned}\langle p^-|\int dt_1 dt_2 I_p^{\mu *}(k_2,t_2)I_p^\nu(k_1,t_1)|p^-\rangle &= \pi^{\mu\nu} = \pi_F^{\mu\nu} + \pi_0^{\mu\nu}\\ \langle p|\int dt_3 dt_4 I_e^{\mu *}(k_2,t_4)I_e^\nu(k_1,t_3)|p\rangle &= i^{\mu\nu} = i_F^{\mu\nu} + i_0^{\mu\nu},\end{aligned} \tag{6.32}$$

where the field dependent parts $i_F^{\mu\nu}$ and $\pi_F^{\mu\nu}$ vanish when the external field vanishes $F_{\mu\nu} = 0$. Then the total probability is

$$w = w_{FF} + w_{F0} + w_{0F} \tag{6.33}$$

In calculating the probabilities w_{F0} and w_{0F} where one of the vertices is regarded as free, the disentangling operation which is basic in our approach is carried out in the usual way (see Secs. 2 and 3).

In the disentanglement of the exponential operator expressions in the calculation of w_{FF} one is faced with a new problem: the appearance of exponential factors which depend on different momenta,

$$e^{ik_1x_1}\cdot e^{-2i(t_2-t_1)\mathcal{H}}e^{ik_2x_2}$$

In formula (6.30) for the probability we write this combination in the form

$$\exp[-i(k_1-k_2,x_1)]\exp(-ik_2x_1)\exp[-2i(t_2-t_1)\mathcal{H}]\exp(ik_2x_2). \tag{6.34}$$

Here we have separated the exponential operator by virtue of the commutativity of the components x_1. For sufficiently large times, the integrations with respect to t_1 and $t_1 - t_2$ can be carried out independently. Integrating with respect to t_1, we verify that $\omega_1 = \omega_2$. The main contribution is given by the region where $k_{1(2)}^2 \leq m^2$; the vectors $\mathbf{k}_{1(2)}$ lie in a narrow cone along the direction of $\mathbf{p}$, so that $|\mathbf{k}_1 - \mathbf{k}_2| \leq m$. For this reason the operator $\exp[-i(k_1 - k_2, x_1)]$ commutes, accurate to terms $\sim m/\varepsilon$, with all the operator combinations contained in the expression for the probability. The disentanglement of the combination

$$\exp(ik_2x_1)\exp[-2i(t_2 - t_1)\mathcal{H}]\exp(ik_2x_1)$$

was carried out in Sec. 3.1:

$$\exp(-ikx_1)\exp(-2iH\tau)\exp(ikx_2)$$
$$= \exp\left\{-\frac{i\mathcal{H}_-}{\omega - \mathcal{H}_-}(k, x_2 - x_1) + \frac{ik^2\tau}{2(\omega - \mathcal{H}_-)}\right\}, \tag{6.35}$$

where $\tau = t_2 - t_1$ and $k = (\omega, \mathbf{k})$. For the photon emission vertex, we obtain similarly

$$\exp(-ikx_4)\exp(ikx_3)$$
$$= \exp\left\{-\frac{i\mathcal{H}}{\mathcal{H} - \omega}(k, x_4 - x_3) + \frac{ik^2\sigma}{2(\mathcal{H} - \omega)}\right\}, \tag{6.36}$$

where $\sigma = t_4 - t_3$. We can now, using the same arguments as in Secs. 2 and 3, change over from operators to classical mean values. While summing over the final states, it is necessary to integrate over all possible trajectories of the produced particle (over its phase space), i.e. to take the integral

$$\sum_F \to \sum_{\text{spin}} \int \frac{d^3x_1 d^3p_-}{(2\pi)^3} \tag{6.37}$$

If it is assumed that in the space-time intervals which are significant for this process the field can be regarded as homogeneous (cf. Sec. 4), then we have

$$\int d^4x_1 \exp[i(k_1 - k_2, x_1)] = (2\pi)^4\delta(k_1 - k_2). \tag{6.38}$$

The block of produced particles in w_{FF} (6.33) is connected with the field dependent part of the polarization operator $\Pi_F^{\mu\nu}$ by the relation

$$2\,\mathrm{Im}\,\Pi_F^{\mu\nu}(k) = ie^2 \int \frac{d^3p_-}{(2\pi)^3} \sum_{s_+,s_-} \times \langle p^-| \int_{-\infty}^{\infty} d\tau I_p^{\mu *}(k) I_p^{\nu}(k)|p^-\rangle\,, \tag{6.39}$$

where in accordance with Eq. (6.32) we represent the polarization operator in the form $\Pi^{\mu\nu} = \Pi_{0R}^{\mu\nu} + \Pi_F^{\mu\nu}$, here $\Pi_{0R}^{\mu\nu}$ is renormalized polarization operator in the absence of a field in the e^2-order. The probability of pair production by a photon in an external field per unit time is

$$W_{ph} = -\frac{1}{2\omega} g^{\mu\nu}\,\mathrm{Im}\,\Pi_{\mu\nu}^F(k)\,. \tag{6.40}$$

Direct calculation gives the following explicit form of $\Pi_F^{\mu\nu}$

$$\Pi_{\mu\nu}^F = G_1\left(g_{\mu\nu} - \frac{k_\mu k_\nu}{k^2}\right) + G_2\frac{\tilde{k}_\mu \tilde{k}_\nu}{\tilde{k}^2} + G_3 k^2 e^2 F^{\sigma\rho} F_\rho^\tau \left(g_{\mu\sigma} - \frac{k_\mu k_\sigma}{k^2}\right)\left(g_{\tau\nu} - \frac{k_\tau k_\nu}{k^2}\right) \tag{6.41}$$

We use the notation $\tilde{k}^\mu = eF^{\mu\nu}k_\nu$ and $\kappa^2(\mu) = -\tilde{k}^2/\mu^6$; μ is the pair-particle mass,

$$G_i^{(s)} = -\frac{\alpha\mu^2}{4\pi}\int_0^\infty d\tau \int_0^\infty \frac{d\xi}{\cosh^2\xi} \times B_\tau g_i^{(s)} \cdot \frac{1}{\tau}\exp\left\{-\frac{4i\cosh^2\xi}{\kappa(\mu)}\left(\tau + \frac{\tau^3}{3}\right) + \frac{ik^2\tau}{\mu^2\kappa(\mu)}\right\}. \tag{6.42}$$

Here the functions $g_i^{(s)}$ for particles with spin 0 and 1/2 have the form

$$g_1^{(\frac{1}{2})} = \frac{2k^2}{\mu^2}\,\frac{1}{\cosh^2\xi} - \frac{16}{3}\tau^2(2\cosh^2\xi + 1)\,,\quad g_2^{(\frac{1}{2})} = 8\tau^2\,,$$
$$g_3^{(\frac{1}{2})} = t\frac{16}{3}\tau^2(2\cosh^2\xi + 1)\,,\quad g_1^{(0)} = \sinh^2\xi\left(\frac{k^2}{\mu^2\cosh^2\xi} - \frac{8}{3}\tau^2\right) \tag{6.43}$$
$$g_2^{(0)} = -4\tau^2\,,\quad g_3^{(0)} = \frac{8}{3}\tau^2\sinh^2\xi\,.$$

We have introduced in Eq. (6.42) the operator B_τ defined by

$$B_\tau \tau^n e^{a\tau^3} = \begin{cases} \tau^n e^{a\tau^3}, & n \geq 0, \\ \tau^n (e^{a\tau^3} - 1), & n = -1, \end{cases} \tag{6.44}$$

which makes it possible to simplify greatly the notation for the results. The polarization operator $\Pi_F^{\mu\nu}$ for a spin $s = 1/2$ coincides with that obtained by Narozhnyi, 1968 for the crossed field.

The probability of the process per unit time is represented in the form

$$W = \frac{dw}{dt} = \int \frac{d^4k}{k^4} J^{\mu\nu} 2\,\mathrm{Im}\,\Pi_{\mu\nu} = W_{FF} + W_{F0} + W_{0F}\,, \tag{6.45}$$

where the tensor $J^{\mu\nu}$, which describes the emission (per unit time) of a polarized photon with mass k^2 from the initial particles, is separated in accordance with (6.32).

We proceed to consider W_{FF}:

$$W_{FF}^{(s_i,s_f)} = \frac{i\alpha m^2}{(2\pi)^2\varepsilon} \int_{-\infty}^{\infty} \frac{dk^2}{k^4} \int_0^{\infty} \frac{du}{(1+u)^2} \sum_i (F_i^{(s_i)} \cdot 2\,\mathrm{Im}\,G_i^{(s_f)})\,. \tag{6.46}$$

Here m is the mass of the initial particle,

$$F_i^{(s)} = \int_{-\infty}^{\infty} d\sigma B_\sigma \frac{f_i^{(s)}}{\sigma} \exp\left\{ -\frac{iu}{\chi}\left(\sigma + \frac{\sigma^3}{3}\right) - \frac{ik^2\sigma}{m^2\kappa(m)} \right\} \tag{6.47}$$

and

$$\begin{aligned} f_1^{(\frac{1}{2})} &= 1 + \frac{k^2}{2m^2} + 2\sigma^2\left(1 + \frac{u^2}{2(1+u)}\right), \quad f_1^{(0)} = 1 - \frac{k^2}{4m^2} + 2\sigma^2\,, \\ f_2^{(\frac{1}{2})} &= \frac{1}{2}\left[\left(3 + \frac{u^2}{1+u}\right)\sigma^2 + 1 + \frac{k^2}{m^2}\left(\frac{2(1+u)}{u^2} + 1\right)\right], \\ f_2^{(0)} &= = \frac{1}{2}\left[1 + 3\sigma^2 + \frac{k^2}{m^2}\frac{(1+u)}{u^2}\right], \quad f_3^{(\frac{1}{2})} = -\frac{k^2(1+u)}{m^2u^2}\,, \\ f_3^{(0)} &= -\frac{k^2(2+u)^2}{4u^2m^2}\,. \end{aligned} \tag{6.48}$$

Taking the integral with respect to k^2 in (6.46) in accordance with the rule (6.23), we obtain the contribution of a virtual photon (index v) to the probability W_{FF} for pair production by a particle in an external field per unit time:

$$W_{FFv}^{(s_i,s_f)} = -\frac{\alpha\mu^2}{8\pi^2\varepsilon}\int_0^\infty \frac{du}{(1+u)^2}\int_0^\infty \frac{d\xi}{\cosh^2\xi}\int_{-\infty}^{\infty} d\sigma \int_{-\infty}^{\infty} d\tau B_\sigma B_\tau$$
$$\times \frac{1}{\tau\sigma}\left[a_1^{(s_i,s_f)}\cdot\delta\left(\sigma-\frac{\mu}{m}\tau\right)+a_2^{(s_i,s_f)}\operatorname{sign}\left(\sigma-\frac{\mu}{m}\tau\right)\right]$$
$$\times \exp\left\{-i\beta\left(\sigma+\frac{\sigma^3}{3}\right)-i\gamma\left(\tau+\frac{\tau^3}{3}\right)\right\}, \tag{6.49}$$

where ε is the energy of the initial particle, $u=\omega/(\varepsilon-\omega)$, $\cosh^2\xi=\omega^2/4\varepsilon_+\varepsilon_-$,

$$\operatorname{sign}(x)=\vartheta(x)-\vartheta(-x),\quad \beta=\frac{u}{\chi},$$

$$\gamma=\left(\frac{\mu}{m}\right)^3\frac{4(1+u)\cosh^2\xi}{\chi u},\quad \chi^2=-\frac{\tilde{p}^2}{m^6},\quad \tilde{p}^\mu=eF^{\mu\nu}p_\nu;$$

$$a_1^{(\frac12,\frac12)}=\frac{2}{\sinh^2\xi}a_1^{(\frac12,0)}=-2a_1^{(0,\frac12)}=-\frac{4}{\sinh^2\xi}a_1^{(0,0)}=\frac{u\chi m^2}{\mu^2(1+u)\cosh^2\xi};$$

$$a_2^{(s_i,s_f)}=(1+2s_f)\left\{\frac{i}{2}\left[\frac{\tau^2}{3u^2}b^{(s_f)}d^{(s_i)}-\frac{m^2}{\mu^2\cosh^2\xi}\right.\right.$$
$$\left.\times\left(1+\frac{\sigma^2 d^{(s_i)}}{1+u}\right)\right]+\frac{(1+u)(\sigma-\frac{\mu}{m}\tau)}{3u\chi}$$
$$\left.\times\left[\left(\frac{b^{(s_f)}d^{(s_i)}}{1+u}+3(-1)^{2s_f}\right)\sigma^2\tau^2+b^{(s_f)}\tau^2\right]\right\};$$

$$b^{(0)}=4\cosh^2\xi-1,\quad d^{(0)}=2(1+u),$$
$$b^{(\frac12)}=8\cosh^2\xi+1,\quad d^{(\frac12)}=2(1+u)+u^2 \tag{6.50}$$

The contribution of a real intermediate photon to the probability of the process per unit time is

$$W_r^{(s_i,s_f)}=\frac{t}{2}\sum_{\lambda=1}^{2}\int\frac{dW_{\gamma\lambda}^{(s_i)}(\varepsilon,\omega)}{d\omega}W_{p\lambda}^{(s_f)}(\omega)d\omega, \tag{6.51}$$

where $dW_{\gamma\lambda}^{(s_i)}/d\omega$ is the spectral density of the radiation probability per unit time and $W_{p\lambda}^{(s_f)}(\omega)$ is the probability of pair production by a photon with the frequency ω; the index λ characterizes the photon polarization. For spin 1/2 particles these probabilities are defined in Eqs. (4.70) and (3.52).

The imaginary part of the polarization operator contained in expression for W_{F0} is well-known for spin 0 and 1/2 particles:

$$\begin{aligned} \operatorname{Im}\Pi_{\mu\nu} &= \frac{\alpha}{3}\left(\frac{k_\mu k_\nu}{k^2} - g_{\mu\nu}\right)\sqrt{1-\frac{4m^2}{k^2}}\varphi^{(s_f)}(k^2)\,, \quad k^2 \geq 4m^2\,; \\ \varphi^{(\frac{1}{2})} &= k^2 + 2m^2\,, \quad \varphi^{(0)} = \frac{k^2}{4} - m^2\,. \end{aligned} \tag{6.52}$$

The threshold value of k^2 for pair production by a virtual photon in the absence of a field is $k^2 = 4\text{m}^2$ Substituting Eq. (6.52) into (6.45), we obtain

$$\begin{aligned} W_{F0}^{(s_i,s_f)} &= \frac{\alpha^2 m^2}{6\sqrt{3}\pi^2\varepsilon}\int_0^\infty \frac{du}{(1+u)^2}\int_1^\infty \frac{dz}{z^2}\sqrt{1-\frac{1}{z}h^{(s_f)}} \\ &\times \left[r_1^{(s_i)}\left(1+\frac{4\mu^2 z(1+u)}{m^2u^2}\right)K_{2/3}(\eta)\right. \\ &\left. - \frac{r_2^{(s_i)}}{2}\int_\eta^\infty K_{1/3}(x)dx\right], \end{aligned} \tag{6.53}$$

where $\text{K}_\nu(x)$ is the Macdonald function,

$$\begin{aligned} &h^{(\frac{1}{2})} = 2(1+2z)\,, \quad h^{(0)} = z - 1\,, \quad r_1^{(0)} = 1\,, \\ &r_1^{(\frac{1}{2})} = 1 + \frac{u^2}{2(1+u)}\,, \quad r_2^{(0)} = 1 - \left(\frac{\mu}{m}\right)^2 z\,, \\ &r_2^{(\frac{1}{2})} = 1 + 2\left(\frac{\mu}{m}\right)^2 z\,, \quad \eta = \frac{2u}{3\chi}\left[1+\frac{4\mu^2(1+u)z}{m^2u^2}\right]^{3/2}. \end{aligned} \tag{6.54}$$

Substituting in Eq. (6.45) the known expression for the current tensor, we obtain for the probability W_{0F}

$$\begin{aligned} W_{0F}^{(s_i,s_f)} &= \frac{2\alpha^2 m^2}{\sqrt{3}\pi^2\varepsilon}\int_0^\infty \frac{du}{(1+u)^2}\int_0^\infty \frac{d\xi}{\cosh^2\xi}\int_{\frac{u^2}{1+u}}^\infty \frac{dy}{y^2} \\ &\times \left[c_1^{(s_i,s_f)}\left(\frac{\mu^2}{m^2}+\frac{y}{4\cosh^2\xi}\right)K_{2/3}(\lambda)\right. \\ &\left. - c_2^{(s_i,s_f)}y\int_\lambda^\infty K_{1/3}(x)dx\right], \end{aligned} \tag{6.55}$$

where

$$c_1^{(s_i,s_f)} = \frac{1+2s_f}{12} \cdot b^{(s_f)} \left(\frac{yd^{(s_i)}}{u^2} - 2\right),$$

$$c_2^{(\frac{1}{2},\frac{1}{2})} = \frac{2c_2^{(\frac{1}{2},0)}}{\sinh^2\xi} = \frac{y-2}{4\cosh^2\xi}, \tag{6.56}$$

$$c_2^{(0,\frac{1}{2})} = \frac{2c_2^{(0,0)}}{\sinh^2\xi} = -\frac{y+4}{8\cosh^2\xi}, \quad \lambda = \frac{8(1+u)\cosh^2\xi}{3u\chi}\left(\frac{\mu^2}{m^2} + \frac{y}{4\cosh^2\xi}\right)^{3/2},$$

and $b^{(s_f)}$ and $d^{(s_f)}$ are given by formula (6.50). The complete expression for the probability is given by formula (6.45).

Let us examine the asymptotic values of the probability W at $\chi \ll 1$. In this case the integral with respect to σ and τ can be obtained by the saddle-point method (we are dealing in fact with the tunnel effect), and the remaining integrals are calculated by the Laplace method. If the difference between the masses μ and m is sufficiently large, the main contribution is given by the probability W_{0F} or W_{F0}, when the block for the light particles is taken in the field, and in the case of equal masses the main contribution is given by the probability W_{FF}. The reason is that the momentum is transferred by the field mainly to the light particles. For $\mu = m$ we have in the extremal point $u = 2$ and $\xi = 0$ this means that the energy is equally divided among the three final particles. When $\mu \ll m$ we have $u \sim \mu/m$ and $\xi = 0$ (the energy of the produced particles is much smaller than the initial energy), and when $\mu \gg m$ we have $u \sim \mu/m$ and $\xi = 0$ (the produced pair carries away almost the entire energy). In the case $\mu = m$ we obtain

$$W^{(s_i,s_f)} = W_{FF}^{(s_i,s_f)} = W_\tau^{(s_i,s_f)} + W_{FFv}^{(s_i,s_f)},$$

$$W_{FF}^{(\frac{1}{2},\frac{1}{2})} = 6W_{FF}^{(0,0)} = \frac{\alpha^2 m^2 \chi}{64\varepsilon}\left(\tau - \sqrt{\frac{8\chi}{3\pi}}\right)\exp\left(-\frac{16}{3\chi}\right), \tag{6.57}$$

$$\tau = tm\frac{H}{H_0} = \frac{t}{t_f} \gg 1.$$

Here t_f is the formation time for photon emission and pair production processes at $\chi \ll 1$. In accordance with the previous analysis the ratio t/t_f is assumed to be large. For $\mu \gg m$, we have

$$
\begin{aligned}
W^{(s_i,s_f)} &= W_{F0}^{(s_i,s_f)}, \quad W_{F0}^{(\frac{1}{2},\frac{1}{2})} = 4W_{F0}^{(0,\frac{1}{2})} \\
&= \frac{16\sqrt{3}}{\chi}\left(\frac{\mu}{m}\right)^2 W_{F0}^{(\frac{1}{2},0)} = \frac{64\sqrt{3}}{\chi}\left(\frac{\mu}{m}\right)^2 W_{F0}^{(0,0)} \\
&= \frac{\alpha^2 m^2 \chi^{5/2}}{2^6 3^{5/4}\sqrt{\pi}\varepsilon}\left(\frac{m}{\mu}\right)^5 \cdot \exp\left[-\frac{4\sqrt{3}}{\chi}\left(\frac{\mu}{m}\right)^2\right]
\end{aligned} \tag{6.58}
$$

In the case $m \gg \mu$ we obtain

$$
\begin{aligned}
W^{(s_i,s_f)} &= W_{0F}^{(s_i,s_f)}, \\
W_{0F}^{(\frac{1}{2},\frac{1}{2})} &= \frac{\alpha^2 m^2 \chi^{5/2}}{8 \cdot 3^{5/4}\sqrt{2\pi}\varepsilon} \cdot \left(\frac{m}{\mu}\right)^4 \exp\left[-\frac{2\sqrt{3}}{\chi}\left(\frac{\mu}{m}\right)^2\right]
\end{aligned} \tag{6.59}
$$

When $\chi \gg \max[(\mu/m)^3, (\mu/m)^{3/2}]$ the main contribution is given by the probabilities W_{FF} and W_{0F} owing to the fact that only small momentum transfers play an important role. The values $u \sim (\mu/m)^3/\chi \ll 1$ contribute and for this reason the interference effects for identical final particles can be neglected. For the probability W_{FF} we have in Eq. (6.49)

$$
\begin{aligned}
&\beta \sim \frac{1}{\chi^2}\left(\frac{\mu}{m}\right)^3, \gamma \sim 1, \text{ i.e. } \sigma \sim \chi^{2/3}\left(\frac{\mu}{m}\right)^2, \ \tau \sim 1; \\
&\delta\left(\sigma - \frac{\mu}{m}\tau\right) \to 0, \ \mathrm{sign}\left(\sigma - \frac{\mu}{m}\tau\right) \to \mathrm{sign}(\sigma).
\end{aligned} \tag{6.60}
$$

Direct calculation of the probabilities W_{FF} and W_{0F} yields

$$
\begin{aligned}
W_{FF}^{(\frac{1}{2},\frac{1}{2})} &= \frac{26}{5} W_{FF}^{(\frac{1}{2},0)} \\
&= -\frac{13\alpha^2 m^3 \chi}{9\sqrt{3}\pi\mu\varepsilon}\left[\ln\left(\frac{\chi\left(\frac{m}{\mu}\right)^{3/2}}{2\sqrt{3}}\right) - C - \frac{16}{39}\right], \\
W_{0F}^{(\frac{1}{2},\frac{1}{2})} &= \frac{26}{5} W_{0F}^{(\frac{1}{2},0)} \\
&= \frac{13\alpha^2 m^3 \chi}{6\sqrt{3}\pi\mu\varepsilon}\left[\ln\left(\frac{\chi}{2\sqrt{3}}\left(\frac{m}{\mu}\right)^2\right) - C - \frac{58}{39}\right], \\
C &= 0.577\ldots, \\
W_{FF(0F)}^{(0,s_f)} &= W_{FF(0F)}^{(\frac{1}{2},s_f)}.
\end{aligned} \tag{6.61}
$$

The latter equality in Eq. (6.61) is provided by the softness of emitted photons. Summing W_{FF} and W_{0F}, we finally obtain

$$\begin{aligned} W^{(\frac{1}{2},\frac{1}{2})} &= \frac{26}{5} W^{(\frac{1}{2},0)} \\ &= \frac{13\alpha^2 m^3 \chi}{18\sqrt{3}\pi\mu\varepsilon}\left[\ln\left(\frac{\chi}{2\sqrt{3}}\left(\frac{m}{\mu}\right)^3\right) - C - \frac{142}{39}\right] \\ &= \frac{13\alpha^2\mu^2\chi(\mu)}{18\sqrt{3}\pi\varepsilon}\left[\ln\frac{\chi(\mu)}{2\sqrt{3}} - C - \frac{142}{39}\right] \end{aligned} \tag{6.62}$$

It should be noted that the probability (6.62) is independent of the initial particle mass. When $\mu \ll m$ and $(\mu/m)^2 \ll \chi \leq (\mu/m)^{3/2}$ the pair production probability is determined by W_{0F} in Eq. (6.61).

The contribution of a real photon to the probability of the process at $\chi \gg 1$ has the form ($s_i = 1/2$)

$$W_r = \frac{3t(\alpha\Gamma(2/3))^2(3\chi)^{4/3}}{14\Gamma^3(1/3)(\gamma\lambda_c)^2}\left[\ln\chi + \frac{7}{4}\left(\ln 3 + \frac{\pi}{\sqrt{3}}\right) - C - \frac{295}{42}\right] \tag{6.63}$$

The probability W_r at arbitrary values of the parameter χ is approximated ($\mu = m, s_f s_i = 1/2$) by the formula

$$W_r = \frac{3\chi}{16} t\left(\frac{\alpha}{\gamma\lambda_c}\right)^2 \ln\left(1 + \frac{\chi}{12}\right)\exp(-16/3\chi)(1 + 0.56\chi + 0.013\chi^2)^{1/6} \tag{6.64}$$

to within 4%. Equations (6.63) and (6.64) were obtained by Baier, Katkov and Strakhovenko, 1991a.

The probabilities (6.62) and (6.63) can be obtained within logarithmic accuracy by the equivalent photon method in the presence of an external field formulated in the previous section. To this end, one should substitute into Eq. (6.28) the pair production probability $W_p(\omega)$ by unpolarized photon with frequency ω instead of $d\sigma_\gamma(\omega)$ and take the integral over ω. Since $W_p(\omega)$ is maximal for $\kappa = \chi(m/\mu)^3 u/(1+u) \sim 1$, the contribution is given by $u \sim \chi^{-1}(\mu/m)^3 = 1/\chi(\mu) \ll 1$. The boundary momentum transfer Δ in Eq. (6.27) is $\Delta = \mu$.

The electroproduction process was stated in this section according to the paper by Baier, Katkov and Strakhovenko, 1971. The results for the crossed field, particularly, Eqs. (6.57) and (6.62) were obtained by Ritus, 1972a, b. Narozhnyi and Nikishov, 1972, considered the pair-production in the combination of the crossed field and Coulomb field with the charge Ze. Their result is reproduced by Eq. (6.61) if we multiply the probability $W_{0F}^{(1/2,1/2)}$ by the factor Z^2.

6.4. Photon Splitting in an Electromagnetic Field

Virtual creation and annihilation of electron-positron pairs is known to induce nonlinear self-action of an electromagnetic field. A characteristic process of nonlinear quantum electrodynamics is the scattering of light by light. In external fields a photon can be split into two ($\gamma \to \gamma_1 + \gamma_2$) as well as deflected (coherently scattered, $\gamma \to \gamma'$). At low photon energies ($\omega \ll m$) the splitting process can be analyzed by using the Heisenberg–Euler effective Largrangian (see, e.g. Berestetskii, Lifshitz and Pitaevskii, 1982, Secs. 129 and 130). For arbitrary photon energies and field intensities, an exact calculation (in terms of the field) is required. Photon splitting was considered in Bialynicki–Birula, 1970 and in Alder *et al.*, 1970, who used an effective Heisenberg–Euler Lagrangian; polarization selection rules, especially with allowance for dispersion, were also obtained by Adler, 1970. Adler, 1971, analyzed in detail the process in an external magnetic field and obtained the allowed-transition amplitude for the general case of an arbitrary photon energy; the expression for this amplitude turned out to be too unwieldy for further use. He used a Green's function for an electron in an external magnetic field in the Schwinger, 1951, proper time representation. Papanyan and Ritus, 1971, 1973, considered photon splitting in a so-called crossed field $\mathbf{E} \perp \mathbf{H}, E = H$, using likewise the electron Green's function in the proper time representation.

The complete solution of the problem in an arbitrary constant field $F_{\mu\nu}$ was obtained by Baier, Milstein and Shaisultanov, 1986, set forth in this section, who used the operator diagrammatic technique developed by Baier, Katkov and Strakhovenko, 1975.

In the general case, the photon splitting probability depends on three dimensionless invariants (see Eq. (B.25) in Appendix B):

$$\begin{aligned} \frac{E}{E_0} &= \frac{e}{m^2}(\sqrt{\mathcal{F}^2 + \mathcal{G}^2} - \mathcal{F})^{1/2} \\ \frac{H}{H_0} &= \frac{e}{m^2}(\sqrt{\mathcal{F}^2 + \mathcal{G}^2} + \mathcal{F})^{1/2}\,, \\ \kappa &= \frac{e}{m^3}\sqrt{|(F_{\mu\nu}k^\nu)^2|}\,, \end{aligned} \tag{6.65}$$

where $\mathcal{F}$ and $\mathcal{G}$ are the field invariants

$$\mathcal{F} = -\frac{1}{4}F_{\mu\nu}F^{\mu\nu} = \frac{1}{2}(\mathbf{E}^2 - \mathbf{H}^2)\,,$$

$$\mathcal{G} = -\frac{1}{4}F^*_{\mu\nu}F^{\mu\nu} = (\mathbf{EH})\,, \tag{6.66}$$

$$F^*_{\mu\nu} = \frac{1}{2}\varepsilon_{\mu\nu\alpha\beta}F^{\alpha\beta}$$

while E_0 and H_0 are the quantum-electrodynamics critical fields (see Eq. (B.20) in Appendix B).The fields E and H in Eq. (6.65) are the electric and magnetic field in a special reference frame in which $\mathbf{E} \parallel \mathbf{H}$. If the invariants H/H_0 and E/E_0 are small compared with unity and with the photon-energy dependent invariant κ, they can be neglected. This is equivalent to transforming the case of a crossed field (or to the constant field approximation, see Appendix B). For the results obtained in a constant and uniform field to be valid, it is necessary that the field changes little over the length (duration) of the process formation. For the process considered, the formation length is (see Eq. (3.21))

$$l_f = \min\left(\frac{E_0}{mE}, \frac{H_0}{mH}\right) \tag{6.67}$$

Photon splitting in a Coulomb (substantially inhomogeneous) field was considered by Baier *et al.*, 1974.

Consider the amplitude of photon splitting ($k \to k_1 + k_2$) in a constant and uniform electromagnetic field $F_{\mu\nu}$ for which $k = k_1 + k_2$. On the mass shell we have $k_1^2 = k_2^2 = k^2 = 0$, $k = (\omega, \mathbf{k})$ etc. Strictly speaking, dispersion takes place in an external field, i.e. the photon acquires a mass that depends on its polarization. This mass is determined from the polarization operator in the given field (see, e.g. Baier, Katkov and Strakhovenko, 1975). The mass acquired, however, turns to be small and is manifested in fact in the polarization selection rules. This question will be analyzed below. By virtue of what has been said, the analysis can be carried out on the mass shell in the approximation called collinear ($\mathbf{k} \parallel \mathbf{k}_1 \parallel \mathbf{k}_2$); it is useful then to introduce the unit vector $\lambda(\lambda^2 = 0, \boldsymbol{\lambda}^2 = 1)$

$$\lambda_\mu = k_\mu/\omega = k_{1\mu}/\omega_1 = k_{2\mu}/\omega_2\,.$$

It is convenient to use for the analysis a special reference frame in which $\mathbf{E} \parallel \mathbf{H}$ (the fields E and H in this frame are given by Eqs. (6.65) and (6.66)). We choose the common direction of these vectors to be the 3-axis of a Cartesian system. The field tensor can then be represented in the form

$$F_{\mu\nu} = C_{\mu\nu}E + B_{\mu\nu}H \tag{6.68}$$

Here

$$\begin{aligned} C_{\mu\nu} &= g^0_\mu g^3_\nu - g^3_\mu g^0_\nu\,, \\ B_{\mu\nu} &= g^2_\mu g^1_\nu - g^1_\mu g^2_\nu\,, \end{aligned} \tag{6.69}$$

where $g_{\mu\nu}$ is the metric tensor.

We use the operator diagrammatic technique developed by Baier, Katkov and Strakhovenko, 1975. In the lowest order of perturbation theory in the interaction with photons, represented by the diagram of Fig. 6.7 where the double line is the electron propagator in an external field, we write the amplitude of the process in the form

$$T_1 = -e^3\,\mathrm{Tr}\left\langle 0|\hat{e}\frac{1}{\hat{P}-m}\hat{e}_1^*\frac{1}{\hat{P}+\hat{k}_1-m}\hat{e}_2^*\frac{1}{\hat{P}+\hat{k}-m}|0\right\rangle\,, \tag{6.70}$$

where $\hat{P} = \gamma^\mu(i\partial_\mu - eA_\mu)$, A_μ is the vector potential of the external field, $e^\mu = e^\mu(k)$; $e^\mu_n = e^\mu(k_n)$ are the photon polarization vectors. It is necessary to add to the amplitude T_1 (6.70) the amplitude $T_2 = T_1(k_1 \leftrightarrow k_2,\ \ e_1 \leftrightarrow e_2)$. The total decay amplitude is $T = T_1 + T_2$. In this approach, the main problem is to calculate the mean value over the states $x = 0 :< 0|\ldots|0>$, which includes an aggregate of noncommuting operators P_μ, and the calculation procedure is based on the completeness of the algebra of the operators $[P_\mu, P_\nu] = -ieF_{\mu\nu}$.

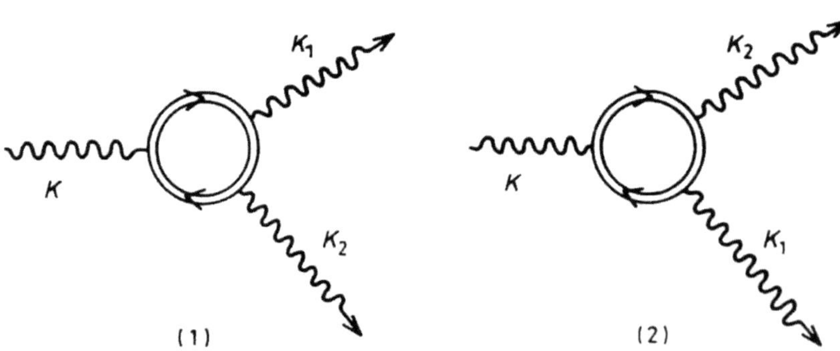

Fig. 6.7. Diagrams for the photon splitting in an external electromagnetic field; the double line represents an electron in an external field.

We proceed now to discuss the dependence of the amplitude of the process on photon polarizations. In the collinear approximation we have in the special reference frame the four-vector λ_μ and vectors $\mathbf{H}$ and $\mathbf{E}$ directed along the z-axis. The photon polarization vectors should be perpendicular to the vector λ_μ and can be expanded in vectors

$$e_{1\mu} = (\lambda B)_\mu, \quad e_{2\mu} = (\lambda C)_\mu \tag{6.71}$$

Using Eq. (6.69) it is easy to verify that $e_{1\mu}\lambda^\mu = e_{2\mu}\lambda^\mu = 0$. If an external field is pure magnetic ($\mathbf{E} = 0$) or pure electric ($\mathbf{H} = 0$), the polarization vectors

$$\begin{aligned} \mathbf{H} = 0: \quad & e_I^\mu = (\lambda C)^\mu, \quad e_{II}^\mu = -(\lambda B)^\mu, \\ \mathbf{E} = 0: \quad & e_I^\mu = (\lambda B)^\mu, \quad e_{II}^\mu = -(\lambda C)^\mu, \end{aligned} \tag{6.72}$$

are the eigenvectors of the Dyson equation which determines the mass $k^2_{(\lambda)}$($\lambda =$ I, II, III, IV; the values $\lambda =$ I and II corresponds to the physical polarizations) acquired by a photon in an external field.

In the collinear approximation the probability of photon splitting per unit time is expressed in terms of its amplitude T as follows (up to the pair production threshold):

$$dW = \frac{1}{32\pi}|T|^2\frac{d\omega_1}{\omega^2}\vartheta\left(\frac{k^2}{\omega} - \frac{k_1^2}{\omega_1} - \frac{k_2^2}{\omega_2}\right). \tag{6.73}$$

The polarization selection rules follow, in particular, from the ϑ-function contained in this expression. It is customary to use a refractive index n, defined by the relation

$$k^2_{(\lambda)} = \omega^2(1 - n^2_{(\lambda)})\,; \tag{6.74}$$

the allowed transitions should then satisfy the relation

$$\omega_1 n_1^2 + \omega_2 n_2^2 - \omega n^2 \geq 0\,. \tag{6.75}$$

Here n_1 and n_2 are the refractive indices for the frequencies ω_1 and ω_2 that depend on the photon polarization.

We proceed now to the important particular case when only a magnetic field ($\mathbf{E} = 0$) or only an electric field ($\mathbf{H} = 0$) is present. Two types of selection rules are known. The first is governed by CP invariance of the electromagnetic interaction. In terms of the polarizations (6.72) at $\mathbf{E} = 0$ or at $\mathbf{H} = 0$ the CP-allowed transitions are

$$e_I \to e_{II}e_{II}, \quad e_I \to e_I e_I, \quad e_{II} \to e_I e_{II} \tag{6.76}$$

Three other transitions (e.g. $e_{II} \to e_I e_I$) are forbidden. These selections rules are valid, naturally, for all energies. The second type entails satisfaction of the condition (6.75). It can be shown that the only CP-allowed transitions (6.76) that satisfy relation (6.75) in the range where the effective Lagrangian can be used is $e_I \to e_{II} e_{II}$. Using the dispersion relation, it was shown by Adler, 1971, that this conclusion is valid* for all $\omega < 2m$. At $\omega > 2m$ a channel is opened for pair production by a photon in an external field (see Sec. 3). The refractive index (see (6.74)) acquires therefore an imaginary part. As a result the spatial component of the photon wave vector acquires a negative imaginary part and the incident wave attenuates as it propagates in the field. The integral with respect to the coordinate along **k** must therefore be taken between finite limits, and the probability $dW(L)$ of photon splitting after propagating a length L in the field becomes meaningful. The expression for $dW(L)$, which must now be used in place of (6.73), is

$$dW(L) = \frac{|T|^2}{32\pi} \frac{d\omega_1}{\omega^2} \rho(L) , \tag{6.77}$$

where

$$\rho(L) = \frac{1}{\pi} \int_0^\infty \frac{1 + e^{-2\gamma L} - 2e^{-\gamma L} \cos((\beta - x)L)}{\gamma^2 + (\beta - x)^2} dx ,$$

$$\beta = \mathrm{Re} \left(\frac{k^2}{\omega} - \frac{k_1^2}{\omega_1} - \frac{k_2^2}{\omega_2} \right) , \tag{6.78}$$

$$\gamma = -\mathrm{Im} \left(\frac{k^2}{\omega} - \frac{k_1^2}{\omega_1} - \frac{k_2^2}{\omega_2} \right)$$

In the limit $\gamma L \ll 1$, $\beta L \gg 1$ (the photon absorption is negligibly small, the effect is considered over large length L), we have $\rho(L) \to L\vartheta(\beta)$, so that $dW(L) = LdW$ (see (6.77) and (6.73)) in accordance with the definition of dW as the probability of the process per unit length (time). In another limiting case $\gamma L \gg 1$ (under conditions of strong absorption),

$$\rho(L) = \frac{1}{\pi\gamma} \left(\frac{\pi}{2} + \arctan \frac{\beta}{\gamma} \right) , \tag{6.79}$$

and in the limit $|\beta/\gamma| \to \infty$ we have asymptotically $\rho(L) \to \vartheta(\beta)/\gamma$. It is clear therefore that only in the limit as $\gamma \to 0$ the function $\rho(L)$ is proportional

*The region $\omega \leq 2m$ was recently analyzed once more, see V. Baier, A. Milstein and R. Shaisultanov, *Phys. Rev. Lett.* **77**, 1691 (1996).

to $\vartheta(\beta)$ and the selection rules based on the inequality (6.75) are valid. In the general case, however, $\rho(L) \neq 0$ regardless of the sign of β so that all the transitions (6.76) are possible.

Expressions for the photon-splitting amplitudes, valid for arbitrary fields $F_{\mu\nu}$ and photon energies ω, are given in the paper by Baier, Milstein and Shaisultanov, 1986. The calculations with these amplitudes are not too complicated. Let us now use them to analyze the photon splitting in single crystals. recognizing that $\mathbf{H} = 0$ and $\mathbf{k} \perp \mathbf{E}$, we have from (6.65)

$$\kappa = \frac{\omega}{m} \cdot \frac{E}{E_0}, \tag{6.80}$$

where E is the electric field at a given distance from the axis (plane). In crystals we have $E/E_0 \ll 1$ and, to reach the region $\kappa \geq 1$ in which the splitting is a maximum, the photon energies must be very high, $\omega \gg m$. We must thus consider the quasiclassical approximation in an electric field. Carrying out the standard expansions in the general expressions for amplitudes we have in the quasiclassical approximation (index Q)

$$\begin{aligned} T_Q &= \frac{e^3}{4\pi^2} m\kappa \int_0^\infty dt \int_0^1 dt_2 \left[\int_0^{t_2} dt_1 e^{i\varphi_Q} \; I_Q + I_Q^{(0)} e^{-it}\right], \\ \varphi &= -\frac{\kappa^2 t^3}{3} [\nu_1 t_1^2 (1-t_1)^2 + \nu_2 t_2^2 (1-t_2)^2 - \nu_1\nu_2 (t_2 - t_1)^2 \\ &\quad \times (t_1 - t_2 + 1)^2] \\ \varphi_Q &= \varphi - t, \end{aligned} \tag{6.81}$$

in which, for the polarizations (6.76),

$$\begin{aligned} I_Q^{I\to II,II} &= i(1+\eta) - 4\kappa^2 t^3 \nu_1\nu_2 (t_2 - t_1)^2 [\nu_1 t_1^2 + \nu_2 (1-t_2)^2], \\ I_Q^{(0)I\to II,II} &= -2it_2^2 e^{i\varphi_0}, \\ I_Q^{I\to I,I} &= i[8t_1(1-t_2) - 1 + \eta] - \frac{\kappa^2 t^3}{2} \{2\nu_1\nu_2 (t_1 - t_2) \\ &\quad \times [1 + (2t_1 - 1)(2t_2 - 1) + 2(t_1 + t_2 - 1)(\eta_1 - \eta_2)] \\ &\quad + (1+\eta)[2\nu_1\nu_2(t_2 - t_1)(t_1 - t_2 + 1) + \eta_1\eta_2]\} \\ I_Q^{(0)I\to I,I} &= -2it_2^2 [e^{i\varphi_0} - \nu_1 e^{i\nu_1^2\varphi_0} - \nu_2 e^{i\nu_2^2\varphi_0}]. \end{aligned} \tag{6.82}$$

In (6.81) and (6.82) we used the notation

$$\begin{aligned} \nu_{1,2} &= \frac{\omega_{1,2}}{\omega}, \\ \eta &= \nu_1(1-2t_1)^2 + \nu_2(1-2t_2)^2, \\ \eta_{1,2} &= 2[\nu_{1,2}(t_1-t_2)(t_1-t_2+1) + t_{2,1}(1-t_{2,1})], \\ \varphi_0 &= -\frac{\kappa^2 t^3}{3} t_2^2(1-t_2)^2 \end{aligned} \tag{6.83}$$

The third amplitude follows from (6.81) and (6.82) by the permutation of the initial and final photons:

$$T_Q^{II\to I,II} = T_Q^{I\to II,II}(\omega \leftrightarrow -\omega_1). \tag{6.84}$$

At $\kappa \ll 1$ we have from Eqs. (6.81)–(6.84) the following expressions

$$\begin{aligned} T_Q^{I\to II,II} &= T_Q^{II\to I,II} \\ &= \frac{13e^3 m}{315\pi^2}\kappa^3 \nu_1\nu_2, \\ T_Q^{I\to I,I} &= \frac{24e^3 m}{315\pi^2}\kappa^3\nu_1\nu_2. \end{aligned} \tag{6.85}$$

The amplitudes (6.81)–(6.84) were calculated numerically in the region $\kappa \geq 1$, in which they reach their maximum values. The quantity $|T^2|/\omega$ is plotted in Fig. 6.8 for the transitions (a) $I \to II, II$, and (b) $I \to I, I$ for which the distribution in ν_1 is symmetric under the substitution $\nu_1 \leftrightarrow \nu_2$, and for the transition $II \to I, II$, for which there is no such symmetry. The value of $|T^2|/\omega$ is measured in units of $\xi = 4\alpha^3 eE/\pi m$. The sets of frequencies for which the plots were drawn are given in the figure captions. Denoting the parameter κ corresponding to the maximum $|T^2|/\omega$ by κ_m, we have $\kappa_m \simeq 80, \simeq 30$, and $\simeq 200$ for $I \to II, II$, $I \to I, I$, and $II \to I, II$, respectively. It can be seen that the probability of the transition $II \to I, II$ is always small with respect to the others, and in the region of its maximum, the dominant transition is $I \to II, II$. We note also that at $\kappa \gg 1$ all the cited amplitudes reach their asymptotes most smoothly, so that the asymptotes can be used only at very large value of the parameter κ. In Fig. 6.9, the functions $d(\nu_1)$ which determine

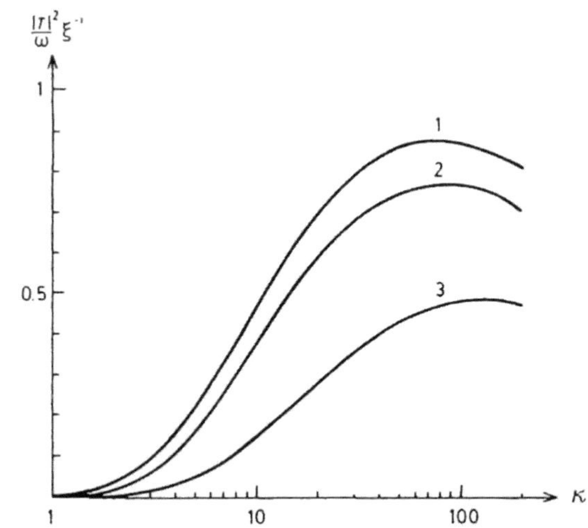

(a)

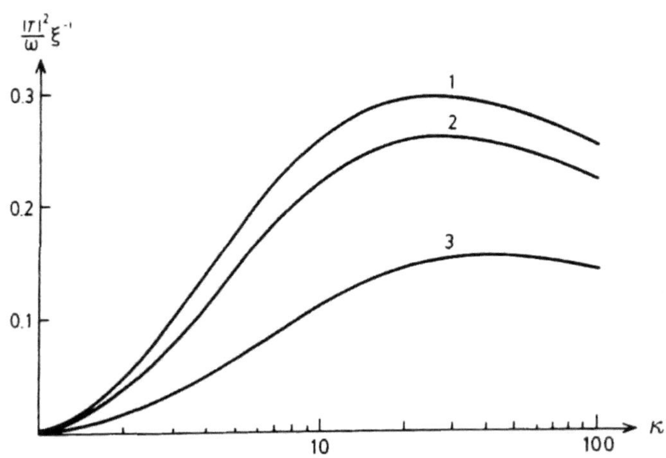

(b)

Fig. 6.8. $|T|^2/\omega$ in units of $\xi = (4\alpha^3/\pi)eE/m$ for transitions:

(a) $I \to II, II$(1 for $\nu_1 = \nu_2 = 0.5$; 2 for $\nu_1 = 0.35$, $\nu_2 = 0.65$; 3 for $\nu_1 = 0.2$, $\nu_2 = 0.8$);

(b) $I \to I, I$ for the same values of $\nu_{1,2}$;

(c) $II \to I, II$(1 for $\nu_1 = 0.35$, $\nu_2 = 0.65$; 2 for $\nu_1 = \nu_2 = 0.5$; 3 for $\nu_1 = 0.65$, $\nu_2 = 0.35$; 4 for $\nu_1 = 0.8$, $\nu_2 = 0.2$).

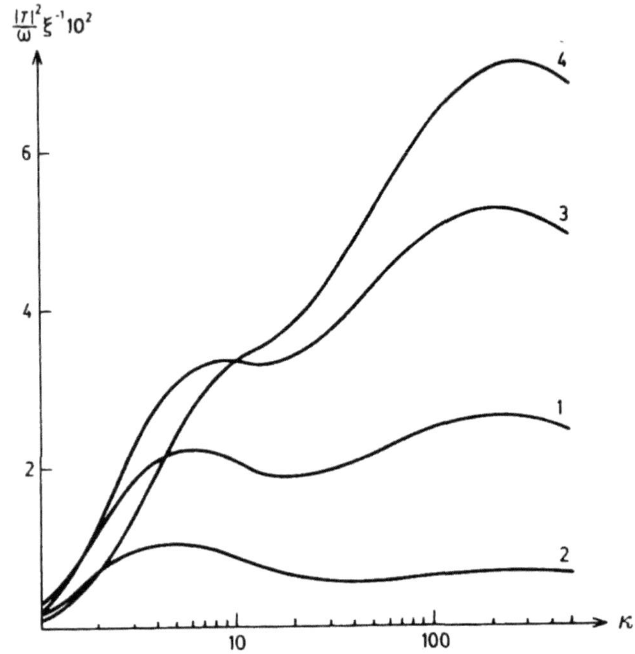

Fig. 6.8 (c) (*Continued*)

the asymptotic ($\kappa \gg 1$) dependence of the amplitudes under discussion on the final photon frequency are present:

$$T_Q = \frac{me^3}{6\pi^2}\Gamma(1/3)(3\kappa)^{1/3}e^{i\pi/3}d(\nu_1)\,.$$

To find the probability of the process we must substitute $|T|^2/\omega$ and the function $\rho(L)$ (6.78) in Eq. (6.77). To find $\rho(L)$ we must first find β and γ (see Eq. (6.78)), i.e. the real and imaginary parts of the refractive index (see Eq. (6.74)). Expressions for the refractive index in the quasiclassical approximation can be found in the book by Baier, Katkov and Fadin, 1973, p.189,

$$n^2_{I,II} = 1 + \frac{\alpha}{12\pi}\left(\frac{E}{E_0}\right)^2 g_{I,II}\,, \tag{6.86}$$

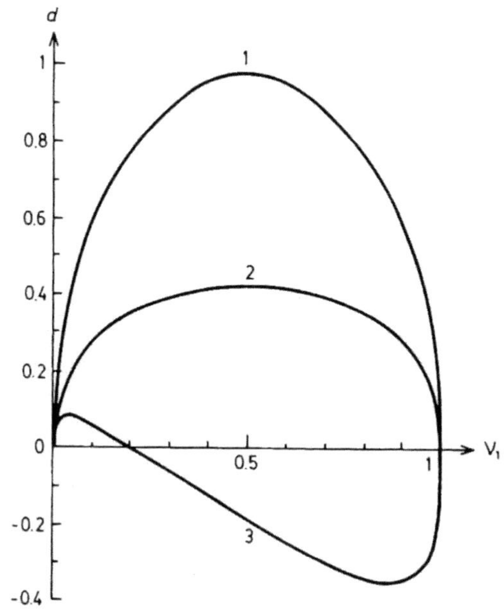

Fig. 6.9. Asymptotic values of the quasiclassical amplitude $T_Q = Dd(\nu_1)$ depending on the photon frequency ν_1 for transitions: $(1) I \to II, II, (2) I \to I, I, (3) II \to I, II$.

where

$$g_{I,II} = -\int_0^1 dv \cdot a_{I,II} \int_0^\infty s \exp\left[-i\left(s + \frac{\kappa^2 f^2 s^3}{48}\right)\right] ds\,,$$
$$a_I = f\left(2 - \frac{f}{2}\right), \quad a_{II} = f(2+f), \quad f = 1 - v^2 \tag{6.87}$$

At $\kappa \ll 1$ we have

$$g_I = \frac{16}{15}, \quad g_{II} = \frac{28}{15}, \tag{6.88}$$

and at $\kappa \gg 1$ the functions g_I and g_{II} decrease:

$$g_{I,II} = c_0 b_{I,II} \kappa^{-4/3} \exp\left(\frac{2}{3} i\pi\right),$$
$$b_I = 2, \quad b_{II} = 3, \quad c_0 = \frac{4 \cdot 6^{2/3}}{7} \frac{\Gamma^2\left(\frac{2}{3}\right)}{\Gamma(7/6)} \sqrt{\pi} \simeq 6.6\,. \tag{6.89}$$

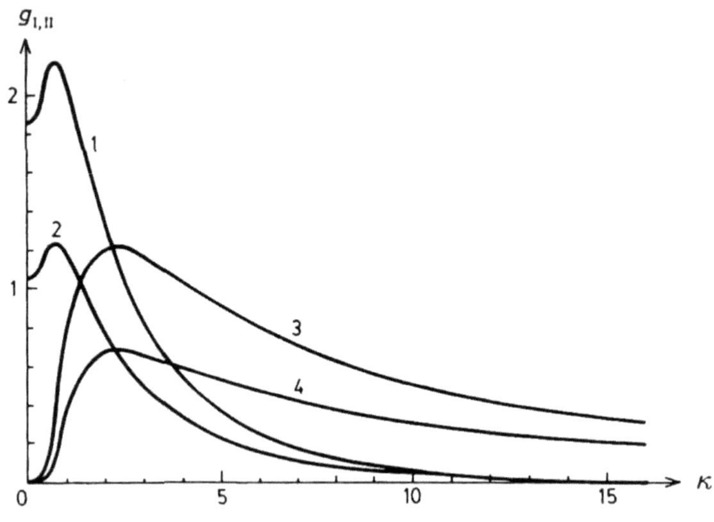

Fig. 6.10. Functions $g_{I,II}(\kappa)$ (Eq. (6.87)); (1) Re g_{II}, (2) Re g_I, (3) Im g_{II}, (4) Im g_I.

Plots of the real and imaginary parts of the functions g_I and g_{II} are shown in Fig. 6.10. Calculation of the refractive index (6.86) yields β and γ (6.78), after which we can obtain $\rho(L)$ at all values of the parameters, whose choice is dictated by the experimental conditions. Note that β and γ depend on $\nu_{1(2)}$.

It is useful to consider the region $\gamma L \ll 1$, from which we can estimate the upper bound of the splitting probability for realistic experimental conditions. In this region $\rho(L)$ is given by the elementary function (6.69). The analysis is particularly simple at $\kappa \gg 1$. In this case

$$\gamma_{I \to II,II} = \frac{\sqrt{3}\alpha c_0}{12\pi}\left(\frac{eE}{m}\right)\kappa^{-\frac{1}{3}}\left[1 + \frac{3}{2}(\nu_1^{-1/3} + \nu_2^{-1/3}\right], \qquad (6.90)$$

and with increasing κ the imaginary part γ decreases: $\gamma \propto \kappa^{-1/3}$ Since $|T^2|/\omega \propto \kappa^{-1/3}$ in this region, the quantity $dW(L)$ (6.77) tends to a finite limit at asymptotically large values $\kappa \gg 1$. This result can be easily understood: actually, the length over which the photons have not yet been converted into e^+e^- pairs increases in proportion to $\kappa^{1/3}$ and the probability $dW(L)$ turns out to be independent of energy in view of this effective lengthening of the region where the photon splitting takes place:

$$
\begin{aligned}
W_{I\to II,II}(L) &\simeq 0.15\,\alpha^2/\pi^2\,,\\
W_{I\to I,I}(L) &\simeq 0.024\,\alpha^2/\pi^2\,.
\end{aligned}
\tag{6.91}
$$

7. PROCESSES WITH HIGH ENERGY PARTICLES MOVING IN A MEDIUM

7.1. Photon Emission in a Macroscopic External Field

The analysis of the quasiclassical method, formulated in Secs. 2 and 4, describing the radiation from relativistic particles was carried out in the case when the spatial size of the field region was much larger than the process formation length. Below we will consider the situation when a radiation formation length exceeds the characteristic size of a region of the field action. The most important example of such field is a Coulomb field (for more details concerning problems discussed in Secs. 7.1–7.3 see Baier, Katkov and Fadin, 1973). In this case, scattering angles are smaller or the same order of magnitude as a characteristic angle of radiation and all the particle trajectories contribute to the process. It is convenient therefore to integrate with respect to time directly in a matrix element.

We proceed now to calculate the matrix element for a photon emission. Let us introduce the final state projection operator

$$
|f\rangle\langle f| = \sum_f \delta(\mathbf{P}-\mathbf{p}_2)|f\rangle\langle f| = \delta(\mathbf{P}-\mathbf{p}_2)\,. \tag{7.1}
$$

Commutating $e^{i\mathbf{kr}}$ with δ-function, we have

$$
\begin{aligned}
e^{i\mathbf{kr}}\delta(\mathbf{P}-\mathbf{p}_2)e^{-i\mathbf{kr}} &= \delta(\mathbf{P}-\mathbf{p}_2-\mathbf{k})\\
\equiv \delta(\mathbf{P}-\mathbf{p}_1+\mathbf{q}) &= \sum_f \delta(\mathbf{P}-\mathbf{p}_1+\mathbf{q})|f\rangle\langle f|\,.
\end{aligned}
\tag{7.2}
$$

Carrying out the disentangling operation (see Sec. 2), we represent the probability of a photon emission in the form

$$
dw = \sum_f \langle i|M^+|f\rangle\langle f|M|i\rangle\,, \tag{7.3}
$$

where

$$
M = \frac{e}{2\pi\sqrt{w}}\int_{-\infty}^{\infty} R(t)\exp\left\{\frac{i\varepsilon kx(t)}{\varepsilon-\omega}\right\}dt\,, \tag{7.4}
$$

here $R(t) = R(\mathbf{P}(t))$ (see Eq. (2.28)), and $\mathbf{P}(t)$ and $x(t)$ are corresponding solutions of the Heisenberg equations of motion in the quasiclassical approximation, when all the involved operators are commutative. The set of states $|f\rangle$ (7.2) characterizes the particle states after the scattering (disregrading the radiation).

The main contribution to the integral in Eq. (7.4) comes from the region where $\varepsilon kx(t)/\varepsilon - \omega) \sim 1$ yielding the estimate for the radiation formation time $t \sim t_c = 2\varepsilon(\varepsilon - \omega/m^2\omega$. Further discussion depends substantially on the value of the momentum transfer q. We can distinguish two characteristic regions of q values, depending on the ratio of the quantities q and q_m, where q_m

$$q_m = \frac{1}{t_c} = \frac{\omega m^2}{2\varepsilon(\varepsilon - \omega)} \tag{7.5}$$

is the minimum momentum transfer. We have in the first region $q \gg q_m$, and in the second $q \sim q_m$. Taking into account the relation $q\rho \sim 1$ (ρ is the impact-parameter), we have in the first region $t_c \gg \rho$, and in the second $t_c \sim \rho$. It clarifies the meaning of this division. Since the scattering time is $t \sim \rho$, and an electron is almost on the mass shell (with an accuracy to terms of order γ^{-2}) the radiation formation time in the first region is much larger than the scattering time, i.e. the scattering in this region is independent of the radiation process. Then the particle trajectory can be represented in the form of an "angle":

$$\begin{aligned} \mathbf{P}(t) &= \mathbf{P}(-\infty)\,, \quad t < 0 \\ \mathbf{P}(t) &= \mathbf{P}(+\infty)\,, \quad t > 0\,. \end{aligned} \tag{7.6}$$

In this case we have from Eq. (7.4)

$$M = \frac{ie(\varepsilon - \omega)}{2\pi\varepsilon\sqrt{\omega}}\left[\frac{R(+\infty)}{kv(+\infty)} - \frac{R(-\infty)}{kv(-\infty)}\right] \tag{7.7}$$

The matrix element in Eq. (7.3) has the form

$$M(\mathbf{q}) = \langle f|M|i\rangle\,, \tag{7.8}$$

where, the final and initial particle states considering the field action within our approximation are

$$
\begin{aligned}
|i\rangle &= U_{\rm in}|\mathbf{p}_i\rangle\,, \\
|f\rangle &= U_{\rm out}|\mathbf{p}_f\rangle\,, \\
U_{\rm in,out} &= \exp\left\{\mp i\int_0^\infty V(\rho,t)dt\right\}
\end{aligned}
\tag{7.9}
$$

Taking into account the relations

$$
\begin{aligned}
\langle f|F(\mathbf{P}(\infty)) &= \langle f|U_{\rm out}F(\mathbf{P})U_{\rm out}^{-1} \\
&= \langle \mathbf{p}_f|F(\mathbf{P})U_{\rm out}^{+} = F(\mathbf{p}_f)\langle f|\,, \\
F(\mathbf{P}(-\infty))|i\rangle &= U_{\rm in}F(\mathbf{P})U_{\rm in}^{-1}|1\rangle \\
&= U_{\rm in}F(\mathbf{P})|\mathbf{p}_i\rangle = F(\mathbf{p}_i)|i\rangle\,,
\end{aligned}
\tag{7.10}
$$

we have

$$
M(\mathbf{q}) = [F(\mathbf{p}_f) - F(\mathbf{p}_i)]\langle f|i\rangle\,,
\tag{7.11}
$$

where the quantity $\langle f|i\rangle$ determines the scattering amplitude. Corrections to the phase and scattering matrix element and the corresponding correction to the bremsstrahlung cross section were calculated by Baier and Katkov, 1976 by means of the operator method.

The factorization of Eq. (7.11) yields the multiplicativity of the cross section for radiation in the region of momentum transfer $q \gg q_m$:

$$
d\sigma_\gamma(\mathbf{k},\mathbf{q}) = W_\gamma(\mathbf{k},\mathbf{q})d^3k d\sigma(\mathbf{q})\,,
\tag{7.12}
$$

where $d\sigma(\mathbf{q})$ is the elastic scattering cross section with the momentum transfer $\mathbf{q}$. The expression for the quantity $W_\gamma(\mathbf{k},\mathbf{q})$ has a universal form regardless of the specific type of field:

$$
W_\gamma(\mathbf{k},\mathbf{q}) = \frac{\alpha}{(2\pi)^2\omega}\frac{\varepsilon-\omega}{\varepsilon}\left|\left[\frac{R(\mathbf{p}_f)}{kv_f} - \frac{R(\mathbf{p}_i)}{kv_i}\right]\right|^2
\tag{7.13}
$$

In the region $q \sim q_m$ we cannot use the trajectory in the form of an "angle" but as far as the momentum transfer in this region is $q \ll m$, we can expand the formula (7.4) in the quantity

$$
\mathbf{q}_\rho(t) = -\frac{\boldsymbol{\rho}}{\rho}\frac{d}{d\rho}\int_{-\infty}^t V(\sqrt{\rho^2+t^2})dt\,,
\tag{7.14}
$$

where $V(r)$ is the potential energy of a particle in the field. Retaining the linear terms of this expansion and integrating by parts in Eq. (7.4), we obtain at $q \ll m$

$$M(\mathbf{q}) = \mathbf{D}\mathbf{A}(\mathbf{q}), \tag{7.15}$$

where

$$\mathbf{D} = \frac{1}{2\pi}\sqrt{\frac{\alpha(\varepsilon-\omega)}{\varepsilon\omega}}\frac{\partial}{\partial \mathbf{p}_i}\left(\frac{R_i}{kv_i}\right),$$

$$\mathbf{A}(\mathbf{q}) = \frac{1}{2\pi i}\int d^2\rho \exp\{i[\chi(\rho)+\mathbf{q}\rho]\} \times \frac{d}{d\boldsymbol{\rho}}\int_{-\infty}^{\infty} V(\sqrt{\rho^2+t^2})e^{iq_{\parallel}t}dt. \tag{7.16}$$

Here $\chi(\rho)$ is the scattering phase in the potential $V(r)$

$$\chi(\rho) = -\int_{-\infty}^{\infty} V(\sqrt{\rho^2+t^2})dt, \quad q_{\parallel} = \frac{\varepsilon kv_i}{\varepsilon-\omega} = \frac{kp_i}{\varepsilon-\omega} \tag{7.17}$$

In the region $m \gg q \gg q_m$ we have

$$\mathbf{A}(\mathbf{q}) = \frac{1}{2\pi}\int d^2\rho e^{i\mathbf{q}\rho}\frac{d}{d\boldsymbol{\rho}}e^{i\chi(\rho)} = \mathbf{q}_{\perp}\langle f|i\rangle, \tag{7.18}$$

and the cross section for radiation takes the form

$$d\sigma_{\gamma}(\mathbf{k},\mathbf{q}) = \frac{\alpha}{(2\pi)^2\omega}\cdot\frac{\varepsilon-\omega}{\varepsilon}\left[\mathbf{q}_{\perp}\frac{\partial}{\partial \mathbf{p}_i}\frac{R_i}{kv_i}\right]^2 d^3k\, d\sigma(\mathbf{q}). \tag{7.19}$$

Thus, taking into account that in this region of momentum transfer q, the expressions (7.19) and (7.12) coincide, we can use the following expression

$$d\sigma_{\gamma}(\mathbf{k},\mathbf{q}) = W_{\gamma}(\mathbf{k},\mathbf{q})d^3k|\mathbf{A}\mathbf{q}_{\perp}|^2\frac{d^2q_{\perp}}{q_{\perp}^4}, \tag{7.20}$$

for arbitrary values of q. In order to find the explicit form of the quantity $W_{\gamma}(\mathbf{k},\mathbf{q})$ we must substitute R in Eq. (7.13) and use the relation

$$kv_f = \frac{\omega}{\varepsilon}[\sqrt{(\mathbf{p}_2+\mathbf{k})^2+m^2} - (\mathbf{n},\mathbf{p}_2+\mathbf{k})] \simeq \varepsilon_2 kp_2/\varepsilon^2, \tag{7.21}$$

where $p_1 = (\varepsilon, \mathbf{p}_1)$ and $p_2 = (\varepsilon - \omega, \mathbf{p}_1 - \mathbf{q} - \mathbf{k})$; $\mathbf{p}_1 \equiv \mathbf{p}_i$. Using the explicit form of the quantity $R(4.3)$ and retaining the main terms of an expansion in γ^{-2}, we obtain (the Coulomb gauge is used)

$$W_\gamma(\mathbf{k}, \mathbf{q}) = \frac{\alpha L}{(2\pi)^2 \varepsilon_1 \varepsilon_2 \omega},$$

$$L = \frac{1}{4\varepsilon_1\varepsilon_2} |\varphi_2^+ \{(\varepsilon_1 + \varepsilon_2)\mathbf{e}^*\mathbf{I} + i(\varepsilon_1 - \varepsilon_2)(\mathbf{e}^*[\sigma\mathbf{I}])\}\varphi_1|^2, \quad (7.22)$$

where

$$\mathbf{I} = \frac{\varepsilon_2[\mathbf{p}_1 - (\varepsilon_1 + m)\mathbf{n}]}{kp_1} - \frac{\varepsilon_1[\mathbf{p}_2 - (\varepsilon_2 + m)\mathbf{n}]}{kp_2} - \mathbf{n}. \quad (7.23)$$

We introduce the variables

$$\mathbf{u} = \frac{\mathbf{p}_1 - \mathbf{n}(\mathbf{n}\mathbf{p}_1)}{m} = \frac{\mathbf{p}_{1\perp}}{m}, \qquad \mathbf{v} = \frac{\mathbf{p}_2 - \mathbf{n}(\mathbf{n}\mathbf{p}_2)}{m} = \frac{\mathbf{p}_{2\perp}}{m}, \quad (7.24)$$

in terms of which the quantity $\mathbf{I}$ takes the form

$$\mathbf{I} = \frac{m}{q_m}\left[\frac{\mathbf{u} - \mathbf{n}}{\zeta} - \frac{\mathbf{v} - \mathbf{n}}{\eta}\right], \quad \mathbf{q}_\perp = m(\mathbf{u} - \mathbf{v}), \quad (7.25)$$

where with relativistic accuracy, we have

$$\zeta = \frac{2\varepsilon_1}{\omega m^2} kp_1 = 1 + u^2, \qquad \eta = \frac{2\varepsilon_2}{\omega m^2} kp_2 = 1 + v^2. \quad (7.26)$$

Summing over polarizations of final particles and averaging over the spin of the initial particle, we obtain

$$L = \mathbf{I}^2\left(1 + \frac{\omega^2}{2\varepsilon_1\varepsilon_2}\right) - (\mathbf{n}\mathbf{I})^2, \qquad \mathbf{I}^2 = \frac{\mathbf{q}_\perp^2}{q_m^2 \zeta\eta}, \quad (\mathbf{n}\mathbf{I})^2 = \frac{(\zeta - \eta)^2 m^2}{(\zeta\eta q_m)^2}. \quad (7.27)$$

The spin and polarization effects are considered in detail in the book by Baier, Katkov and Fadin, 1973.

7.2. Bremsstrahlung Cross Section in an Unscreened Coulomb Field

For Coulomb potential of a nucleus we have

$$V(r) = \xi/r\,, \quad \xi = Z\alpha\,,$$

$$\chi(\rho) \simeq -\int_{-a}^{a} V(\sqrt{\rho^2+t^2})dt \simeq 2\xi \ln\frac{\rho}{a}\,, \tag{7.28}$$

where a is the regularization parameter, which is assumed to be much larger than any characteristic dimension. The cross section turns out to be independent of this parameter.

As it was noted, the factor $W_\gamma(\mathbf{k},\mathbf{q})$ in the general expression for the cross section (7.20) is independent of the field, and therefore we must calculate only the quantity $\mathbf{A}\mathbf{q}_\perp/q_\perp^2$ (see Eq. (7.16)). Taking into account the relation

$$\begin{aligned}\int_{-\infty}^{\infty} e^{iq_\parallel t}\frac{d}{d\rho}V(\sqrt{\rho^2+t^2})dt &= -\xi\int_{-\infty}^{\infty}\frac{\rho e^{iq_\parallel t}}{(\rho^2+t^2)^{3/2}}dt \\ &= -2\xi q_\parallel K_1(q_\parallel \rho)\,,\end{aligned} \tag{7.29}$$

where K_1 is the Macdonald function and using the explicit form of the scattering phase in a Coulomb field, we obtain (see Gradshteyn and Ryzhik, 1965)

$$\begin{aligned}\frac{1}{q_\perp^2}\mathbf{A}\mathbf{q}_\perp &= -\frac{2\xi q_\parallel}{q_\perp}\int_0^\infty \left(\frac{\rho}{a}\right)^{2i\xi} J_1(q_\perp\rho)K_1(q_\parallel\rho)\rho d\rho \\ &= -\frac{2\xi}{q^2}\left(\frac{2}{aq}\right)^{2i\xi}(1+i\xi)\Gamma^2(1+i\xi) \\ &\quad \times F(-i\xi, 1+i\xi; 2; q_\perp^2/q^2)\,,\end{aligned} \tag{7.30}$$

where $q^2 = q_\perp^2 + q_\parallel^2$; $F(a,b,c;z)$ is the hypergeometric function. We used the transformation of the hypergeometric function from an argument $1-y^{-1}$ to an argument $1-y$:

$$F(2+i\xi, 1+i\xi; 2; 1-y^{-1}) = y^{1+2i\xi}F(-i\xi, 1+i\xi; 2; 1-y)\,. \tag{7.31}$$

For this function we have

$$(1+i\xi)F(-i\xi, 1+i\xi; 2; 1-y) = V(y) + i\xi W(y)\,, \tag{7.32}$$

where

$$V(y) = F(i\xi, -i\xi; 1; 1-y) = V^*(y);$$

$$W(y) = \frac{1}{\xi^2}\frac{dV(y)}{dy} \tag{7.33}$$

It is convenient to use the variable $y = (kp_1)(kp_2)/\varepsilon_1\varepsilon_2 q^2$ which at $q \sim q_m$ is approximately equal to

$$y \simeq \frac{q_\parallel^2}{q^2} = 1 - \frac{q_\perp^2}{q^2} \tag{7.34}$$

Hence we have in this region

$$\frac{1}{q_\perp^4}|\mathbf{A}\mathbf{q}_\perp|^2 = \frac{4\xi^2}{q^4}\Phi(y),$$
$$\Phi(y) = V^{-2}(0)[V^2(y) + \xi^2 y^2 W^2(y)], \tag{7.35}$$

where $V^{-2}(0) = \Gamma^2(1-i\xi)\Gamma^2(1+i\xi) = (\pi\xi/\sinh\pi\xi)^2$ We have $\Phi(y) \simeq 1$ for $q_\perp \gg q_\parallel$, i.e. for $y \ll 1$.

To calculate the integral cross sections, in particular, the spectrum of radiated photons it is convenient to use the variables y, ζ, η in terms of which the cross section takes the form

$$d\sigma_\gamma = \frac{2\alpha\xi^2\varepsilon_2 d\omega dy d\zeta d\eta}{\pi m^2\varepsilon_1\omega S\zeta\eta}\Phi(y)$$
$$\times\left[\left(\frac{\varepsilon_1}{\varepsilon_2} + \frac{\varepsilon_2}{\varepsilon_1}\right)\left(\frac{1}{y} - 1\right) - \frac{2(\zeta-\eta)^2}{\zeta^2\eta^2\delta^2}\right], \tag{7.36}$$

where

$$S = [2\lambda\eta\zeta(\zeta+\eta-2) - (\lambda\eta\zeta)^2 - (\zeta-\eta)^2]^{1/2},$$
$$\lambda = \delta^2\left(\frac{1}{y} - 1\right), \quad \delta = \frac{q_m}{m} = \frac{\omega m}{2\varepsilon_1\varepsilon_2}. \tag{7.37}$$

The integration with respect to η is carried out between the zeros of the function S. The integrals involved in the considered cross section are:

$$\int d\eta(S\eta)^{-1} = \pi/\zeta,$$

$$\int d\eta S^{-1}\eta^{-2} = \pi[1 + \lambda(\zeta-2)]\zeta^{-2}, \tag{7.38}$$

$$\int d\eta S^{-1}\eta^{-3} = \pi[(1-\lambda\zeta)^2 - 6\lambda(1-\lambda)(\zeta-1)]\zeta^{-3}.$$

The integrals with respect to y have the form

$$\int_{\delta^2}^{1} \Phi(y)\frac{\lambda}{\delta^2}dy = \int_{\delta^2}^{1} \Phi(y)\frac{1-y}{y}dy\,, \quad \int_{\delta^2}^{1} \Phi(y)\left(\frac{\lambda}{\delta}\right)^2 dy\,.$$

Using the hypergeometric equation, it can be easily shown that

$$\int_{\delta^2}^{1} \Phi(y)\frac{1-y}{y}dy = V^{-2}(0)(1-y)[VW + V^2 - y(1-y)\xi^2 W^2]_{\delta^2}^{1}$$

$$\simeq -2\left(\ln\delta + \frac{1}{2} + f(\xi)\right)\,, \tag{7.39}$$

where

$$f(\xi) = \frac{1}{2}[\psi(1+i\xi) + \psi(1-i\xi) - 2\psi(1)]$$

$$= \sum_{n=1}^{\infty}(-1)^{n+1}\zeta(2n+1)\xi^{2n} = \xi^2\sum_{n=1}^{\infty}\frac{1}{n(n^2+\xi^2)} \tag{7.40}$$

Here $\zeta(2n+1)$ is the Riman zeta-function, $\psi(x) = (\ln\Gamma(x))'$ Only small values of $y \sim \delta^2$ contribute to the other integral, so that

$$\int_{\delta^2}^{1} \Phi(y)\left(\frac{\lambda}{\delta}\right)^2 dy \simeq \delta^2\int_{\delta^2}^{1}\frac{dy}{y^2} \simeq 1\,. \tag{7.41}$$

Taking trivial integrals with respect to ζ, we finally obtain the bremsstrahlung spectrum at electron scattering on nucleus

$$d\sigma_\gamma = \frac{4Z^2\alpha^3\varepsilon_2 d\omega}{m^3\varepsilon_1\omega}\left(\frac{\varepsilon_1}{\varepsilon_2} + \frac{\varepsilon_2}{\varepsilon_1} - \frac{2}{3}\right)\left(\ln\frac{1}{\delta} - \frac{1}{2} - f(\xi)\right) \tag{7.42}$$

For $\xi \to 0$, $f(\xi) \to 0$ and we obtain the radiation spectrum in the Born approximation. For $\xi \gg 1$ we have $f(\xi) \simeq \ln\xi + C$ ($C = 0.577\ldots$ is the Euler constant). In this limit a purely classical description of the scattering is valid ($Z\alpha = Ze^2/\hbar c \gg 1$ at $\hbar \to 0$).

7.3. Bremsstrahlung in a Screened Coulomb Field

It can be shown (Olsen and Maximon, 1959) that the difference between the bremsstrahlung cross section that is exact in ξ and that calculated in the Born approximation is independent of the screening. In other words, the screening and higher order corrections can be taken into account independently. In the presence of a screening the factor $(1-F(q))/q^2$ (where $F(q)$ is the form-factor of atomic electrons) appear in place of the quantity q^{-2} in the scattering amplitude. This yields the factor $(1-F(q))^2/q^4$ instead of q^{-4} in the bremsstrahlung cross section. Hence, the difference of screened and unscreened cross sections which we have to calculate in the Born approximation, contains the factor $\{(1-F(q))^2-1\}/q^4$. This quantity is substantially different from zero only in the region of small momentum transfers q, whereas q can be represented in the form of $q = m\delta\zeta y^{-1/2}$ since $\zeta \sim \eta$ in this region. The integration with respect to η is therefore carried out as in (7.38). The following integral appears instead of the integral (7.39):

$$\begin{aligned}\mathcal{F}(\zeta\delta) &= \frac{1}{2}\int_0^1 \frac{1-y}{y}\left[\left(1-F\left(\frac{m\zeta\delta}{\sqrt{y}}\right)\right)^2-1\right]dy \\ &= \int_{\zeta\delta}^{\infty}[(1-F(mx))^2-1]\frac{x^2-(\zeta\delta)^2}{x^3}dx\,.\end{aligned} \tag{7.43}$$

Integrating the term $\{1-F(mx)\}^2$, we can assume the lower limit to be zero, since $\{1-F(mx)\} \propto x$ and the integral converges at its lower limit.

To calculate explicitly the quantity $\mathcal{F}(\zeta\delta)$, we use the atomic form-factor in the Moliere approximation:

$$\frac{1-F(q)}{q^2} = \sum_{i=1}^{3}\frac{\alpha_i}{\beta_i^2+q^2}\,, \tag{7.44}$$

where

$$\begin{aligned}&\alpha_1 = 0.10\,, \quad \alpha_2 = 0.55\,, \quad \alpha_3 = 0.35\,; \\ &\beta_i = 8.25\cdot 10^{-3}Z^{1/3}b_i m\,; \\ &b_1 = 6.0\,, \quad b_2 = 1.2\,, \quad b_3 = 0.3\,.\end{aligned} \tag{7.45}$$

Substituting (7.44) in (7.43), we obtain after simple calculations

$$\begin{aligned}\mathcal{F}(\zeta\delta) = &-\frac{1}{2}\sum_{i=1}^{3}\alpha_i^2(1+B_i) \\ &+\sum_{i\neq j}^{3}\alpha_i\alpha_j\left\{[(1+B_j)/(B_i-B_j)]\ln(1+B_j)+\frac{1}{2}\right\}\,,\end{aligned} \tag{7.46}$$

where $B_i = (\beta_i/\zeta\delta\, m)^2$. In the case of a full screening ($B_i \gg 1$) which means that the screening radius is much smaller than the distance corresponding to a minimum momentum transfer, we obtain from Eq. (7.46) the following expression:

$$\mathcal{F}(\zeta\delta) = \ln\,(111\, Z^{-1/3}\zeta\delta)\,. \tag{7.47}$$

At last, for the bremsstrahlung spectrum in a Coulomb field at arbitrary screening and taking into account the higher-order Coulomb corrections, we have

$$d\sigma_\gamma = \frac{4Z^2\alpha^3}{m^2\varepsilon_1^2}\frac{d\omega}{\omega}\left[(\varepsilon_1^2+\varepsilon_2^2)\psi_1 - \frac{2}{3}\varepsilon_1\varepsilon_2\psi_2\right], \tag{7.48}$$

where

$$\begin{aligned}\psi_1 &= \ln\frac{1}{\delta} - \frac{1}{2} - f(\xi) + \int_1^\infty \frac{\mathcal{F}(\zeta\delta)}{\zeta^2}d\zeta\,,\\ \psi_2 &= \ln\frac{1}{\delta} - \frac{1}{2} - f(\xi) + 6\int_1^\infty \frac{\mathcal{F}(\zeta\delta)}{\zeta^4}(\zeta-1)d\zeta\,,\\ \delta &= \frac{\omega m}{2\varepsilon_1\varepsilon_2}\,.\end{aligned} \tag{7.49}$$

Substituting in Eqs. (7.49) and (7.48) the quantity $\mathcal{F}(\zeta\delta)$ (7.47), we obtain in the case of a full screening

$$\begin{aligned}\psi_1 &= \ln\,(111\, Z^{-1/3}) + \frac{1}{2} - f(\xi) = \ln\,(183\, Z^{-1/3}) - f(\xi)\,,\\ \psi_2 &= \ln\,(111\, Z^{-1/3}) + \frac{1}{3} - f(\xi) = \psi_1 - \frac{1}{6}\end{aligned} \tag{7.50}$$

The spectrum of radiated photons in this case is

$$d\sigma_\gamma^{sc} = \frac{4Z^2\alpha^3}{m^2\varepsilon_1^2}\cdot\frac{d\omega}{\omega}\cdot\left\{\left(\varepsilon_1^2+\varepsilon_2^2-\frac{2}{3}\varepsilon_1\varepsilon_2\right)\left[\ln\,(183\, Z^{-1/3}) - f(\xi)\right] + \frac{1}{9}\varepsilon_1\varepsilon_2\right\} \tag{7.51}$$

Mean energy losses per unit length of an electron travelling in a medium of atom density n are characterized by the radiation length $L_{\rm rad}$

$$L_{\rm rad}^{-1} = n\int_0^{\varepsilon_1}\frac{\omega}{\varepsilon_1}\cdot\frac{d\sigma_\gamma}{d\omega}\cdot d\omega\,. \tag{7.52}$$

In the absence of a screening we find, using Eq. (7.42)

$$L_{\rm rad}^{-1} = \frac{4Z^2\alpha^3 n}{m^2}\left(\ln\frac{2\varepsilon_1}{m} - \frac{1}{3} - f(\xi)\right), \tag{7.53}$$

and in the case of a full screening we obtain from Eq. (7.51)

$$L_{\rm rad}^{-1} = \frac{4Z^2\alpha^3 n}{m^2}\left[\ln\left(183\, Z^{-1/3}\right) - f(\xi) + \frac{1}{18}\right] \tag{7.54}$$

7.4. Radiation from Relativistic Particles Colliding in a Medium in the Presence of an External Field

At very high energies, the formation length of basic processes of quantum electrodynamics turns out to be very large, and the processes come to depend on the medium in which they are played out, as well as on external fields. Landau and Pomeranchuk, 1953, were the first to draw attention to the fact that if the formation length for bremsstrahlung becomes comparable to the distance over which multiple scattering becomes important, the bremsstrahlung will be suppressed. Migdal, 1955, 1957, developed a quantitative theory of this phenomenon. As it was shown by Baier and Katkov, 1972, external fields can affect the bremsstrahlung as well. The indicated effects become especially apparent in two physical phenomena under active study at present. One of them is the radiation from high energy electrons traversing aligned single crystals, and the second is the radiation from the collision of e^+e^- beams in the so-called linear colliders (at particle energies of hundreds of GeV or more). In this section the theoretical description of the bremsstrahlung taking into account a medium and an external field action is stated according to Baier, Katkov and Strakhovenko, 1987c.

Consider the effect of an external field on the bremsstrahlung process. This effect is associated with a reduction in the formation length of a photon (either real or virtual) due to the relatively large rotation of particle velocity over this length, and the corresponding increase of the radiation angle's cone. If a photon of frequency Ω is emitted by an electron (position) of energy ε at an angle ϑ to its velocity, the formation length of such a photon will be given by (see Eq. (6.1))

$$l_\Omega \sim \frac{\varepsilon - \Omega}{\varepsilon\Omega\left(\frac{1}{\gamma^2} + \vartheta^2\right)} = \frac{\gamma \lambda_c}{u_\Omega(1+\gamma^2\vartheta^2)}\,, \tag{7.55}$$

where $u_\Omega \equiv \Omega/(\varepsilon - \Omega)$. The characteristic radiation angles in weak fields are $\vartheta \sim 1/\gamma$, and we can neglect the influence of the external field if

$$\dot{v} l_\Omega = \frac{eH}{\varepsilon} l_\Omega \ll \frac{1}{\gamma}\,. \tag{7.56}$$

Substituting (7.55) into (7.56), we have the criterion for a field to be weak:

$$\gamma\frac{eH}{\varepsilon}\cdot\frac{\gamma^2}{\varepsilon u_\Omega}=\frac{\chi}{u_\Omega}\ll 1\,. \tag{7.57}$$

In strong fields, where $\chi/u \gg 1$, with characteristic radiation angles $\vartheta \gg 1/\gamma$, the effective radiation angle ϑ_{ef} is determined by a self-consistency argument: the deviation angle of the particle in the field over the photon formation length must not exceed ϑ_{ef}, i.e. (cf. Eq. (6.2))

$$\dot{v}l_\Omega(\vartheta_{ef})\sim\vartheta_{ef}\,,$$

$$\vartheta_{ef}\sim\frac{1}{\gamma}\left(\frac{\chi}{u_\Omega}\right)^{1/3}=\left(\frac{eH}{\varepsilon^2u_\Omega}\right)^{1/3}, \tag{7.58}$$

$$l_\Omega(\vartheta_{ef})\sim\frac{\gamma}{mu_\Omega}\left(\frac{u_\Omega}{\chi}\right)^{2/3}=\left(\frac{\varepsilon}{u_\Omega(eH)^2}\right)^{1/3}.$$

It can be seen from (7.58) that when $\chi/u \gg 1$ neither the characteristic radiation angle nor the photon formation length depends on the mass of radiating particle.

It follows from (7.58) that, for $\chi/u_\Omega \gg 1$, the characteristic photon formation length is reduced by a factor $(\chi/u_\Omega)^{2/3}$, and the radiation angles are increased by a factor $(\chi/u_\Omega)^{1/3}$. Since the relevant parameter χ/u_Ω depends on the photon frequency, this effect is manifested earlier for soft photons. This is the reason why the field has a significant effect on virtual photon emission in e^+e^- collisions ($\Omega\simeq q_0'=\omega m^2/4\varepsilon(\varepsilon-\omega)$, where ω is the frequency of the real photon) even for moderate fields and at relatively low particle energies, when the parameter $\chi(\varepsilon-\omega)/\omega\equiv\chi/u$ is small, and $\chi(\varepsilon-q_0')/q_0'\simeq 4\gamma^2\chi/u$ is large. In this case, the features of a real photon emission do not change directly, but a significant change takes place in the virtual photon spectrum at momentum transfers $|q|\le q_{\rm min}(4\gamma^2\chi/u)^{1/3}$, increasing the lower bound on the effective momentum transfer and resulting in a corresponding decrease of the cross section. It follows from Eq. (7.57), that the radiation block describing a virtual photon absorption and a real photon emission changes significantly at $\chi/u\ge 1$. Since the formation length of a real photon l_ω falls off as $(u/\chi)^{2/3}$ for $\chi/u \gg 1$ (see Eq. (7.58)), the bremsstrahlung cross section falls off in just the same way. In further calculations, we neglect the polarization of a medium, which is important only for the emission of very soft photons, and which, moreover, is diminished by a factor $(u/\chi)^{2/3}$ when $\chi/u \gg 1$.

We start out with the quasiclassical theory of radiation, developed in Secs. 2 and 4. In this theory the emission probability is

$$dw = \frac{\alpha}{(2\pi)^2}\frac{d^3k}{\omega}\int dt_1 dt_2 R^*(t_2)\,R(t_1)$$
$$\times \exp\left\{-\frac{i\varepsilon}{\varepsilon-\omega}[kx(t_2)-kx(t_1)]\right\},$$

where $k=(\omega,\mathbf{k})$ is the four-momentum of the photon, $k^2=0, x(t)=(t,\mathbf{r}(t)), t$ is the time, and $\mathbf{r}(t)$ is the particle coordinate on a classical trajectory. For 1/2 spin particles, we have to relativistic accuracy

$$\begin{aligned}
R(t) &= \varphi_f^+(A+i\boldsymbol{\sigma}\mathbf{B})\varphi_i\,,\\
A &= \frac{1}{2}\left(1+\frac{\varepsilon}{\varepsilon-\omega}\right)\mathbf{ev} \simeq \frac{1}{2}\left(1+\frac{\varepsilon}{\varepsilon-\omega}\right)\mathbf{e}\boldsymbol{\vartheta}\,,\\
\mathbf{B} &= \frac{\omega}{2(\varepsilon-\omega)}[\mathbf{eb}]\,,\\
\mathbf{b} &= \mathbf{n}-\mathbf{v}+\frac{\mathbf{n}}{\gamma}\simeq -\boldsymbol{\vartheta}+\frac{\mathbf{n}}{\gamma}\,,
\end{aligned} \tag{7.59}$$

where the angle $\boldsymbol{\vartheta}=(\mathbf{v}-\mathbf{n}(\mathbf{nv}))/v\simeq \mathbf{v}_\perp$, and $\mathbf{v}_\perp$ is the component of particle velocity perpendicular to the vector $\mathbf{n}=\mathbf{k}/\omega$. If we are not interested in the initial and final particle polarizations, then

$$\begin{aligned}
R^*(t_2)R(t_1) &\to \frac{1}{2(\varepsilon-\omega)^2}\left[\frac{\omega^2}{\gamma^2}+(\varepsilon^2+(\varepsilon-\omega)^2)\boldsymbol{\vartheta}\boldsymbol{\vartheta}'\right]\\
&\equiv \frac{1}{2}\mathcal{L}(\boldsymbol{\vartheta}',\boldsymbol{\vartheta})\,,
\end{aligned} \tag{7.60}$$

where the notation is introduced $\boldsymbol{\vartheta}'=\boldsymbol{\vartheta}(t_2), \boldsymbol{\vartheta}=\boldsymbol{\vartheta}(t_1)$. If the particle moves along a definite trajectory, then by substituting the classical values of $\mathbf{r}(t)$ and $\mathbf{v}(t)$ into Eqs. (7.58) – (7.59), we obtain the desired probability for this process. If the scattering occurs, Eq. (7.58) must be averaged over all possible particle trajectories. This operation is performed with the aid of the distribution function, averaged over atomic positions in the scattering medium and satisfying the kinetic equation with the external field (particle acceleration) taken into account. The emission probability per unit time is then

$$dW \equiv \left\langle \frac{dw}{dt} \right\rangle = \frac{\alpha}{(2\pi)^2} \frac{d^3k}{\omega} \operatorname{Re} \int_0^\infty d\tau \exp\left(-i\frac{\varepsilon}{\varepsilon'}\omega\tau\right)$$
$$\times \int d^3v d^3v' d^3r d^3r' \mathcal{L}(\vartheta', \vartheta) F_i(\mathbf{r}, \mathbf{v}, t)$$
$$\times F_f(\mathbf{r}', \mathbf{v}', \tau; \mathbf{r}, \mathbf{v}) \cdot \exp\left\{i\frac{\varepsilon}{\varepsilon'}(\mathbf{k}, \mathbf{r}' - \mathbf{r})\right\},$$
$$\varepsilon' = \varepsilon - \omega\,. \tag{7.61}$$

The distribution function F in (7.61) satisfies the kinetic equation

$$\frac{\partial F(\mathbf{r}, \mathbf{v}, t)}{\partial t} + \mathbf{v}\frac{\partial F(\mathbf{r}, \mathbf{v}, t)}{\partial \mathbf{r}} + \mathbf{w}\frac{\partial F(\mathbf{r}, \mathbf{v}, t)}{\partial \mathbf{v}}$$
$$= n \int \sigma(\mathbf{v}, \mathbf{v}')[F(\mathbf{r}, \mathbf{v}', t) - F(\mathbf{r}, \mathbf{v}, t)] d^3v'\,, \tag{7.62}$$

where n is the density of atoms in the medium, and $\sigma(\mathbf{v}, \mathbf{v}')$ is the scattering cross section. The normalization condition

$$\int d^3r d^3v F(\mathbf{r}, \mathbf{v}, t) = 1 \tag{7.63}$$

should also be satisfied, as well as the intial condition for F_f

$$F_f(\mathbf{r}', \mathbf{v}', 0; \mathbf{r}, \mathbf{v}) = \delta(\mathbf{r} - \mathbf{r}')\delta(\mathbf{v} - \mathbf{v}')\,.$$

In Eq. (7.61), we integrate over $d^3r d^3(r' - r)$, taking advantage of the fact that $F_f(\mathbf{r}', \mathbf{v}', \tau; \mathbf{r}, \mathbf{v})$ can only depend on the coordinate difference $\mathbf{r}' - \mathbf{r}$:

$$dW = \frac{\alpha}{(2\pi)^2} \frac{d^3k}{\omega} \operatorname{Re} \int_0^\infty d\tau \int d^3v d^3v'$$
$$\times \mathcal{L}(\vartheta', \vartheta) F_i(\mathbf{v}, t) F_k(\mathbf{v}', \tau; \mathbf{v})\,, \tag{7.64}$$

where

$$F_i(\mathbf{v}, t) = \int d^3r F_i(\mathbf{r}, \mathbf{v}, t)\,;$$

$$F_k(\mathbf{v}', \tau; \mathbf{v}) = \exp\left(-\frac{i\varepsilon\omega\tau}{\varepsilon'}\right) \int d^3r' \exp\left\{i\frac{\varepsilon}{\varepsilon'}(\mathbf{k}, \mathbf{r}' - \mathbf{r})\right\}$$
$$\times F_f(\mathbf{r}', \mathbf{v}', \tau; \mathbf{r}, \mathbf{v})\,. \tag{7.65}$$

Here

$$F_k(\mathbf{v}', \tau; \mathbf{v}) = U(\boldsymbol{\vartheta}', \boldsymbol{\vartheta}; \tau)\delta(|\mathbf{v}'| - |\mathbf{v}|),$$
$$U(\boldsymbol{\vartheta}, \boldsymbol{\vartheta}; 0) = \delta(\boldsymbol{\vartheta}' - \boldsymbol{\vartheta}), \tag{7.66}$$

if it is assumed that in ultrarelativistic limit $\mathbf{wv} \sim 0(\gamma^{-3}), \mathbf{w} \simeq \mathbf{w}_\perp \simeq \mathbf{w} - \mathbf{v}(\mathbf{vw})$, and the scattering cross section is

$$\sigma(\mathbf{v}, \mathbf{v}') = \delta(|\mathbf{v}'| - |\mathbf{v}|)\sigma(\boldsymbol{\vartheta}, \boldsymbol{\vartheta}'). \tag{7.67}$$

Using Eq. (7.62) for F, we obtain the following equation for $U(\boldsymbol{\vartheta}', \boldsymbol{\vartheta}; \tau)$

$$\frac{\partial U}{\partial \tau} + i\frac{\omega\varepsilon}{2\varepsilon'}\left(\frac{1}{\gamma^2} + \vartheta'^2\right)U + \mathbf{w}\frac{\partial U}{\partial \boldsymbol{\vartheta}'}$$
$$= n\int d^2\vartheta''\sigma(\boldsymbol{\vartheta}', \boldsymbol{\vartheta}'')[U(\boldsymbol{\vartheta}'', \boldsymbol{\vartheta}; \tau) - cU(\boldsymbol{\vartheta}', \boldsymbol{\vartheta}; \tau)]. \tag{7.68}$$

A similar equation for the $\boldsymbol{\vartheta}$-dependence of U can be obtained by letting $\boldsymbol{\vartheta}' \to \boldsymbol{\vartheta}$ and $\mathbf{w} \to -\mathbf{w}$. If the final state of the charged particle is of no interest, the probability (7.64) must be integrated over d^3v' The resulting emission probability per unit time, normalized to a single particle moving at a speed $\mathbf{v}$ is then

$$dW = \frac{\alpha}{(2\pi)^2}\frac{d^3k}{\omega}\,\mathrm{Re}\int_0^\infty d\tau \exp\left(-\frac{ia\tau}{2}\right)$$
$$\times\left[\left(\frac{\omega}{\gamma\varepsilon'}\right)^2 V_0(\boldsymbol{\vartheta}, \tau) + \left(1 + \frac{\varepsilon^2}{\varepsilon'^2}\right)\boldsymbol{\vartheta}\mathbf{V}(\boldsymbol{\vartheta}, \tau)\right], \tag{7.69}$$

where $a = \omega m^2/\varepsilon\varepsilon'$, and $V_\mu(\boldsymbol{\vartheta}, \tau)$ satisfies the equation

$$\frac{\partial V_\mu}{\partial \tau} + i\frac{b\vartheta^2}{2}V_\mu - \mathbf{w}\frac{\partial V_\mu}{\partial \boldsymbol{\vartheta}} = n\int d^2\vartheta'\sigma(\boldsymbol{\vartheta}, \boldsymbol{\vartheta}')[V_\mu(\boldsymbol{\vartheta}', \tau) - V_\mu(\boldsymbol{\vartheta}, \tau)]. \tag{7.70}$$

Here $b = \omega\varepsilon/\varepsilon'$; the initial conditions for V_μ are

$$V_0(\boldsymbol{\vartheta}, 0) = 1, \quad \mathbf{V}(\boldsymbol{\vartheta}, 0) = \boldsymbol{\vartheta} \tag{7.71}$$

If the scattering cross section $\sigma(\boldsymbol{\vartheta}, \boldsymbol{\vartheta}')$ depends solely on the angle difference $\boldsymbol{\vartheta} - \boldsymbol{\vartheta}'$, then Eq. (7.70) is most conveniently solved by Fourier transforming with respect to the variables $\boldsymbol{\vartheta}$:

$$\begin{aligned} \varphi_\mu(\mathbf{x},\tau) &= \frac{1}{(2\pi)^2}\int d^2\vartheta e^{i\boldsymbol{\vartheta}\mathbf{x}} V_\mu(\boldsymbol{\vartheta},\tau)\,, \\ V_\mu(\boldsymbol{\vartheta},\tau) &= \int d^2x e^{-i\boldsymbol{\vartheta}\mathbf{x}} \varphi_\mu(\mathbf{x},\tau)\,. \end{aligned} \tag{7.72}$$

From Eq. (7.70), with initial conditions (7.71), we have

$$\begin{aligned} &\frac{\partial\varphi_\mu(\mathbf{x},\tau)}{\partial\tau} - \frac{ib}{2}\Delta_x\varphi_\mu(\mathbf{x},\tau) + i\mathbf{w}\mathbf{x}\varphi_\mu(\mathbf{x},\tau) \\ &= n\int d^2\vartheta(e^{i\boldsymbol{\vartheta}\mathbf{x}} - 1)\sigma(\vartheta)\varphi_\mu(\mathbf{x},\tau) \\ &\equiv (2\pi)^2 n[\Sigma(x) - \Sigma(0)]\varphi_\mu(\mathbf{x},\tau)\,, \\ &\quad \varphi_0(\mathbf{x},0) = \delta(\mathbf{x})\,, \quad \boldsymbol{\varphi}(\mathbf{x},0) = -i\boldsymbol{\nabla}\delta(\mathbf{x})\,. \end{aligned} \tag{7.73}$$

If the angular dependence of the radiation distribution is of no interest, Eq. (7.69) must be integrated over the photon emission angles $\boldsymbol{\vartheta}$. Bearing in mind that $d^3k = \omega^2 d\omega d^2\vartheta$, and that

$$\begin{aligned} \frac{1}{(2\pi)^2}\int V_0(\boldsymbol{\vartheta},\tau)d^2\vartheta &= \varphi_0(0,\tau)\,, \\ \frac{1}{(2\pi)^2}\int \boldsymbol{\vartheta}\mathbf{V}(\boldsymbol{\vartheta},\tau)d^2\vartheta &= -i\boldsymbol{\nabla}\boldsymbol{\varphi}(\mathbf{x},\tau)|_{\mathbf{x}=0}\,, \end{aligned} \tag{7.74}$$

we obtain the following expression for the spectral distribution of emission probability per unit time:

$$\begin{aligned} \frac{dW}{d\omega} = \alpha\omega\,\mathrm{Re}\int_0^\infty d\tau \exp\left(-i\frac{a\tau}{2}\right)\Bigg[\left(\frac{\omega}{\gamma\varepsilon'}\right)^2\varphi_0(0,\tau) \\ - i\left(1 + \frac{\varepsilon^2}{\varepsilon'^2}\right)\boldsymbol{\nabla}\boldsymbol{\varphi}(0,\tau)\Bigg] \end{aligned} \tag{7.75}$$

We now introduce the function

$$\phi_\mu(\mathbf{x}) = \int_0^\infty d\tau \exp\left(-i\frac{a\tau}{2}\right)\varphi_\mu(\mathbf{x},\tau)\,, \tag{7.76}$$

which satisfies the equation

$$\begin{aligned} &\frac{i}{2}(a - b\Delta)\phi_\mu(\mathbf{x}) + i\mathbf{w}\mathbf{x}\phi_\mu(\mathbf{x}) \\ &\quad = (2\pi)^2 n[\Sigma(x) - \Sigma(0)]\phi_\mu(\mathbf{x}) - \varphi_\mu(\mathbf{x},0)\,. \end{aligned} \tag{7.77}$$

Making use of the solution (7.77), the emission probability spectrum may be put in the form

$$\frac{dW}{d\omega} = \alpha\omega \,\mathrm{Re}\left[\left(\frac{\omega}{\gamma\varepsilon'}\right)^2 \phi_0(0) - i\left(1+\frac{\varepsilon^2}{\varepsilon'^2}\right)\nabla\phi(0)\right]. \tag{7.78}$$

7.5. Influence of an External Field on the Radiation Process in a Screened Coulomb Potential

If a diffraction by the nucleus, which is significant at angles $\vartheta_{ef} \geq \vartheta_2 \simeq (Z^{1/3}\alpha\gamma)^{-1}$, is not taken into account, the scattering cross section in a screened Coulomb potential takes the form

$$\sigma(\vartheta) = \frac{4Z^2\alpha^2}{\varepsilon^2(\vartheta^2+\vartheta_1^2)^2}, \tag{7.79}$$

where $\vartheta_1 = \kappa/\varepsilon$, $\kappa = 1/R$, R is the screening radius, and its Fourier transform may be expressed in terms of the Macdonald function

$$\Sigma(x) = \frac{1}{(2\pi)^2}\int e^{i\vartheta\mathbf{x}}\frac{4Z^2\alpha^2}{\varepsilon^2(\vartheta^2+\vartheta_1^2)^2}d^2\vartheta$$
$$= \frac{1}{\pi}\left(\frac{Z\alpha}{\varepsilon}\right)^2 \cdot \frac{x}{\vartheta_1}K_1(x\vartheta_1). \tag{7.80}$$

Bearing in mind that contributions to the cross section come from $x \sim 1/\vartheta_{ef} \ll \vartheta_1^{-1}$, and expanding $K_1(x\vartheta_1)$ in powers of $x\vartheta_1$, we obtain from Eq. (7.73), the following equation for $\varphi_\mu(x,\tau)$:

$$\frac{\partial\varphi_\mu}{\partial\tau} - i\frac{b}{2}\Delta\varphi_\mu + i\mathbf{wx}\varphi_\mu$$
$$= 2\pi n\left(\frac{Z\alpha x}{\varepsilon}\right)^2\left(\ln\frac{x\vartheta_1}{2}+C-\frac{1}{2}\right)\varphi_\mu, \tag{7.81}$$

where $C = 0.577\ldots$ is Euler's constant. When the characteristic emission angles $\vartheta_{ef} \gg \vartheta_2 \simeq (Z^{1/3}\alpha\gamma)^{-1}$, no further "tuning" of the scattering angle to the emission angle is possible, and Eq. (7.81) simplifies considerably:

$$\frac{\partial\varphi_\mu}{\partial\tau} - i\frac{b}{2}\Delta\varphi_\mu + i\mathbf{wx}\varphi_\mu = 2\pi n\left(\frac{Z\alpha x}{\varepsilon}\right)^2\cdot\varphi_\mu\cdot\ln\frac{\vartheta_1}{\vartheta_2}. \tag{7.82}$$

Equation (7.81) can be expressed in the same form if we wish to obtain the emission probability to logarithmic accuracy. We must then take as ϑ_2 the characteristic emission angle ϑ_{ef}, which in weak fields and with no multiple scattering is $\sim 1/\gamma$. In strong fields $(\chi/u \gg 1)$, $\vartheta_{ef} \sim \gamma^{-1}(\chi/u)^{1/3}$, so that for estimates we can use the approximate formula

$$\vartheta_{ef} \sim \frac{1}{\gamma}\left[1+\left(\frac{\chi}{u}\right)^{1/3}\right], \tag{7.83}$$

and the equation for φ_μ then takes the form

$$\frac{\partial \varphi_\mu}{\partial \tau} - i\frac{b}{2}\Delta\varphi_\mu + i\mathbf{w}\mathbf{x}\varphi_\mu = -qx^2\varphi_\mu \tag{7.84}$$

where

$$q = 2\pi n\left(\frac{Z\alpha}{\varepsilon}\right)^2 \ln\frac{\vartheta_{ef}}{\vartheta_1} \tag{7.85}$$

And subsequent estimates of ϑ_{ef} require that a broadening of emission angles be taken into account, because of multiple scattering. This effect (the Landau–Pomeranchuk effect) can become dominant in a weak field, where $\chi/u \ll 1$. Let us introduce the variable $\zeta = 1 + \gamma^2\vartheta^2$ and estimate the change in the square of the particle deviation angle over the photon formation length $l \sim \varepsilon(\varepsilon-\omega)/m^2\omega\zeta = \gamma/mu\zeta$ owing to the external field and multiple scattering. We obtain as a result the following estimate for ζ_{ef}

$$\left(\frac{\chi}{u\zeta}\right)^2 + \frac{Z\alpha\omega_0^2\gamma}{m^2u\zeta}\ln\frac{\zeta}{\gamma^2\vartheta_1^2} \lesssim \zeta, \tag{7.86}$$

where $\omega_0^2 = 4\pi nZ\alpha/m$.

Equation (7.84) is of the same form as the Schrödinger equation for an oscillator in an external field, and with initial conditions (7.73) its solution takes the form of a Gaussian distribution. We therefore seek a solution of (7.84) in the form

$$\begin{aligned} \varphi_0 &= \exp[\alpha(\tau) + \beta(\tau)x^2 + \gamma(\tau)\mathbf{w}\mathbf{x}], \\ \varphi &= [A(\tau)\mathbf{x} + B(\tau)\mathbf{w}]\varphi_0 . \end{aligned} \tag{7.87}$$

Substituting this expression into Eq. (7.84) and solving the system of equations thus obtained with initial conditions (7.73), we have

$$\varphi_0 = \frac{qe^{f(\tau)}}{\pi\nu\sinh\nu\tau}, \quad \nu = \sqrt{2ibq}\,;$$

$$\varphi = \left(\frac{\nu\mathbf{x}}{b\sinh\nu\tau} + \frac{\mathbf{w}}{\nu}\tanh\frac{\nu\tau}{2}\right)\varphi_0\,; \tag{7.88}$$

$$f(\tau) = -\frac{qx^2}{\nu}\coth\nu\tau - \frac{w^2\tau}{4q} + \frac{w^2}{2\nu q}\tanh\frac{\nu\tau}{2} - \frac{i\mathbf{wx}}{\nu}\tanh\frac{\nu\tau}{2}$$

The solution (7.88) and Eq. (7.75) give the following expression for the emission probability spectrum per unit time:

$$\frac{dW}{d\omega} = \frac{\alpha}{\pi}\mathrm{Re}\frac{q\omega}{\nu}\int_0^\infty \frac{d\tau}{\sinh\nu\tau}\left[\left(\frac{\omega}{\gamma\varepsilon'}\right)^2 - \left(1+\frac{\varepsilon^2}{\varepsilon'^2}\right)\left(\frac{2i\nu}{b\sinh\nu\tau}\right.\right.$$
$$\left.\left. + \frac{w^2}{\nu^2}\tanh^2\frac{\nu\tau}{2}\right)\right]\exp\left(-\frac{ia\tau}{2} - \frac{w^2\tau}{4q} + \frac{w^2}{2\nu q}\tanh\frac{\nu\tau}{2}\right). \tag{7.89}$$

Integrating the term containing $\sinh^{-2}\nu\tau$ by parts, we have, to logarithmic accuracy

$$\frac{dW}{d\omega} = -\frac{\alpha}{\pi}\mathrm{Im}\,\nu\int_0^\infty d\tanh\left[\frac{1}{\gamma^2\sinh\nu\tau} + \frac{1}{2}\left(\frac{\varepsilon}{\varepsilon'} + \frac{\varepsilon'}{\varepsilon}\right)\right.$$
$$\left.\times\left(\frac{1}{\gamma^2} + \frac{w^2}{\nu^2}\right)\tanh\frac{\nu\tau}{2}\right]\exp\left(-\frac{ia\tau}{2} - \frac{w^2\tau}{4q} + \frac{w^2}{2\nu q}\tanh\frac{\nu\tau}{2}\right) \tag{7.90}$$

One should take into account that in formulae (7.75), (7.78), (7.89) and (7.90) the subtraction is assumed: $\frac{dW}{d\omega} \to \frac{dW(w,\nu)}{d\omega} - \frac{dW(0,0)}{d\omega}$

This probability was obtained in a somewhat different form by Baryshevskii and Tikhomirov, 1986 by the functional integration method. For $\mathbf{w} = 0$, Eq. (7.90) becomes the probability derived by Migdal, 1957. For weak multiple scattering, where the second term in Eq. (7.86) can be neglected, we can expand (7.90) in powers of $\nu\tau$. The principal term in this expansion gives the emission probability in a constant field in the absence of scattering, and a correction $\propto q$ gives, to logarithmic accuracy, the emission cross section for scattering by a screened Coulomb potential in the presence of an external field:

$$\frac{d\sigma}{d\omega} = \frac{4Z^2\alpha^3}{15m^2\omega}\ln\left[\left(1+\left(\frac{\chi}{u}\right)^{1/3}\right)\frac{1}{\gamma\vartheta_1}\right]\cdot\left[\frac{\omega^2}{\varepsilon^2}(x^4\Upsilon - 3x^2\Upsilon' - x^3) + \left(1+\frac{\varepsilon'^2}{\varepsilon^2}\right)(x^4\Upsilon + 3x\Upsilon - 5x^2\Upsilon' - x^3)\right], \quad (7.91)$$

where $\vartheta_1 = q_1/\varepsilon$, q_1 is determined by the parameters of a potential, and we have

$$\Upsilon(x) = \int_0^\infty \sin\left(x\tau + \frac{\tau^3}{3}\right)d\tau\,, \quad x = \left(\frac{u}{\chi}\right)^{2/3} \quad (7.92)$$

The logarithm in Eq. (7.91) is multiplied by the function obtained by Borisov and Zhukovskii, 1975. It should be noted that since the lower bound on the effective momentum transfer increases as $(\chi/u)^{2/3}$ in strong fields, a situation in which there is full screening in the absence of an external field may not be so when the fields are sufficiently strong.

We now return to Eq. (7.73) and introduce the notation

$$\begin{aligned} V(x) &= (2\pi)^2 n[\Sigma(x) - \Sigma(0)] \\ &= n\int (e^{i\boldsymbol{\vartheta}\mathbf{x}} - 1)\sigma(\vartheta)d^2\vartheta\,. \end{aligned} \quad (7.93)$$

We will solve this equation in the Born approximation for scattering by $V(x)$, with no restrictions on the magnitude of the external field. Assuming $V = 0$, we obtain in the zeroth approximation

$$\frac{\partial\varphi_\mu^{(0)}}{\partial\tau} - \frac{ib}{2}\Delta\varphi_\mu^{(0)} + i\mathbf{wx}\,\varphi_\mu^{(0)} = 0\,. \quad (7.94)$$

We can obtain the solution of this equation from (7.88), letting $q \to 0$:

$$\begin{aligned} \varphi^{(0)} &= \frac{\mathbf{y}}{b}\varphi_0^{(0)}\,; \quad \mathbf{y} = \frac{\mathbf{x}}{\tau} + \frac{b\tau}{2}\mathbf{w}\,, \\ \varphi_0^{(0)}(\mathbf{x},\tau) &= \frac{1}{2i\pi b\tau}\exp\left(-\frac{x^2}{2ib\tau} - \frac{i\tau}{2}\mathbf{wx} - \frac{ibw^2\tau^3}{24}\right) \end{aligned} \quad (7.95)$$

Substituting the solution (7.95) into Eq. (7.75), we obtain the well-known emission probability in a constant external field. In the first Born approximation, we have

$$\varphi_\mu^{(1)} = \varphi_\mu^{(0)} + \psi_\mu\varphi_0^{(0)}\,. \quad (7.96)$$

Transforming then to variables $\mathbf{y}$, $s = 1/\tau$, we find that the function $\psi_\mu(\mathbf{y}, s)$ satisfies the equation

$$\frac{\partial \psi_\mu}{\partial s}(\mathbf{y}, s) + \frac{ib}{2}\Delta\psi_\mu(\mathbf{y}, s) = -\frac{y_\mu}{bs^2} V\left(\frac{\mathbf{y}}{s} - \frac{b\mathbf{w}}{2s^2}\right), \tag{7.97}$$

where we have put $y_0 = b$. We seek a solution of Eq. (7.97) using the Green's function $G(\mathbf{y}, s)$ which satisfies the equation

$$\frac{\partial G}{\partial s} + i\frac{b}{2}\Delta G = -\delta(\mathbf{y})\delta(s), \tag{7.98}$$

which we can solve. for example, via a Fourier transformation:

$$G(\mathbf{y} - \mathbf{y}', s - s') = \frac{\vartheta(s' - s)}{2i\pi b(s' - s)} \exp\left[-\frac{(\mathbf{y} - \mathbf{y}')^2}{2ib(s' - s)}\right] \tag{7.99}$$

Using Eq. (7.99), we obtain a solution of the equation for $\psi_\mu(\mathbf{y}, s)$ in the form

$$\begin{aligned}
\psi_\mu(\mathbf{y}, s) &= \int G(\mathbf{y} - \mathbf{y}', s - s') \frac{y'_\mu}{bs'^2} V\left(\frac{\mathbf{y}'}{s'} - \frac{b}{2s'^2}\mathbf{w}\right) d^2y' ds' \\
&= \frac{1}{2i\pi b^2} \int_s^\infty \frac{ds'}{s' - s} \int d^2y' Q_\mu \\
&\quad \times V\left(\mathbf{y}' - \frac{b}{2s'^2}\mathbf{w}\right) \exp\left[-\frac{(s'\mathbf{y}' - \mathbf{y})^2}{2ib(s' - s)}\right], \\
Q_0 &= y_0\,, \quad \mathbf{Q} = s'\mathbf{y}'.
\end{aligned} \tag{7.100}$$

Transforming variables to $s' = (\xi + 1)s$ and $\mathbf{y}' = \boldsymbol{\eta} + \mathbf{x}/(\xi + 1) + b\tau^2\mathbf{w}/2(\xi + 1)$, and reverting to $\tau = s^{-1}$, $\mathbf{x} = \mathbf{y}\tau - b\tau^2\mathbf{w}/2$, we have

$$\begin{aligned}
\psi_0(\mathbf{x}, \tau) &= \frac{1}{2i\pi b} \int_0^\infty \frac{d\xi}{\xi} \int \exp\left[-\frac{\eta^2(1 + \xi)^2}{2ib\xi\tau}\right] \\
&\quad \times V\left(\boldsymbol{\eta} + \frac{\mathbf{x}}{1 + \xi} + \frac{b\xi\tau^2}{2(1 + \xi)^2}\mathbf{w}\right) d^2\eta\,, \\
\boldsymbol{\psi}(\mathbf{x}, \tau) &= \frac{1}{2i\pi b^2\tau} \int_0^\infty \frac{d\xi}{\xi} \int d^2\eta \exp\left[-\frac{\eta^2(1 + \xi)^2}{2ib\xi\tau}\right] \\
&\quad \times V\left(\boldsymbol{\eta} + \frac{\mathbf{x}}{1 + \xi} + \frac{b\xi\tau^2}{2(1 + \xi)^2}\mathbf{w}\right) \\
&\quad \times \left[\boldsymbol{\eta}(1 + \xi) + \mathbf{x} + \frac{b\tau^2}{2}\mathbf{w}\right].
\end{aligned} \tag{7.101}$$

Substituting the solution (7.101) into (7.96) and then into (7.75), we obtain in the first Born approximation

$$\frac{dW^{(1)}}{d\omega} = \frac{\alpha\omega}{(2\pi b)^2} \operatorname{Re} \int_0^\infty \frac{d\tau}{\tau} \exp\left(-i\frac{a\tau}{2} - \frac{ibw^2\tau^3}{24}\right)$$
$$\times \int_0^\infty \frac{d\xi}{\xi} \int d^2\eta\, V\left(\eta + \frac{b\xi\tau^2}{2(1+\xi)^2}\mathbf{w}\right)$$
$$\times \exp\left\{i\eta^2\frac{(1+\xi)^2}{2b\xi\tau}\right\} \left[\left(1+\frac{\varepsilon^2}{\varepsilon'^2}\right)\left(\frac{\eta^2(1+\xi)^2}{\xi(b\tau)^2}\right.\right.$$
$$\left.\left. + \frac{(1+\xi)^2}{2b\xi}\mathbf{w}\eta + \frac{w^2\tau^2}{4}\right) - \left(\frac{\omega}{\gamma\varepsilon'}\right)^2\right]. \tag{7.102}$$

It is convenient to transform variables in (7.102), letting $\tau \to 2\tau, \xi = e^{2z}$, whereupon the emission probability spectrum takes the form

$$\frac{dW^{(1)}}{d\omega} = \frac{\alpha\omega}{(\pi b)^2} \operatorname{Re} \int_0^\infty \frac{d\tau}{\tau} e^{-\rho(\tau)} \int_0^\infty dz \exp\left(i\eta^2\frac{\cosh^2 z}{b\tau}\right)$$
$$\times V\left(\eta + \frac{b\tau^2}{2\cosh^2 z}\mathbf{w}\right)\left\{-\left(\frac{\omega}{\gamma\varepsilon'}\right)^2\right.$$
$$\left. + \left(1+\frac{\varepsilon^2}{\varepsilon'^2}\right)\left[w^2\tau^2 + \left(\frac{\eta^2}{b^2\tau^2} + \frac{2}{b}\mathbf{w}\eta\right)\cosh^2 z\right]\right\},$$
$$\rho(\tau) = i\left(a\tau + \frac{1}{3}bw^2\tau^3\right) \tag{7.103}$$

This can be expressed directly in terms of the scattering cross section. In order to do so, we must use the definition (7.93) and carry out the appropriate Gaussian integrals over η in (7.103). We then obtain

$$\frac{dW^{(1)}}{d\omega} = \frac{\alpha\omega n}{\pi b} \operatorname{Im} \int_0^\infty d\tau e^{-\rho(\tau)} \int_0^\infty \frac{dz}{\cosh^2 z}$$
$$\times \int d^2\vartheta\,\sigma(\vartheta)[F(0) - F(\vartheta)], \tag{7.104}$$

where

$$F(\vartheta) = \left[\left(1+\frac{\varepsilon^2}{\varepsilon'^2}\right)\left(\frac{i}{b\tau}+\frac{\vartheta^2}{4\cosh^2 z}-\mathbf{w}\vartheta\tau+\mathbf{w}^2\tau^2\right)-\frac{\omega^2}{\gamma^2\varepsilon'^2}\right]$$
$$\times\exp\left\{-\frac{ib\tau}{4\cosh^2 z}(\vartheta^2-2\mathbf{w}\vartheta\tau)\right\}. \qquad (7.105)$$

Equations (7.103)–(7.105) describe the photon emission probability to quasi-classical (relativistic) accuracy, upon particle scattering in an external field when the scattering cross section is of the form of $\sigma(\boldsymbol{\vartheta},\boldsymbol{\vartheta}')=\sigma(\boldsymbol{\vartheta}-\boldsymbol{\vartheta}')$. This subsumes scattering by a screened Coulomb potential (see Eq. (7.79)).

In weak fields we can expand the functions which enter into (7.103)–(7.105) in power of $w\tau$. Keeping the principal terms of this expansion and terms $\sim w^2\tau^2$, we obtain the following expression for the emission cross section in a weak field ($\chi/u \ll 1$) for a screened Coulomb potential:

$$d\sigma_{sc}^{(1)} = \frac{4Z^2\alpha^3 d\omega}{3m^2\omega}\left\{\frac{\omega^2}{\varepsilon^2}\left(\ln\frac{1}{\gamma\vartheta_1}+\frac{1}{3}\right)\right.$$
$$+2\left(1+\frac{\varepsilon'^2}{\varepsilon^2}\right)\left(\ln\frac{1}{\gamma\vartheta_1}+\frac{7}{12}\right)+\frac{2}{5}\left(\frac{\chi}{u}\right)^2\left[\left(1+\frac{\varepsilon'^2}{\varepsilon^2}\right)\right.$$
$$\left.\left.\times\left(43\ln\frac{1}{\gamma\vartheta_1}-\frac{251}{15}\right)+\frac{\omega^2}{\varepsilon^2}\left(32\ln\frac{1}{\gamma\vartheta_1}-\frac{199}{15}\right)\right]\right. \qquad (7.106)$$

In the absence of a field, the cross section (7.106) is the same as the standard cross section in a screened potential (7.50) and (7.51), if we put $f(\xi)=0$, which corresponds to the first Born approximation ($\gamma\vartheta_1 = Z^{1/3}/111$).

In strong fields ($\chi/u \gg 1$), contributions to the integrals (7.103) and (7.104) come from $w^2\tau^2 \sim \eta^{-2} \sim \vartheta^2 \sim \gamma^{-2}(\chi/u)^{2/3} \gg \gamma^{-2}$ Therefore, up to terms $\sim (u/\chi)^{2/3}$, we can neglect the term linear in τ which appears in the argument $\rho(\tau)$ of the exponential ($\rho(\tau) \simeq ibw^2\tau^2/3$), and we can omit the term $\omega^2/\gamma^2\varepsilon'^2$ in brackets in (7.103) and (7.105). We can then rotate the integration contour over τ by an angle $-\pi/6$, so that $\tau \to e^{-i\pi/6}\tau$ ($i\tau^3 \to \tau^3$), and transform to the variable ρ. The integrals over η and ϑ in (7.103) and (7.104) are conveniently calculated via an exponential parametrization. With $\chi/u \gg 1$, we find after some fairly tedious calculations that

$$\frac{d\sigma_{sc}^{(1)}}{d\omega} = \frac{2Z^2\alpha^3}{5m^2\omega}\Gamma\left(\frac{1}{3}\right)\left(\frac{u}{3\chi}\right)^{2/3}$$
$$\times\left(1+\frac{\varepsilon'^2}{\varepsilon^2}\right)\left[\ln\left(\frac{1}{\gamma\vartheta_1}\left(\frac{\chi}{u}\right)^{1/3}\right)+D_{sc}\right], \tag{7.107}$$

where $D_{sc} = 2.3008$. The emission inhibition factor $(u/\chi)^{2/3}$ in (7.107) occurs, as was shown in Sec. 7.4, because of a decrease in the photon formation length. The factor $(\chi/u)^{1/3}$ in the argument of the logarithm is associated with the increase in effective emission angles $\vartheta_{ef} \sim \gamma^{-1}(\chi/u)^{1/3}$ Note that since the quantum parameter $\chi \propto 1/m^3$, the cross section (7.107) is independent of the mass of the radiating particle.

7.6. Influence of an External Field on the Radiation in an Unscreened Coulomb Potential

In a purely Coulomb radiation problem, the scattering cross section for momentum transfer close to the minimum depends not only on the angular difference $\boldsymbol{\vartheta}' - \boldsymbol{\vartheta} = \boldsymbol{\theta}$ (on the transverse momentum transfer $\mathbf{q}_\perp \simeq \varepsilon\boldsymbol{\theta}$), but on the emission angle itself as well (on the longitudinal momentum transfer $q_\parallel \simeq u\varepsilon(\vartheta^2 + \gamma^{-2}/2)$. This is connected with the fact that for such values of momentum transfer, the emission of a photon influences the particle scattering (see Sec. 7.1). In this case, the scattering cross section is conveniently given in the form

$$\begin{aligned}\sigma(\boldsymbol{\vartheta}', \boldsymbol{\vartheta}) &= \sigma(\boldsymbol{\vartheta}, \boldsymbol{\vartheta}') \\ &= \sigma\left(\boldsymbol{\theta}, \frac{\boldsymbol{\vartheta}+\boldsymbol{\vartheta}'}{2}\right) \\ &= \sigma(\boldsymbol{\theta}, \vartheta_1) + \tilde{\sigma}\,,\end{aligned} \tag{7.108}$$

where

$$\tilde{\sigma} = \sigma\left(\boldsymbol{\theta}, \boldsymbol{\vartheta} + \frac{\boldsymbol{\theta}}{2}\right) - \sigma(\boldsymbol{\theta}, \vartheta_1)\,. \tag{7.109}$$

Here $\sigma(\boldsymbol{\theta}, \vartheta_1)$ is the scattering cross section in the screened potential (7.79). We assume that ϑ_1 is small enough that $\vartheta_1 \ll \vartheta_{ef}$ (7.83), and large enough that when $\theta \geq \vartheta_1$, we can neglect the longitudinal momentum transfer: $q_\perp \simeq \varepsilon\theta \gg q_\parallel$

$$\vartheta_1 \ll \vartheta_{ef} \simeq \frac{1}{\gamma}\left[1 + \left(\frac{\chi}{u}\right)^{1/3}\right],$$

$$\vartheta_1 \gg \frac{1}{\varepsilon} q_\parallel(\vartheta_{ef}) \simeq \frac{u}{2}\left(\frac{1}{\gamma^2} + \vartheta_{ef}^2\right). \tag{7.110}$$

The right-hand (collision) side of Eq. (7.70) may be represented by a sum of a screened cross section and $\tilde{\sigma}$. In the first Born approximation, these terms contribute linearly and independently to the emission cross section. We calculate the corresponding cross section in a screened potential above. In the remainder with $\tilde{\sigma}$, we make use of the fact that when $\theta > \vartheta_1, \tilde{\sigma}$ falls off as ϑ_1^2/θ^2, and we may expand the contributing terms in powers of $\theta/\vartheta \sim \theta/\vartheta_{ef}$:

$$V_\mu(\boldsymbol{\vartheta}') \simeq V_\mu(\boldsymbol{\vartheta}) + \boldsymbol{\theta}\frac{\partial V_\mu}{\partial \boldsymbol{\vartheta}} + \frac{1}{2}\theta_i\theta_j\frac{\partial^2 V_\mu}{\partial\vartheta_i\partial\vartheta_j} + \cdots$$

$$\tilde{\sigma} \simeq \tilde{\sigma}(\boldsymbol{\theta}, \boldsymbol{\vartheta}) + \frac{1}{2}\boldsymbol{\theta}\frac{\partial\tilde{\sigma}(\boldsymbol{\theta}, \boldsymbol{\vartheta})}{\partial\boldsymbol{\vartheta}} + \cdots \tag{7.111}$$

Substituting this expansion into Eq. (7.70), we obtain a Fokker–Planck type of equation:

$$\frac{\partial V_\mu}{\partial \tau} + ib\frac{\vartheta^2}{2}V_\mu - \mathbf{w}\frac{\partial V_\mu}{\partial \boldsymbol{\vartheta}} = \frac{1}{4}\frac{\partial}{\partial \boldsymbol{\vartheta}}\left(\frac{\partial V_\mu}{\partial \boldsymbol{\vartheta}}\tilde{\theta}^2(\vartheta)\right),$$

where

$$\tilde{\theta}^2(\vartheta) = n\int d^2\theta \cdot \theta^2\tilde{\sigma}(\theta^2, \vartheta). \tag{7.112}$$

For a nuclear potential, $\tilde{\theta}^2(\vartheta)$ is of the form ($|q^2| = \varepsilon^2\theta^2 + q_\parallel^2$)

$$\tilde{\theta}^2(\vartheta) = 4\left(\frac{Z\alpha}{\varepsilon}\right)^2 \pi n\int d^2\theta \cdot \theta^2\left[\frac{1}{(\theta^2 + \delta^2\zeta^2)^2} - \frac{1}{(\theta^2 + \vartheta_1^2)^2}\right]$$

$$= 8\pi n\left(\frac{Z\alpha}{\varepsilon}\right)^2 \ln\frac{\varepsilon_1}{\delta\zeta}, \delta = \frac{u}{2\gamma^2},$$

$$\zeta = 1 + (\gamma\vartheta)^2. \tag{7.113}$$

Let $\boldsymbol{\nu} = \boldsymbol{\vartheta} + \mathbf{w}\tau$, and

$$2\pi n \left(\frac{Z\alpha}{\varepsilon}\right)^2 \ln\frac{\vartheta_1}{\delta\zeta} \equiv q \ln\frac{\vartheta_1}{\delta\zeta} \equiv qL(\boldsymbol{\vartheta}), \tag{7.114}$$

Then Eq. (7.112) takes the form

$$\frac{\partial V_\nu(\boldsymbol{\nu},\tau)}{\partial\tau} + \frac{ib}{2}\vartheta^2(\tau)V_\mu(\boldsymbol{\nu},\tau) = q\frac{\partial}{\partial\boldsymbol{\nu}}\left(\frac{\partial V_\mu}{\partial\boldsymbol{\nu}}L(\boldsymbol{\vartheta}(\tau))\right) \tag{7.115}$$

We next find the zeroth approximation to the solution of (7.115):

$$\frac{\partial V_\mu^{(0)}(\boldsymbol{\nu},\tau)}{\partial\tau} + \frac{ib}{2}\vartheta^2(\tau)V_\mu^{(0)}(\boldsymbol{\nu},\tau) = 0,$$

$$V_0^{(0)}(\boldsymbol{\nu},\tau) = \phi(\boldsymbol{\nu},\tau) = \exp\left[-\frac{ib}{2}\int_0^\tau d\tau'\vartheta^2(\tau')\right]$$

$$= \exp\left[-\frac{ib\tau}{2}\left(\boldsymbol{\nu}^2 - \boldsymbol{\nu}\mathbf{w}\tau + \frac{w^2\tau^2}{3}\right)\right],$$

$$\mathbf{V}^{(0)} = \boldsymbol{\nu}\phi(\boldsymbol{\nu},\tau). \tag{7.116}$$

Substitution of this solution into Eq. (7.69) gives the spectral and angular distribution of the emission probability in a constant field. In the first Born approximation, we seek a solution of Eq. (7.115) in the form

$$\begin{aligned} V_\mu(\boldsymbol{\nu},\tau) &= V_\mu^{(0)}(\boldsymbol{\nu},\tau) + V_\mu^{(1)}(\boldsymbol{\nu},\tau), \\ V_\mu^{(1)}(\boldsymbol{\nu},\tau) &= v_\mu(\boldsymbol{\nu},\tau)\phi(\boldsymbol{\nu},\tau). \end{aligned} \tag{7.117}$$

The equation for v_μ then takes the form

$$\frac{\partial v_\mu}{\partial\tau} = q\phi^{-1}\frac{\partial}{\partial\boldsymbol{\nu}}\left(\frac{\partial V_\mu^{(0)}}{\partial\boldsymbol{\nu}}L(\boldsymbol{\nu} - \mathbf{w}\tau)\right), \tag{7.118}$$

and, with the initial data $v_\mu(\boldsymbol{\vartheta},0) = 0$, we obtain its solution simply by integrating the right-hand side of this equation:

$$v_\mu(\boldsymbol{\nu},\tau) = q\int_0^\tau d\tau'\phi^{-1}(\boldsymbol{\nu},\tau') \times \frac{\partial}{\partial\boldsymbol{\nu}}\left(\frac{\partial V_\mu^{(0)}(\boldsymbol{\nu},\tau')}{\partial\boldsymbol{\nu}}L(\boldsymbol{\nu} - \mathbf{w}\tau')\right) \tag{7.119}$$

Substituting the solution (7.119) first into (7.117) and then into (7.69), we obtain the following expression

$$\frac{d\widetilde{W}}{d\omega} = \frac{\alpha q\omega}{(2\pi)^2} \operatorname{Re} \int_0^\infty d\tau e^{-i\frac{a\tau}{2}} \int_0^\tau d\tau' \phi^{-1}(\boldsymbol{\nu}, \tau')$$
$$\times \left[\left(\frac{\omega}{\gamma\varepsilon'}\right)^2 \frac{\partial}{\partial\boldsymbol{\nu}} \left(\frac{\partial\phi(\boldsymbol{\nu}, \tau')}{\partial\boldsymbol{\nu}} L(\boldsymbol{\nu} - \mathbf{w}\tau') \right) \right.$$
$$+ \left(1 + \frac{\varepsilon^2}{\varepsilon'^2}\right) (\boldsymbol{\nu} - \mathbf{w}\tau) \frac{\partial}{\partial\nu_i} \left(L(\boldsymbol{\nu} - \mathbf{w}\tau') \right.$$
$$\left. \left. \times \frac{\partial}{\partial\nu_i} \boldsymbol{\nu}\phi(\boldsymbol{\nu}, \tau') \right) \right] . \tag{7.120}$$

Note that

$$\phi(\boldsymbol{\nu}, \tau)\phi^{-1}(\boldsymbol{\nu}, \tau') = \phi(\boldsymbol{\nu} - \mathbf{w}\tau', \tau - \tau') . \tag{7.121}$$

Integrating in Eq. (7.120) by parts and making use of (7.121), we may write (7.120) in the symmetric form

$$\frac{d\widetilde{W}}{d\omega} = -\frac{\alpha q\omega}{(2\pi)^2} \operatorname{Re} \int_0^\infty d\tau e^{-i\frac{a\tau}{2}} \int d^2\boldsymbol{\vartheta} L(\boldsymbol{\vartheta})$$
$$\times \int_0^\tau d\tau' \left[\left(\frac{\omega}{\gamma\varepsilon'}\right)^2 \left(\frac{\partial}{\partial\boldsymbol{\vartheta}} \phi(\boldsymbol{\vartheta}, \tau - \tau') \right) \right.$$
$$\times \left(\frac{\partial}{\partial\boldsymbol{\vartheta}} \phi(\boldsymbol{\vartheta} + \mathbf{w}\tau', \tau') \right) + \left(1 + \frac{\varepsilon^2}{\varepsilon'^2}\right) \left(\frac{\partial}{\partial\vartheta_i} (\boldsymbol{\vartheta} - \mathbf{w}(\tau - \tau')) \right.$$
$$\left. \left. \times \phi(\boldsymbol{\vartheta}, \tau - \tau') \right) \left(\frac{\partial}{\partial\vartheta_i} (\boldsymbol{\vartheta} + \mathbf{w}\tau') \phi(\boldsymbol{\vartheta} + \mathbf{w}\tau', \tau') \right) \right] . \tag{7.122}$$

Making the change of variables $\tau - \tau' = (1-\mu)\tau/2, \tau' = (1+\mu)\tau/2$, and letting $\tau \to 2\tau$, and $\boldsymbol{\vartheta} = \boldsymbol{\theta} - \mu\mathbf{w}\tau$, we obtain a correction to the emission probability spectrum of the form

$$\frac{d\widetilde{W}}{d\omega} = -\frac{\alpha q\omega}{\pi^2} \operatorname{Re} \int_0^\infty d\tau \tau e^{-\rho(\tau)}$$
$$\times \int_0^1 d\mu \int d^2\theta e^{-ib\tau\theta^2} \ln \frac{\delta(1 + \gamma^2(\boldsymbol{\theta} - \mathbf{w}\tau)^2)}{\boldsymbol{\vartheta}_1}$$

$$\times \{b^2\tau^2(1-\mu^2)[R_1 + R_2(\theta^2 - w^2\tau^2)]$$

$$\times \left(\theta^2 - \mathbf{w}\boldsymbol{\theta}\mu\tau + \frac{\mu^2-1}{4}w^2\tau^2\right) - 2R_2$$

$$+ ib\tau R_2[2\boldsymbol{\theta}(\boldsymbol{\theta} + \mathbf{w}\mu\tau) + (1-\mu^2)w^2\tau^2]\}, \qquad (7.123)$$

where

$$\rho(\tau) = i\left(a\tau + \frac{1}{3}bw^2\tau^3\right),$$

$$R_1 = \frac{\omega^2}{\varepsilon'^2\gamma^2}, \quad R_2 = 1 + \frac{\varepsilon^2}{\varepsilon'^2}.$$

Expanding the expressions in (7.123) in powers of $w\tau$ at $w \to 0$, and carrying out some straightforward but tedious calculations, we obtain the following correction to the emission cross section for the waek field ($\chi/u \ll 1$) case:

$$\frac{d\tilde{\sigma}}{d\omega} = \frac{4Z^2\alpha^3}{3m^2\omega}\left\{\frac{\omega^2}{\varepsilon^2}\left(\ln\frac{\vartheta_1}{\delta} - \frac{5}{6}\right) + 2\left(1 + \frac{\varepsilon'^2}{\varepsilon^2}\right)\left(\ln\frac{\vartheta_1}{\delta} - \frac{13}{12}\right)\right.$$
$$\left. + \frac{2}{5}\left(\frac{\chi}{u}\right)^2\left[\frac{\omega^2}{\varepsilon^2}\left(32\ln\frac{\vartheta_1}{\delta} - \frac{662}{105}\right) + \left(1 + \frac{\varepsilon'^2}{\varepsilon^2}\right)\left(43\ln\frac{\vartheta_1}{\delta} - \frac{3967}{420}\right)\right]\right\}$$
$$(7.124)$$

Combining this result with the cross section (7.106) for the screened potential, we obtain the emission cross section at a nucleus, with $\chi/u \ll 1$, and taking corrections $\sim (\chi/u)^2$ into account (cf. (7.42))

$$\frac{d\sigma}{d\omega} = \frac{4Z^2\alpha^3}{3m^2\varepsilon^2\omega}\left\{[2(\varepsilon^2 + \varepsilon'^2) + \omega^2]\left(\ln\frac{2\gamma}{u} - \frac{1}{2}\right)\right.$$
$$+ \frac{2}{5}\left(\frac{\chi}{u}\right)^2\left[\omega^2\left(32\ln\frac{2\gamma}{u} - \frac{137}{7}\right)\right.$$
$$\left.\left. + (\varepsilon^2 + \varepsilon'^2)\left(43\ln\frac{2\gamma}{u} - \frac{733}{28}\right)\right]\right\}. \qquad (7.125)$$

In a strong external field, the calculation is the same as in the screened case (see the discussion preceding Eq. (7.107)). For $\chi/u \gg 1$, the correction $d\tilde{\sigma}/d\omega$ to the emission cross section takes the form

$$\frac{d\tilde{\sigma}}{d\omega}=\frac{2Z^2\alpha^3}{5m^2\omega}\Gamma\left(\frac{1}{3}\right)\left(\frac{u}{3\chi}\right)^{2/3}\left(1+\frac{\varepsilon'^2}{\varepsilon^2}\right)\left[\ln\left(\frac{\vartheta_1}{\delta}\left(\frac{u}{\chi}\right)^{2/3}\right)+\widetilde{D}\right], \quad (7.126)$$

where $\widetilde{D}=-3.0131$. The argument of the logarithm in (7.126) determines the limit of a full screening of the potential during photon emission. When there is a full screening, we must have

$$\left(\frac{\chi}{u}\right)^{2/3}\times\frac{\delta}{\vartheta_1}=\left(\frac{\chi}{u}\right)^{2/3}a\times q_{\min}\ll 1, \quad (7.127)$$

where $q_{\min}=\varepsilon\delta=\omega\, m^2/2\varepsilon(\varepsilon-\omega)$, and a is the characteristic radius of the potential. For example, in an amorphous substance under standard conditions, $a\sim a_0Z^{-1/3}$ (where a_0 is the Bohr radius), and in oriented single crystals, this length is determined by the amplitude of thermal vibrations. Equation (7.127) implies that a fully screened potential in the absence of an external field potential ($q_{\min}\,a\ll 1$) can become unscreened in the presence of a strong field. Note that the condition (7.127) depends slightly on the photon frequency ($\omega^{1/3}$), and is independent of the particle mass. Combining (7.126) with the asymptotic expression (7.107) for the emission cross section in a screened potential, we obtain for the emission cross section at a nucleus in an external field, when $\chi/u\gg 1$,

$$\frac{d\sigma_{\rm nosc}}{d\omega}=\frac{2Z^2\alpha^3}{5m^2\omega}\left(\frac{u}{3\chi}\right)^{2/3}\left(1+\frac{\varepsilon'^2}{\varepsilon^2}\right)\left[\ln\left(\frac{2\gamma}{u}\left(\frac{u}{\chi}\right)^{1/3}\right)+D_{\rm nosc}\right], \quad (7.128)$$

where $D_{\rm nosc}=D_{sc}+\widetilde{D}=-0.7123$. The cross section (7.128), like (7.107), is independent of the mass of the radiating particle. Making use of Eq. (7.52) which characterizes mean particle energy losses per unit length in a medium, and spectral distributions of emission cross sections, we can calculate the corresponding radiation length. So, in the limit of a strong field ($\chi\gg 1$), we obtain, in the case of a full screening

$$L_{sc}^{-1}(\chi)=\frac{232Z^2\alpha^3\pi n\Gamma\left(\frac{1}{3}\right)}{405\sqrt{3}m^3(3\chi)^{2/3}}\left(\ln\frac{\chi^{1/3}}{\gamma\vartheta_1}+1.2824\right) \quad (7.129)$$

In the unscreened case, we have

$$L_{\rm nosc}^{-1}(\chi) = \frac{232Z^2\alpha^3\pi n\Gamma\left(\frac{1}{3}\right)}{405\sqrt{3}m^2(3\chi)^{2/3}}\left(\ln\frac{2\gamma}{\chi^{1/3}} - 2.7491\right) \tag{7.130}$$

7.7. Photoproduction of Charged Particles in a Medium

It is known that in quantum electrodynamics there is a relation among differential probabilities of the photon emission by a charged particle and the pair production by a photon. In general, such relations are missing for the cross sections integrated over outgoing angles of final particles. However, it can be shown (see Olsen, 1955) that, in the case under consideration, this relation holds for the spectrum of final particles as well, owing to the smallness of all the characteristic angles in the problem. The spectral probability of a photoproduction process can be obtained therefore from the corresponding formulae describing the radiation process by the substitution rules:

$$\omega \to -\omega\,, \quad \varepsilon \to -\varepsilon\,, \quad d\omega \to -(\varepsilon/\omega)^2 d\varepsilon\,.$$

For example, we obtain from Eq. (7.48) for the distribution over the energy of the created e^+e^- pair at arbitrary screening:

$$d\sigma_p = \frac{4Z^2\alpha^3}{m^2}\frac{d\varepsilon_1}{\omega}\left[(\varepsilon_1^2+\varepsilon_2^2)\psi_1 + \frac{2}{3}\varepsilon_1\varepsilon_2\psi_2\right]. \tag{7.131}$$

In the unscreened case $\psi_1 = \psi_2 = \ln\delta^{-1} - 1/2 + f(\xi)$, and the total cross section of the pair production by a photon in a Coulomb field takes the form

$$\sigma_p = \int_0^\omega \frac{d\sigma_p}{d\varepsilon_1}d\varepsilon_1 = \frac{28Z^2\alpha^3}{9m^2}\left[\ln\frac{2\omega}{m} - f(\xi) - \frac{109}{42}\right]. \tag{7.132}$$

In the case of a full screening the quantities $\psi_{1,2}$ are determined by Eq. (7.50), then we have

$$\sigma_p^{sc} = \frac{28Z^2\alpha^3}{9m^2}\left[\ln(183Z^{-1/3}) - f(\xi) - \frac{1}{42}\right]. \tag{7.133}$$

Using the substitution rules, we can find the spectral properties of the photo-production process in the presence of an external field as well. For the characteristic parameters of the problem, we have

$$\begin{aligned}
&\frac{\chi}{u} = \frac{\kappa\varepsilon}{\omega u} \to \frac{\kappa\varepsilon_1\varepsilon_2}{\omega^2}, \quad \kappa = \frac{H}{H_0}\cdot\frac{\omega}{m}, \quad a = \frac{\omega m^2}{\varepsilon_1\varepsilon_2} = 2\delta m\,, \\
&\Delta = \frac{\vartheta_{ef}}{\vartheta_1} = \frac{1}{\gamma\vartheta_1}\left[1+\left(\frac{\chi}{u}\right)^{1/3}\right] \to \frac{m}{q_1}\left[1+\left(\frac{\kappa\varepsilon_1\varepsilon_2}{\omega^2}\right)^{1/3}\right], \\
&\nu = \sqrt{2ibq} = \sqrt{2iaq_0} = 2\sqrt{i\delta m q_0}\,, \\
&q_0 = \gamma^2 q = 2\pi n\left(\frac{Z\alpha}{m}\right)^2 \ln\Delta\,, \quad \frac{w^2}{q} = \frac{w_0^2}{q_0}, \quad w_0 = \frac{eH}{m}\,.
\end{aligned} \tag{7.134}$$

Making use of the transformations (7.134), we can represent the spectrum of the created pair in a symmetric form. For example, we obtain from Eq. (7.90) the spectral probability of the photoproduction process to logarithmic accuracy:

$$\begin{aligned}
\frac{dw}{d\varepsilon_1} = \frac{\alpha}{\pi}\left(\frac{m}{\omega}\right)^2 \operatorname{Im}\nu\int_0^\infty d\tau\left[\frac{1}{\sinh\nu\tau} - \frac{1}{2}\left(\frac{\varepsilon_1}{\varepsilon_2}+\frac{\varepsilon_2}{\varepsilon_1}\right)\left(1+\frac{w_0^2}{\nu^2}\right)\tanh\frac{\nu\tau}{2}\right] \\
\times\exp\left(-\frac{ia\tau}{2} - \frac{w_0^2\tau}{4q_0} + \frac{w_0^2}{4\nu q_0}\tanh\frac{\nu\tau}{2}\right).
\end{aligned} \tag{7.135}$$

For a weak multiple scattering ($\nu\tau \ll 1$), we have (cf. Eq. (7.91)):

$$\begin{aligned}
\frac{d\sigma}{d\varepsilon_1} = \frac{4Z^2\alpha^2}{15m^2\omega}(\ln\Delta)\Big[x^4\Upsilon - 3\,x^2\Upsilon' - x^3 \\
+\left(\frac{\varepsilon_1^2+\varepsilon_2^2}{\omega^2}\right)(x^4\Upsilon + 3x\Upsilon - 5x^2\Upsilon' - x^3)\Big],
\end{aligned} \tag{7.136}$$

$$x = \left(\frac{\omega^2}{\kappa\varepsilon_1\varepsilon_2}\right)^{2/3}$$

Similar transformations of the corresponding probabilities may be easily performed in other cases as well. For example, carrying out the indicated transformations in Eq. (7.106) and integrating over the energy of one of the created particles, we obtain the total photoproduction cross section in a screened Coulomb potential for a weak external field

$$\sigma_p^{sc}(\kappa \ll 1) = \frac{28Z^2\alpha^3}{9m^2}\left[\ln\frac{m}{q_1} + \frac{10}{21} + \frac{\kappa^2}{1225}\left(396\ln\frac{m}{q_1} - \frac{799}{5}\right)\right]. \tag{7.137}$$

Comparing this expression with Eq. (7.133), we note that, as it follows from Eq. (7.50), $\ln(183Z^{-1/3}) = \ln(111Z^{-1/3}) + 1/2 \equiv \ln(m/q_1) + 1/2$. In the limiting case of a strong external field ($\kappa \gg 1$), we have

$$\sigma_p^{sc} = \frac{8Z^2\alpha^3\Gamma^3\left(\frac{1}{3}\right)}{25m^2(3\kappa)^{2/3}\Gamma\left(\frac{2}{3}\right)}\left(\ln\frac{m\kappa^{1/3}}{q_1} + 0.9416\right). \tag{7.138}$$

In the absence of screening, we find for the weak field case ($\kappa \ll 1$) the following expression for the total photoproduction cross section:

$$\sigma_p = \frac{28Z^2\alpha^3}{9m^2}\left[\ln\frac{2\omega}{m} - \frac{109}{42} + \frac{\kappa^2}{1225}\left(396\ln\frac{2\omega}{m} - \frac{30389}{35}\right)\right] \tag{7.139}$$

For the strong field case ($\kappa \gg 1$) the total photoproduction cross section takes the form

$$\sigma_p = \frac{8Z^2\alpha^3\Gamma^3\left(\frac{1}{3}\right)}{25m^2(3\kappa)^{2/3}\Gamma\left(\frac{2}{3}\right)}\left(\ln\frac{2\omega}{m\kappa^{1/3}} - 3.4307\right) \tag{7.140}$$

Note that the cross sections (7.138) and (7.140) are independent of the created particle mass.

Chapter 2

FAST PARTICLES PASSAGE THROUGH SINGLE CRYSTALS

8. OUTLINE OF CRYSTALLOGRAPHY

8.1. Crystal Lattice, Planes and Axes

The description of crystal lattice types is contained in some special guide books and in many books concerning a theory of solids (see, e.g. Animalu, 1977; Kittel, 1956). Here, for the sake of the reader's convenience, some brief information is given.

An ideal crystal is built up by the repetition of a structural motif—some particular atomic group by translation in three dimensions. The nature of the pattern can be studied if each repetition of the motif is replaced by a representative point. The infinite three-dimensional array of points (all of which have identical environments in similar orientation) thus derived is a space lattice. There could be only fourteen distinct lattices, that was first proved rigidly by Bravais, and they are frequently called Bravais lattices. They may be determined by a set of space-lattice vectors:

$$\mathbf{l} = l_1\mathbf{a}_1 + l_2\mathbf{a}_2 + l_3\mathbf{a}_3 \, , \tag{8.1}$$

where $l_1, l_2, l_3 = 0, \pm 1, \pm 2, \ldots$. The unit cell of atomic structure may be enclosed by the parallelepiped generated by the different translations corresponding to the base vectors $\mathbf{a}_1, \mathbf{a}_2$ and $\mathbf{a}_3$. If there are s atoms with

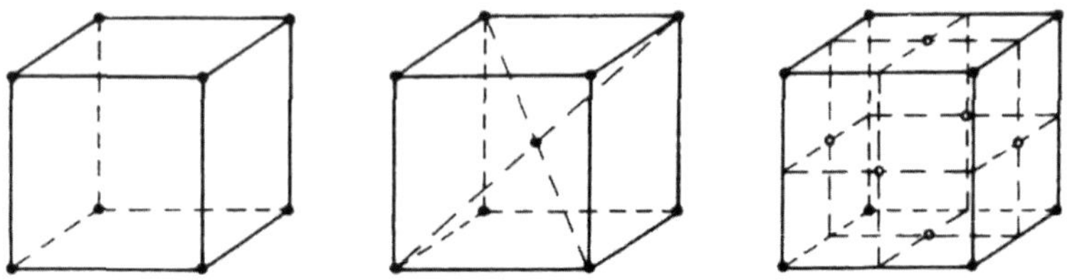

Fig. 8.1. Three types of cubic lattices: a) primitive, b) body-centered, c) all-face centered.

coordinates $\mathbf{r}_a$ in the unit cell then the distribution of atoms in a crystal is described by the function

$$\rho(\mathbf{r}) = \sum_{\mathbf{l}} \sum_{a=1}^{s} \delta(\mathbf{r} - \mathbf{r}_a - \mathbf{l}) , \tag{8.2}$$

being, obviously, periodic:

$$\rho(\mathbf{r} + \mathbf{l}) = \rho(\mathbf{r}) . \tag{8.3}$$

If we represent atoms as rigid spheres (assuming that a crystal consists of identical atoms), then all the possible crystal structures correspond to more or less close packing of such spheres in the box having crystal sizes. The quantative measure of packing is the packing coefficient f, which is the ratio of the volume occupied by atoms (rigid spheres) in the unit cell to its volume.

In further applications, crystals which have the cubic (isometric) lattice will be considered. The simplest lattice, termed a primitive lattice (P), has a lattice point at each corner of the cube (see Fig. 8.1(a)). Two other cubic lattices are body-centered (bcc) (see Fig. 8.1(b)) and all face-centered (fcc) (see Fig. 8.1(c)). Some characteristics of these cubic lattices are given in Table 8.1.

The orientation of crystal planes may be determined by giving the coordinates of three atoms lying in the plane and belonging to different crystallographic axes. The Miller indices are conventionally used. These are integers h, k and l (enclosed in parentheses (hkl)) connected with the orientation of a plane (see Fig. 8.2) relative to crystal axes as follows: the plane (hkl) in the cubic lattice crystal intersects the axes $\mathbf{a}_1$, $\mathbf{a}_2$ and $\mathbf{a}_3$ correspondingly at the distances $1/h, 1/k$ and $1/l$ from the origin, so that the equation of this plane has the form:

$$hx + ky + lz = 1 . \tag{8.4}$$

Table 8.1. Characteristics of cubic lattices.

Lattice type	Number of atoms per unit cell	Number of nearest neighbors	Distances between atoms along axes (in units of a)			Distances between adjacent planes (in units of a)		
			$\langle 100\rangle$	$\langle 110\rangle$	$\langle 111\rangle$	(100)	(110)	(111)
Primitive	1	6	1	$\sqrt{2}$	$\sqrt{3}$	1	$1/\sqrt{2}$	$1/\sqrt{3}$
bcc	2	8	1	$\sqrt{2}$	$\sqrt{3}/2$	1/2	$1/\sqrt{2}$	$1/2\sqrt{3}$
fcc	4	12	1	$1/\sqrt{2}$	$\sqrt{3}$	1/2	$1/2\sqrt{2}$	$1/\sqrt{3}$
fcc(d)	8	4	1	$1/\sqrt{2}$	$\sqrt{3}/4$ $3\sqrt{3}/4$	1/4	$1/2\sqrt{2}$	$\sqrt{3}/4$ $1/4\sqrt{3}$

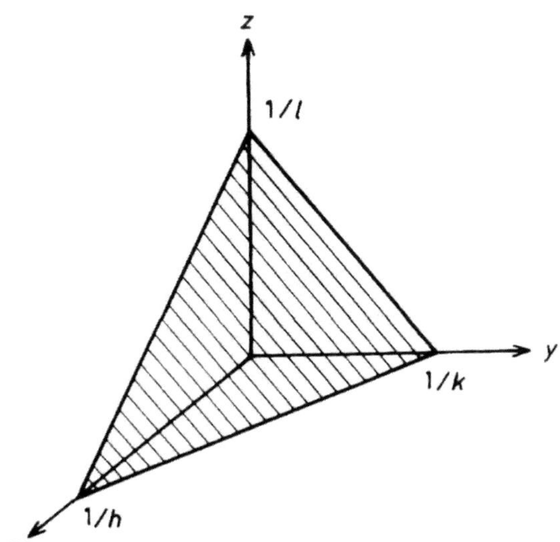

Fig. 8.2. The plane having Miller indices (hkl).

The main planes and their Miller indices are shown in Fig. 8.3 for a cubic lattice crystal. For negative intercepts, a corresponding index acquires a streak, e.g. $(\bar{h}kl)$. Every specific set of indices (hkl) generates a family of parallel planes. Owing to a crystal symmetry, different sets turn out to be equivalent.

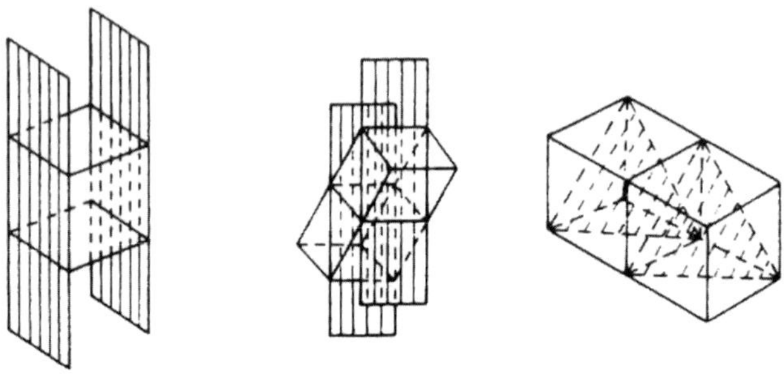

Fig. 8.3. The main planes and their Miller indices for a cubic crystal.

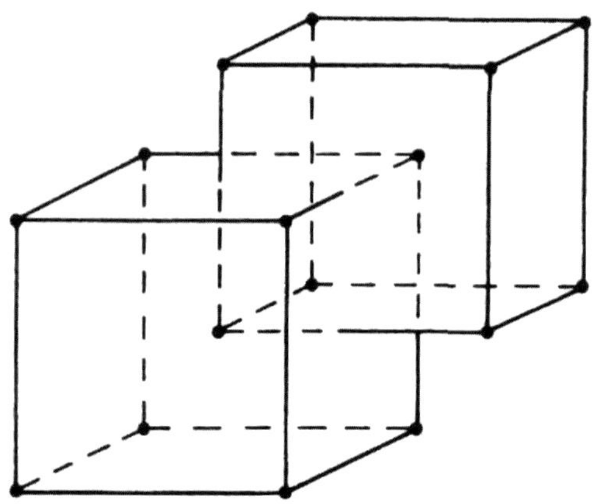

Fig. 8.4. Body-centered cubic lattice as a combination of two primitive lattices.

For example, the planes (100), (010) and (001) are equivalent for cubic lattice crystals, as well as the planes (110), (101) and (011). The totality of all equivalent planes is conventionally denoted as $\{hkl\}$. With increasing order of Miller indices the distance between corresponding planes decreases, e.g., in Fig. 8.3, these distances (in units of lattice constant) are 1 for (100) plane, $2^{-1/2}$ for (110) plane and $3^{-1/2}$ for (111) plane. Note that in some cases the more complicated structures may be regarded as a sum of more simple ones. For example, the (*bcc*)-type lattice can be represented as two primitive lattices shifted in a parallel way along the diagonal to the middle of a cube (see Fig. 8.4).

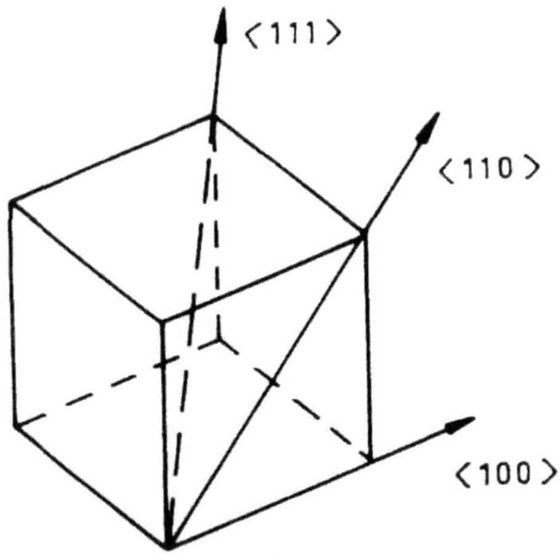

Fig. 8.5. The main crystallographic axes for a primitive cubic lattice.

Side by side with planes, a crystal lattice is characterized by straight lines, called axes, passing through lattice points. Consider an axis passing through the origin with the nearest lattice point having the numbers l_1, l_2 and l_3 (see Eq. (8.1)), correspondingly h, k and l, then the symbol $[hkl]$ denotes a family of parallel axes with identical properties. The main axes of a primitive lattice are shown in Fig. 8.5. The symbol $\langle hkl\rangle$ denotes the totality of axes families, which are equivalent owing to a crystal symmetry. For example, the axes of cubic crystals [100], [010], [001], $[\bar{1}00]$, $[0\bar{1}0]$ and $[00\bar{1}]$ form the family of axes denoted by the symbol ⟨100⟩. Similarly, the totality of axes [111], $[\bar{1}11]$ etc. is denoted by the symbol ⟨111⟩. For cubic crystals, the axes $\langle hkl\rangle$ are perpendicular to the planes $\{hkl\}$. Consider now several examples.

1. Primitive lattice (see Fig. 8.1(a)).

There is one atom per unit cell, which is a cube with the edge length (lattice constant) a. The close packing of rigid spheres implies that there is only one sphere with the radius $R = a/2$ at each corner of the cube, i.e. the packing coefficient is

$$f = \frac{V_s}{a^3} = \frac{4\pi}{3}\left(\frac{a}{2}\right)^3\frac{1}{a^3} = \frac{\pi}{6} \simeq 0.524\,. \tag{8.5}$$

2. Body-centered cubic lattice (see Fig. 8.1(b)).

One additional atom (comparatively to a primitive lattice) is situated at the center of the cube, so that there are two atoms per unit cell. The close packing now implies rigid spheres at each corner of the cube and one at the center, i.e. the radius $R = 3^{1/2}a/4$, and the packing coefficient is

$$f = \frac{2V_s}{a^3} = \frac{\pi\sqrt{3}}{8} \simeq 0.680\,. \tag{8.6}$$

3. Face-centered cubic lattice (see Fig. 8.1(c)).

Here there are atoms at each corner of the cube and at the center of each face, i.e. the unit cell contains four atoms. The radius of a rigid sphere at close packing is $R = 2^{-3/2}a$, and we obtain for the packing coefficient:

$$f = \frac{4\,V_s}{a^3} = \frac{\pi\sqrt{2}}{6} \simeq 0.740\,. \tag{8.7}$$

Note that this packing is, side by side with the hexagonal one, the most compact packing of rigid spheres.

4. Diamond like lattice $fcc_{(d)}$.

Many crystals (diamond, Si, Ge), widely used, possess such lattices, every atom is situated at the center of a tetrahedron and the four nearest atoms are at each vertex of this tetrahedron (Fig. 8.6). For the diamond like lattice,

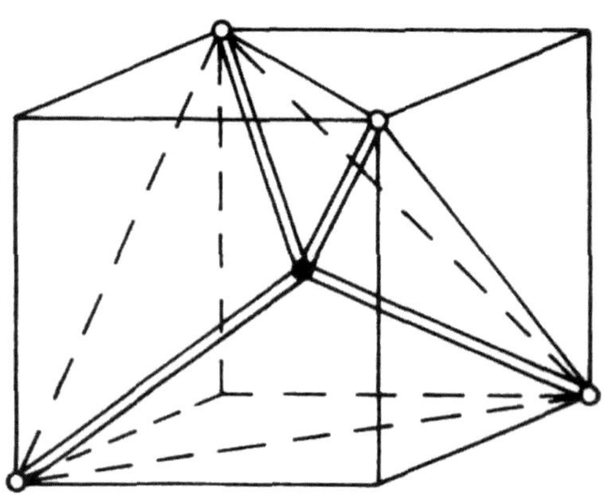

Fig. 8.6. Tetrahedral binding in diamond.

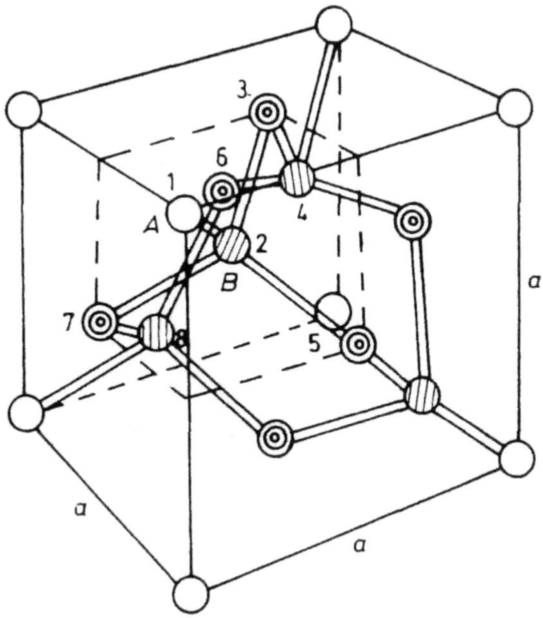

Fig. 8.7. Cubic cell in the diamond lattice.

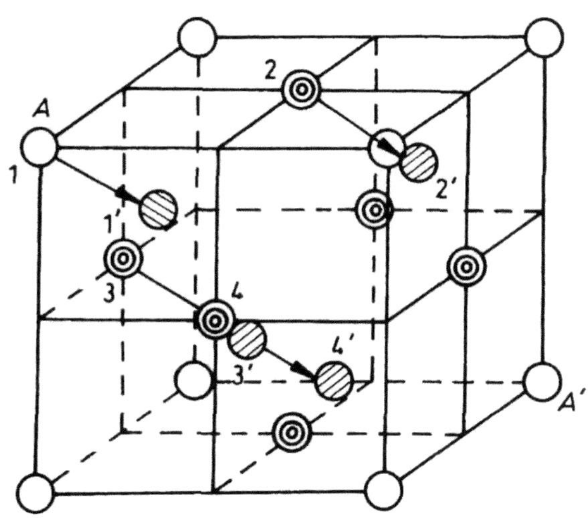

Fig. 8.8. Construction of the diamond-type lattice.

we can select a group of 18 atoms which build a cubic cell as shown in Fig. 8.7. To represent the location of atoms in this group, we divide the unit cell of (fcc) lattices into 8 equal cubes (see Fig. 8.8). Putting additional atoms (shaded circles in Fig. 8.8) into centers of four of these cubes, we obtain diamond like lattices. Making use of Fig. 8.8, we find that there are 8 atoms per unit cell in the diamond like lattice. This lattice can be represented as a sum of two (fcc)-type lattices shifted parallel to the diagonal (AA') of the cube at one quarter of its length, which is clearly seen in Fig. 8.8. Under this shift the atoms indexed by 1, 2, 3 and 4 will take places indexed by $1'$, $2'$, $3'$ and $4'$, whilst the other atoms of primary fcc lattices (open and concentric circles) take places outside the cube. Considering the packing of rigid spheres, we must take into account that the shortest distance between atoms is 1–$1'$ distance, i.e. the corresponding radius $R = 3^{1/2}\, a/8$, and we find for the packing coefficient:

$$f = \frac{8\, V_s}{a^3} = \frac{\pi\sqrt{3}}{16} \simeq 0.340\,. \tag{8.8}$$

Evidently, the packing coefficient is small, so that there are empty channels, especially, along $\langle 110\rangle$ direction. It is clearly seen in Fig. 8.9, that we could imbed as many atoms into empty channels as the lattice already contains.

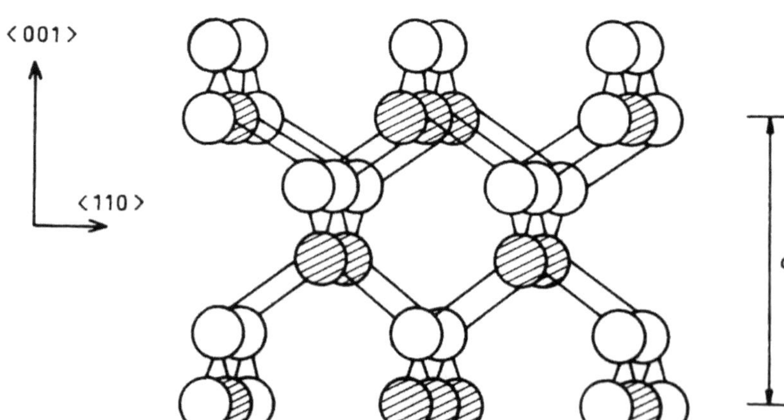

Fig. 8.9. Diamond lattice along $\langle 110\rangle$ axis. Overlapping circles show atoms lying on $\langle 110\rangle$ axis. Shaded circles correspond to atoms belonging to (fcc)-lattice.

8.2. Reciprocal Lattice

A crystal potential is periodic (see Eq. 8.3)

$$U(\mathbf{r}+\mathbf{l}) = U(\mathbf{r})\,. \tag{8.9}$$

The periodic function $U(\mathbf{r})$ can be represented as Fourier series

$$U(\mathbf{r}) = \sum_{\mathbf{q}} G(\mathbf{q}) e^{-i\mathbf{qr}}\,, \tag{8.10}$$

where the vectors $\mathbf{q}$, over which the summation is carried out, satisfy the condition $e^{-i\mathbf{ql}} = 1$ in accordance with Eq. (8.9). Then

$$\mathbf{ql} = 2\pi N\,, \quad N \text{ is an integer}\,. \tag{8.11}$$

It follows from Eq. (8.11) that the vectors $\mathbf{q}$ can be represented in the form

$$\mathbf{q} = 2\pi(n_1\mathbf{b}_1 + n_2\mathbf{b}_2 + n_3\mathbf{b}_3)\,, \tag{8.12}$$

where n_1, n_2 and n_3 are integers, and the vectors $\mathbf{b}_i$, called the reciprocal lattice vectors, are chosen to satisfy the relation

$$\mathbf{a}_i\mathbf{b}_j = \delta_{ij}\,. \tag{8.13}$$

The explicit form of $\mathbf{b}_i$ is

$$\mathbf{b}_1 = [\mathbf{a}_2 \times \mathbf{a}_3]\Omega_0^{-1}, \quad \mathbf{b}_2 = [\mathbf{a}_3 \times \mathbf{a}_1]\Omega_0^{-1}, \quad \mathbf{b}_3 = [\mathbf{a}_1 \times \mathbf{a}_2]\Omega_0^{-1}\,, \tag{8.14}$$

where $\Omega_0 = (\mathbf{a}_1(\mathbf{a}_2 \times \mathbf{a}_3))$ is the volume of the unit cell. For example, the reciprocal lattice for primitive cubic lattice with edge length a will be of the same type with edge length $b = 1/a$. The reciprocal lattice can be regarded as the Fourier transform of the direct lattice and vice versa. If we multiply the vector $\mathbf{q}$ (8.12) by arbitrary direct lattice vector $\mathbf{r} = m_1\mathbf{a}_1 + m_2\mathbf{a}_2 + m_3\mathbf{a}_3$, then the equation

$$\mathbf{qr} = 2\pi(m_1n_1 + m_2n_2 + m_3n_3) = 2\pi N\,, \tag{8.15}$$

determines a family of parallel crystal planes. If the integers n_1, n_2 and n_3 are relatively prime, they are Miller indices of these planes (otherwise, they must be divided by their greatest common divisor).

8.3. Geometric Structure Factor

According to Eq. (8.3), the density $\rho(\mathbf{r})$ can be represented as Fourier series

$$\rho(\mathbf{r}) = \Omega_0^{-1} \sum_{\mathbf{q}} S(\mathbf{q}) e^{-i\mathbf{q}\mathbf{r}}, \tag{8.16}$$

where $\Omega_0 = l^3$ for a cubic lattice, l is the lattice constant (the length of the cube edge), and Fourier component $S(\mathbf{q})$ is determined by the formula:

$$S(\mathbf{q}) = \int \rho(\mathbf{r}) e^{i\mathbf{q}\mathbf{r}} d^3 r \tag{8.17}$$

The integration in (8.17) is carried out over one unit cell if we represent the density $\rho(\mathbf{r})$ within the unit cell in the form (cf. Eq. (8.2)) of $\rho(\mathbf{r}) = \sum_{a=1}^{s} \delta(\mathbf{r} - \mathbf{r}_a - \mathbf{l})$, we obtain

$$S(\mathbf{q}) = \sum_{a=1}^{s} e^{i\mathbf{q}\mathbf{r}_a}. \tag{8.18}$$

The quantity $S(\mathbf{q})$ is called a geometric structure factor or simply a structure factor. It appears in the description of the X-ray diffraction, pair production and photon emission processes in crystal. Let us give several examples.

(A) Body-centered cubic lattice.

There are two atoms in the unit cell with coordinates (see Fig. 8.1(b) and Table 8.1)

$$\mathbf{r}_1 = (0,0,0), \quad \mathbf{r}_2 = l\left(\frac{1}{2}, \frac{1}{2}, \frac{1}{2}\right). \tag{8.19}$$

Substituting these $\mathbf{r}_a$ into Eq. (8.18) and using the representation of reciprocal lattice vectors (8.12), we obtain for (*bcc*) structure factor

$$S = 1 + \exp[i\pi(n_1 + n_2 + n_3)] = 1 + \cos\pi(n_1 + n_2 + n_3). \tag{8.20}$$

(B) Face-centered cubic lattice.

There are four atoms in the unit cell (see Fig. 8.1(c) and Table 8.1) with coordinates

$$\begin{aligned} \mathbf{r}_1 &= (0,0,0), \quad \mathbf{r}_2 = l\left(\frac{1}{2}, \frac{1}{2}, 0\right), \\ \mathbf{r}_3 &= l\left(0, \frac{1}{2}, \frac{1}{2}\right), \quad \mathbf{r}_4 = l\left(\frac{1}{2}, 0, \frac{1}{2}\right) \end{aligned} \tag{8.21}$$

Substituting these $\mathbf{r}_a$ into Eq. (8.18), we find for (fcc) structure factor

$$S = 1 + \exp[i\pi(n_1 + n_2)] + \exp[i\pi(n_1 + n_3)] + \exp[i\pi(n_2 + n_3)] = 1 + \cos\pi(n_1 + n_2) + \cos\pi(n_1 + n_3) + \cos\pi(n_2 + n_3)\,. \tag{8.22}$$

(C) Diamond like lattice *fcc(d)*.

As noted above, the diamond like lattice can be represented as a sum of two fcc-lattices, each containing four atoms and shifted along the diagonal of the cube at one fourth of its length (see Fig. 8.7). Then the atoms of the first fcc-lattice are situated at the points

$$\begin{aligned}
\mathbf{r}_1 &= (0,0,0) = \left(\frac{1}{8},\frac{1}{8},\frac{1}{8}\right) - \mathbf{r}_0 = \mathbf{r}_1' - \mathbf{r}_0\,,\\
\mathbf{r}_2 &= \left(0,\frac{1}{2},\frac{1}{2}\right) = \left(\frac{1}{8},\frac{5}{8},\frac{5}{8}\right) - \mathbf{r}_0 = \mathbf{r}_2' - \mathbf{r}_0\,,\\
\mathbf{r}_3 &= \left(\frac{1}{2},0,\frac{1}{2}\right) = \left(\frac{5}{8},\frac{1}{8},\frac{5}{8}\right) - \mathbf{r}_0 = \mathbf{r}_3' - \mathbf{r}_0\,,\\
\mathbf{r}_4 &= \left(\frac{1}{2},\frac{1}{2},0\right) = \left(\frac{5}{8},\frac{5}{8},\frac{1}{8}\right) - \mathbf{r}_0 = \mathbf{r}_4' - \mathbf{r}_0\,,
\end{aligned} \tag{8.23}$$

where the auxiliary vector $\mathbf{r}_0 = (1/8, 1/8, 1/8)$ is introduced; all the vectors in Eq. (8.23) (as well as in Eq. (8.24)) are given in units of the lattice constant l. For the atoms of the second fcc-lattice, we have

$$\begin{aligned}
\mathbf{r}_5 &= \left(\frac{1}{4},\frac{1}{4},\frac{1}{4}\right) = \mathbf{r}_1' + \mathbf{r}_0, \mathbf{r}_6 = \left(\frac{1}{4},\frac{3}{4},\frac{3}{4}\right) = \mathbf{r}_2' + \mathbf{r}_0\,,\\
\mathbf{r}_7 &= \left(\frac{3}{4},\frac{1}{4},\frac{3}{4}\right) = \mathbf{r}_3' + \mathbf{r}_0, \mathbf{r}_8 = \left(\frac{3}{4},\frac{3}{4},\frac{1}{4}\right) = \mathbf{r}_4' + \mathbf{r}_0\,.
\end{aligned} \tag{8.24}$$

The vectors $\mathbf{q}$ (8.12) are expressed by reciprocal lattice vectors $\mathbf{b}_i$ which, according to Eq. (8.14), have the form:

$$\begin{aligned}
\mathbf{b}_1 &= l^2\left(\left(\frac{1}{2},0,\frac{1}{2}\right) \times \left(\frac{1}{2},\frac{1}{2},0\right)\right) \cdot \frac{4}{l^3}\\
&= \frac{1}{l}(-1,1,1), \mathbf{b}_2 = \frac{1}{l}(1,-1,1), \mathbf{b}_3 = \frac{1}{l}(1,1,-1)\,,
\end{aligned} \tag{8.25}$$

then the vectors $\mathbf{q}$ (8.12) are

$$\mathbf{q} = \frac{2\pi}{l}(-n_1 + n_2 + n_3, n_1 - n_2 + n_3, n_1 + n_2 - n_3)\,. \tag{8.26}$$

Substituting Eqs. (8.23)–(8.26) into Eq. (8.18), we have for the structure factor of diamond like lattice:

$$\begin{aligned} S &= \sum_{a=1}^{8} \exp(i\mathbf{q}\mathbf{r}_a) = 2\cos(\mathbf{q}\mathbf{r}_0) \sum_{a=1}^{4} \exp(i\mathbf{q}\mathbf{r}'_a) \\ &= 2\cos(\mathbf{q}\mathbf{r}_0) e^{i\mathbf{q}\mathbf{r}_0} S_{fcc}\,, \end{aligned} \tag{8.27}$$

where the quantity S_{fcc} is given by Eq. (8.22). Taking into account that $\mathbf{q}\mathbf{r}_0 = \pi(n_1 + n_2 + n_3)/4$, we have, finally

$$\begin{aligned} S &= \left\{1 + \exp\left[i\frac{\pi}{2}(n_1 + n_2 + n_3)\right]\right\} \cdot [1 + \cos\pi(n_1 + n_2) \\ &\quad + \cos\pi(n_1 + n_3) + \cos\pi(n_2 + n_3)] \\ &= \left\{1 + \exp\left[i\frac{\pi}{2}(n_1 + n_2 + n_3)\right]\right\}(\cos\pi n_1 + \cos\pi n_3) \\ &\quad \times (\cos\pi n_1 + \cos\pi n_2)\,. \end{aligned} \tag{8.28}$$

9. RELATIVISTIC PARTICLE MOTION IN A CRYSTAL POTENTIAL

9.1. Crystal Potential with Regard for Lattice Vibrations

We represent the potential of an individual atom of the crystal lattice as the Fourier integral

$$\varphi_a(\mathbf{r}) = \frac{1}{(2\pi)^3} \int g_a(\mathbf{q}) e^{-i\mathbf{q}\mathbf{r}} d^3q\,. \tag{9.1}$$

Atoms deviate from their equilibrium locations in crystal lattice owing to thermal (zero) vibrations. The probability $w(\mathbf{x})$ of a displacement $\mathbf{x}$ has the Gaussian form

$$w(\mathbf{x}) = \frac{1}{(2\pi u_1^2)^{3/2}} \exp\left(-\frac{x^2}{2u_1^2}\right), \quad \int w(\mathbf{x}) d^3x = 1\,, \tag{9.2}$$

where u_1 is the amplitude of thermal vibrations. The potential of an individual atom, averaged over thermal displacements, has the form

$$\varphi(\mathbf{r}) = \int \varphi_a(\mathbf{r} - \mathbf{x}) w(\mathbf{x}) d^3x\,. \tag{9.3}$$

Using Eqs. (9.1) and (9.2), we find for the Fourier transform of the potential (9.3):

$$\begin{aligned} g(\mathbf{q}) &= \int \varphi(\mathbf{r}) e^{i\mathbf{q}\mathbf{r}} d^3 r \\ &= \int \varphi_a(\mathbf{r}-\mathbf{x}) w(\mathbf{x}) e^{i\mathbf{q}\mathbf{r}} d^3 x d^3 r \\ &= \int \varphi_a(\mathbf{r}) e^{i\mathbf{q}\mathbf{r}} d^3 r \int w(\mathbf{x}) e^{i\mathbf{q}\mathbf{x}} d^3 x \\ &= g_a(\mathbf{q}) \exp(-u_1^2 q^2/2) \,. \end{aligned} \tag{9.4}$$

The crystal potential is obtained by summing up the potentials of individual atoms with regard for their locations in crystal lattice:

$$\begin{aligned} U(\mathbf{r}) = \sum_i \varphi(\mathbf{r}-\mathbf{r}_i) &= \frac{1}{(2\pi)^3} \int d^3 p g(\mathbf{p}) \\ \times \sum_i \exp[i(\mathbf{p}, \mathbf{r}_i - \mathbf{r})] &= \frac{1}{l^3} \int d^3 p g(\mathbf{p}) S(\mathbf{p}) e^{-i\mathbf{p}\mathbf{r}} \\ \times \sum_{\mathbf{q}} \delta(\mathbf{p}-\mathbf{q}) &= \frac{1}{l^3} \sum_{\mathbf{q}} S(\mathbf{q}) g(\mathbf{q}) e^{-i\mathbf{q}\mathbf{r}} \\ \equiv \sum_{\mathbf{q}} G(\mathbf{q}) e^{-i\mathbf{q}\mathbf{r}} \,, \end{aligned} \tag{9.5}$$

where $\mathbf{q} = 2\pi(n_1, n_2, n_3)/l$; l is the lattice constant; $S(\mathbf{q})$ is the structure factor (see Eqs. (8.18), (8.20), (8.22), and (8.28)).

The Moliere approximation for the individual atom potential (see Eq. (7.44)) will be used in particular calculations when the quantity $g_a(\mathbf{q})$ has the form

$$g_a(\mathbf{q}) = 4\pi Z\alpha \sum_{i=1}^{3} \frac{\alpha_i}{q^2 + \beta_i^2} \,, \tag{9.6}$$

then we obtain from Eqs. (9.4) and (9.5)

$$G(\mathbf{q}) = \frac{4\pi Z\alpha}{l^3} \exp\left(-\frac{u_1^2 q^2}{2}\right) S(\mathbf{q}) \sum_{i=1}^{3} \frac{\alpha_i}{q^2 + \beta_i^2} \,. \tag{9.7}$$

Side by side with the Moliere approximation of the atomic potential, other approximations are used, e.g. Doyle–Turner (see Doyle and Turner, 1968) and

Cox-Bonham (see Cox and Bonham, 1967) approximations. Crystal potentials calculated using these approximations (see Bazylev and Zhevago, 1987) turn out to be sufficiently similar in their shape and magnitude. For example, the difference of the potential well depth is less than 15%.

The motion of relativistic particles travelling along the family of crystal planes $\{hkm\}$ is mainly governed by the continuum potential of these planes, which can be obtained from Eq. (9.5) by letting in (9.5)

$$\mathbf{q} = 2\pi n(h,k,m)/l\,. \tag{9.8}$$

Then the sum (over n) in (9.5) becomes one-dimensional, giving the corresponding one-dimensional periodic continuum potential. Similarly, while travelling along the axes $\langle hkm\rangle$, relativistic particles experience, mainly, the continuum potential of these axes. The latter is reproduced by Eq. (9.5), if the summation is restricted by the values of n_1, n_2 and n_3, satisfying the condition

$$n_1 h + n_2 k + n_3 m = 0\,. \tag{9.9}$$

The potential of an isolated atomic chain which is dominant at small distances $r_\perp = \rho \ll l$ to this chain, can be obtained from Eq. (9.5) with regard for condition (9.9) if we integrate over $\mathbf{q}_\perp$ instead of the corresponding summation:

$$\sum_{\mathbf{q}} \to \frac{s}{(2\pi)^2}\int d^2 q_\perp\,;$$

$$U_{st}(\rho) = \frac{1}{(2\pi)^2 d}\int d^2 q_\perp \exp\left(-\frac{u_1^2\mathbf{q}_\perp^2}{2} - i\mathbf{q}_\perp\rho\right) g_a(\mathbf{q}_\perp)\,, \tag{9.10}$$

where s is the area per one chain, d is the mean distance between atoms in this chain. Using Eq. (9.6) and an exponential parametrization, we can take the Gaussian integral with respect to $\mathbf{q}_\perp$ Simple transformations yield the following integral representation for the potential of an isolated atomic chain (Avakyan, Zhevago and Jan Shi, 1982):

$$U_{st}(\rho) = \frac{Ze^2}{d}\sum_{i=1}^{3}\alpha_i \exp\left(\frac{\beta_i^2 u_1^2}{2}\right)$$
$$\times \int_0^1 \frac{dt}{t}\exp\left(-\frac{\rho^2 t}{2u_1^2} - \frac{\beta_i^2 u_1^2}{2t}\right) \tag{9.11}$$

In the limiting case $u_1 \to 0$, the integral over t can be extended to ∞, and we have, as a result

$$U_{st}(\rho) = \frac{2Ze^2}{d} \sum_{i=1}^{3} \alpha_i K_0(\beta_i \rho) \,. \tag{9.12}$$

This formula has meaning only in the region $u_1 \ll \rho \ll l$.

The expressions (9.11) and (9.12) are useful at investigating the characteristic features of the continuum axes potential very close to a chain. In applications, the continuum axes potential can be conveniently represented in the whole area per one axis by a simple function which reproduces the behavior (9.11) close to the axis. Below, we shall use the following form of the continuum axes potential:

$$\begin{gathered} U(x) = V_0 \left[\ln \left(1 + \frac{1}{x + \eta} \right) - \ln \left(1 + \frac{1}{x_0 + \eta} \right) \right] , \\ x = \frac{\rho^2}{a_s^2} , \quad x_0^{-1} = \pi n d a_s^2 = \frac{\pi a_s^2}{s} , \end{gathered} \tag{9.13}$$

where n is the atom density in the crystal, a_s is the effective screening radius of the potential. For estimates, we can assume that $V_0 \simeq Ze^2/d, \eta \simeq 2u_1^2/a_s^2$. Actually, these parameters were determined by means of a fitting procedure, using Eqs. (9.5) and (9.9). They are given in Table 9.1.

Table 9.1. Parameters of the axial potential (9.13).

Crystal	Axis	T(K)	u_1 10^{-8}cm	V_0(eV)	η	u_s 10^{-8}cm	x_0	U_0(eV)
C	111	293	0.040	29	0.025	0.326	5.5	103
Si	111	293	0.075	54	0.150	0.299	15.1	106
Si	110	293	0.075	70	0.145	0.324	15.8	140
V	111	293	0.082	135	0.135	0.306	18.0	280
Cr	111	293	0.061	165	0.122	0.272	20.7	358
Fe	111	293	0.068	180	0.145	0.276	19.8	363
Ge	111	293	0.085	91	0.130	0.300	16.3	191
Ge	110	280	0.083	110	0.115	0.337	15.8	243
Ge	110	100	0.054	114.5	0.063	0.302	19.8	318
W	111	293	0.050	417	0.115	0.215	39.7	937
W	111	77	0.030	348	0.027	0.228	35.3	1255

The parameters of planes continuum potential were determined by a similar procedure. This potential was chosen in the form of

$$U(x) = V[\cosh(\delta(\sqrt{1+\eta^2} - \sqrt{y^2+\eta^2})) - 1], \tag{9.14}$$

where $y = 2x/d_{pl}, |y| \leq 1$. Note that the parameter η is proportional to u_1^2 and $\delta \sim d_{pl}/a_s$. For some crystals, the values of these parameters are given in Table 9.2. It should be noted that the evaluation of a one-dimensional sum over n in Eq. (9.5) is not a very difficult task, so this form of planes continuum potential can be adequately used in numerical calculations.

Table 9.2. Parameters of the potential (9.14) for the (110) plane of various crystals at room temperature.

Crystal	V(eV)	δ	η^2	d_{pl} (10^{-8}cm)	U_0(eV)
C	7.7	2.2	0.005	1.26	23
Si	3.5	2.9	0.0045	1.92	23
Ge	4.5	3.2	0.0052	2.00	40
Fe	8.0	3.1	0.002	1.017	70
W	6.4	3.85	0.0007	1.119	130

9.2. Charged Particle Motion in a Crystal

Consider the motion of a changed particle in the potential $U(\mathbf{r})$. If the vector of initial velocity of a relativistic particle is not too close to some axial or planar direction, its motion, in accordance with the estimate of the potential well depth V_0 (9.13), can be described in terms of the perturbation theory in crystal potential. The first order of such an approach is the rectilinear trajectory approximation. Remember that the acceleration of the relativistic particles (see Eqs. (1.29)–(1.31)) is almost transverse

$$\dot{\mathbf{v}} \simeq \dot{\mathbf{v}}_\perp = -\frac{1}{\varepsilon}\frac{\partial U(\mathbf{r})}{\partial \mathbf{r}_\perp}. \tag{9.15}$$

By substituting into the right-hand side of Eq. (9.15) the coordinate $\mathbf{r}$ in the form $\mathbf{r} = \mathbf{r}_0 + \mathbf{v}_0 t$, making use of Eq. (9.5) and integrating with respect to time t, we obtain

$$\mathbf{v}(t) = \mathbf{v}_0 - \frac{1}{\varepsilon}\sum_{\mathbf{q}} \frac{\mathbf{q}_\perp}{q_\parallel} G(\mathbf{q}) e^{-i\mathbf{q}\mathbf{r}_0 - iq_\parallel t}, \tag{9.16}$$

where $q_{\|} = \mathbf{q}\mathbf{v}_0, \mathbf{q}_{\perp} = \mathbf{q} - \mathbf{v}_0(\mathbf{q}\mathbf{v}_0), 1 - |\mathbf{v}_0| \ll 1$. Equation (9.16) describes the transverse particle motion characterized by the set of frequencies $\omega_0(\mathbf{q}) \equiv q_{\|}(\mathbf{q})$ (see Eq. (9.5) for notations). If the angle ϑ_0 particle velocity $\mathbf{v}_0$ to some crystal axis $\mathbf{e}_z$ is small, we have for vectors $\mathbf{q}$ satisfying the condition $\mathbf{q}\mathbf{e}_z = 0$

$$q_{\|} \simeq \boldsymbol{\vartheta}_0 \mathbf{q}_{\perp}, \tag{9.17}$$

where $\boldsymbol{\vartheta}_0 = \mathbf{v}_0 - \mathbf{e}_z(\mathbf{v}_0\mathbf{e}_z)$, whilst for the remaining vectors $\mathbf{q}$, we have $q_{\|} \sim q_{\perp}$, and their contribution to the sum in Eq. (9.16) is negligible. Then from Eq. (9.16) we obtain ($G(\mathbf{q}) \sim V_0$) an estimate of the variation of transverse particle velocity

$$|\Delta\mathbf{v}| \sim \frac{V_0}{\varepsilon}\left|\frac{\mathbf{q}_{\perp}}{\boldsymbol{\vartheta}_0\mathbf{q}_{\perp}}\right| \tag{9.18}$$

When $\boldsymbol{\vartheta}_0\mathbf{q}_{\perp} \sim \vartheta_0 q_{\perp}$ (at axial orientation) we have $|\Delta\mathbf{v}| \sim V_0/\varepsilon\vartheta_0$. The rectilinear trajectory approximation as well as Eq. (9.16) are valid, if the condition $|\Delta\mathbf{v}| \ll \vartheta_0$ is fulfilled, and we obtain

$$\frac{|\Delta\mathbf{v}|}{\vartheta_0} \sim \frac{V_0}{\varepsilon\vartheta_0^2} \ll 1\,. \tag{9.19}$$

If the opposite inequality holds, the special type of motion, called channeling, takes place. It is determined by the continuum axes potential. If the condition (9.19) is fulfilled, but the direction $\boldsymbol{\vartheta}_0$ turns out to be close to some crystal axes ($\mathbf{e}_y$) lying in the plane perpendicular to $\mathbf{e}_z$, then the vectors $\mathbf{q}$ contributing to Eq. (9.16) must satisfy the additional condition $\mathbf{q}\mathbf{e}_y = 0$. This means that a particle moves in the continuum planes potential, which depends on the x-coordinate. In the planar case, we consider the angle ϑ_0 in Eq. (9.19) to be the particle velocity component v_{0x} perpendicular to the corresponding plane. The rectilinear trajectory approximation holds, if $V_0/\varepsilon v_{0x}^2 \ll 1$, and planar channeling takes place in the opposite case.

9.3. Particle Motion at Planar Channeling

A great many papers and reviews deal with the particle motion at channeling conditions. Much of the basic theory of the channeling process can be found in the review published by Lindhard, 1965. Later on reviews by Gemmel, 1974, and Andersen *et al.*, 1977 and a book by Ohtsuki, 1983 were published. An extensive bibliography concerning channeling problems can be found in the mentioned works.

The motion at planar channeling takes place in one-dimensional potential $U(x)$, and can be therefore described in the most simple way. The Hamiltonian corresponding to the motion along the axis $\mathbf{e}_x$ has a non-relativistic form, where the energy ε plays a role of the particle mass.

$$\varepsilon \simeq \sqrt{p_z^2 + p_y^2 + m^2}\,, \quad \mathcal{H} = \frac{p_x^2}{2\varepsilon} + U(x)\,. \tag{9.20}$$

It is well known, that a one-dimensional motion can be described in quadratures for any potential:

$$t = \int \frac{dx}{v_\perp(x)} = \sqrt{\frac{\varepsilon}{2}} \int \frac{dx}{\sqrt{\varepsilon_\perp - U(x)}}\,, \qquad \varepsilon_\perp = \mathcal{H}\,, \tag{9.21}$$

where $\varepsilon_\perp$ is the integral of motion. If the denominator $v_\perp(x)$ in (9.21) vanishes at so-called turning points, the motion in the x-direction is finite. Then a positron is moving between two adjacent planes in symmetric potential $U(x) = U(-x)$, $U(0) = 0$ and $U_0 = U(d/2)$ is the potential well depth; d is the interplanar distance. A channeled electron is moving around one plane with $U(x) \to -U(x)$. Obviously, for particles having $\varepsilon_\perp < U_0$, the motion is finite. They will be termed channeled particles (or particles in a channel). In the opposite case $\varepsilon_\perp > U_0$, the motion is infinite, and particles with such values of $\varepsilon_\perp$ will be termed non-channeled (or over-barrier) particles.

Assuming the thickness of the crystal transition layer to be small comparatively to the particle's path during the period of its oscillation in the potential $U(x)$, we easily obtain the distribution over particle transverse energy $\varepsilon_\perp$ after entering into crystal. Taking into account that the relativistic character of the particle motion and the smallness of the ratio $p_\perp/p_\parallel$, we have for the potential $U(z,x) = \vartheta(z)U(x)$

$$\varepsilon_\perp(z = -0, x_0) - \varepsilon_\perp(z = +0, x_0) = U(x_0)$$

$$\varepsilon_\parallel(z = -0, x_0) - \varepsilon_\parallel(z = +0, x_0) = -U(x_0)$$

so that, just after entering into crystal, the particle transverse energy is

$$\varepsilon_\perp = \varepsilon_0 + U(x_0)\,, \quad \varepsilon_0 = \frac{\varepsilon\vartheta_0^2}{2}\,, \tag{9.22}$$

where x_0 is the entry point. For $\vartheta_0 = 0 (\varepsilon_0 = 0)$, all the particles are channeled, since $\varepsilon_\perp = U(x_0) \le U_0$. For $\varepsilon_0 \ge U_0$ all the particles are moving above the potential barrier, since $\varepsilon_\perp \geq U_0$. The angle ϑ_c, satisfying the condition $\varepsilon_0(\vartheta_c) = U_0$, is called the critical Lindhard angle:

$$\vartheta_c = \sqrt{\frac{2U_0}{\varepsilon}} \tag{9.23}$$

Under realistic conditions, the spatial width of a beam is much larger than interplanar distances, and therefore we can consider the uniform distribution of particles over transverse coordinate x outside of a crystal. Then we find

$$\begin{aligned}\varphi(\varepsilon_\perp) &= \frac{dN(\varepsilon_\perp)}{d\varepsilon_\perp} = \frac{1}{d}\left|\frac{d\varepsilon_\perp}{dx_0}\right|^{-1} \\ &= \frac{1}{d}\sum_i \left|\frac{dU(x_0)}{dx_{0i}}\right|^{-1},\end{aligned} \tag{9.24}$$

where x_{0i} is a solution of Eq. (9.22). The following consideration will be restricted to the case of potentials having two symmetric turning points for arbitrary $\varepsilon_\perp \leq U_0$. If the distribution (9.24) holds, the crystal will be termed thin, contrarily to thick crystal where some processes change the function $\varphi(\varepsilon_\perp)$. This change is mainly determined by multiple scattering and the radiation energy losses of particles. The particle distribution over transverse energy in thick crystals will be considered in the next section.

Use of a distribution function over phase space is the serious alternative to the detailed investigation of individual particle trajectories. The theory of radiation in a medium in the presence of an external field formulated in Sec. 7 may serve as an example of such an approach. Let us consider some simple cases. If we know the distribution function $\varphi(\varepsilon_\perp)$, the distribution in the phase space can be easily obtained. Indeed, for definite value $\varepsilon_\perp$, the averaged coordinate distribution is

$$dw_{\varepsilon_\perp}(x) = \frac{dt(x)}{\tau(\varepsilon_\perp)} = \frac{dx}{v_\perp(x)\tau(\varepsilon_\perp)}, \tag{9.25}$$

where $\tau(\varepsilon_\perp) = T_c(\varepsilon_\perp)/2$ for channeled particles and $\tau(\varepsilon_\perp) = T_{nc}(\varepsilon_\perp)$ for over-barrier particles. Here $T(\varepsilon_\perp)$ is the period of particle oscillation. We find from Eq. (9.21):

$$\tau(\varepsilon_\perp) = \sqrt{\frac{\varepsilon}{2}}\int_{-x_c}^{x_c} \frac{dx}{\sqrt{\varepsilon_\perp - U(x)}}, \tag{9.26}$$

where x_c is determined by the condition

$$\varepsilon_\perp - U(x_c) = 0\,, \quad x_{nc} = \frac{d}{2} \tag{9.27}$$

Taking into account the distribution over transverse energy, we have

$$dN(\varepsilon_\perp, x) = \frac{\varphi(\varepsilon_\perp)d\varepsilon_\perp dx}{\tau(\varepsilon_\perp)v_\perp(x)}\,. \tag{9.28}$$

Going over to the momentum distribution by means of the relation

$$\varepsilon_\perp = \frac{p_x^2}{2\varepsilon} + U(x)\,, \tag{9.29}$$

we obtain

$$dN(p_x, x) = \widetilde{\varphi}\left(\frac{p_x^2}{2x} + U(x)\right) dp_x dx\,, \tag{9.30}$$

where

$$\widetilde{\varphi}(\varepsilon_\perp) = \frac{\varphi(\varepsilon_\perp)}{\tau(\varepsilon_\perp)}\,. \tag{9.31}$$

The expressions (9.30) and (9.31) imply the uniform distribution of particles over phase space for the one-dimensional problem, if the function $\varphi(\varepsilon_\perp)$ is proportional to the oscillation period $T(\varepsilon_\perp)$. Integrating (9.28) with respect to $\varepsilon_\perp$, we find the resulting distribution over x

$$\frac{dN(x)}{dx} = \sqrt{\frac{\varepsilon}{2}} \int \frac{\widetilde{\varphi}(\varepsilon_\perp)d\varepsilon_\perp}{\sqrt{\varepsilon_\perp - U(x)}}\,. \tag{9.32}$$

In particular, by substituting (9.24) into Eqs. (9.31) and (9.32), we obtain the coordinate distribution in a thin crystal.

Let us now investigate specific properties of a particle trajectory. Our main problem is to calculate the frequencies and to carry out the Fourier analysis of a particle motion, i.e. to find

$$v_m = \int_0^T dt v_x(t) e^{im\omega_0 t}\,, \quad m = 1, 2, 3 \ldots\,,$$

since the radiation characteristics at quasi-periodic motion (see Secs. 1.8–1.12 and 5) are directly determined by these quantities.

Consider the particle motion in the planar potential (9.14), neglecting lattice vibrations ($u_1 = 0$):

$$U(x) = V(\cosh y - 1)\,, \quad y = \frac{x}{a_s}, -\frac{d}{2} \leq x \leq \frac{d}{2}\,,$$
$$U_0 = V(\cosh\delta - 1)\,, \quad \delta = \frac{d}{2a_s}\,. \tag{9.33}$$

Such approximation for planar potential leads mainly to a change of its shape near to crystal planes and therefore affects the motion of a relatively small fraction of channeled particles. For positrons, the form (9.33) can be used directly, since the minimum of the potential is situated in the middle between the adjacent planes ($V > 0$). Then we find for channeled positrons, using Eqs. (9.21):

$$t = \sqrt{\frac{\varepsilon}{2V}}a_s \int_{-y_0}^{y} \frac{dy}{\sqrt{\cosh y_0 - \cosh y}}\,, \tag{9.34}$$

where $y = x/a_s$, the turning point y_0 is determined by the relation $\cosh(y_0) = 1 + 2z\sinh^2(\delta/2)$, and $z = \varepsilon_\perp/U_0$. By letting $\sin\psi = \{(\cosh y_0 - \cosh\ y)/(\cosh y_0 - 1)\}^{1/2}$, the integral (9.34) is reduced to

$$t = \sqrt{\frac{\varepsilon}{V}}a_s \frac{F(\varphi,k)}{\cosh\frac{y_0}{2}}\,, \quad F(\varphi,k) = \int_0^{\varphi} \frac{d\varphi}{\sqrt{1 - k^2\sin^2\psi}}\,. \tag{9.35}$$

Here

$$k^2 = z\,\sinh^2\left(\frac{\delta}{2}\right) \Big/ \left(1 + z\,\sinh^2\left(\frac{\delta}{2}\right)\right)\,,$$

$$\varphi = \begin{cases} \beta, & y \leq 0\,; \\ \pi - \beta, & y \geq 0\,; \end{cases} \quad \beta = \text{arc}\sin\sqrt{\frac{\cosh y_0 - \cosh y}{\cosh y_0 - 1}}\,, \tag{9.36}$$

the function $F(\varphi, k)$ is the elliptic integral of the first kind as determined by Gradshteyn and Ryzhik, 1965. For the periods of channeled particles, we have

$$T_c = T_0 k' K(k)\,, \tag{9.37}$$

where $T_0 = 4a_s(\varepsilon/V_0)^{1/2}, k' = (1 - k^2)^{1/2}$, and $K(k) \equiv F(\pi/2, k)$ is the complete elliptic integral of the first kind. Note that with increasing transverse energy $\varepsilon_\perp(z)$ the period T_c decreases monotonically from the value $\pi T_0/2$, for

$\varepsilon_\perp = 0$, to the value $T_0 K(\tanh\delta/2)/\cosh\delta/2$, for $\varepsilon_\perp = U_0$. For example, the latter quantity is $T_c = 1.115 T_0$ for $\delta = 2.5$ (the (110) Si plane).

In the considered case the velocity Fourier transform has the form

$$\int_0^{T/2} dt v_x(t) e^{i\tilde{\omega}t} = 2ia_s \int_0^{y_0} dy \sin\left[\frac{F(\beta,k)}{K(k)}\pi\left(l-\frac{1}{2}\right)\right], \tag{9.38}$$

where the relation $\tilde{\omega}T = 2\pi(2l-1)$ is taken into account which is valid for channeled particles. Going over to the variable $u = F(\beta,k)$ and making use of the expansion (see Gradshetyn and Ryzhik, 1965)

$$\mathrm{sn}u = \frac{\pi}{kK}\sum_{n=1}^{\infty}\frac{\sin\left[(n-\frac{1}{2})\frac{\pi u}{K}\right]}{\sinh\left[(n-\frac{1}{2})\frac{\pi K'}{K}\right]}, \quad K' = K(k'),$$

we find, finally

$$v_m^c = (1-e^{i\pi m})\frac{2\pi i a_s}{\sinh\left[\frac{\pi m K'}{2K}\right]}. \tag{9.39}$$

For over-barrier particles ($z = \varepsilon_\perp/U_0 > 1$), the dependence $t(y)$ is given by the integral (9.34), if we put the quantity $-\delta$ as the lower limit (in place of $-y_0$). Thus, this dependence is also determined by the function $F(\beta,k)$, where the quantities $\beta(y)$ and k were introduced in Eq. (9.36). In this case we have for the period

$$T_{nc} = \frac{T_0}{2}k'[K(k) - F(\alpha,k)], \tag{9.40}$$

where $\alpha = \beta(y=\delta) = \arcsin(1-1/z)$. Note that at $z \to 1$, the ratio T_c/T_{nc} tends to 2. This is connected to approximately equal velocities (at $z \simeq 1$) of channeled and over-barrier particles, whilst channeled particles pass the distance d twice during a period. As it will be seen below, such sharp distinction of periods leads to the clear separation of the contributions to the radiation spectrum from these particle groups.

For v_m^{nc}, we have correspondingly

$$v_m^{nc} = 4\pi a_s \sum_{n=1}^{\infty}\frac{g_n\cos(\lambda g_n)}{\sinh\left(g_n\frac{K'}{K}\right)\left[g_n^2 - (\frac{\pi m}{1-\lambda})^2\right]}, \tag{9.41}$$

where $g_n = \pi(n-1/2), \lambda = F(\alpha,k)/K$. Note that at $\pi m/(1-\lambda) \to g_n, \cos(\lambda g_n) \to 0$ and the ratio $\cos(\lambda g_n)/(g_n - \pi m/(1-\lambda))$ remains finite.

We use the potential (9.33) to describe the motion of electrons. It is convenient to choose $U_e(0) = 0$:

$$U_e(x) = -V\left[\cosh\left(\left|\frac{x}{a_s}\right| - \delta\right) - \cosh\delta\right], \quad -\frac{d}{2} \le x \le \frac{d}{2} \tag{9.42}$$

For electrons channeled in this potential

$$t = a_s\sqrt{\frac{\varepsilon}{2V}}\int_{-y_0}^{y_0}\frac{dy}{\sqrt{\cosh(\delta-|y|)-\cosh(\delta-y_0)}}, \tag{9.43}$$

where now $\cosh(\delta - y_0) = 1 + 2(1-z)\sinh^2(\delta/2)$ and the remaining notations are as in Eq. (9.34). The expression analogous to (9.38) reads now

$$\int_0^{T/2} dt v_x(t) e^{i\tilde{\omega}t} = 2ia_s\int_0^{y_0} dy \sin\left[\pi\left(l-\frac{1}{2}\right)\frac{F(\rho,k)}{F(\rho_0,k)}\right], \tag{9.44}$$

where

$$\rho = \arcsin\sqrt{1-\frac{(1-z)\sinh^2\left(\frac{\delta}{2}\right)}{\sinh^2\left(\frac{\delta-y}{2}\right)}}, \quad \rho_0 = \rho|_{y=0} = \arcsin\sqrt{z}\,.$$

Letting in Eq. (9.44) $u = F(\rho, k)$ when $dy = -du\ 2k'\ \mathrm{sn}(u)/\mathrm{cn}(u)$ and using the expansion of the quantity $\mathrm{sn}(u)/\mathrm{cn}(u)$ (Gradshteyn and Ryzhik, 1965):

$$\frac{\mathrm{sn}u}{\mathrm{cn}u} = \frac{\pi}{k'K}\sum_{n=1}^{\infty}(-1)^{n+1}\sin\left(\frac{\pi nu}{K}\right)\tanh\left(\frac{\pi nK'}{K}\right), \tag{9.45}$$

we can take the integral (9.44) and, by the same token, find the quantity v_m^c:

$$\begin{aligned}
T_e(z<1) &= T_0 kF(\rho_0,k)\,, \quad k^2 = \frac{\tilde{z}-1}{\tilde{z}-z}\,;\\
T_e(1\le z\le\tilde{z}) &= \frac{T_0}{2}[K - F(\rho_1,k)]\,, \quad k^2 = \frac{\tilde{z}-z}{\tilde{z}-1}\,;\\
T_e(z\ge\tilde{z}) &= \frac{T_0}{2}k'F\left(\frac{\pi}{2}-\rho_1\,,k\right), \quad k^2 = \frac{z-\tilde{z}}{z-1}\,,
\end{aligned} \tag{9.46}$$

where $\tilde{z} = \coth^2(\delta/2)$, the quantity $\rho_0(z)$ is determined in (9.44), and $\rho_1 = \arcsin(1/\cosh(\delta/2))$. With z increasing from 0 to 1, the period T_e increases. Its behavior in the neighborhood of the point $z = 1$ is described by the function

$\ln(1/|1-z|)$, and at further increase of z, it decreases monotonically (for $z \gg 1$, $T_e/T_0 \propto z^{-1/2}$). Here two circumstances should be noted: (1) the periods of channeled and over-barrier particles run the same values; (2) the interval of these values is infinite.

The main features of the electron motion at planar channeling can be easily observed by means of the simple potential model called "inverted parabola" (see Baier, Katkov and Strakhovenko, 1979b):

$$U(x) = U_0 \left[1 - \left(1 - \frac{2|x|}{d}\right)^2\right], \quad -\frac{d}{2} \le x \le \frac{d}{2}. \tag{9.47}$$

Within this model, we have

$$\begin{aligned} v_c(t) &= \frac{\kappa d}{2\cosh f}\sinh(f - \kappa|t|), -\frac{T_c}{2} \le t \le \frac{T_c}{2}; \\ v_{nc}(t) &= \frac{\kappa d}{2\sinh f}\cosh(f - \kappa|t|), \quad 2|t| \le T_{nc}, \end{aligned} \tag{9.48}$$

where

$$\begin{aligned} &\kappa = \frac{2}{d}\sqrt{\frac{2U_0}{\varepsilon}}, \quad f = \frac{1}{2}\ln\left|\frac{1+\sqrt{z}}{1-\sqrt{z}}\right|, \quad z = \frac{\varepsilon_\perp}{U_0}, \\ &T_c = T_0 f(z), \quad T_0 = \frac{4}{\kappa}, \quad T_{nc} = \frac{T_0}{2} f(z). \end{aligned} \tag{9.49}$$

The direct calculation of velocity Fourier components yields

$$\begin{aligned} v_m^c &= (1 - e^{i\pi m})\frac{d}{1 + \left(\frac{2\pi m}{T_c\kappa}\right)^2}, \\ v_m^{nc} &= \frac{d}{1 + \left(\frac{2\pi m}{T_{nc}\kappa}\right)^2}. \end{aligned} \tag{9.50}$$

With a view to references, we give some characteristics of the positron motion in the parabolic potential

$$U(x) = U_0 \left(\frac{2x}{d}\right)^2, \quad |x| \le \frac{d}{2} \tag{9.51}$$

The expression describing the velocity is well known:

$$v(t) = \frac{\kappa d}{2}\sqrt{z}\cos\kappa t, \quad |t| \le \frac{T}{2}, \tag{9.52}$$

where

$$T_c = \frac{\pi}{2} T_0 \,, \quad T_{nc} = \frac{T_0}{2} \arcsin \frac{1}{\sqrt{z}} \,. \tag{9.53}$$

For velocity Fourier components, we have

$$\begin{aligned} v_m^c &= \frac{\pi d}{2} \sqrt{z} \delta_{m,1} \,; \\ v_m^{nc} &= (-1)^{m+1} \frac{d}{\left(\frac{2\pi m}{\kappa T_{nc}}\right)^2 - 1} \,. \end{aligned} \tag{9.54}$$

At last, consider the conditions providing the classical description of the motion at planar channeling. According to quantum mechanics, such description is valid for large values of quantum numbers. The quasiclassical estimate of a number of levels in the one-dimensional potential well at planar channeling yields

$$N \sim \frac{d}{\lambda_c} \sqrt{\frac{\varepsilon U_0}{m^2}} \sim \sqrt{\gamma} \gg 1 \,, \quad \lambda_c = \frac{\hbar}{mc} \tag{9.55}$$

It is taken into account in (9.55) that $U_0 \sim 10 - 100$ eV, $d \sim 1 - 3$ A. Note that, owing to a specific shape of a potential well (cf. Eqs. (9.47) and (9.51)), the condition (9.55) is more restrictive for electrons than for positrons. The correspondence of classical (9.54) and quantum description of energy level population was considered by Andersen *et al.* 1977 and also by the authors Baier, Kathov and Strakhovenko, 1980. The quantum mechanical description of particle motion by means of wave functions, when the number of energy levels (9.55) is relatively small, can be found in a book by Ohtsuki, 1983.

9.4. Particle Motion at Axial Channeling

Because of very complicated potential shapes in a transverse plane, the analytical solution of Eq. (9.15) cannot be found. And what is more, such a solution does not exist for any realistic potential of an isolated atomic chain. The adequate approach here is the statistical description of particle distribution in the phase space. In some important cases, e.g., calculating the total intensity and radiation spectra in magnetic bremsstrahlung limit, when the indicated radiation characteristics depend only on the local values of particle transverse coordinates (see Sec. 4), it is sufficient to know only the coordinate distribution of particles. Owing to the periodicity of this distribution, we shall consider it in the area per one axis. The transverse particle energy

$$\varepsilon_\perp = \frac{p_\perp^2}{2\varepsilon} + U(\boldsymbol{\rho}) \tag{9.56}$$

is an integral of motion, therefore, as in the planar case, we introduce the distribution function $\varphi(\varepsilon_\perp)$. Then (cf. Eq. (9.28)) we have

$$dN(\varepsilon_\perp, \boldsymbol{\rho}) = \varphi(\varepsilon_\perp) f(\varepsilon_\perp, \boldsymbol{\rho}) d\varepsilon_\perp d^2\rho\,, \tag{9.57}$$

where

$$\begin{aligned} f(\varepsilon_\perp, \boldsymbol{\rho}) &= \frac{1}{s(\varepsilon_\perp)} \vartheta(\varepsilon_\perp - U(\boldsymbol{\rho}))\,; \\ \vartheta(x) &= \begin{cases} 1, & x > 0\,, \\ 0, & x < 0\,. \end{cases} \end{aligned} \tag{9.58}$$

Here $s(\varepsilon_\perp)$ is the accessible area of a motion around one axis (see Lindhard, 1965). The distribution of particles in the phase space has the form (cf., Eq. (9.30))

$$dN(\mathbf{p}_\perp, \boldsymbol{\rho}) = \tilde{\varphi}\left(\frac{\mathbf{p}_\perp^2}{2\varepsilon} + U(\boldsymbol{\rho})\right) d^2p_\perp d^2\rho\,, \tag{9.59}$$

where (cf. Eq. (9.31))

$$\tilde{\varphi}(\varepsilon_\perp) = \frac{\varphi(\varepsilon_\perp)}{2\pi\varepsilon s(\varepsilon_\perp)} \tag{9.60}$$

It follows from the definition (9.60) and Eq. (9.59), that the distribution of particles in the transverse phase space is uniform, if the function $\varphi(\varepsilon_\perp)$ is proportional to the accessible area of motion $s(\varepsilon_\perp)$.

Integrating (9.57) with respect to $\varepsilon_\perp$, we obtain the coordinate distribution of particles in the transverse plane

$$dN(\boldsymbol{\rho}) = \left[\int d\varepsilon_\perp \vartheta(\varepsilon_\perp - U(\boldsymbol{\rho})) \frac{\varphi(\varepsilon_\perp)}{s(\varepsilon_\perp)}\right] d^2\rho\,. \tag{9.61}$$

We emphasize the important property of axial distributions which substantially differ from planar ones: the distributions (9.58) and (9.61) become independent of the coordinate for $\varepsilon_\perp \geq U_0$. In particular, it takes place for angles of incidence ϑ_0 larger than the critical Lindhard angle $\vartheta_c = (2U_0/\varepsilon)^{1/2}$.

At axial channeling in a thin crystal the function $\vartheta(\varepsilon_\perp)$ has the form (cf. Eq. (9.24))

$$\begin{aligned}\varphi_0(\varepsilon_\perp) &= n_\perp d^2\rho_0 \\ &= n_\perp \left|\frac{\partial^2 U(x,y)}{\partial x \partial y}\right|^{-1}, \\ \boldsymbol{\rho}_0 &= (x,y)\,,\end{aligned} \tag{9.62}$$

where the coordinate $\boldsymbol{\rho}_0(\varepsilon_\perp, \boldsymbol{\vartheta}_0)$ is a solution to the equation

$$\varepsilon_\perp = \frac{\varepsilon\vartheta_0^2}{2} + U(\boldsymbol{\rho}_0)\,. \tag{9.63}$$

Here $n_\perp$ is the density of axes in the transverse plane ($n_\perp = 1/s, s$ is the area per one axis). If we assume the potential near an axis to be axially symmetric, then we have

$$\varphi_0(\varepsilon_\perp) = 2\pi n_\perp \rho_0(\varepsilon_\perp, \vartheta_0)\left|\frac{dU(\rho_0)}{d\rho_0}\right|^{-1}. \tag{9.64}$$

A sharp change of the axis potential takes place in the region $\rho \leq a_s$ having an area much smaller than the total area per one axis (see Eq. (9.13))

$$\pi a_s^2/s = \pi n_\perp a_s^2 = \pi n d a_s^2 = \frac{1}{x_0} \ll 1\,,$$

which leads to a very different population of energy levels by electrons and positrons. While almost all the positrons have an energy $\varepsilon_\perp \ll U_0$ for $\vartheta_0 = 0$, electrons in this case have an energy $\varepsilon_\perp \simeq U_0$. For example, for $\vartheta_0 \simeq \vartheta_c/2$, about 80% of the electrons have $\varepsilon_\perp \geq U_0$ and therefore are uniformly distributed over transverse coordinates. The remaining electrons (captured into a channel) have an energy $\varepsilon_\perp \simeq U_0$ and their distribution over ρ is also nearly uniform. On the contrary, positrons exhibit very smooth dependence of energy levels population on incident angle ϑ_0 up to angles of the order of the critical angle ϑ_c, after which all the particles are in over-barrier states.

Let us find the electron coordinate distribution at $\vartheta_0 = 0$. This can be conveniently carried out, if we directly integrate in (9.61) with respect to ρ_0, using Eq. (9.64):

$$\begin{aligned}dN_-(\boldsymbol{\rho}) &= \left[\int \frac{d^2\rho_0}{\pi\rho_0^2}\vartheta(\rho_0 - \rho)\right]\frac{d^2\rho}{s} \\ &= \frac{d^2\rho}{s}\ln\frac{s}{\pi\rho^2}\,.\end{aligned} \tag{9.65}$$

For positrons at $\vartheta_0 = 0$, we have

$$dN_+ = \frac{d^2\rho}{s} \ln\left(\frac{1}{1 - \pi\rho^2/s}\right) . \tag{9.66}$$

At last, note that the quantum features of the motion at axial channeling become unessential for lower energies compared to the planar case. This is connected with both the large value of the potential well depth U_0 and the two-dimensional character of the problem.

10. DECHANNELING

10.1. Diffusion at Planar Channeling in Thick Crystals

As it was noted, the distribution function (DF) of particles over transverse energy $\varphi(\varepsilon_\perp)$ is an important characteristic of motion. In thin crystals, this function is entirely determined by initial conditions. In thick crystals, some processes, mainly, multiple scattering and radiation energy losses change the distribution $\varphi(\varepsilon_\perp)$. While the radiation process decreases to some extent, the particle transverse energy, multiple scattering, on the average, increases it. As a result, particles can leave a channel (dechanneling).

We introduce the critical dechanneling length owing to multiple scattering l_d over which the root mean square angle of scattering becomes equal to the critical channeling angle ϑ_c (the Lindhard angle):

$$l_d = \frac{\alpha\rho_0}{2\pi} L_{\mathrm{rad}} , \quad \rho_0 = \frac{2U_0\varepsilon}{m^2} , \tag{10.1}$$

where $\alpha = 1/137, L_{\mathrm{rad}}$ is the radiation length in an amorphous medium. One should bear in mind that for channeled positrons multiple scattering is suppressed providing an increase of their dechanneling length by several times compared to l_d. Radiation energy losses are mainly determined by two mechanisms. The first of them is bremsstrahlung at collisions with individual nuclei and electrons. Under channeling, this process is modified compared to the amorphous medium case. The second is properly the radiation at channeling.

For energies satisfying $\rho_0 \ll 1$ (see Eq. (10.1)), the critical dechanneling length l_d is substantially smaller than the characteristic radiation length connected to energy losses. So that in this case multiple scattering is the main cause influencing the kinetics of the DF.

Let us consider the evolution of the DF. We introduce the DF which depends on the transverse energy $\varepsilon_\perp$ and the coordinate x at depth l:

$$F(\varepsilon_\perp, x, l) = \varphi(\varepsilon_\perp, l) f(\varepsilon_\perp, x), \tag{10.2}$$

where (cf. Eq. (9.58))

$$\begin{aligned} f(\varepsilon_\perp, x) &= \frac{1}{T}\sqrt{\frac{\varepsilon}{2(\varepsilon_\perp - U(x))}}, \\ f dx &= \frac{dt}{T} \end{aligned} \tag{10.3}$$

Here T is the period of particles moving above the barrier and the half-period for channeled particles. The normalization conditions have the form

$$\int F d\varepsilon_\perp dx = \int \varphi(\varepsilon_\perp, l) d\varepsilon_\perp = \int f dx = 1. \tag{10.4}$$

Variation of the DF owing to multiple scattering of particles is described by a kinetic equation

$$\frac{\partial F}{\partial l} = \int d\sigma [w(\varepsilon_\perp - \sigma, \sigma) F(\varepsilon_\perp - \sigma) - w(\varepsilon_\perp, \sigma) F(\varepsilon_\perp)], \tag{10.5}$$

where $w(\varepsilon_\perp, \sigma)$ is the probability of scattering when the transverse energy changes from $\varepsilon_\perp$ to $\varepsilon'_\perp = \varepsilon_\perp + \sigma$. Assuming this change to be small and expanding in powers of σ, we obtain the Fokker–Planck equation

$$\begin{aligned} \frac{\partial F}{\partial l} &= \int d\sigma \left[-\sigma \frac{\partial (wF)}{\partial \varepsilon_\perp} + \frac{\sigma^2}{2} \frac{\partial^2 (wF)}{\partial \varepsilon_\perp^2} \right] \\ &= -\frac{\partial}{\partial \varepsilon_\perp} \left(\frac{\Delta \varepsilon_\perp}{\Delta l} F \right) + \frac{1}{2} \frac{\partial^2}{\partial \varepsilon_\perp^2} \left(\frac{(\Delta \varepsilon_\perp)^2}{\Delta l} F \right), \end{aligned} \tag{10.6}$$

where

$$\frac{\Delta \varepsilon_\perp}{\Delta l} = \int d\sigma \cdot \sigma w(\varepsilon_\perp, \sigma), \quad \frac{(\Delta \varepsilon_\perp)^2}{\Delta l} = \int d\sigma \sigma^2 w(\varepsilon_\perp, \sigma) \tag{10.7}$$

are the mean and mean-square variations of the transverse energy per unit length. Using Eq. (9.20), we have

$$\begin{aligned} \Delta \varepsilon_\perp &= \frac{1}{2} \varepsilon (\Delta \vartheta_x)^2, \\ (\Delta \varepsilon_\perp)^2 &= \varepsilon^2 v_x^2 (\Delta \vartheta_x)^2 = 4(\varepsilon_\perp - U(x)) \Delta \varepsilon_\perp. \end{aligned} \tag{10.8}$$

Here ϑ_x is the x-component of the electron multiple scattering angle. Note that, generally speaking, the quantity $\Delta\varepsilon_\perp$ depends on the x-coordinate. Integrating in Eq. (10.6) with respect to x, we obtain an equation for the DF $\varphi(\varepsilon_\perp, l)$:

$$\frac{\partial\varphi}{\partial l} = -\frac{\partial}{\partial\varepsilon_\perp}\left(\left\langle\frac{\Delta\varepsilon_\perp}{\Delta l}\right\rangle\varphi\right) + 2\frac{\partial^2}{\partial\varepsilon_\perp^2}\left(\left\langle\frac{\Delta\varepsilon_\perp}{\Delta l}(\varepsilon_\perp - U(x))\right\rangle\varphi\right), \tag{10.9}$$

where the notation for the mean value is introduced

$$\langle\phi\rangle \equiv \int dx \cdot f(x)\phi = \frac{1}{T}\int dx \cdot \phi\sqrt{\frac{\varepsilon}{2(\varepsilon_\perp - U(x))}} \tag{10.10}$$

Equation (10.9) can be represented as follows:

$$\frac{\partial\varphi}{\partial l} = 2\frac{\partial}{\partial\varepsilon_\perp}D\frac{\partial}{\partial\varepsilon_\perp}\cdot\frac{\varphi}{T}, \tag{10.11}$$

where

$$D = \sqrt{\frac{\varepsilon}{2}}\int dx\left(\frac{\Delta\varepsilon_\perp}{\Delta l}\right)\sqrt{\varepsilon_\perp - U(x)}. \tag{10.12}$$

The magnitude of $(\Delta\vartheta_x)^2/\Delta l$ is large enough only in the narrow interval $x \sim u_1 \ll d$ (u_1 is the amplitude of thermal vibrations), where $U(x) \ll U_0$, and therefore the coefficient D in the case of electron planar channeling can be represented in the form

$$D = \left(\frac{\varepsilon}{2}\right)^{3/2}\sqrt{\varepsilon_\perp}d\cdot\left(\frac{\Delta\vartheta_x^2}{\Delta l}\right)_{am}, \tag{10.13}$$

where $(\Delta\vartheta_x^2/\Delta l)_{am}$ is the rate of change of the mean square scattering angle in an amorphous medium:

$$\frac{\Delta\vartheta_x^2}{\Delta l} \simeq \frac{2\pi}{\alpha\gamma^2}L_{\rm rad}^{-1}. \tag{10.14}$$

Letting $\tau = l/l_d$ (see Eq. (10.1) for l_d), $\xi = (2\varepsilon_\perp/\varepsilon)^{1/2}/\vartheta_c = (\varepsilon_\perp/U_0)^{1/2}$ and introducing the function $g = \varphi\, d/\vartheta_c T$, we can rewrite Eq. (10.11) with regard for Eqs. (10.13) and (10.14):

$$\frac{\partial g(\xi,\tau)}{\partial\tau} = \frac{1}{2\kappa(\xi)}\frac{\partial^2}{\partial\xi^2}g(\xi,\tau). \tag{10.15}$$

Here

$$\kappa(\xi) = \frac{\xi}{d} \int \frac{dx}{\sqrt{\xi^2 - U(x)/U_0}} \,. \tag{10.16}$$

Consider the DF at the depth $l \gg l_d (\tau \gg 1)$. We are interested in the region of particle motion in the channel or not high above the barrier where $\xi \leq 1$, i.e. $\xi^2 \ll \tau$. In this region, as it follows from Eq. (10.15), the function $g(\xi, \tau)$ is independent of ξ. This assertion may prove to be invalid for $|\xi - 1| \ll 1$, where $\kappa(\xi)$ is comparable with τ. Then the diffusion over $\xi(\varepsilon_\perp)$ caused by multiple scattering is suppressed, however, particles may easily go through this narrow region of ξ owing to single scattering; the details of such a passage turn out to be unessential.

In the region $\tau^{1/2} \gg \xi \gg 1$, where $\kappa(\xi) = 1$, the solution of Eq. (10.15) has the form

$$g(\xi, \tau) = \frac{1}{\sqrt{2\pi\tau}} \int d\xi_0 g(\xi_0, 0) \exp\left[-\frac{(\xi - \xi_0)^2}{2\tau}\right] \,, \tag{10.17}$$

where $g(\xi_0, 0)$ is the initial particle distribution. If $g(\xi_0, 0)$ vanishes for $\xi_0 \geq 1$, then we find from Eq. (10.17) for $\xi^2 \ll \tau$:

$$g(\xi, \tau) \simeq \frac{1}{\sqrt{2\pi\tau}} \int g(\xi_0, 0) d\xi_0 = \frac{1}{\sqrt{2\pi\tau}} \,. \tag{10.18}$$

Since this expression is independent of ξ, it can be extended to the region $\xi \leq 1$ as well, so that the asymptotic form of the solution to Eq. (10.11) is

$$\varphi(\xi^2, \tau) = \frac{\vartheta_c T(\xi)}{d\sqrt{2\pi\tau}} \,. \tag{10.19}$$

According to Eqs. (10.2), (10.3) and (10.19), the number of particles in a given interval of ξ and x, at a fixed value of τ, is

$$\begin{aligned} dN(\xi, x, \tau) &= F(\xi^2, x, \tau) d\xi^2 dx \\ &= \frac{d\xi^2}{\sqrt{2\pi\tau}} \frac{dx}{d\sqrt{\xi^2 - U(x)/U_0}} \\ &= 2\frac{dx}{d} \frac{d(\sqrt{\xi^2 - U(x)/U_0})}{\sqrt{2\pi\tau}} = \frac{dx d|\tilde{v}_x|}{d\sqrt{2\pi\tau}} \,, \end{aligned} \tag{10.20}$$

where $\tilde{v}_x = \pm(\xi^2 - U(x)/U_0)^{1/2} = v_x/\vartheta_c$, v_x is the classical velocity of a particle at point x. It is seen from Eq. (10.19) that the distribution is uniform over both the coordinate x and the velocity v_x, which corresponds to the analysis carried out in Sec. 9. If we are also interested in the distribution with respect to the velocity component v_y (the plane (x, y) is perpendicular to the planes forming the channel), a similar analysis yields the uniform distribution in two-dimensional phase space ($v_y \leq \vartheta_c$):

$$dN(v_x, v_y, x, \tau) = \frac{d\tilde{v}_x dv_y}{2\pi\tau}\frac{dx}{d} \tag{10.21}$$

We emphasize that the obtained result (10.21) is valid for arbitrary dependence of the diffusion coefficient D on transverse energy $\varepsilon_\perp$. The magnitude of the coefficient D only determines a scale of depth, where the uniform distribution is established. Actually, only the fact that, for $\varepsilon_\perp \gg U_0$, the coefficient D for both electrons and positrons takes the form (10.13), was used in derivation of Eq. (10.21).

10.2. Diffusion at Axial Channeling in Thick Crystals

In this case the dechanneling length l_d^{ax} is defined by the expression

$$l_d^{ax} = \frac{\alpha}{4\pi}\rho_0 L_{\text{rad}} \,. \tag{10.22}$$

For particles moving along a crystal axis, the DF has the form (9.57):

$$\begin{aligned} F(\varepsilon_\perp, \rho, l) &= \varphi(\varepsilon_\perp, l) f(\varepsilon_\perp, \rho)\,, \\ f(\varepsilon_\perp, \rho) &= \frac{\vartheta(\varepsilon_\perp - U(\rho))}{s(\varepsilon_\perp)} \end{aligned} \tag{10.23}$$

Transformations, similar to those developed above, yield the kinetic equation in the case of axial channeling (cf. Eq. (10.11))

$$\frac{\partial\varphi}{\partial l} = \frac{\partial}{\partial\varepsilon_\perp}\mathcal{D}\frac{\partial}{\partial\varepsilon_\perp}\cdot\frac{\varphi}{s(\varepsilon_\perp)}\,, \tag{10.24}$$

where

$$\begin{aligned} \mathcal{D} &= \int d^2\rho\left(\frac{\Delta\varepsilon_\perp}{\Delta l}\right)(\varepsilon_\perp - U(\rho))\vartheta(\varepsilon_\perp - U(\rho))\,, \\ \Delta\varepsilon_\perp &\simeq \frac{\varepsilon}{2}(\Delta\vartheta)^2\,, \quad \frac{\Delta\varepsilon}{\Delta l} = \frac{\varepsilon}{2}\cdot\frac{(\Delta\vartheta)^2}{\Delta l} \end{aligned} \tag{10.25}$$

Here $(\Delta\vartheta)^2/\Delta l$ is the rate of change of the mean square scattering angle at a given impact parameter ρ. The right-hand side of Eq. (10.24) is of the order of φ/l_d for $\varepsilon_\perp \sim U_0$, and we can put $\partial\varphi/\partial l = 0$ for $l \gg l_d$. Then the solution to Eq. (10.24) takes the form

$$\varphi = \text{const} \quad s(\varepsilon_\perp)\,, \tag{10.26}$$

and the particle distribution in the phase space is uniform in accordance with the analysis developed in Sec. 9.4 (see Eq. (9.59)). This distribution can be represented in the form (cf. Eq. (10.21)):

$$dN(\mathbf{v}_\perp, \rho, l) = \frac{d^2\rho}{s} \cdot \frac{d^2 v_\perp}{\pi\vartheta_s^2(l)}\,. \tag{10.27}$$

Sometimes, e.g., for positrons at zero angle of incidence, when multiple scattering is noticeably suppressed, the dechanneling length is much larger than l_d (10.22). Then the asymptotic solutions (10.26) and (10.27) are established at substantially larger depths. The distribution (10.27) is valid, if the condition

$$2\,\frac{\varepsilon_\perp}{\varepsilon} \ll \vartheta_s^2(l) = \frac{4\pi}{\alpha L_{\text{rad}}} \cdot \int \frac{dl}{\gamma^2(l)} = \int dl\dot{\vartheta}_s^2(l) \tag{10.28}$$

is satisfied. A possible change of the total particle energy is taken into account in Eq. (10.28) in the adiabatic approximation.

The presentation in Secs. (10.1) and (10.2) is developed according to papers by Baier, Katkov and Strakhovenko, 1983b, 1984b. In the second paper (1984b), the detailed analysis was carried out of the electron distribution with regard to the conservation of the orbital momentum projection on the axis direction. Equations for the DF were considered also in Beloshitskii and Kumakhov, 1973, 1982; Lindhard, 1965.

10.3. Electron Density Near Crystal Axes

As it was noted in Sec. 9, at the penetration of charged particles into a crystal, a redistribution of particles flux takes place. So that, the particle density becomes a function of the distance ρ from the axis (we consider the area per one axis). In this section we shall investigate the electron density in the region where the fluctuations of the crystal potential are substantial, i.e. at distances $\rho \le u_1$. At such values of ρ the gradient of this density is considerably different from zero only for the angles of incidence $\vartheta_0 \simeq 0$ and only at a very small depth.

Really, the gradient of the distribution $F(\rho)$ at $\rho \leq u_1$ is determined initially by the population of the lowest states of transverse energy. Since the effective nuclear density for electrons in these states is two to three orders of magnitude higher than the average nuclear density in an amorphous medium, the diffusion owing to multiple scattering increases by the same factor. This leads to the rapid establishment of a homogeneous distribution of electrons in the region $\rho \leq u_1$. The corresponding length l_u is much shorter than the characteristic length l_d:

$$l_u = \frac{\pi u_1^2}{s} l_d \,. \tag{10.29}$$

Here s is the area per one axis, l_d is given by Eq. (10.22). In all considered cases, the crystal thickness $L \gg l_u$, and therefore we can assume the electron density in the region of interest $\rho \leq u_1$ to be constant.

We proceed to the consideration of the particular form of distribution $F(\rho)$ for electrons incident strictly along a crystal axis, when $\vartheta_0 = 0$ (actually we assume $\vartheta_0 \ll \vartheta_c = (2U_0/\varepsilon)^{1/2}$). Then the electron transverse energy is determined by the value of the potential at the entry point $\rho_0 : \varepsilon_\perp = U(\rho_0)$, and we have for an arbitrary, axially symmetric potential

$$d\,N(\varepsilon_\perp(\rho_0), \rho) = \frac{d^2\rho_0}{s} \frac{d^2\rho}{\pi\rho_0^2} \vartheta(\rho_0 - \rho) \,. \tag{10.30}$$

Since, for given ρ_0, the effective nuclear density is $s/\pi\rho_0^2$ times higher than in a corresponding amorphous medium, the dechanneling length l_{ρ_0} reads

$$l_{\rho_0} = \frac{\pi\rho_0^2}{s} l_d \,. \tag{10.31}$$

As it was shown in the previous section, for $l \gg l_d$ the channeled electrons are uniformly distributed in the transverse phase space including the coordinate ρ. On the basis of this result, we assume the electron distribution to be uniform at $l = l_{\rho_0}$ for $\rho < \rho_0$, and to be coincident with the initial distribution for $\rho > \rho_0$. Then the electron distribution at the depth l has the form

$$\begin{aligned} dN(\zeta, \xi) &= F(\xi, \zeta) d\xi \,, \\ F(\xi, \zeta) &= \vartheta(1 - \zeta)[(1 - \ln\zeta)\vartheta(\zeta - \xi) \\ &\quad - \ln\xi \cdot \vartheta(\xi - \zeta)] + \vartheta(\zeta - 1) \,, \quad \zeta = \frac{l}{l_d} \,, \\ \xi &= \frac{\pi\rho^2}{s} \,. \end{aligned} \tag{10.32}$$

To calculate the radiation characteristics, we have to know the electron distribution averaged over the whole thickness l_0 of a crystal:

$$\bar{F}(\xi) = \frac{1}{\zeta_0}\int_0^{\zeta_0} F(\xi,\zeta)d\zeta = \left[\left(2 - \ln\zeta_0 - 2\frac{\xi}{\zeta_0}\right)\vartheta(\zeta_0 - \xi) - \ln\xi\vartheta(\xi - \zeta_0)\right]\vartheta(1 - \zeta_0) + \vartheta(\zeta_0 - 1) \times \left[1 + \frac{1}{\zeta_0}(1 - 2\xi)\right], \tag{10.33}$$

where $\zeta_0 = l_0/l_d$. According to what has been said above, the characteristics of the incoherent radiation are determined by the value of the distribution function at $\rho = 0$:

$$\bar{F}(0) = (2 - \ln\zeta_0)\vartheta(1 - \zeta_0) + \left(1 + \frac{1}{\zeta_0}\right)\vartheta(\zeta_0 - 1). \tag{10.34}$$

For the angles of incidence $\vartheta_0 \neq 0$ ($\vartheta_0 \geq \vartheta_c$) and $l_0 \ll l_d(\zeta_0 \ll 1)$, the distribution function is not changed by multiple scattering, then the function $F(0, \vartheta_0)$ has the form (see Eqs. (9.61)–(9.63))

$$F(0,\vartheta_0) = \int d^2\rho_0 \left[\frac{1}{s(\rho_0,\vartheta_0)}\vartheta\left(U_0 - \frac{\varepsilon\vartheta_0^2}{2} - U(\rho_0)\right) + \frac{1}{s}\vartheta\left(\frac{\varepsilon\vartheta_0^2}{2} + U(\rho_0) - U_0\right)\right], \tag{10.35}$$

where

$$s(\rho_0,\vartheta_0) = \int d^2\rho\vartheta\left(\frac{\varepsilon\vartheta_0^2}{2} + U(\rho_0) - U(\rho)\right).$$

In the general case of arbitrary crystal thickness, the value $F(0, \vartheta_0)$ can be obtained by substituting a solution of the kinetic equation (10.24) into (10.23).

Chapter 3

PAIR PRODUCTION IN ALIGNED SINGLE CRYSTAL

11. GENERAL EXPRESSIONS FOR THE PROBABILITY OF ELECTRON-POSITRON PAIRS PHOTO-PRODUCTION IN CRYSTALS

11.1. Basic Features of Photo-Production Process

When high-energy photons interact with a crystal the mechanism by which electron-positron pairs are produced is considerably modified in comparison with an amorphous medium. Even in the 1950s it was established that when the photon momentum **k** lies near directions of axes or planes, there is an interference of the contributions to pair production from different centers. For certain values of the photon angles of incidence and its energy, the probability of this mechanism, called the coherent pair production (see, e.g., Ter-Mikaelyan, 1972; Diambrini Palazzi; 1968, and references cited therein), differs substantially from the probability of independent pair production at separate centers, which occurs in an amorphous medium (the Bethe–Heitler mechanism, see Sec. 7). The coherent character of the process is connected with the quasiperiodic character of produced particles motion in a crystal potential (see Sec. 5), which follows from the quasiclassical theory of pair production formulated in Sec. 2. According to this theory, relativistic particles produced by a photon with the frequency $\omega \gg m$ move approximately along **k** within a narrow cone of angle $\sim m/\omega$. If this direction turns out to be close to some crystal axes or planes, the oscillation amplitude of the particle's velocity increases and so does the probability of the process.

Let ϑ_0 be the photon angle of incidence with respect to a chosen axis and V_0 be the scale of the corresponding continuum potential. For $\vartheta_0 \gg (V_0/m)^{1/2}$, produced particles move high above the potential barrier, that can be described in terms of rectilinear trajectory (see Eq. (9.16)). In the rest system of a crystal there is only an electric field $\mathbf{E}$. In the system of reference moving with a relativistic velocity $\mathbf{v} = \mathbf{n}v(\mathbf{n} = \mathbf{k}/\omega, \kappa^2 = \omega^2 - \mathbf{k}^2 = 0)$ along the photon momentum direction, a magnetic field $\mathbf{H} = \gamma_v[\mathbf{E}\mathbf{v}]$ arises $(\gamma_v = (1-\mathbf{v}^2)^{-1/2} \gg 1)$ and, as is well known, the resultant field in this system can be represented with relativistic accuracy in the form of plane waves with frequencies $\gamma_v|q_{\parallel}|v(q_{\parallel} = \mathbf{qn})$. The equivalent photon flux averaged over time and over transverse coordinates is the sum of partial contributions $\mathbf{J}_q$. The latter quantity has the form

$$\mathbf{J}_q = -\mathbf{n}\frac{\gamma_v}{4\pi e^2}\frac{|G(\mathbf{q})|^2}{|q_{\parallel}|}\mathbf{q}_{\perp}^2\,, \tag{11.1}$$

where $\mathbf{q}_{\perp} = \mathbf{q} - \mathbf{n}(\mathbf{qn})$. In the interaction region, the transverse size of which is of the order $\lambda\!\!\!^{-}_c = 1/m(\hbar/mc)$, and the longitudinal size is the process formation length, which in the c.m.s. system of the incident and equivalent photons is of the order $2\pi/\gamma_v|q_{\parallel}|$, there are $N_q \simeq 2\pi\lambda\!\!\!^{-}{}_c^2|\mathbf{J}_q/q_{\parallel}\gamma_v|$ photons. The effective strength of interaction is characterized by the parameter

$$\alpha N_{\text{ph}} = \alpha\sum_{\mathbf{q}} N_{\mathbf{q}} \simeq \sum_{\mathbf{q}}\frac{|G(\mathbf{q})|^2}{m^2 q_{\parallel}^2}\mathbf{q}_{\perp}^2\,. \tag{11.2}$$

This parameter is purely classical (it does not contain Planck's constant $\hbar$) and always arises in problems with an external electromagnetic field: the parameter ξ^2 in the theory of processes in an intense plane wave (see Appendix B) has the same meaning; the same is true of the parameter ρ in the problem of radiation at quasiperiodic motion (see Sec. 5). For $\alpha N_{\text{ph}} \ll 1$, the external field can be taken into account in perturbation theory. In this case the process can be considered as the pair production by two colliding photons: a real photon of the frequency ω and a quasireal $(\omega_1^2 - \mathbf{k}_1^2 \simeq \omega_1^2/\gamma_v^2)$ photon with the frequency ω_1. Kinematics of the two-photon process yields

$$kk_1 = \omega|q_{\parallel}| = m^2 + p_+p_- \geq 2m^2\,. \tag{11.3}$$

Taking into account that the momenta of produced particles are collinear to the accuracy $l/\gamma(p_+p_- \sim m^2)$ and making use of estimates given in Sec. 9 ($|G(\mathbf{q})| \sim V_0, q_\parallel \sim q_\perp\vartheta_0, q_\perp \sim 2\pi/d$), we find that the probability of coherent pair production is maximum for the angles of incidence

$$\vartheta_0 \sim \vartheta_m = \frac{m^2\Delta}{\pi\omega}, \tag{11.4}$$

where Δ is the typical distance between axes or planes. Making use of mentioned estimates, we obtain from Eq. (11.2)

$$\alpha N_{\text{ph}} \sim \left(\frac{V_0}{m\vartheta_0}\right)^2, \tag{11.5}$$

i.e. the perturbation theory, the first approximation of which is just the coherent pair production theory, can be applied for angles of incidence satisfying the condition

$$\vartheta_0 \gg \vartheta_v \equiv V_0/m\,. \tag{11.6}$$

The location of the maximum (11.4) shifts with increase of the energy toward smaller angles. When a value $\vartheta_m \sim \vartheta_v$ is reached, the theory of coherent pair production becomes inapplicable in the region of its maximum. In this sense this theory is completely inadequate in the high energy region.

For $\vartheta_0 \ll \vartheta_v$, the parameter ρ is large:

$$\rho = 2\gamma^2\langle(\Delta\mathbf{v})^2\rangle \simeq \alpha N_{\text{ph}} \gg 1 \tag{11.7}$$

and the transverse motion of the particles produced is ultrarelativistic ($p_\perp \gg m$). Then the process formation time τ_f is that of the particle velocity turn over the angle ϑ_γ, which is much shorter than a period of transverse oscillations T:

$$\tau_f \sim T\rho^{-1/2} \ll T \tag{11.8}$$

In this case the field strength almost does not change during the formation time, and the constant field limit (see Sec. 3) for the probability of pair production becomes applicable.

Generally speaking, Eq. (9.16) becomes inapplicable for photon angles of incidence $\vartheta_0 \sim \vartheta_c$ when produced particles are channeled or move not high above the potential barrier. In this connection, we have

$$\rho \sim \frac{V_0^2}{m^2\vartheta_c^2} \sim \rho_e \equiv \frac{\omega V_0}{m^2} \tag{11.9}$$

When the parameter ρ_c is not very large ($\rho_c \leq 1$), the pair production probability is strongly suppressed in the region $\vartheta_0 \leq \vartheta_c$. In the opposite case $\rho_c \gg 1$, the constant field limit is valid for $\vartheta_0 \leq \vartheta_c$, when characteristics of the process depend only on local values of the particle acceleration and can be obtained by means of Eq. (9.16). Thus, the rectilinear trajectory approximation can be used for the quasiclassical description of the pair production in crystal regardless of the value of the parameter ρ_c. However, in the following presentation based on papers by Baier, Katkov and Strakhovenko, 1984; 1985a, b; 1986a, b, (see also Baier, Katkov and Strakhovenko, 1989a, b; 1991b) it will be assumed that

$$\rho_c = \frac{\omega V_0}{m^2} \gg 1 \tag{11.10}$$

when the physical picture of the phenomenon cannot be reduced to the coherent pair production. The mechanism of pair production in a field of atomic chains was also considered in the papers by Kimball and Cue, 1985; Belkacem, Cue and Kimball, 1985; Kimball, Cue and Belkacem, 1986; Baryshevskii and Tikhomirov, 1985. The assumption that capture of the produced particles into the channeling regime has an important role, previously stated in Baryshevskii and Tikhomirov, 1982, 1983; Kimball *et al.*, 1983, has not been confirmed.

11.2. Basic Formulae

General expressions for the probability of pair production by a photon in an external field are derived in Sec. 3 within quasiclassical approximations. For unpolarized photons, the probability of the process summed up over polarizations of final particles is given by Eqs. (3.27)–(3.31). Substituting the value (9.16) of the transverse particle velocity evaluated on a rectilinear trajectory in the potential (9.5) into Eqs. (3.30)–(3.31) and taking corresponding integrals, we find for the pair production prabability:

$$w_e = \frac{i\alpha m^2}{2\pi\omega^2} \int \frac{d^3 r_0}{V} \int_0^{\omega} d\varepsilon \cdot \int dt \int \frac{d\tau}{\tau + i0} \left\{ 1 + \frac{\varepsilon^2 + \varepsilon'^2}{\varepsilon\varepsilon'} \right.$$
$$\left. \times \left(\sum_{\mathbf{q}} G(\mathbf{q}) \frac{\mathbf{q}_\perp}{m q_\parallel} \exp[-i(\mathbf{q}\mathbf{r}_0 + q_\parallel t)] \cdot \sin\frac{q_\parallel \tau}{2} \right)^2 \right\} e^{i\widetilde{A}}, \quad \varepsilon' = \omega - \varepsilon\,, \tag{11.11}$$

where

$$\widetilde{A} = \frac{m^2\omega\tau}{2\varepsilon\varepsilon'}\left\{1 + \sum_{\mathbf{q},\,\mathbf{q}'} \frac{G(\mathbf{q})G(\mathbf{q}\prime)}{m^2}\cdot\right.$$

$$\times \exp\{-i[(\mathbf{q}+\mathbf{q}')\mathbf{r}_0 + (q_{\parallel} + q'_{\parallel})t]\}\frac{\mathbf{q}_{\perp}\mathbf{q}'_{\perp}}{q_{\parallel}q'_{\parallel}}$$

$$\left.\times\left[\frac{2\sin\left(\frac{q_{\parallel}+q'_{\parallel}}{2}\tau\right)}{(q_{\parallel}+q'_{\parallel})\tau} - \frac{4\sin\frac{q_{\parallel}\tau}{2}\sin\frac{q'_{\parallel}\tau}{2}}{q_{\parallel}q'_{\parallel}\tau^2}\right]\right\} \tag{11.12}$$

In this expression the time t appears only in the combination $r_{0\parallel} + t$, and after the substitution $r_{0\parallel} + t \to r_{0\parallel}$ the dependence on time t remains only in the limits of integration over $r_{0\parallel}$. This dependence can be neglected if the crystal thickness (in the longitudinal direction) is considerably greater than the process formation length l_f. This condition is essentially always satisfied, since the length l_f is about (3–60) 10^{-6} cm for $\omega \sim 100$ GeV depending on the crystal and axis types and increases very slowly with the increase of ω (asymptotically $l_f \propto \omega^{1/3}$). Then we can omit the integration over t and extend the region of integration over τ to infinity, since the integrals converge rapidly. As a result we obtain for the probability of electron-positron pair production per unit time:

$$\frac{dw_e}{dt} \equiv W_e = \frac{i\alpha m^2}{2\pi\omega^2}\int\frac{d^3r_0}{V}\int^{\omega} d\varepsilon$$

$$\times\int_{-\infty}^{\infty}\frac{d\tau}{\tau+i0}\left\{1 + \frac{\varepsilon^2+\varepsilon'^2}{\varepsilon\varepsilon'}\left(\sum_{\mathbf{q}}\frac{G(\mathbf{q})\mathbf{q}_{\perp}}{mq_{\parallel}}\cdot e^{i\mathbf{q}\mathbf{r}_0}\sin(q_{\parallel}\tau)\right)^2\right\}e^{iA}, \tag{11.13}$$

$$A = \frac{m^2\omega\tau}{\varepsilon\varepsilon'}\left\{1 + \sum_{\mathbf{q},\,\mathbf{q}'}\frac{G(\mathbf{q})G(\mathbf{q}')}{m^2q_{\parallel}q'_{\parallel}}(\mathbf{q}_{\perp}\mathbf{q}'_{\perp})e^{-i(\mathbf{q}+\mathbf{q}',\mathbf{r}_0)}\Psi(q_{\parallel},q'_{\parallel},\tau)\right\},$$

$$\Psi(q_{\parallel},q'_{\parallel},\tau) = \frac{\sin[(q_{\parallel}+q'_{\parallel})\tau]}{(q_{\parallel}+q'_{\parallel})\tau} - \frac{\sin(q_{\parallel}\tau)}{q_{\parallel}\tau}\,\frac{\sin(q'_{\parallel}\tau)}{q'_{\parallel}\tau}.$$

As it was explained above, the probability (11.13), formally obtained for $\vartheta_0 \gg \vartheta_c$, is valid right down to $\vartheta_0 = 0$ being a general expression of the pair production probability for arbitrary energies and directions of incident photons. For relatively large angles ϑ_0 the probability (11.13) goes over to the well known expression of coherent pair production theory, and for vanishing angles ϑ_0 it reproduces the result of a constant field approximation. These transitions will be traced in detail below. In the intermediate region of angles ϑ_0 it is necessary to use Eq. (11.13) directly. Making use of Eqs. (3.12) and (3.23), we can similarly obtain the probabilities of pair production for polarized initial and final particles. We give here the corresponding probability (per unit time) for polarized initial photons (see Eq. (3.25)):

$$W_e = \frac{i\alpha m^2}{2\pi\omega^2} \int \frac{d^3r_0}{V} \int d\varepsilon \int_{-\infty}^{\infty} \frac{d\tau}{\tau + i0} e^{iA} \left\{ (\mathbf{e}^*\mathbf{e}) \left[1 + \varphi(\varepsilon) \right.\right.$$

$$\times \left. \left(\sum_{\mathbf{q}} \frac{G(\mathbf{q})\mathbf{q}_\perp}{mq_\parallel} e^{-i\mathbf{q}\mathbf{r}_0} \sin q_\parallel \tau \right)^2 \right] + \sum_{\mathbf{q},\, \mathbf{q}'} \frac{G(\mathbf{q})G(\mathbf{q}')}{m^2 q_\parallel q'_\parallel} \cdot e^{-i(\mathbf{q}+\mathbf{q}', \mathbf{r}_0)}$$

$$\times \left[\frac{i\varphi(\varepsilon)}{2} ((\mathbf{e}^*\mathbf{q}_\perp)(\mathbf{e}\mathbf{q}'_\perp) - (\mathbf{e}^*\mathbf{q}'_\perp)(\mathbf{e}\mathbf{q}_\perp))(g(q_\parallel\tau)\sin q'_\parallel\tau - g(q'_\parallel\tau)\sin q_\parallel\tau) \right.$$

$$+ ((\mathbf{e}^*\mathbf{e})(\mathbf{q}_\perp\mathbf{q}'_\perp) - (\mathbf{e}^*\mathbf{q}_\perp)(\mathbf{e}\mathbf{q}'_\perp) - (\mathbf{e}^*\mathbf{q}'_\perp)(\mathbf{e}\mathbf{q}_\perp))(g(q_\parallel\tau)g(q'_\parallel\tau)$$

$$\left.\left. + \sin q_\parallel\tau \cdot \sin q'_\parallel\tau) \right] \right\}, \qquad (11.14)$$

$$g(x) = \frac{\sin x}{x} - \cos x\,, \quad \varphi(\varepsilon) = \varepsilon/\varepsilon' + \varepsilon'/\varepsilon\,.$$

12. PROBABILITY OF THE PHOTO-PRODUCTION IN CRYSTAL IN THE CONSTANT FIELD APPROXIMATION

12.1. Probability of the Process at $\vartheta_0 \ll \vartheta_V$. Corrections to the Constant Field Approximation

Let us turn to analysis of the general expression (11.13). The behavior of the probability (11.13) for various angles and energies of incident photons is determined by the dependence on them of phase A in (11.13). Now we shall estimate it for $\vartheta_0 \ll V_0/m$. Consider, for definiteness the case when the momentum of incident photon lies near to a crystal axis whose direction we take as the z-axis of the coordinate system. For estimates we assume that $G(\mathbf{q}) \sim V_0$, and therefore the order of magnitude of the double sum in A is $(V_0/m)^2(q_\perp/q_\parallel)^2\Psi(q_\parallel, q'_\parallel, \tau)$, where the function Ψ is defined in (11.13). We introduce the notation $\mathbf{q}_t$ for the vectors $\mathbf{q}$ lying in the plane (x, y). For them $q_z = 0, q_\parallel \sim \vartheta_0 q_\perp, q_\perp \sim q$, and for all remaining vectors $q_\parallel \sim q_\perp \sim q$. Then the contribution to the sum of terms with $q_z, q'_z \neq 0$ is $(V_0/m)^2\Psi \leq (V_0/m)^2$, since $|\Psi| \leq 1$ for any values of the arguments. Since $(V_0/m)^2 \ll 1$, this contribution can be neglected. Thus, we keep in the sum only terms with $\mathbf{q}_t$, for which its value is $\sim (V_0/\vartheta_0 m)^2\Psi$. The large value of the phase A leads to an exponential suppression of the probability and therefore the characteristic values of the variable τ in the integral (11.13), which have the meaning of the process formation time (length), are adjusted in such a way that the large factor $(V_0/\vartheta_0 m)^2$ is compensated by the quantity Ψ, i.e. the contribution at small angles of incidence is from $q_\parallel \tau, q'_\parallel \tau \ll 1$. Expanding the function Ψ accordingly, we obtain

$$A = \frac{m^2\omega\tau}{\varepsilon\varepsilon'}\left\{1 - \frac{\tau^2}{3}\sum_{\mathbf{q}_t, \mathbf{q}'_t}\frac{G(\mathbf{q}_t)G(\mathbf{q}'_t)}{m^2}(\mathbf{q}_t\mathbf{q}'_t)\cdot\exp[-i(\mathbf{q}_t + \mathbf{q}'_t, \boldsymbol{\rho}_0)]\right.$$
$$\left.\times\left[1 - \frac{\tau^2}{10}((\mathbf{n}\mathbf{q}_t)^2 + (\mathbf{n}\mathbf{q}'_t)^2 + \frac{2}{3}(\mathbf{n}\mathbf{q}_t)(\mathbf{n}\mathbf{q}'_t))\right]\right\}, \tag{12.1}$$

where $\boldsymbol{\rho}_0$ denotes the component of the vector $\mathbf{r}_0$ lying in the (x, y) plane. We can rewrite Eq. (12.1) in terms of a potential, which depends only on the coordinate transverse to the axis (see Eq. (9.9))

$$U(\boldsymbol{\rho}_0) = \sum_{\mathbf{q}_t} G(\mathbf{q}_t)e^{-i\mathbf{q}_t\boldsymbol{\rho}_0}, \tag{12.2}$$

namely

$$A = \frac{m^2\omega\tau}{\varepsilon\varepsilon'}\left\{1 + \frac{\tau^2}{3}\mathbf{b}^2 + \frac{\tau^4}{15}\left[(\mathbf{b}(\mathbf{n}\boldsymbol{\nabla})^2\mathbf{b}) + \frac{1}{3}\left((\mathbf{n}\boldsymbol{\nabla})\mathbf{b}\right)^2\right]\right\}, \tag{12.3}$$

where $\mathbf{b} = \boldsymbol{\nabla}U(\boldsymbol{\rho}_0)/m, \boldsymbol{\nabla} = \partial/\partial\boldsymbol{\rho}_0$. After a similar procedure, the expression multiplying the exponential in the integrand of (11.13) takes the form

$$1 - \frac{\varepsilon^2 + \varepsilon'^2}{\varepsilon\varepsilon'}\tau^2\left[\mathbf{b}^2 + \frac{\tau^2}{3}(\mathbf{b}(\mathbf{n}\boldsymbol{\nabla})^2\mathbf{b})\right]. \tag{12.4}$$

Note that $\mathbf{b}$ is expressed in terms of the particle acceleration $\mathbf{b} = -\gamma \mathbf{w}(\boldsymbol{\rho}_0)$, so that the entire expression is dependent on the local characteristics of the motion, which is inherent in pair production by a photon in a constant field. The terms in (12.3) which contain $\mathbf{n}$ have a relative magnitude $\sim (m\vartheta_0/V_0)^2$, and we can expand e^{iA} in these terms. After substituting (12.3) and (12.4) into (11.13) and performing this expansion, we take the integrals over τ by means of well-known relations

$$\begin{aligned} \int_{-\infty}^{\infty} d\tau \exp\left\{-i\lambda\left(\tau + \frac{\tau^3}{3}\right)\right\} &= \frac{2}{\sqrt{3}} K_{1/3}\left(\frac{2\lambda}{3}\right), \\ \int_{-\infty}^{\infty} d\tau \tau \exp\left\{-i\lambda\left(\tau + \frac{\tau^3}{3}\right)\right\} &= -\frac{2i}{\sqrt{3}} K_{2/3}\left(\frac{2\lambda}{3}\right), \end{aligned} \tag{12.5}$$

and the recurrence relations for the MacDonald functions $K_\nu(z)$. Carrying out this calculation, we obtain for the pair production probability in the limit $\vartheta_0 \ll V_0/m$

$$\begin{aligned} W_e^F &= \frac{\alpha m^2}{\pi\sqrt{3}\omega^2} \int_0^\omega d\varepsilon \int \frac{d^2\rho_0}{s} \left\{\varphi(\varepsilon) K_{2/3}(\lambda) + \int_\lambda^\infty dy K_{1/3}(y) - \frac{1}{3}\varphi(\varepsilon)\right. \\ &\times \frac{(\mathbf{b}(\mathbf{n}\nabla)^2\mathbf{b})}{\mathbf{b}^4}\left[K_{2/3}(\lambda) - \frac{2}{3\lambda}K_{1/3}(\lambda)\right] - \frac{\lambda}{30\mathbf{b}^4}[((\mathbf{n}\nabla)\mathbf{b})^2 \\ &+ 3(\mathbf{b}(\mathbf{n}\nabla)^2\mathbf{b})]\left[K_{1/3}(\lambda) - \frac{4}{3\lambda}K_{2/3}(\lambda) - \varphi(\varepsilon)\right. \\ &\times \left.\left.\left(\frac{4}{\lambda}K_{2/3}(\lambda) - K_{1/3}(\lambda) - \frac{16}{9\lambda^2}K_{1/3}(\lambda)\right)\right]\right\}, \end{aligned} \tag{12.6}$$

where $\varphi(\varepsilon) = \varepsilon/\varepsilon' + \varepsilon'/\varepsilon, \lambda = 2m^2\omega/3\varepsilon\varepsilon'|\mathbf{b}|$. Since the integrand in (12.6) is independent of the coordinate z, we have $\int d^3r_0/V \to \int d^2\rho_0/S$ where S is the area of the crystal cross-section perpendicular to the chosen axis. In view of the periodicity of a crystal, the averaging over $\boldsymbol{\rho}_0$ can be carried out over the area s per one axis (see Eq. (9.13)). The first two terms in (12.6), which are independent of $\mathbf{n}$, give the constant (over the pair formation length) field limit of the probability obtained in Sec. 3, and the remaining terms are a correction $\propto \vartheta_0^2$.

Further simplifications of Eq. (12.6) are possible if the potential around the axis can be assumed to be axisymmetric. In some cases, for example, for the $\langle 111 \rangle$ axes in crystal structures of the $fcc(d)$ and bcc types, the potential is very accurately axisymmetric about the chosen axis over the entire region s associated with this axis. However, even if the potential is not axisymmetric over the entire region s, it in any case has this property for $|\boldsymbol{\rho}_0| \leq a_s$, where a_s is the effective screening radius of the potential of the string and $\boldsymbol{\rho}_0$ is measured from the axis. The strength of the electric field in which pair production occurs is maximal just in the region $|\boldsymbol{\rho}_0| \leq a_s$. Assuming that $U = U(\boldsymbol{\rho}_0^2)$, we can carry out the integration over angles of the vector $\boldsymbol{\rho}_0$ in Eq. (12.6):

$$\begin{aligned} W_e^F &= \frac{\alpha m^2}{\pi\sqrt{3}\omega^2} \int_0^{x_0} \frac{dx}{x_0} \int_0^{\omega} d\varepsilon \Bigg\{ \varphi(\varepsilon) K_{2/3}(\lambda) + \int_{\lambda}^{\infty} dy K_{1/3}(y) \\ &\quad - \frac{1}{6}\left(\frac{m\vartheta_0}{V_0}\right)^2 \Bigg[\varphi(\varepsilon) \cdot \frac{xg'' + 2g'}{xg^3} \left(K_{2/3}(\lambda) - \frac{2}{3\lambda} K_{1/3}(\lambda) \right) \\ &\quad + \frac{1}{20} \frac{1}{g^4 x^2} (2x^2 g'^2 + g^2 + 14gg'x + 6x^2 gg'') \cdot \Bigg(\lambda K_{1/3}(\lambda) \\ &\quad - \frac{4}{3} K_{2/3}(\lambda) - \varphi(\varepsilon) \left(4K_{2/3}(\lambda) - (\lambda + 16/9\lambda) K_{1/3}(\lambda) \right) \Bigg) \Bigg] \Bigg\}, \end{aligned} \tag{12.7}$$

where we have gone over to the variable $x = \rho_0^2/a_s^2; x_0^{-1} = \pi a_s^2 nd; d$ is the average distance between atoms along the axis, n is the density of atoms in the crystal, and the notation $U'(x) = V_0 g(x)$ has been used; then we have

$$\lambda = \frac{m^3 \omega a_s}{3\varepsilon\varepsilon' V_0 \sqrt{x}|g|} = \frac{1}{3} \frac{\omega^2}{\varepsilon\varepsilon'} \frac{1}{\kappa(a_s)} \frac{1}{\sqrt{x}|g|} \tag{12.8}$$

Equation (12.7) is applicable for any shape of an axisymmetric potential. In specific calculations we shall use an explicit form of the axis potential given by Eq. (9.13). For this potential, the magnitudes of the electric field and the parameter κ are

$$\begin{aligned} eE &= -\frac{\partial U}{\partial \rho} = \frac{2V_0}{a_s} \sqrt{x} g(x), \\ g(x) &= \frac{1}{x+\eta} - \frac{1}{x+\eta+1}, \\ \kappa(x) &= \frac{2V_0 \omega}{a_s m^3} \sqrt{x} \cdot g(x). \end{aligned} \tag{12.9}$$

The electric field (12.9) vanishes at $x = 0$ as a result of thermal vibrations, reaches a maximum at $x = x_m = [(1 + 16\eta(1+\eta))^{1/2} - 1 - 2\eta]/6$ and then falls off. In what follows we shall use $\kappa_m = \kappa(x_m)$ and $\kappa_s = V_0\omega/a_s m^3$ Note that the spectrum $dW_e^F/d\varepsilon$ of one of the produced particles follows from Eq. (12.7), if we omit the integration with respect to ε.

Equations (12.6) and (12.7) were obtained from the probability (11.13) in the region $\vartheta_0 \gg \vartheta_c$ where the latter is rigorously applicable, but as the angle ϑ_0 decreases further, the derived probabilities change insignificantly, which obviously follows from their explicit form, so that, as it was explained above, Eqs. (12.6) and (12.7) and also Eq. (11.13) are valid down to $\vartheta_0 = 0$.

The pair production probability is shown in Fig. 12.1 as a function of photon energy for several materials. The spectrum of produced electrons (positrons) for various photon energies is shown in Fig. 12.2 for the $\langle 111 \rangle$ axis in tungsten. It can be seen that the peak at $\varepsilon = \omega/2$ for relatively low energies gradually

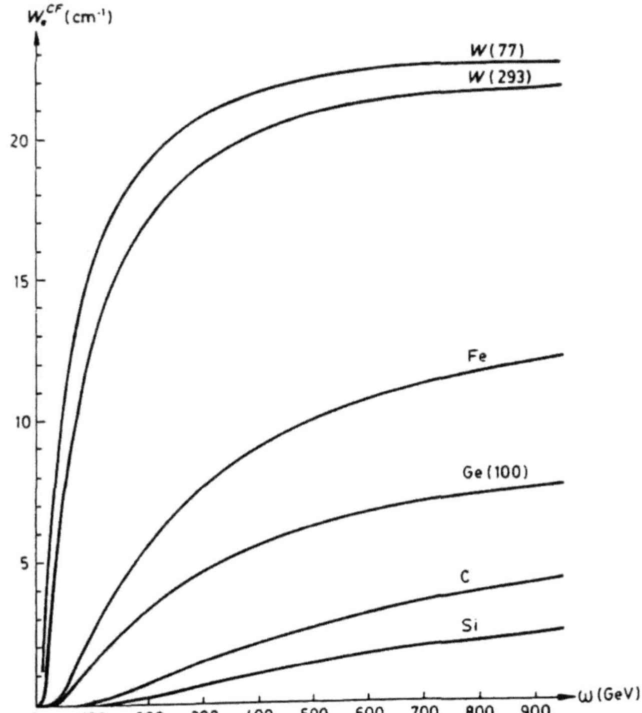

Fig. 12.1. Probability of production of a pair by a photon for incidence angle $\vartheta_0 = 0$ with respect to the $\langle 111 \rangle$ axis (for Ge with respect to the $\langle 110 \rangle$ axis). The numbers in parentheses denote the temperature of the crystal; where they are not given, $T = 293$ K.

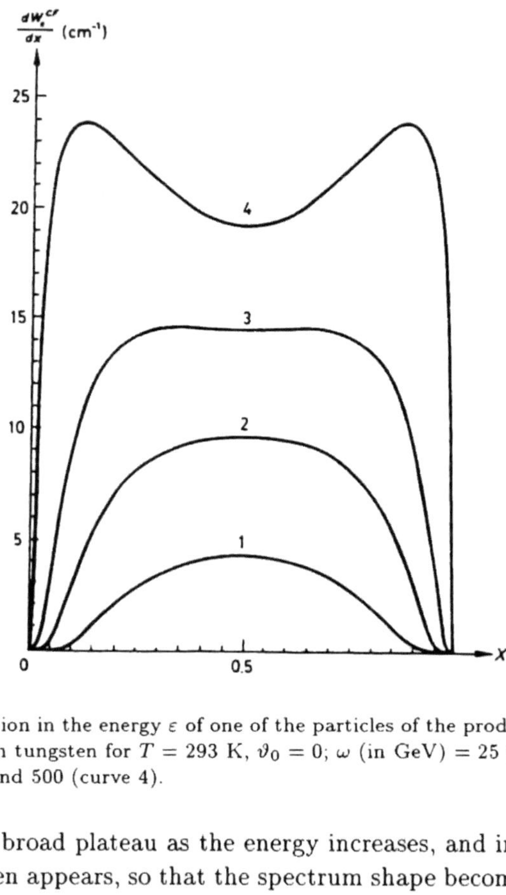

Fig. 12.2. Distribution in the energy ε of one of the particles of the produced pair ($x = \varepsilon/\omega$) for the $\langle 111 \rangle$ axis in tungsten for $T = 293$ K, $\vartheta_0 = 0$; ω (in GeV) = 25 (curve 1), 50 (curve 2), 100 (curve 3), and 500 (curve 4).

goes over into a broad plateau as the energy increases, and in the middle of a plateau a dip then appears, so that the spectrum shape becomes rather similar to the Bethe–Heitler spectrum. The probability (12.7) has the form of

$$W_e^F = F_1 + \left(\frac{m\vartheta_o}{V_o}\right)^2 F_2 \,. \tag{12.10}$$

The ratio F_2/F_1 of the functions in Eq. (12.10) is shown in Fig. 12.3 as a function of energy for various materials. At those values of ω for which $F_2 > 0$, the probability (12.10) has a minimum at $\vartheta_0 = 0$. For $\kappa_s \simeq (5.1\text{–}5.3)\eta^{1/2}$ the function F_2 changes sign and for larger values of κ_s the probability W_e has a maximum at $\vartheta_0 = 0$. As $\omega(\kappa_s)$ continues to increase, the height of this maximum increases and its width begins to narrow, since for $\kappa_s \gg 1$ the function F_2 increases faster than F_1. The maximum value of the probability

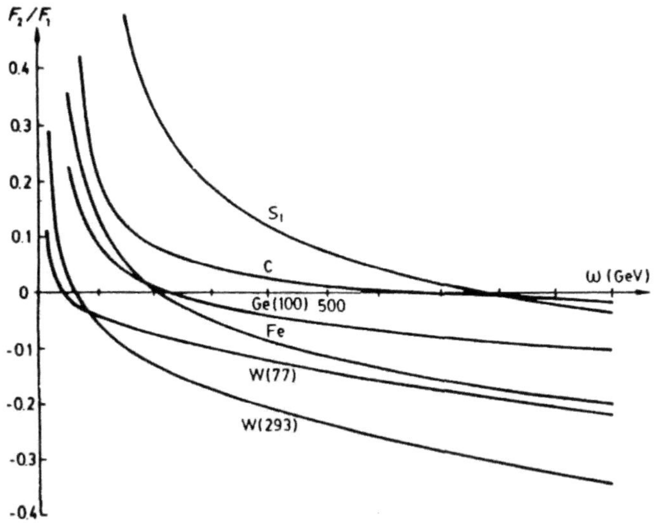

Fig. 12.3. The ratio F_2/F_1 characterizing the magnitude of the correction to the constant field limit. The notation is the same as in Fig. 12.1.

W_e at ϑ_0 is reached for $\kappa_s \gg 1$, when an explicit asymptotic expression for W_e^F can be obtained. The corresponding analysis will be carried out in Sec. 12.3.

12.2. Pair Production for Small Values of κ_s

In a constant field the probability of pair production for small values of the parameter κ is determined by Eq. (3.58) applicable within good accuracy up to $\kappa = 2$. If the maximum value $\kappa(x_m)$ is small (see Eq. (12.9)), we can use the Laplace method in integration over transverse coordinate in Eq. (12.7). Expanding the quantity $\kappa(x)$ at $x = x_m$ and taking corresponding integrals, we obtain for the probability per unit time retaining the principal term of the expansion:

$$W_e^F = \frac{9\alpha V_0\sqrt{\pi\eta}}{16\sqrt{2}mx_0a_s}\varphi(x_m),$$
$$\varphi(x_m) = \left(\frac{\kappa_m\psi^3(x_m)}{4\eta^2|\psi''(x_m)|}\right)^{1/2} e^{-\frac{8}{3\kappa_m}}, \qquad (12.11)$$

where

$$\kappa_m = \frac{\kappa_s}{\sqrt{\eta}}\psi(x_m), \quad \kappa_s = \frac{V_0\omega}{m^3 a_s},$$

$$\psi(x) = 2\sqrt{\eta x}g(x) = \frac{2\sqrt{\eta x}}{(\eta + x)(\eta + x + 1)}, \tag{12.12}$$

$$x_m = \frac{1}{6}\{\sqrt{1 + 16\eta(1+\eta)} - 1 - 2\eta\}.$$

The relative error of Eq. (12.11) at $\kappa_m = 2$ is $\leq 25\%$.

The exponential factor in (12.11) inherent for pair production in a constant field varies most rapidly with photon energy ω. For small values of the parameter κ_m, the probability of e^+e^- pair production by a photon in an axis falls off so rapidly that the process can be characterized by the effective threshold ω_t. The specific value ω_t can be obtained from the condition $\kappa_m(\omega_t) \sim 1$. We shall define the quantity ω_t more rigorously as the photon energy for which the probability W_e^F equals the probability W_{BH} of Bethe–Heitler mechanism in the corresponding amorphous medium. A rough estimate of the energy ω_t is

$$\omega_t \sim \frac{m}{Z\alpha} \cdot \frac{u_1 d}{\lambda_c}, \tag{12.13}$$

from which it is evident that the effect appears earliest with increase of the photon energy in materials with large Z and small u_1 and d. In Table 12.1 calculated values of the energy ω_t are given for some crystals.

Table 12.1. Characteristics of pair production process.

Crystal	Axis	T (K)	κ_s at ω 100 GeV	κ_m/κ_s	ω_t (GeV)	ω_m (TeV)	r_{max}
C	111	293	0.13	6.0	90	10	156
Si	111	293	0.27	2.0	150	14	68
Si	110	293	0.32	2.1	120	12	78
Ge	111	293	0.45	2.2	100	7.9	25
Ge	110	280	0.48	2.4	70	7	29
Ge	110	100	0.56	3.5	50	4	29
W	111	293	2.9	2.4	22	1.4	10.5
W	111	77	2.3	5.6	13	1.1	10.8

To calculate the spectral probability of the process in the near-threshold region, we omit the integration over ε in Eq. (12.7), neglecting correction terms $\propto \vartheta_0^2$ and using the asymptotic representation of K_ν functions for large values of arguments ($K_\nu(z) \simeq e^{-z}(\pi/2z)^{1/2}$). Integrating by the Laplace method, we find

$$\frac{dW_e^F}{d\varepsilon} = \frac{\sqrt{3}\alpha V_0}{mx_0\omega a_s}\left[\frac{\psi^3(x_m)}{4\eta|\psi''(x_m)|}\right]^{1/2} \times \left[1 - \frac{\varepsilon}{\omega}\left(1 - \frac{\varepsilon}{\omega}\right)\right] \exp\left\{-\frac{2\omega^2}{3\kappa_m\varepsilon(\omega - \varepsilon)}\right\}. \qquad (12.14)$$

The spectral distribution (12.14) is symmetric with respect to the substitution $\varepsilon \to \omega - \varepsilon(\varepsilon^- \leftrightarrow \varepsilon^+)$ and has a sharp peak at $\varepsilon^\pm = \omega/2$. For $\kappa_m \leq 1$ the spectral distribution of the process is essentially different from the Bethe–Heitler spectrum which depends on photon energy rather smoothly.

Let us elucidate, finally, conditions of the applicability of the quasiclassical approach in the near-threshold region. This is necessary in view of the fact that, for such energies, electrons are produced very close ($\rho \sim u_1$) to atomic chain being at the bottom of a potential well. The number of levels for electrons with transverse energy $\varepsilon_\perp \leq U(x_m)$ can be easily estimated:

$$n \sim \frac{1}{5}\kappa_m\left(\frac{u_1}{\lambda_c}\right)^3 \qquad (12.15)$$

Taking into account that for all the crystals $u_1/\lambda_c \geq 10$, we find that for those values of κ_m at which the probability (12.11) is not too small, the number of levels is large $n \gg 1$, providing the applicability of the quasiclassical approach.

12.3. Pair Production at $\kappa_s \gg 1$

Let us divide the integration region in Eq. (12.7) into three parts: (1) $0 \leq x \leq x_1 \ll 1$, (2) $x_1 \leq x \leq x_2$, (3) $1 \ll x_2 \leq x \leq x_0$, assuming $x_0 \gg 1, \lambda(x_1) \ll 1$ and $\lambda(x_2) \ll 1$. It is necessary also to take into account that the integral over energy gets its main contribution from the region in which $\omega^2 \sim 4\varepsilon\varepsilon'$ Then in regions 1 and 3 we can replace $x^{1/2}g(x)$ by approximate expressions for small and large x correspondingly and go over to integration over λ, while in region 2 we can use the expansion of the functions $K_\nu(\lambda)$ for $\lambda \ll 1$. Extending the integration over x to infinity and matching the probabilities from different regions, we find after rather tedious calculations

$$W_e^F(\kappa_s \gg 1) \simeq \frac{\alpha}{3\pi\sqrt{3}} \cdot \frac{V_0}{m x_0 a_s}\left\{c_1 \kappa_s^{-1/3}[\ln \kappa_s + B_1(\eta)] - \left(\frac{m\vartheta_0}{V_0}\right)^2 c_2 \kappa_s^{1/3}(1+2\eta)[\ln \kappa_s + B_2(\eta)]\right\}, \quad (12.16)$$

where

$$c_1 = \frac{45}{7}\frac{\Gamma^3\left(\frac{2}{3}\right)}{\Gamma\left(\frac{1}{3}\right)}\left(\frac{4}{3}\right)^{1/3} \simeq 6.558\,,$$

$$c_2 = \frac{7\Gamma^3\left(\frac{1}{3}\right)}{15.495\Gamma\left(\frac{2}{3}\right)}\left(\frac{3}{4}\right)^{1/3} \simeq 0.097\,;$$

$$B_1(\eta) = B_1(0) - \frac{3}{2}\int_0^\infty \frac{dx}{x}[(1+x)^{-2/3} - (1+\eta x)^{-2/3} \times (1+(1+\eta)x)^{-2/3}]\,,$$

$$B_2(\eta) = B_2(0) - \frac{3}{2}\int_0^\infty \frac{dx}{x}[(1+x)^{-1/3} - (1+\eta x)^{-1/3}(1+(1+\eta)x)^{-1/3}] + \frac{9}{16} - \frac{3}{16(1+2\eta)}\int_0^\infty \frac{dx}{(1+\eta x)^{4/3}(1+(1+\eta)x)^{4/3}}\,, \quad (12.17)$$

$$B_1(0) = \frac{\pi\sqrt{3}}{2} + \frac{5}{2}\ln 3 + \ln 2 - \frac{417}{70} - C \simeq -0.374\,,$$

$$B_2(0) = \frac{\pi\sqrt{3}}{2} + \frac{5}{2}\ln 3 + \ln 2 + \frac{10959}{3080} - C \simeq 3.700\,.$$

Here $C = 0.577\ldots$ is Euler's constant. The integrals in B_1 and B_2 can be expressed in terms of hypergeometric functions. Taking into account that usually $\eta \ll 1$, we give the following approximate expressions

$$B_1(\eta) = -0.374 - 3.975\beta^{2/3}\left(1 + \frac{8}{15}\beta + \frac{7}{18}\beta^2\right) + \beta\left(\frac{3}{2} + \frac{9}{8}\beta + \frac{13}{14}\beta^2\right),$$

$$B_2(\eta) = 3.700 - 2.31\beta^{1/3}\left(1 + \frac{2}{3}\beta - \frac{19}{63}\beta^2\right) - \frac{\beta}{160}\left(30 + 144\beta - 29\beta^2\right), \quad (12.18)$$

where $\beta = \eta/(1+\eta)$.

In deriving (12.16) the region of integration over x was extended to infinity. Since the main (logarithmic) contribution to (12.16) is from the region $x \sim \kappa_s^{2/3}$ (region 3), it is clear that the asymptotic behavior (12.16) is valid only for $\kappa_s^{2/3} < x_0$, i.e. for photon energies less than a few TeV. At such energies when $\kappa_s^{2/3} \sim x_0$, the main contribution is from large distances from the axis, where the axial symmetry of the potential can be destroyed and it is necessary to take into account the influence of several axes. Then to find the probability of the process for small angles of incidence, we should substitute the potential (12.2) into Eq. (12.6).

The function $F_1(\kappa_s \gg 1) \equiv F_1^{as}$ in (12.10) and (12.16) has a maximum at the point $\kappa_s^0 = \exp[3 - B_1(\eta)]$, near which the function F_1^{as} varies very smoothly. Therefore the energy value ω^0 obtained from κ_s^0 is only an order of magnitude estimate. However, the value $W_e^{F\max}$ is given by Eq. (12.16) quite satisfactorily and is in a good agreement with the results of a numerical calculation according to Eq. (12.7):

$$W_e^{F\max} \simeq k \frac{\alpha}{\sqrt{3}\pi} \frac{V_0}{m a_s x_0}, \tag{12.19}$$

where the coefficient $k = c_1 \exp(B_1(\eta)/3 - 1)$ depends on the material and, for the materials usually used, varies in the range 1.59–1.93 when η varies from 0.15 (silicon) to 0.025 (diamond).

It is of interest to determine the maximum value of the increase of the probability for pair production in the field of axes over the pair production probability in the corresponding amorphous medium (via the Bethe–Heitler mechanism). Taking (12.19) and W_{BH} (cf., for example, with Eq. (7.133)), we obtain

$$r^{\max} \equiv \left(\frac{W_e^F}{W_{BH}}\right)_{\max} \simeq \frac{3\sqrt{3}k}{28} \frac{V_0 m a_s d}{(Z\alpha)^2 \ln(183 Z^{-1/3})}$$
$$\sim \frac{1}{3} \frac{m a_s}{Z\alpha \ln(183 Z^{-1/3})}, \tag{12.20}$$

where the second (simplified) estimate is relatively crude. It can be seen from this that $r^{\max}$ is greater, the lower is Z and the larger is a_s. The greatest gain (among the materials used) is achieved in diamond ($r^{\max} \sim 160$); values of $r^{\max}$ are given in Table 12.1. It is seen from Eq. (12.13) that the effect exhibits primarily for those materials for which the gain $r^{\max}$ is minimal.

13. MODIFIED THEORY OF COHERENT PAIR PRODUCTION

13.1. Coherent Pair Production Theory

Let us consider now the region of relatively large photon angles of incidence. We shall show first that in this region the results of the standard theory of coherent pair production follow from Eq. (11.13). From the estimates made at the beginning of Sec. 12.1 it follows that the order of magnitude of the double sum in A (11.13) is $(V_0/m\vartheta_0)^2$. Therefore for $\vartheta_0 \gg V_0/m$ the term with the double sum in A is small, and $\exp(iA)$ can be expanded accordingly. As a result the probability (11.13) takes the form

$$W_e^{\text{coh}} = \frac{i\alpha m^2}{2\pi\omega^2}\int^{\omega} d\varepsilon \int_{-\infty}^{\infty} \frac{d\tau}{\tau + i0} e^{i\frac{m^2\omega\tau}{\varepsilon\varepsilon'}}$$
$$\times \sum_{\mathbf{q},\,\mathbf{q}'} \frac{G(\mathbf{q})G(\mathbf{q}')}{m^2 q_\parallel q'_\parallel}(\mathbf{q}_\perp \mathbf{q}'_\perp)\left\{\frac{\varepsilon^2 + \varepsilon'^2}{\varepsilon\varepsilon'} \sin q_\parallel\tau \sin q'_\parallel\tau\right.$$
$$\left. + \frac{im^2\omega\tau}{\varepsilon\varepsilon'}\left[\frac{\sin(q_\parallel + q'_\parallel)\tau}{(q_\parallel + q'_\parallel)\tau} - \frac{\sin q_\parallel\tau \sin q'_\parallel\tau}{q_\parallel q'_\parallel \tau^2}\right]\right\}$$
$$\times \int \frac{d^3 r_0}{V} e^{-i(\mathbf{q}+\mathbf{q}')r_0} \qquad (13.1)$$

Integration over $\mathbf{r}_0$ in (13.1) is elementary and gives $\delta_{\mathbf{q}+\mathbf{q}',0}$ ($\mathbf{q}$ is a discrete variable), after which the sum over $\mathbf{q}'$ is calculated and the integrals over τ are easily evaluated by means of the theory of residues. Finally, we obtain

$$W_e^{\text{coh}} = \frac{\alpha}{2\omega^2}\int_0^{\omega} d\varepsilon \sum_{\mathbf{q}} \frac{|G(\mathbf{q})|^2}{q_\parallel^2} q_\perp^2 \left\{\frac{\varepsilon^2 + \varepsilon'^2}{\varepsilon\varepsilon'} + \frac{2\omega m^2}{q_\parallel^2 \varepsilon\varepsilon'}\right.$$
$$\left.\times \left(q_\parallel - \frac{\omega m^2}{2\varepsilon\varepsilon'}\right)\right\}\vartheta\left(q_\parallel - \frac{\omega m^2}{2\varepsilon\varepsilon'}\right). \qquad (13.2)$$

Carrying out the integration over ε in Eq. (13.2), we find

$$W_e^{\text{coh}} = \frac{\alpha}{\omega}\sum_{\mathbf{q}} \frac{|G(\mathbf{q})|^2 \mathbf{q}_\perp^2}{q_\parallel^2}\vartheta\left(q_\parallel - \frac{2m^2}{\omega}\right) f\left(\frac{2m^2}{\omega q_\parallel}\right), \qquad (13.3)$$

where

$$f(x) = \left(1 + x - \frac{x^2}{2}\right)\ln\frac{1 + \sqrt{1-x}}{1 - \sqrt{1-x}} - (1+x)\sqrt{1-x}\,. \qquad (13.4)$$

The probabilities (13.2) and (13.3) coincide with the results of the standard theory of coherent pair production (see, e.g. Ter-Mikaelyan, 1972; Diambrini-Palazzi, 1969). Equation (13.3) can be obtained from the well-known expression for the cross-section of pair production by two photons (see, e.g. Berestetskii, Lifshitz and Pitaevskii, 1982)

$$\sigma_{\gamma\gamma} = \frac{\pi\alpha^2}{m^2} y f(y) \vartheta(1-y)\,, \quad y = \frac{2m^2}{kq}\,, \tag{13.5}$$

where the θ-function corresponds to the threshold of the reaction $\gamma + \gamma \to e^- + e^+$. Making use of the equivalent photon flux $\mathbf{J}_q$ (11.1), we obtain for the desired probability per unit time

$$W = \sum_{\mathbf{q}} \sigma_{\gamma\gamma}(k, \mathbf{q}) |\mathbf{J}_q|\,. \tag{13.6}$$

The probability (13.3) depends substantially on the photon angle of incidence with respect to the direction of an axis. For small angles of incidence, we have

$$\begin{aligned} q_{\parallel} &\simeq \frac{2\pi}{d} k + (\mathbf{q}_\perp \mathbf{n}_\perp)\vartheta_0\,, \quad |\mathbf{n}_\perp| = 1\,, \\ k &= 0, 1, 2, \ldots, \quad \vartheta_0 \ll 1\,. \end{aligned} \tag{13.7}$$

In this case the main contribution to the sum is given by $k = 0$, and the probability (13.3) takes the form ($\cos\varphi(\mathbf{q}_\perp) = (\mathbf{q}_\perp \mathbf{n}_\perp)/|\mathbf{q}_\perp||\mathbf{n}_\perp|$)

$$\begin{aligned} W_e^{\mathrm{coh}}(\vartheta_0 \ll 1) &= \frac{\alpha}{\omega\vartheta_0^2} \sum_{\mathbf{q}_\perp} \frac{|G(\mathbf{q}_\perp)|^2}{\cos^2\varphi(\mathbf{q}_\perp)} \vartheta(1-x) f(x)\,, \\ x &= \frac{2m^2}{\omega q_\perp \vartheta_0 \cos\varphi} \end{aligned} \tag{13.8}$$

As it was noted in Sec. 11.1 (see Eq. (11.4)), the probability (13.8) has a maximum at the angle of incidence $\vartheta_0 \sim \vartheta_m = m^2\Delta/\pi\omega \sim m^2 a_s/\omega$, which decreases with increasing photon energy. Taking into account that the region of applicability of Eqs. (13.3) and (13.8) is bounded below by the angle $\vartheta_v = V_0/m(\vartheta_0 \gg \vartheta_v)$, we find that, for sufficiently large photon energy $\omega \geq \omega_b$, the peak of the probability obtained within coherent pair production theory is situated in the region where this theory is inapplicable

$$\omega_b = \frac{m^2 a_s}{\vartheta_v} = \frac{m^3 a_s}{V_0}\,. \tag{13.9}$$

By the definition (12.9), $\kappa_s(\omega) = V_0\omega/m^3 a_s$, and we obtained $\kappa_s(\omega_b) = 1$. Thus, starting from energies for which the pair production in the constant field of an axis becomes essential, the coherent theory of the process is inapplicable. Note that even at such energies this theory is valid for the tail ($\vartheta_0 \gg \vartheta_v$) of the probability distribution $W(\vartheta_0)$. The magnitude of the effect exceeds the amorphous value up to the angles ϑ_0 of the order of

$$\vartheta_0 \sim \sqrt{\frac{\alpha}{\omega W_{BH}}} V_0 . \tag{13.10}$$

13.2. Generalization of Coherent Pair Production Theory

For $\kappa_s \gg 1$ we can obtain from the general formula (11.13) approximate expressions for the probabilities, which region of applicability is considerably broader than that of the standard theory of coherent pair production. For this purpose it is necessary to take into account that the second term in the double sum in the function A in (11.13) has an order of magnitude $(V_0/m\vartheta_0)^4/\kappa_s^2$, while the first term contains contributions of two types: for $q_{\parallel} + q'_{\parallel} = 0$ of order $(V_0/m\vartheta_0)^2$, and for $q_{\parallel} + q'_{\parallel} \neq 0$ of order $(V_0/m\vartheta_0)^3/\kappa_s$. Contributions of order $(V_0/m\vartheta_0)^3/\kappa_s$ (and even more so for those of order $(V_0/m\vartheta_0)^4/\kappa_s^2$) can turn out to be small even for angles $\vartheta_0 \leq V_0/m$. Assuming these contributions to be small, we shall carry out the corresponding expansions in e^{iA}, while the terms with $q_{\parallel} + q'_{\parallel} = 0$ in the first term in the double sum in A will be retained in the exponential. As a result we obtain an expression which coincides in form with (13.1) except that in it we must make the substitution

$$\begin{aligned} \exp\left(i\frac{m^2\omega\tau}{\varepsilon\varepsilon'}\right) &\to \exp\left(\frac{im_*^2\omega\tau}{\varepsilon\varepsilon'}\right) , \\ m_*^2 &= m^2(1+\rho/2) , \end{aligned} \tag{13.11}$$

and in the term $\sin(q_{\parallel} + q'_{\parallel})\tau/(q_{\parallel} + q'_{\parallel})\tau$ it is necessary to assume $q_{\parallel} + q'_{\parallel}, \neq 0$. Here (see Eqs. (11.2) and (11.7))

$$\begin{aligned} \frac{\rho}{2} &= \gamma^2[\langle \mathbf{v}^2\rangle - \langle \mathbf{v}\rangle^2] \\ &= \sum_{\mathbf{q},\, \mathbf{q}'} \frac{G(\mathbf{q})G(\mathbf{q}')}{m^2 q_{\parallel} q'_{\parallel}} (\mathbf{q}_{\perp}\mathbf{q}'_{\perp}) \cdot [\delta_{q_{\parallel}+q'_{\parallel},0} - \delta_{q_{\parallel},0}\delta_{q'_{\parallel},0}] \\ &= \sum_{\mathbf{q}, q_{\parallel}\neq 0} \frac{|G(\mathbf{q})|^2 \mathbf{q}_{\perp}^2}{m^2 q_{\parallel}^2} \end{aligned} \tag{13.12}$$

The transition $m \to m_*$ corresponds to the transition to the effective mass in the field of a wave (see Appendix B).

The remaining calculations are carried out in the same way as in transition from Eq. (13.1) to Eq. (13.2). Finally, we obtain

$$W_e^{\rm mcoh} = \frac{\alpha}{2\omega^2}\int_0^\omega d\varepsilon \sum_{\mathbf{q}} \frac{|G(\mathbf{q})|^2 \mathbf{q}_\perp^2}{q_\parallel^2}$$

$$\times \left[\frac{\varepsilon^2+\varepsilon'^2}{\varepsilon\varepsilon'} + \frac{2\omega m^2}{q_\parallel^2 \varepsilon\varepsilon'}\left(q_\parallel - \frac{\omega m_*^2}{2\varepsilon\varepsilon'}\right)\right] \vartheta\left(q_\parallel - \frac{\omega m_*^2}{2\varepsilon\varepsilon'}\right) \quad (13.13)$$

and after integration over ε we have

$$W_e^{\rm mcoh} = \frac{\alpha}{\omega}\sum_{\mathbf{q}} \frac{|G(\mathbf{q})|^2}{q_\parallel^2}\mathbf{q}_\perp^2 \vartheta\left(q_\parallel - \frac{2m_*^2}{\omega}\right)\tilde{f}(\tilde{x}), \quad (13.14)$$

where

$$\tilde{f}(z) = \left(1 + \frac{2z - z^2}{2+\rho}\right)\ln\frac{1+\sqrt{1-z}}{1-\sqrt{1-z}} - \left(1 + \frac{2z}{2+\rho}\right)\sqrt{1-z},$$

$\tilde{x} = 2m_*^2/\omega q_\parallel$. Equations (13.13) and (13.14) are no more complicated than Eqs. (13.2) and (13.3) but have a broader range of applicability. For $\vartheta_0 \gg V_0/m (\rho \ll 1)$, Eqs. (13.13) and (13.14) go over respectively into Eqs. (13.2) and (13.3).

Making use of Eq. (11.14) and carrying out similar calculations, we obtain the probability of e^+e^- pair production by a polarized photon within modified theory. This probability (per unit time) has the form

$$W_e^{\rm mcoh}(\mathbf{e}) = \frac{\alpha}{2\omega^2}\int_0^\omega d\varepsilon \sum_{\mathbf{q}} \frac{|G(\mathbf{q})|^2}{q_\parallel^2}\vartheta\left(q_\parallel - \frac{\omega m_*^2}{2\varepsilon\varepsilon'}\right)$$

$$\times \left\{(\mathbf{e}^*\mathbf{e})\mathbf{q}_\perp^2 \left[\varphi(\varepsilon) + \frac{2\omega m^2}{\varepsilon\varepsilon' q_\parallel^2}\left(q_\parallel - \frac{\omega m_*^2}{2\varepsilon\varepsilon'}\right)\right]\right.$$

$$\left. + \frac{1}{2}\left(\frac{\omega m_*^2}{\varepsilon\varepsilon' q_\parallel}\right)^2 \left[(\mathbf{e}^*\mathbf{e})\mathbf{q}_\perp^2 - 2|(\mathbf{q}_\perp\mathbf{e})|^2\right]\right\}, \quad (13.15)$$

where m_*^2 is defined in Eq. (13.11). At $\rho \to 0 (\vartheta_0 \gg V_0/m)$ the probability (13.15) goes over into the probability of the process with polarized initial photons obtained within the standard approach (see Ter-Mikaelyan, 1972).

14. ORIENTATION DEPENDENCE OF THE PAIR PRODUCTION PROCESS IN CRYSTALS

14.1. Orientation Dependence of the Probability

Let us consider the dependence of the pair production probability on the photon angle of incidence ϑ_0. The behavior of an orientation curve substantially depends on the magnitude of the parameter κ_s. For relatively small photon energies, when this parameter is $\kappa_s \ll 1$, the pair production is suppressed for $\vartheta_0 \ll V_0/m$, whilst in the range of angles ϑ_0 where the probability of the process is noticeably different from zero, the orientation dependence is described by the coherent theory of pair production (see Sec. 13.1). At increasing photon energies when the parameter κ_s becomes of order of unity, in the first place, the coherent theory of pair production is inapplicable near the peak of an orientation curve, and, secondly, the mechanism of pair production by the averaged field of axes is turned on. The position of the peak is shifted toward smaller angles ϑ_0, and when the sign of the correction in Eq. (12.10) changes, this position achieves the point $\vartheta_0 = 0$. With a further increase of the photon energy the peak stays at $\vartheta_0 = 0$, while its angular width narrows.

The calculation of the orientation dependence of the pair production probability by means of the complete formula (11.13) is a fairly complicated problem, while the calculation of the probability using the formulae for the constant field with the corrections (12.7) and the modified coherent pair production formula (13.14) is rather simple. Taking into account that the numerical calculation is most laborious just in the regions of applicability of asymptotic expressions, the complete formula (11.13) can be reasonably used for numerical calculations only in the intermediate region. Moreover, the analysis shows that for $\kappa_s \gg 1$ the interval where both Eqs. (12.7) and (13.14) are inapplicable is relatively narrow ($\Delta\vartheta_0 \sim \kappa_s^{-1/3} V_0/m$), while for $\kappa_s \sim 1$, the dependence on ϑ_0 is sufficiently smooth. This enables one to use an interpolation procedure in the intermediate region. The character of such an interpolation can be studied on the model of pair production in the field of a plane wave because of the analogy noted above of pair production by a photon in a crystal and in a plane wave field. In the latter case, owing to the relative simplicity of the corresponding formulae, we can make exact calculations (see Appendix D).

As an illustration, the probability $W_e(\vartheta_0)$ is shown in Fig. 14.1(a) for the $\langle 110 \rangle$ axis of cooled ($T = 100$ K) germanium crystal and in Fig. 14.1(b) for the $\langle 111 \rangle$ axis in tungsten at room temperature. The change of ϑ_0 corresponds to the rotation of the incident photon momentum in the plane containing the

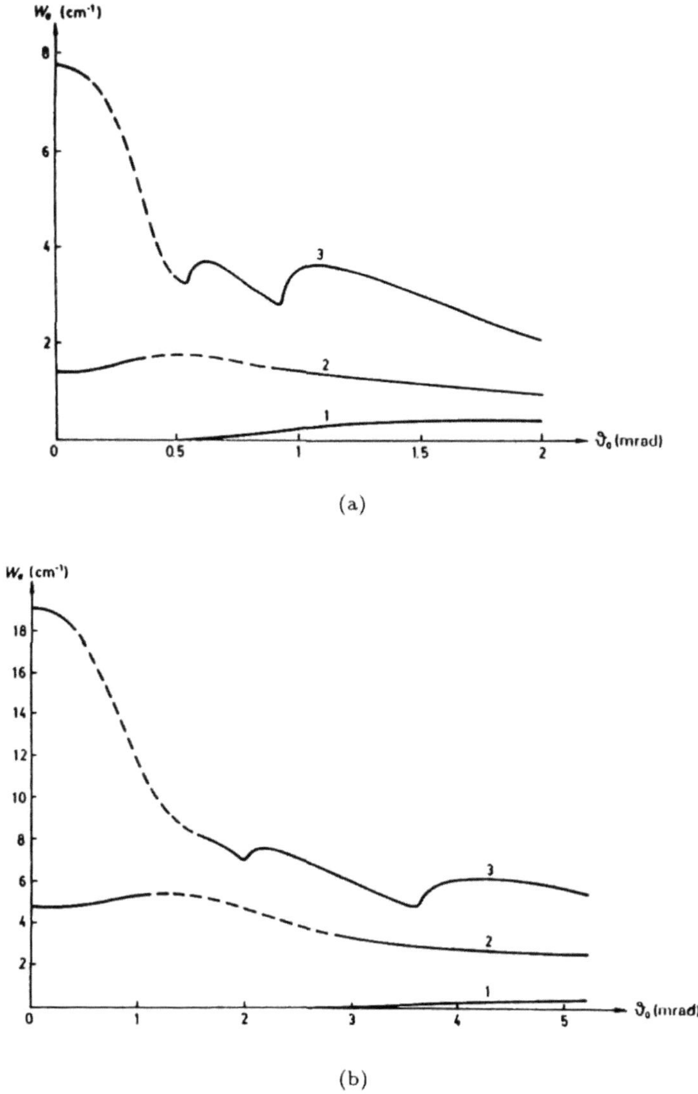

Fig. 14.1. Orientation dependence of the probability of pair production by a photon in (a) $\langle 110 \rangle$ axis of Ge at $T = 100$ K for ω (in GeV) = 30 (curve 1), 100 (curve 2), and 1000 (curve 3). The values of κ_m correspondingly are 0.6, 2, and 20. (b) $\langle 111 \rangle$ axis of W at $T = 293$ K for ω (in GeV) = 5 (curve 1), 35 (curve 2), and 300 (curve 3). The values of κ_m are 0.35, 2.4, and 21.

chosen axis and forming the angle φ_0 with the $(\bar{1}10)$ plane. To avoid the main planes, the value $\tan(\varphi_0) \simeq 0.092$ was chosen in the case of tungsten crystal. Under the conditions indicated above, the calculation yields $\rho/2 = 1.87(V_0/m\vartheta_0)^2$ for Ge and $\rho/2 = 1.04(V_0/m\vartheta_0)^2$ for W.

Curve 1 in Figs. 14.1(a), (b) demonstrates the situation at $\kappa_m < 1$ when the constant field effects are small and the specific nature of the crystal exhibits itself in the coherent pair production mechanism. The effect exists only for such ϑ_0 which correspond to values $\rho \ll 1$, so that probabilities are described by the standard formula (13.3). Curve 2 in these figures corresponds to the case when, as the energy grows, the mechanism of pair production in a constant field is already switched on. In accordance with the results of Sec. 12.1, the probability $W_e(\vartheta)$ is a minimum at $\vartheta_0 = 0(\kappa_m < 4)$. With further increase of the energy, this minimum goes over into a maximum and, then, as the energy increases, the width of this maximum narrows. Curve 3 in Figs. 14.1(a), (b) corresponds to this case. Note that the quantity $W_e(0) = W_e^F$ continues to grow as long as $\kappa_s < 25$, and the maximum value $W_e^{F\max}$ with good accuracy is given by Eq. (12.19).

14.2. Comparison of Theory with Experiment

Side by side with the photo-production of e^+e^- pairs in the crystal potential averaged over thermal vibrations, the incoherent pair production on the potential fluctuations connected with these vibrations takes place in a crystal. The theory of incoherent processes in crystals is stated in detail in Sec. 21. Here we note only that, since the contribution of incoherent mechanism is essential for $\kappa_s \leq 1$, it must be taken into account in comparison of the theory with experiment in this region of photon energies. However, the influence of the average crystal field on the incoherent process can be neglected for $\kappa_s \leq 1$ with good accuracy, and only its modification (as compared to the amorphous case) owing to the periodicity of crystal lattice must be taken into account.

In experiments (see Belkacem *et al.*, 1987; Bak *et al.*, 1988a) the probabilities $W_e(\omega)$ and $W_e(\vartheta_0)$ were measured for $\langle 110 \rangle$ axis of Ge crystal at $T = 100$ K. In this situation $W_e^F = W_{BH}$ at $\omega = \omega_t = 50$ GeV and a rearrangement of the angular dependence due to the change in sign of the correction in Eq. (12.10) occurs at $\omega \simeq 230$ GeV. The function $W_e(\omega)$ at $\vartheta_0 = 0$ is shown in Fig. 12.1. In Fig. 14.2 the same curve is shown with the addition of a modified incoherent contribution $W_{BH}^m = 0.28$ cm^{-1}, whereas $W_{BH} = 0.32$ cm^{-1} The experimental points are from (Belkacem *et al.*, 1987;

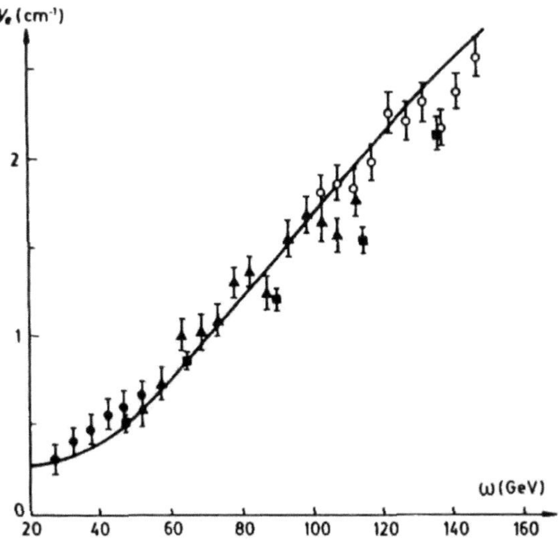

Fig. 14.2. Total probability of pair production in Ge $\langle 110 \rangle (T = 100$ K, $\vartheta_0 = 0)$ with regard for the incoherent contribution. The experimental data are from Belkacem *et al.*, 1987 (except the squares), the squares are from Bak *et al.*, 1988a.

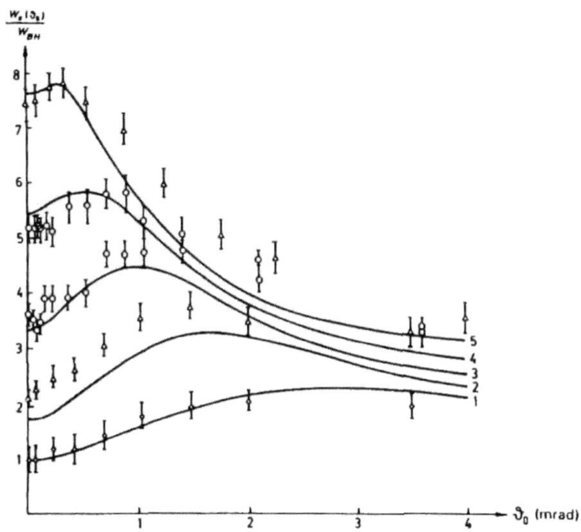

Fig. 14.3. Orientation dependence of the total probability of pair production in Ge $\langle 110 \rangle$ $(T = 100$ K$)$ averaged over photon energy intervals according to the experimental conditions of Belkacem *et al.*, 1987.

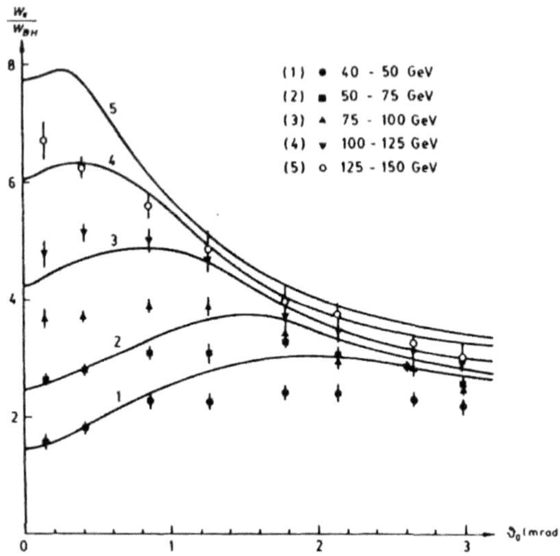

Fig. 14.4. The same as Fig. 14.3 but for other photon energy intervals (data of Bak *et al.*, 1988a).

Bak *et al.*, 1988a). In Figs. 14.3 and 14.4 there are given curves of the orientation dependence, additionally averaged over photon energy intervals in accordance with the conditions of indicated experiments. On the whole, it is seen that the theory quite satisfactorily describes the experimental data.

15. PAIR PRODUCTION BY PHOTONS INCIDENT NEAR CRYSTALLINE PLANES

15.1. Pair Production in the Continuum Plane Potential

In Secs. 11 and 13, a theory was developed of e^+e^- pair production by photons in crystals, which is valid for any energies and directions of incidence of the photon. Proceeding from general expressions, a comprehensive analysis was carried out in Sec. 12 of the case when the photon momentum $\mathbf{k}$ is in nearly the same direction as the crystalline axis. In the present section we deal with a quantitative study of the situation when the photon propagates near crystalline planes. The angle between $\mathbf{k}$ and the nearest axis is assumed to be so large that the coherent axial effects (see Eq. (13.10)) do not exceed the amorphous values.

In the planar case polarization effects can manifest themselves. In order to take them into account within the constant field approximation, we use Eq. (11.14), expanding all the quantities as it was done in the beginning of Sec. 12.1. Taking the corresponding integrals, we obtain the pair production probability by a polarized photon in the constant field approximation

$$W_e = \frac{\alpha m^2}{\pi\sqrt{3}\omega^2}\int d\varepsilon \int \frac{d^2\rho_0}{s}\left\{(\mathbf{e}^*\mathbf{e})[\varphi(\varepsilon)K_{2/3}(\lambda)\right.$$
$$\left. + \int_\lambda^\infty dy K_{1/3}(y) - \left[\frac{2}{\mathbf{b}^2}(\mathbf{e}^*\mathbf{b})(\mathbf{e}\mathbf{b}) - (\mathbf{e}^*\mathbf{e})\right]K_{2/3}(\lambda)\right\}, \quad (15.1)$$

here $\lambda = 2m^2\omega/3\varepsilon\varepsilon'|\mathbf{b}|$. In the axial case $\boldsymbol{\nabla} = \partial/\partial\boldsymbol{\rho}_0, \mathbf{b} = \boldsymbol{\nabla} U(\boldsymbol{\rho}_0)/m$; the potential $U(\boldsymbol{\rho}_0)$ is defined by Eq. (12.2). In the planar case $\int d^2\rho_0/s \to \int dx/d_{pl}$ where d_{pl} is the distance between the adjacent planes;

$$\boldsymbol{\nabla} \to d/dx\,, \quad \mathbf{b} \to m^{-1}dU(x)/dx, \quad U(x) = \sum_{q_x} G(q_x)\exp(-ixq_x)\,.$$

The x-coordinate is perpendicular to the plane; $\mathbf{n}\boldsymbol{\nabla} \to \vartheta_0 d/dx$, where ϑ_0 is the angle between the direction of incidence of the photon and the plane.

Below we will use both the plane potential as the sum over q_x and its analytical approximation in the form of (9.14). If the amplitude of thermal vibrations, u_1, goes to zero, $\eta^2 \to 0$ and expression (9.14) is transformed into the potential (9.33). In the potential (9.14) the local values of the quantum parameter κ, determining the behavior of the probability of pair production by a photon in a constant field, will be

$$\kappa(x) = \frac{E(x)}{E_0}\frac{\omega}{m} = \kappa_p \psi(y)\,,$$
$$\kappa_p = \frac{2V\delta\omega}{m^3 d_{pl}} \equiv \frac{V\omega}{a_s m^3}\,,$$
$$\psi(y) = \sqrt{\frac{y^2}{y^2+\eta^2}}\sinh\delta(\sqrt{1+\eta^2} - \sqrt{y^2+\eta^2})\,, \quad (15.2)$$
$$E_0 = \frac{m^2}{e} \simeq 1.32\cdot 10^{16}\text{V /cm}\,.$$

The electric field $E(x)$ and the parameter $\kappa(x)$ vanish on the plane (because of thermal vibrations) and in the middle between the planes (because of the symmetry) and reach a maximum at the point y_0 where $\psi'(y_0) = 0$.

Let us first discuss the behavior of the probability of the process at a planar orientation in the constant field approximation. Integrating Eq. (15.1) by parts over ε, we reduce it to the form

$$\begin{aligned} W_{eF}^{pl} &= \frac{2\alpha m^2}{3\sqrt{3}\pi\omega^2}\int_0^\omega d\varepsilon \int_0^1 dy\, \{|e_x|^2 \\ &\times (\varphi(\varepsilon)+1) + |e_y|^2 (\varphi(\varepsilon)+4)\} K_{2/3}(\lambda) \\ &\equiv |e_x|^2 W_\perp + |e_y|^2 W_\parallel\,, \end{aligned} \tag{15.3}$$

where $\lambda = 2\omega^2/\varepsilon\varepsilon'\kappa_p\psi(y)$. Expression (15.3) coincides (with an accuracy up to coordinate averaging) with formula (3.55) if in the latter the external field is considered to be purely electric and is directed along the x-axis. From Eq. (15.3) it can be seen that in the case of linear polarization of the photon along the x-axis (perpendicularly to the plane), the corresponding probability $W_\perp$ is less than $W_\parallel$, the probability of the process at linear polarization of the photon along the y-axis (in parallel to the plane).

At small values of the parameter κ the quantity λ in Eq. (15.3) proves to be large at any y and we can use for $K_{2/3}(\lambda)$ its asymptotic expression at $\lambda \gg 1 : K_\nu(\lambda) \simeq e^{-\lambda}(\pi/2\lambda)^{1/2}$. Under these conditions the main contribution comes from the range of values of y near the point y_0 where the field has its maximum value, and near $\varepsilon \simeq \varepsilon' \simeq \omega/2$ with respect to the energy, both resulting in a minimum (but large) value of λ

$$\begin{aligned} W_{eF}^{pl}(\kappa \ll 1) \simeq & \frac{3\alpha m^2\kappa_m^2}{16\omega}\left(|e_x|^2 + 2|e_y|^2\right) \\ & \times \left(\frac{\pi}{2\kappa_p|\psi''(y_0)|}\right)^{1/2} \exp\left(-\frac{8}{3\kappa_m}\right) \end{aligned} \tag{15.4}$$

Here κ_m is the maximum value of the parameter $\kappa(x) : \kappa_m = \kappa_p\psi(y_0)$. Within this limit, just as in the constant field, we have $W_\parallel = 2W_\perp$, at $\kappa \ll 1$. At small enough κ_m the probability (15.4) is exponentially small and in this respect one may talk of the energy threshold of the process under consideration. The threshold photon energy ω_t can be defined more strictly by the relation: $\overline{W}_{eF}(\omega = \omega_t) = (W_\parallel + W_\perp)/2 = W_{BH}$, the probability in a corresponding amorphous medium. For some substances the values of ω_t are given in Table 15.1.

Table 15.1. Characteristics of pair production in the (110) plane of various crystals at room temperature.

Crystal	ω_t GeV	$\left[\frac{\langle E^2\rangle}{E_0^2}\right]^{1/2} \times 10^7$	$l_0 W_{BH} \frac{\omega}{m} \times 10^{-5}$	$l_{\min}\left(\frac{1}{2}\right)$ (mm)
C	550	3.73	3.0	4.2
Si	920	2.59	8.8	6.4
Ge	700	4.41	12.9	3.9
Fe	350	7.28	6.2	2.2
W	240	12.1	11.3	1.3

The probability (15.3) reaches a maximum at high energies (at $4 \div 30$ TeV for the (110) plane) when $\kappa_m \gg 1$. Of interest is therefore the asymptotic of the expression (15.3) at high values of κ. It can be conveniently obtained by taking advantage of the integral representation of the function $K_{2/3}(\lambda)$ (see Gradshteyn and Ryzhik, 1965)

$$K_{2/3}(\lambda) = \frac{1}{4\pi i}\int_{-c-i\infty}^{-c+i\infty} ds\Gamma(-s)\Gamma\left(-s-\frac{2}{3}\right)\left(\frac{\lambda}{2}\right)^{2s+\frac{2}{3}}, \qquad c > \frac{2}{3} \tag{15.5}$$

Substituting Eq. (15.5) into Eq. (15.3), integrating over ε and closing the integration contour over s to right, we obtain at $\kappa \gg 1$

$$W_{eF}^{pl} = \frac{2}{3}\frac{\alpha m^2}{\omega}\left[|e_x|^2(2A-1) + |e_y|^2(3A-1)\right],$$
$$A = \frac{3}{7}\left[\frac{\Gamma\left(\frac{2}{3}\right)}{\Gamma\left(\frac{1}{3}\right)}\right]^2 \langle(3\kappa(x))^{2/3}\rangle, \tag{15.6}$$

where

$$\langle f(x)\rangle = \int_{-d_{pl}/2}^{d_{pl}/2} dx \cdot f(x)/d_{pl}.$$

Note that $\kappa(x)$ has been defined in Eq. (15.2). Expressions (15.6) for $W_{\|}$ and $W_{\perp}$, and for the unpolarized case $\overline{W}_{eF} = (W_{\|} + W_{\perp})/2$, have a maximum at $\omega = \omega_0$. However, because of a smooth dependence of these probabilities on ω in the region of the maximum, the values of ω_0 obtained from Eq. (15.6) may serve merely as an estimation. Meanwhile, the maximum values themselves $W^{\max}$ are defined from Eq. (15.6) with good accuracy; it is confirmed by the numerical calculation. From Eq. (15.6) we find

$$W^{\max} = \frac{4}{3}\frac{\alpha m^2}{\omega_0}, \tag{15.7}$$

where

$$\begin{aligned} \omega_0^{\parallel} &= (\widetilde{A})^{-3/2}, \quad \omega_0^{\perp} = \left(\frac{3}{2}\right)^{3/2} \cdot \omega_0^{\parallel}, \\ \overline{\omega}_0 &= \left(\frac{6}{5}\right)^{3/2} \omega_0^{\parallel}, \end{aligned} \tag{15.8}$$

here $\widetilde{A} = A\omega^{-2/3}$; the quantity A is defined in Eq. (15.6) and $\widetilde{A}$ is independent of the photon frequency.

Figure 15.1 illustrates the behavior of the probability (15.3) versus the photon energy. It is seen that in the frequencies range $\omega \geq \omega_t$ a rapid growth of the probability is observed, which slows down in the region of a maximum; having passed through it the probability drops very slowly. The maximum value of $W_{\parallel}^{\max}$ turns out to be comparable with the maximum probability in the axial case. Since the field of the plane is weaker than that of the axis, at planar orientations, the effect manifests itself and reaches maximum values at higher energies in comparison with the axial orientation (cf. the corresponding values of ω_t).

Corrections ($\propto \vartheta_0^2$) to the constant field approximation are obtained in Baier, Katkov and Strakhovenko, 1987. These corrections determine the region of applicability of the constant field approximation and are also useful in finding a tendency of the ϑ_0-dependence of W_e. The relative magnitude of the correction in comparison with the leading term (15.6) is found in the mentioned paper to be parametrically $\sim (\vartheta_0 m/V)^2 \kappa_p^{2/3}$ This parameter is also obtained from a direct estimation of the applicability condition for the constant field approximation: the variation of the field is negligible over the length of pair formation. One should, however, bear in mind that, because of the smallness of the numerical coefficient in the correction, information contained in it is reduced, since the following terms of the expansion in ϑ_0^2 may turn out to be larger. In particular, taking into account the correction $\propto \vartheta_0^4$ shows that the probabilities $W_{\parallel}$ and $W_{\parallel}$, for large κ, decrease as the angle ϑ_0 grows from zero, while taking into account the same for the first correction alone does not allow such a conclusion to be drawn.

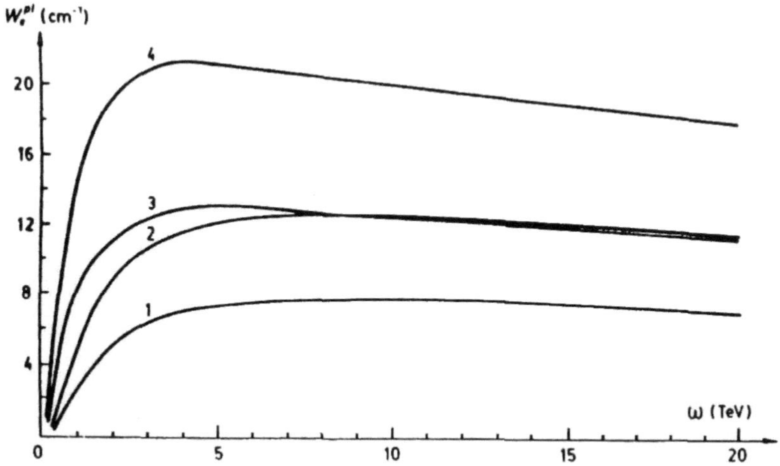

(a)

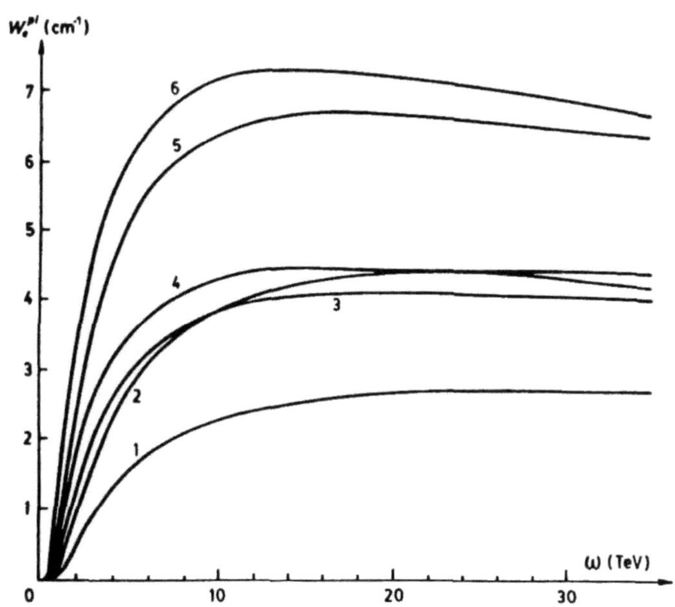

(b)

Fig. 15.1. Dependence of the probabilities $W_\perp$ and $W_{||}$ on photon energy obtained acoording to (15.3) for the (110) plane; (a) for Fe (curves 1 and 2) and W (curves 3 and 4) (b) for Si (curves 1 and 2), diamond (curves 3 and 5) and for Ge (curves 4 and 6).

15.2. The Process at $\vartheta_0 > V/m$

For relatively large angles of incidence, $\vartheta_0 > V/m$, we can use Eq. (13.15). At planar orientation $\sum_{\mathbf{q}}$ in Eq. (13.15) becomes one-dimensional:

$$\frac{dW_{pl}^{\text{mcoh}}}{d\varepsilon} = \frac{\alpha}{2(\omega\vartheta_0)^2}\sum_{k=1}^{\infty}|U(k)|^2\vartheta(1-z)\left\{|e_x|^2\left[\varphi(\varepsilon) + \frac{4z(1-z)}{1+\rho/2} - 2z^2\right] + |e_y|^2\left[\varphi(\varepsilon) + \frac{4z(1-z)}{1+\rho/2} + 2z^2\right]\right\}, \quad (15.9)$$

$$z = \frac{\omega m_*^2 d_{pl}}{4\pi\varepsilon\varepsilon' k\vartheta_0}.$$

Here $U(k)$ is a Fourier-component of the plane potential

$$U(k) = \frac{1}{d_{pl}}\int_{-d_{pl}/2}^{d_{pl}/2} dx U(x)\exp[2i\pi kx/d_{pl}],$$

$$m_*^2 = m^2(1+\rho/2), \quad (15.10)$$

$$\frac{\rho}{2} = \frac{2}{(m\vartheta_0)^2}\sum_{k=1}^{\infty}|U(k)|^2 = \frac{\langle U^2\rangle - \langle U\rangle^2}{(m\vartheta_0)^2}$$

The quantity $U(k)$ is clearly connected with $G(\mathbf{q})$. For instance, if the Moliere potential is used for the atom, then for the (110) plane of the structures $fcc_{(d)}$ (diamond, Si, Ge) and bcc (tungsten, iron) we find according to Eqs. (9.5)–(9.8), (8.20) and (8.28):

$$U(k) = \frac{4\pi Ze^2}{ld_{pl}^2}\exp\left(-\frac{u_1^2q_x^2}{2}\right)\sum_{i=1}^{3}\frac{\alpha_i}{\beta_i^2+q_x^2},$$
$$q_x = \frac{2\pi k}{d_{pl}} \quad (15.11)$$

while for the (100) plane of these structures, the right-hand side of the expression (15.11) should be divided by two. After the spectral distribution (15.9) has been integrated over ε, we find for the total probability:

$$W_{pl}^{\text{mcoh}} = \frac{\alpha}{\omega\vartheta_0^2}\sum_{k=1}^{\infty}|U(k)|^2\vartheta(1-\mu)\left\{\left[1+\frac{\mu(2-\mu)}{2+\rho} \pm \frac{\mu^2}{4}\right]\times\ln\frac{1+\sqrt{1-\mu}}{1-\sqrt{1-\mu}} - \sqrt{1-\mu}\left[1+\frac{2\mu}{2-\rho} \mp \frac{\mu}{2}\right]\right\}, \quad (15.12)$$

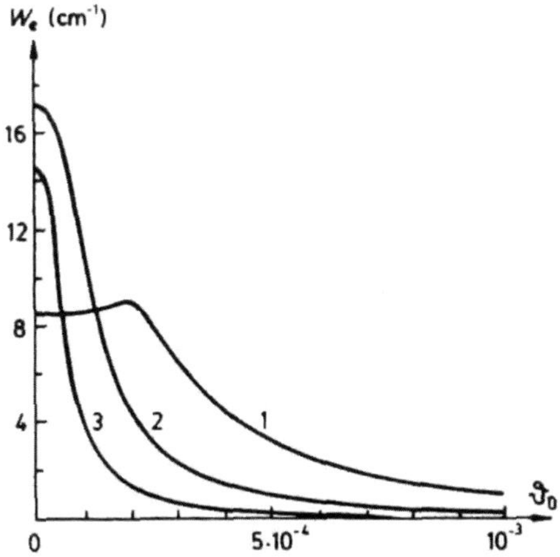

(a)

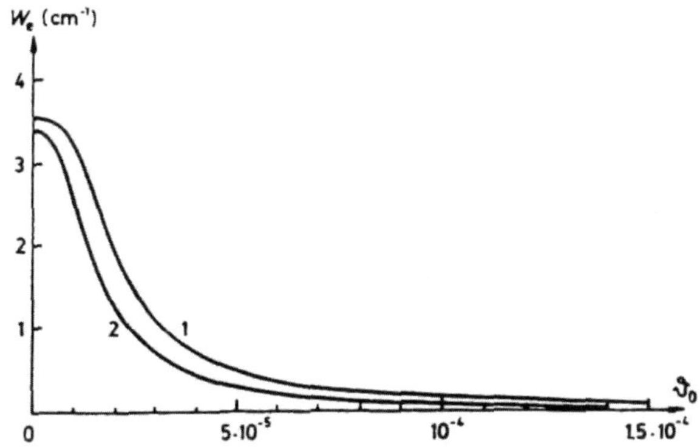

(b)

Fig. 15.2. Dependence of the probability of pair production by an unpolarized photon incident at the angle ϑ_0 with respect to the (110) plane; (a) in W at ω (in TeV) = 0.7 (1), 4 (2), and 10 (3); (b) in Si at ω (in TeV) = 25 (1) and 50 (2).

where $\mu = m_*^2 d_{pl}/(\pi\omega k\vartheta_0)$; the upper signs in Eq. (15.12) give $W_{\parallel}$ and the lower ones give $W_{\perp}$. It is seen that $W_{\parallel} > W_{\perp}$ in the coherent limit too.

The orientation dependence of the probability for the process with the incoherent contribution neglected) is shown in Fig. 15.2. Under these conditions the magnitude of ρ proves to be $\rho_{\rm W} \simeq (U_0/m\vartheta_0)^2/6$ and $\rho_{\rm Si} \simeq$

$(5.2)^{-1}(U_0/m\vartheta_0)^2$, according to Eqs. (15.10) and (15.11); U_0 is the depth of the potential well. The quantity U_0/m is $\sim 2.5 \times 10^{-4}$ in W and $\sim 4.5 \times 10^{-5}$ in Si. It is the quantity that determines the order of the width of the orientation curve at $\kappa \sim 1$. In accordance with the above consideration, for sufficiently high photon energy, these curves have a maximum at $\vartheta_0 = 0$ while their angular width narrows with further increase of the energy.

15.3. Polarization Effects

To describe the propagation of an electromagnetic wave in a crystal, we will take advantage of the method given in Baier, Katkov and Fadin, 1973, Sec. 13. Let us introduce the refractive index $\tilde{n}$ by the relation $\tilde{n} = |\mathbf{k}|/\omega$ where $\mathbf{k}$ is the wave vector and ω the wave frequency. The imaginary part of $\tilde{n}^2$ is connected with the photon absorption probability in a known manner: $\mathrm{Im}\,\tilde{n}^2 = W_e/\omega$, and the real one may be reconstructed using the dispersion relation

$$\mathrm{Re}\,\tilde{n}^2 = 1 - \frac{\mathcal{P}}{\pi}\int_{-\infty}^{\infty} \frac{d\omega'\,\mathrm{Im}\,\tilde{n}^2(\omega')}{\omega' - \omega} \tag{15.13}$$

Now we take into account that W_e in Eqs. (11.13) and (11.14) is of the form

$$W_e = \frac{1}{\omega}\int_{-\infty}^{\infty} \frac{d\tau}{\tau + i0} R(\tau)\exp[i\tau S(\tau)/\omega]\,, \tag{15.14}$$

where the function $S(\tau)$ is positive at any real value of τ, and the ω-dependence is represented explicitly. Substituting Eq. (15.14) into Eq. (15.13) and making allowances for a connection between $\mathrm{Im}\,\tilde{n}^2$ and W_e, we find

$$\tilde{n}^2 = 1 + \frac{i}{\omega^2}\int_{-\infty}^{\infty} \frac{d\tau}{\tau + i0} R(\tau)\,[1 + \mathrm{sgn}\,(\tau)] \times \exp\left[i\tau S(\tau)/\omega\right]\,. \tag{15.15}$$

Here $\mathrm{sgn}(\tau)$ is 1 for $\tau > 0$ and is -1 for $\tau < 0$. As has been shown in the above-mentioned book (see pp.190 and191), the relation (15.15) is one of the representation of the dispersion equation whose solutions determine the types of waves which may propagate in a crystal. In the planar case, these are waves linearly polarized along the x-axis (perpendicularly to the plane, index$_\perp$) and along the y-axis (index$_\parallel$). The function $R(\tau)$ in Eq. (15.15) may be written as $R(\tau) = |e_x|^2 R_\perp(\tau) + |e_y|^2 R_\parallel(\tau)$; an explicit form of the functions $R_\perp$ and $R_\parallel$

is obtained through a comparison of the expression (15.14) with Eqs. (11.13) and (11.14). In terms of the functions $R_\perp$ and $R_\parallel$, we have for the frequencies of the eigensolutions of the dispersion equation:

$$\begin{aligned} \mathrm{Re}\,\omega_{\parallel,\perp} &= |\mathbf{k}|\left\{1-\frac{i}{2\omega^2}\int_{-\infty}^{\infty}\frac{d\tau\,\mathrm{sgn}\,(\tau)}{\tau+i0}R_{\parallel,\perp}(\tau)\exp\left[\frac{i\tau S(\tau)}{\omega}\right]\right\}, \\ \mathrm{Im}\,\omega_{\parallel,\perp} &= -\frac{|\mathbf{k}|}{2\omega^2}\int_{-\infty}^{\infty}\frac{d\tau}{\tau+i0}R_{\parallel,\perp}(\tau)\exp\left[\frac{i\tau S(\tau)}{\omega}\right] = -\frac{W_{\parallel,\perp}}{2}. \end{aligned} \tag{15.16}$$

For example, in the constant field limit ($\vartheta_0 \to 0$) we obtain

$$\begin{aligned} \omega_\parallel &= |\mathbf{k}|\left\{1-\frac{\alpha m^2}{3\pi\omega^3}\int_0^\omega d\varepsilon\int_0^1 dy\,[\varphi(\varepsilon)+4]\right. \\ &\qquad \left.\times\left[\frac{i}{\sqrt{3}}K_{2/3}(\lambda)-L_{2/3}(\lambda)\right]\right\}, \\ \omega_\perp &= |\mathbf{k}|\left\{1-\frac{\alpha m^2}{3\pi\omega^3}\int_0^\omega d\varepsilon\int_0^1 dy\,[\varphi(\varepsilon)+1]\right. \\ &\qquad \left.\times\left[\frac{i}{\sqrt{3}}K_{2/3}(\lambda)-L_{2/3}(\lambda)\right]\right\}, \end{aligned} \tag{15.17}$$

where $\lambda = 2\omega^2/3\varepsilon\varepsilon'\kappa(x), y = 2x/d_{pl}$, while the function $L_{2/3}$ is defined by the following equation (compare with (7.92))

$$L_{2/3}(\lambda) = \int_0^\infty dt\cdot t\cdot\cos\left[\frac{3}{2}\lambda\left(t+\frac{t^3}{3}\right)\right]. \tag{15.18}$$

At $\kappa \gg 1$ the contribution comes from $\lambda \ll 1$. In this case $L_{2/3}(\lambda) \simeq K_{2/3}(\lambda)/3$ and the real part of the correction in $\omega_\parallel, \omega_\perp$ differs from the imaginary part by a factor $-(3)^{-1/2}$.

Propagation of an arbitrary polarized wave can be found through its representation as the sum of its eigensolutions. If at the entrance to the crystal $\mathbf{E}_0 = \mathbf{e}_x E_{10} + \mathbf{e}_y E_{20}$, then at the depth l we have $\mathbf{E}(l) = \mathbf{e}_x E_{10}\exp(-i\omega_\perp l) + \mathbf{e}_y E_{20}\exp(-i\omega_\parallel l)$. Forming a quadratic combination $E_i(l)E_j^*(l)$ and averaging over parameters on which the quantities E_{10} and E_{20} may depend in the case of partial polarization of the incident beam, we obtain for the density matrix of the photon beam at depth l

$$
\begin{aligned}
\rho_{ij}(e) &= \frac{1}{2}N(0)P_{ij}(l)\,, \\
P_{11} &= [1+\eta_3(0)]e^{-W_\perp l}\,, \quad P_{22} = [1-\eta_3(0)]\,e^{-W_\parallel l}\,, \\
P_{12} &= P_{21}^{*} = [\eta_1(0) - i\eta_2(0)]\exp\{-\overline{W}l - il\,\mathrm{Re}\,(\omega_\perp - \omega_\parallel)\}\,,
\end{aligned}
\tag{15.19}
$$

where $\eta_i(0)$ are the values of the Stokes parameters of an incident beam and $N(0)$ is the number of photons incident on a crystal. Comparing Eq. (15.19) with the standard expression for the density matrix, we can find the number of photons and the magnitude of the Stokes parameters at a depth l in a crystal

$$
\begin{aligned}
N(l) &= N(0)\,[\cosh(al) + \eta_3(0)\sinh(al)]\,e^{-\overline{W}l}\,, \\
\eta_1(l) &= \frac{\eta_1(0)\cos(bl) - \eta_2(0)\sin(bl)}{\cosh(al) + \eta_3(0)\,\sinh(al)}\,, \\
\eta_2(l) &= \frac{\eta_1(0)\sin(bl) + \eta_2(0)\cos(bl)}{\cosh(al) + \eta_3(0)\,\sinh(al)}\,, \\
\eta_3(l) &= \frac{\eta_3(0) + \tanh(al)}{1 + \eta_3(0)\,\tanh(al)}\,, \quad a = \mathrm{Im}\,(\omega_\perp - \omega_\parallel) \\
&= \frac{1}{2}(W_\parallel - W_\perp), \quad b = \mathrm{Re}(\omega_\perp - \omega_\parallel), \quad \overline{W} = \frac{1}{2}(W_\parallel + w_\perp)\,.
\end{aligned}
\tag{15.20}
$$

It is seen from the derived solution that as l increases, the number of photons $N(l)$ and the parameters $\eta_{1,2}(l)$ damp exponentially, while $\eta_3(l) \to 1$, i.e. at a sufficiently large depth, regardless of the initial polarization, the beam proves to be linearly polarized perpendicularly to the plane.

Let us treat the possibility of using a crystal to obtain and to measure the polarization of a very energetic photon beam. As long as the parameter $\kappa_m \ll 1$, the probability of pair production by a photon in the plane field is exponentially small and we can put $a = 0$ in Eq. (15.20). Then we have

$$
\begin{aligned}
N(l) &\simeq N(0)e^{-W_{BH}l}\,, \\
\eta_1(l) &= \eta_1(0)\cos(bl) - \eta_2(0)\sin(bl)\,, \\
\eta_2(l) &= \eta_2(0)\cos(bl) + \eta_1(0)\sin(bl)\,, \\
\eta_3(l) &= \eta_3(0)\,,
\end{aligned}
\tag{15.21}
$$

i.e. a polarization does not arise, but the linear polarization may be transformed into a circular one. This problem was analyzed in Cabibo *et al.*, 1962b in terms of the coherent pair production theory.

We want to explain the appearance of the factor $\exp(W_{BH}l)$ in the expression for the number of photons in Eq. (15.20). Generally speaking, the probability of pair production by a photon in a crystal is a sum of the coherent term (11.14) and an incoherent part, $W_{\rm ncoh}$, which may be different from the value in an amorphous medium, W_{BH} (see, e.g. Ter-Mikaelyan, 1972 for the coherent approach). In the expressions for a and b in Eq. (15.20) the quantity $W_{\rm ncoh}$ cancels, but it remains in the half-sum $(W_\perp + W_\parallel)/2$.

Let the initial beam be completely linearly polarized at an angle $\pi/4$ to the x-axis, i.e. in Eq. (15.21) $\eta_1(0) = 1$ and $\eta_2(0) = \eta_3(0) = 0$. Then $\eta_2(l) = \sin(bl)$ and at a depth $l_0 = \pi/2b$, polarization becomes circular. Let us estimate the length l_0 in the considered case $\kappa_m \ll 1$. Taking into account that $\lambda \gg 1$ in Eq. (15.17) and that the function $L_{2/3}(\lambda) \simeq -(2/3\lambda)^2$, we derive for b the expression

$$b(\kappa_m \ll 1) = \mathrm{Re}\,(\omega_\perp - \omega_\parallel)\big|_{\kappa_m \ll 1} = \frac{\alpha m^2}{30\pi\omega}\langle\kappa^2(x)\rangle$$
$$= \frac{\alpha\omega}{30\pi}\left\langle\left(\frac{E}{E_0}\right)^2\right\rangle, \qquad (15.22)$$

which is well known (with an accuracy up to averaging) from the theory of photon propagation in a constant field (see e.g. Baier, Katkov and Fadin, 1973, p. 192). The average values of the ratio of a crystal plane field to the critical field E_0 (see Eq. (15.2)), calculated for the potential (9.14) are listed in Table 15.1. One should bear in mind (for example, to estimate the quantity κ_m) that the maximum values of $(E/E_0)_{\rm max}$ for the crystals in this table are roughly three times larger than $\langle(E/E_0)^2\rangle^{1/2}$. We also present the values of the product l_0W_{BH},derived using Eq. (15.22) where W_{BH} was found according to formula (19.17) in Baier, Katkov and Fadin, 1973. The quantity $l_0W_{BH} =$ constant $10^5\ m/\omega$ where the constant varies from 3 to 13. The upper boundary of applicability for expression (15.22) is determined by the condition $\omega \ll \omega_t$; note that at $\omega \leq \omega_t$ it may be used as a rough estimation. The lower boundary of frequencies for which the linear-to-circular-polarization transformation effect is really usable is determined by permissible losses of photons. As ω decreases the required thickness of the crystal and the quantity l_0W_{BH} grows, since

$l_0 \propto \omega^{-1}$. For example, in silicon, at $\omega = 500$ GeV, $l_0 = 12$ cm and $\simeq 40\%$ of photons remains, which is quite acceptable. On the contrary, in tungsten there is practically no region where the effect applies because of a relatively small value of ω_t.

Let us now consider the effect of polarization transformation at very high photon energies when $\kappa \gg 1$. For the same initial conditions as above, we have from Eq. (15.20)

$$\eta_1(l) = \frac{\cos(bl)}{\cosh(al)}, \quad \eta_2(l) = \frac{\sin(bl)}{\cosh(al)}, \quad \eta_3(l) = \tanh(al). \tag{15.23}$$

Taking into account that at $\kappa \gg 1, b = -3^{-1/2}a$, we find the maximum values $\eta_2(l_0) = -0.35$ and $l_0 = 0.63/|b|$ (rather than $\pi/2b$, as at $\kappa \ll 1$). If we neglect the factor $\exp(-W_{BH}l)$ in the expression for $N(l)$ and recollect that at $k \gg 1, W_{\parallel} = 1.5W_{\perp}$, then for $N(l_0)$ we obtain $N(l_0)/N(0) \simeq 7 \times 10^{-3}$ So, at $\kappa \gg 1$ there is no complete transformation of the linear polarization into a circular one and it achieves a maximum value at large losses of photons. A similar situation occurs also at lower κ as long as $a \sim |b|$. Moreover, since the quantity b alters its sign (cf. Eq. (15.22)), there is a range of frequencies for which it is close to zero.

As far as the polarization capability of the crystal is concerned, for $\vartheta_0 = 0$ it becomes significant only at $\omega > \omega_t$ when the mechanism of pair production in a plane field is "switched on" For an initially unpolarized beam we have from Eq. (15.20)

$$N(l) = \frac{1}{2}N(0)(e^{-W_{\perp}l} + e^{-W_{\parallel}l}), \qquad \eta_1 = \eta_2 = 0, \quad \eta_3 = \tanh(al). \tag{15.24}$$

Using the expressions found for the probabilities $W_{\perp,\parallel}$ in the plane field (15.3) and taking the Bethe–Heitler contribution into account, we can analyze the behavior of the quantities l and $N(l)/N(0)$ for a given polarization $\eta_3(l)$ at the exit from the crystal for substances presented in Table 15.1. With increasing ω, the quantity l first decreases, achieves a minimum in the region where the probabilities $W_{\perp,\parallel}$ are maximum, and then increases very slowly. For $\eta_3(l) = 0.5$ one can find the minimum values of the quantity l in the same table. The ratio $N(l)/N(0)$ increases fast, reaches a maximum in the 1–2 TeV region and then decreases gradually. For $\eta_3(l) = 0.5$, the ratio $[N(l)/N(0)]_{\max} \simeq 0.1$–$0.15$, while for $\eta_3(l) = 0.3$ it is 0.27–0.33. At $\omega \sim \omega_t$ the beam polarization is accompanied by large losses. For example, for $\omega = 200$ GeV in tungsten the value of $\eta_3(l) = 0.5$ is achieved on the length $l = 1.8$ cm when $N(l)/N(0) \simeq 4 \times 10^{-3}$ In the asymptotic region when $\kappa \gg 1$, neglecting the incoherent contribution, we find a simple relation between the number of photons and the degree of beam polarization:

$$\left(\frac{N(l)}{N(0)}\right)_{as} = \frac{[1-\eta_3(l)]^2}{[1+\eta_3(l)]^3} \tag{15.25}$$

From the point of view of the orientation dependence, polarization of the photon beam by a crystal occurs most efficiently in the angular range where the probabilities $W_{\perp,\parallel}(\vartheta_0)$ are maximum. For $\omega \gg \omega_t$ this situation takes place at $\vartheta_0 = 0$, i.e. at planar orientation.

Concerning the possibility of using the crystal in an analysis of the polarization of a photon beam at high energy with respect to the numbers of pairs produced in a simple of definite thickness, from the expression for $N(l)$ in Eq. (15.20) it can be seen that (1) no conclusions may be drawn on the circular polarization; (2) performing measurements at different positions of the crystal, which are distinguished by a rotation around the direction of incidence of the beam, one can find merely the ratio between the Stokes parameters $\eta_1(0)$ and $\eta_3(0)$, whereas if the theoretical value of the quantity la is used, one can find their absolute magnitudes.

Chapter 4

EMISSION OF RADIATION BY HIGH-ENERGY PARTICLES IN ORIENTED SINGLE CRYSTALS

16. GENERAL THEORY OF RADIATION FROM HIGH-ENERGY PARTICLES

16.1. Qualitative Consideration of Radiation for Various Angles and Energies of Incident Particles

The emission of a photon by a charged particle accompanied by creation of an electron-positron pair by a photon in the case when the initial particle moves at a small angle ϑ_0 relative to the direction of an axis or a plane in a single crystal is quite different from the corresponding process in an amorphous medium. In Chap. 3 we developed an approach for the description of the creation of pairs by high-energy photons in single crystals which is valid at any energies and angles of incidence of the photon. This approach is based on a general semiclassical formalism for the description of phenomena in external fields (Secs. 2, 3). A similar approach will be applied below to the problem of radiation emission. Several new aspects have to be faced and, in particular, it is necessary to allow for a redistribution of a flux of charged particles in a crystal.

We shall consider qualitative features of the emission of radiation in an oriented single crystal. It is well known (see Sec. 1.8), that the features of this process depend strongly on the ratio of the characteristic emission angle $\vartheta_\gamma \equiv$

here $\varphi(u) = 1 + u + (1+u)^{-1}$, $C = 0.577\ldots$—is the Euler constant. The functions $B_1(\eta)$ and $B_2(\eta)$ occur in the asymptotic representation of the probability of creation of pairs by a photon in the case when $\kappa_s \gg 1$ (see Eq. (12.17)). For $\eta \ll 1$ we have obtained the following asymptotic expression of the functions $l_{1,2}(\eta)$

$$l_1(\eta) = 3.975\beta^{2/3}\left(1 + \frac{8}{15}\beta + \frac{7}{18}\beta^2\right) - \beta\left(\frac{3}{2} + \frac{9}{8}\beta + \frac{13}{14}\beta^2\right),$$
$$l_2(\eta) = 2.31\beta^{1/3}\left(1 + \frac{2}{3}\beta - \frac{19}{63}\beta^2\right) + \frac{1}{160}\beta(30 + 144\beta - 29\beta^2), \tag{17.13}$$

where $\beta = \eta/(1+\eta)$.

In the derivation of Eq. (17.11) the integration with respect to x is carried out between infinite limits. Since the main (logarithmic) contribution to Eq. (17.11) comes from the range $x \sim (\chi_s/u)^{2/3}$ (range 3), it is clear that the asymptotic expression (17.11) is valid only if $(\chi_s/u)^{2/3} < x_0$.

We must bear in mind that Eq. (17.11) is not valid at low frequencies in the limit $u \to 0$. The physical reason for the inadmissibility of the magnetic bremsstrahlung approach for the distribution of the low-frequency radiation is that at low frequencies the radiation is formed throughout the particle transit trajectory along the axis. We shall now analyze this topic. The magnetic bremsstrahlung approach is valid if the transverse momentum corresponding to an impact parameter $x^{1/2}a_s$ exceeds considerably the mass:

$$\frac{\Delta p_\perp}{m} \simeq \frac{2V_0\sqrt{x}}{ma_s}\int_{-\infty}^{\infty} dt\left(x + \frac{t^2\vartheta_0^2}{a_s^2}\right)^{-2} = \frac{\pi V_0}{mx\vartheta_0} \gg 1\,.$$

Substituting here $x = (\chi_s/u)^{2/3}$, we can find the lower limit to the frequency range in which the magnetic bremsstrahlung description is valid (see Eqs. (17.6), (17.7) and (17.11))

$$\left(\frac{\chi_s}{u}\right)^{4/3}\frac{1}{\rho(\vartheta_0)} \ll 1, \text{ or } u \gg u_{\min} = \frac{\chi_s}{\rho^{3/4}(\vartheta_0)} \tag{17.14}$$

In connection with this criterion we must draw attention to the anomalously small (by a factor $(\chi_s/u)^{2/3}$) first correction in the expansion of (17.11). It should be pointed out that if $\chi_s \ll 1$, then the asymptotic expression (17.11) describes the spectrum outside the range of fundamental frequencies (Eq. (17.11)) applies when $u \ll \chi_s$, whereas the main contribution to the radiation intensity comes from $u \sim \chi_s$), whereas for $\chi_s \gg 1$ the condition $u \ll \chi_s$ is satisfied by frequencies with $u \sim 1$, which dominate the intensity.

the region of the continuous potential. It then follows from Eq. (16.4) that $u \sim (1+1/\rho)^{1/2}\chi_s$, where the parameter

$$\chi_s = V_0\varepsilon/(m^3 a_s) \tag{16.5}$$

represents quantum effects in the course of emission of radiation. In fact, our estimate (valid up to $\chi_s \sim 1$) shows that the ratio $\omega/\varepsilon \ll 1$ for characteristic emission frequencies, which allows us to ignore the recoil and use the classical theory of the emission of radiation, may be satisfied only if $\chi_s \ll 1$. However, in this case the validity of the classical description of the emission of radiation breaks down for $\rho^{1/2} \leq \chi_s$, i.e. for $\vartheta_0 \geq V_0/(m\chi_s) = ma_s/\gamma \equiv \vartheta_m$. It should be pointed out that a typical value for crystals is $ma_s \sim 10^2 \gg 1$.

At sufficiently high angles of incidence ϑ_0 we can apply the theory of coherent bremsstrahlung—CBS (see Ref. Uberall, 1956; Ferretti, 1950, Chaps. 1 and 2 in Ref. Ter-Mikaelyan, 1969 and Chap. 8 in Ref. Kalashnikov, 1981; the criterion of validity of the theory of CBS is discussed in Ref. Akhiezer, Boldyshev, Shul'ga, 1979), which represents essentially the Born approximation to the potential of a crystal and its validity requires (in our terms) that the condition of the emission of dipole radiation is satisfied simultaneously with the condition of validity of the rectilinear trajectory approximation.

We shall now consider the dependence of the emission pattern on the particle energy. At low energies (up to several megaelectron-volts in the axial case and up to several tens of megaelectron-volts in the planar case) the radiation is dominated by the transition between energy levels in the relevant potential well. We shall consider the range of energies when the number of these levels is high and the particle motion can be described classically.

As long as condition $\rho_c \ll 1$ is satisfied (we then have $\vartheta_v \ll \vartheta_c \ll \vartheta_\gamma \ll \vartheta_m$) the radiation is of a dipole nature for all angles of incidence of a particle on a crystal. The parameter χ is small; $\chi_s = \rho_c/(2ma_s) \ll \rho_c \ll 1$ and $\chi_s/\sqrt{\rho_c} \ll 1$, i.e. the classical description of radiation is valid and, although it does break down for angles $\vartheta_0 \sim \vartheta_m$, at these angles we can already apply the CBS theory which is then valid if $\vartheta_0 \gg \vartheta_c$. At higher energies the condition $\rho_c \sim 1$ begins to apply, i.e. we now have $\vartheta_v \sim \vartheta_c \sim \vartheta_\gamma \ll \vartheta_m$, but all that we have said so far about the nature of emitted radiation remains valid except that at low angles of incidence $\vartheta_0 \leq \vartheta_c$ the radiation is no longer of the dipole nature. Finally, when $\rho_c \gg 1$, which corresponds to $\vartheta_\gamma \ll \vartheta_c \ll \vartheta_v$, at angles of incidence $\vartheta_0 \ll \vartheta_v$, the magnetic bremsstrahlung description applies ($\rho \gg 1$) and the CBS theory can be used in the opposite case when $\vartheta_0 \gg \vartheta_v$ [if $\vartheta_0 \sim \vartheta_v$, we find from Eq. (16.2) that $\rho \sim 1$ and the dipole emission condition is no longer obeyed]. At energies of this kind the parameter χ_s can no longer be regarded as small (for example, already for $\chi_s = 0.1$ the classically calculated total intensity in a constant field is approximately 1.5 times greater than the correct result) and it becomes essential to allow exactly for the quantum recoil in the course of emission of radiation.

We shall obtain one further estimate of the length of formation l_f for angles of incidence $\vartheta_0 \leq \vartheta_c$. Since $l_f \propto \varepsilon/(m^2 u)$, we find from Eq. (16.4) that if $\rho_c \ll 1$, then $l_f \propto l_0 \rho_c^{1/2} \propto 1/\omega_0$, where $l_0 = a_s/\vartheta_v$ and the length l_f rises proportionally to $\varepsilon^{1/2}$; when the condition $\rho_c \sim 1$ begins to be satisfied, this rise ceases and as long as $u \sim \chi_s$, i.e. up to $\rho_c \sim 10^2$, the length of formation of the radiation remains practically constant: $l_f \sim l_0$; on further increase in the energy we have $u \sim 1$ and $l_f \propto \varepsilon/m^2$; at energies $\varepsilon \sim 10$ TeV or higher, this law changes to a slower one: $l_f \propto l_0 \chi^{1/3}$ The value of l_0 is greater for crystals of lighter elements and, for example, along the $\langle 111 \rangle$ axis it varies from 5.7×10^{-5}cm for diamond and 2.6×10^{-6}cm for tungsten. When the energy is $\varepsilon \sim 10$ TeV, the length of formation of radiation becomes of the same order of magnitude for different substances: $l_f \sim 10^{-3}$cm. We shall assume that the thickness of a crystal L satisfies the condition $L \gg l_f$.

In Secs. 16–18 we shall assume that our crystal is thin, i.e. we shall ignore the change in the distribution function in the transverse phase space during the passage of a particle through the crystal. This approximation is justified if (for a discussion see the beginning of Sec. 10.1) the thickness of the crystal is less than the characteristic energy loss length L_{ch}

$$L_{\text{ch}}^{-1} = I(\varepsilon)/\varepsilon \,, \tag{16.6}$$

where $I(\varepsilon)$ is the total radiation intensity. An analysis (see Secs. 10, 17) demonstrates that the range of thickness in which a crystal can be regarded as thin and the condition $L \gg l_f$ be still satisfied can be found at any energy. On the other hand, in the range of energies under discussion ($\rho_c \gg 1$), for a thick crystal satisfying the condition $\vartheta_0 \leq \vartheta_v$ there is always a specific electromagnetic shower. Then the description of the emmission of radiation in terms of one particle is no longer satisfactory and it is necessary to solve the appropriate equation of the cascade theory in which the kernels are the expressions obtained below for the description of the emission of radiation and the expressions describing the creation of a pair by photon obtained in Chap. 3. These problems will be detailed in Sec. 20.

In Secs. 16–18 a theory of the emission of radiation by electrons and positrons in thin crystals will be developed for the range $\rho_c \gg 1$ ($\varepsilon \gg m^2/V_0$) and we shall allow for the quantum effects. The results obtained will then be valid for any angle of incidence ϑ_0. In the range $\vartheta_0 \ll \vartheta_v$ these results reduce to the theory of emission in axial fields in the magnetic bremsstrahlung limit (see Sec. 17). For $\vartheta_0 \geq \vartheta_v$, the general formulas yield an expression for the probability representing a modification of the standard theory of CBS, whereas for $\vartheta_0 \gg \vartheta_v$, we obtained the standard theory of CBS rather than its modification (Sec. 18). The general expression obtained for the description of the orientation dependence of the total intensity of the emitted radiation is found to be valid at any energies at which we can apply the semiclassical theory of the emission of radiation (i.e. it is valid when $\rho_c \leq 1$).

16.2. Semiclassical Description of Photons Emission by Charged Particles in a Single Crystal

The most satisfactory approach to the problem of emission of radiation by relativistic particles is the formalism utilizing the semiclassical operator method developed in Sec. 2, because it is valid for all types of external fields, including inhomogeneous and alternating fields. For the majority of quantum numbers of motion (corresponding to the semiclassical condition) this method can be used to go over from the exact quantum expressions by a series of transformations to quantities on a classical trajectory of a particle when the recoil due to the emission of a photon can be allowed for exactly. The general expression for the energy of the emitted radiation has been obtained in Sec. 2 (see Eqs. (2.27), (2.37)). These equations describe all the characteristics of the emitted radiation, including the spin and polarization phenomena. The formulas for unpolarized electrons, but allowing for the photon polarization can be found in Sec. 4.3 (see Eqs. (4.58), (4.59)). The energy distribution of the photon, emitted by the particle on a given trajectory, is described by equations

$$dE = e^2 \frac{d^3k}{(2\pi)^2} \int dt_1 \int dt_2 \, L(t_1, t_2) \exp[ik'(x_1 - x_2)] \,, \tag{16.7}$$

where

$$k'_\mu = k_\mu \frac{\varepsilon}{\varepsilon'} \,, \quad \varepsilon' = \varepsilon - \omega \,, \quad k_\mu = (\omega, \mathbf{k})$$

$$L(t_1, t_2) = \frac{(\varepsilon + \varepsilon')^2}{4\varepsilon'^2} (\mathbf{e}^* \mathbf{v}_1)(\mathbf{e}\mathbf{v}_2) - \frac{\omega^2}{4\varepsilon'^2} \times [(\mathbf{e}^*\mathbf{v}_2)(\mathbf{e}\mathbf{v}_1) - (\mathbf{e}\mathbf{e}^*)(\mathbf{v}_1\mathbf{v}_2 - 1 + 1/\gamma^2)] \,. \tag{16.8}$$

Here $x_{1,2} \equiv x(t_{1,2}), \mathbf{v}_{1,2} \equiv \mathbf{v}(t_{1,2})$. Summing over the photon polarization, we find that

$$\bar{L} = \frac{\varepsilon^2 + \varepsilon'^2}{2\varepsilon'^2}(\mathbf{v}_1\mathbf{v}_2 - 1) + \frac{\omega^2}{2\gamma^2\varepsilon'^2} \tag{16.9}$$

Below, we shall generally assume that a crystal is thin and the condition $\rho_c \gg 1$ is satisfied. In this case the extremely difficult task of averaging Eqs. (16.7) and (16.9)—derived for a given trajectory—over all possible trajectories of a particle in a crystal simplifies radically. In fact, if $\rho_c \gg 1$ then in the range where the trajectories are essentially rectilinear ($\vartheta_0 \le \vartheta_c, v_\perp \sim \vartheta_c$) the mechanism of emission of radiation is of the magnetic bremsstrahlung nature and the characteristics of the radiation can be expressed in terms of the local parameters of motion. The averaging procedure can then be carried out simply if we know the distribution function in the transverse phase space $dN(\rho, \mathbf{v}_\perp)$, which for a thin crystal is defined directly by the initial conditions of incidence of particles on a crystal. For a given incidence angle ϑ_0 (which is the angle between the momentum of the incident particle and the crystal axis or the plane), we have $dN/N = d^3rF(\mathbf{r}, \vartheta_0)/V$, where V is the volume of a crystal and N is the total number of particles. In the axial case the function $F(\mathbf{r}, \vartheta_0)$ is of the form

$$F_{ax}(\rho, \vartheta_0) = \int \frac{d^2\rho_0}{S(\varepsilon_\perp(\rho_0))}\vartheta(\varepsilon_\perp(\rho_0) - U(\rho)), \tag{16.10}$$

where $U(\rho)$ is the continuous axial potential dependent on the transverse coordinate ρ normalised so that $U(\rho) = 0$ at the boundary of a cell;

$$S(\varepsilon_\perp(\rho_0)) = \int d^2\rho\vartheta(\varepsilon_\perp(\rho_0) - U(\rho_0)), \qquad \varepsilon_\perp(\rho_0) = \varepsilon\vartheta_0^2/2 + U(\rho_0));$$

$\vartheta(x)$ is the Heaviside function: $\vartheta(x) = 0$ when $x < 0$ and $\vartheta(\mathrm{x}) = 1$ when $x > 0$. In the planar case, we have (see Sec. 9.4)

$$F_{pl}(x, \vartheta_0) = \int \frac{dx_0\vartheta(\varepsilon_{\perp 0} - U(x))}{v(\varepsilon_{\perp 0}, x)}\left[\int \frac{dy\vartheta(\varepsilon_{\perp 0} - U(y))}{v(\varepsilon_{\perp 0}, y)}\right]^{-1}, \tag{16.11}$$

where $U(x)$ is the continuous potential of a plane dependent on the transverse coordinate $x, v(\varepsilon_{\perp 0}, x) = [2(\varepsilon_{\perp 0} - U(x))/\varepsilon]^{1/2}$ is the transverse velocity of the particle, and $\varepsilon_{\perp 0} = \varepsilon\vartheta_0^2/2 + U(x_0)$. It should be noted that in the case when $\varepsilon_\perp(\rho_0) > U_0$ (above-barrier particles) the distribution (16.10) becomes uniform, whereas the distribution (16.11) becomes uniform only if $\varepsilon_{\perp 0} \gg U_0$ (U_0 is the depth of the potential well).

Substituting Eq. (16.9) into Eq. (16.7) and summing over the trajectories of particles in a crystal with allowance for the redistribution of the flux [Eq. (16.10), (16.11)], we find that simple transformations yield an expression for the energy loss of the particle

$$E = -\frac{am^2}{4\pi^2}\int \frac{d^3k}{\varepsilon\varepsilon'}\int \frac{d^3r}{V}F(\mathbf{r},\vartheta_0)\int e^{-iA}\left[1+\frac{\varepsilon^2\varepsilon'^2}{4\varepsilon\varepsilon'}\gamma^2(\mathbf{v}_1-\mathbf{v}_2)^2\right]dt_1 dt_2\,, \tag{16.12}$$

$$A = \frac{\omega\varepsilon}{2\varepsilon'}\int_{t_1}^{t_2} dt\left[1/\gamma^2+(\mathbf{n}-\mathbf{v}(t))^2\right],\quad \mathbf{n}=\mathbf{k}/\omega, \alpha = e^2 = 1/137$$

The particle velocity can be represented in the form $\mathbf{v}(t) = \mathbf{v}_0 + \Delta\mathbf{v}(t,\mathbf{v}_0)$. Then the integrand in phase A in Eq. (16.12) can be written in the form

$$1/\gamma^2 + \mathbf{u}^2 - 2\mathbf{u}\Delta\mathbf{v} + (\Delta\mathbf{v})^2\,, \tag{16.13}$$

where

$$\mathbf{u} = \mathbf{n} - \mathbf{v}_0 \simeq \mathbf{n} - \mathbf{v}_0(\mathbf{n}\mathbf{v}_0) \simeq \mathbf{n}_\perp\,.$$

Thus it can be assumed that vector $\mathbf{u}$ is lying perpendicular to the vector $\mathbf{v}_0$ plane. The main contribution to the loss of energy (16.12) is given by the regions of trajectory where the vector $\mathbf{n}$ is directed close to vector $\mathbf{v}_0$: $|\mathbf{u} - \Delta\mathbf{v}(t,\mathbf{v}_0)| \sim 1/\gamma, |\Delta\mathbf{v}| \sim \sqrt{\rho}/\gamma$. The parameter ρ was defined in Eqs. (16.1)–(16.3). Integrating in Eq. (16.12) over $\mathbf{u}(d^3k = \omega^2 d\omega d\mathbf{u})$, we find (compare with (5.31))

$$\begin{aligned} E &= \frac{i\alpha m^2}{2\pi\varepsilon^2}\int \omega d\omega\int\frac{d^3r}{V}F(\mathbf{r},\vartheta_0)\int\frac{dt d\tau}{\tau - i0}\left[1+\frac{\varepsilon^2+\varepsilon'^2}{4\varepsilon\varepsilon'}\gamma^2\right.\\ &\quad\times\left.\left(\Delta\mathbf{v}\left(t+\frac{\tau}{2}\right)-\Delta\mathbf{v}\left(t-\frac{\tau}{2}\right)\right)^2\right]\exp(-iA_1)\,,\\ A_1 &= \frac{\omega\varepsilon\tau}{2\varepsilon'}\left[\frac{1}{\gamma^2}+\frac{1}{\tau}\int_{-\tau/2}^{\tau/2}ds(\Delta\mathbf{v}(\tau+s))^2\right.\\ &\quad\left.-\left(\frac{1}{\tau}\int_{-\tau/2}^{\tau/2}ds\Delta\mathbf{v}(t+s)\right)^2\right], \end{aligned} \tag{16.14}$$

where we turned to variables $t,\tau : t_1 = t - \tau/2, t_2 = t + \tau/2$.

If $\vartheta_0 \gg \vartheta_c$, we find $\Delta\mathbf{v}(t)$ using the rectilinear trajectory approximation:

$$\Delta\mathbf{v}(t) = -\frac{1}{\varepsilon}\sum \frac{G(\mathbf{q})}{q_{\parallel}}\mathbf{q}_{\perp}\exp[-i(q_{\parallel}t + \mathbf{qr})], \tag{16.15}$$

where $q_{\parallel} = \mathbf{qv}_0, \mathbf{q}_{\perp} = \mathbf{q} - \mathbf{v}(\mathbf{qv}_0)$. If $\rho_c \gg 1$, there is a range of angles ϑ_0 which satisfies $\vartheta_c \ll \vartheta_0 \ll \vartheta_v$, i.e. when both the magnetic bremsstrahlung description and the rectilinear trajectory approximation can be applied. However, since the magnetic bremsstrahlung approach provides the same description for all angles $\vartheta_0 \ll \vartheta_v$, the formulas obtained in the range of ϑ_0 remain valid (see below) right up to $\vartheta_0 = 0$. In the range $\vartheta_0 \geq \vartheta_v$, the rectilinear trajectory approximation is known to be valid for $\rho_c \gg 1$. Substituting Eq. (16.15) into Eq. (16.14), we find after simple transformations a general expression for the intensity of radiation valid for any angle of incidence ϑ_0:

$$I \equiv \frac{dE}{dt} = \frac{i\alpha m^2}{2\pi\varepsilon^2}\int \omega d\omega \int \frac{d^3r}{V}F(\mathbf{r},\vartheta_0)\int \frac{d\tau}{\tau - i0}\left[1 - \varphi(\varepsilon)\right.$$
$$\left.\times\left(\sum_{\mathbf{q}} \frac{G(\mathbf{q})\mathbf{q}_{\perp}}{mq_{\parallel}}\sin(q_{\parallel}\tau)e^{-i\mathbf{qr}}\right)^2\right]\exp(-iA_2), \quad \varphi(\varepsilon) = \varepsilon/\varepsilon' + \varepsilon'/\varepsilon,$$
$$A_2 = \frac{m^2\omega\tau}{\varepsilon\varepsilon'}\left[1 + \sum_{\mathbf{q},\mathbf{q}'} \frac{G(\mathbf{q})G(\mathbf{q}')}{m^2 q_{\parallel}q'_{\parallel}}(\mathbf{q}_{\perp}\mathbf{q}'_{\perp})\left(\frac{\sin(q_{\parallel} + q'_{\parallel})\tau}{(q_{\parallel} + q'_{\parallel})\tau}\right.\right.$$
$$\left.\left.- \frac{\sin q_{\parallel}\tau}{q_{\parallel}\tau}\frac{\sin q'_{\parallel}\tau}{q'_{\parallel}\tau}\right)\exp\{-i(\mathbf{q}+\mathbf{q}')\mathbf{r}\}\right] \tag{16.16}$$

Similar expressions are obtained also for the quantities describing the polarization of the emitted radiation (see Eq. (16.8)). In this case, integrating over the angles of emission of a photon, we find that the spectral distribution of the polarization density matrix is

$$dE_{ij} = \frac{i\alpha}{4\pi}\omega d\omega \int \frac{dtd\tau}{\tau - i0}\exp(-iA_1)T_{ij}, \tag{16.17}$$

where

$$T_{ij} = \delta_{ij}\left[\frac{1}{\gamma^2} + \frac{\varphi(\varepsilon)}{4}(\mathbf{a}_2 - \mathbf{a}_1)^2\right] + \delta_{ij}(\mathbf{a}_1\mathbf{a}_2) - a_{1i}a_{2j} - a_{2i}a_{1j}$$
$$- \frac{\varphi(\varepsilon)}{2}(a_{1i}a_{2j} - a_{2i}a_{1j}). \tag{16.18}$$

Here A_1 is given by Eq. (16.14), $\mathbf{a}_{1,2} = \Delta\mathbf{v}(t_{1,2}) - \mathbf{Q}, \mathbf{Q} = \frac{1}{\tau}\int_{t_1}^{t_2} ds\Delta\mathbf{v}(s)$. As before, $\mathbf{v}(t) = \mathbf{v}_0 + \Delta\mathbf{v}(t)$, where $\mathbf{v}_0$ is the average velocity.

The Stokes parameters $\xi^{(i)}$ can be directly obtained from (16.17) using formulae (A.41). We emphasize that Eqs. (16.17) and (16.18) describe the spectral distribution of the polarization density matrix for any given particle trajectory and have the same generality as formula (2.46). The latter can be obtained from (16.17) by taking trace. It should be noted that the expressions for A_1 and T_{ij} are note affected by the substitution $\Delta\mathbf{v} \to \Delta\mathbf{v} + \mathbf{b}$, where $\mathbf{b}$ is any vector independent of time.

Substituting Eq. (16.15) into Eq. (16.18) and reinserting the summation over the trajectories of particles in a crystal with allowance for the redistribution of the flux [Eqs. (16.10) and (16.11)], we find a general expression for the quantities describing the polarization of the emitted radiation. For example, if $\xi^{(2)} = I^{(2)}/I$, we find that

$$I^{(2)} = \frac{i\alpha}{4\pi\varepsilon^2} \int d\omega\omega\varphi(\varepsilon) \int \frac{d^3r}{V} F(\mathbf{r}, \vartheta_0) \int_{-\infty}^{\infty} \frac{d\tau}{\tau - i0} \exp(-iA_2)$$
$$\times \sum_{\mathbf{q},\mathbf{q}'} \frac{G(\mathbf{q})G(\mathbf{q}')}{q_{\parallel}q_{\parallel}} ((\mathbf{q} \times \mathbf{q}')\mathbf{v}_0) \left[\sin q_{\parallel}\tau \left(\cos q'_{\parallel}\tau - \frac{\sin q'_{\parallel}\tau}{q'_{\parallel}\tau}\right)\right.$$
$$\left. \times \sin q'_{\parallel}\tau \left(\cos q_{\parallel}\tau - \frac{\sin q_{\parallel}\tau}{q_{\parallel}\tau}\right)\right] \exp[-i(\mathbf{q} + \mathbf{q}')\mathbf{r}] \,. \qquad (16.19)$$

In Eq. (16.16) and (16.19) we have made the substitution $\tau \to 2\tau$. We shall assume that the thickness of a crystal is considerably greater than the length of formation of radiation l_f, so that we can consider a description in terms of the intensity.

It is known that there is a relationship between the differential probabilities of the emission of a photon by a charged particles and the creation of a pair by a photon in quantum electrodynamics. The probability integrated over the angles of emission of the final particles do not generally exhibit such a relationship. However, we can show that in view of the smallness of all characteristic angles of the problem, such a relationship applies in our case also to the spectrum of the final particles. Therefore, Eq. (16.16) can be obtained from Eq. (11.13) by means of the substitutions $\varepsilon \to -\varepsilon, \omega \to -\omega, \varepsilon' \to \varepsilon', \varepsilon^2 d\varepsilon \to -\omega^2 d\omega$ allowing for a redistribution of a flux of charged particles and using the self-evident relationship between the intensity and the probability $dI = \omega dW$

Equation (16.16) describes the spectral properties of the radiation (these properties are obtained by dropping the integral with respect to ω) for the range $\vartheta_0 \gg \vartheta_c$ and any value of ρ_c, whereas if $\vartheta_0 \leq \vartheta_c$, such a distribution is obtained if the condition $\rho_c \gg 1$ is obeyed. However in the case of the total (integral) intensity of radiation, the range of validity of Eq. (16.16) is wider and it is the same as that of the semiclassical approximation. This situation arises because at energies when $\rho_c \leq 1$ and the magnetic bremsstrahlung description ceases to be valid, the classical theory of the emission of radiation can be used for angles $\vartheta_0 \leq \vartheta_c$ and the total intensity then depends only on the local characteristics of the motion of a particle.

We shall show below that if $\vartheta_0 \ll \vartheta_v$, then Eq. (16.16) describes the emission of radiation in the magnetic bremsstrahlung limit, whereas for $\vartheta_0 \gg \vartheta_v$ we obtain the familiar expression for coherent bremsstrahlung. In the intermediate range of angles one should use Eq. (16.16) directly.

17. EMISSION OF RADIATION FOR SMALL ENTRY ANGLES (MAGNETIC BREMSSTRAHLUNG LIMIT)

17.1. General Equations for the Intensity of Radiation

The behavior of the intensity I (16.16) for various entry angles and energies of the particles is determined by the dependence on them of the phase A_2 given there. We shall estimate it for $\vartheta_0 \ll V_0/m \equiv \vartheta_v$. For definiteness we shall consider the case in which the momentum of the incident particles lies near a crystallographic axis whose direction we shall take as the z-axis of the coordinate system. The order of magnitude of the double sum in A_2 is* $(G/m)^2(q_\perp/q_\parallel)^2\Psi(q_\parallel, q'_\parallel, \tau)$. We shall introduce the notation $\mathbf{q}_t$ for the vector $\mathbf{q}$ lying in the plane(x, y). For them $q_z = 0$ and there the quantities in Eq. (16.16) can be estimated in the following ways:

$$G(\mathbf{q}) \sim V_0, \quad q_\perp \sim 1/a_s\,, \quad q_\parallel \sim \vartheta_0/a_s\,. \tag{17.1}$$

For all remaining vectors $q_\parallel \sim q_\perp \sim 1/a_s$. Then the contribution to the sum of terms with $q_z \neq 0$ will be $\sim (V_0/m)^2\Psi \leq (V_0/m)^2$. Since $(V_0/m)^2 \ll 1$, this contribution can be neglected. Thus, we keep in the sum only terms with $\mathbf{q}_t$, for which its value is $\sim (V_0/m\vartheta_0)^2\Psi$. The large value of the phase A_2 leads to an exponential suppression of the intensity I, therefore the characteristic value of the variable τ in the integral (16.16) which have the meaning of the length (time) of formation of the process, are adjusted in such a way that

---

*Here $\Psi(q_\parallel, q'_\parallel, \tau) = \frac{\sin[(q_\parallel + q'_\parallel)\tau]}{(q_\parallel + q'_\parallel)\tau} - \frac{\sin(q_\parallel\tau)}{q_\parallel\tau}\frac{\sin(q'_\parallel\tau)}{q'_\parallel\tau}$.

the large factor $(V_0/m\vartheta_0)^2$ is compensated for by the quantity $\Psi(q_{\parallel}\tau)$, i.e. the contribution at small entry angles is from $q_{\parallel}\tau \ll 1$. Expanding the phase A_2 accordingly, we find an approximate expression for it for $\vartheta_0 \ll \vartheta_v$:

$$A_2 \simeq \frac{m^2\omega\tau}{\varepsilon\varepsilon'}\left\{1 - \frac{\tau^2}{3}\sum_{\mathbf{q}_t,\mathbf{q}'_t} G(\mathbf{q}_t)G(\mathbf{q}'_t)\frac{(\mathbf{q}_t\mathbf{q}'_t)}{m^2}\exp[-i(\mathbf{q}_t+\mathbf{q}'_t)\boldsymbol{\rho}]\right.$$
$$\left.\times\left[1-\frac{\tau^2}{10}\left((\mathbf{q}_t\boldsymbol{\nu})^2+(\mathbf{q}'_t\boldsymbol{\nu})^2+\frac{2}{3}(\mathbf{q}_t\boldsymbol{\nu})(\mathbf{q}'_t\boldsymbol{\nu})\right)\right]\right\}, \qquad (17.2)$$

here $\boldsymbol{\nu} = \mathbf{v}_0/v_0$ is the direction of entry of the initial particle, $\boldsymbol{\rho} = \mathbf{r}_t$. We can rewrite (17.2) in terms of the average potential of the atomic strings of the crystal $U(\boldsymbol{\rho})$ (see Eq. (12.2)):

$$A_2 = \frac{m^2\omega\tau}{\varepsilon\varepsilon'}\left\{1+\frac{1}{3}\mathbf{b}^2\tau^2+\frac{\tau^4}{15}[(\mathbf{b}(\boldsymbol{\nu}\boldsymbol{\nabla})^2\mathbf{b})+((\boldsymbol{\nu}\boldsymbol{\nabla})\mathbf{b})^2/3]\right\}, \qquad (17.3)$$

where the quantity $\mathbf{b}$ is in proportion to the acceleration of the particle

$$\mathbf{b} = -\gamma\mathbf{w}(\boldsymbol{\rho}) = \boldsymbol{\nabla}U(\boldsymbol{\rho})/m, \quad \boldsymbol{\nabla} = \partial/\partial\boldsymbol{\rho}. \qquad (17.4)$$

We find correspondingly for the pre-exponential factor in (16.16)

$$\{\cdots\} \simeq 1+\varphi(\varepsilon)\tau^2\left[\mathbf{b}^2+\frac{\tau^2}{3}(\mathbf{b}(\boldsymbol{\nu}\boldsymbol{\nabla})\mathbf{b})\right]. \qquad (17.5)$$

The terms containing τ^5 in A_2 are small, of the order $(m\vartheta_0/V_0)^2$, and they can be expanded into the pre-exponential factor. Then taking integral over τ we have obtained the following expression for spectrum distribution of the intensity of radiation $dI^F(\omega)$ for $\vartheta_0 \ll V/m$.

$$dI^F = \frac{\alpha m^2\omega d\omega}{\sqrt{3}\pi\varepsilon^2}\int\frac{d^2\rho}{s}\left\{\int\frac{d^2\rho_0}{s(\varepsilon_\perp(\boldsymbol{\rho}_0))}\vartheta(\varepsilon_\perp(\boldsymbol{\rho}_0)-U(\boldsymbol{\rho}))\right.$$
$$\times\left[\varphi(\varepsilon)K_{2/3}(\lambda)-\int_\lambda^\infty dyK_{1/3}(y)\right]$$
$$-\frac{1}{3\mathbf{b}^4}(\mathbf{b}(\boldsymbol{\nu}\boldsymbol{\nabla})^2\mathbf{b})\varphi(\varepsilon)\left[K_{2/3}(\lambda)-\frac{2}{3\lambda}K_{1/3}(\lambda)\right]$$
$$+\frac{\lambda}{30\mathbf{b}^4}[((\boldsymbol{\nu}\boldsymbol{\nabla})\mathbf{b})^2+3(\mathbf{b}(\boldsymbol{\nu}\boldsymbol{\nabla})^2\mathbf{b})]\left[K_{1/3}(\lambda)-\frac{4}{3\lambda}K_{2/3}(\lambda)\right.$$
$$\left.\left.+\varphi(\varepsilon)\left(\frac{4}{\lambda}K_{2/3}(\lambda)-K_{1/3}(\lambda)\left(1+\frac{16}{9\lambda^2}\right)\right)\right]\right\}, \qquad (17.6)$$

where $\lambda = 2m^2\omega/3\varepsilon\varepsilon'|\mathbf{b}|$. Since the expression for dI^F is independent of z, it follows that $\int d^3r/V \to \int d^2\rho/s$, where s is the cross-sectional area per axis. The first two terms in Eq. (17.6) independent of $\boldsymbol{\nu}$ represent the spectral distribution of the intensity in the magnetic bremsstrahlung limit (see Eq. (4.26)) if we allow for a redistribution of the flux of charged particles; the other terms represent a correction proportional to ϑ_0^2 (compare with Eq. (12.6)). Equation (17.6) allows for the fact that in the range $\vartheta_0 \sim \vartheta_c$ we have $(m\vartheta_0/V)^2 \sim 1/\rho_c$ and retention of terms of this kind in Eq. (17.6) would represent unnecessary precision, because these terms are dropped in the derivation of Eq. (16.16). Therefore, corrections proportional to $(m\vartheta_0/V_0)^2$ need be included only if $\varepsilon > U_0$, when we have $F(\mathbf{r}, \vartheta_0) = 1$.

The further approximations in Eq. (17.6) are possible where the potential of an axis can be regarded as axially symmetric. Sometimes, for example, for the $\langle 111\rangle$ axis in crystal structures of the type $fcc_{(d)}$ and bcc, the real potential is with high accuracy axially symmetric with respect to the distinguished axis over the entire region s and in any case has this property at distances from the axis $\leq a_s$ where the electric field strength has its greatest value. Believing $U = U(\rho^2)$ the integration with respect to the angles of the vector $\boldsymbol{\rho}$ would be carried out in Eq. (17.6), then

$$dI^F = \frac{\alpha m^2\omega d\omega}{\sqrt{3}\pi\varepsilon^2}\int_0^{x_0}\frac{dx}{x_0}\left\{\int_0^{x_0}\frac{dy}{y_0(\varepsilon_\perp(y))}\vartheta(\varepsilon_\perp(y) - U(x))\right.$$
$$\times\left[\varphi(\varepsilon)K_{2/3}(\lambda) - \int_\lambda^\infty dy K_{1/3}(y)\right]$$
$$-\frac{1}{6}\left(\frac{m\vartheta_0}{V_0}\right)^2\left[\varphi(\varepsilon)\frac{xg'' + 2g'}{xg^3}\left(K_{2/3}(\lambda) - \frac{2}{3\lambda}\cdot K_{1/3}(\lambda)\right)\right.$$
$$-\frac{1}{20g^4x^2}(2x^2g'^2 + g^2 + 14gg'x + 6x^2gg'')\left(\lambda K_{1/3} - \frac{4}{3}\ K_{2/3}(\lambda)\right.$$
$$\left.\left.\left.+\varphi(\varepsilon)(4K_{2/3}(\lambda) - (\lambda + 16/9\lambda)K_{1/3}(\lambda))\right)\right]\right\}, \qquad (17.7)$$

where we have adopted a new variable $x = \rho^2/a_s^2; x_0^{-1} = \pi a_s^2 nd$; d is the average distance between the atoms of a chain forming the axis; n is the density of atoms in a crystal;

$$y_0(\varepsilon_\perp(y)) = \int_0^{x_0} dx\vartheta(\varepsilon_\perp(y) - U(x)), \quad \varepsilon_\perp(y) = \varepsilon\vartheta_0^2/2 + U(y). \qquad (17.8)$$

The notation $U'(x) = V_0 g(x)$ is used in Eq. (17.8), so that

$$\lambda = \frac{m^3 \omega a_s}{3\varepsilon\varepsilon' V_0 g\sqrt{x}} = \frac{u}{3\chi_s g\sqrt{x}}, \quad \chi_s = \frac{V_0 \varepsilon}{a_s m^3}, \quad u = \frac{\omega}{\varepsilon'}. \tag{17.9}$$

The above expression (17.7) represents the spectral distribution of the radiation intensity (after integration with respect to ω, it gives the total radiation intensity) for an arbitrary form of an axially symmetric potential of the selected axis. In specific calculations we shall use a potential (9.13). For this potential

$$\begin{aligned} g(x) &= \frac{1}{x+\eta} - \frac{1}{x+\eta+1} = \frac{1}{(x+\eta)(x+\eta+1)}, \\ \chi(x) &= \frac{eE(x)\gamma}{m^2} = \frac{2V_0\varepsilon}{m^3 a_s}\sqrt{x}g(x) = 2\chi_s\sqrt{x}g(x), \end{aligned} \tag{17.10}$$

where $\chi(x)$ is the local value of the quantum parameter χ (at a distance of $x = \rho^2/a_s^2$ from the axis).

Equations (17.6) and (17.7) are derived from the intensity equation (16.16) within the range of validity of the latter ($\vartheta_0 \gg \vartheta_c$). However, the intensity

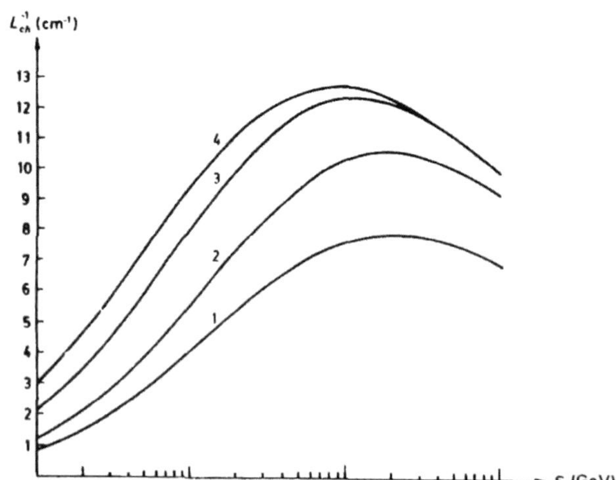

Fig. 17.1. Energy dependence of the reciprocal of the characteristic energy loss for Si ⟨110⟩ at $T = 293$ K(1), for diamond ⟨111⟩ at $T = 293$ K(2), for Ge ⟨110⟩ at $T = 280$ K(3), and for Ge ⟨110⟩ at $T = 100$ K(4).

changes only slightly on further reduction in the angle ϑ_0, as demonstrated by Eq. (17.7). Hence, it follows that Eqs. (17.6) and (16.7) and, therefore, Eq. (16.13) are all valid up to $\vartheta_0 = 0$.

In this section, in all specific calculations and estimates we shall assume, for the sake of simplicity, that the distribution is uniform along the transverse coordinates (which is true in the case of large angles of incidence $\vartheta_0 > \vartheta_c$ and approximately correct in the case of beams with a large angular spread $\Delta\vartheta_0 \sim \vartheta_c$). Numerical calculations were carried out for the potential of Eqs. (9.13) and (17.10). The influence of the redistribution of a flux on the orientation dependence will be discussed in Sec. 19.

The results are illustrated in Figs. 17.1 and 17.2. Figure 17.1 gives the reciprocal of the characteristic length of the loss of energy by a particle in the magnetic bremsstrahlung limit $L_{\text{ch}}^{-1} = -\varepsilon^{-1}d\varepsilon/dl = I^M(\varepsilon)/\varepsilon$ as a function of the electron energy for some crystals. We can see that if $\varepsilon \approx 1$ TeV, the value of L_{ch}^{-1} reaches its maximum, near which the energy dependence of L_{ch}^{-1} is fairly weak and the value of L_{ch} is a few tens of times less than the radiation wavelength in the corresponding amorphous substance (for example, in the case of Si $\langle 110\rangle$, we have $(L_{\text{rad}}/L_{\text{ch}}|_{\max} \approx 75)$. Figure 17.2 shows the spectral distribution of the intensity $dI^M/d\omega$ (in the magnetic bremsstrahlung limit) of

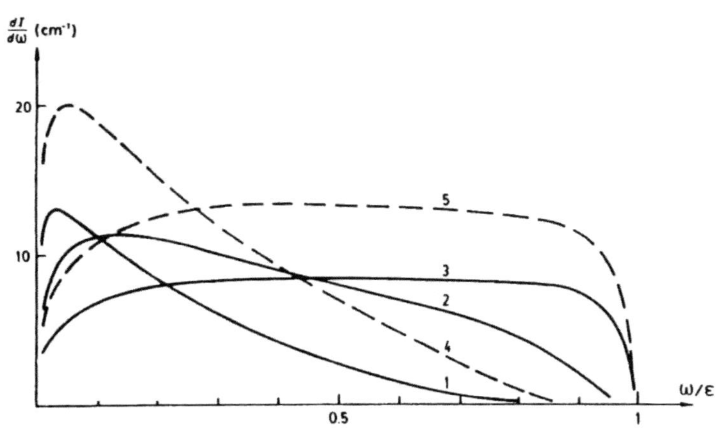

Fig. 17.2. Spectral intensity of radiation obtained for a given energy in the case of Si $\langle 110\rangle$ at $T = 293$ K : $\varepsilon = 100$ GeV (1), $\varepsilon = 700$ GeV (2), $\varepsilon = 5$ TeV (3). The dashed curves give the corresponding results for Ge $\langle 110\rangle$ at $T = 280$ K : $\varepsilon = 100$ GeV (4), and $\varepsilon = 3$ TeV (5).

the emitted photons calculated for different electrons energies along the $\langle 110\rangle$ axis in silicon and germanium. We can see that if the energy is $\varepsilon = 100$ GeV, then the value of $dI^M/d\omega$ has a maximum at $\omega/\varepsilon \sim 1$ and it falls rapidly upon increase in the frequency. However, as the energy is increased, the distribution becomes more uniform over the whole spectrum right up to $\omega \approx \varepsilon$.

17.2. Behavior of the Spectral Intensity of Radiation for Ultrahigh Energies

The spectral characteristics of the radiation depend very strongly on the relationship between the values of $u = \omega/\varepsilon'$ and χ. If $u \gg \chi_s\eta^{-1/2}$, the emission is exponentially suppressed, so that the emission spectrum consists mainly of the frequencies characterized by $u \leq \chi_s\eta^{-1/2}$. If $u \ll \chi_s$, we can obtain an explicit asymptotic expression for the spectral distribution of the radiation intensity, which is particularly convenient for the analysis in the range of high energies when $\chi_s \gg 1$. The range of integration with respect to x in Eq. (17.7) can be split into three parts: (1) $0 \leq x \leq x_1 \leq 1$; (2) $x_1 \leq x \leq x_2 \gg 1$; (3) $x_2 \leq x \leq x_0$, if we assume that $x_0 \gg x_2 \gg 1$, where $\lambda(x_1) \ll 1$ and $\lambda(x_2) \ll 1$. Then, in the ranges 1 and 3 we can replace $g(x)$ with approximate expressions at low and high values of x, and we can then integrate with respect to λ, whereas in range 2 we can use the expansion of $K_\nu(\lambda)$ valid in the range $\lambda \ll 1$. Extending integral over x in region (3) *ad infinitum* we have found as a result of fairly cumbersome calculations:

$$\frac{dI^F}{\omega d\omega} = \frac{\alpha m^2}{3\sqrt{3}\pi x_0\varepsilon^2}\left[\Phi_1 - \left(\frac{m\vartheta_0}{V_0}\right)^2\Phi_2\right], \tag{17.11}$$

where

$$\Phi_1 = \Gamma(2/3)\left(\frac{6\chi_s}{u}\right)^{2/3}\varphi(u)\left[\ln\frac{18\sqrt{3}\chi_s}{u} - \frac{\pi}{2\sqrt{3}} - C - \frac{3}{2\varphi(u)}l_1(\eta)\right],$$

$$\Phi_2 = t\frac{4}{405}\Gamma(1/3)(1+2\eta)\left(\frac{6\chi_s}{u}\right)^{4/3}\varphi(u)\left[\ln\frac{18\sqrt{3}\chi_s}{u} + \frac{\pi}{2\sqrt{3}} - C + \frac{3}{8} - \frac{21}{4\varphi(u)} - l_2(\eta)\right]; \tag{17.12}$$

$$l_1(\eta) = -B_1(\eta) + B_1(0),$$

$$l_2(\eta) = -B_2(\eta) + B_2(0),$$

here $\varphi(u) = 1 + u + (1+u)^{-1}$, $C = 0.577\ldots$—is the Euler constant. The functions $B_1(\eta)$ and $B_2(\eta)$ occur in the asymptotic representation of the probability of creation of pairs by a photon in the case when $\kappa_s \gg 1$ (see Eq. (12.17)). For $\eta \ll 1$ we have obtained the following asymptotic expression of the functions $l_{1,2}(\eta)$

$$\begin{aligned} l_1(\eta) &= 3.975\beta^{2/3}\left(1 + \frac{8}{15}\beta + \frac{7}{18}\beta^2\right) - \beta\left(\frac{3}{2} + \frac{9}{8}\beta + \frac{13}{14}\beta^2\right), \\ l_2(\eta) &= 2.31\beta^{1/3}\left(1 + \frac{2}{3}\beta - \frac{19}{63}\beta^2\right) + \frac{1}{160}\beta(30 + 144\beta - 29\beta^2), \end{aligned} \tag{17.13}$$

where $\beta = \eta/(1+\eta)$.

In the derivation of Eq. (17.11) the integration with respect to x is carried out between infinite limits. Since the main (logarithmic) contribution to Eq. (17.11) comes from the range $x \sim (\chi_s/u)^{2/3}$ (range 3), it is clear that the asymptotic expression (17.11) is valid only if $(\chi_s/u)^{2/3} < x_0$.

We must bear in mind that Eq. (17.11) is not valid at low frequencies in the limit $u \to 0$. The physical reason for the inadmissibility of the magnetic bremsstrahlung approach for the distribution of the low-frequency radiation is that at low frequencies the radiation is formed throughout the particle transit trajectory along the axis. We shall now analyze this topic. The magnetic bremsstrahlung approach is valid if the transverse momentum corresponding to an impact parameter $x^{1/2}a_s$ exceeds considerably the mass:

$$\frac{\Delta p_\perp}{m} \simeq \frac{2V_0\sqrt{x}}{ma_s}\int_{-\infty}^{\infty} dt\left(x + \frac{t^2\vartheta_0^2}{a_s^2}\right)^{-2} = \frac{\pi V_0}{mx\vartheta_0} \gg 1\,.$$

Substituting here $x = (\chi_s/u)^{2/3}$, we can find the lower limit to the frequency range in which the magnetic bremsstrahlung description is valid (see Eqs. (17.6), (17.7) and (17.11))

$$\left(\frac{\chi_s}{u}\right)^{4/3}\frac{1}{\rho(\vartheta_0)} \ll 1, \text{ or } u \gg u_{\min} = \frac{\chi_s}{\rho^{3/4}(\vartheta_0)} \tag{17.14}$$

In connection with this criterion we must draw attention to the anomalously small (by a factor $(\chi_s/u)^{2/3}$) first correction in the expansion of (17.11). It should be pointed out that if $\chi_s \ll 1$, then the asymptotic expression (17.11) describes the spectrum outside the range of fundamental frequencies (Eq. (17.11)) applies when $u \ll \chi_s$, whereas the main contribution to the radiation intensity comes from $u \sim \chi_s$), whereas for $\chi_s \gg 1$ the condition $u \ll \chi_s$ is satisfied by frequencies with $u \sim 1$, which dominate the intensity.

We shall now discuss the qualitative character of the spectral distribution of radiation intensity depending on the particle energy. For the relative low energies, when $\chi_s \ll 1$ (but $\rho_c = 2V_0\varepsilon/m^2 \gg 1$, this is the condition for the magnetic bremsstrahlung limit) the main contribution to the radiation intensity gives the region $\eta^{1/2} \leq x \leq 1 (u_1 \leq \rho \leq a_s)$ and the maximum of the spectral distribution is found for frequencies $\omega \sim \chi_s\varepsilon \ll \varepsilon$. On increasing $\chi_s(\varepsilon)$ the fraction contributed by frequencies $\omega \sim \varepsilon$ becomes significant ($\chi_s \sim 1$) and for $\chi_s \gg 1$ the frequencies $\omega \sim \varepsilon$ dominate the emission. At such high electron energies the spectral intensity of the radiation depends weakly on the frequency of the emitted photon. This follows from Eq. (17.11) and is illustrated in Fig. 17.2. In the range under discussion the value of L_{ch}^{-1} depend weakly on the energy (Fig. 1), so that the situation is on the whole similar to the properties of the bremsstrahlung emitted on a single nuclei (Bethe–Heitler mechanism), but the radiation lengths are now a few tens of times less than the corresponding radiation lengths in the amorphous medium. In the same range of energies an analogous situation occurs in the process of creation of an electron-positron pair by a photon (see Sec. 12.3). Hence, it follows that at low angles of incidence of a photon (or an electron) of sufficiently high energy on a single crystal, a special electron-photon shower develops (over distances much shorter than in the corresponding amorphous substance). The evolution of such showers is in many respects different from the Bethe–Heitler mechanism that will be considered in Sec. 20.

17.3. Total Radiation Intensity

We shall now consider the calculation of the total radiation intensity. It is convenient to carry out the integration of Eq. (17.7) by parts with respect to the variable u (and then in some terms the integration by parts with respect to the variable x), so that the final result is

$$
\begin{aligned}
I^F &= i_1(\chi_s) - \frac{1}{\rho(\vartheta_0)} i_2(\chi_s) = \frac{\alpha m^2}{3\sqrt{3}\pi x_0} \int_0^\infty \frac{u\,du}{(1+u)^4} \\
&\times \int_0^{x_0} dx \left\{ \int d^2\rho_0 \frac{\vartheta(\varepsilon_\perp(\rho_0) - U(\rho))}{s(\varepsilon_\perp(\rho_0))} (4u^2 + 5u + 4) K_{2/3}(\lambda) \right. \\
&\left. - \left(\frac{m\vartheta_0}{V_0}\right)^2 \frac{3\chi_s\sqrt{x}}{10(1+u)^2} \cdot K_{1/3}(\lambda) \left[5A(u) + \frac{1+2\eta}{x} B(u) + \frac{g(x)}{2} C(u)\right] \right\},
\end{aligned}
\tag{17.15}
$$

where

$$\rho(\vartheta_0) = (2V/m\vartheta_0)^2, \quad A(u) = 4u^3 - u^2 + 14u - 1,$$
$$B(u) = -4u^3 + 27u^2 - 2u + 27, \quad C(u) = 16u^3 - 43u^2 + 38u - 43.$$

As pointed out already, the range of validity of Eq. (17.15) is the same as that of the semiclassical approximation. At low energies (in the classical limit) it is well known that $I(x) \propto \chi^2(x)$ (see Eq. (1.39)) irrespective of ϑ_0 and this remains valid right up to the angles where the emission of frequencies $\omega \sim \gamma^2\vartheta_0/a_s \sim \varepsilon$ becomes significant and the classical theory ceases to be valid. Therefore, at relatively low energies when $\chi_s \ll 1$, we can have $i_2/i_1 \ll 1$ and the classical description is valid for the orientational dependence of radiation intensity. Thus, the term with i_2 represents a quantum correction. In fact, it follows from Eq. (17.15) that if $\chi_s \ll 1$, we find for functions i_1, i_2

$$\begin{aligned} i_1 &= \frac{8}{3x_0}\alpha m^2\chi_s^2\left[(1+2\eta)\ln\frac{1+\eta}{\eta} - 2\right] \\ i_2 &= \frac{\alpha m^2\sqrt{3}\pi\chi_s^3[87(1+2\eta) + 218\sqrt{\eta(1+\eta)}]}{40x_0\sqrt{\eta(1+\eta)}(\sqrt{\eta}+\sqrt{1+\eta})^3} \end{aligned} \tag{17.16}$$

We can see from these equations that $i_2/i_1 \propto \chi_s\eta^{-1/2}(\eta \ll 1)$. It is because the contribution in i_2 is given on the whole by region $\chi(x)$ near its maximum value $\chi_{\max} \sim \chi_s/\eta^{-1/2}$.

At high energies ($\chi_s \gg 1$), the expression for $L_{\text{ch}}^{-1} = I^F(\varepsilon)/\varepsilon$ becomes

$$L_{\text{ch}}^{-1} \simeq \frac{\alpha V_0}{mx_0a_s}\left[F(\chi_s) - \frac{8\chi_s}{9\rho(\vartheta_0)}\right], \quad F(\chi_s) = \frac{g_1}{\chi_s^{1/3}}(\ln\chi_s + g_0), \tag{17.17}$$

where

$$g_1 = (2/3)^6 6^{2/3}\Gamma(2/3) \simeq 0.3925,$$
$$g_0 = \frac{\pi}{2\sqrt{3}} + \frac{5}{2}\ln 3 + \ln 2 - C - \frac{99}{32} - l_1(\eta) \simeq 0.6756 - l_1(\eta).$$

It is clear from Eq. (17.17) that if $\chi_s \gg 1$, then the magnetic bremsstrahlung description is valid in the range where $\chi_s^{4/3}/\rho(\vartheta_0) \ll 1$. This is in agreement with Eq. (17.14) if we bear in mind that for these values of χ_s the main contribution to the intensity comes from $u \sim 1$.

The quantity $F(\chi_s)$ in Eq. (17.17) reaches its maximum at the point $\chi_s^0 = \exp(3 - g_0)$, near which it varies extremely smoothly. Since $\chi_s^0 \geq e^3 \gg 1$, this justifies the use of Eq. (17.17) in the region of the maximum. In view of the smooth variation of $F(\chi_s)$ near the maximum, the energy ε_0 obtained from χ_s^0 represents only an estimate. However, the value of $(L_{\text{ch}}^{-1})_{\text{max}}$ given by Eq. (17.17) is quite satisfactory and it is in good agreement with the results of a numerical calculation carried out on the basis of Eq. (17.7)

$$\max L_{\text{ch}}^{-1} \simeq 0.543 \frac{\alpha V_0}{m x_0 a_s} \exp\left[-\frac{l_1(\eta)}{3}\right] \tag{17.18}$$

It is interesting to compare Eq. (17.18) with the maximum probability of creation of pairs by a photon [Eq. (12.19)]. We have the relationship $W_e^{\max} L_{\text{ch}}^{\max} \simeq 0.72$; however, we must bear in mind that the characteristics of pair creation and emission of radiation have maxima at different energies. At very high energies the values of $W_e L_{\text{ch}} \simeq 5 \cdot 3^6 \Gamma^3(2/3)\,(7 \cdot 2^7 \pi^2)^{-1} \simeq 1.023$. Although the product of these quantities is, as in Bethe–Heitler case, a constant (the constants are different: 1.023 and 0.778), the quantities W_e and L_{ch}^{-1} themselves depend on the energy, in contrast to the Bethe–Heitler case. It should also be pointed out that the factor $\exp(-l_1(\eta)/3)$ depends weakly on the medium and it varies (for the usually investigated substances) in the range 0.90–0.75 as η changes from 0.025 for diamond to 0.15 for silicon.

Table 17.1. Some values characterizing radiation process.

Crystal	Axis	$T(K)$	χ_s (ε = 1 TeV)	ρ_c (ε = 10 GeV)	$r_\gamma^{\max}$
C	111	293	1.3	2.2	168
Si	111	293	2.7	4.1	71
Si	110	293	3.2	5.4	81
Ge	111	293	4.5	7.0	26
Ge	110	280	4.8	8.4	30
Ge	110	100	5.6	8.8	30
W	111	293	29	32	11

The maximum excess of $L_{\rm rad}$ above $L_{\rm ch}$ in the corresponding amorphous substance is of considerable interest (compare with Eq. (12.20))

$$r_\gamma^{\max} = \left(\frac{L_{\rm rad}}{L_{\rm ch}}\right)_{\max} \simeq 0.426 \frac{V_0 m a_s d \exp\left[-l_1(\eta)/3\right]}{(Z\alpha)^2 \ln\left(183 Z^{-1/3}\right)} \sim \frac{m a_S}{3Z\alpha \ln\left(183Z^{-1/3}\right)}, \tag{17.19}$$

where the second (simplified) estimate is fairly rough. Hence, we can see that $r_\gamma^{\max}$ increases upon reduction of Z and upon increase of a_s. The greatest enhancement (among the investigated substances) is achieved for diamond ($r_\gamma^{\max} \sim 160$); the values of $r_\gamma^{\max}$ are listed in Table 17.1.

17.4. Total Probability of Emission of Radiation

We shall now consider calculation of the total radiation probability which determines the total number of emitted photons and we shall do this in the magnetic bremsstrahlung approximation. Then, using Eq. (17.9)–(17.10), we obtain for the total probability for the case when $\chi_s \ll 1$

$$\begin{aligned} W_1^M = (\chi_s \ll 1) &= \frac{5\alpha m^2}{2\sqrt{3}x_0\varepsilon} \int_0^{x_0} dx \chi(x) = \frac{10\alpha V_0}{\sqrt{3}a_S m x_0} \\ &\times \left[\sqrt{1+\eta}\arctan\left(\frac{x_0}{1+\eta}\right)^{1/2} - \sqrt{\eta}\arctan\left(\frac{x_0}{\eta}\right)^{1/2}\right] \\ &\simeq \frac{5\alpha\pi V_0}{\sqrt{3}a_s m x_0}\left(\sqrt{1+\eta} - \sqrt{\eta} - \frac{2}{\pi\sqrt{x_0}}\right). \end{aligned} \tag{17.20}$$

In the other limiting case of $\chi_s \gg 1$, integrating Eq. (17.11) and taking into account that $d\omega = \varepsilon(1+u)^{-2}du$, we find

$$W_1^M(\chi_s \gg 1) = \frac{28 \cdot 6^{2/3}\Gamma(2/3)\alpha V_0}{81 m x_0 a_s \chi_s^{1/3}}\left[\ln \chi_s + D(\eta)\right], \tag{17.21}$$

where

$$D(\eta) = \ln 2 + \frac{5}{2}\ln 3 + \frac{\pi}{2\sqrt{3}} - C + \frac{3}{28} - l_1(\eta) \simeq 3.8765 - l_1(\eta).$$

Using Eqs. (17.16), (17.17), (17.20) and (17.21) one can find the mean value of the emitted photon frequency

$$\begin{aligned} \bar{\omega}(\chi_s \ll 1) &= \frac{8\chi_s\varepsilon}{5\sqrt{3}\pi}\frac{(1+2\eta)\ln(1+\eta^{-1})-2}{\sqrt{1+\eta}-\sqrt{\eta}-2(\pi\sqrt{x_0})^{-1}} \\ \bar{\omega}(\chi_s \gg 1) &= \frac{16\varepsilon}{63}\frac{\ln\chi_s+0.676-l_1(\eta)}{\ln\chi_s+3.877-l_1(\eta)}\,. \end{aligned} \qquad (17.22)$$

An analysis of the above expressions for the intensity and the total probability shows that a large number of soft photons is emitted and they do not significantly affect the loss of energy by the particle. At high particle energies ($\chi_s \geq 1$) these photons can nevertheless manifest their presence on experiments. In particular, they can form (as a result of the Bethe–Heitler mechanism) secondary pairs of charged particles, the number of which is determined by the probability described by Eq. (17.21). Figure 17.3 shows the dependence of the total probability W_1^M on the energy of particles in different substances. We can see that the probability on the left-hand side of the graph exhibits a plateau (when the probability of emission in external field is practically independent of the energy for $\chi \ll 1$). This is followed by a region of logarithmic fall where the falling curve can be approximated satisfactorily by a straight line. Figure 17.4 shows the probability of the emission of a photon integrated over the frequencies in the range $\omega_0 \div \varepsilon$ for different values of ε considered as a function of ω_0. The fact that even for $\omega_0 \sim \varepsilon$ this probability obeys $W_1^M(\omega_0 \div \varepsilon) \sim L_{\text{ch}}^{-1}$ confirms the conclusions reached above on the contribution of the soft photons.

We shall conclude this section with a theoretical analysis of the case when the contribution of the range $x \gg 1$ becomes important. In the above approximation we used the potential of Eq. (9.13) characterized by $g(x) \propto x^{-2}$ when $x \gg 1$. We shall first assume that the law describing the fall of the potential in the range $x \gg 1$ is different: $g(x) \propto x^{-(\mu+1)}$, so that $\chi(x) \propto \chi_s x^{-(\mu+1/2)}$ (see Eq. (17.10)). If $\chi_s \gg 1$, the range $\chi(x) \geq 1$ lies to the left of the point $x \sim \chi^{2/(2\mu+1)} \equiv x_1$. The contribution to the total intensity from the range $x \sim x_1$ is $I \sim \chi_s^{2/3}$ Therefore, in the case of the potential which falls faster than $x^{-1}(\mu > 1)$, the radiation intensity is governed by the range $x \sim 1$. In this case the characteristics of the radiation behave in the same way as in the magnetic bremsstrahlung approach and, in particular, the total intensity obeys $I \propto \chi_s^{2/3}$, where the criterion of the validity of the approach is $\vartheta_0 \gg \vartheta_v \chi_s^{-1/3}$. However,

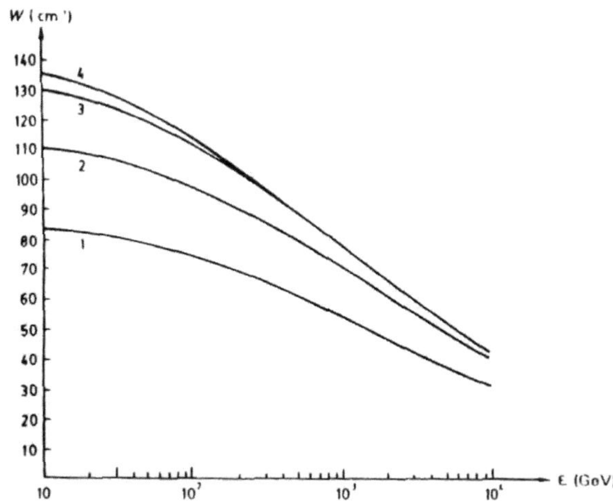

Fig. 17.3. Energy dependence of the total probability of emission of radiation in Si $\langle 110\rangle$ at $T = 293$ K; (1), in diamond $\langle 111\rangle$ at $T = 293$ K (2), in Ge $\langle 110\rangle$ at $T = 280$ K (3), and in Ge $\langle 110\rangle$ at $T = 100$ K (4).

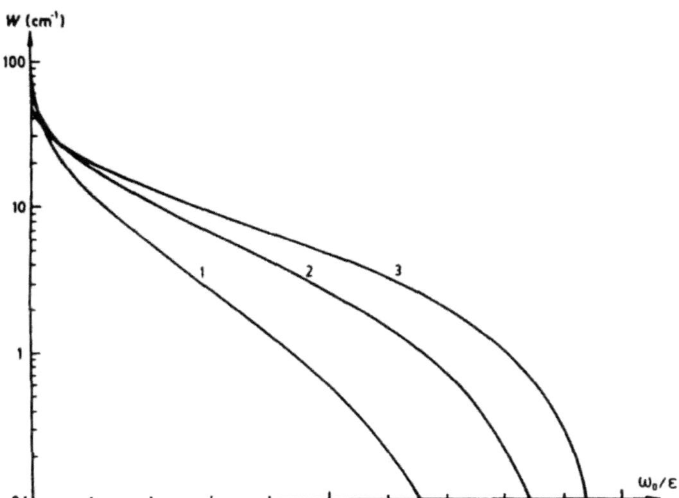

Fig. 17.4. Dependence of the probability of photon emission $W(\omega_0) = \varepsilon \int_{\omega_0}^{\varepsilon} (dW/d\omega) d\omega$ of the energy $\omega \geq \omega_0$ on ω_0 for Si $\langle 110\rangle$ at $T + 293$ K : $\varepsilon = 100$ GeV (1), $\varepsilon = 400$ GeV (2), $\varepsilon = 2$ TeV (3).

if $\mu < 1$, then for $\chi_s \gg 1$ the contribution of the range $x \sim x_1$ becomes predominant and the criterion of validity of the magnetic bremsstrahlung description is then $\vartheta_0 \ll \vartheta_v \chi_s^{-2\mu/(2\mu+1)}$. As the energy is increased, the value of x_1 becomes compared with x_0 (it becomes a boundary of the area per one axis) and there is a change in the physical situation. If $x_1 \gg x_0(\chi(x) \gg 1$ for all values of x), we have $I \propto \chi_s^{2/3}$ and the criterion of validity becomes $\vartheta_0 \gg \vartheta_v \chi_s^{-1/3} x_0^{-(2\mu-1/2)/3}$. For the potential of Eq. (9.13) we have $\mu = 1$ and if $\chi_s \gg 1$, then $I \propto \chi_s^{2/3} \ln \chi_s$. The appearance of the logarithm is due to the fact that the whole range $1 \leq x \leq x_1$ contributes. In this case the change in the physical situation is that in the total intensity we now have $\ln \chi_s \rightarrow 3 \ln x_0/2$. Moreover, if $x_1 \geq x_0$, the potential on the axis can no longer be regarded as axially symmetric and instead of Eq. (17.7) we have to use Eq. (17.6). It should be pointed out that for the potential of Eq. (9.13) the value of x_1 becomes comparable with x_0 at energy $\varepsilon \sim$ 10–20 TeV, which depends on the substance.

17.5. Emission of Radiation with Regard to the Particle Energy Loss

As stated above (see Sec. 10.1) the multiple scattering and emission of radiation are the main processes affecting the particle distribution function in thick crystals. The variation of this function in the region of middle energy ($\rho_c = 2V_0\varepsilon/m^2 \leq 1$) is given mainly by multiple scattering and it was considered in Sec. 10. With the increase of particles energy, the angle of multiple scattering decreases ($\vartheta_s^2 \propto \varepsilon^{-2}$) and the energy losses per unit of length becomes larger, so that for $\chi_s \geq 1 (\chi_s = \rho_c/2ma_s \sim 10^{-2}\rho_c)$ the emission of radiation dominates in the variation of the distribution function. Since for high energies ($\chi_s \geq 1$) the emission of radiation depends on the relation between transverse velocity $v_\perp$ and angle $\vartheta_v = V_0/m$, we define the length l_v so that a squared multiple scattering angle for it is equal to ϑ_v^2 in an amorphous medium.

$$l_v = \frac{\vartheta_v^2}{\dot{\vartheta}_s^2} = \frac{\alpha}{4\pi} L_{\rm rad} \left(\gamma \frac{V_0}{m}\right)^2 = L_{\rm rad} \frac{\alpha}{4\pi} \frac{a_s^2}{\lambda_c^2} \chi_s^2 \geq L_{\rm rad} \chi_s^2 . \tag{17.23}$$

Since for $\chi_s \geq 1$ it is valid that $L_{\rm rad} \gg L_{\rm ch}$ it follows from Eq. (17.23) the validity of the magnetic bremsstrahlung approach for the small incident angles ($\vartheta_0 \ll V_0/m$) if a particle does not lose almost the whole of the energy.

The other parameter defining a variation of the distribution function is the length of dechanneling l_d (see Sec. 10.2)

$$l_d = \frac{\alpha}{4\pi}\rho_c L_{\rm rad} = L_{\rm rad}\frac{\alpha}{2\pi}\frac{a_s}{\lambda_c}\chi_s \sim \frac{1}{10}L_{\rm rad}\chi_s \,. \tag{17.24}$$

Taking into account for $\chi_s \geq 1$ the ratio $r = L_{\rm rad}/L_{\rm ch} \sim 10 \div 100$ it follows $l_d > L_{\rm ch}$ but in this case the value $l_d/L_{\rm ch}$ is not so large as $l_v/L_{\rm ch}$. However we have to remember (see Sec. 10.2) that the dechanneling length of positrons can be sometimes greater than l_d in Eq. (17.24) due to the weakening of the scattering of positrons moving in a channel.

In condition of the validity of the magnetic bremsstrahlung limit if the particle distribution function $\varphi(\varepsilon, \boldsymbol{\rho}, l)$ at the depth l is known, the spectral distribution of the radiation intensity of a non-polarized beam of particles at the depth l is of the form

$$\frac{dI(l)}{d\omega} = \int \frac{dI(\boldsymbol{\rho},\varepsilon)}{d\omega}\varphi(\varepsilon,\boldsymbol{\rho},l)d^2\rho d\varepsilon \,, \tag{17.25}$$

where $dI(\boldsymbol{\rho},\omega)/d\omega$ is given by the following expression (see e.g. formula (17.6))

$$\frac{dI(\boldsymbol{\rho},\varepsilon)}{d\omega} = \frac{\alpha m^2\omega}{\sqrt{3}\pi\varepsilon^2}\left[\int_\lambda^\infty dy K_{5/3}(y) + \frac{\omega^2}{\varepsilon(\varepsilon-\omega)}K_{2/3}(\lambda)\right] \,, \tag{17.26}$$

here

$$\lambda = \frac{2m^3\omega}{3\varepsilon(\varepsilon-\omega)|\boldsymbol{\nabla} U(\boldsymbol{\rho})|} \equiv \frac{2\omega}{3(\varepsilon-\omega)\chi(\boldsymbol{\rho})} \,.$$

Generally speaking, finding the distribution function $\varphi(\varepsilon, \boldsymbol{\rho}, l)$ in a concrete situation is a quite complicated problem. Here we shall use a uniform distribution with respect to the transverse coordinate, i.e.

$$\varphi(\varepsilon,\boldsymbol{\rho},l) = \frac{1}{s}g(\varepsilon,l) \,, \tag{17.27}$$

where s is the area per one axis. This is valid, in any case, when the interval of the angles of incidence ϑ_0 is not too narrow, $\Delta\vartheta_0 \leq \vartheta_c = (2V_0/\varepsilon)^{1/2}$. In this case, from the very beginning the above-barrier states for which the distribution is uniform are mainly occupied. For example, under the experimental condition of Belkacem *et al.*, 1985 where the angular width of the incident beam was ∓ 30 μrad, we have 80% of the particles in the above-barrier states, and by taking into account as much mosaic structures as possible (10 μrad), this number increases to 88%. Hence, even with the kinetics neglected, we have a uniform distribution for most of the electrons under the conditions of this experiment.

Making allowances for Eq. (17.27), one can perform, in Eq. (17.25) the integration over $d^2\rho$ for any given potential $U(\rho)$. The concrete calculations have been made for the potential $U = U(x)(x = \rho^2/a_s^2, a_s$ is the screening radius) in the form (9.13).

For the variation of the average energy of an electron at depth l, we have an equation

$$\frac{d\langle\varepsilon(l)\rangle}{dl} = -\langle Q(\varepsilon)\rangle\,, \tag{17.28}$$

where

$$Q(\varepsilon) = \int \frac{d^2\rho}{s} I(\chi(\rho))\,, \quad I(\chi(\rho)) = \int \frac{dI(\rho,\varepsilon)}{d\omega} d\omega$$

is the total radiation intensity; the notation $\langle f(\varepsilon)\rangle = \int d\varepsilon g(\varepsilon, l) f(\varepsilon)$ denotes the averaging over the energy distribution at a given depth. In the given potential the quantity $Q(\varepsilon)$ is calculated rather easily. It turns out that Eq. (17.28) may be solved with sufficient accuracy and without the use of explicit form of the function $g(\varepsilon, l)$. The method is based on an approximation of the found function $Q(\varepsilon)$ in the energy interval we investigated in by a simpler function. So if the average energy of the particles when they leave the crystal is high enough, then $Q(\varepsilon) = a\varepsilon(1 + b\varepsilon/\varepsilon_0)$ with a high degree of accuracy where a and b are the fitting parameters. Such a form is determined by the properties of the radiation intensity: for $\chi > 1$ the ratio $I(\chi)/\chi$ depends weakly on χ. For example, within the range of the value of χ from 1 to 15 the quantity $I(\chi)/\chi \simeq 2\alpha m^2/15$ and its relative change is less than 10% (see Fig. 5.4). In the appropriate energy interval, the radiation length remains constant with the same accuracy $L_{\text{ch}} \simeq L_{\text{ch}}^{(0)}$

$$\frac{1}{L_{\text{ch}}^{(0)}} \equiv \frac{Q(\varepsilon_0)}{\varepsilon_0} \simeq \frac{\alpha V_0}{2a_s x_0 m}\,. \tag{17.29}$$

The rate of the energy losses in comparison with the corresponding amorphous media may be roughly estimated as $r \sim ma_s/[3Z\alpha \ln(183Z^{-1/3})]$ (see Eq. (17.19)).

For the given approximation $Q(\varepsilon)$, Eq. (17.28) takes the form

$$\frac{d\langle\varepsilon(l)\rangle}{\varepsilon_0 dl} = -\frac{a}{\varepsilon_0}\left[\langle\varepsilon(l)\rangle + \frac{b}{\varepsilon_0}\langle\varepsilon^2(l)\rangle\right]$$
$$= -a\left[y + b(y^2 + \Delta^2)\right] , \qquad (17.30)$$

where

$$y = \frac{\langle\varepsilon(l)\rangle}{\varepsilon_0}, \quad \Delta^2(l) = \frac{\langle\varepsilon^2(l)\rangle - \langle\varepsilon(l)\rangle^2}{\varepsilon_0^2}$$

is the dispersion of the distribution with respect to ε, and $y(0) = 1$. For the monochromatic initial beam of particles, $\Delta^2(0) = 0$, while the quantity $\Delta^2(l)$ increases as l grows, and $\Delta^2(l) \to 0$ at $y(l) \to 0$. In addition, we would like to remark that for the uniform distribution (this is the maximum estimate) $\Delta^2 = 1/12$.

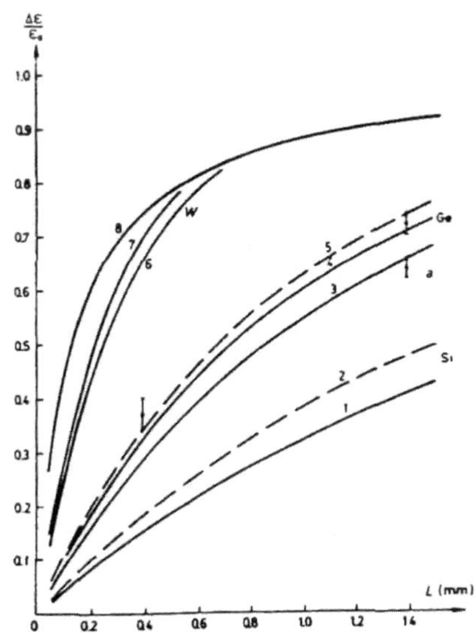

Fig. 17.5. Average relative energy losses of electrons in dependence on crystal thickness: 1. Si⟨111⟩, T = 293 K, ε_0 = 200 GeV; 2. Si⟨110⟩, T = 293 K, ε_0 = 200 GeV; 3. Ge⟨110⟩, T = 280 K, ε_0 = 150 GeV; 4. Ge⟨110⟩, T = 100 K, ε_0 = 150 GeV; 5. Ge⟨110⟩, T = 100 K, ε_0 = 200 GeV; 6. W⟨111⟩, T = 293 K, ε_0 = 100 GeV; 7. W⟨111⟩, T = 77 K, ε_0 = 100 GeV; 8. the curve, calculated with using the classical formula, for crystal Ge ⟨110⟩, T = 100 K, ε_0 = 150 GeV. The experimental results are taken from Belkacem *et al.*, 1985: a: at T = 280 K, the other at T = 100 K.

The analysis performed has shown that the explicit form of the energy distribution influences weakly on the average relative energy loss $\Delta\varepsilon(L)/\varepsilon_0 = 1 - y(L)$ calculated according to Eq. (17.30). For example, under the experimental condition by Belkacem *et al.*, 1985 the result changes by less than 2–3%. Within this accuracy, it is possible to omit Δ^2 in Eq. (17.30) and after that it can be solved simply. Figure 17.5 presents the value of $\Delta\varepsilon(L)/\varepsilon_0$ for different crystals, axis and energies versus the crystal thickness L. These values have been calculated at $\Delta^2 = 0$. In accordance with the above (cf. Eq. (17.29)), the losses depend weakly on the energy (cf. curves 4 and 5) and the difference in energy losses for different media and axes is mainly determined by different values of $L^0_{\rm ch}$ (see Eq. (17.29)). The temperature dependence proves to be different in the classical ($\chi \ll 1$) and quantum ($\chi \geq 1$) regions. This is due to the fact that $I(\chi) \propto \chi^2$ in the classical region, i.e. it is proportional to the squared field; then the region $\rho \sim u_1$(u_1 is the termal vibration amplitude) and $\rho \sim a_s$ contributes approximately equally when integrating over $d^2\rho$. As has already been mentioned, $I(\chi) \propto \chi$ for $\chi \geq 1$ and the contribution of $\rho \sim u_1$ to $\int d^2\rho$ turns out to be small. This is illustrated in Fig. 17.5: curves 6 and 7 are closer to each other (larger χ) than curves 3 and 4. Curve 8 has been obtained under the same conditions as curve 4 (the condition of the experiment by Belkacem *et al.*, 1985) but according to the classical formula. A sharp distinction between them demonstrates the magnitude of the quantum effects in this case. On the other hand, curves 3 and 4 are in good agreement with the experiment and this indicates the necessity to use quantum electrodynamics when describing the process. As far as we know, this is the first direct observation of quantum effects in radiation in a strong external field. The agreement also confirms the assumption of a uniform distribution of the particle flow.

We now proceed to the discussion of the spectral distribution of the radiation intensity $dI/d\omega$. Under the accepted assumptions, averaged over the transverse coordinates, the quantity $dI/d\omega$ is of the form, for a given electron energy,

$$\frac{dI}{d\omega} = \frac{\alpha m^2\omega}{\pi\sqrt{3}\varepsilon^2 x_0}\int_0^{x_0} dx\left[1 + \frac{\omega^2}{\varepsilon\varepsilon'} + \frac{3x^2 + (1+2\eta)x - \eta(1+\eta)}{3(x+\eta)(1+x+\eta)}\right] K_{2/3}(\lambda)\,, \tag{17.31}$$

where

$$\lambda = \frac{\omega}{3\varepsilon'\chi_s}\frac{(x+\eta)(1+x+\eta)}{\sqrt{x}}\,, \quad \varepsilon' = \varepsilon - \omega\,.$$

The results of the calculation according to this formula are presented in Fig. 17.6 (curves 1, 4, 5 and 6). The calculated average energies of emitted photons ω are very consistent with Eq. (17.22). The energy dependence of the shape of the spectrum is rather weak, as is seen from curves 4 and 5 in Fig. 17.6. For them energy differs by a factor of 2.

The energy losses of an electron lead to varying the spectrum $dI/d\omega$. To find it at a given depth, it is necessary to average $dI/d\omega$ (see Eq. (17.31)

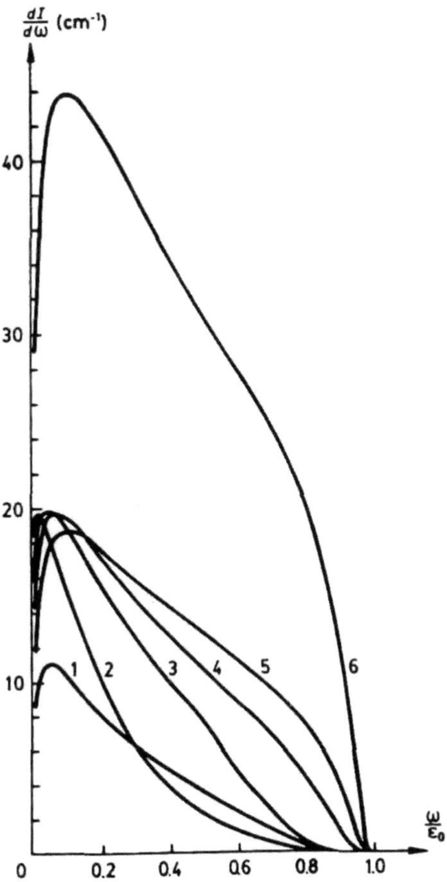

Fig. 17.6. Spectral intensity of radiation at fixed energy: 1. Si$\langle 111\rangle, T = 293$ K$, \varepsilon_0 =$ 200 GeV; 4. Ge$\langle 110\rangle, T = 100$ K$, \varepsilon_0 = 150$ GeV; 5. Ge$\langle 110\rangle, T = 100$ K$, \varepsilon_0 = 300$ GeV; 6. $W\langle 111\rangle, T = 293$ K$, \varepsilon_0 = 100$ GeV. Spectral intensity averaged over crystal thickness in Ge $\langle 110\rangle, T = 100$ K$, \varepsilon_0 = 150$ GeV; 2: for $L = 1.4$ mm, 3: for $L = 0.4$ mm.

with the distribution function $g(\varepsilon, l)$ (Eq. (17.31)). As has been mentioned above, taking into account the dispersion of the distribution with respect to ε exerts weak influence on the quantity $\Delta\varepsilon(l)/\varepsilon_0$. A similar assertion is not so evident for the spectral distribution. Nevertheless, we use it here as a first (and simple!) approximation. Then $dI(l)/d\omega$ can be obtained from Eq. (17.31) by the substitution $\varepsilon \to \langle\varepsilon(l)\rangle$, where $\langle\varepsilon(l)\rangle$ is the solution of Eq. (17.28). The observed spectrum reads

$$\frac{1}{L}\int_0^L dl \frac{dI(l)}{d\omega}.$$

The results calculated for the experimental condition are given in Fig. 17.6. It is seen that the spectrum narrows with increasing thickness and the maximum in distribution shifts to the left (cf. curves 4, 3 and 2). The average number of the emitted photons grows as well.

In this section, based on Baier, Katkov, Strakhovenko, 1986d, multiple character of the emission of photons has been taken into account by the dependence on depth l of the average energy. The results of calculations of electron-positron showers development (see Sec. 20) will show consistency of this approach.

18. MODIFIED THEORY OF COHERENT BREMSSTRAHLUNG

18.1. Theory of Coherent Bremsstrahlung (CBS)

The estimates of phase A_2 made at the beginning of Sec. 17.1 remain valid also for $\vartheta_0 \geq \vartheta_v$, except that now the factor in the double sum is $(\vartheta_v/\vartheta_0)^2 \leq 1$, so that values $|q_\parallel\tau| \sim 1$ contribute. If $\vartheta_0 \gg \vartheta_v$, then this factor is small, the term with the double sum from A_2 can be moved in Eq. (16.16) into the pre-exponential part. As a result the spectral distribution of the radiation intensity is of the form

$$\begin{aligned} dI^{\text{coh}} &= -\frac{i\alpha m^2}{2\pi\varepsilon^2}\omega d\omega \int_{-\infty}^{\infty} \frac{d\tau}{\tau - i0} \exp\left(-i\frac{m^2\omega\tau}{\varepsilon\varepsilon'}\right) \sum_{\mathbf{q},\mathbf{q}'} \frac{G(\mathbf{q})G(\mathbf{q}')}{m^2 q_\parallel q'_\parallel} \\ &\times (\mathbf{q}_\perp \mathbf{q}'_\perp) \cdot \left\{\varphi(\varepsilon)\, \sin(q_\parallel\tau)\sin(q'_\parallel\tau) + i\frac{m^2\omega\tau}{\varepsilon\varepsilon'}\left[\frac{\sin(q_\parallel + q'_\parallel)\tau}{(q_\parallel + q'_\parallel)\tau}\right.\right. \\ &\left.\left. - \frac{\sin q_\parallel\tau}{q_\parallel\tau} \cdot \frac{\sin q'_\parallel\tau}{q'_\parallel\tau}\right]\right\} \int \frac{d^3r}{V} \exp[-i(\mathbf{q}+\mathbf{q}')\mathbf{r}]\,. \end{aligned} \tag{18.1}$$

Integration over coordinates $\mathbf{r}$ in (18.1) is elementary and gives $\delta_{\mathbf{q}+\mathbf{q}',0}$ ($\mathbf{q}$ is a discrete variable), after that the sum over $\mathbf{q}'$ is calculated and the integrals over τ are easily evaluated by means of the theory of residues. Finally we obtain

$$dI^{coh} = \frac{\alpha\omega d\omega}{4\varepsilon^2}\sum_{\mathbf{q}}|G(\mathbf{q})|^2\frac{\mathbf{q}_\perp^2}{q_\parallel^2}\left\{\varphi(\varepsilon) - \frac{2m^2\omega}{\varepsilon\varepsilon' q_\parallel^2}\cdot\left(|q_\parallel| - \frac{m^2\omega}{2\varepsilon\varepsilon'}\right)\right\}$$
$$\times\,\vartheta\left(|q_\parallel| - \frac{m^2\omega}{2\varepsilon\varepsilon'}\right) \qquad (18.2)$$

Going over to the variable $u = \omega/\varepsilon'$ and carrying out the integration we find

$$I^{coh} = \frac{\alpha}{4}\sum_{\mathbf{q}}|G(\mathbf{q})|^2\frac{\mathbf{q}_\perp^2}{q_\parallel^2}\Phi(\bar{z});\ \bar{z} = \frac{2\varepsilon|q_\parallel|}{m^2}, \qquad (18.3)$$

where

$$\Phi(z) = \left[\ln(1+z) - \frac{z(2+3z)}{2(1+z)^2}\right]\left[1 - \frac{4(3+z)}{z^2}\right]$$
$$+\frac{z}{(1+z)^2}\left[4 + \frac{z(3+2z)}{3(1+z)}\right] \qquad (18.4)$$

The spectral distribution (18.2) coincides with the results of the standard theory of coherent bremsstrahlung (see, for example, Ter-Mikaelian, 1969).

Using the Eqs. (16.8), (16.17) and (16.18), the spectral distribution of polarized photons can be also found

$$dI^{coh}(\mathbf{e}) = \frac{\alpha\omega d\omega}{8\varepsilon^2}\sum_{\mathbf{q}}|G(\mathbf{q})|^2\vartheta\left(|q_\parallel| - \frac{m^2\omega}{2\varepsilon\varepsilon'}\right)\frac{1}{q_\parallel^2}$$
$$\left\{(\mathbf{e}\mathbf{e}^*)\mathbf{q}_\perp^2\left[\varphi(\varepsilon) - \frac{2m^2\omega}{\varepsilon\varepsilon' q_\parallel^2}\left(|q_\parallel| - \frac{m^2\omega}{2\varepsilon\varepsilon'}\right)\right]\right.$$
$$\left.+\frac{1}{2}\left(\frac{m^2\omega}{\varepsilon\varepsilon' q_\parallel}\right)^2[2|\mathbf{q}_\perp\mathbf{e})|^2 - (\mathbf{e}\mathbf{e}^*)\mathbf{q}_\perp^2]\right\} \qquad (18.5)$$

The spectral intensity (18.2) can be much higher than the Bethe–Heitler bremsstrahlung intensity for small angles of incidence ϑ_0 with respect to the selected axis. In this case ($\vartheta_0 \ll 1$) the quantity $q_{\parallel}$ can be represented in the form (see Eq. (13.7))

$$q_{\parallel} \simeq \frac{2\pi}{d} n + \mathbf{q}_{\perp} \mathbf{v}_{0\perp} \tag{18.6}$$

The main contribution to the sum (18.2) for small ϑ_0 is given by $\mathbf{q}$ with $n = 0$, then

$$q_{\parallel} \simeq \left(\frac{2\pi}{f} k \cos\varphi + \frac{2\pi}{h} l \sin\varphi \right) \vartheta_0 \,, \tag{18.7}$$

where f and h are the characteristic periods of the potential in the transverse plane, $k, l = 0, \mp 1, \mp 3 \ldots$. It follows from Eq. (18.2) that the maximum emission of radiation will be observed for the angles (cf. Eq. (11.4))

$$\vartheta_0 \sim \vartheta_m(\omega) \sim \frac{m^2 a_s \omega}{\varepsilon(\varepsilon - \omega)} = \frac{u}{\gamma} \frac{a_s}{\lambda_c} \sim 10^2 \frac{u}{\gamma} \,, \qquad u = \frac{\omega}{\varepsilon - \omega} \tag{18.8}$$

In the spectral distribution of the radiation intensity one can observe the typical oscillations connected with the set of frequencies defined by the harmonics of particle quasiperiodic motion with the frequency $\omega_0 = q_{\parallel}$ (see Eq. (16.5)). From Eq. (5.7) for $\rho \ll 1$ in the dipole approximation we find

$$\omega' = \frac{\omega\varepsilon}{\varepsilon - \omega} = u\varepsilon = \frac{2 q_{\parallel} \gamma^2}{1 + \gamma^2 \vartheta^2} \,. \tag{18.9}$$

Using Eq. (16.5) for a particle velocity and formulae obtained in Sec. 5.2, one can find the spectral-angular intensity distribution for polarized photons. The angular distribution and the polarization problems has been considered in Sec. 10 of the Ter-Mikaelian book 1969.

18.2. Extension of the CBS Theory

For $\chi_s \gg 1$, one can obtain from the general expression (16.16) approximate expressions for the spectral distribution of radiation, the region of applicability of which is broader than that of the standard CBS theory. For this purpose it is necessary to take into account that the phase A_2 in (16.16) has for $q_{\parallel} + q'_{\parallel} \neq 0$ terms of the order of $(V_0/m\vartheta_0)^3/\chi_s$ and $(V_0/m\vartheta_0)^4/\chi_s^2)$ which can be small even for $\vartheta_0 \leq V_0/m$ if $\chi_s \gg 1$. Therefore, assuming that these contributions are small, we shall carry out the corresponding expansion of $\exp(iA_2)$, while the term with $q_{\parallel} + q'_{\parallel} = 0$ in the first term of the double sum in A_2 will be retained in the exponential. As a result we obtain an expression which coincides in the form with (18.1) where we must make the substitution

$$\exp\left(-\frac{im^2\omega\tau}{\varepsilon\varepsilon'}\right) \to \exp\left(-\frac{im_*^2\omega\tau}{\varepsilon\varepsilon'}\right), \quad m_*^2 = m^2(1+\rho/2) \tag{18.10}$$

and in the term $\dfrac{\sin(q_{\parallel} + q'_{\parallel})\tau}{(q_{\parallel} + q'_{\parallel})\tau}$ it is necessary to assume $q_{\parallel} + q'_{\parallel} \neq 0$. Here

$$\begin{aligned}\frac{\rho}{2} &= \gamma^2[\langle \mathbf{v}^2\rangle - \langle \mathbf{v}\rangle^2] = \frac{1}{m^2}\sum_{\mathbf{q},q'} G(\mathbf{q})G(\mathbf{q}')\frac{(\mathbf{q}_{\perp}\mathbf{q}'_{\perp})}{q_{\parallel}q'_{\parallel}} \\ &\times [\delta_{q_{\parallel}+q'_{\parallel},0} - \delta_{q'_{\parallel}0}\delta_{q'_{\parallel},0}] = \sum_{\substack{\mathbf{q},\mathbf{q}_{\parallel}\\ \neq 0}} |G(\mathbf{q})|^2\frac{\mathbf{q}_{\perp}^2}{m^2 q_{\parallel}^2}\end{aligned} \tag{18.11}$$

The remaining calculations are carried out in the same way as in the transition from (18.1) to (18.2). We finally obtain

$$\begin{aligned}dI^{\text{mcoh}} &= \frac{\alpha\omega d\omega}{4\varepsilon^2}\sum_{\mathbf{q}} |G(\mathbf{q})|^2\frac{\mathbf{q}_{\perp}^2}{q_{\parallel}^2}\left[\varphi(\varepsilon) - \frac{2m^2\omega}{\varepsilon\varepsilon' q_{\parallel}^2}\left(|q_{\parallel}| - \frac{m_*^2\omega}{2\varepsilon\varepsilon'}\right)\right] \\ &\times \vartheta(|q_{\parallel}| - m_*^2\omega/(2\varepsilon\varepsilon')).\end{aligned} \tag{18.12}$$

After integration over ω in (18.12) we have for the total intensity of radiation

$$I^{\text{mcoh}} = \frac{\alpha}{4}\sum_{\mathbf{q}} |G(\mathbf{q})|^2\frac{\mathbf{q}_{\perp}^2}{q_{\parallel}^2}F(z), \quad z = \frac{2\varepsilon|q_{\parallel}|}{m_*^2},$$

where

$$\begin{aligned}F(z) &= \left[\ln(1+z) - \frac{z(2+3z)}{2(1+z)^2}\right]\left[1 - \frac{8(3+z)}{(2+\rho)z^2}\right] \\ &+ \frac{z}{(1+z)^2}\left[\frac{8}{2+\rho} + \frac{z(3+2z)}{3(1+z)}\right]\end{aligned} \tag{18.13}$$

Equations (18.12) and (18.13) are not more complicated than (18.2) and (18.3) but have a significantly broader range of applicability. For $\vartheta_0 \gg V_0/m (\rho \ll 1)$, Eqs. (18.12) and (18.13) go over respectively into (18.2) and (18.3).

The transition $m \to m_*$ corresponds to the transition to the effective mass of a particle. It means that at ultrahigh energies when the parameter χ_s is large, the basic harmonic is still radiated for the angles of incidence ϑ_0 where the condition $\rho(\vartheta_0) < \chi_s^{2/3}$ is valid. It means that Compton scattering of only one equivalent photon dominates. On the other hand for $\rho \geq 1$ the momentum of a particle transverse to the mean velocity is comparable with the mass of the particle. For these conditions Compton scattering takes place in the bound states of the transverse motion and the process depends mainly on the mean value of the electron velocity $\mathbf{V}$. This velocity in turn depends on the particle energy ε and effective mass m_* so as in the usual kinematics a particle velocity depends on its energy and mass, i.e.

$$\mathbf{V}^2 = 1 - 1/\gamma_*^2 , \quad \gamma_*^2 = \varepsilon^2/m_*^2 .$$

This description ceases to be valid when the contribution of other harmonics becomes essential. Then $\rho(\vartheta_0) > \chi_s^{2/3}$ and several equivalent photons are scattered by electron at once. In this case the modified theory of CBS is inapplicable.

Using this analysis and the general expressions (16.17) and (16.18) the spectral distribution of the polarized photons in the modified theory of CBS has been found

$$\begin{aligned} dI^{\mathrm{mcoh}}(\mathbf{e}) = \frac{\alpha\omega d\omega}{8\varepsilon^2} \sum_{\mathbf{q}} |G(\mathbf{q})|^2 \frac{1}{q_\parallel^2} \vartheta\left(|q_\parallel| - \frac{m_*^2\omega}{2\varepsilon\varepsilon'}\right) \Bigg\{ &(\mathbf{e}\mathbf{e}*)\mathbf{q}_\perp^2 \\ \times \left[\varphi(\varepsilon) - \frac{2m^2\omega}{\varepsilon\varepsilon' q_\parallel^2}\left(|q_\parallel| - \frac{m_*^2\omega}{2\varepsilon\varepsilon' q_\parallel}\right)\right] & \\ + \frac{1}{2}\left(\frac{m_*^2\omega}{\varepsilon\varepsilon' q_\parallel}\right)[2|(\mathbf{q}_\perp\mathbf{e})|^2 - (\mathbf{e}\mathbf{e}^*)\mathbf{q}_\perp^2]\Bigg\} . & \end{aligned} \tag{18.14}$$

It is easily convinced that for $\rho \to 0 (m_*^2 \to m^2)$ Eq. (18.14) turns into Eq. (18.5). When calculations are carried out using Eqs. (18.11)–(18.14), special attention must be given for a satisfactory allowance for the contributions of higher-order planes when the vector $\boldsymbol{\nu} = \mathbf{v}/|\mathbf{v}_0|$ is accidentally in one of these planes.

At relatively low particle energies and not too large angles of incidence, when $z \ll 1$, we find that the expansion of Eq. (18.13) in terms of z and retention of the leading term (proportional to z^2) gives a classical expression for the total radiation intensity. The higher terms of the expansion represent the quantum corrections and the values of these corrections can be used to estimate the range of validity of the classical expression. In estimating these corrections we shall bear in mind that the terms with additional powers of z are proportional to $q_\parallel$ and contain in the sum of Eq. (18.13), $q_\parallel$ in the numerator. Consequently, the contribution to these sums is generally made by higher harmonics (when summation in Eq. (18.13) can be replaced approximately by integration) and is truncated at the amplitude of thermal vibrations $u_1(|\mathbf{q}|u_1 \leq 1)$, so that these corrections depend effectively on $\chi_u = \chi_s a_s/u_1$ and the relative values of the correction terms are determined by the parameter

$$\lambda_u = \frac{\varepsilon}{m^2}\frac{\vartheta_0}{u_1} = \frac{m}{V_0}\vartheta_0\chi_u\,. \tag{18.15}$$

Let us consider the spectral intensity of the radiation in the extreme quantum limit, when the parameter in question has the value $\lambda = 2\varepsilon|q_\parallel|_{\min}/m^2 \sim \varepsilon\vartheta_0/m^2 a_s \gg 1$. In this case the maximum of the intensity of the coherent radiation is attained at such values of ϑ_0 where the theory of coherent emission becomes invalid. Bearing in mind that if $\lambda \gg 1$ and $\vartheta_0 \sim V_0/m$, then $\chi_s \sim \lambda \gg 1$, we can conveniently use a modified theory of coherent emission when the spectral distribution becomes

$$\frac{dI^{\rm mcoh}}{d\omega} \simeq \frac{\alpha\omega}{4\varepsilon^2}\left(\frac{\varepsilon}{\varepsilon'} + \frac{\varepsilon'}{\varepsilon}\right)\sum_{\mathbf{q}} |G(\mathbf{q})|^2 \frac{\mathbf{q}_\perp^2}{q_\parallel^2}\vartheta(u_0 - u)\,, \tag{18.16}$$

where

$$u_0 = \frac{2\varepsilon|q_\parallel|}{m^2(1+\rho/2)}\,, \qquad u = \frac{\omega}{\varepsilon'}$$

The spectral intensity described by Eq. (18.16) has a sharp maximum near the end of the spectrum at $\omega \simeq \varepsilon\lambda(1+\lambda)^{-1}$ with a relatively small (in terms of λ^{-1}) width $\Delta\omega \sim \varepsilon(1+\rho/2)/\lambda = m^2(1+\rho/2)/2|q_\parallel|_{\min}$:

$$\left(\frac{dI}{d\omega}\right)_{\max} = \frac{\alpha\rho|q_\parallel|_{\min}}{2(2+\rho)} \tag{18.17}$$

We can see that both the amplitude of the maximum of the spectral curve and its effective width are independent of the energy. However, in calculating the total radiation intensity we find, as it follows from Eq. (18.13), that there is a

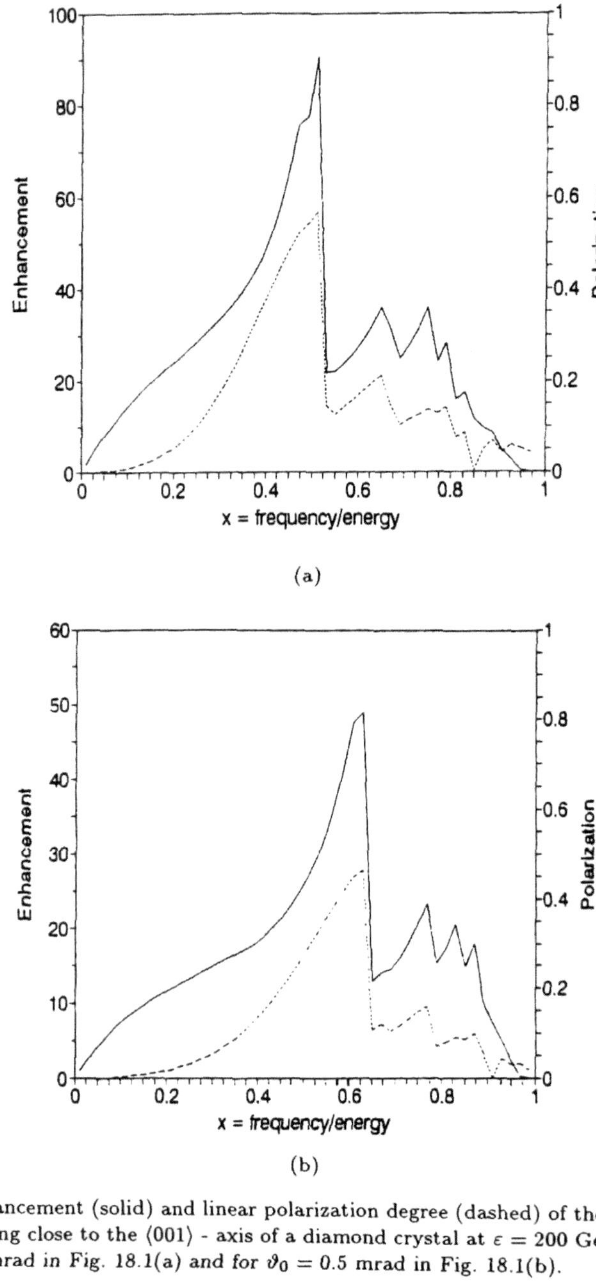

Fig. 18.1 Enhancement (solid) and linear polarization degree (dashed) of the radiation from particles moving close to the $\langle 001 \rangle$ - axis of a diamond crystal at $\varepsilon = 200$ GeV, $\sin\phi = 0.25$; for $\vartheta_0 = 0.3$ mrad in Fig. 18.1(a) and for $\vartheta_0 = 0.5$ mrad in Fig. 18.1(b).

weak (logarithmic) dependence of the intensity on the particle energy:

$$I \simeq \frac{\alpha}{8} m^2 \rho \ln \lambda \,. \tag{18.18}$$

This behavior of the spectral and total intensities of the emitted radiation is typical in the theory of undulator radiation in the extreme quantum case and is in full agreement with the results discussed above (see Subsec. 5.2). These features of the emission spectrum are associated with the dominant contribution of the lowest harmonics to the spectrum of equivalent photons in the case when $\lambda \gg 1$ and then its discreteness becomes important. The behavior of the spectrum of equivalent photons in crystals differs radically from the corresponding spectrum in the Bethe–Heitler case, when this spectrum is continuous and extends without any boundary to the arbitrarily low frequencies. The spectral curves (Fig. 18.1) for various values of λ illustrate these features of the spectral distribution of the radiation.

18.3. Hard Photon Emission from High Energy Electrons and Positrons in Single Crystals

In this section, a theory is stated, which describes hard photon emission from charged particles in single crystals taking into account both the change of the effective electron mass in the plane wave field and the influence of the field caused by the continuous potential of crystal planes. The results presented have been obtained by Baier, Katkov, Strakhovenko, 1992. The photon emission process can be represented as the Compton scattering of equivalent photons on a charged particle. The theory of coherent bremsstrahlung (CBS) is exhausted by this picture both qualitatively and quantitatively (see e.g. Eq. (4.8) in Baier, Katkov, Strakhovenko, 1989). The 4-momentum conservation in Compton effect yields:

$$\frac{\omega}{\varepsilon} = \frac{s}{1 + s + (\gamma \vartheta_{ph})^2}\,, \tag{18.19}$$

here $\gamma = \varepsilon/m$; ϑ_{ph} is the angle between the momentum of outgoing photon and the initial particle momentum $\mathbf{p}$; $s = 2(qp)/m^2, (qp) = q_0\varepsilon - \mathbf{qp}$. The momentum q_μ of an initial (equivalent) photon is in the crystal case one of reciprocal lattice vectors $\mathbf{q}$ taken in a corresponding frame of reference. In the crystal c.m. frame one has: $s = 2\varepsilon|q_\parallel|/m^2 \equiv \lambda$ (see above), $q_\parallel = (\mathbf{qp})/|\mathbf{p}|$. The CBS intensity spectrum has the peak situated just at the boundary (according to Eq. (18.19)) frequency $\omega_b = \varepsilon s/(1+s)$. For different angles of incidence a crystal is characterized by a certain set of waves, every one having its own value of the quantity $q_{\parallel}$. The position of the most soft peak in the CBS intensity spectrum is determined by the minimal value $|q_{\parallel}|_{\min}$ dependent only on the geometric factors. It turns out that at a given angle of incidence to the axes ϑ_0 the parameter $|q_{\parallel}|_{\min}$ takes a minimal value, when the particle momentum $\mathbf{p}$ is in the $(1\bar{1}0)$ plane forming the angle $\vartheta_0 \ll 1$ to the axes $\langle 001\rangle$ for $fcc(d)$-structure (C,Si,Ge) and to the axes $\langle 110\rangle$ for bcc-structure (W, Fe). Thus, other things being equal, one gets the most hard CBS spectrum when particles are moving under plane channeling conditions or not high above the plane potential barrier. In this connection, the influence of the field corresponding to the continuous plane potential on the radiation process should be taken into account. This field in the first place leads to an ordinary radiation at planar channeling giving a contribution to the soft part of the spectrum and secondly affects the emission of hard quanta.

The influence of an external field on the Compton effect was considered in Sec. 5.6. In this section a straightforward derivation of the formulae is given for the radiation spectrum taking into account an action of an external field and the change of the charged particle mass in the plane wave field.

Let us start from the general formula (16.16) derived in Sec. 16 for the radiation spectrum at sufficiently high energy for arbitrary crystal orientation

$$\frac{dW_\gamma}{d\omega} = -\frac{i\alpha m^2}{2\pi\varepsilon^2}\int \frac{d^3r_0}{V}F(\mathbf{r}_0,\vartheta_0)\int_{-\infty}^{\infty}\frac{d\tau}{\tau - i0}$$
$$\times\left[\left(\sum_{\mathbf{q}}\frac{G(\mathbf{q})\mathbf{q}_\perp}{mq_\parallel}\sin(q_\parallel\tau)e^{-i\mathbf{q}\mathbf{r}(t)}\right)^2\beta - 1\right]e^{-iA}, \qquad (18.20)$$

here $\beta = \varepsilon/\varepsilon' + \varepsilon'/\varepsilon$, $\varepsilon' = \varepsilon - \omega; q_\parallel = (\mathbf{q}\mathbf{v}_0)$, $\mathbf{v}_0$ is the mean velocity of the particle traversing a crystal; $\mathbf{r}(t) = \mathbf{r} + t\mathbf{v}_0; \alpha = 1/137$ is the fine structure constant; the crystal potential is used in the form of $U(\mathbf{r}) = \sum G(\mathbf{q})e^{-i\mathbf{q}\mathbf{r}}$;

$$A = \frac{m^2\omega\tau}{\varepsilon\varepsilon'}\left\{1 + \sum_{\mathbf{q},\mathbf{q}'}\frac{G(\mathbf{q})G(\mathbf{q}')}{m^2 q_\parallel q'_\parallel}(\mathbf{q}_\perp\mathbf{q}'_\perp)\left[\frac{\sin(q_\parallel + q'_\parallel)\tau}{(q_\parallel + q'_\parallel)\tau}\right.\right.$$
$$\left.\left. - \frac{\sin(q_\parallel\tau)}{q_\parallel\tau}\,\frac{\sin(q'_\parallel\tau)}{q'_\parallel\tau}\right]e^{-i(\mathbf{q}+\mathbf{q}',\mathbf{r}(t))}\right\}. \qquad (18.21)$$

The function $F(\mathbf{r},\vartheta_0)$ in Eq. (18.20) gives the coordinate distribution depending on the parameters of the incident beam and generally speaking on time.

Below the crystal is assumed to be thin i.e. the kinetics of this distribution is neglected. As it was shown in Sec. 5.6, the problem of a photon emission by a charged particle in simultaneously acting a plane wave field and an external one can be solved analytically only if the radiation in a pure external field has the magnetic bremsstrahlung nature. In the case under consideration the parameter $\rho_c = 2U_0\varepsilon/m^2$ where U_0 is the potential well depth should be large: $\rho_c \gg 1$, and the angle Ψ of the particle velocity $\mathbf{v}_0$ to the plane should satisfy the condition $\Psi \leq U_0/m$. When the angle ϑ_0 of the vector $\mathbf{v}_0$ to the chosen axis is small, the contribution to the sum $\sum_q$ is given by the vectors $\mathbf{q}_t$ situated in the plane perpendicular to this axis, i.e. all the sums become two-dimensional. The continuous planar potential corresponds to the terms in the sum with $\mathbf{q}_t$ perpendicular to the plane. To isolate this contribution we split all the sums into two parts: $\sum_q \to \sum_{q_t}^{(F)} + \sum_{q_t}^{(W)}$, where the sum $\sum_{q_t}^{(F)}$ contains $\mathbf{q}_t$ for which $q_{\parallel} \to 0$, when $\Psi \to 0$, and the sum $\sum^{(W)}$ contain all the other $\mathbf{q}_t$. The functions contained in the sum $\sum^{(F)}$ can be expanded in the powers of $|q_{\parallel}|\tau \ll 1$. After that this sum is expressed in terms of a continuous planar potential $U(y) = \sum_{\mathbf{q}_y} G(\mathbf{q}_y)e^{-iq_y y}$, where y is the coordinate perpendicular to the plane. Carrying out these transformations and changing $\tau \to \tau\varepsilon/m^2$ we get from Eq. (18.20)

$$\frac{dW_\gamma}{dx} = \frac{i\alpha m^2}{2\pi\varepsilon}\int \frac{d^2\rho_0}{S_0}F(\rho_0, \vartheta_0)\int_{-\infty}^{\infty}\frac{d\tau}{\tau - i0}[1 + \beta\tau^2(\chi + \mathbf{C})^2]$$
$$\times \exp\{-iu\tau[1 + \tau^2\chi^2/3 - 4(\chi\mathbf{A}) - (\mathbf{B})^2 + \rho/2 + \tilde{\Sigma}]\}, \quad (18.22)$$

here ρ_0 is the coordinate vector in the plane perpendicular to the chosen axis, S_0 is the area per one axis, over which coordinate averaging is carried out, the quantity β is determined in Eq. (18.20), $u = \omega/\varepsilon', x = \omega/\varepsilon$,

$$\begin{Bmatrix}\mathbf{A}\\ \mathbf{B}\\ \mathbf{C}\end{Bmatrix} = \sum_{\mathbf{q}_t}^{(W)} \frac{G(\mathbf{q}_t)\mathbf{q}_t}{mq_{\parallel}}e^{-i\mathbf{q}_t\mathbf{r}(t)}\begin{Bmatrix} if_2(s)/s\\ f_1(s)\\ isf_1(s)/2\end{Bmatrix}, s = 2q_{\parallel}\varepsilon/m^2, s' = s(q'_{\parallel})$$

$$f_1(s) = 2\sin(s\tau/2)/s\tau, \quad f_2(s) = \cos(s\tau/2) - f_1(s), \quad (18.23)$$

$$\chi = \frac{i\varepsilon}{m^3}\sum_{\mathbf{q}_y} G(\mathbf{q}_y)\mathbf{q}_y e^{-iq_y y}, \rho/2 = \sum_{\mathbf{q}_t, q_{\parallel}\neq 0}^{(W)} |G(\mathbf{q}_t)/m|^2(\mathbf{q}_t/q_{\parallel})^2$$

$$\tilde{\sum} = \sum_{\mathbf{q}_t, \mathbf{q}'_t}^{(W)} \frac{G(\mathbf{q}_t)G(\mathbf{q}'_t)}{m^2 q_{\parallel}q'_{\parallel}}(\mathbf{q}_t\mathbf{q}'_t)[f_1(s + s') - \delta_{\mathbf{q}_t+\mathbf{q}'_t,0}]e^{-i(\mathbf{r}(t), \mathbf{q}_t+\mathbf{q}'_t)}$$

Note, that $|\chi| = \varepsilon m^{-3}|dU/dy| = \varepsilon m^{-3}|eE(y)| = \varepsilon|E(y)|/(mE_0)$, where $E(y)$ is a magnitude of the electric field of the plane at distance y from it, $E_0 = m^2/e \cong 1.32 \cdot 10^{16}$V/cm is the critical field. This coincides with the standard definition of the quantum parameter χ characterizing the properties of magnetic bremsstrahlung. If we denote time averaging by $\langle \cdots \rangle$, then the parameter $\rho/2$ characterizing in particular the intensity of an electromagnetic wave can be written in the form of

$$\rho/2 = \left\langle \sum_{\mathbf{q}_t, \mathbf{q}'_t}^{(W)} \frac{G(\mathbf{q}_t)G(\mathbf{q}'_t)}{m^2 q_\parallel q'_\parallel}(\mathbf{q}_t\mathbf{q}'_t) f_1(s+s') e^{-i(\mathbf{r}(t), \mathbf{q}_t + \mathbf{q}'_{\| lower(1.5pt)t})} \right\rangle,$$

i.e. the average value of the first term in the double sum $\Sigma^{(W)}$ in Eq. (18.21) has been isolated in Eq. (18.22). Note that the average value of the quantity $\tilde{\Sigma}$ is equal to zero. Equation (18.22) completely corresponds to Eq. (5.77), if we substitute in the latter a set of crystal waves in place of one wave and let $\zeta = 0$, since in Eq. (18.22) the initial particle is assumed to be unpolarized. The exponential function in Eq. (18.22) can be expanded in the wave fields (terms containing $\Sigma^{(W)}$). At $\vartheta_0 \gg \vartheta_v$ such expansion is valid (see the discussion in Sec. 4 of Ref. 1) at arbitrary photon frequency. When $|s| \gg 1$ it is valid in a neighborhood of the hard peak position for $\vartheta_0 \geq \vartheta_v$ as well. Carrying out this expansion, we find that after time averaging:

$$\begin{aligned} \frac{dW_\gamma}{dx} &= \frac{i\alpha m^2}{2\pi\varepsilon} \int \frac{dy}{d_{pl}} F(y, \vartheta_0) \int_{-\infty}^{\infty} \frac{d\tau}{\tau - i0} \\ &\quad \times \exp[-iu\tau(1 + \rho/2 + \tau^2\chi^2/3)] \cdot \left\{ 1 + \beta\tau^2\chi^2 + \sum_{\mathbf{q}_t}^{(W)} \frac{|G(\mathbf{q}_t)|^2}{m^2 q_\parallel^2} \mathbf{q}_t^2 \right. \\ &\quad \times [(1 + \beta\tau^2\chi^2) \cdot (iu\tau f_1^2(s) - 4\Lambda_1(f_2(s)u\tau\mu)^2) \\ &\quad \left. + \beta(\sin(s\tau/2) + 4iu\Lambda_1\tau^2\mu^2 s f_2(s)) \cdot \sin(s\tau/2)] \right\}, \end{aligned} \qquad (18.24)$$

here d_{pl} is the distance between the planes, near which the motion occurs; $\Lambda_1 = 2(\chi\mathbf{q}_t)^2/\chi^2\mathbf{q}_t^2, \mu = \chi/|s|$. In what follows we assume the parameter s to be positive: $s \to |s| = 2\varepsilon|q_\parallel|/m^2$, since the integrand in Eq. (18.24) is an even function of s. In the case of one wave considered in Sec. 5.6, the parameter Λ_1 has the following form: $\Lambda_1 = 1 + \lambda_3\cos(2\phi_1)$, where ϕ_1 is the angle of the wave polarization vector to the external field direction, and λ_3 is the degree of wave linear polarization. A comparison of this expression for Λ_1 with one obtained in a crystal case proves, that equivalent photons in a crystal are completely polarized ($\lambda_3 = 1$) along the direction $\mathbf{q}_t$. The parameter μ characterizes (see Sec. 5.6) the strength of an external field in the sense of its influence on the Compton scattering of wave photons (on the coherent radiation in a crystal). The order of magnitude of the parameter μ is the ratio of the momentum transferred by the external field to a charged particle on the wavelength to its mass.

By integrating in Eq. (18.24) over τ it is convenient to go over to the variable $z = |s|\tau$ and to carry out the integration by parts of terms containing high powers of τ. All the integrals are taken by means of the formula:

$$\frac{1}{2\pi i}\int_{-\infty}^{\infty}\frac{dz}{z-i0}e^{-i(zt+\lambda^2 z^3/3)} = \frac{\vartheta(t)}{\pi\sqrt{3}}\int_a^{\infty} dx K_{\frac{1}{3}}(x) + \vartheta(-t)$$

$$\times\left[1 - \frac{1}{3}\int_a^{\infty} dx(J_{\frac{1}{3}}(x) + J_{-\frac{1}{3}}(x))\right] \quad (18.25)$$

and relations obtained from it by differentiating with respect to t. In Eq. (18.25) J_ν are the Bessel functions of the first kind, and K_ν are the Bessel functions of the third kind; $a = 2t^{3/2}/3\lambda$. The integration over τ in Eq. (18.24) yields:

$$\frac{dW_\gamma}{dx} = \frac{\alpha m^2}{\pi\varepsilon\sqrt{3}}\int\frac{dy}{d_{pl}}F(y,\vartheta_0)\Bigg\{(1+\rho/2)\cdot\beta\cdot K_{\frac{2}{3}}(z_0)$$

$$-\int_{z_0}^{\infty} dx K_{\frac{1}{3}}(x) + \sum_{\mathbf{q}_t}^{(W)}\frac{|G(\mathbf{q}_t)|^2}{m^2 q_{\|}^2}\mathbf{q}_t^2$$

$$\times\Bigg\{(g_1+\nu)\int_{z_+}^{\infty} dx K_{\frac{1}{3}}(x) - 2g_1\int_{z_0}^{\infty} dx K_{\frac{1}{3}}(x)$$

$$+(g_1-\nu)\Bigg[\vartheta(u-u_0)\int_{z_-}^{\infty} dx K_{\frac{1}{3}}(x) + \pi\sqrt{3}\vartheta(u_0-u)$$

$$\times\left(1-\frac{1}{3}\int_{z_-}^{\infty} dx(J_{\frac{1}{3}}(x) + J_{-\frac{1}{3}}(x))\right)\Bigg]$$

$$+\mu\nu(3\chi/2u)^{1/3}\Bigg[(g_4+4\nu g_2)z_+^{1/3}K_{\frac{1}{3}}(z_+) - 2\beta(1-\Lambda_1)z_0^{1/3}$$

$$\times K_{\frac{1}{3}}(z_0) + (g_4 - 4\nu g_2)\Bigg(\vartheta(u-u_0)K_{\frac{1}{3}}(z_-)$$

$$+\frac{\pi}{\sqrt{3}}\vartheta(u_0-u)(J_{\frac{1}{3}}(z_-)+J_{-\frac{1}{3}}(z_-))\Bigg)\ z_-^{1/3}\Bigg]$$

$$+\nu^2(3\chi/2u)^{2/3}\Bigg[(g_2-g_3)z_+^{2/3}K_{\frac{2}{3}}(z_+)+2(g_2+g_3)z_0^{2/3}K_{\frac{2}{3}}(z_0)$$

$$+(g_2-g_3)z_-^{2/3}\Bigg(\vartheta(u-u_0)K_{\frac{2}{3}}(z_-)+\frac{\pi}{\sqrt{3}}\vartheta(u_0-u)$$

$$\times(J_{-\frac{2}{3}}(z_-)-J_{\frac{2}{3}}(z_-))\Bigg)\Bigg]\Bigg\}\Bigg\}, \tag{18.26}$$

here $\vartheta(z)$ is the Heaviside function: $\vartheta(z)=1$ for $z>0$ and $\vartheta(z)=0$ for $z<0$, the notations are introduced:

$$z_0=(2u/3\chi)(1+\rho/2)^{3/2}\,,\quad z_\pm=z_0|1\pm u_0/u|^{3/2}\,,\quad u_0=|s|/(1+\rho/2)\,,$$
$$g_1=\beta/4+\nu^2(1+\rho/2-4\Lambda_1\mu^2)\,,\quad g_2=\Lambda_1[\beta(1+\rho/2)-1]\,, \tag{18.27}$$
$$g_3=1-4\Lambda_1\beta\mu^2\,,\quad g_4=\beta(1+\Lambda_1)\,,\quad \nu=u/|s|\,.$$

The terms in Eq. (18.26) outside of the sum $\Sigma^{(W)}$ represent the usual magnetic bremsstrahlung in the field of a plane slightly corrected owing to the presence of waves. In the argument z_0 this correction is reduced to the substitution: $m^2 \to m_*^2 = m^2(1+\rho/2)$ since the parameter $\chi \sim m^{-3}$ The same correction happens in the parameter u_0, determining a boundary of the Compton spectrum at a given frequency of the initial photon (at a given value of the quantity $q_\parallel$ in our case). A transition from Eq. (18.26) to the case of one wave is provided by the substitution:

$$\Sigma^{(W)}|G(\mathbf{q}_t)\mathbf{q}_t|^2/m^2q_\parallel^2 \to \rho/2=\xi_0^2\,,$$

where ξ_0^2 is the intensity of this wave, and if the change of the mass m to the effective value m_* is not taken into account, by the additional expansion of the terms describing the magnetic bremsstrahlung in $\rho/2$. An expression obtained in this way is consistent with the results of Sec. 5.6. In the case of a strong external field ($\mu=\chi/s\gg 1$) the Eqs. (18.22) and (18.26) are reduced to the formulae of the magnetic bremsstrahlung, where one should add together vectors of accelerations induced by an external field and a wave field, to determine the parameter χ.

In the case of a weak field ($\mu \ll 1$), which is practically more interesting, it influences most strongly on the coherent radiation spectrum near the intensity peak of the latter, i.e. at $u \simeq u_0$. Equation (18.26) implies at $\mu \ll 1$ the spectrum variation of the order of unity for photon frequencies satisfying the condition $|u - u_0| \leq \mu^{2/3}$. This is physically stipulated by the fact, that the length of the coherent radiation formation $l_w \sim |u - u_0|^{-1}$ (see Sec. 5.6). Then the length l_w becomes larger than the corresponding formation length in an external field $l_f \sim \chi^{-1}(1 + \chi/u)^{1/3}$ at sufficiently small distances from the boundary frequency. This means the change of the radiation mechanism within the mentioned range of frequencies. The considered change of the intensity spectrum is illustrated by Fig. 5.4 in one wave and weak external field case. Then the probability (first Born approximation in a wave) can be represented as:

$$\frac{dW_\gamma}{dx} = \frac{dW_\gamma^{(F)}}{dx} + \xi_0^2 \frac{dW_\gamma^{(W)}}{dx}, \tag{18.28}$$

where the first term describes the magnetic bremsstrahlung induced by an external field. Note, that Eq. (18.26) has the same form, but in contrast to Eq. (18.28) a set of waves is inserted, and as it was explained above, the term $dW_\gamma^{(F)}/dx$ is slightly changed. The part of the probability proportional to ξ_0^2 in Eq. (18.28) can be interpreted by the usual rules in terms of a cross section

$$d\sigma = \frac{8\pi\alpha\varepsilon}{m^4 s} dW_\gamma^{(W)}, \tag{18.29}$$

i.e. we can discuss a change of the Compton scattering in an external field. Equation (18.26) implies that this cross section at μ tending to zero is reduced to the ordinary Compton formula multiplied by the factor

$$f(u) = \frac{\vartheta(u - u_0)}{\pi\sqrt{3}} \int_{z_-}^{\infty} dx K_{\frac{1}{3}}(x) + \vartheta(u_0 - u) \times \left[1 - \frac{1}{3} \int_{z_-}^{\infty} dx (J_{\frac{1}{3}}(x) + J_{-\frac{1}{3}}(x))\right],$$

which noticeably differs from unity in the narrowing (at $\mu \to 0$) neighborhood of the point $u = u_0$, where $f(u_0) = 1/3$. The function $I^{(W)} = \xi_0^2 \cdot x dW_\gamma^{(W)}/dx$

is shown in Fig. 5.4 for two values of the parameter $\mu = \chi/s$. It was calculated taking into account all the terms in the probability $dW_\gamma^{(W)}$. It is seen, that oscillations arise in the monotone (at $\chi = 0$) spectral curve, the peak is lowering and moving to the left from the point $u = s$, and the tail of the distribution appears in the kinematically prohibited (at $\chi = 0$) frequency domain $u > s$. The scale of these alterations increases with an increase of the parameter μ.

According to what has been said in the beginning of Sec. 18.3, we consider the radiation process for initial particle velocity $\mathbf{v}_0$ being nearly aligned with the axis $\langle 001 \rangle$ for the crystal structure $fcc(d)$. Let ϑ_0 be the angle of the vector $\mathbf{v}_0$ to this axis, and ϕ_0 be the angle of the projection $\mathbf{v}_0$ onto the plane (001) with respect to the plane $(1\bar{1}0)$. Let us remind, that l is the lattice constant and $\mathbf{q} = 2\pi/l \cdot (m, n, k)$, where m, n and k are integers. For the axis $\langle 001 \rangle$ the quantities $\mathbf{q}_t$ and $q_{\parallel}$ have the form

$$\mathbf{q}_t = \frac{2\pi}{l}(m, n, 0)\,, \quad q_{\parallel} = \frac{\pi\sqrt{2}}{l} \sin\vartheta_0[(m+n)\cos\phi_0 + (m-n)\sin\phi_0]\,.$$

Similarly, in the case of the crystal structure (bcc) we consider the radiation from particles nearly aligned with the $\langle 110 \rangle$ axis. Then ϑ_0 is the angle of velocity $\mathbf{v}_0$ to this axis, and ϕ_0 is the angle of the projection $\mathbf{v}_0$ onto the plane (110) to the plane $(1\bar{1}0)$, and

$$\mathbf{q}_t = \frac{2\pi}{l}(m, -m, k)\,, \quad q_{\parallel} = \frac{2\pi}{l} \sin\vartheta_0[k \cdot \cos\phi_0 + m\sqrt{2}\sin\phi_0]\,.$$

Let us start with the ordinary situation, when the influence of the plane field on the radiation process can be neglected. In general, for $\vartheta \ll 1$ there is always a plane containing the chosen axis and having an angle ψ to the velocity $\mathbf{v}_0$ less than the corresponding critical angle of planar channeling ψ_c. However, it is possible to avoid the main planes. For example, at $\sin\phi_0 = 0.25$ (see Fig. 18.1) the angle ψ to the plane $(\bar{3}50)$ for $\vartheta_0 = 0.3$ mrad is only 2.34 μrad being less than the corresponding ψ_0 value. But since this plane is very weak ($U_0 \simeq 1.2$ eV, $\chi_{\max} \simeq 0.037$), its field influence on coherent radiation is small. It is seen in Fig. 18.1, that the peak becomes harder, and their heights diminish, when the angle ϑ_0 is increased. The peaks shift to the right due to the increase of the parameter s which is proportional to ϑ_0 (see Eq. (18.19)). The diminution of their heights is owing to the decreasing strength of interaction (decreasing value of the parameter $\rho/2$). The arising radiation has an essential linear polarization. The spectral distribution of the polarization degree is also shown in Fig. 18.1. Note that the maxima of enhancements (peaks) are

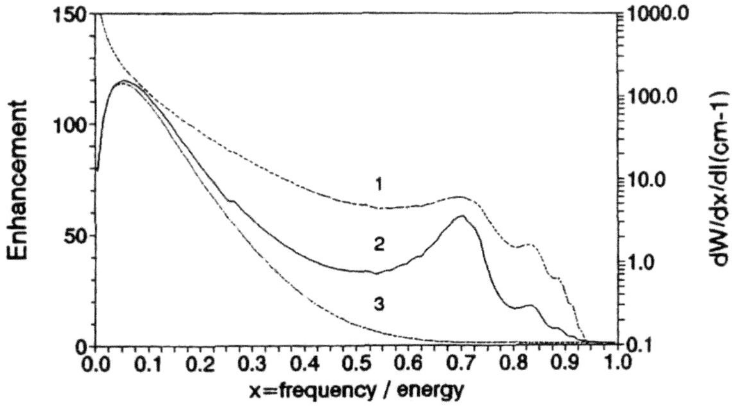

(a)

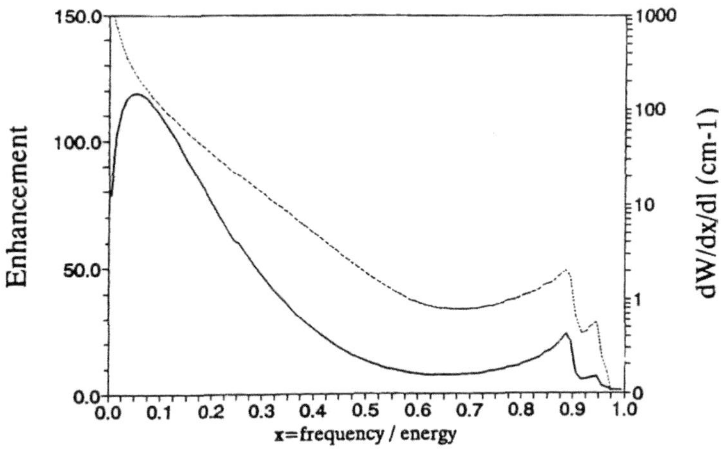

(b)

Fig. 18.2. Probability (dashed (1)), enhancement (solid (2)) and constant field contribution (dotted (3) in Fig. 18.2(a) for the radiation from particles moving close to the $\langle 001 \rangle$ axis and the $(1\bar{1}0)$ plane of a diamond crystal at $\varepsilon = 200$ GeV, $\psi \geq 15$ mrad; for $\vartheta_0 = 0.2$ mrad in Fig. 18.2(a) and $\vartheta_0 = 0.6$ mrad in Fig. 18.2(b).

situated at the same frequencies as those of the polarization degree. The enhancement means the ratio of the spectrum calculated for the crystal to the theoretical value of the spectrum in the corresponding amorphous material. The similar change in shape of the hard part of the spectrum occurs when the angle ϑ_0 increases for particles moving close to the strong $(1\bar{1}0)$-plane. This can be seen in Fig. 18.2. The calculation of curves presented in this figure was carried out by Eq. (18.26) for uniform distribution, i.e. letting $F(y, \vartheta_0) = 1$. The particles moving over a potential barrier have approximately a distribution like that. The angle ψ of the vector $\mathbf{v}_0$ to the plane $(1\bar{1}0)$ is expressed by means of the angles ϑ_0 and $\phi_0 : \psi \simeq \vartheta_0 \sin\phi_0$. The calculation shows that the shape of the spectrum is practically not changed by the variation of the angle ϕ_0, keeping the inequality $\psi_c < \psi < U_0/m$ valid. On the other hand for $\psi < \psi_c$ one should take into account the particle flux redistribution ($F(y, \vartheta_0) \neq 1$), while for $\psi > U_0/m$ the radiation induced by the plane field necessarily loses its magnetic bremsstrahlung nature, and the developed description of the field influence on coherent bremsstrahlung becomes inapplicable. The hard peak in Fig. 18.2 even though for $\vartheta_0 = 0.2$ mrad is situated to the right of that in Fig. 18.1 for $\vartheta_0 = 0.5$ mrad. This obviously illustrates the assertion about the maximal hardness of the radiation at the motion close to the $(1\bar{1}0)$ plane. For $x = \omega/\varepsilon < 0.5$ the radiation induced by the plane field gives the main contribution to the intensity. This is seen in comparison of the summary curve 2 in Fig. 18.2(a) with the curve 3 representing the contribution of the magnetic bremsstrahlung in Eq. (18.26) and the Bethe–Heitler (incoherent) spectrum.

The inclusion of the latter explains, why the enhancement given by the curve 2 is tending not to zero but to unity at increasing photon frequency. Since the parameter s is proportional to ε, the hard peak according to Eq. (18.19) moves to the right proportionally to $s/(1+s)$ at increasing energy. But since the value of the parameter s is already not small ($s_{\min} \simeq 3$ for $\varepsilon = 200$ GeV), this shift turns out to be non-linear with respect to ε. On the contrary, the soft peak, as it should be at magnetic bremsstrahlung, moves to the right linearly with respect to ε and becomes wider. These changes of the spectrum shape are seen in Figs. 18.2(a) and 18.3. We emphasize, that the heights of the both soft and hard peaks are practically independent of the energy ε. This assertion holds at least up to the energy of order of a few TeV

It is well known, that the radiation from channeled electrons and positrons turned out to be different, since they move in different potentials. Channeled electrons are focused around the planes (flux redistribution) moving on the average in the stronger field. On the contrary the coherent radiation from electrons and positrons is usually the same. In the case under consideration the field of the plane causes a difference in the radiation from e^+ and e^- in the hard part of the spectrum as well. The spectra from e^+ and e^- are shown

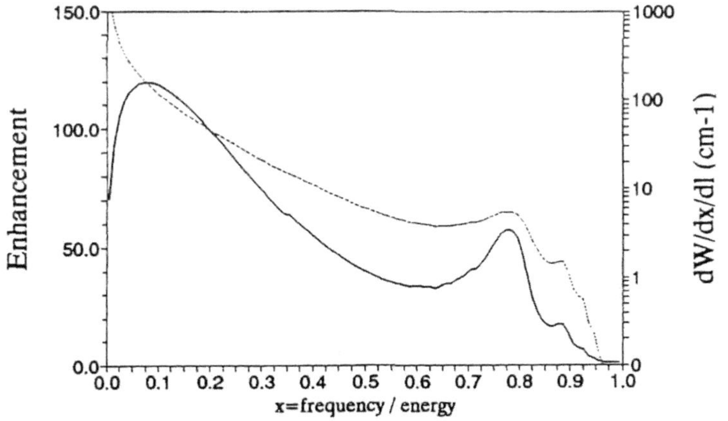

(a)

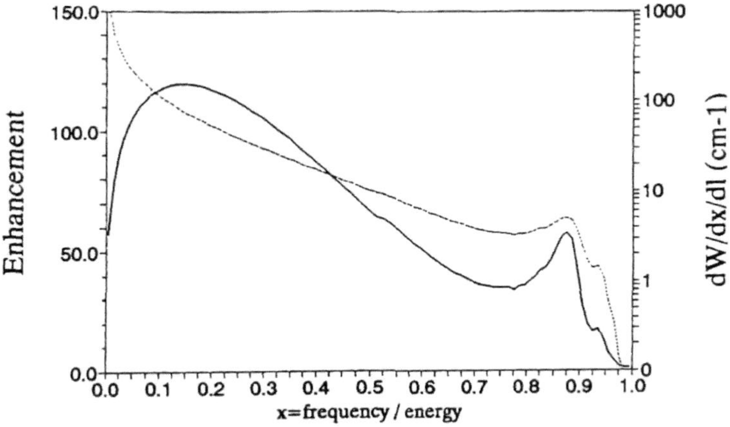

(b)

Fig. 18.3. Probability (dashed) and enhancement (solid) under conditions of Fig. 18.2(a); at $\varepsilon = 300$ GeV (a) and 600 GeV (b).

in Fig. 18.4 for Ge and W crystals at $\phi_0 = 0(\psi = 0)$. The difference of the two spectra (from e^+ and e^-) is about $10 - 15\%$ in the vicinity of the hard peak position being maximal for W crystal. The magnetic bremsstrahlung contribution is negligibly small in this frequency region, therefore the mentioned difference is completely stipulated by the influence of the field on the coherent radiation process, which is actually Compton scattering of equivalent

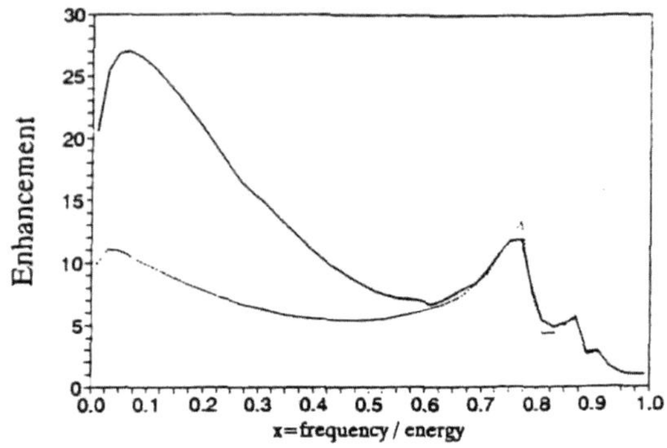

(a)

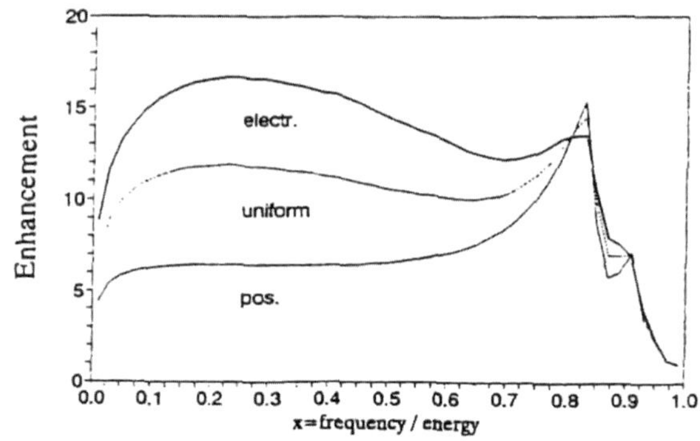

(b)

Fig. 18.4. (a) Enhancement for electrons (solid) and positrons (dashed) channeled in Ge $(1\bar{1}0)$ plane close to $\langle 001 \rangle$ axis at $\varepsilon = 200$ GeV, $\vartheta_0 = 0.4$ mrad; (b) enhancement for electrons and positrons channeled in $W(1\bar{1}0)$ plane close to $\langle 110 \rangle$ axis at $\varepsilon = 200$ GeV, $\vartheta_0 = 0.5$ mrad, the dashed curve is calculated for uniform distribution.

photons on particles of the incident beam. The experimental observation of this difference could verify the possibility of the external field influence on one of the fundamental QED-processes, what has not been still succeeded. High crystal quality is necessary for the experiment of such kind, in particular the mosaic spread and the accuracy of the angle ψ measurement should be a few times less than the critical angle ψ_c. The requirements are not so rigid for an observation of the spectra like these in Figs. 18.2 and 18.3, but in every case the

Table 18.1. Parameters characterizing continuous potential of the $(1\bar{1}0)$ plane and some properties of the radiation from particles moving close to this plane at $\varepsilon = 200$ GeV; ψ_c is the critical angle of planar channeling, U_0 is the depth of the potential well, χ_{max} is the maximal value of the parameter χ, ϑ_0 is the angle of the incident particle momentum with respect to axes ($\langle 001 \rangle$ for fcc (d) structure and $\langle 110 \rangle$ for bcc structure), L_{ch} is the radiation length under indicated conditions, $x_{max} = \omega_{max}/\varepsilon$ is the hard peak position, r is the magnitude of the enhancement at $x = x_{max}$, W is the magnitude of the probability $d^2W/dx\,dl$ at $x = x_{max}$.

Crystal type	ψ_c μrad	U_0/m μrad	χ_{max}	ϑ_0 mrad	L_{ch}^{-1} cm^{-1}	x_{max}	r	W cm^{-1}
C *fcc(d)*	15.3	46	0.24	0.2	4.3	0.71	58.4	6.1
Si *fcc(d)*	14.6	42	0.17	0.3	2.1	0.72	22.0	2.9
Ge *fcc(d)*	18.2	65	0.28	0.4	4.4	0.76	11.8	6.1
Fe *bcc*	26.2	135	0.55	0.35	11.8	0.77	20.3	13.5
W *bcc*	36.5	250	1.28	0.5	29.9	0.81	14.4	46.2

mosaic spread should not exceed the value U_0/m. The values of the angles ψ_c and U_0/m are listed in Table 18.1 for the $(1\bar{1}0)$ plane of some crystals. Some characteristics of radiation calculated for uniform distribution ($F(y, \vartheta_0) = 1$) are also given in Table 18.1.

An increase of the nucleus charge Z leads for identical crystal structures to decreasing enhancement for all the frequencies, so that the maximal enhancement is achieved in diamond. The magnitude of the radiation probability in the vicinity of the peak may serve as a more objective characteristic of a crystal fitness for hard photons production. It is seen in Table 18.1, that this probability does not have monotone behavior with respect to Z being maximal for tungsten. The maximal values of the parameter χ depending on a distance from the plane are also given in Table 18.1. It is clear, that the effective χ value at averaging over this distance is less than χ_{max}. Note, that the field of the plane accounts for about two-thirds of the quantity L_{ch}^{-1} characterizing the rate of the particle energy loss for all examples in Table 18.1.

Thus, the most preferable situation for hard photons production is realized, when incident particles move close to the $(1\bar{1}0)$ plane of a crystal. If the soft part of the spectrum arising under these conditions is undesirable, the crystal orientation can be chosen like that by obtaining Fig. 18.1, i.e. one should avoid the main planes.

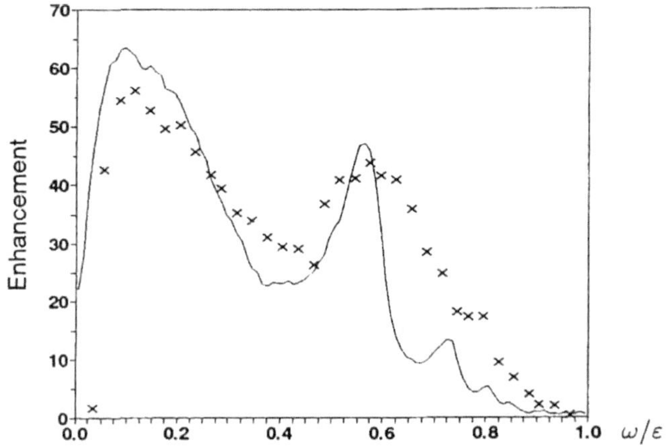

(a)

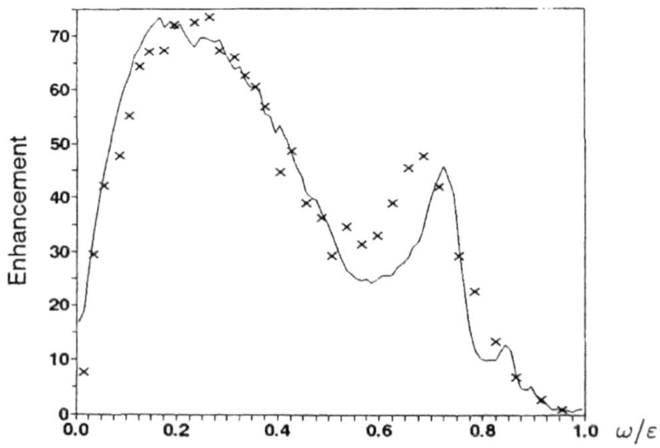

(b)

Fig. 18.5. Enhancement versus relative photon energy for 70 GeV (a), 150 GeV (b) and 240 GeV (c) electrons incident on a 500 ηm thick diamond crystal at an angle of 0.3 mrad to the $\langle 100 \rangle$ axis. Angular regions around (110) planes are 0–11 μrad (a), 0–8 μrad (b), and 0–5 μrad (c).

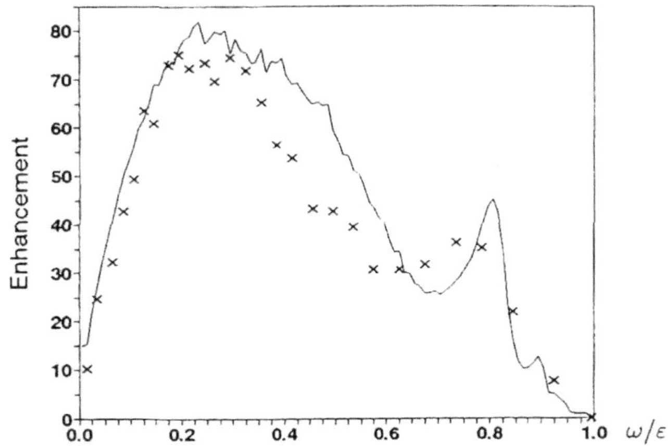

(c)

Fig. 18.5. (*Continued*)

For the first time very pronounced high-energy photon peaks were observed in Medenwaldt *et al.*, 1992, where the radiation from 70, 150 and 240 GeV electrons incident at 0.1–1.0 mrad to the axis in diamond and Si crystals were investigated. The energy of the photons in the peaks was found to be 0.7–0.8 times the initial particle energy with yields of 50 times the Bethe–Heitler one (in diamond). The peaks consist of single photons. Some experimental results from this paper and the corresponding theoretical curves are presented in Fig. 18.5.

19. ORIENTATIONAL DEPENDENCE OF THE RADIATION INTENSITY

19.1. Discussion of Results

The orientational dependence of the spectral distribution and the total intensity of the radiation is given by Eq. (16.16). The calculation carried out using this formula is a fairly complex computational task. On the other hand, it is relatively simple if we use approximate expressions (17.6), (17.7) and (17.15) derived from Eq. (16.16) for small angles of incidence, in the constant field approximation with a correction, and for large angles, correspondingly Eqs. (18.12) and (18.13) obtained in the modified theory of CBS. Therefore, the general formula (16.16) should be used only in the intermediate range of angles $\vartheta_0 \sim V_0/m$. Moreover, using the interpolation procedure described in Appendix D we can, in the first approximation, confine ourself by these simple formulae to find the orientational dependence of the emitted radiation.

We shall consider this problem in detail. As pointed out already, the range of validity of Eq. (16.16) is independent of the parameter ρ_c and it is the same as the range of validity of the semiclassical approximation. If $\vartheta_0 \leq \vartheta_c$, the orientational dependence for a thin crystal is related to a redistribution of the flux of incident particles, i.e., it is related to a change in the distribution function $F(\mathbf{r}, \vartheta_0)$ (see Eq. (16.10) versus the angle ϑ_0. This distribution is different for electrons $(-)$ and positrons $(+)$, so that the emission of the radiation in the range $\vartheta \leq \vartheta_c$ depends on the sign of the particle charge. If $\vartheta_0 = 0$, we find from Eq. (16.10) for the arbitrary axially symmetric potential (see Eqs. (9.65) and (9.66))

$$\begin{gathered} F_{ax}^{(-)}(\rho^2, 0) = \ln(x_0/x)\,, \quad F_{ax}^{(+)}(\rho^2, 0) = \ln[x_0/(x_0 - x)]\,, \\ x = \rho^2/a_s^2, \end{gathered} \tag{19.1}$$

If $\chi_s \leq 1$, the contribution to the total intensity comes from the region $\eta \leq x \leq 1$, where $F_{ax}^{(-)} \geq \ln x_0$ and $F_{ax}^{(+)} \sim 1/x_0$, from which it follows that as compared with the case of the uniform distribution over the transverse coordinate, the intensity $I^{(-)}$ is enhanced at $\vartheta_0 = 0$ by a factor of approximately $\ln x_0$ and the intensity $I^{(+)}$ is weakened approximately by a factor $1/x_0$. When the angle ϑ_0 is increased, the distribution $F_{ax}^{(-)}(\rho^2, \vartheta_0)$ rapidly becomes uniform, whereas $F_{ax}^{(+)}(\rho^2, \vartheta_0)$ varies quite slowly. The behavior of the distribution $F(\mathbf{r}, \vartheta_0)$ becomes important when we discuss the emission of radiation from a beam of electrons characterized by a finite angular width $\Delta\vartheta_0$, since already for $\Delta\vartheta_0 \simeq \vartheta_c/2$ approximately 80% of the electrons are in above-barrier states and, naturally, are distributed uniformly over the transverse coordinates. Moreover, we should bear in mind that in view of the mentioned redistribution of the particle flux, multiple scattering of electrons is enhanced and that of positrons is weakened in the range $\vartheta_0 \leq \vartheta_c$, so that we can have a situation when a crystal is thin for positrons but not for electrons.

If $\vartheta_0 \geq \vartheta_c$, the distribution of particles (16.10) is uniform for electrons as well as for positrons. In this case when parameter χ_s is small ($\chi_s \ll 1$), the radiation intensity has a classical form (the term i_1^M in Eq. (17.16)) and does not depend on ϑ_0 (there is a plateau on the curve of the orientational dependence) up to the angles when the quantum effects in the emission of radiation appear (see the discussion in Sec. 18.2). As soon as the angles of incidence reach the value $\vartheta_0 \geq V_0/m\chi_s$, a drop of the intensity described by Eq. (18.3) starts. The intensity of the radiation emitted by electrons and positrons is the same for $\vartheta_0 > \vartheta_c$. The regions of validity of Eqs. (17.15) and (18.3) for $\chi_s \ll 1$ overlap, so that for $\chi_s \ll 1$ the orientational dependence of the radiation intensity is described completely by these simple formulae. As the parameter χ_s increases, this plateau narrows, and disappears completely when $\chi_s \sim 1$. If $\chi_s \geq 1$, the intensity of electron radiation has a maximum, as previously, at $\vartheta_0 = 0$ and decreases monotonously as ϑ_0 grows. For positrons, the intensity $I^{(+)}$ has a minimum as before at $\vartheta_0 = 0$; then with increasing ϑ_0 the intensity $I^{(+)}$ grows, achieves its maximum at $\vartheta_0 \simeq \vartheta_c$ and coincides with that for electrons when $\vartheta_0 > \vartheta_c$. It should be pointed out that with an increase in χ_s the contribution to the intensity for $\vartheta_0 < \vartheta_c$ comes from increasing values of x. Consequently, the difference between the intensities of the radiation emitted by electrons and positrons decreases and for sufficiently high values of $\chi_s \sim x_0^{3/2} \gg 1$ the difference practically disappears. Figure 19.1

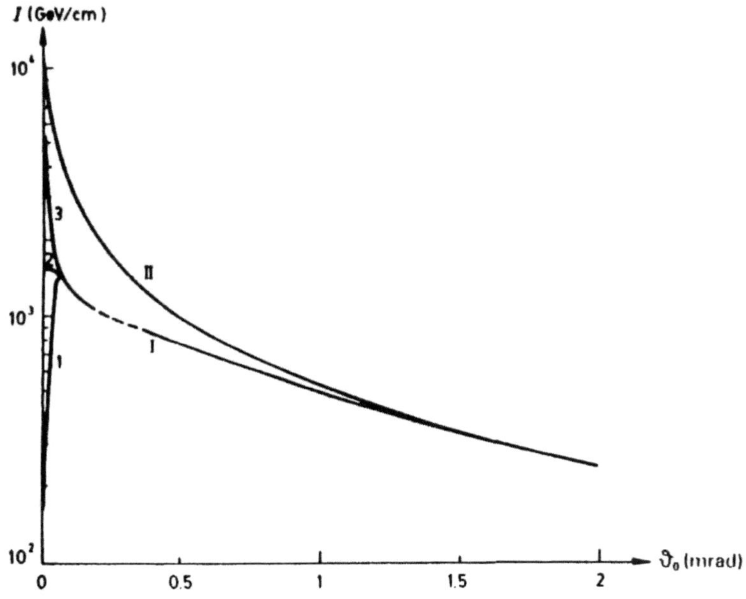

Fig. 19.1. Orientational dependence of the radiation intensity in Ge $\langle 110 \rangle$ at $T = 100$ K and $\varepsilon = 150$ GeV.

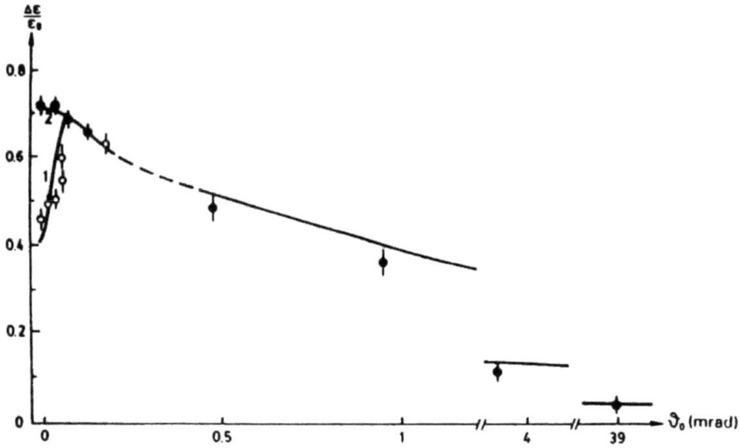

Fig. 19.2. Relative energy losses versus angle of incidence with respect to the axis $\langle 110 \rangle$ of Ge crystal ($T = 100$ K, $\varepsilon_0 = 150$ GeV, $L = 1.4$ mm).

demonstrates the typical behavior of the angular dependence of the radiation intensity in a thin crystal, for $\chi_s \sim 1$. It is seen that for $\vartheta_0 < \vartheta_c$ the intensity depends considerably on the charge of the particles and on the angular width of the beam $\Delta\vartheta_0$: curves 1 and 3 represent, correspondingly, the radiation intensity from positrons and electrons for $\Delta\vartheta_0 = 0$; curve 2 is obtained for the uniform distribution over particle coordinates corresponding to a wide angular spread of the incident beam ($\Delta\vartheta_0 \sim \vartheta_c$). All these curves coincide at $\vartheta_0 > \vartheta_c$. Curve I has been obtained according to formulae of the present chapter, curve II is calculated according to the CBS theory. The comparison of curve I and curve II gives the region of validity the CBS theory. The curves in Fig. 19.2 have been obtained from calculations based on the derived results. The experimental data are taken from Belkacem *et al.*, 1985 where the orientation dependence of a relative energy loss $\Delta\varepsilon/\varepsilon_0$ has been measured. The crystal used in the experiment can not be considered to be thin for electrons because in an appropriate amorphous medium the angle of multiple scattering $\langle\vartheta_s\rangle \simeq 35\mu$rad $\simeq \vartheta_c/2$. Furthermore, the angular width of the incident beam was $\Delta\vartheta_0 \simeq \mp 30\,\mu$rad and the quality of the crystal could correspond to the overall curvature less than 10 μrad. Since for the indicated effective divergence $\Delta\vartheta_0 \sim$ 30–40 μrad most of the electrons immediately populate the above-barrier states, and the intensified multiple scattering forces the rest of electrons to leave the channel, the electron distribution was taken as uniform.

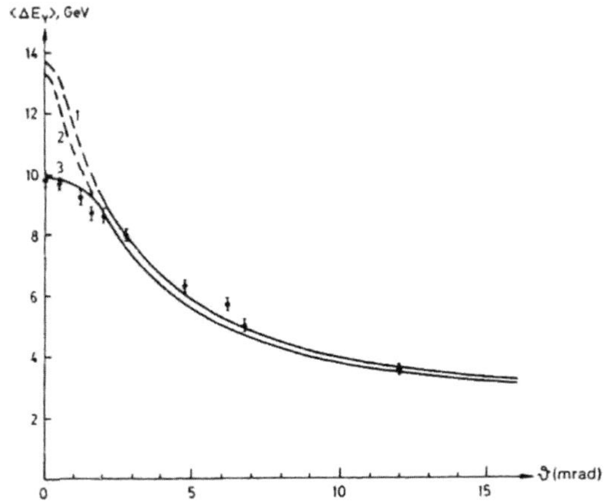

Fig. 19.3. Energy losses of electrons with the initial energy $\varepsilon_0 = 28$ GeV traversing a 0.3 mm W-crystal near ⟨111⟩ axis direction.

On the other hand, for positrons in the first approximation one can neglect multiple scattering and the calculation has been performed for a thin crystal with the effective width of the incident beam taken into account. Note that in thinner crystals and for a smaller angular divergence of the incident beam the orientation dependence of the intensity at $\vartheta_0 < \vartheta_c$ can be observed for electrons as well. Further calculations of the energy losses have been performed as in Sec. 17.5. The solid parts of the curves are calculated according to formulae (17.15) and (18.13). The dashed part is the result of interpolation.

The orientation dependence of energy losses from 28 GeV electrons traversing a W-crystal was investigated in Baier *et al.*, 1989, 1992c. Experimental data from Baier *et al.*, 1992c and results of theoretical calculations are shown in Fig. 19.3. It is seen that curve 3 calculated with regard for experimental conditions: divergence and multiple scattering of initial electron beam, absorption of created photons is rather different as compared with the curve 2 which takes into account the absorption of photons only and with the curve 1 calculated for the ideal case.

19.2. A Simple Model of Kinetic Phenomena for the Description of Radiation in Single Crystals

An additional interest in the radiative process was stimulated by the observation of a hard peak in the electron energy loss distribution (Belkacem *et al.*, 1986a,b). The origin of this peak is due to multiple photon emission by an electron traversing a crystal of sufficient thickness and to the actual addition of photon energies by the detector. This explanation was given in Baier, Katkov, Strakhovenko, 1988e and then confirmed by the reanalysis of the experimental data by Belkacem *et al.*, 1988.

The radiation characteristics contain indirect information about the kinetics of electrons traversing a crystal. The kinetics is mainly determined by the multiple scattering and the radiation processes. The magnitude of the multiple scattering and the radiation energy losses can be characterized by an energy dependent dechanneling length $l_d(\varepsilon)$ and a radiation length $L_{\text{ch}}(\varepsilon)$. At some energy ε_{cr} these two lengths become equal (e.g. for Ge, $\varepsilon_{cr} \simeq$ 40–60 GeV depending on the axis and the temperature). At $\varepsilon < \varepsilon_{cr}$, the multiple scattering dominates in the kinetic, but at $\varepsilon > \varepsilon_{cr}$, the role of the radiation becomes more and more important. In turn, a correct description of the radiation is possible only with an adequate allowance for the kinetics. This is especially important for crystals of intermediate thickness, where the radiation from particles initially captured into a channel is essential. There is a group of electrons, which while traversing a crystal, remains all the time close to the axis. In this region of distance ρ from the axis both multiple scattering and the radiation (including bremsstrahlung) are enhanced by a factor of about 10 compared to the above-barrier values. Multiple scattering results on the average in an increase of the transverse energy $\varepsilon_\perp$ in contrast to the radiation effects decreasing it. These processes together cause a strong intermixing of electrons belonging to this group over $\varepsilon_\perp$, leading to a uniform distribution over the transverse coordinate ρ for them.

In this section a simple model, based on the approach developed in Baier, Katkov, Strakhovenko, 1988e, 1991c is presented, which permits one in particular to predict the energy loss spectra of electrons traversing a crystal of intermediate (and large) thickness. As in previous sections, we use an axially symmetric approximation $U(x)$ of a crystal potential in the region around a single axis with an area $S_0 = \pi r_0^2 = (nd)^{-1}$, where n is the density of atoms in the crystal, d is the mean distance between them along the axis. The explicit form of the potential $U(x)$ is given by Eq. (9.13). A generalization to an arbitrary potential $U(\rho)$ is straightforward. We remind that for electrons

$$U(x) = V_0 \ln[(1 + 1/\eta)/(1 + 1/(x + \eta))]\,, \tag{19.1}$$

where $x = \rho^2/a_s^2$; V_0, a_s, η are fitting parameters (see Table 9.1).

Table 19.1. Characteristic thicknesses of the phenomenon. All quantities were calculated at room temperature, thickness L_0, L_1 are given in mm.

Crystal	Axis	ε_{cr} [GeV]	$\chi_s(\varepsilon_{cr})$	$\varepsilon_0 = \varepsilon_{cr}$	
				L_0	L_1
C	$\langle 111\rangle$	48	0.06	0.31	1.5
Si	$\langle 110\rangle$	53	0.17	0.14	1.9
Ge	$\langle 110\rangle$	60	0.29	0.06	0.9
W	$\langle 111\rangle$	30	0.86	0.007	0.25

		$\varepsilon_0 = 150$ GeV		$\varepsilon_0 = 250$ GeV		$\varepsilon_0 = 500$ GeV	
		L_0	L_1	L_0	L_1	L_0	L_1
C	$\langle 111\rangle$	0.46	2.1	0.5	2.3	0.58	2.6
Si	$\langle 110\rangle$	0.21	2.7	0.23	2.9	0.27	3.3
Ge	$\langle 110\rangle$	0.1	1.3	0.11	1.4	0.14	1.7
W	$\langle 111\rangle$	0.016	0.5	0.019	0.6	0.025	0.76

For a given $\varepsilon_\perp$ the scattering magnitude can be characterized by the parameter $t_d(\varepsilon_\perp, \varepsilon)$:

$$t_d(\varepsilon_\perp, \varepsilon) = l_d(\varepsilon)[\vartheta(\varepsilon_\perp - U_0) + \vartheta(U_0 - \varepsilon_\perp)(x + \eta)/(x_0 + \eta)]\,. \qquad (19.2)$$

Here $l_d(\varepsilon) = (\alpha/2\pi)U_0\varepsilon L_{\text{rad}}/m^2$ is the conventional dechanneling length, $U_0 \equiv U(x_0)$ is the potential well depth, $x_0 = r_0^2/a_{\text{a}}^2$ is the boundary value of x, L_{rad} is the radiation length in a corresponding amorphous medium. The dependence of x in Eq. (19.2) on $\varepsilon_\perp$ is defined by the equation $U(x) = \varepsilon_\perp$. Relation (19.2) means that for channeled ($\varepsilon_\perp < U_0$) particles the length $t_d(\varepsilon_\perp, \varepsilon)$ decreases in practice ($\eta \ll 1$) in proportion to the area $S(\varepsilon_\perp)/S_0$. A similar behavior holds for the radiation length in the continuous potential $T_{\text{ch}}(\varepsilon_\perp, \varepsilon) = \varepsilon/I(\varepsilon, \varepsilon_\perp)$, where $I(\varepsilon, \varepsilon_\perp)$ is the radiation intensity from an electron with total energy ε and transverse energy $\varepsilon_\perp$:

$$T_{\text{ch}}(\varepsilon_\perp, \varepsilon) = L_{\text{ch}}(\varepsilon)[\vartheta(\varepsilon_\perp - U_0) + \vartheta(U_0 - \varepsilon_\perp)(2x + 1)/(2x_0 + 1)]\,. \qquad (19.3)$$

Here $L_{\text{ch}}(\varepsilon)$ is the radiation length for above-barrier ($\varepsilon_\perp > U_0$) particles.

Let us separate all the particles into two groups using an initial distribution over $\varepsilon_\perp$. The particles having initial value of $\varepsilon_\perp$ smaller than some boundary value $\varepsilon_\perp^b$ belong to the first group (I), all others, including above-barrier ones, belong to the second group (II). The radiation from electrons of the second group is described by the simulation of a cascade under the conditions of above-barrier motion, where, within the constant field approximation, the probability (and the intensity, see Eq. (19.3)) of photon emission is independent of $\varepsilon_\perp$. In view of the strong intermixing in the transverse phase space of the particles belonging to the first group, photon emission from them is described by the simulation of a cascade at some intermediate value of $\varepsilon_\perp = \varepsilon_\perp^{ef}$ for all the particles of the first group. Then radiation characteristics (energy loss spectra, photon multiplicity, etc.) are presented as a sum of the contributions from the two groups weighted by the fraction of particles in each of them

$$F = aF_I + (1-a)F_{11}\,. \tag{19.4}$$

The constant field approximation is valid for incident angles ϑ_0 smaller than $\vartheta_v = V_0/m$. Then the orientation dependence of the yield F in this interval of ϑ_0 is entirely due to that of the fraction a of electrons belonging to the first group, since the parameters $\varepsilon_\perp^b$ and $\varepsilon_\perp^{ef}$ depend only on the initial electron energy ε_0, the type and thickness of the crystal.

Let us formulate the prescription.

(1) The parameter $\varepsilon_\perp^{ef}$ is given by the solution of the following equation

$$\exp[L/T_{\text{ch}}(\varepsilon_\perp^{ef},\varepsilon_0)] = t_d(\varepsilon_\perp^{ef},\varepsilon_0)/L\,, \tag{19.5}$$

where L is the crystal thickness, the functions $t_d(\varepsilon_\perp,\varepsilon), T_{\text{ch}}(\varepsilon_\perp,\varepsilon)$ are defined by Eqs. (19.2) and (19.3).

(2) The equation for the parameter $\varepsilon_\perp^b$ reads

$$\frac{T_{\text{ch}}(\varepsilon_\perp^b,\varepsilon_0)}{L}\{\exp[L/T_{\text{ch}}(\varepsilon_\perp^b,\varepsilon_0)]-1\} = 1 + \frac{t_d(\varepsilon_\perp^b,\varepsilon_0)V_0^2}{2T_{\text{ch}}(\varepsilon_\perp^b,\varepsilon_0)U_0^2} \tag{19.6}$$

It is important, that only the above-barrier values of the radiation $L_{\text{ch}}(\varepsilon_0)$ and the dechanneling $l_d(\varepsilon_0)$ lengths at the initial electron energy ε_0 are sufficient for calculation of the parameters $\varepsilon_\perp^{ef}$ and $\varepsilon_\perp^b$. The meaning of Eq. (19.5) is in a self-consistent determination of the $\varepsilon_\perp$ domain boundary, below which the transverse phase space distribution becomes uniform for a given crystal thickness. A detailed analysis of this problem was performed in Sec. 10.2 without allowance for the radiation action.

In this case as it was shown at the end of Sec. 10.2 the uniform distribution is formed for $\varepsilon_\perp \le \varepsilon_\perp^0$, where $\varepsilon_\perp^0$ is determined by the equation

$$t_d(\varepsilon_\perp^0,\varepsilon_0) = L\,. \tag{19.7}$$

Taking into account radiation energy losses which can be estimated for the energy region considered by a simple formula

$$\varepsilon(L) = \varepsilon_0 \exp(-L/T_{\rm ch}(\varepsilon_\perp^0, \varepsilon_0)) \tag{19.8}$$

using the fact that $t_d \propto \varepsilon$ (see Eq. (19.2)), and setting $\varepsilon_\perp^0 \equiv \varepsilon_\perp^{ef}$ one immediately gets Eq. (19.5).

It is the point, that as a consequence of radiation and multiple scattering, particles having initially the $\varepsilon_\perp$-value larger than $\varepsilon_\perp^{ef}$ are also involved in the first group. This "capture" occurs mainly because the increasing width $\Delta\varepsilon_\perp$ of some initial $\varepsilon_\perp$ level. In turn, large energy losses of particles belonging to the first group allows one to estimate this width, taking into account multiple scattering and Eq. (19.8):

$$d(\Delta\varepsilon_\perp)^2/dl = 2\varepsilon_\perp U_0/t_d(\varepsilon_\perp, \varepsilon) \simeq 2U_0^2 \exp(l/T_{\rm ch})/t_d(\varepsilon_\perp, \varepsilon_0)\,. \tag{19.9}$$

It is assumed, that $\varepsilon_\perp$-states, for which the mean value $\langle(\Delta\varepsilon_\perp)^2\rangle = \int_0^L dl(\Delta\varepsilon_\perp)^2/L$ is of the order of the characteristic potential well scale $\langle(\Delta\varepsilon_\perp)^2\rangle \simeq V_0^2$, are captured into the first group. This condition leads to Eq. (19.6).

For fixed ε_0 a peak like that observed in Belkacem *et al.*, 1986 appears in the electron energy loss distribution for crystal thickness L larger than L_0 defined by the equation $x_{ef}(L_0) = 1(U(x_{ef}) = \varepsilon_\perp^{ef}$, see Eq. (19.5)). Note, that the thickness L_0 under the conditions of the mentioned paper estimated in Baier, Katkov, Strakhovenko, 1988e by using similar arguments was then confirmed experimentally. For $L < L_0$ a separation of all particles into two groups becomes insufficient. Indeed, for very small thicknesses $L \ll L_0$ no intermixing occurs and one has to use a continuous initial distribution over $\varepsilon_\perp$ corresponding in our terms to an infinite number of groups. Thus, the proposed model is valid for $L > L_0$. Beginning with the thickness L_1, defined by the equation $x_{ef}(L_1) = x_0$, all particles belong to the second group.

In Table 19.1 are listed for some crystals thicknesses L_0 and L_1 which determine the interval, where the kinetics of initially channeled electrons is important leading to the appearance of a peak in the energy loss spectra. We expect the most pronounced peak at a crystal thickness of $L \simeq 2L_0$. The phenomenon exists for $\varepsilon > \varepsilon_{cr}$, whose values are tabulated as well. In addition the electron energy should be sufficiently high to preserve the essentially quantum character of the radiation process, determined by the parameter χ_s. In tungsten, the peak may be observed at relatively low energy $\varepsilon \simeq \varepsilon_{cr} = 30$ GeV, for which all necessary conditions are fulfilled. In Fig. 19.4 the dependence of the thickness L_0 on the initial energy is given. One can see the interrelation between L_0 in different crystals.

Let us compare the energy loss spectra calculated in this model to experimental data on radiation from 150 GeV electrons traversing aligned Ge $\langle 110 \rangle$ crystal. For $L = 0.14$ cm we have to put $a = 0$ as follows from Eq. (19.5). The orientation dependence of the total energy loss for this thickness was discussed above (see Fig. 19.2). The results for the energy loss spectra are presented in Fig. 19.5. A satisfactory agreement of the theory and data is seen. The

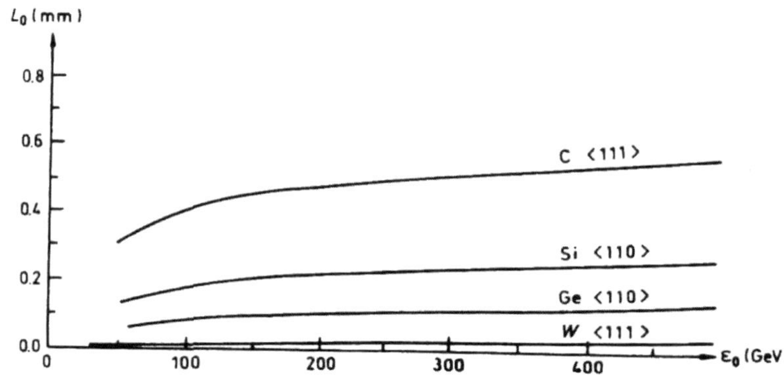

Fig. 19.4. Dependence of thickness L_0 on the initial particle energy.

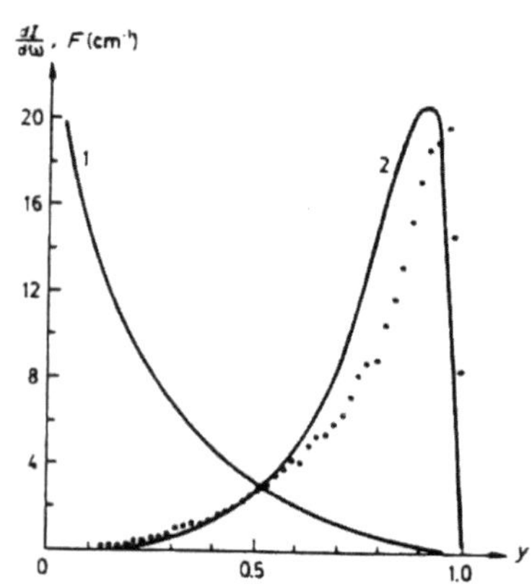

Fig. 19.5. Energy loss spectra (curve 2) for $a = 0$ and true photon spectra (curve 1) for 150 GeV electrons traversing 1.4 mm Ge $\langle 110 \rangle$ crystal, $T = 100$ K.

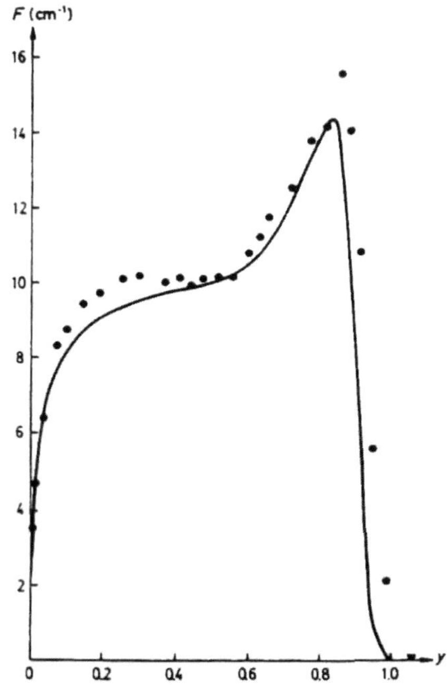

Fig. 19.6. Energy loss spectrum for 150 GeV electrons traversing a 0.185 mm Ge ⟨110⟩ crystal, $T = 100$ K. The Doyle–Turner potential was used.

drastic difference of the true photon spectra (curve 1) and the energy loss spectra (curve 2) is due to the high multiplicity rate.

In Fig. 19.6 the calculated energy loss spectrum in Ge ($T = 100$ K, ⟨110⟩ axis) crystal for $L = 0.185$ mm and experimental data are shown (Artru *et al.*, 1989). The Doyle–Terner potential without any additional approximations was used. By using our standard potential the region of $y < 0.5$ rises by about 20%. The main contribution to this region is made by particles of the second group.

In the same experiment the Ge crystal of thickness $L = 0.07$ mm was used which is smallar than the critical thickness L_0. Figure 19.7 shows the absence of the peak in the energy loss distribution.

The results of comparing the energy loss spectra calculated within the developed model to experimental data (Medenwaldt *et al.*, 1989a) on radiation from 150 GeV electrons traversing aligned Ge (axis $\langle 110\rangle, L = 0.205$ mm) and Si ($(\langle 110\rangle, L = 0.5$ mm) crystals are shown in Figs. 19.8 and 19.9. Our standard

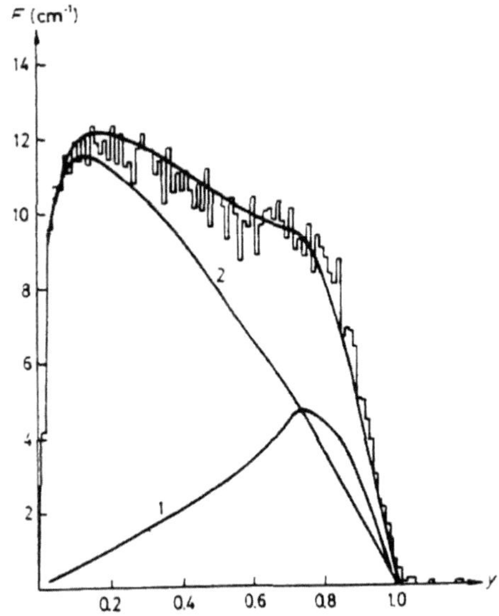

Fig. 19.7. Energy loss spectrum for 150 GeV electrons traversing a 0.07 mm Ge ⟨110⟩ crystal, $T = 100$ K. The curves 1 and 2 represent correspondingly contributions from electrons of the first and second group.

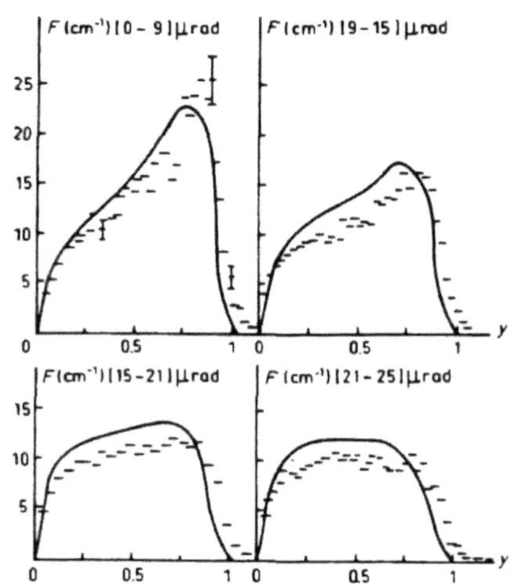

Fig. 19.8. Energy loss spectra for 150 GeV electrons traversing a 0.205 mm Ge ⟨110⟩ crystal at indicated incident angles.

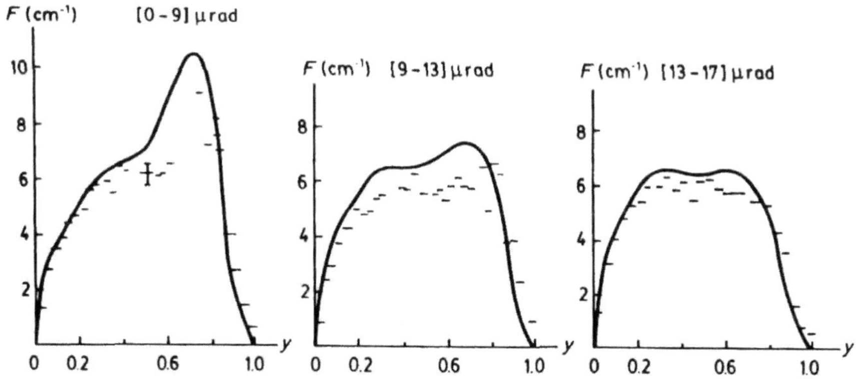

Fig. 19.9. As Fig. 19.8, but for a 0.5 mm Si ⟨110⟩ crystal.

potential was used in the calculations. Note that here the change in the shape of the energy loss spectra is completely due to the change of the fraction of particles captured into the first group (the quantity a in Eq. (19.4)).

In conclusion, it can be stated that the presented model gives not only a qualitative, but also quite a satisfactory quantitative description of the phenomena. The model permits one to find regions of thicknesses and energies where the peak appears and to predict the shape of the spectra without any fitting parameters. It allows one to estimate the physical situation, in particular for planning new experiments.

20. CASCADE PROCESSES IN THE AXES OF THE ALIGNED SINGLE CRYSTALS

20.1. Kinetic Equations

In Sec. 17.2 we have pointed out the existence of particular electron-photon showers in aligned single crystals. At small angles of incidence ϑ_0 of the initial photons or electrons, $\vartheta_0 \ll V_0/m$(V_0 is the scale of the axis or plane potential), a shower takes place in the fields of the separate axis (planes). Of significance is that at energies considerably exceeding the creation threshold of an electron–positron pair, the shower can generate over lengths (depending upon the substance and orientation) one or two orders of magnitude shorter than the shower in an appropriate amorphous substance, which is due to the standard interaction mechanism with individual nuclei (Bethe–Heitler (BH) mechanism). Note that this particular mechanism in question is the most effective mechanism of electromagnetic energy losses. In Chapters 3 and 4 the theory has been developed for pair creation and radiation which is valid at any angles of incidence and offers the possibility of tracing the dynamics of the showers in aligned single crystals at any ϑ_0. However, the largest effects occur at small ϑ_0 and in the fields of the axes so we will deal with this case.

The theory of cascade showers in an amorphous substance has been formulated and developed by Bhabha and Heitler, 1937; Carlson and Oppenheimer, 1937; Landau and Rumer, 1938. In these papers, kinetic equations have been derived to describe the evolution of a cascade and an analytical solution for these equations have been found in the case when allowance is made for the radiation and pair creation only (Landau and Rumer, 1938). Note that the probability of the latter have been found by Bethe and Heitler (see Secs. 7.3, 7.7). Subsequently, a great number of papers developing this theory have been published (for a review, see for example monograph (Rossi, 1952)).

Let us discuss the distinctive features of the shower in aligned single crystals.

1. As has already been mentioned the characteristic lengths of radiation and pair creation here may be essentially shorter than in an appropriate amorphous substance (in a nonaligned crystal).

2. There is a sharp boundary (threshold) ω_t with respect to photon energies below which the probability of pair photoproduction in the axis field falls off exponentially. We choose this in such a way that the photoproduction probability in the field is $W_e^F(\omega_t) = W_e^{BH}$. The values of ω_t are given in Table 12.1. At the same time, if the energy of the particle is $\varepsilon \sim \omega_t$ it emits intensely on account of the mechanism under discussion.

3. Within the $\omega > \omega_t$ region, both the pair creation probability $W_e^F(\omega)$ and the characteristic length of the energy losses $L_{\text{ch}}(\varepsilon) = \varepsilon/I(\varepsilon)(I(\varepsilon)$ is the radiation intensity) are energy-dependent, while for the Bethe–Heitler process the corresponding quantities W_e^{BH} and L_{rad} are constant.

4. At $\varepsilon \leq \omega_t$ the total radiation probability is roughly one order of magnitude higher than the inverse characteristic length of the energy losses caused by the radiation $L_{\text{ch}}^{-1}(\varepsilon)$ (note that this probability can be found using the magnetic-bremsstrahlung limit). This means that a great number of relatively soft quanta that exert no influence on the energy losses of the particle radiating them are emitted. Nevertheless, these photons can produce pairs due to

the Bethe–Heitler mechanism (i.e. with relative low probability). As a result, the particular shower evolves due to a mixed mechanism for which a large number of photons can produce a noticeable number of pairs, despite a low probability for the photoproduction process. It is then natural that the number of photons in the shower will be considerably greater than the number of charged particles. In what follows we will distinguish "hard" showers when the particle energy is $\varepsilon \gg \omega_t$ and "soft" showers when the particles energy is $\varepsilon \lesssim \omega_t$.

The indicated peculiarities of the evolution of the shower in an aligned crystal make it far different from the standard cascade shower in an amorphous substance. In this case, it is impossible to use both the results derived for this case and the method of solving a kinetic equation. Our analysis involves an analytical solution for a hard cascade and numerical results based on a Monte Carlo simulation of the process. Such calculations have been made for the hard and soft showers using the characteristics of the radiation and pair creation which were derived in Secs. 12 and 17.

An evolving electron-photon shower when passing through matter is characterized by the number of photons $N_\gamma(\omega, t)$ and of charged particles $N_e(\varepsilon, t)$ with energies $\omega(\varepsilon)$, at a given depth t. The functions N_γ and N_e satisfy the following kinetic equations (Landau and Rumer, 1938)

$$\begin{aligned}\frac{\partial N_\gamma(\omega,t)}{\partial t} &= -W(\omega)N_\gamma(\omega,t) + \int_\omega^\infty W_\gamma(\varepsilon,\omega)N_e(\varepsilon,t)d\varepsilon\,,\\ \frac{\partial N_e(\varepsilon,t)}{\partial t} &= -W_\gamma(\varepsilon)N_e(\varepsilon,t)\\ &\quad + \int_\varepsilon^\infty W_\gamma(\varepsilon',\varepsilon'-\varepsilon)N_e(\varepsilon',t)d\varepsilon'\\ &\quad + 2\int_\varepsilon^\infty W(\omega,\varepsilon)N_\gamma(\omega,t)d\omega\,, \end{aligned} \tag{20.1}$$

where $W(\omega, \varepsilon)$ is the differential probability for electron–positron pair creation by a photon with energy ω when the energy of one of the pair particles is ε (while that of the other is $\varepsilon' = \varepsilon - \omega$), $W(\omega)$ is the total pair creation probability, $W_\gamma(\varepsilon, \omega)$ is the differential emission probability of a photon with energy ω by a particle with energy ε, $W_\gamma(\varepsilon)$ is the total radiation probability. It is convenient to introduce new functions $\nu(\varepsilon, t)$ and $N_0(\varepsilon'; \varepsilon, t)$ defined as follows:

$$\frac{\partial N_0(\varepsilon';\varepsilon,t)}{\partial t} = -W_\gamma(\varepsilon)N_0(\varepsilon';\varepsilon,t) + \int_\varepsilon^\infty W_\gamma(\varepsilon'',\varepsilon''-\varepsilon)N_0(\varepsilon';\varepsilon'',t)d\varepsilon'', \tag{20.2}$$

$$N_0(\varepsilon';\varepsilon,0) = \delta(\varepsilon-\varepsilon'),$$
$$N_e(\varepsilon,t) = \int_0^t d\tau \int_\varepsilon^\infty \nu(\varepsilon',t-\tau)N_0(\varepsilon';\varepsilon,\tau)d\varepsilon'. \tag{20.3}$$

It is seen from the definitions (20.2) and (20.3), that the distribution function $N_0(\varepsilon';\varepsilon,t)$ is only connected with the radiation kinetics, while $\nu(\varepsilon,t)$ characterizes the source of the electron–positron pair. Note that it follows from (20.2) that

$$\int_0^{\varepsilon'} N_0(\varepsilon';\varepsilon,t)d\varepsilon = 1$$

with (20.2) and (20.3) substituted into (20.1) we find

$$\frac{\partial N_\gamma(\omega,t)}{\partial t} = -W(\omega)N_\gamma(\omega,t) + \int_\omega^\infty d\varepsilon \int_0^t d\tau \times \nu(\varepsilon,t-\tau)W_\gamma(\varepsilon,\omega,\tau), \tag{20.4}$$
$$\nu(\varepsilon,t) = 2\int_\varepsilon^\infty W(\omega,\varepsilon)N_\gamma(\omega,t)d\omega,$$

where

$$W_\gamma(\varepsilon,\omega,\tau) = \int_\omega^\varepsilon W_\gamma(\varepsilon',\omega)N_0(\varepsilon;\varepsilon',\tau)d\varepsilon' \tag{20.5}$$

Substituting the second of Eqs. (20.4) into the first, we have the final equation for $N_\gamma(\omega,t)$

$$\frac{\partial N_\gamma(\omega,t)}{\partial t} = -W(\omega)N_\gamma(\omega,t) + 2\int_\omega^\infty d\varepsilon \int_\varepsilon^\infty d\omega' \times \int_0^t d\tau W(\omega',\varepsilon)W_\gamma(\varepsilon,\omega,\tau)N_\gamma(\omega',t-\tau). \tag{20.6}$$

Equation (20.6) is exact. We will solve it making certain simplifications. An important step consists in the following substitution

$$W_\gamma(\varepsilon,\omega,\tau) = W_\gamma(\varepsilon(\tau),\omega)\,,$$
$$\frac{d\varepsilon(\tau)}{d\tau} = -I(\varepsilon(\tau))\,, \tag{20.7}$$

where $I(\varepsilon)$ is the radiation intensity of a particle with energy ε. In this case we neglect the energy dispersion of the electrons upon radiation. Strictly speaking, this is prohibitive in the calculation of the energy distribution of electrons itself in the course of the radiation. When calculating the energy losses the substitutions (20.7) however is quite justified[a] (see Sec. 17.5). Equations (20.6) and (20.7) are far simpler than the initial Eq. (20.1) and this makes it possible for further progress in their solution for the case when the radiation and pair creation probabilities are considerably different from the Bethe–Heitler ones.

The next step is to simplify the expressions for the differential radiation and pair creation probabilities. We define these using the equalities

$$W(\omega,\varepsilon) = W(\omega)\frac{\vartheta(\omega-\varepsilon)}{\omega}\,,$$
$$W_\gamma(\varepsilon,\omega) = L_{\text{ch}}^{-1}(\varepsilon)\frac{\vartheta(\varepsilon-\omega)}{\omega}\,. \tag{20.8}$$

Here $L_{\text{ch}}^{-1}(\varepsilon) = (-1/\varepsilon)\partial\varepsilon/\partial t$. The characteristic radiation length $L_{\text{ch}}(\varepsilon)$ and the pair production probability $W(\omega)$ are presented in Secs. 12 and 17. Below we will restrict ourselves to the situation where the constant field limit is applicable to these quantities. This limit is true under the condition that the angle of particle motion is $\vartheta \le V_0/m$, with respect to the axis. At $\varepsilon,\omega \ge \omega_t$ it suffices if the angle of incidence ϑ_0 of the initial particle satisfies this condition since the characteristic angles of radiation and pair creation are $\sim m/\varepsilon \ll V_0/m$ for this case. The analysis made for the Bethe–Heitler cascade shows that the replacement (20.8) also has little influence on the characteristics of the shower (the error constitutes a few percent). As a result, the substitutions (20.7) and (20.8) give rise for the Bethe–Heitler cascade to a 10% discrepancy with Landau–Rumer's solution. In the case of a hard shower in single crystals when $\omega,\varepsilon \gg \omega_t$, the probabilities $W_\gamma(\varepsilon,\omega)$ and $W(\omega,\varepsilon)$ are similar, in form, to the Bethe–Heitler ones (see Secs. 12 and 17) and, hence, the substitution (20.8) is admissible, with the indicated accuracy.

[a]For the standard Bethe–Heitler shower, this replacement results in an error of a few percent (cf., for example, ref Rossi, 1952).

Let us introduce the following variables and functions:

$$\eta = \ln\frac{\omega_0}{\omega}, \zeta = \ln\frac{\omega_0}{\varepsilon},$$
$$a(\zeta) = L_{\text{ch}}^{-1}(\omega_0 e^{-\zeta}), b(\eta) = W(\omega_0 e^{-\eta}), \tag{20.9}$$
$$N(\omega, t) = \frac{1}{\omega_0}N(\eta, t), n(\eta, t) = e^{-\eta}N(\eta, t),$$

where ω_0 is the energy of the initial particle, for example, of a photon. With Eqs. (20.7)–(20.9) substituted into (20.6), we get

$$n_t + b(\eta)n = 2\int_0^t d\tau \int_0^\eta d\zeta \int_0^\zeta d\eta' e^{\eta'-\zeta}$$
$$\times a(\zeta(\tau))b(\eta')n(\eta', t-\tau)\vartheta(\eta - \zeta(\tau)), \tag{20.10}$$

where $n_t \equiv \partial n/\partial t$. Of interest is the region $t \gg L_{\text{ch}}(\varepsilon_0)$ wherein the cascade has the time to evolve to a considerable extent. Here the characteristic values of the quantities are $\eta, \zeta \gg 1$. It is the region we will consider below. In the energy range where a hard cascade occurs, the functions $a(\zeta)$ and $b(\eta)$ are very smooth (Secs. 12 and 17).

Let us now transform Eq. (20.10) acting on it by the operator $\partial/\partial t + a(\eta)\partial/\partial\eta$ and bearing in mind that the main contribution to the right-hand side comes from $\eta - \zeta, \eta - \eta' \sim 1$. The terms with derivatives $a'(\zeta)[b'(\eta)]$ are of the order $1/\zeta[1/\eta]$ and may be omitted. Nevertheless, the dependence of $a(\zeta)[b(\eta)]$ on $\zeta(\eta)$ should be taken into account and this is actually done in the adiabatic approximation. With the above assumptions taken into account, we derive the final form of the kinetic equation:

$$n_{tt} + a(\eta)(n_{\eta t} + b(\eta)n_\eta) + b(\eta)n_t$$
$$= 2b(\eta)a(\eta)\int_0^\eta d\zeta \int_0^\zeta d\eta' e^{\eta'-\zeta}n(\eta', t)$$
$$= 2b(\eta)a(\eta)\int_0^\eta dy(1 - e^{-y})n(\eta - y, t). \tag{20.11}$$

20.2. Solution of the Kinetic Equation with Constant Coefficients

Let us find the solution of Eq. (20.11) in the case when the functions a and b are constant. This situation takes place, for example, where a shower evolves in an amorphous medium. The solution will be found using the Laplace transform with respect to the variable η:

$$\nu(\lambda, t) = L(n(\eta, t)) = \int_0^{\infty} e^{-\lambda\eta} n(\eta, t) d\eta \,. \tag{20.12}$$

Applying the Laplace transform to Eq. (20.11) we take into account that

$$\begin{gathered} L(n_\eta(\eta, t)) = \lambda\nu(\lambda, t) - n(0, t) \,, \\ L\left(\int_0^{\eta} n_1(y) n_2(\eta - y) dy\right) = L(n_1(\eta)) L(n_2(\eta)) = \nu_1(\lambda)\nu_2(\lambda) \,. \end{gathered} \tag{20.13}$$

Not uniform term in the equation for function $\eta(\lambda, t)$, being the result of the Laplace transform, is equal zero

$$n_t(0, t) + bn(0, t) = 0 \,. \tag{20.14}$$

The last equation means that an initial photon produces the electron–positron pair with the probability b:

$$n(0, t) = n(0, 0) \exp(-bt) \,.$$

As a result the equation for $\eta(\lambda, t)$ is the form

$$\lambda(\lambda + 1) \left(\frac{\partial}{\partial t} + b\right) \left(\frac{\partial}{\partial t} + a\lambda\right) \nu(\lambda, t) = 2ab\nu(\lambda, t) \,. \tag{20.15}$$

Equation (20.15) with respect to the variable t is a linear differential equation of the second-order with constant coefficients. The solution of this equation will be found in the form $\exp(\mu t)$, then we obtain for μ the following square equation and its solution

$$\begin{gathered} \lambda(\lambda + 1)(\mu + b)(\mu + a\lambda) = 2ab \,, \\ \mu_{1,2} = -\frac{a\lambda + b}{2} \pm \sqrt{\frac{(a\lambda - b)^2}{4} + \frac{2ab}{\lambda(\lambda + 1)}} \end{gathered} \tag{20.16}$$

The total solution of Eq. (20.15) is the form

$$\nu(\lambda, t) = c_1(\lambda) \exp(\mu_1 t) + c_2(\lambda) \exp(\mu_2 t) \,. \tag{20.17}$$

As a result of the reverse Laplace transform we find the total form of the solution for $n(\eta, t)$

$$n(\eta,t) = \frac{1}{2\pi i}\int_{c-i\infty}^{c+i\infty} d\lambda e^{\lambda\eta} \cdot \left[c_1(\lambda)e^{\mu_1 t} + c_2(\lambda)e^{\mu_2 t}\right]. \tag{20.18}$$

Using the initial conditions we find the expressions for $c_1(\lambda)$ and $c_2(\lambda)$. When a photon is an initial particle these conditions are as follows

$$n(\eta,0) = \delta(\eta)\,, \quad n'(\eta,0) = -b\delta(\eta)\,. \tag{20.19}$$

Substituting the solution for $n(\eta,t)$ in form (20.18) to Eq. (20.19), we obtain the following set of equations for c_1 and c_2 and its solution

$$\begin{aligned} c_1 + c_2 = 1, &\quad c_1\mu_1 + c_2\mu_2 = -b\,; \\ c_1 = (\mu_2 + b)/(\mu_2 - \mu_1), &\quad c_2(\mu_1 + b)/(\mu_1 - \mu_2)\,. \end{aligned} \tag{20.20}$$

For definiteness we put Re $\mu_1 >$ Re μ_2 and introduce the notation $\mu_1 \equiv \mu$. Then we obtain for the exponentially increasing part of the solution (20.18):

$$n^{(\gamma)}(\eta,t) = \frac{1}{2\pi i}\int_{c-i\infty}^{c+i\infty} d\lambda \frac{\mu + \lambda a}{2\mu + \lambda a + b} e^{\mu t + \lambda\eta} \tag{20.21}$$

When an electron is an initial particle the initial conditions are as follows

$$\begin{aligned} n(\eta,0) = 0, &\quad n'(\eta,0) = a\,; \\ c_1 + c_2 = 0, &\quad c_1\mu_1 + c_2\mu_2 = a\,, \end{aligned} \tag{20.22}$$

for this case

$$c_1 = -c_2 = a/(\mu_1 - \mu_2) = a/(2\mu + \lambda a + b)\,. \tag{20.23}$$

20.3. Solution of the Kinetic Equation in the Adiabatic Approximation

As it follows from Eq. (20.21), the number of particles in a cascade grows exponentially as it evolves in a substance. In view of this, a solution of Eq. (20.11) will be sought in the form $n(\eta,t) = A(\eta,t)\exp[f(t)]$. Introducing the notation $\partial f/\partial t \equiv f_t = \mu$ and $\partial f/\partial \eta \equiv f_\eta = \lambda$, neglecting the derivatives of A and the second ones of f (these terms are $\sim 1/\eta$) expending the exponent in $n(\eta - y, t)$

with respect to y and, finally, taking the corresponding integral in (20.11), we obtain for μ and λ the following equation (compare with (20.16))

$$\lambda(\lambda+1)(\mu+b)(\mu+a\lambda) = 2ab\,,$$

$$\mu(\lambda) = -\frac{a\lambda+b}{2} + \sqrt{\frac{(a\lambda-b)^2}{4} + \frac{2ab}{\lambda(\lambda+1)}}\,. \tag{20.24}$$

The solution of this partial differential equation of the first-order is of the form

$$f(\eta,t) = F(\mu,\eta,t) = \mu t + \int_0^\eta \lambda(\mu, a(y), b(y))dy\,,$$

where $\mu = \mu(\eta,t)$ is found from the condition $\partial F/\partial\mu = 0$. The corresponding expression for $n(\eta,t)$ is convenient for representing as a contour integral which can be evaluated, using the saddle-point method:

$$n(\eta,t) = \frac{1}{2\pi i}\int_{c-i\infty}^{c+i\infty} d\mu C(\mu,\lambda,a(\eta),b(\eta)) \times \exp\left[\mu t + \int_0^\eta \lambda(\mu,a(y),b(y))dy\right] \tag{20.25}$$

According to what has been said above, the derivatives of $a(\eta)$ and $b(\eta)$ will be neglected when substituting (20.25) into Eq. (20.11). To solve Eq. (20.11) it is necessary now to find an explicit form of the function C in (20.25) in accordance with the initial conditions. To do this, it suffices to solve the problem with constant a and b. In the case when the initial particle is a photon, the part of the solution that increases exponentially at large t is given by Eq. (20.21). From the comparison of (20.25) and (20.21) we find that

$$C^{(\gamma)}(\mu,\lambda,a(\eta),b(\eta)) = -\frac{\lambda_\mu(\mu+\lambda a)}{2\mu+\lambda a+b}\,, \qquad \lambda^\mu \equiv \partial\lambda|\lambda\mu\,. \tag{20.26}$$

At large $t \gg a^{-1} = L_{\text{ch}}$ the integral (20.25) is taken using the saddle-point method. As a result, we have for the number of photons at a large depth t:

$$N_\gamma^{(\gamma)}(\eta,t) = -\frac{\lambda_\mu(\mu+\lambda a)}{\sqrt{2\pi d}(2\mu+\lambda a+b)} \times \exp\left[\mu t + \eta + \int_0^\eta \lambda(\mu,a(y),b(y))dy\right]\,, \tag{20.27}$$

where

$$t + \int_0^\eta \lambda_\mu(\mu, a(y), b(y))dy = 0\,,$$
$$d = \int_0^\eta \lambda_{\mu\mu}(\mu, a(y), b(y))dy\,. \tag{20.28}$$

The first of the equalities (20.28) defines μ as a function of t and $\eta(\mu = \mu(\eta, t))$. Using (20.4) and (20.3) and taking (20.7) into account ($N_0(\varepsilon'; \varepsilon, \tau) = \delta(\varepsilon'(\tau) - \varepsilon)$) we have, with the same accuracy, the expression for the number of charged particles:

$$N_e(\eta, t) = \frac{2b}{(\lambda+1)(\mu+\lambda a)} N_\gamma(\eta, t) = \frac{\lambda}{a}(\mu + b)N_\gamma(\eta, t)\,. \tag{20.29}$$

If the initial particle is an electron (positron) with energy ε_0, the number of photons with an energy $\omega(\eta = \ln(\varepsilon_0/\omega))$ is found by the comparison of (20.25) with (20.23) and (20.18)

$$N_\gamma^{(e)}(\eta, t) = -\frac{\lambda_\mu a}{\sqrt{2\pi d}\lambda(2\mu + \lambda a + b)} \times \exp\left[\mu t + \eta + \int_0^\eta \lambda(\mu, a(y), b(y))dy\right]. \tag{20.30}$$

Of course relation (20.29) holds in this case.

Of great interest is the total number of particles $\Pi(\eta_f, t)$ with energy $\omega, \varepsilon \geq \varepsilon_f; \eta_f = \ln(\varepsilon_0/\varepsilon_f)$. To find $\Pi(\eta_f, t)$ the function $N(\eta', t)$ should be integrated over η' with the weight $e^{-\eta'}(-d\omega = \varepsilon_0 d\eta' e^{-\eta'})$. With the accepted accuracy the calculation can be made by expanding the exponent argument in (20.27) near the point $\eta' = \eta$ and by changing to the integration variable $\eta' - \eta$. Taking the first of the equalities (20.28) into account, we obtain:

$$\Pi(\eta_f, t) = \int_0^{\eta_f} N(\eta', t)e^{-\eta'}d\eta' = \frac{1}{\lambda}e^{-\eta_f}N(\eta_f, t)\,. \tag{20.31}$$

Making an allowance for (20.28) it is easy to see that a maximum number of particles, at fixed η, is achieved at $t = t_{op}$ defined from the condition $\mu(\eta, t) = 0$. From Eq. (20.24) we get $\lambda = 1$ at $\mu = 0$ for any a and b. Calculating λ_μ and $\lambda_{\mu\mu}$ at the point $\lambda = 1$ and $\mu = 0$ we find

$$\lambda_\mu(\mu = 0) = -\frac{2(a+b)}{5ab}\,, \quad \lambda_{\mu\mu}(\mu = 0) = \frac{4}{125}\left(\frac{7}{a^2} - \frac{1}{ab} + \frac{17}{b^2}\right). \tag{20.32}$$

For t_{op} we then obtain from (20.28):

$$t_{op} = \frac{2}{5}\int_0^\eta dy\left(\frac{1}{a(y)} + \frac{1}{b(y)}\right) \tag{20.33}$$

At t close to t_{op}, the main contribution to the integral over μ comes from the vicinity of the point $\mu = 0$. Expanding the exponent argument in Eq. (20.25) with due regard to the relation (20.32) and taking the integrals, we derive explicit expressions for the distribution functions $N_\gamma(\eta, t)$ and $N_e(\eta, t)$ which are of the form of Gaussian distributions in time (depth):

$$\begin{aligned} N_\gamma(\eta, t) &= \frac{2e^{2\eta}}{5\sqrt{2\pi d}b(\eta)}\exp\left[-\frac{(t - t_{op})^2}{2d(\eta)}\right], \\ N_e(\eta, t) &= b(\eta)N_\gamma(\eta, t)/a(\eta), \\ d(\eta) &= \frac{4}{125}\int_0^\eta dy\left(\frac{7}{a^2(y)} - \frac{1}{a(y)b(y)} + \frac{17}{b^2(y)}\right) \end{aligned} \tag{20.34}$$

For electron-photon showers in an amorphous substance ($b/a = 7/9$, $a^{-1} = L_{\text{rad}}$) we have

$$\begin{aligned} N_\gamma(\eta, t_{op}) &= \frac{0.197}{\sqrt{\eta}}e^{2\eta}, \\ N_e(\eta, t_{op}) &= \frac{0.153}{\sqrt{\eta}}e^{2\eta};\ t_{op} = \frac{32}{35}\eta L_{\text{rad}}\,. \end{aligned} \tag{20.35}$$

This is in good agreement (the difference is $\approx 10\%$) with the standard shower theory. Note that our computer simulation of the hard cascade is consistent with the results of the preceding analytical calculation.

20.4. Results

Besides the hard cascade (when the particle and photon energy is $\varepsilon, \omega \gg \omega_t$) whose evolution is analytically analyzed in the foregoing section, a situation when the lower energy boundary is $\varepsilon_f, \omega_f \ll \omega_t$ is possible. In this case, at a photon energy $\omega \leq \omega_t$, the Bethe–Heitler mechanism of pair creation basically works, while the particle radiation in the field of an aligned single crystal still remains rather intense ($b \simeq W_e^{BH}, a \gg 1/L_{\text{rad}}$).

It is obvious that the major part of the photon is emitted over the characteristic length $l \sim \int_0^\eta dy/a(y) = \int_0^\eta dy L_{\text{ch}}(\varepsilon_0\, e^{-y}) \ll t_{op}$ and there is a wide enough interval wherein the number of photons remains practically constant (smoothly reducing on account of the absorption) to a length $t \sim 1/W^{BH}$ at which a noticeable photon absorption starts due to a transformation into the e^+e^- pairs. In the situation under analysis, two essentially different cases can occur. In the first case, the initial energy is $\omega_0(\varepsilon_0) \gg \omega_t$ and a mixed cascade develops, while in the second one $\omega_0(\varepsilon_0) \sim \omega_t$ and a soft cascade develops.

We would like to start with the results of the computer simulation for a mixed cascade. The simulation has been performed for a silicon crystal along the $\langle 110 \rangle$ axis and at $T = 293$ K. Figure 20.1 shows the number of charged particles and photons versus the depth at which the shower evolves in the crystal, for various energies of the initial photon. The lower boundary of the photon energy in the cascade is taken to be $\omega_f = 100$ MeV and it

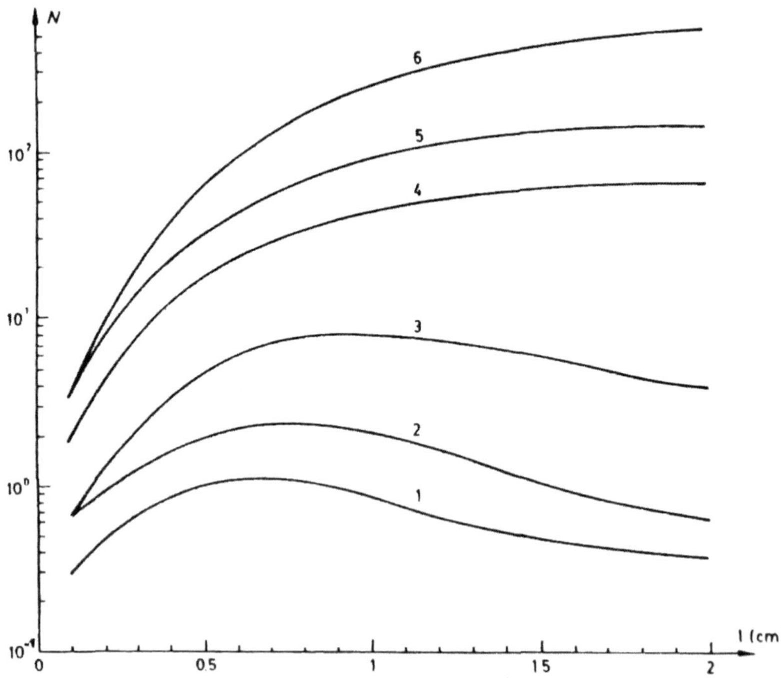

Fig. 20.1. The number of charged particles with $\varepsilon > 10$ GeV in the Si crystal (the $\langle 110 \rangle$ axis, $T = 293$ K) in the case when the photon with $\omega_0 = 0.4$ TeV is incident on the crystal (curve 1), $\omega_0 = 1$ TeV (curve 2) and $\omega_0 = 4$ TeV (curve 3). For the same conditions, curves 4, 5 and 6 show the number of photons with $\omega > 100$ MeV.

corresponds roughly to the effective threshold of electron–positron pair creation in a material. It can be seen from Fig. 20.1 that the number of photons reaches a plateau, in agreement with the discussion above. The lower boundary for the charged-particle energy is taken equal to $\varepsilon_f = 10$ GeV. On the other hand, this choice ensures the applicability of the magnetic bremsstrahlung approximation employed, and, on the other hand, at $\varepsilon < 10$ GeV the photons emitted are basically those ones whose energy is $\omega < 100$ MeV. The results presented allow one to optimize the crystal thickness in the crystal-based ultrahigh-energy electron and photon detector or converter.

Figure 20.2 illustrates the number N_γ of photons with $\omega > 100$ MeV and the total number of charged particles N_e versus the energy ω_0 of the initial photon incident on a Si crystal of thickness $L = 1$ cm at a small angle $\vartheta_0 \ll V_0/m$ to the $\langle 110 \rangle$ axis. It is interesting that the ratio of the number of photons N_γ to that of charged particles N_e does not in practice depend on the initial energy and is $N_\gamma/N_e \simeq 11$. The curves in Fig. 20.2 permit one to find with good accuracy (Baier, Katkov, and Strakhovenko, 1986) the energy of superhard photons upon detection. As the photon energy increases it turns out to be sufficient for this purpose to detect the charged particles only, without additional devices for the photon-into-e^+e^--pair conversion.

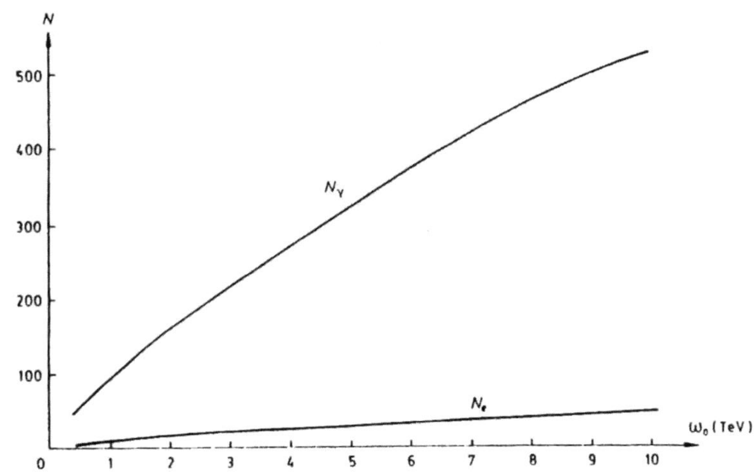

Fig. 20.2. The total number N_e of charged particles and photons with $\omega > 100$ MeV (N_γ) at a depth of 1 cm in the aligned Si single crystal (the $\langle 110 \rangle$ axis, $T = 293$ K) depending on the initial photon energy ω_0.

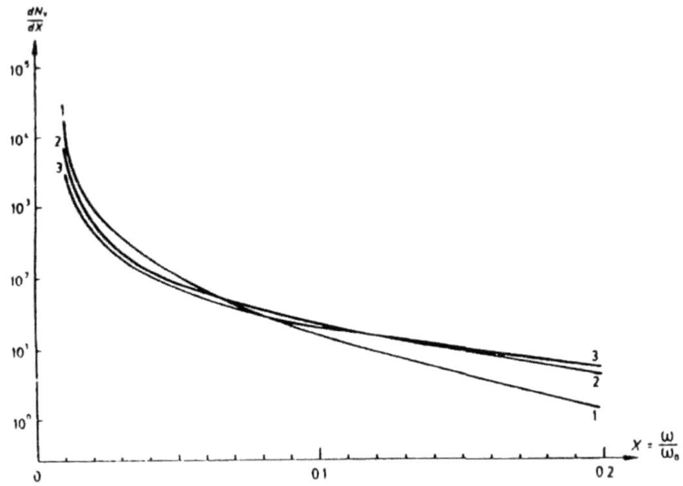

Fig. 20.3. The energy distribution of photons at a 1 cm depth in the Si aligned single crystal (the $\langle 110 \rangle$ axis, $T = 293$ K) for the initial-photon energy $\omega_0 = 0.4$ TeV (curve 1, $\omega_0 = 1$ TeV (curve 2) and $\omega_0 = 4$ TeV (curve 3).

Figure 20.3 demonstrates the energy distributions of the photons for various energies of the initial photons.

In the case when the initial particle (photon) energy is of the order of the threshold energy ω_t the soft cascade develops wherein, as already mentioned, the emission of a great number of photons leads to the creation of a noticeable number of e^-e^+ pairs despite the quantity W_e^{BH} being small. In this case, for $l \le L_{\text{ch}} \ll L_{\text{rad}}$ the method of successive approximations is suited for an analytical calculation inasmuch as the relative amount of the secondary particles is small. For example, within the framework of the accepted approach we find the number of photons at depth l, for the initial electron

$$N_\gamma^{(e)}(l) = \int_0^l W_\gamma(t)dt = \int_{\varepsilon(l)}^{\varepsilon_0} W_\gamma(\varepsilon) \cdot L_{\text{ch}}(\varepsilon)\frac{d\varepsilon}{\varepsilon}$$

$$= \int_0^\eta \frac{dy}{a(y)} W_\gamma(\varepsilon_0 e^{-y}) \simeq \overline{W}_\gamma \ell, \eta = \ln\frac{\varepsilon_0}{\varepsilon(l)} \qquad (20.36)$$

The latter equality in Eq. (20.36) is satisfied with good accuracy due to the weak dependence of the total radiation probability W_γ on ε in the energy range under consideration. In this range the magnetic bremsstrahlung description is valid, and the quantum effects for the total probability are still relatively small. Correspondingly, for the number of secondary charged particles we obtain

$$\begin{aligned} N_e^{(e)}(l) &= 2\int_0^l W_e^{BH} N_\gamma(t)dt \\ &= 2W_e^{BH}\int_0^l W_\gamma(t)(l-t)dt \\ &\simeq W_e^{BH}\overline{W}_\gamma \ell^2 \simeq W_e^{BH} N_\gamma^{(e)}(l)l\,. \end{aligned} \tag{20.37}$$

One can in principle calculate the terms of the next approximations as well. At $l \ll L_{\rm ch}$ these terms are however small, whereas at $l \geq L_{\rm ch}$ the secondary particles are produced mainly with an energy lower than the threshold ε_f.

The experiments have been made on electron radiation and pair creation by photons in a single crystal at a high enough particle energy. In these experiments (Belkacem *et al.*, 1985, 1986) Ge single crystals of thickness $L = 0.4$ mm and 1.4 mm, cooled down to 100 K were used. The initial electrons with energy $\varepsilon_0 = 150$ GeV and photons with energies $\omega_0 < 155$ GeV were incident on a single crystal in a direction near to the $\langle 110\rangle$ axis. Bearing in mind that for these conditions $\omega_t = 50$ GeV, this experimental situation can be referred to as the case of a soft cascade. We would like to discuss at first the case when the initial particle is an electron. A first estimate of the mean energy losses caused by the radiation has been obtained in Sec. 17.5. However, no allowance was made there for the creation of secondary particles whose numbers can be estimated according to Eqs. (20.36) and (20.37). For $L = 1.4$ mm, one has $\overline{W}_\gamma = 120\ \text{cm}^{-1}$ and $W_e^{BH} = 0.28\ \text{cm}^{-1}$ (the last value is taken with regard to thermal vibration). We then have from (20.36) and (20.37)

$$N_\gamma^{(e)}(L) \simeq 16.8; \quad N_e^{(e)}(L) \simeq 0.67\,. \tag{20.38}$$

Taking into account that for $L = 0.4$ mm, $\overline{W}_\gamma = 111\ \text{cm}^{-1}$, we find the number of photons and secondary electrons for this thickness:

$$N_\gamma^{(e)}(L) \simeq 4.4; \quad N_e^{(e)}(L) \simeq 1/20\,. \tag{20.39}$$

We have made a mumerical calculation of the cascade for the experimental conditions indicated above. For the number of photons with energy higher than

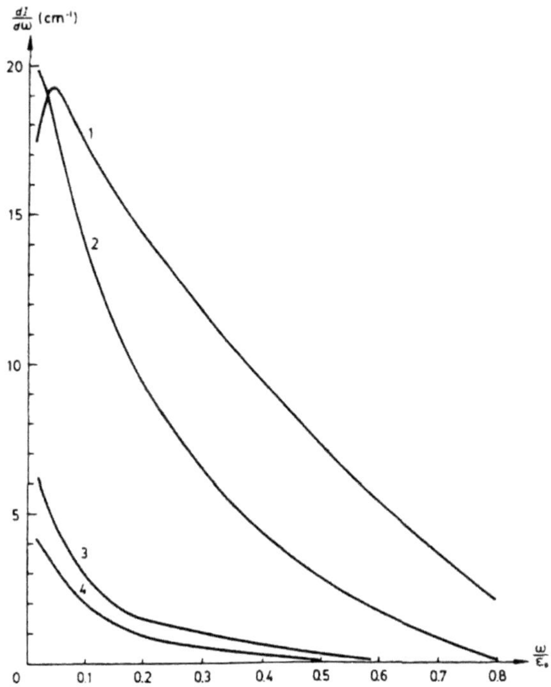

Fig. 20.4. The average intensity: $dI/d\omega = (\omega/L)\, dN(\omega)/d\omega$, where $N_\gamma(\omega)$ is the number of photons with energy ω leaving the Ge crystal ($\langle 110\rangle$ axis, $T = 100$ K). Curve 1 is for the initial electrons with $\varepsilon_0 = 150$ GeV ($L = 0.4$ mm) and curve 2 is for $L = 1.4$ mm. Curve 3 is for the initial photons with $\omega_0 = 150$ GeV and $L = 1.4$ mm and curve 4 is for photons with $\omega_0 = 100$ GeV and $L = 1.4$ mm.

$\omega_f = 100$ MeV we have obtained N_γ (1.4 mm) $\simeq 13.9$ and N_γ (0.4 mm) $\simeq 3.9$, while for the total number of photons the results are N_γ (1.4 mm) $\simeq 17.3$ and N_γ (0.4 mm) $\simeq 4.4$; the results obtained are well consistent with the estimates (20.38) and (20.39). The agreement is also good for the number of secondary particles.

Figure 20.4 illustrates the averaged (over crystal thickness) spectral radiation intensity obtained from the computer simulation; it is in good agreement with the results of Sec. 17.5. Thus, the use of $N_0(\varepsilon; \varepsilon', \tau)$ as $\delta(\varepsilon'(\tau) - \varepsilon)$ is now justified. Figure 20.5 presents the computation results on the energy distribution of charged particles for $\varepsilon > 5$ GeV. Note that despite a considerable

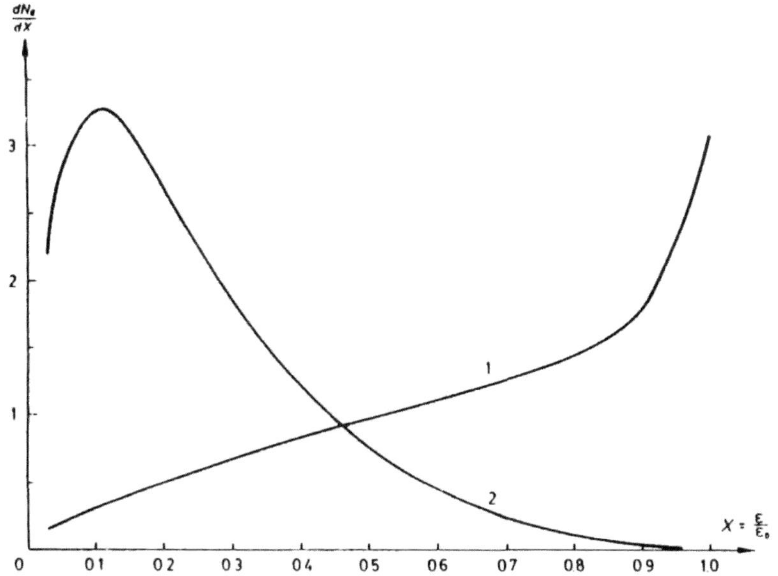

Fig. 20.5. The energy distribution of charged particles leaving the Ge crystal for the initial electrons with $\varepsilon_0 = 150$ GeV ($\langle 110 \rangle$ axis, $T = 100$ K); the thickness is $L = 0.4$ mm (curve 1) and $L = 1.4$ mm (curve 2).

number of secondary particles (see Eq. (20.38)), most of them have an energy of $\varepsilon < 5$ GeV and hence, they have no influence in practice on the spectrum for $\varepsilon > 5$ GeV. In view of this, the results in Fig. 20.5 can be directly compared with the distribution in Fig. 2 of the paper by Belkacem *et al.*, 1985 (with the substitution $x \to 1 - x$). It is seen that the results of our calculation are well consistent with this experiment. With the distribution found, we obtain for the average energy of electrons with $\varepsilon > 5$ GeV, at the exit from the crystal $\bar{\varepsilon} = 0.26\varepsilon_0$ at $L = 1.4$ mm and $\bar{\varepsilon} = 0.66\varepsilon_0$ at $L = 0.4$ mm. These values are nearly the same as those in Sec. 17.5 and the experimental data.

We have also analyzed the case in which a photon with energy ω_0 is incident on a crystal and the electron–positron pair creation has been studied under the conditions of the experiment (Belkacem *et al.*, 1986). Curves 1 and 2 in Fig. 20.6 show the spectral distribution of electrons (positrons) produced by a photon with $\omega_0 = 150$ GeV and $\omega_0 = 100$ GeV, with further radiation neglected. The calculation of this distribution within the framework of

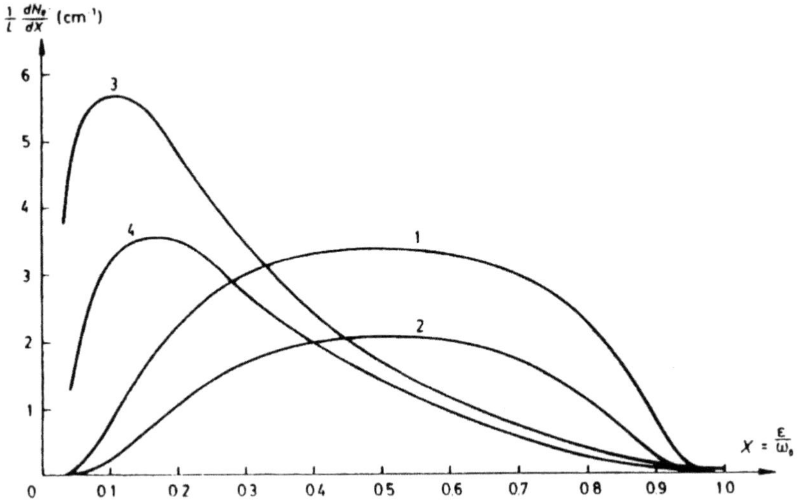

Fig. 20.6. The energy distribution of electrons at the exit from the Ge single crystal ($\langle 110 \rangle$ axis, $T = 100$ K, $L = 1.4$ mm) for the initial photon energy $\omega_0 = 150$ GeV (curve 3) and $\omega_0 = 100$ GeV (curve 4). For comparison, the same distribution is given in a direct e^+e^- pair creation process for $\omega_0 = 150$ GeV (curve 1) and $\omega_0 = 100$ GeV (curve 2).

the cascade theory for the same initial photons (curves 3 and 4) demonstrate a dramatic difference of the spectra of the particles produced in a crystal of thickness $L = 1.4$ mm in comparison with the spectrum which would be observed in a very thin crystal (curves 1 and 2). For emitted photons with $\omega > 100$ MeV, at $\omega_0 = 150$ GeV we have $N_\gamma \simeq 4.8$, while for $\omega_0 = 100$ GeV we have $N_\gamma = 3.4$. The total number of photons is 6 and 4.4, respectively.

The development of electromagnetic cascades along crystalline axes was measured for the first time in Medenwaldt *et al.*, 1989b. Some results of these measurements are presented in Fig. 20.7. We have here examples of soft and mixed cascades with electrons as initial particles and therefore the angular dependence of the measured multiplicity results from the electron radiation intensity. Let us estimate the number of created charged particles at axial alignment (at a tilting angle equal to zero in Fig. 20.7). The crystal thickness L in the experiment was large: $L \gg L_{\rm ch}$ and therefore the estimate (20.37) for the number of created charge particles N_e is inapplicable. Assuming that pairs in a soft cascade are produced by Bethe–Heitler mechanism, we find evident

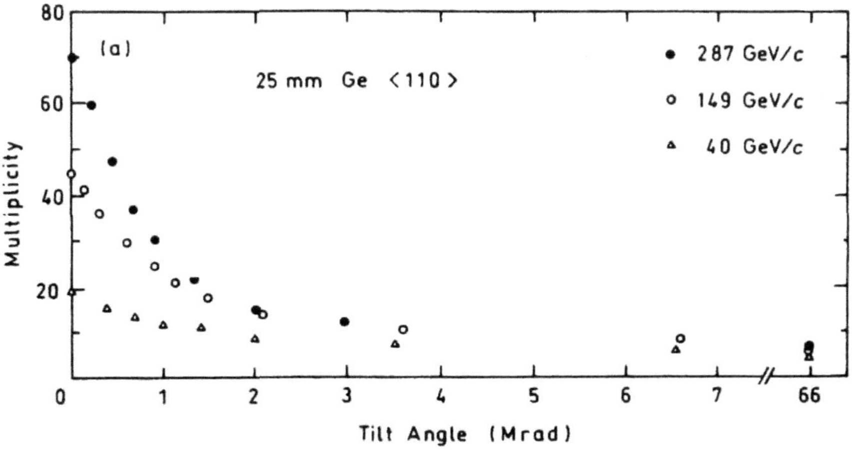

Fig. 20.7.

generalization of Eq. (20.37) to the case $L \gg L_{\text{ch}}$:

$$N_{1e}^{(e)} = 2\int d\omega \int_0^L dz \frac{dW_\gamma(\varepsilon(z), \omega)}{d\omega}[1 - \exp(-W_e^{BH}(L - z))] . \quad (20.40)$$

The quantity $\varepsilon(z)$ is defined in Eq. (20.7). In the interval of energies 10 GeV $< \varepsilon <$ 300 GeV which corresponds to experimental conditions, the quantity $L_{\text{ch}}^{-1} = I(\varepsilon)/\varepsilon$ can be fitted to a good accuracy by a simple expression

$$L_{\text{ch}}^{-1}(\varepsilon) = A + B\ln(\varepsilon/20 \text{ GeV}) , \quad (20.41)$$

where $A = 3.22$ cm^{-1}, $B = 2.57$ cm^{-1}. Using this approximation and properties of a soft part of the radiation spectrum, we obtain from Eq. (20.40) $N_{1e}^{(e)} = 2\overline{W}_\gamma Z_{ef}(\varepsilon_0, \omega_0)(1 - \exp(-W_e^{BH} L))$, where

$$Z_{ef}(\varepsilon_0, \omega_0) = B^{-1} \ln\left[1 - \frac{1}{2} B L_{\text{ch}}(\varepsilon_0) \ln \frac{\varepsilon_0 \chi_s(\varepsilon_0)}{2\omega_0}\right] \quad (20.42)$$

Here ω_0 is a lower boundary of effective pair production by the Bethe–Heitler mechanism and we put below $\omega_0 = 1$ GeV. For experimental conditions we obtain $N_{1e}^{(e)}(\varepsilon_0 = 40 \text{ GeV}) = 20$, $N_{1e}^{(e)}(\varepsilon_0 = 149 \text{ GeV}) = 44$ and $N_{1e}^{(e)}(\varepsilon_0 = 287 \text{ GeV}) = 53$. If the initial energy ε_0 is large enough the second generation of charged particles $N_{2e}^{(e)}$ can arise. To contribute to this process, photons emitted by initial particles must have energy $\omega > \omega_c$. The boundary value ω_c is consistently defined by the equation $Z_{ef}(\omega_c/2, \omega_0) = 0$ and we obtain

$$N_{2e}^{(e)} = (2\overline{W}_\gamma)^2 B^{-1}[1 - (1 + W_e^{BH} L)\exp(-W_e^{BH} L)] \int_0^{Z_{ef}(\varepsilon_0, \omega_c)} dz$$

$$\times \int_{f(z)}^1 dy \ln[1 + 3BA^{-1}\ln(y/f(z)], \qquad (20.43)$$

$$f(z) = \exp\left[\frac{2}{3}\frac{e^{-BZ}ef - e^{-Bz}}{BL_{\text{ch}}(\varepsilon_0)}\right]$$

From Eq. (20.43) we obtain $N_{2e}^{(e)}(\varepsilon_0 = 287 \text{ GeV}) = 22$ and $N_{2e}^{(e)}(\varepsilon_0 = 149 \text{ GeV}) < 0.1$. So the total number of created particles $N_e^{(e)} = N_{1e}^{(e)} + N_{2e}^{(e)}$ at $\varepsilon_0 =$ 287 GeV is $N_e^{(e)} = 75$. The obtained values for multiplicities are in good agreement with the experimental data confirming theoretical description of the process.

21. INCOHERENT RADIATION AND PAIR CREATION IN CRYSTALS

21.1. Main Features of the Incoherent Processes

In Secs. 11–20 the theory of electromagnetic processes was developed for particle motion in the potential of an aligned single crystal, averaged over thermal vibrations of atoms. Besides the given motion of particles in the external field of an aligned crystal the particles are scattered due to the potential fluctuations associated with atomic oscillations in the crystalline lattice. Such scattering is accompanied by processes in crystals usually called incoherent. Let us consider the characteristic peculiarities of incoherent radiation in crystals differing from bremsstrahlung in amorphous matter. In the aligned-crystal potential, the charged particles follow curvilinear trajectories, that directly leads to a redistribution of the flux of incident particles. The role of the nonuniformity of the colliding-particles distribution in bremsstrahlung has been considered by Baier, Katkov and Strakhovenko, 1982a. Similar problems for crystals has been discussed by Bazilev, Golovisnin and Demura, 1987. In the case of the particle motion in crystals, the electron and nuclear density is nonuniform. In addition, the bremsstrahlung process is directly influenced by this trajectory curving at fairly high energies. This influence is connected with a shortening of the photon formation length on account of a relatively large change of the particle velocity on this length and the appropriate widening of the cone of radiation angles. For analysis we shall use the results of Sec. 7. The account follows the authors' paper [Baier, Katkov, Strakhovenko, 1987e].

In Sec. 7.4 it has been shown that the value $\chi = E\gamma/E_0$ is the parameter characterizing the immediate influence of the external field on the bremsstrahlung process. For $\chi \ll 1$ this influence can be neglected. In strong fields when $\chi/u \gg 1 (u = \omega/(\varepsilon - \omega)$ the characteristic angles of radiation are $\vartheta_{\text{eff}} \sim (1/\gamma)(\chi/u)^{1/3}$ and the photon formation length $l_f \sim (\gamma \lambda_c/u)(u/\chi)^{2/3}$. Since this length at large χ/u falls off as $(u/\chi)^{2/3}$, the bremsstrahlung cross section reduces in the same manner.

Because of the high density of nuclei near the axis where incoherent processes are essential, the Landau–Pomeranchuk effect might manifest itself at high electron energies. Therefore, for ϑ_{eff} to be estimated consistently, a widening of the radiation angles, caused by multiple scattering, should be taken into account as well. Making allowance for a change of the angle of deflection of a particle on the photon formation length, both on account of the external field and multiple scattering, we obtain from the self-consistency condition that (see Eq. (7.86))

$$\frac{\chi^2}{u^2\zeta^2} + \frac{4\pi Z^2\alpha^2\gamma n}{m^3 u\zeta} \ln \frac{\zeta}{\gamma^2\vartheta_u^2} \lesssim \zeta, \quad \zeta = 1 + \gamma^2\vartheta^2, \tag{21.1}$$

where Z is the nuclear charge, n is the nuclear density of a medium, $\vartheta_u = 1/\varepsilon u_1$ is the angle corresponding to the lower boundary of the momentum transfer $q_{\text{min}} = 1/u_1$ where u_1 is the amplitude of thermal vibrations. Allowing for the fact that

$$\chi(u_1) \sim \frac{V_0\gamma}{m^2 u_1} \sim \frac{Z\alpha\gamma}{m^2 d u_1}, \quad n \sim \frac{1}{4\pi u_1^2 d}, \tag{21.2}$$

where V_0 is the scale of the axis potential, d is the mean distance between the atoms in the chain forming the axis, we have for the ratio of the contributions of multiple scattering and of the external field to the left side of Eq. (21.1)

$$k = \frac{\dot{\vartheta}_s^2 l_f}{(w l_f)^2} \sim \frac{u\zeta}{\gamma}\frac{d}{\lambda_c} \ln \frac{\zeta}{\gamma^2\vartheta_u^2}. \tag{21.3}$$

At high energies when the parameter $\chi \geq 1$ this ratio is always small since it is small at $\chi(u_1) \sim 1$, while at $\chi(u_1) \gg 1$ the quantity $\zeta_{\text{eff}} \propto \gamma^{2/3}$. The maximum estimate for k yields $(\chi(u_1) \simeq 1, u \simeq 1)$

$$k_{\text{max}} \sim \frac{Z\alpha\lambda_c}{u_1} \ln \left(\frac{u_1}{\lambda_c}\right)^2 \ll 1. \tag{21.4}$$

The condition $k \ll 1$ allows the perturbation theory for the scattering description to be used and the Landau–Pomeranchuk effect to be neglected in the calculation of the probabilities of the incoherent processes.

21.2. Role of the Inhomogeneous Particle Density

Most of the bremsstrahlung generated by high-energy particles is formed at very low momentum transfer. In a space-time picture this means the process proceeds at long distances. In this case, particle radiation at large impact parameters may prove to be strongly dependent on the shapes of the colliding beams. In the case we are interested in, the electron state is not a plane wave but a wave packet in which the distribution along the transverse coordinate is not homogeneous. In addition, the distribution of the atoms in the chain (for definiteness, the axial case is considered) is also characterized by a definite density in the transverse plane, which is connected with the amplitude of the thermal vibrations. Under such conditions, it is impossible to use the standard bremsstrahlung theory. The bremsstrahlung theory involving the nonuniformity of the particle distribution has been developed by the authors with the help of the space-time approach which naturally takes into account the nonhomogeneity of the particle density. In application to our problem the major result of Baier, Katkov, Strakhovenko, 1982 may be represented as follows:

$$d\sigma_\gamma(\omega,\zeta) = \Phi(0)d\sigma_{\gamma 0}(\omega,\zeta) + d\sigma_{\gamma 1}(\omega,\zeta)\,, \tag{21.5}$$

where $d\sigma_{\gamma 0}(\omega,\zeta)$ is the standard radiation cross section for the homogeneous distribution of particles in the beams and

$$d\sigma_{\gamma 1}(\omega,\zeta) = \frac{Z^2\alpha^3}{m^2\pi}\frac{\varepsilon-\omega}{\varepsilon}\frac{d\omega}{\omega}\cdot U(\zeta)F(\beta,\zeta)d\zeta\,, \tag{21.6}$$

here

$$\begin{aligned} U(\zeta) &= \frac{\varepsilon-\omega}{\varepsilon} + \frac{\varepsilon}{\varepsilon-\omega} - \frac{4(\zeta-1)}{\zeta^2}\,, \\ F(\beta,\zeta) &= \frac{\beta^2}{\zeta^2}\left\{\int d^2\rho\rho[K_0(\beta\rho)K_2(\beta\rho) - K_1^2(\beta\rho)]\frac{d\Phi(\rho)}{d\rho}\right. \\ &\quad \left. -2\int d^2x\left[\int d^2\rho n_a(\mathbf{x}-\boldsymbol{\rho})\frac{\boldsymbol{\rho}}{\rho}K_1(\beta\rho)\right]^2 n_e(\mathbf{x})\right\}\,; \\ \beta &= \sqrt{a^{-2}+\zeta^2 q_{\min}^2}\,, \quad q_{\min} = \frac{m^2\omega}{2\varepsilon(\varepsilon-\omega)} = \frac{mu}{2\gamma} \end{aligned} \tag{21.7}$$

In Eq. (21.7), a is the screening radius, n_a the atomic density in the transverse plane, n_e the ratio of the electron (positron) density to that given by the homogeneous distribution, and K_n are the Macdonald functions,

$$\begin{aligned}\Phi(\boldsymbol{\rho}) &= \int d^2x n_e(\boldsymbol{\rho}+\mathbf{x})n_a(\mathbf{x})\,,\\ n_a(\mathbf{x}) &= \frac{1}{2\pi u_1^2}e^{-x^2/2u_1^2}\,,\end{aligned} \tag{21.8}$$

where u_1 is the amplitude of thermal vibrations. The electron (positron) density depends on the particular situation. The analysis of the electron distribution when they pass through a crystal has been made is Sec. 10. Let us consider the integral

$$\mathbf{Q}(\mathbf{x}) = \int d^2\rho n_a(\mathbf{x}-\boldsymbol{\rho})\frac{\boldsymbol{\rho}}{\rho}K_1(\beta\rho) \tag{21.9}$$

being a part of Eq. (21.7). If we represent $\mathbf{Q}(\mathbf{x})$ as the Fourier integral, then

$$\mathbf{Q}(\mathbf{x}) = \frac{1}{(2\pi)^2}\int d^2q_\perp e^{i\mathbf{q}_\perp\mathbf{x}}n_a(q_\perp)\boldsymbol{\varphi}(\mathbf{q}_\perp)\,, \tag{21.10}$$

where

$$\begin{aligned}n_a(q_\perp) &= \frac{1}{2\pi u_1^2}\int d^2\rho\exp\left(-\frac{\rho^2}{2u_1^2}+i\mathbf{q}_\perp\boldsymbol{\rho}\right)\\ &= e^{-q_\perp^2u_1^2/2}\,,\\ \times\boldsymbol{\varphi}(\mathbf{q}_\perp) &= 2\pi i\frac{\mathbf{q}_\perp}{q_\perp}\int\rho d\rho J_1(q_\perp\rho)K_1(\beta\rho) = \frac{2\pi i}{\beta}\frac{\mathbf{q}_\perp}{q_\perp^2+\beta^2}\,.\end{aligned} \tag{21.11}$$

Substituting (21.11) into (21.10) we have

$$\mathbf{Q}(\mathbf{x}) = \frac{i}{2\pi\beta}\mathbf{q}_\perp(\mathbf{x})\,;\quad \mathbf{q}_\perp(\mathbf{x}) = \int\frac{\exp(-q_\perp^2u_1^2/2)}{q_\perp^2+\beta^2}\mathbf{q}_\perp e^{-i\mathbf{q}_\perp\mathbf{x}}d^2q_\perp\,. \tag{21.12}$$

Note that $q_\perp^2+\beta^2 = q_\perp^2+q_\parallel^2+a^{-2} = q^2+a^{-2}$. The value of $\mathbf{q}_\perp(\mathbf{x})$ corresponds to the momentum transfer averaged over the position of the atoms when an electron is at distance $\mathbf{x}$ from the axis. Hence, the subtraction of the second term in (21.7) is associated with the extraction of pure fluctuation processes.

In the simplest case of a homogeneous distribution of electrons, $n_e(\mathbf{x}) = 1$, $\Phi(\rho) = 1$ and the second term in (21.7) contains

$$G \equiv \int d^2x \mathbf{Q}^2(\mathbf{x}) = \frac{1}{\beta^2} \int d^2q_\perp \frac{q_\perp^2 e^{-q_\perp^2 u_1^2}}{(q_\perp^2 + \beta^2)^2} . \tag{21.13}$$

Making in (21.13) a trivial integration over φ and going over to the integration variable $x = q_\perp^2/\beta^2$, we get

$$\begin{aligned} G(\delta) &= \frac{2\pi}{\beta^2} g(\delta) , \\ g(\delta) &= \frac{1}{2} \int_0^\infty dx \frac{xe^{-\delta x}}{(1+x)^2} \\ &= -\frac{1}{2}[e^\delta(1+\delta)Ei(-\delta) + 1] ; \quad \delta = u_1^2\beta^2 . \end{aligned} \tag{21.14}$$

Here $Ei(-x)$ is the integral exponential function for which it is convenient to use the following series representation at $x \leq 1$

$$Ei(-x) = C + \ln x + \sum_{k=1}^{\infty} \frac{(-x)^k}{kk!} . \tag{21.15}$$

For the electron energies under consideration $q_{\min} a \ll 1$, then if we allow for the fact that ~ 1 mainly contributes, we have $\beta = 1/a$ and $\delta = \delta_0 = u_1^2/a^2$. In the Moliere potential $a \simeq 111 Z^{-1/3}\lambda_c = 0.043 Z^{-1/3} nm$, and the amplitude of thermal vibrations is, generally speaking, comparable with the screening radius ($\delta \leq 1$), so that the function $g(\delta)$ should be calculated with sufficient accuracy. The exception is diamond for which $\delta \simeq 1/35$.

Substituting (21.14) into (21.5)–(21.7), we obtain after integration over ς ($d\sigma_{\gamma 0}$ is given by Eq. (7.51))

$$\begin{aligned} d\sigma_\gamma(\omega) = \frac{4Z^2\alpha^3}{m^2} \frac{\varepsilon - \omega}{\varepsilon} \frac{d\omega}{\omega} \Bigg\{ & \left(\frac{\varepsilon - \omega}{\varepsilon} + \frac{\varepsilon}{\varepsilon - \omega} - \frac{2}{3} \right) \\ & \times [\ln(183 Z^{-1/3}) - g(\delta_0) - f(Z\alpha)] + \frac{1}{9} \Bigg\} , \end{aligned} \tag{21.16}$$

where the function $f(Z\alpha)$ determines the Coulomb corrections (see Eq. (7.40)). If the electron distribution is not homogeneous but the density gradients are essential only at $\rho \gg u_1$, the cross section (21.16) should be multiplied by the factor $n_e(0)$. The behavior of the electron density near the axis has been discussed in Sec. 10.3.

21.3. Influence of the Effective Crystalline Fields on Incoherent Radiation

At sufficiently high energies of the particles, when the quantum parameter $\chi(\rho) = \gamma E(\rho)/E_0$ becomes comparable with unity, the external field begins to have an influence on the bremsstrahlung process. Since the incoherent radiation occurs mainly at distance ρ, where the nuclear density differs considerably from zero, $\rho \leq u_1$, we introduce the parameter $\chi_1 = \varepsilon V_0/m^3 u_1$ (V_0 is the characteristic scale of the potential). Note that since the electric fields of the axes are maximum at $\rho \simeq u_1$, χ_1 is a maximum estimate for quantum effects in the effective external field of the crystal. Let us also introduce a boundary energy ε_1 at which the parameter χ_1 is equal to unity ($\chi_1(\varepsilon_1) = 1$).

So, for $\varepsilon \ll \varepsilon_1$ the influence of the field on the incoherent radiation may be neglected, while at $\varepsilon \gg \varepsilon_1$ there appears the incoherent radiation suppression factor $\sim (u/\chi_1)^{2/3}$ due to a decrease of photon formation length. It is necessary to note that in the region where the field has a great influence on incoherent processes the radiation emission and the pair creation probabilities in effective fields of the atomic chains are one or two orders of magnitude higher than the probabilities of the corresponding incoherent (Bethe–Heitler) processes. Consequently, at energies $\varepsilon > \varepsilon_1$ there is no necessity to calculate cross sections with high accuracy and it is sufficient to confine oneself to a logarithmic accuracy.

The logarithmic accuracy of the analysis in the region of noticeable influence of the field on incoherent processes makes the picture of electromagnetic processes local in the transverse-to axis plane and enables the results obtained for the constant external field to be applied. Indeed, when calculating bremsstrahlung on separate nuclei with this accuracy we can consider the region of impact parameters $q_{max}^{-1} < \rho < q_{min}^{-1} = u_1$, not including their boundaries. We can consider also the particle and nuclear densities as well as the effective fields as constant because their inhomogeneity manifests itself only at distances $\rho \geq u_1$. Let us note also that, as the analysis in Sec. 21.1 shows, the ratio of the angle of multiple scattering to that of the particle turn in the effective potential of the chain on the photon formation length at $\varepsilon \geq \varepsilon_1$ is small ($k \ll 1$, see Eqs. (21.3) and (21.4)), and the incoherent processes can be studied using the perturbation theory with respect to scattering.

With such approach to the problem, advantage can be taken directly of the results of Sec. 7. In Sec. 7.5, the radiation cross section in a screened potential was calculated in the presence of an external field, using the Fourier transform of the scattering cross section (see Eq. (7.80)). To be more precise, the right-hand side of Eq. (7.73) includes the quantity $\Sigma(\mathbf{x}) - \Sigma(0)$. For the atomic distribution in the chain (21.8), this quantity may be represented as follows ($\theta_u = 1/\varepsilon u_1, \vartheta_1 = 1/\varepsilon a$)

$$\Sigma(\mathbf{x}) - \Sigma(0) = \left(\frac{Z\alpha}{\pi\varepsilon}\right)^2 \cdot \int \frac{d^2\theta}{(\theta^2 + \vartheta_1^2)^2}(1 - e^{-\theta^2/\theta_u^2})(e^{i\boldsymbol{\theta}\mathbf{x}} - 1) . \tag{21.17}$$

Calculating this integral under the assumption accepted in Sec. 7.5, $x\theta_u \ll 1$, we get

$$\Sigma(\mathbf{x}) - \Sigma(0) \simeq \frac{1}{2\pi}\left(\frac{Z\alpha}{\varepsilon}\right)^2 x^2 \left[\ln\frac{x\vartheta_1}{2} + C - \frac{1}{2} - g(\delta_0)\right] , \tag{21.18}$$

where $g(\delta_0)$ is given by Eq. (21.14). Comparing the result obtained with Eq. (7.81), we conclude that it is necessary simply to redefine the constant which is added to the logarithm on the right-hand side of Eq. (7.73). At middle energies, when $\chi_1 \ll 1$, this change contributes noticeably ($\sim 10\%$) to the total radiation intensity. A concrete calculation of the radiation probability using the results of Sec. 7 leads one to Eq. (21.16). At high energies when $\chi_1 \gg 1$ and the effective electric field has a substantial influence on the bremsstrahlung process we make use of the local picture, with logarithmic accuracy, and make the following choice

$$\begin{aligned} q_{\max} &= m\left(\frac{\chi_1}{u}\right)^{1/3} = \left(\frac{V_0\varepsilon}{u_1 u}\right)^{1/3} , \quad q_{\min} = \frac{1}{u_1} \\ q_{\max}/q_{\min} &= mu_1(\chi_1/u)^{1/3} = (V_0\varepsilon u_1^2/u)^{1/3} . \end{aligned} \tag{21.19}$$

Taking into consideration that the characteristic x in Eq. (21.18) is $x \sim \vartheta_{\text{eff}}^{-1} \sim \varepsilon/q_{\max}$, this formula can be represented, with logarithmic accuracy, as follows

$$\Sigma(\mathbf{x}) - \Sigma(0) \simeq -\frac{1}{2\pi}\left(\frac{Z\alpha}{\varepsilon}\right)^2 x^2 \ln\frac{q_{\max}}{q_{\min}} . \tag{21.20}$$

In addition, at a given distance $\mathbf{r}_\perp$ from the axis, we will consider, in accordance with the local picture, the process with constant densities $n_e(\mathbf{r}_\perp) \simeq n_e(0)$ (see Sec. 10.3) and $n_a(\mathbf{r}_\perp) = (1/2\pi u_1^2)\exp(-\mathbf{r}_\perp^2/2u_1^2)$ and in the field $e\mathbf{E}(\mathbf{r}_\perp) = -\boldsymbol{\nabla}_\perp U(\mathbf{r}_\perp)$ where $U(\rho)$ is the effective potential of the atomic chain. The particular form of this potential is given by Eq. (9.13). For the parabolic shape of the Fourier transform of the cross section (21.20) the radiation problem at scattering in the presence of an external field is solved exactly in Sec. 7, and the spectral density of the radiation probability is given by Eq. (7.90). However, by virtue of the condition $k \ll 1$ (see (21.3) and (21.4) there is no necessity to use this exact solution and it is possible to expand it in terms of the parameter $q(\rho)$ characterising the magnitude of multiple scattering. The leading term of this expansion gives the radiation probability in the constant external field in the absence of scattering, while the corrections $\sim q$ determine, with logarithmic accuracy, the local cross section of incoherent radiation on nuclei in the effective field of the chains (see Eq. (7.91))

$$\frac{d\sigma_\gamma(\rho)}{d\omega} = \frac{4Z^2\alpha^3}{15m^2\omega} d^2\rho n_e(\rho) n_a(\rho) f(\rho) \ln\left[mu_1\left(1+\left(\frac{\chi_1}{u}\right)^{1/3}\right)\right], \quad (21.21)$$

where

$$f(\rho) = \frac{\omega^2}{\varepsilon^2}(z^4\Upsilon - 3z^2\Upsilon' - z^3) + \left(1+\frac{(\varepsilon-\omega)^2}{\varepsilon^2}\right)(z^4\Upsilon + 3z\Upsilon - 5z^2\Upsilon' - z^3).$$

Here

$$\Upsilon(z) = \int_0^\infty d\tau \sin\left(z\tau + \frac{\tau^3}{3}\right), \quad z \equiv z(\rho) = \left(\frac{u}{\chi(\rho)}\right)^{2/3}. \quad (21.22)$$

At $\chi_1 \gg 1$ advantage may be taken, with the same (logarithmic) accuracy of Eq. (7.107) and after integration over ρ we get

$$\frac{d\sigma_\gamma}{d\omega} = \mathcal{A}_\gamma \frac{2Z^2\alpha^3\Gamma\left(\frac{1}{3}\right)}{5m^2\omega}\left(\frac{u}{3\chi_1}\right)^{2/3} \ln\left[mu_1\left(\frac{\chi_1}{u}\right)^{1/3}\right]\left(1+\frac{(\varepsilon-\omega)^2}{\varepsilon^2}\right), \quad (21.23)$$

where the constant $\mathcal{A}_\gamma$ is given by the expression

$$\mathcal{A}_\gamma = \int d^2\rho n_e(\rho) n_a(\rho)\left(\frac{\chi_1}{\chi(\rho)}\right)^{2/3} \quad (21.24)$$

For the particular form of the chain potential $U(x)$ (9.13), the parameter $\chi(\rho)$ at $\rho \le u_1 (x \le \eta)$ is of the form ($\eta \ll 1$)

$$\chi(\rho) \simeq 2\chi_1 \frac{\rho u_1}{\rho^2 + 2u_1^2}, \quad \chi_1 = \frac{V_0\gamma}{m^2 u_1} \simeq \frac{Z\alpha\gamma}{m^2 d u_1}. \quad (21.25)$$

Substituting (21.25) into (21.24) and making allowances for the homogeneity of the electron density at $\rho \leq u_1$, we obtain the following approximate expression for $\mathcal{A}_\gamma$

$$\mathcal{A}_\gamma \simeq n_e(0)2^{-\frac{1}{3}} \int_0^\infty dx e^{-x} \cdot x^{-\frac{1}{3}}(1+x)^{2/3} \simeq 1.508\, n_e(0)\,. \qquad (21.26)$$

Multiplying (21.23) by ω, we obtain after integration over ω with logarithmic accuracy the contribution to the radiation intensity of the incoherent process

$$\frac{I^{\text{ncoh}}}{\varepsilon} = \frac{8.29\pi\Gamma(\frac{1}{3})Z^2\alpha^3 N}{5 \cdot 3^5 3^{1/6} m^2 \chi_1^{2/3}} \cdot \mathcal{A}_\gamma \ln(g(m u_1 \chi_1^{1/3})\,, \qquad (21.27)$$

where N is the average atomic density in the crystal. Let us compare the intensity obtained with that in the field of a separate chain (Eq. (17.17)) for a homogeneous distribution of electrons ($n_e = 1$) at large $\chi_s \gg 1(\chi_s/\chi_1 = u_1/a_s < 1)$

$$\frac{I^{\text{ncoh}}}{I^F} = 3.41 \frac{\mathcal{A}_\gamma Z^2\alpha^2 N^2 x_0 a_s}{V_0 m} \frac{\chi_s^{1/3}}{\chi_1^{2/3}}$$

$$\times \frac{\ln(m u_1 \chi_1^{1/3})}{\ln \chi_s} \simeq 1.64 \frac{\lambda_c}{a_s}\left(\frac{u_1}{a_s}\right)^{1/3}$$

$$\times \frac{Z\alpha}{\chi_1^{1/3}} \frac{\ln(m u_1 \chi_1^{1/3})}{\ln \chi_s} \lesssim 10^{-2} \frac{Z\alpha}{\chi_1^{1/3}} \frac{\ln(m u_1 \chi_1^{1/3})}{\ln \chi_s}\,. \qquad (21.28)$$

It follows from this equation that at high electron energies, when $\chi_s \gg 1$, the contribution of incoherent radiation may be neglected.

21.4. Incoherent Photoproduction of Electron–Positron Pairs

The probability of pair creation by a photon at high particle energies is derived from the radiation probability using the substitutions $\varepsilon \to -\varepsilon, \omega \to -\omega$. Besides, a change in the phase volume $\omega^2 d\omega \to -\varepsilon^2 d\varepsilon$ should be taken into account. Since the distribution of the photons passing through crystals remains the same in the used order of the perturbation theory, at such substitution one needs to use the expressions for the radiation probability with $n_e(\rho) = 1$. Then, for middle photon energies, when the parameter $\kappa_1 = V_0\omega/m^3 u_1$ is small ($\kappa_1 \ll 1$) and photoproduction in the effective potential of the chain is exponentially suppressed, the photoproduction cross section is derived from Eq. (21.16) using the substitutions indicated above

$$d\sigma_p(\varepsilon_1) = \frac{4Z^2\alpha^3}{m^3}\frac{d\varepsilon_1}{\omega}\left[\left(\frac{\varepsilon_1^2+\varepsilon_2^2}{\omega^2}+\frac{2\varepsilon_1\varepsilon_2}{3\omega^2}\right)\right.$$

$$\left.\times\left(\ln 183Z^{-1/3} - g(\delta_0) - f(Z\alpha)\right) - \frac{\varepsilon_1\varepsilon_2}{9\omega^2}\right], \qquad (21.29)$$

$$\varepsilon_1 + \varepsilon_2 = \omega .$$

After integration of (21.29) over ε_1, we obtain for the total cross section of the pair creation in an aligned crystal

$$\sigma_p = \frac{28Z^2\alpha^3}{9m^2}\left[\ln 183Z^{-1/3} - \frac{1}{42} - f(Z\alpha) - g(\delta_0)\right] \qquad (21.30)$$

At photon energies $\omega \geq \omega_1(\kappa(\omega_1) = 1)$ the effective field of the chain changes the process of incoherent pair creation. Using Eqs. (21.21) and (21.22), with $n_e(\rho) = 1$, and the appropriate substitutions taken into account, we get

$$\frac{d\sigma_p(\rho)}{dx} = \frac{4Z^2\alpha^3}{15m^2}n_a(\rho)f_p(\rho)d^2\rho\,\ln\left[mu_1(1+(\kappa_1 x(1-x))^{1/3})\right], \qquad (21.31)$$

where $x = \varepsilon_1/\omega, f_p(\rho) = (\varepsilon/\omega)^2 f(z_p(\rho))$ and the function $f(z)$ defined in Eq. (21.21),

$$z_p(\rho) = (\kappa(\rho)x(1-x))^{-\frac{2}{3}} \qquad (21.32)$$

At photon energies $\omega \gg \omega_1(\kappa_1 \gg 1)$, Eq. (21.23) is applicable and we then obtain

$$\frac{d\sigma_p}{dx} = \mathcal{A}_p\frac{2Z^2\alpha^3\Gamma(1/3)}{5m^2}(\varepsilon\kappa_1 x(1-x))^{-\frac{2}{3}}$$

$$\times\,(x^2+(1-x)^2)\ln\left(mu_1\kappa_1^{1/3}\right), \qquad (21.33)$$

where (see Eq. 21.26))

$$\mathcal{A}_p = \int d^2\rho n_a(\rho)\left(\frac{\kappa_1}{\kappa(\rho)}\right)^{\frac{2}{3}} \simeq 1.508 .$$

Integrating (21.33) over x we obtain, with logarithmic accuracy, the following expression for the total cross section of a pair creation on fluctuations of the potential

$$\sigma_p = A_p \frac{8Z^2\alpha^3\Gamma^3\left(\frac{1}{3}\right)\ln\left(mu_1\kappa_1^{1/3}\right)}{25m^2\Gamma\left(\frac{2}{3}\right)(3\kappa_1)^{2/3}}. \tag{21.34}$$

It is interesting to compare the probability of the incoherent pair creation with that of the pair creation in the effective potential of the chain at high $\kappa_s \gg 1(\kappa_s/\kappa_1 = u_1/a_s < 1)$

$$\begin{aligned} R_p &= \frac{W_p^{\text{ncoh}}}{W_p^F} = \frac{\sigma_p^N}{W_p^F} \simeq A_p \frac{28 \cdot 2^{1/3}}{125 \cdot 3^{5/6}} \cdot \left(\frac{\Gamma(1/3)}{\Gamma(2/3)}\right)^4 \\ &\times Z\alpha \frac{\lambda_c}{a_s}\frac{\kappa_s^{1/3}}{\kappa_1^{2/3}}\frac{\ln\left(mu_1\kappa_1^{1/3}\right)}{\ln\kappa_s} \simeq 2.5Z\alpha\frac{\lambda_c}{a_s}\left(\frac{u_1}{a_s}\right)^{1/3} \\ &\times \frac{\ln\left(mu_1\,\kappa_1^{1/3}\right)}{\kappa_1^{1/3}\ln\kappa_s} \lesssim 10^{-2}\frac{Z\alpha\,\ln\left(mu_1\,\kappa_1^{1/3}\right)}{\kappa_1^{1/3}\,\ln\,\kappa_s} \end{aligned} \tag{21.35}$$

Thus, at photon energies when $\kappa_s \gg 1$ the contribution of the incoherent pair creation may be omitted.

21.5. Total Picture of the Radiation Processes in Crystals

Let us consider the qualitative picture of the electromagnetic processes in aligned crystals, involving incoherent effects, versus the particle energy. The quasiclassical radiation theory is assumed to be valid, i.e. particle energies $\varepsilon > 100$ MeV will be considered. The low-energy range is the range where the parameter $\rho_c = 2V_0\varepsilon/m^2$ is small ($\rho_c \ll 1$), that means nonrelativistic transverse motion. The middle energies are characterized by larger values of the parameter $\rho_c \geq 1$ and, at the same time, by small values of the quantum parameter $\chi_s = V_0\varepsilon/m^3a_s(\rho_c/\chi_s = 2a_s/\lambda_c \sim 10^2)$. Finally, in the range of high energies $\chi_s \geq 1$.

At low energies the classical radiation theory for particles moving in the effective fields of atomic chains (planes) holds since soft frequencies are emitted in a relatively narrow interval of frequencies, $\omega \sim \omega_c \ll \varepsilon$. Radiation is of a dipole nature and its total intensity is less than that of the radiation caused by the Bethe–Heitler mechanism whose spectrum extends up to $\omega \sim \varepsilon$.

In the range of middle energies, the energy losses in the field of the axis are larger than those due to the Bethe–Heitler mechanism. The emitted photons are still soft ($\omega \ll \varepsilon$). At $\chi_s \sim 0.1$ the quantum effects start to manifest themselves.

Both at low and middle energies, owing to a redistribution of the charged-particle flux at angles of incidence $\vartheta_0 < \vartheta_c = (2U_0/\varepsilon)^{1/2}$ the density of the electrons near the chains $n_e(\rho \leq u_1)$ can be considerably higher and that of the positrons considerably lower than the average (initial) particle density, what varies the probability of incoherent radiation to the same extent. One more effect in incoherent processes manifests itself directly on account of ordered positions of the atoms in a crystal thereby leading to the cut-off of a minimum momentum transfer connected with the amplitude of thermal vibrations rather than with the screening radius, as in the Bethe–Heitler mechanism. Generally speaking, as seen from the above calculations, both the screening and the ordered atomic arrangement contribute. This reduces the relevant cross sections by 10% to 20%.

In the high-energy range the radiation intensity in constant (on the photon formation length) fields of the crystalline axes is one to two orders of magnitude higher than the corresponding intensity of the Bethe–Heitler mechanism. In this case, the characteristic frequencies of the emitted photons are comparable with the particle energy ($\omega \sim \varepsilon$). In addition, at photon energies when $\kappa_s \geq 1$ a mechanism of pair creation by photons manifests itself in a constant external field. At low and middle energies this mechanism is exponentially suppressed. The probability of electron–positron pair photoproduction by this mechanism in the maximum is also one to two orders of magnitude larger than the Bethe–Heitler probability.

Because of a considerable curving of the trajectory in the crystal-axis field and of the resulting shortening of the photon formation length during the bremsstrahlung at $\chi_s \sim 1$, incoherent processes begin to be suppressed and at the energies when this parameter is large ($\chi_s \gg 1$) their contribution becomes negligible. While in the high-energy region the factors mentioned above hinder an experimental separation of the contribution of incoherent processes, at low and middle energies the emission of hard photons ($\omega \sim \varepsilon$) is completely governed by the incoherent process. According to the above analysis, measuring the hard part of the spectrum at such energies gives direct information on the electron and positron densities near the atomic chains, which is necessary to create an adequate kinetic picture of the motion of charged particles in aligned single crystals. For this purpose, it is desirable to have experimental data for a sufficiency large set of crystal thicknesses.

In the experiment (Ganenko *et al.*, 1987) measurements have been made of the yield of hard 700 MeV photons emitted by 800 MeV electrons passing through silicon and diamond single crystals at axial orientation. This experiment corresponds to the range of middle energies. We use the experimental data for a non-hard collimation of a photon beam ($\vartheta_k = 2/\gamma$) at $\vartheta_0 = 0$ when the azimuthal scattering can be omitted and the peculiarities of incoherent radiation manifest themselves in pure form. Experimentally, a radiation yield under such conditions exceeds that in a nonaligned crystal roughly by 2.3 ± 0.1 for a Si crystal (thickness is 30 μm $\langle 111 \rangle$ axis) and 1.3 ± 0.1 for a diamond one (thickness is 300 μm $\langle 100 \rangle$ axis). In accordance with the results obtained above this enhancement is given by the factor $n_e(0)$ since $l_d(\mathrm{Si}) = 39$ μm, $l_d(c) = 54$ μm and one cannot consider these crystals as very thin. Making use of Eq. (10.34), ($n_e(0) = \bar{F}(0)$) and taking into account that ζ_0^{-1} (Si) $= 1.3$ and $\zeta_0^{-1}(\mathrm{C}) = 0.18$, we obtain $n_e(0)(\mathrm{Si}) = 2.26$ and $n_e(0)(C) = 1.18$. As it is seen, these results are consistent with the experimental data within its error limits. This agreement is particularly good for Si crystals where the enhancement under discussion is rather large.

Chapter 5

ELECTROMAGNETIC RADIATION AT CHANNELING

In this chapter the radiation from charged particles, whose energy is not higher than several tens GeV, will be under consideration. This problem has been discussed by a number of theoretical groups. The most complete results are contained in their books and reviews: Kumakhov (1986); Bazylev and Zhevago (1987); Akhiezer and Shul'ga (1982); Baryshevskii (1982); Vorob'ev (1984). The wide circle of bibliograhies are also contained there. The consideration of the present chapter follows the authors' papers: Baier, Katkov, Strakhovenko (1979a, b; 1980; 1981a, b, d; 1983a–d; 1984a–d).

22. RADIATION IN THIN CRYSTALS

22.1. Radiation from Electrons and Positrons at Planar Channeling. Dipole Case

For a description of the radiation in a crystal the problem of summing the contributions of all possible trajectories whose form is essentially dependent on the point of incidence x_0 for a given angle of incidence ϑ_0 arises. This situation differs from one in an ideal undulator. In other words it is necessary to find the particle distribution over transverse energy $\varepsilon_\perp = \varepsilon\vartheta_0^2/2 + U(x_0)$ and make the appropriate averages. In a thin crystal by definition, we neglect the alteration of the distribution function (DF) and use the initial distribution (9.24) for DF. By this reason, the description of the radiation in thin crystals, when the crystal thickness $L \ll \ell_d$, becomes more simple. The experiment using thin crystals is convenient for the quantitative comparison with the theory not only for the reason of the comparable simplicity of the theory consideration, or else the radiation process is extracted in a pure form without the mixing of the rather complicated kinetic picture. On the other hand the crystal thickness L must be not too small, allowing a particle to have a number of oscillations while moving in the potential well:

$$d(\varepsilon/U_0)^{1/2} \ll L \ll \ell_d\,. \tag{22.1}$$

Below the last condition is assumed to be valid.

At planar channeling, the value of $\varepsilon_\perp$ fixes the trajectory type. The radiation characteristics in the dipole approximation for a given trajectory have been obtained in Sec. 1.9. Here we shall use these results for the particle motion in the potential (9.33). In particular, for the description of the radiation polarization, Eqs. (1.142) and (1.143) may be used. For the case of the planar channeling one has to put $\beta = 0$ (a linear polarization exists only) and make the corresponding averages. Let us remind that the dipole approximation is valid when the parameter $\rho_c = 2U_0\varepsilon/m^2$ (compare with Eq. (16.3)) is small $\rho_c \ll 1$. For the planar case, this condition is valid in the energy region $\varepsilon \leq 1$ GeV, when the photons are emitted with the frequency $\omega \ll \varepsilon$ and a classical description of radiation may be used.

For positrons, substituting the Fourier component of the velocity v_m (9.39) and (9.41) into the formula

$$w_m = -2\pi m i v_m/T\,, \tag{22.2}$$

we have the Fourier component of the acceleration w_m, T is the period of the particle motion. Substituting w_m into formula (1.132) and integrating over the photon-emission angles up to $\vartheta = \vartheta_{\max}(y_{\max} = [1 + (\gamma\vartheta_{\max})^2])$, we find the spectral distribution of the radiation intensity from positrons for the given value of $\varepsilon_\perp$

$$\frac{dI}{d\xi} = I_0\frac{\pi^4\xi}{f^2}\sum_{m=1}^{\infty} B^2(m)\left(\eta_m^2 - \eta_m + \frac{1}{2}\right) \times \vartheta(1-\eta_m)\cdot\vartheta(2\eta_m y_{\max} - 1)\,, \tag{22.3}$$

where

$$I_0 = 8e^2V_0^2\gamma^2/a_s^2m^2\,, \quad \xi = \omega/\omega_c\,, \quad \omega_c = 4\pi\gamma^2/T_0$$

is the characteristic frequency of the radiation, and

$$\eta_m^c = \xi f^c/(2m-1), \quad f^c = T_c/T_0; \quad \eta_m^{nc} = \xi f^{nc}/2m, \quad f^{nc} = 2T_{nc}/T_0 \,. \tag{22.4}$$

The quantity $B(m)$ in (22.3) has the form (see (9.39) and (9.41))

$$B(m) = \begin{cases} \frac{1}{2}\left[\sinh\left(g_m \frac{K'}{K}\right)\right]^{-1}, & z < 1 \\ \sum\limits_{n=1}^{\infty} \frac{g_n \cos(\lambda g_n)}{\sinh\left(g_n \frac{K'}{K}\right)\left[g_n^2 - \left(\frac{\pi m}{1-\lambda}\right)^2\right]}, & z > 1\,. \end{cases} \tag{22.5}$$

In calculations it is convenient to use the expression for f that is valid for any $z = \varepsilon_\perp/U_0$

$$f = [\sinh(\delta/2)]^{-1} \int_{A(z)}^{\sqrt{z}} dy[\psi(y)]^{-1}\,, \tag{22.6}$$

where

$$\begin{aligned} A(z) &= \vartheta(z-1)\,\sqrt{z-1}\,, \\ \psi(y) &= \sqrt{(z-y^2)(\widetilde{z}-1+z-y^2)} \\ \widetilde{z} &= \coth^2(\delta/2)\,. \end{aligned} \tag{22.7}$$

For the total intensity, not taking into account a collimation of radiation, we have

$$I = I_0 \frac{\sinh^3(\delta/2)}{3f} \int_{A(z)}^{\sqrt{z}} dy\,\psi(y)\,. \tag{22.8}$$

This quantity can be expressed in terms of the elliptic integrals

$$I = \frac{I_0}{g(1-k^2)} \left\{ \frac{E(k)-E(\alpha,k)}{K(k)-F(\alpha,k)} \cdot \frac{1+k^2}{1-k^2} \right.$$
$$\left. -1 - \frac{k^2 \sin 2\alpha}{2(1-k^2)} \cdot \frac{\sqrt{1-k^2\sin^2\alpha}}{K(k)-F(\alpha,k)} \right\}, \tag{22.9}$$

where $\sin\alpha = \vartheta(z-1)(1-z^{-1})^{1/2}$, $E(\alpha,k)$ is the elliptic integral of the second kind.

The total intensity (22.8) grows with the increase of z in the interval (0, 1), and at the further increase of z begins to fall so that

$$I\Big|_{z\to\infty} = \frac{I_0}{24}\left(\frac{\sinh 2\delta}{2\delta} - 1\right)\,. \tag{22.10}$$

For particles, moving in the channel ($\sin\alpha = 0$), formula (22.9) is simplified:

$$I^c = \frac{I_o}{g(1-k^2)}\left[\frac{E(k)}{K(k)}\cdot\frac{1+k^2}{1-k^2} - 1\right]. \qquad (22.11)$$

The obtained quantities are to be averaged over states with various $\varepsilon_\perp$ For a classical description the particle density in the interval $d\varepsilon_\perp$ is given by Eq. (9.24). Using the explicit form of the potential (9.33), we have for the averaged value of any radiation characteristic $\varphi(z)$

$$\langle\varphi\rangle = \delta^{-1}\int_b^{1+b} dz\ \varphi(z)[\psi(y=\sqrt{b})]^{-1}, \qquad (22.12)$$

where $b = \varepsilon_0/U_0 = (\vartheta_0/\vartheta_c)^2$ The value $\langle\varphi\rangle$ depends on the incident angle ϑ_0 via b. In Fig. 22.1 the functions $\langle I\rangle/I_0$ depending on b are shown ($\langle I^c\rangle$ — curve 1, $\langle I^{nc}\rangle$ — curve 2 and $\langle I\rangle = \langle I^c\rangle + \langle I^{nc}\rangle$ — curve 3) according to Eqs. (22.12) and (22.8). When the angle of incidence (b) increases the total number of particles in a channel decreases slowly but the averaged intensity $\langle I^c\rangle$ grows and reaches its maximum at $b = 0.69$. It is connected with the dominant contribution to the radiation intensity from particles having $z \simeq 1$. For $b > 1$ only the above-barrier particles contribute to the radiation intensity $\langle I\rangle$ which decreases when the angle of incidence increases and tends to the value given by Eq. (22.10). In Fig. 22.2, the spectral intensity (see Eqs. (22.30) and (22.13)) is shown normalized to unity $[(d\langle I\rangle/d\xi)/\langle I\rangle]$ for different angles of

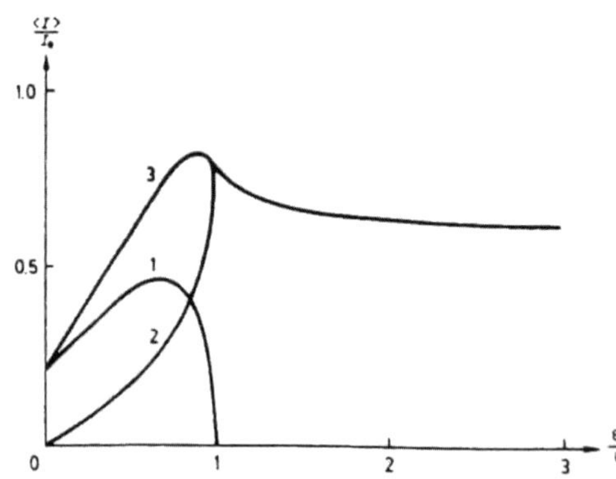

Fig. 22.1. Radiation intensity from positrons depending on ε_0/U_0 averaged over $x_0, \delta = 2.5$.

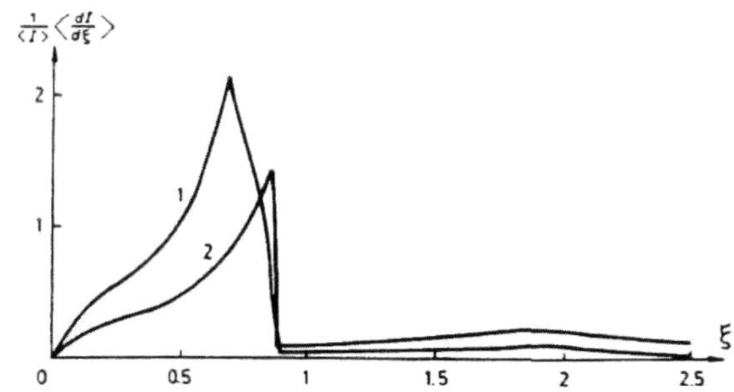

Fig. 22.2. Spectral distribution of the radiation intensity averaged over x_0 (summary contribution).

incidence ($b = 0.2$ (curve 1) and $b = 0.9$ (curve 2)). For the particles moving in the channel, the first harmonic dominates (the sum of the other ones is less than 1%) and the break in the point $\xi = 1/f^c_{\min} \simeq 0.9$ corresponds to the boundary frequency of this harmonic. The peaks in Fig. 22.2, are in the points $\xi = 1/f(\varepsilon_\perp = \varepsilon_0)$ so their positions are determined by the incident angle only. For $\xi > 0.9$, the spectrum is practically determined by the above-barrier particles (the wide maximum in the region $\xi \simeq 1.8$ corresponds to the first harmonic radiation). The clear separation of the contribution to the intensity spectra from above-barrier particles and from ones moving in a channel is connected with the cardinal changing of the period of particle motion at the transition to the above-barrier state. Above we have put the angle of incidence fixed. If the spread of the incident angles $\Delta\vartheta_0$ is compared with ϑ_c then it is necessary to average the function $\langle\varphi\rangle$ in Eq. (22.12) with the distribution function $\chi(\vartheta_0/\vartheta_c)$

$$\langle\varphi\rangle_{\vartheta_0} = \int d\vartheta_0 \langle\varphi\rangle \chi(\vartheta_0/\vartheta_c) \Big/ \int d\vartheta_0 \chi(\vartheta_0/\vartheta_c)\,. \tag{22.13}$$

If χ is an even function then taking into account Eq. (22.12), formula (22.13) may be presented in the form

$$\langle\varphi\rangle_{\vartheta_0} = \left(\int_0^1 dz \int_0^{\sqrt{z}} dy + \int_1^\infty dz \int_{\sqrt{z-1}}^{\sqrt{z}} dy\right) \frac{\varphi(z)}{\psi(y)} \chi(y)/\delta \int_0^\infty dy \chi(y)\,, \tag{22.14}$$

where we have changed the order of integration over ϑ_0 and z; the function $\psi(y)$ has been defined in Eq. (22.7). The contributions of the above-barrier particles

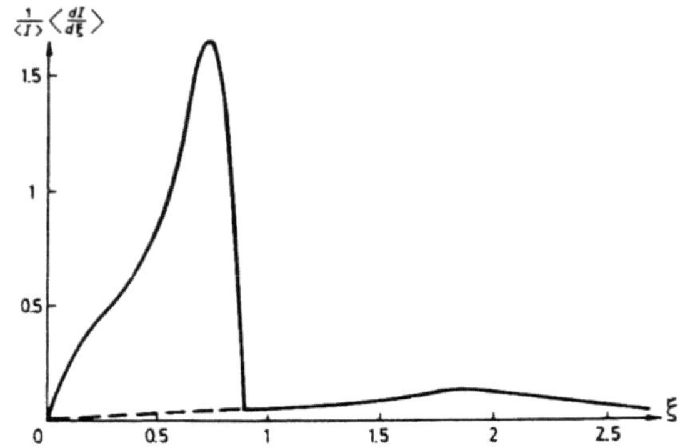

Fig. 22.3. Spectral distribution of the radiation intensity from positrons averaged over x_0 and ϑ_0, $\delta = 2.5$.

and those ones moving in a channel are separated explicitly and given by the second and first terms in parentheses accordingly. The numerical calculation has been carried out for the cases where $\chi(y) \propto \vartheta(1-y)$ (uniform distribution in the interval $|\vartheta_0| \leq \vartheta_c$). In this case Eq. (22.14) has the form

$$\langle\varphi\rangle_{\vartheta_0} = \frac{\sinh(\delta/2)}{\delta}\left\{\int_0^1 dz\,\varphi(z) + \int_1^2 dz\,\varphi(z)\left[F\left(\frac{\pi}{2}-\alpha,k\right)-F(\alpha,k)\right]k'\right\}. \tag{22.15}$$

In Fig. 22.3 the result of calculations of the quantity $\langle dI/d\xi\rangle_{\vartheta_0}/\langle I\rangle_{\vartheta_0}$ is presented according to Eqs. (22.15) and (22.3). For the total intensty we have

$$\langle I\rangle_{\vartheta_0} = 0.46\, I_0\,, \quad \langle I^c\rangle_{\vartheta_0} = 0.32\, I_0\,. \tag{22.16}$$

Note finally that a collimation of the radiation under consideration has no such influence as in an undulator. This is connected with the spread of the particle motion periods.

Let us consider the radiation from electrons. For electrons, the spectral distribution of the radiation intensity for a fixed value $\varepsilon_\perp(z)$ can be presented in the form (22.3), where one has to make substitutions $f \to f_e$ and $B(m) \to B_e(m)$. The value f_e is obtained from Eq. (22.6) by the substitution $\psi(y) \to \psi_e(y) = [(y^2 - z + 1)(y^2 - z + \tilde{z})]^{1/2}$. We give the expression for k^2 together with $B_e(m)$ (in the case of positrons there is the single expression for k^2 (see

Eq. (9.36)). Thus

$$B_e(m) = \begin{cases} \sum\limits_{n=1}^{\infty} \frac{g_n}{\left(\frac{kK'}{f} g_m\right)^2 + g_n^2} \cdot \frac{\cosh\left(\frac{f g_n}{kK'}\right)}{\sinh\left(\frac{K}{K'} g_n\right)}, & k^2 = \frac{\tilde{z}-1}{\tilde{z}-z}, \quad z < 1 \\ \sum\limits_{n=1}^{\infty} \frac{g_n}{\left(\pi m \frac{K'}{f}\right)^2 + g_n^2} \cdot \frac{\sinh\left(g_n \frac{f}{K'}\right)}{\cosh\left(\frac{K}{K'} g_n\right)}, & k^2 = \frac{\tilde{z}-z}{\tilde{z}-1}, \quad 1 \le z \le \tilde{z} \\ \sum\limits_{n=1}^{\infty} \frac{\pi n}{\left(\frac{\pi m k' K'}{f}\right)^2 + (\pi n)^2} \cdot \frac{\sinh\left(\frac{\pi n f}{k' K'}\right)}{\cosh\left(\pi n \frac{K}{K'}\right)}, & k^2 = \frac{z-\tilde{z}}{z-1}, \quad z \ge \tilde{z}, \end{cases} \tag{22.17}$$

where $\mathfrak{g}_m = \pi(m - 1/2)$.

For electrons the radiation intensity for a fixed $\varepsilon_\perp$ is given by Eq. (22.8) with the substitution $\psi(y) \to \psi_e(y)$. The averaging over x_0 is made using Eq. (22.12) where we have to change $\varphi(z) \to \varphi_e(z)$ $\psi(y = b^{1/2}) \to \psi_e(y = b^{1/2})$. It is necessary to do this for the averaging over angle ϑ_0 (see Eqs. (22.13) and (22.14)).

The functions f_e and I_e can be presented by the elliptic integrals (see, for example, Eq. (9.46) for the period). Let us remind that (1) the functions f_e (period) pass through the same values for electrons moving in a channel and for above-barrier states; (2) the interval of these values is infinite. The first property leads to the essential (practically full) overlapping of the contributions to the spectra from these two groups of particles, the second one gives the rather large width of the spectral distribution (with respect to the radiation spectra from positrons).

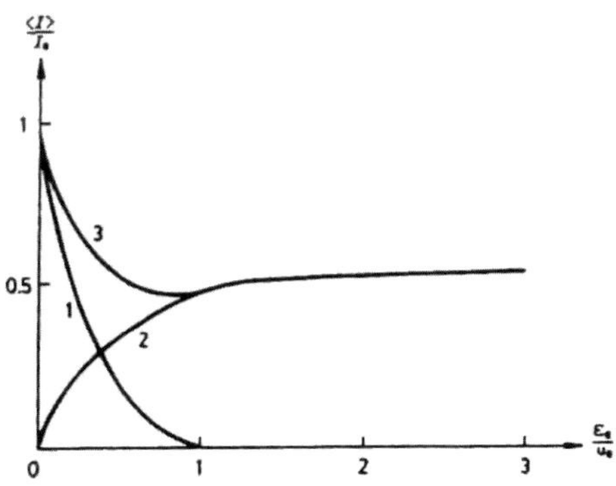

Fig. 22.4. Radiation intensity from electrons averaged over x_0.

The intensity I_e^c monotonically falls down to zero for z changing from 0 to 1, and I_e^{nc} grows tending to its asymptotic value (22.10) for $z \gg 1$. The form of the average distribution is determined by these reasons as well as by the population of states for different z. In Fig. 22.4 the function I_e averaged over x_0 is shown depending on the variable $b = \varepsilon_0/U_0 = (\vartheta_0/\vartheta_c)^2(\langle I_e^c\rangle$ — curve 1, $\langle I_e^{nc}\rangle$ — curve 2, and $\langle I_e\rangle = \langle I_e^c\rangle + \langle I_e^{nc}\rangle$ — curve 3). At first the value $\langle I_e\rangle$ decreases, reaches its minimum at $b = 0.83$ and then slowly grows tending to the asymptotic value defined by Eq. (22.10).

In Fig. 22.5 the spectral intensity is shown depending on ξ for various values of the parameter b. As for the case of positrons there are the breaks on the curves in the points $\xi = (2n-1)/f(z=b)$ and $\xi = 2n/f(z=1+b)$ connected with the boundary frequency of the n-th harmonic for electrons moving in a channel and above-barrier electrons, respectively. One can clearly see the overlapping of the contributions to the spectra from these particle groups. For example, the breaks on the curve for $b = 0.4$ in the points $\xi \simeq 2$ and $\xi \simeq 4$ correspond to the limits of the first and second harmonics for above-barrier particles and the break at $\xi = 2.3$ corresponds to the first harmonic for particles in the channel. For the case $b = 0.7$ the limits of the first and second harmonics lie at $\xi \simeq 1.2$ and 3.7 for particles in the channel and at $\xi \simeq 2.5$ and 4.9 for particles moving above the barrier. A further difference between the radiation spectra of electrons and positrons lies in the fact that in the case of electrons not one but the first few harmonics contribute. These features are clearly illustrated by the behavior of the integrated spectrum shown in Fig. 22.6,

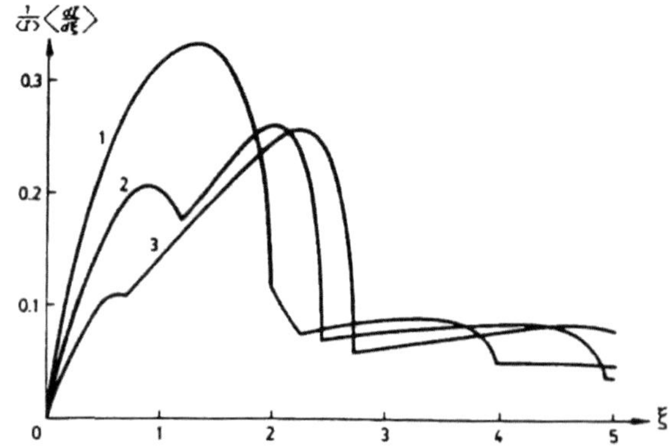

Fig. 22.5. Spectral distribution of the radiation intensity from electrons averaged over x_0 (summary contribution) for various angles of incidence, $\delta = 2.5$.

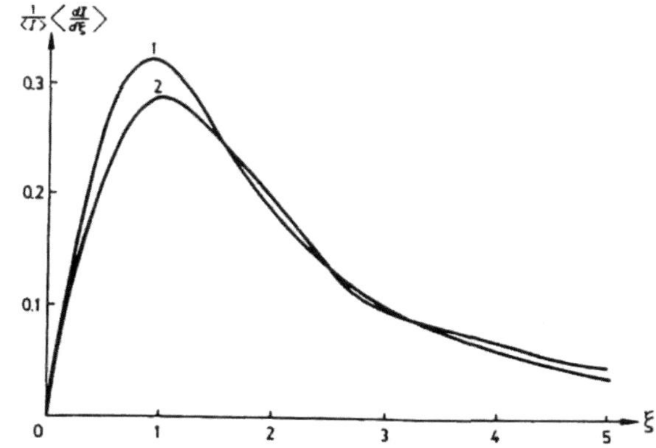

Fig. 22.6. Spectral distribution of the radiation intensity from electrons in channel (curve 1) and summary contribution (curve 2), $\delta = 2.5$.

where we present the graph of $(\langle dI_e \rangle_\vartheta / d\xi)/\langle I_e \rangle_\vartheta$ the total spectrum normalized to unity (curve 2) and the radiation spectrum of the particles moving in the channel, also normalized to unity (curve 1). The averaging over ϑ was done for a uniform distribution $|\vartheta_0| \leq \vartheta_c$. We see that the position of the maximum and the width of the spectral distribution are very close for the two cases. For the integrated intensities, the calculation gives

$$\langle I_e^c \rangle_{\vartheta_0} = 0.45\, I_0, \; \langle I_e \rangle_{\vartheta_0} = \langle I_e^c \rangle_{\vartheta_0} + \langle I_e^{nc} \rangle_{\vartheta_0} = 0.68\, I_0 \,. \tag{22.18}$$

The results obtained in this section are in complete qualitative and fairly good quantitative agreement with those of Baier, Katkov, Strakhovenko, 1979*b* where the radiation at electron channeling was studied for the more simple shape of potentials (an inverted parabola).

The ratio of the radiation intensities from positrons and electrons averaged over x_0 shown in Fig. 22.7 has the steep decrease for small angles and the maximum at $\vartheta_0 \simeq \vartheta_c$. The position of this maximum varies weakly over a wide range of the parameter δ.

Thus, there is an essential difference between the spectral distributions of the radiation intensity from electrons and positrons. For positrons, as noted above, the peak in $dI/d\omega$ is relatively narrow and the radiation spectra of the particles moving in the channel and above the barrier are actually different, while for electrons this peak is considerably wider and the radiation spectra from particles above and below the barrier overlap considerably.

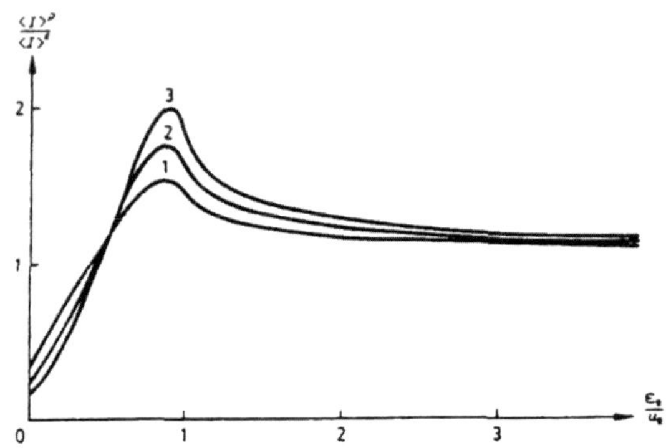

Fig. 22.7. Ratio of the radiation intensity from positrons and electrons averaged over x_0 depending on ε_0/U_0 for various values of the parameter δ.

The ratio of the spectral distributions of the radiation intensity from electrons and positrons in channeling and the corresponding characteristic of the bremsstrahlung (for an amorphous medium) in the vicinity of the peak in the radiation spectrum of the channeled particles can be roughly estimated by the formula

$$\left.\frac{d\langle I\rangle/d\omega}{dI_{\rm br}/d\omega}\right|_{\omega\simeq\omega_c} \sim \frac{\langle I\rangle}{\omega_c}L_{\rm rad} \sim \alpha\left(\frac{U_0}{m}\right)^{3/2}\frac{L_{\rm rad}}{d}\sqrt{\gamma} \tag{22.19}$$

For example, for silicon (the (110) plane) at the energy $\varepsilon = 1$ GeV this ratio is $\sim 10^2$ The measurement of the ratio given by Eq. (22.19) is convenient for distinguishing the radiation at channeling on a background bremsstrahlung.

22.2. Radiation Intensity at Axial Channeling in the Classical Region

In the case of an axial channeling particles move in two-dimensional transverse space. Then it is impossible to obtain analytically even the trajectory of a particle and more so to make a very cumbersome analysis of the radiation from actual trajectories. However, in the classical limit, the total radiation intensity is the local characteristic of the particle motion and the problem is simplified substantially for this case. Really, if one is interested in characteristics of the radiation that depend on the instantaneous values of the coordinates and the momenta of a particle only, then one needs to know only the distribution of particles in the transverse to the direction of axes phase space. This distribution in thin crystals is given by Eqs. (9.61) and (9.62).

For ultrarelativistic particles in the potential $U(\rho)$ the instantaneous radiation intensity is determined by the well known equation (1.18):

$$I(\rho) = \frac{2}{3}e^2 w^2 \gamma^4 = \frac{2}{3}\frac{e^2}{m^2}\gamma^2(\nabla U)^2 \equiv A(\nabla U)^2\,, \qquad (22.20)$$

where w is the instantaneous acceleration, $\gamma = \varepsilon/m$ is the Lorentz factor. Using the distribution function (9.61), one obtains for the radiation intensity from particles during axial channeling in a single crystal

$$I(l) = A\int d\varepsilon_\perp \varphi(\varepsilon_\perp) S^{-1}(\varepsilon_\perp)\int d^2\rho(\nabla U)^2\vartheta(\varepsilon_\perp - U(\rho))\,. \qquad (22.21)$$

The integrals over $d^2\rho$ in (22.21) are taken within one elementary cell. The contribution to the radiation from particles moving above a barrier (when $\varepsilon_\perp \geq U_0$, $U_0 = \max U$, $U \geq 0$) is given by

$$I_{nc} = A_{n_\perp}\int d^2\rho(\nabla U)^2 \int_{\varepsilon_\perp \geq U_0} d\varepsilon_\perp \varphi(\varepsilon_\perp)\,. \qquad (22.22)$$

Since at $\varepsilon_\perp \geq U_0$ the whole transverse coordinate space is accessible then $s(\varepsilon_\perp) = 1/n_\perp$. In the case $\varepsilon_0 \geq U_0$ all the particles are moving above the barrier, then using the normalization condition for the function $\varphi(\varepsilon_\perp)$ one has for the radiation intensity

$$I_{as} = An_\perp \int d^2\rho(\nabla U)^2\,. \qquad (22.23)$$

Thus, the radiation intensity at $\varepsilon_0 \geq U_0$ ($\vartheta_0 \geq \vartheta_c$, $\vartheta_c = (2U_0/\varepsilon)^{1/2}$ is the Lindhard angle) does not depend on the incident angle or the sign of the charge of the particle for any potential. Note that the validity of the classical description breaks for the rather large angles of incidence. This situation differs from the case of planar channeling, where the difference between radiation intensities of the particles with positive and negative charge and their dependence on ε_0 vanish in the limit $\varepsilon_0 \gg U_0$ only.

Let us discuss some general properties of the radiation intensity for axial channeling. The main contribution to integrals (22.21)–(22.23) is given by the region where the gradient of the potential $U(\rho)$ is large. This region has a characteristic scale a_s which is a screening radius. Inside this region, with good accuracy the potential is axially symmetric. Let us represent the elementary cell as a circle with a radius $r_0 = (\pi n_\perp)^{-1/2}$. For real crystal $r_0 \gg a_s$. This means, that one can extend to infinity the upper limit of the integration over ρ in Eq. (22.23), then

$$I_{as} \simeq 2\pi A n_\perp \int_0^\infty d\rho\rho \left(\frac{dU}{d\rho}\right)^2 = 2\pi A n_\perp U_0^2 \int_0^\infty d\rho \quad \rho \left[f'(\rho)\right]^2, \qquad (22.24)$$

where $f(\rho) = U(\rho)/U_0$. If f is a function of ρ/a_s, then after the substitution $\rho \to \rho a_s$ one can see, that I_{as} does not depend on the specific value of the screening radius and depend on a shape of the potential only (for the real crystal it is necessary to consider the more complicated case, when there are actually two characteristic scales). This result differs essentially from the case of planar channeling, where the intensity of the radiation increases when $a_s \to 0$. For an axially symmetric potential we obatin the radiation intensity substituting (9.64) in Eq. (22.21)

$$I(\vartheta_0) = 4\pi^2 A n_\perp \int_0^{r_0} d\rho_0 \frac{\rho_0}{S(\varepsilon_\perp(\rho_0))} \int d\rho \quad \rho \cdot \left(\frac{dU}{d\rho}\right)^2 \vartheta(\varepsilon_\perp(\rho_0) - U(\rho)), \qquad (22.25)$$

where $S(\varepsilon_\perp(\rho_0)) = 2\pi \int \rho d\rho\, \vartheta(\varepsilon_\perp(\rho_0) - U(\rho)), \varepsilon_\perp(\rho_0) = U(\rho_0) + \varepsilon\vartheta_0^2/2$. Changing to the variables $x = \rho^2/a_s^2, y = \rho_0^2/a_s^2$ we have, after some manipulations, for the radiation intensity from positrons $I^+(\vartheta_0)$

$$I^+(\vartheta_0) = 4\pi A n_\perp \left\{ Q \left[\vartheta(\varepsilon_0 - U_0) + \frac{y_+}{x_0}\vartheta(U_0 - \varepsilon_0)\right] \right.$$
$$\left. + \vartheta(U_0 - \varepsilon_0) \int_{y_+}^{x_0} \frac{dy}{x_0 - x_+(y)} \cdot \int_{x_+(y)}^{x_0} dx \cdot x U'^2(x) \right\}, \qquad (22.26)$$

where $\varepsilon_0 = \varepsilon\vartheta_0^2/2$, $x_0 = r_0^2/a_s^2$, $U_0 = U(0) \geq 0, U(x_0) = 0$; quantities y_+, $x_+(y)$ are determined by the equations

$$U(x_+(y)) = U(y) + \varepsilon_0,\ U(y_+) = U_0 - \varepsilon_0,\ Q = \int_0^{x_0} dx\, x[U'(x)]^2. \qquad (22.27)$$

For the radiation intensity from electrons $I^-(\vartheta_0)$ we find

$$I^-(\vartheta_0) = 4\pi A n_\perp \left\{ Q\left[1 - \frac{y_-}{x_0}\vartheta(U_0 - \varepsilon_0)\right] + \vartheta(U_0 - \varepsilon_0)\int_0^{y_-} \frac{dy}{x_-(y)} \int_0^{x_-(y)} dx \cdot x U'^2(x) \right\}, \tag{22.28}$$

where

$$U(x_-(y)) = U(y) - \varepsilon_0, \quad U(y_-) = \varepsilon_0 . \tag{22.29}$$

The obtained Eqs. (22.26) and (22.28) give the orientation dependence of the radiation intensity in thin crystals for any form of the axially symmetric potential. If we use the potential (9.13) dependent on two characteristic scales: the amplitude of thermal vibrations u_1 and the screening radius a_s, then we find for the total radiation intensity from positrons depending on ϑ_0

$$\begin{aligned} I^+(\vartheta_0) = I_0 \Bigg\{ & \vartheta(\varepsilon_0 - U_0)\, \varphi(\eta) + \vartheta(U_0 - \varepsilon_0)\Bigg[(1 + 2\eta)\ \ln\ \lambda \\ & + (1-\lambda)\left(\eta + \frac{1+\eta}{\lambda}\right) + \frac{1}{x_0}\Bigg[\left(\lambda\eta - \frac{1+2\eta}{1-\lambda}\right)\ \ln\ \lambda \\ & - \eta(1 + \lambda + 2\eta)\ \ln\left(1 + \frac{1}{\eta}\right) - \frac{1-\lambda}{\lambda}(1 + \eta(1-\lambda)) \\ & \times\left(\ln\ \frac{x_0[1 + \eta(1-\lambda)]}{1+\eta} + 1\right)\Bigg]\Bigg]\Bigg\}, \end{aligned} \tag{22.30}$$

where

$$\begin{aligned} I_0 &= \frac{8\pi e^2 V_0^2}{3m^2}\gamma^2 n_\perp , \\ \lambda &= \exp(\varepsilon_0/V_0), \\ \varepsilon_0 &= \varepsilon\vartheta_0^2/2, \\ \varphi(\eta) &= (1 + 2\eta)\ \ln\left(1 + \tfrac{1}{\eta}\right) - 2 . \end{aligned} \tag{22.31}$$

We retain, in Eq. (22.30), the term proportional to $1/x_0$, since at $\varepsilon_0 = 0 (\lambda = 1)$ all other terms vanish. The quantity $I^+(\vartheta_0)$ increases monotonically from a small value at $\varepsilon_0 = 0$

$$I^+(0) = \frac{I_0}{x_0}\left[1 + 2\eta - 2\eta(1+\eta)\ \ln\left(1 + \frac{1}{\eta}\right)\right] \tag{22.32}$$

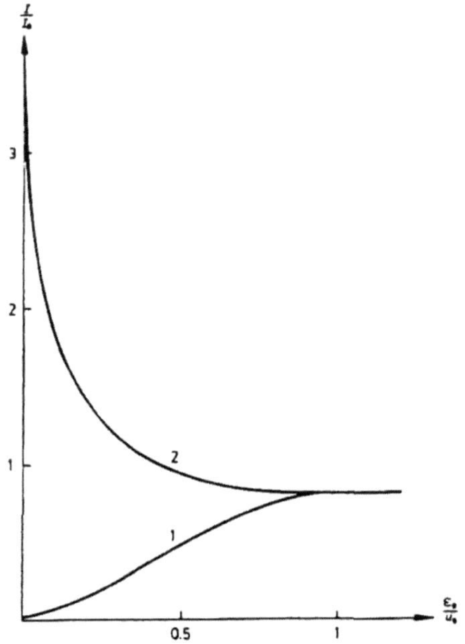

Fig. 22.8. Radiation intensity dependence on $\varepsilon_0/U_0 = (\vartheta_0/\vartheta_c)^2$ at axial channeling of positrons (curve 1) and electrons (curve 2) in tungsten ($\langle 100\rangle$ axis). The value I_0 is given by Eq. (22.31).

and reaches its maximum value at $\varepsilon_0 = U_0$

$$I^+(\varepsilon_0 = U_0) = I_{as} \equiv I_0 \cdot \varphi(\eta) \tag{22.33}$$

and for the further increase of angle ϑ_0, it remains constant. Note that $dI^+/d\varepsilon_0(\varepsilon_0 = U_0) = 0$. The function $I^+(\vartheta_0)$ for W is shown in Fig. 22.8 (curve 1) and for Si in Fig. 22.9a (curve 1).

Substituting the potential (9.13) in Eq. (22.28), we obtain the total radiation intensity at axial channeling of electrons and its dependence on ϑ_0:

$$I^-(\vartheta_0) = I_0\left\{\vartheta(\varepsilon_0 - U_0)\ \varphi(\eta) + \vartheta(U_0 - \varepsilon_0)\left[\frac{\ln\ \lambda}{\lambda} - \mu + \mu(1+\lambda)\right.\right.$$

$$\times\left(\ln(\mu(1+\eta)) - \eta\right) + \frac{1}{2\lambda}\ \ln\left[\frac{1}{\eta(1+\eta)(\mu+\frac{1}{x_0})^2}\right]$$

$$\times \left(\frac{(1+2\eta)\ln(1+\frac{1}{\eta})}{(1+\eta\mu)^2} - 2 \right) + \frac{1+2\eta}{(1+\mu\eta)^2} \left[\mu(1+\eta\mu) \right.$$

$$\times \left((1+\eta)\ln\left(1+\frac{1}{\eta}\right) - \ln(\mu(1+\eta)) + \eta \ln \lambda \right)$$

$$\left. \left. \left. - \frac{\ln \lambda}{\lambda} \left(\frac{1}{2} \ln \lambda + \ln(\mu(1+\eta)) \right) \right] \right] \right\} , \qquad (22.34)$$

where $\mu = 1 - 1/\lambda$ and the remaining definitions are given in Eq. (22.30). The quantity $I^-(\vartheta_0)$ reaches its maximum value at $\vartheta_0 = 0$

$$I^-(0) = I_{as} \cdot \ln \frac{x_0}{\sqrt{\eta(1+\eta)}} \qquad (22.35)$$

and when ε_0 increases from 0 to $U_0, I^-(\vartheta_0)$ decreases monotonically from $I^-(0)$ to I_{as} at $\varepsilon_0 = U_0$ and then remains constant. The function $I^-(\vartheta_0)$ for W is shown in Fig. 22.8 (curve 2) and for Si in Fig. 22.9b (curve 1).

The results (22.32) and (22.35) agree with the estimate performed in Sec. 19 according to which the radiation intensity from electrons is essentially larger than the value I_{as} for $\vartheta_0 = 0$ and from positrons is suppressed as compared with this value. Remember, that the value of intensity I_{as} corresponds to the uniform distribution over transverse coordinates. These differences for electrons and positrons are connected with the redistribution of a particle flux. In the Coulomb type potential, the problem of the particle motion has an exact solution. Therefore the attempt to use this potential (see Avakyan, Zhevago, Jan Shi, 1982) for the description of radiation at axial channeling is quite natural. But it seems unsatisfactory to use a potential of the form

$$U(\rho) = \frac{\alpha}{\rho}\vartheta(\rho - u_1) + \frac{\alpha}{u_1}\vartheta(u_1 - \rho) \qquad (22.36)$$

for the description of the radiation. For example, let us obtain, according to Eqs. (22.26) and (22.28), the radiation intensity for the potential (22.36) at $\vartheta_0 = 0$

$$\begin{aligned} I^+_{(1)} &= \frac{u_1^2}{r_0^2}\left(\ln \frac{r_0^2}{u_1^2} + 1\right) \cdot I^{(1)}_{as} \\ I^-_{(1)} &= \left(\ln \frac{r_0^2}{u_1^2} - 1\right) \cdot I^{(1)}_{as} \\ I^{(1)}_{as} &= \frac{2\pi e^2 U_0^2}{3m^2}\gamma^2 n_\perp . \end{aligned} \qquad (22.37)$$

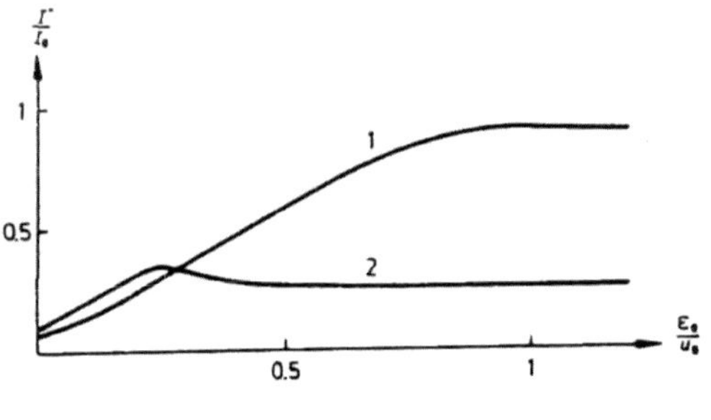

(a)

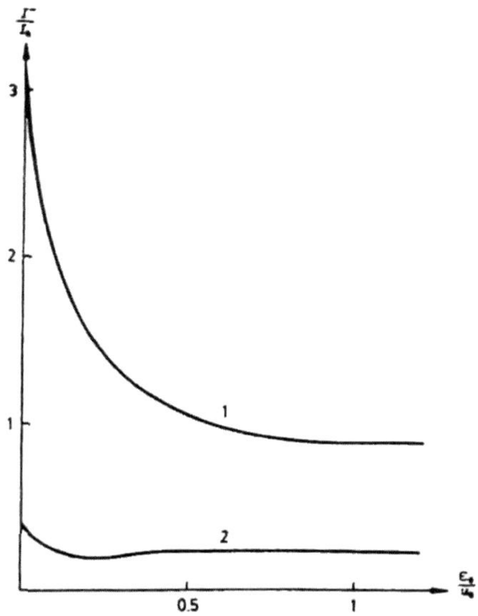

(b)

Fig. 22.9. Radiation intensity dependence on ε_0/U_0 in Si at axial (1) ($\langle 100\rangle$ axis) and planar (2) ((110) plane) channeling of positrons (a) and electrons (b). Here we use the axial channeling units.

Since the potential (22.36) depends on one characteristic scale only (thermal vibration amplitude u_1) then for particles moving above the barrier the radiation intensity is independent of u_1. Then one should bear in mind, that in a realistic potential for a chain of atoms, the region of integral convergence is $u_1 \leq \rho \leq a_s$ but $\rho \sim u_1$ in the potential (22.36). This means that Eq. (22.37), in particular, diminishes substantially the radiation intensity from positrons since $a_s^2/u_1^2 \geq 10$. Moreover, one can obtain an erroneous notion of a radiation spectrum using the potential (22.36), because the main contribution in this case is given by the trajectories with $\rho \sim u_1$, while actually the whole interval from u_1 to a_s contributes. It is obviously interesting to compare the radiation intensity obtained for axial channeling with the radiation intensity for planar channeling. The radiation intensity from positrons in Si is shown in Fig. 22.9a as a function of the incident angle ϑ_0 with respect to the axis $\langle 100 \rangle$ (curve 1) for axial channeling and with respect to the plane (110) (curve 2) for planar channeling. The quantities U_0 and I_0 being taken for the axial case. In Fig. 22.9b the same is plotted for electrons. One can see in the figures that the radiation intensity from electrons for axial channeling $I_{as}^-(\vartheta_0)$ appreciably exceeds that for planar channeling $I_{pl}^-(\vartheta_0)$ for all angles ϑ_0. For positrons, the ratio between $I_{ax}^+(\vartheta_0)$ depends on the incident angle ϑ_0 and only for $\vartheta_0 \geq \vartheta_c$, $I_{ax}^+(\vartheta_0)$ appreciably exceeds $I_{pl}^+(\vartheta_0)$. For wide enough (in ϑ_c units) beams the relation between axial and planar channeling can be described by the ratio of the asymptotic intensities $r_{as} = I_{as}^{ax}/I_{pl}^{ax}$ (see Eqs. (22.10) and (22.33)). This ratio r_{as} is 3.3 for C, 3.8 for Si, 3.7 for Ge and ~ 7 for W as compared the axis $\langle 100 \rangle$ with the plane (110). One can see that although the quantities I^{ax} and I^{pl} are proportional to the square of the corresponding U_0 and for axial channeling the value of U_0 is several times larger than for planar channeling, actually in a rough estimate $r_{as} \sim U_0^{ax}/U_0^{pl}$, i.e. the gain r_{as} increases only linearly. The reason is that for the above-barrier motion during planar channeling particles are in the region of a large gradient of the potential a relatively longer time than during axial channeling.

For radiation from electrons the gain has a maximum at $\vartheta_0 = 0$, i.e. for beams with a small angular spread. One can obtain an estimate of $r(\vartheta_0 = 0)$ taking into account the fact that $I_{pl}^-(0)$ for all crystals is nearly twice as much as I_{as}^{-pl} and for axial channeling it is necessary to use Eq. (22.35).

For the spectral distribution of the radiation we shall give here a simple estimation of the characteristic frequencies only. As has already been mentioned the characteristic radius of the region giving the main contribution to the radiation is the screening radius a_s. The transverse velocity of an electron in the channel is $\vartheta_\perp \leq \vartheta_c$; therefore, the corresponding frequency of the motion is ϑ_c/a_s, and the characteristic radiation frequency, with the Doppler effect taken into account, is equal to

$$\omega_{\text{ch}} \simeq 2\gamma^2\vartheta_c/a_s\,. \tag{22.38}$$

Since we have taken the upper limit of velocity $\vartheta_\perp$ in a channel, the frequency $\omega_{\rm ch}$ is expected to be not far from the maximum of the spectral distribution on its right slope. With the help of (22.38), we can estimate the number of emitted photons as $N_{\rm ph} \simeq E_{\rm ph}/\omega_{\rm ch}$. The comparison of these estimates with all, known to us, experiments dealing with electron radiation spectrum measurements for the axial case with electron energies ranging from 28 MeV to 10 GeV shows the agreement between them (see Aideshvili *et al.*, 1984; Bochek *et al.*, 1983; Swent *et al.*, 1979; Aganyantz *et al.*, 1983; Bulgakov *et al.*, 1983). If we are interested in certain intervals of the radiation frequencies, Eq. (22.38) is suitable for the choice of an appropriate crystal.

Let us consider the possibility of using the synchrotron approximation for the radiation spectrum at the axial channeling. It should be pointed out that even for $\rho_c \gg 1$ in the case $\chi_s \ll 1$, the maximum of the spectral intensity occurs at frequencies $u \sim u_{\rm min}$ (see Eq. (17.14)) where the synchrotron approximation becomes inapplicable. Nevertheless, it can be used at frequencies in the range $\omega \geq \varepsilon\chi_s$. Such situation occurs, for example, in the experimental results of Bulgakov *et al.*, 1986, where silicon crystal near the $\langle 111 \rangle$ axis was used and the energy of the incident particles was $\varepsilon = 10$ GeV. Using Eqs. (16.4) and (17.14) we found that the basic frequency is $\omega \sim \varepsilon\chi_s \simeq 270$ MeV and also that $\omega_{\rm min} \sim \varepsilon u_{\rm min} \simeq 90$ MeV; $\rho_c \simeq 4.1$. Fig. 22.10 shows the results of a calculation

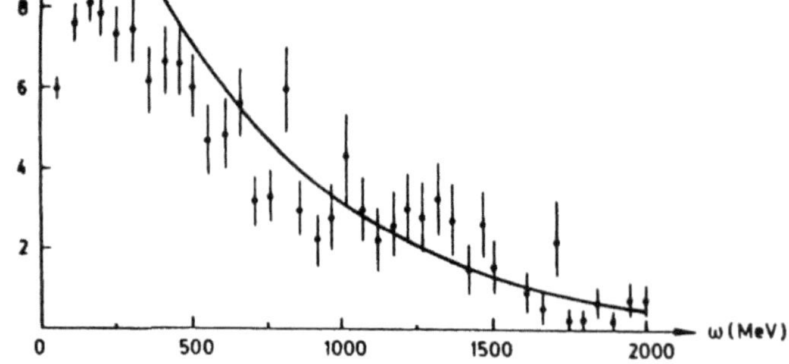

Fig. 22.10. Spectral intensity of radiation in Si $\langle 111 \rangle$ $\varepsilon = 10$ GeV.

of the spectrum carried out in the synchrotron approximation allowing for the effective divergence of the beam and the corresponding experimental conditions. One can see that the calculated curve describes quite satisfactorily the experimental data at frequencies $\omega > 300$ MeV. Further experimental results with improved statistics in a wider range of energies were obtained by Bak *et al.*, 1985, p. 575. Detailed investigation of radiation from electrons and positrons at axial channeling was made by Bak *et al.*, 1988b, where a series of spectra are presented in a broad range of incidence angles in Si and Ge crystals.

22.3. Quantum Effects in Channeling Radiation

In this section the quantum effects in the total intensity of radiation at channeling are considered. With the particle energy increase it appears that the recoil at radiation becomes important and one has to take into account the quantum effects. This situation is controlled by the parameter χ_s and at $\chi_s \ll 1$ the classical theory is valid. Because of $\rho/\chi_s \sim 2ma_s \sim 10^2$, at $\chi_s \sim 1$, when the radiation is essentially quantum, one has $\rho \sim 10^2$ so that the radiation is of the synchrotron nature.

However, one should keep in mind that the quantum effects in the synchrotron radiation are "turned on" rather early. For example, already at $\chi = 0.1$ the classical value of the total intensity is around 1.5 times larger than the correct result. Even at $\chi \sim 10^{-2} (\rho \sim 1)$ quantum corrections are getting quite noticeable. Generally speaking, in the quantum region the simple formula of the type (22.20) for the total intensity does not exist. But in synchrotron radiation limit ($\rho \gg 1$) the description of the radiation is again getting local. So, for us the interval $1 \leq \rho_c \ll 10^2$ is most interesting, because at $\rho \leq 1$ the quantum corrections are very small and classical electrodynamics is applicable and at $\rho_c \geq 10^2 (\chi_s \geq 1)$ the constant field (local) description is valid. This means that the most strong effects of nonlocality in the total intensity of radiation becomes apparent just inside discussed interval of ρ, i.e. in the region where one can put $\chi \ll 1$. For an analysis of this problem we shall use the results obtained in Sec. 5.4. Using the potential (9.13) we have for the current values of the parameters χ and ρ:

$$\chi(x) = \chi_s 2\sqrt{x} g(x), \;\; \rho(x) = 4V_0 \varepsilon x \; g(x)/m^2, \;\; g(x) = [(x+\eta)(1+x+\eta)]^{-1} .$$

After the averaging of Eq. (5.62) with the uniform particle distribution over coordinates $\boldsymbol{\rho}$, which corresponds to particles moving above barrier or in a thick crystal, we obtain

$$\bar{I} = \frac{8e^2m^2\chi_s^2}{3x_0}\left[S(\eta) - \frac{55\sqrt{3}}{16}\chi_s R(\eta) - \frac{57\sqrt{3\pi}}{160ma_s}A(\eta) + 48\chi_s^2 T(\eta)\right], \quad (22.39)$$

where $x_0 = r_0^2/a_s^2$, $r_0^2 = (\pi n_\perp)^{-1}$, $n_\perp$ is the density of the axes,

$$\begin{aligned}
&S(\eta) = (1+2\eta)\ln((1+\eta)/\eta) - 2\,,\\
&R(\eta) = \frac{3\pi}{4\sqrt{\eta(1+\eta)}}\frac{1}{\left(\sqrt{\pi}+\sqrt{1+\eta}\right)^5}\,,\\
&A(\eta) = 1/\sqrt{\eta} + 1/\sqrt{1+\eta} + 4(\sqrt{1+\eta} - \sqrt{\eta})\,,\\
&T(\eta) = 4\left[(1/\eta + 1/(1+\eta))/3 - 4\ \ln(1+1/\eta)(5\eta(1+\eta)+1) + 20\eta + 10\right]
\end{aligned} \quad (22.40)$$

The terms with functions S, R and T are written with the same coefficients as in the decomposition (5.62), the linearly increasing with the energy parameter χ_s depends on the field at the screening radius. These functions are presented in Fig. 22.11 in the interval of η within which are all the used crystals. It's seen

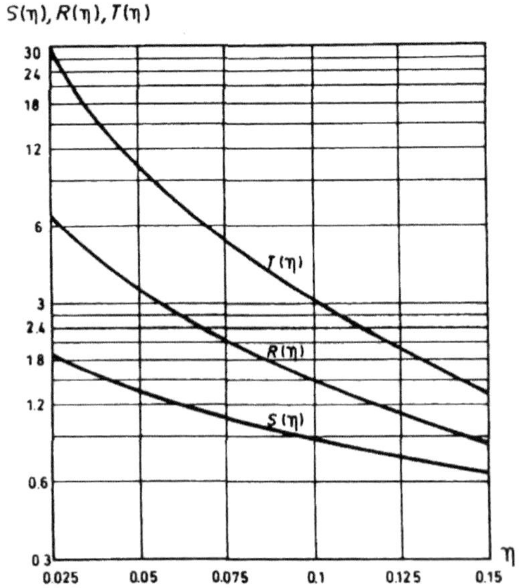

Fig. 22.11. Functions $S(\eta)$, $R(\eta)$ and $T(\eta)$ in the interval of η which includes all the used crystals parameters.

that the functions R and T are appreciably larger than S particularly for small amplitudes of thermal vibrations (η). This means that the distances smaller than the screening radius give noticeable contribution. From the above it follows that the validity region of the expansion (22.39), which is quite satisfactory if the corrections do not exceed 10–15%, reduces in the crystal with a small amplitude of thermal vibrations. The correction $\propto 1/\rho$ gives the energy independent term with the function $A(\eta)$. As compared with the main term ($S(\eta)$) it is (at room temperature): in diamond 4.7%, in tungsten 4.5% (for axis $\langle 111 \rangle$), in silicon 3%, and in germanium 2.8% (for axis $\langle 110 \rangle$). Let us be reminded that according to what has been said in Sec. 5.4 the expansion over $1/\rho$ is valid up to $\rho = 1$. Thus, in the validity region of this expansion the contribution of the term $\propto 1/\rho$ is smaller than 5% and is independent of energy. As a result one can use the synchrotron radiation description of total radiation intensity and the $1/\rho$ term can be estimated.

With an energy increase, the quantum corrections become large and expansion (5.62) becomes invalid. In this situation one should use the complete

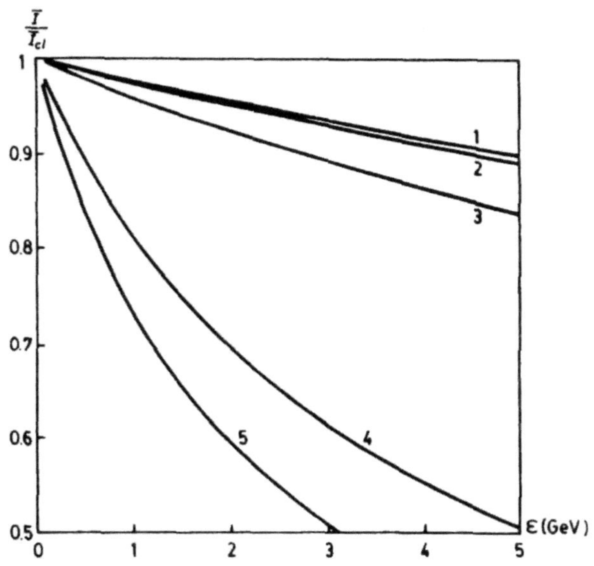

Fig. 22.12. Mean radiation intensity at channeling divided by the mean classical intensity versus electron energy at room temperature in single crystals: curve 1 is for Si ($\langle 110 \rangle$ axis), curve 2 is for diamond ($\langle 111 \rangle$ axis), curve 3 is for Ge ($\langle 110 \rangle$ axis). The curve 4 is for tungsten (axis $\langle 111 \rangle$) at room temperature and curve 5 at $T = 77$ K.

expression for the radiation intensity valid at arbitrary χ. One can also use the approximate expression for the radiation intensity (the accuracy is better than 2% for any χ, see Eq. (4.57))

$$I(\chi) = (2/3)e^2m^2\chi^2[1 + 4.8(1+\chi)\ln(1+1.7\chi) + 2.44\chi^2]^{-2/3}. \qquad (22.41)$$

The result of averaging this expression is shown in Fig. 22.12 where the ratio $\bar{I}/\bar{I}_{cl}$ is presented, where $\bar{I}_{cl}$ is the mean classical intensity, as the function of the electron energy. It is seen, that the quantum corrections exceed 10% at the energy near 5 GeV in Si and diamond and at the energy near 3 GeV in Ge. So, for the energy up to 5 GeV the quantum corrections in these crystals are rather small. The experimental data for Si and diamond (see Potylitzin *et al.*, 1987 and Antipenko *et al.*, 1990) are known for the energy near 1 GeV, where the corrections are a few percent what is less than the accuracy of both theory and data.

The situation is different for tungsten where already at the energy of 1 GeV (axis $\langle 111 \rangle$, room temperature) the energy dependent correction is 19% (the correction $\propto 1/\rho$ is up to 4.5%). As a result, at the energy 1 GeV the radiation intensity from electrons in W for these conditions is 77% of the classical value. For energies 2 GeV and 5 GeV one has 66% and 47%, respectively. Thus in W the quantum effects in the radiation are very strong. They are still more strong at lower temperature (axis $\langle 111 \rangle$, $T = 77$ K, curve 5), where the energy dependent correction is 27% at 1 GeV (the correction $\propto 1/\rho$ is about 6.5%). The experimental situation in tungsten is still uncertain.

23. RADIATION IN THICK CRYSTALS WITH ALLOWANCES FOR DECHANNELING AND COLLIMATION

23.1. Radiation Yield in Thick Crystals

For a theoretical description of the specific radiation which occurs when high-energy particles are moving near the crystalline axes or planes in single crystals, a relation between the dechanneling length l_d (see Sec. 10) and the characteristic length of energy loss L_{ch} is very important. As far as the value of l_d increases and L_{ch} decreases with the increase of the particle energy, these lengths become equal at some energy $\varepsilon = \varepsilon_{cr} : l_d(\varepsilon_{cr}) = L_{\text{ch}}(\varepsilon_{cr})$. Then for $\varepsilon < \varepsilon_{cr}$ the distribution function (DF) in a transverse phase space varies mainly due to multiple scattering and for $\varepsilon > \varepsilon_{cr}$ it varies mainly due to the radiation energy losses. According to estimates by Baier, Katkov, Strakhovenko, 1983 the quantity ε_{cr} is about tens of GeV in the axial and hundreds of GeV in the planar cases. To evaluate the possibility of using the effect under discussion as a source of hard and directed radiation concentrated within a comparatively narrow frequency range, one should know the total radiation intensity into a given solid angle. In this case, the interesting question is to find an optimal thickness of the crystal, for which the energy of the radiation yield, at a given collimation of the photon beam, has a maximum. The consideration essentially depends on a ratio $\varepsilon/\varepsilon_{cr}$. For $\varepsilon > \varepsilon_{cr}$, $L_{\rm ch} < l_d$ and a particle loses the main part of its energy already in a thin crystal and in this case all radiation concentrates in the angle $\sim \vartheta_c$. The peculiarities of radiation at such high energies have been considered in Sec. 4.

Practically the case $\varepsilon \ll \varepsilon_{cr}$ is interesting when the radiation is classical and the optimal thickness L_0 is much larger than the dechanneling length l_d. Really, if we do not take into account a photon absorption, then the radiation yield grows with thickness until a particle emits the noticeable part of its energy. For ultrarelativistic particles it takes place over the length $\sim \min(L_{\rm rad}, L_{\rm ch})$. On the other hand, the photon absorption becomes substantial on the length $\sim L_{\rm rad}$. Since, in the region of energy $\varepsilon \ll \varepsilon_{cr}$, when $l_d/L_{\rm ch} \ll 1$ and $l_d/L_{\rm rad} \ll 1$ we have $L_0 \gg l_d$ so that it is necessary to consider the radiation formation in thick crystals.

The classical intensity of radiation depends only on the local values of the particles coordinates and momenta and to find it, one has to know DF in the transverse phase space only. (The validity of classical description has been discussed at the end of the previous section). We use the asymptotic form of DF obtained in Sec. 10 for thick crystals. Taking into account the gaussian form of the particle distribution over velocities for high lying states, we present DF in the form

$$\begin{aligned} dF^{pl}(v_\perp, x, l) &= \frac{dx}{d} \cdot \frac{d^2 v_\perp}{\pi g(l)} \cdot \exp\{-v_\perp^2/g(l)\} \\ dF^{ax}(v_\perp, \rho, l) &= \frac{d^2\rho}{S} \cdot \frac{d^2 v_\perp}{\pi g(l)} \cdot \exp\{-v_\perp^2/g(l)\}\,, \end{aligned} \tag{23.1}$$

where d is the distance between neighboring planes, S is the area of a cell per one atomic chain,

$$g(l) = \Delta^2 + \int_0^l dt\; \dot{\vartheta}^2(t)\,. \tag{23.2}$$

The first term in $g(l)$ reflects the character of the angular (velocity) distribution established at the initial stage of the electron motion in a crystal which depends on the angular (over ϑ_0) distribution in an incident beam. Even with the angle of incidence $\vartheta_0 = 0$, the angular spread of particles in the crystal proves to be roughly equal to the Lindhard angle $\vartheta_c = (2U_0/\varepsilon)^{1/2}$; therefore, assuming the incident beam to be rather narrow, $\vartheta_0 \leq \vartheta_c$, we can put $\Delta^2 = c_1\vartheta_c^2$ where c_1 is a coefficient of the order of unity. The magnitude of c_1 can be influenced also by the imperfection of the crystal, for example, its mosaic structure. The uncertainty in the choice of the value of c_1 determines the calculation accuracy which could be estimated as $(\ln L/l_d)^{-1}$, where L is the thickness of a crystal. In what follows, according to the definition of the value l_d in Sec. 10, we set $c_1^{pl} = 2$, $c_1^{ax} = 1$.

The specific features of the radiation under consideration, which occurs for electrons moving in a continuous potential of axes (planes), are dictated by the particles which are in the channel or not too high above the barrier. The contribution of these particles can be enhanced by a photon collimator. For this purpose, the collimation angle ϑ_{col} (we count it from the axis oriented to the center of the collimator) should be chosen approximately equal to the Lindhard angle $\vartheta_{col} \sim \vartheta_c$. On the other hand, if $\vartheta_{col} < m/\varepsilon = 1/\gamma$, we shall omit a fraction of the radiation of interest. For energies under consideration from several hundreds of MeV to several GeV, we have $\vartheta_s \gg 1/\gamma$ and $\vartheta_s \gg \vartheta_c$ for the crystal thickness $L \sim L_0$. Thus, the condition

$$\max(\gamma^{-1},\ \vartheta_c) \leq \vartheta_{col} \ll \vartheta_s \tag{23.3}$$

is assumed to be satisfied and then the factor $\exp(-\vartheta_\perp^2/g(l))$ can be substituted for unity, and the distribution proves to be uniform in the transverse phase space. In the case when $\vartheta_s \sim 1/\gamma$ (assuming $\vartheta_s \gg \vartheta_c$, as previously), this situation can take place at relatively low energies; any collimator will be reached by the radiation from all the particles and the distribution (23.1) needs to be used. The exponential factor in (23.1) should be kept also when detecting photons with the outgoing angle $\vartheta_{\text{ph}} \sim \vartheta_s$, for example, at $\vartheta_{col} \sim \vartheta_s$. As a consequence of the uniform distribution, for the intensity of the radiation under consideration (labelled by "ch") at the depth l we have

$$\begin{aligned} I_{\text{ch}}(\Delta, l) &= \int dF^{ax}(\mathbf{v}_\perp, \boldsymbol{\rho}, l) \int_\Delta d^2\vartheta I(\boldsymbol{\vartheta} - \mathbf{v}_\perp, \boldsymbol{\rho}) \\ &= \frac{1}{\pi g(l)} \int_\Delta d^2\vartheta \int \frac{d^2\rho}{S} I(\boldsymbol{\rho}) = \frac{\Omega}{\pi g(l)} \langle I \rangle\,, \end{aligned} \tag{23.4}$$

where $\boldsymbol{\Delta}(\Delta_x, \Delta_y)$ is the collimator angle, Ω is the solid angle of the collimator.

In the case of a slit collimator for the plane $|\vartheta_x| \le \Delta_x$, we find

$$\begin{aligned} I_{\text{ch}}(\Delta_x, l) &= \int_{-\Delta_x}^{\Delta_x} d\vartheta_x \int_{-\infty}^{\infty} d\vartheta_y \int dF^{pl}(\mathbf{v}_\perp, x, l) I(\boldsymbol{\vartheta} - \mathbf{v}_\perp, x) \\ &= \int_{-\Delta_x}^{\Delta_x} d\vartheta_x \int_{-\infty}^{\infty} d\vartheta_y \, \exp\{-\vartheta_y^2/g(l)\} \int \frac{dx}{d} I(x) \\ &= \frac{2\Delta_x}{\sqrt{\pi g(l)}} \langle I \rangle \, . \end{aligned} \tag{23.5}$$

For photon collimation in two directions, both expressions for the planar and axial cases can be presented in the form

$$dI_{\text{ch}}/d\Omega = \langle I \rangle / \pi g(l)) \, , \tag{23.6}$$

where $\langle I \rangle = \int dx \; I(x)/d$ for the plane and $\langle I \rangle = \int d^2\rho \; I(\rho)/S$ for the axis. It is worth noting that $\langle I \rangle$ coincides with I_{as}, the asymptotic value achieved by the intensity at the incident angles $\vartheta_0 \ge \vartheta_c$ for the axial case and $\vartheta_0 \gg \vartheta_c$ for the planar case. We emphasize that (23.6) incorporates the radiation from all the particles rather than from the above-barrier only. The appearance of I_{as} in (23.6) is due to the uniformity of the distribution in transverse phase space (under the indicated assumptions, see Eq. (23.3)), while the uniformity of the distribution over $\mathbf{v}_\perp$ resulted in (23.6) in a uniform distribution with respect to the solid angle of emitted photons for the angles $\vartheta_{\text{ph}} \ll \vartheta_s$.

The values $\langle I \rangle = I_{as}$ are readily calculated for any potential of the axes, or planes. For the planar potential used (9.33), the expression for I_{as} is given by formulae (22.10) and for the axis potential (9.13) it does so by Eq. (22.33).

Besides the radiation under study, the bremsstrahlung (below with the subscript "br") also contributes to the intensity. Generally speaking, the bremsstrahlung can change in comparison with a amorphous medium, under channeling conditions, Assuming the condition (23.3) to be satisfied, for the bremsstrahlung intensity at depth l we find

$$\frac{dI_{\text{br}}}{d\Omega} = \frac{I_{\text{br}}}{\pi g(l)} = \frac{1}{\pi g(l)} \cdot \frac{\varepsilon}{L_{\text{rad}}} \tag{23.7}$$

Because of $L_{\text{ch}}, L_{\text{rad}} \ge l_d$ the energy variation will be taken into account in an adiabatic approximation considering both mechanisms of radiation losses

$$d\varepsilon/ds = -\varepsilon(1 + R) \, , \tag{23.8}$$

where $R = R(\varepsilon) = I_{as}(\varepsilon)L_{\text{rad}}/\varepsilon$, $s = l/L_{\text{rad}}$. The solution of this equation has the form

$$\frac{\varepsilon(s)}{\varepsilon(0)} = \frac{e^{-s}}{1 + R(\varepsilon_0)(1 - e^{-s})} \equiv Q(s) \tag{23.9}$$

The value $R(\varepsilon)$ is the ratio of the radiation intensity under discussion and bremsstrahlung intensity. Substituting $\varepsilon(s)$ into (23.6) and (23.7), we find

$$\begin{aligned} \frac{dI_{\text{ch}}}{d\Omega} &= \frac{\alpha}{(2\pi)^2} \cdot \frac{\varepsilon_0^3 R(\varepsilon_0)}{m^2 L_{\text{rad}}} \cdot Q^2(s) \cdot f(s) \\ \frac{dI_{\text{br}}}{d\Omega} &= \frac{\alpha}{(2\pi)^2} \cdot \frac{\varepsilon_0^3}{m^2 L_{\text{rad}}} \cdot Q(s) f(s)\,, \end{aligned} \tag{23.10}$$

where

$$f(s) = \left[\frac{1}{2}(1+R)^2(e^{2s} - 1) - 2R(1+R)(e^s - 1) + R^2 s + l_d/L_{\text{rad}}\right]^{-1} \tag{23.11}$$

Here $\varepsilon_0 = \varepsilon(0)$, $R = R(\varepsilon_0)$, l_d is defined in (10.1) and (10.22), and $Q(s)$ is given by Eq. (23.9).

The photon absorption, which should be taken into consideration at rather large thickness of a crystal, can be characterized using the average absorption length L_{ph}. This length is determined by the frequency interval of emitted photons we are interested in; being dependent on the energy of the initial particles and on the media.

With the absorption taken into account, the radiation brightness, which is by definition the emitted energy $E_{\text{ch}}(s)$ or $E_{\text{br}}(s)$ per unit solid angle after passage of the depth l, is

$$\begin{aligned} \frac{dE_{\text{ch}}(s)}{d\Omega} &= \frac{\alpha}{(2\pi)^2} \cdot \frac{\varepsilon_0^3 R(\varepsilon_0)}{m^2} \cdot \phi(s) \\ \frac{dE_{\text{br}}(s)}{d\Omega}(s) &= \frac{\alpha}{(2\pi)^2} \cdot \frac{\varepsilon_0^3}{m^2} \cdot \Psi(s)\,, \end{aligned} \tag{23.12}$$

where

$$\begin{aligned} \phi(s) &= e^{-cs} \int_0^s dx\; e^{cx} Q^2(x) f(x) \\ \psi(s) &= e^{-cs} \int_0^s dx\; e^{cx} Q(x) f(x)\,, \end{aligned} \tag{23.13}$$

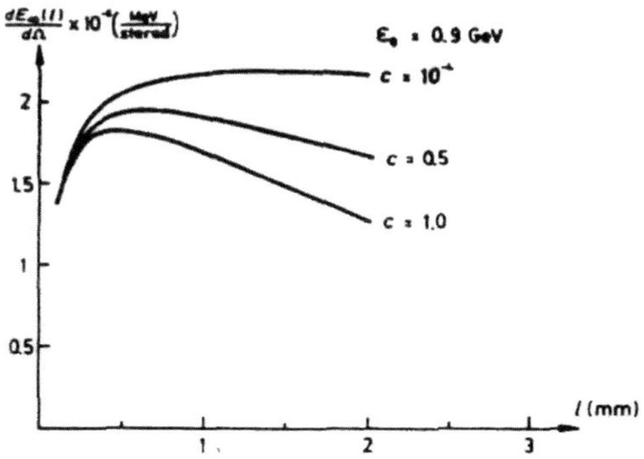

(a)

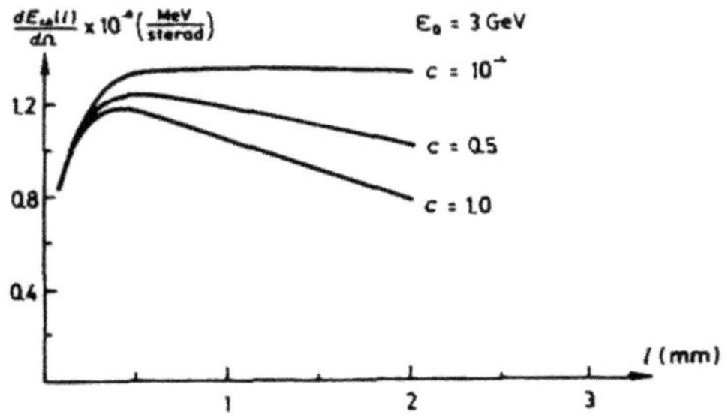

(b)

Fig. 23.1. Radiation brightness $dE_{ch}/d\Omega$ upon motion near the $\langle 111 \rangle$ axis in tungsten for the initial electron energies (a) $\varepsilon_0 = 0.9$ GeV and (b) $\varepsilon_0 = 3$ GeV versus the crystal thickness l for various values of the electron absorption coefficient $c = L_{rad}/L_{ph}$.

$c = L_{rad}/L_{ph}$. Let us define the radiation yield as ratio of the energy emitted into a given solid angle to the initial energy of an electron: $\eta = E(s)/\varepsilon_0$.

We start our consideration with the dependence of the brightness and optimal thickness on the photon absorption coefficient. We have calculated this coefficient assuming the photon energy to be within the 5 to 100 MeV interval with Compton scattering and electron-positron pair creation by photon taken into account. Note that the Compton cross section is large at comparatively low photon energies when the scattered photons do not reach the collimator. As the photon energy grows, the Compton cross section reduces rapidly, and in the region where the scattered photons can reach the collimator this cross section is very small. The total cross section for photon absorption in the

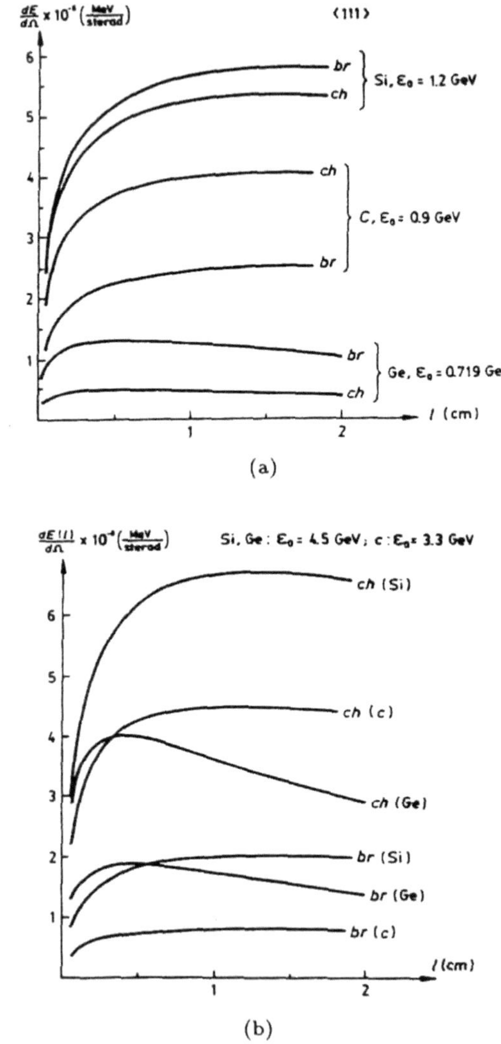

Fig. 23.2. Radiation brightness $dE/d\Omega$ upon motion near the $\langle 111 \rangle$ axis (ch) and bremsstrahlung brightness (br) in (a) Si ($\varepsilon_0 = 1.2$ GeV), Ge ($\varepsilon_0 = 0.719$ GeV) and diamond ($\varepsilon_0 = 0.9$ GeV); in (b) Si and Ge ($\varepsilon_0 = 4.5$ GeV) and diamond ($\varepsilon_0 = 3.3$ GeV).

energy region under consideration depends only slightly on the photon energy. This enables one to replace the corresponding cross section by its average value in the indicated interval of frequencies when calculating L_{ph} and to proceed without an exact knowledge of the spectrum of emitted photons. The procedure described above gives, for different substances, $c = L_{\rm rad}/L_{\rm ph} \simeq 0.5$ to 0.6. Note that the inaccuracy in the value of c has little influence on the value of the optimal yield $\eta(L_0)$. This assertion is illustrated by Fig. 23.1. It is clear that the doubling of c (evidently, this exceeds the uncertainty in c determination) decreases the magnitude of the optimal thickness by not more than 6%. It is also clear from Fig. 23.1 that both the optimal thickness and the yield increases with a decreasing factor c. If we did not make allowance for the absorption at all, the yield would be saturated in the same range of the optimal thickness values.

Figures 23.2 and 23.3 show the radiation brightness in diamond, silicon, germanium and tungsten as a function of the thickness at various energies of

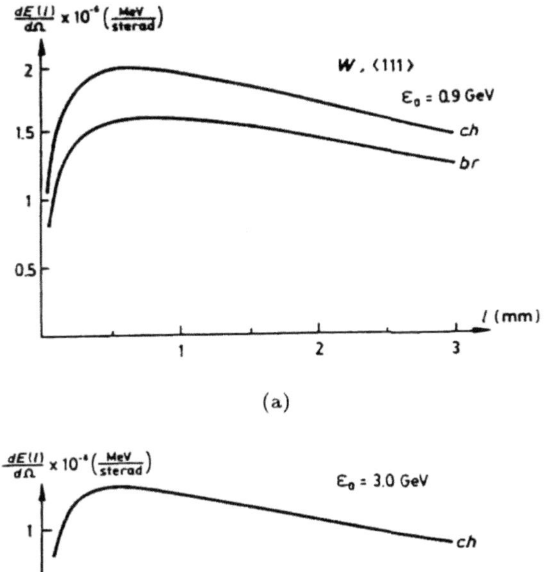

(a)

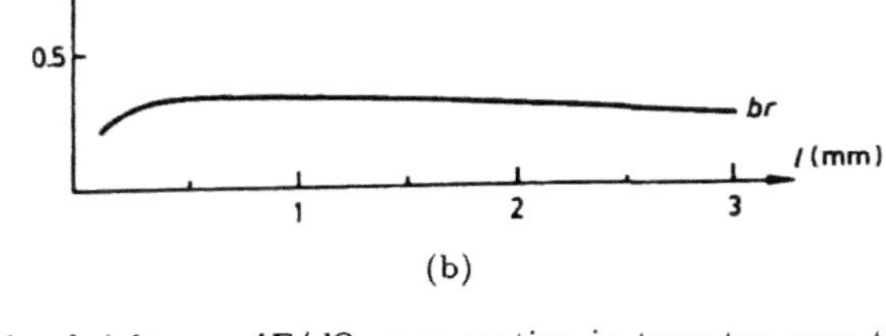

(b)

Fig. 23.3. Radiation brightness $dE/d\Omega$ upon motion in tungsten near the $\langle 111 \rangle$ axis (ch) and bremsstrahlung brightness at (a) $\varepsilon_0 = 0.9$ GeV and (b) $\varepsilon_0 = 3.0$ GeV.

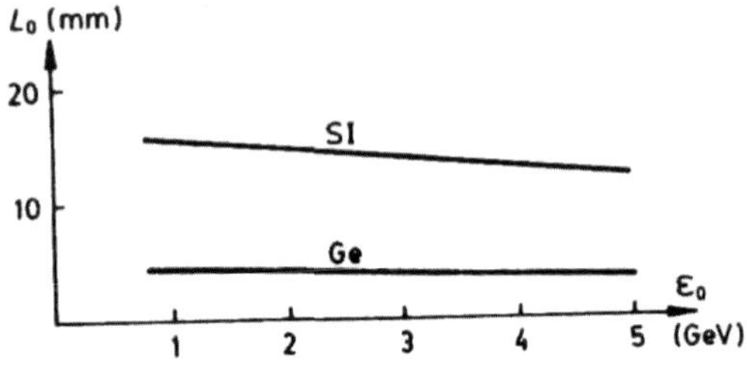

(a)

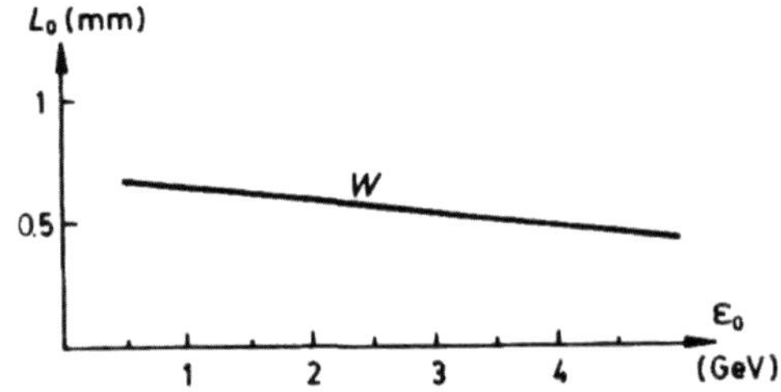

(b)

Fig. 23.4. Optimal thickness versus the initial electron energy in Si, Ge, and W.

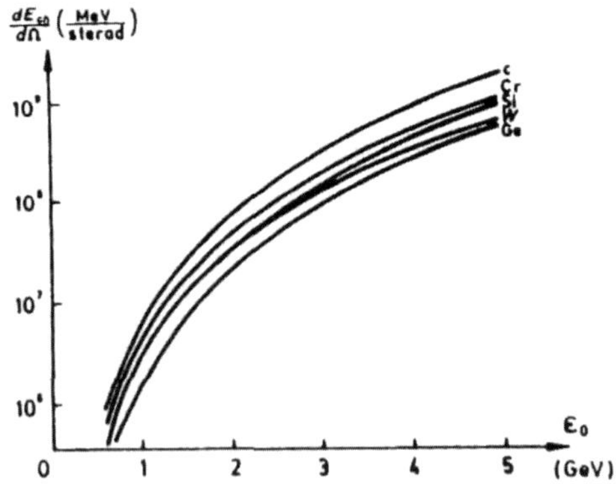

Fig. 23.5. Brightness at optimal thickness in Ge, W, Cr, and diamond as a function of the initial electron energy.

incident electrons which have been calculated according to Eq. (23.12). It is seen that the brightness first grows with thickness, then achieves a maximum, and, finally, begins reducing slowly, so that there is a wide plateau in the vicinity of the optimal thickness. This means that the brightness is close to an optimal one in a large range of thickness. For instance, at $\varepsilon_0 = 1.2$ GeV the optimal thickness $L_0 = 1.6$ cm in Si; for $L = 0.8$ cm, the brightness decreases only by 2%. Thus, the thickness of a crystal needs to be chosen in the region where the sharp increase in brightness stops. Further increase of the thickness has no practical effect. It turns out that the optimal thickness reduces insignificantly with energy (in the range of energies under consideration). This is illustrated by Fig. 23.4.

It is interesing that the maximum brightness $dE_{\rm ch}(L_0)/d\Omega$ at a given energy proves to be the same within a factor of about 3 for different single crystals. This is clearly seen in Fig. 23.5 wherein the brightness at the optimal thickness for a diamond, silicon, tungsten, chromium, and germanium crystal is presented as a function of energy.

Let us estimate the brightness for the potential (9.13)

$$\left.\frac{dE_{\rm ch}}{d\Omega}\right|_{L\sim L_0} \simeq \text{const.}\ \varepsilon_0^4 \left(\frac{V_0}{Z}\right)^2 d\cdot\varphi(\eta)\ln\left(\frac{m^2}{\alpha U_0\varepsilon_0}\right), \tag{23.14}$$

where the function $\varphi(\eta)$ is defined by Eq. (22.31). In this estimate, the quantities dependent on the type of single crystal are explicitly written. Since $V_0 \sim Z/d$ in a first approximation, then the basic dependence on the atomic number Z cancels. When comparing the brightness of different crystals (at a given energy), the dependence on the characteristics of a particular crystal proves to be a logarithmic one (hence, weak), except for the explicitly entering factor $1/d$. However, for different substances, the values of the latter differ by not more than a factor 2 (see Table 4 in the review by Gemmel, 1974.) Thus, one can draw the conclusion that in various single crystals (not only in those given in Fig. 23.5) the maximum brightness (at optimal thickness) is approximately the same (within the factor of scale 3) and lie near the curves in Fig. 23.5. In addition, formula (23.14) allows a purposeful search for crystals with maximal brightness: since the r.h.s. of (23.14) $\sim 1/d$, one should take the crystals with the lowest d. The quantity $\eta \propto u_1^2$ (u_1 is the amplitude of thermal vibrations); it is seen from (22.31) that crystals with minimal u_1 are needed. And, finally, in the argument of the logarithm in (23.14) there is U_0 in the denominator; in order that the logarithm is large, minimal $U_0 (U_0 \sim Z)$ are required. Following these arguments, one can see that among those usually using crystals the diamond will have the most intense brightness.

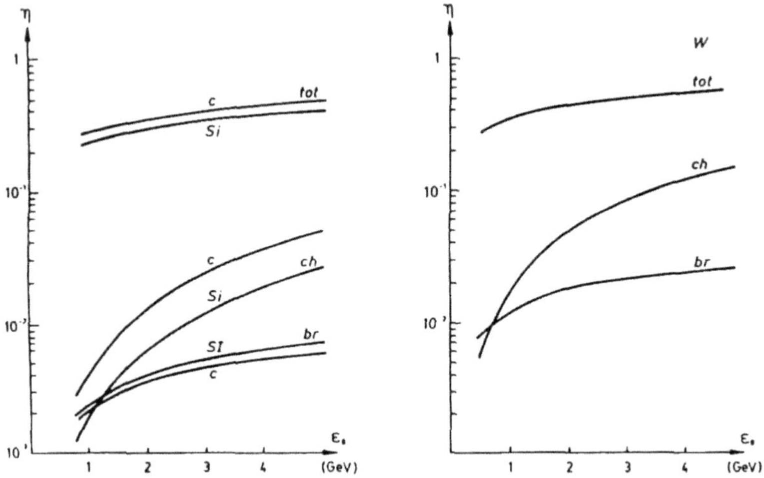

(a) (b)

Fig. 23.6. Radiation yield $\eta(L_0) = E(L_0)/\varepsilon_0$ at optimal thickness in (a) Si and diamond and (b) in tungsten at the angle of collimation ϑ_c.

A high brightness is also inherent in Cr, Fe, and V (the curve for iron would lie between the curves for Cr and Si in Fig. 23.5). Fig. 23.6 gives the radiation yield from the thickness L_0 into the solid angle $\pi\vartheta_c^2$ (to the cone whose opening angle is the Lindhard angle ϑ_c). It is seen that $\eta(L_0)$ achieves 10% in W at $\varepsilon_0 = 3.5$ GeV. This figure also presents the relative values of the total radiation energy losses of a particle (due to both radiation mechanism); $\eta_{tot} = [\varepsilon_0 - \varepsilon(L_0)]/\varepsilon_0$. The values of η_{tot} are large and attain 50% in W at $\varepsilon_0 = 4$ GeV and 40% in Si at $\varepsilon_0 = 5$ GeV. We would like to emphasize that the collimator (the Lindhard angle for this case) is reached only by a fraction of this energy $\eta_{tot} > \eta_{\rm ch} + \eta_{\rm br}$.

The radiation from electrons incident along the crystal axis ⟨111⟩ in thick single crystals has been studied experimentally. The studies have been made in Si at the energy $\varepsilon_0 = 1.2$ GeV (Adeyshvili *et al.*, 1984) and in Ge at $\varepsilon_0 = 0.719$ GeV (Bochek *et al.*, 1983). The measured brightness and radiation yield into a given solid angle, as well as the specific characteristic features of each experiment are given in Table 23.1. In addition, this table presents theoretical values of these quantities calculated by means of Eq. (23.12). It is seen that the experimental data are consistent with theory.

Table 23.1. Radiation yield in thick crystals.

Crystal	ε_0 (MeV)	L (mm)	$10^4/\gamma$	$10^4\vartheta_c$	$10^4\vartheta_{col}$	$10^4\vartheta_g$	$dE_{ch}/d\Omega$ (10^6 MeV/sr)
Si	1200	1.6	4.26	4.20	8.37	22	3.83
Ge	719	2.0	7.11	7.29	7.25	87	0.45

Crystal	$dE_{br}/d\Omega$ (10^6 MeV/sr)	$10^2\eta_{ch}$ theor.	$10^2\eta_{br}$ theor.	$10^2(\eta_{ch}+\eta_{br})$ theor.	$10^2\eta$ exper.
Si	4.02	0.89	0.15	1.04	0.6
Ge	1.21	0.13	0.35	0.48	0.43

For Si, the discrepancy is about 40% and for Ge it is about 10%. For Ge, the measurements have been made for different solid angles satisfying the inequalities (23.3). With an accuracy better than 20%, the brightness turns out to be independent of the angle of collimation; this can be regarded as an experimental confirmation of the uniform character of the distribution with respect to $\mathbf{v}_\perp$ (see the discussion after formula (23.6)). In the experiment for Si ($\varepsilon_0 = 1.2$ GeV), the 12 to 160 MeV photons were detected. In comparison, we have assumed that the whole radiation, which is due to the motion in a continuous potential, is detected (upper estimate). The value of η_{br}, indicated in Table 23.1 for this experiment, was obtained by integration of the standard spectral distribution of bremsstrahlung within the indicated range of frequencies. The main conclusions of the theory under consideration have been verified in the experimental work of Vorobev *et al.*, 1986, where the total yield as well as the spectral distribution of radiation have been studied for electron energies $\varepsilon_0 = 900$ MeV in thick crystals of diamond, silicon and tungsten.

With temperature variation of the crystal, the potential (9.13) will change as well as the kinetics of approaching the asymptotic DF. In the range of validity of Eq. (23.1), the effects of temperature variation reduce to a change of the potential parameters. For the $\langle 111\rangle$ axis in W, at $T = 77$ K, these parameters are quoted in Table 9.1. It is seen that the screening radius a_s slightly varies when going from the normal room temperature to the boiling point of nitrogen, while the depth of the potential well U_0 and η vary strongly. As a result, the radiation brightness $dE_{ch}/d\Omega$ increases somewhat. This is illustrated by Fig. 23.7. Since the rate of energy losses grows (R grows), l_d

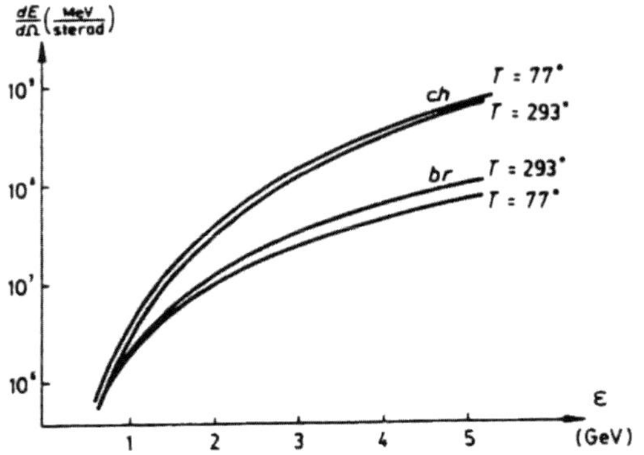

Fig. 23.7. Radiation brightness $dE_{\rm ch}/d\Omega$ and $dE_{\rm br}/d\Omega$ for the $\langle 111\rangle$ axis in tungsten at the optimal thickness for $T = 293$ K and 77 K, respectively.

increases, and L_0 decreases, the magnitude of $dE_{\rm br}/d\Omega$ reduces somewhat. The scale of the effect for brightness is about 20 to 30%. For the radiation yield into the Lindhard angle ($\vartheta_{col} = \vartheta_c$), the effect is more noticeable because $\vartheta_c^2 \propto U_0$.

23.2. Spectral Distribution of Dipole Radiation at Planar Channeling of Electrons with Allowance for Dechanneling and Collimation

When the dipole approximation is valid, to allow for the collimation of the photon beam it is necessary to integrate the angular distribution of the radiation (formula (1.131)) over the solid angle of the collimator. As a result we obtain

$$dI = \frac{16e^2}{T^4}\gamma^6\pi\frac{d\xi}{\xi}\sum_{k=1}^{\infty}|kv_k|^2\int_\Omega dn_x dn_y \cdot \delta(1+s^2+t^2-k/\xi)$$
$$\times\,[(1+s^2+t^2)^2 - 4s^2]/(1+s^2+t^2)^4\,. \tag{23.15}$$

Here T is the period of the motion, depending on the transverse energy $\varepsilon_\perp$, $\xi = \omega T/4\pi\gamma^2$, $\gamma = \varepsilon/m$, $\varepsilon(m)$ is the energy (mass) of the particles, ω is the frequency of the photon, and v_k is the Fourier transform of the z-component of the velocity; the x coordinate is perpendicular to the planes that form the channel, the z axis is parallel to these planes, and the region Ω is defined by the condition that the photon hits the collimator. The variables s and t are associated with the angular variables n_x and n_y of the photon and with the components of the average velocity V_x and V_y of the particle in the following way:

$$s = \gamma(n_x - V_x), \quad t = \gamma(n_y - V_y). \tag{23.16}$$

The expression (23.15) is intermediate and it should be averaged, taking into account the distribution of the particles over the transverse energy $\varepsilon_\perp$ and V_y. This distribution is obtained by integrating over crystal thickness l, in the interval $0 < l < L$, the corresponding distribution, which depends on the depth l. Since the form of the last has not been obtained yet, we shall start from a simplified model. We assume that a beam with a Gaussian distribution in the variables ϑ_x and ϑ_y, with width Δ_0, is incident on the crystal. According to the results of Sec. 9 one can find the initial distribution. In this distribution we then set $\Delta_0^2 \to \Delta_0^2 + \vartheta_s^2(l)/2$, where $\vartheta_s^2(l)$ is the mean square multiple-scattering angle in the corresponding amorphous medium at depth l. The distribution thus obtained has the form

$$\frac{dN(\varepsilon_\perp, V_y, l)}{d\varepsilon_\perp dV_y} = \frac{1}{2\pi\Delta^2(l)} \int_{-\infty}^{\infty} d\vartheta_0 \frac{dN(\varepsilon_\perp, \vartheta_0)}{d\varepsilon_\perp} \exp\left\{-\frac{\vartheta_0^2 + V_y^2}{2\Delta^2(l)}\right\}, \tag{23.17}$$

where $dN(\varepsilon_\perp, \vartheta_0)/d\varepsilon_\perp$ is the initial distribution over $\varepsilon_\perp$, defined by Eq. (9.24), at a fixed value of incident angle ϑ_0. If a change of the particle energy can be neglected, then one can put $\Delta^2(l) = \Delta_0^2 + \dot{\vartheta}_s^2 l/2$. Since the distribution (23.17) goes over into the initial distribution for $l = 0$, and it is in agreement with the asymptotic expression (23.1) for $1 \gg l_d$, these are all reasons to hope that the formula (23.17) describes a behavior of the distribution function in thick crystals rather well for all important values of l. Note that it is valid for electrons only since their multiple scattering near crystal planes differs weakly from that in the corresponding amorphous medium.

If for $U(x)$ we use the inverted-parabola type of potential defined by Eq. (9.47), then after integration over the path of the particle through the crystal we obtain the distribution with respect to $\varepsilon_\perp$ and $v = V_y/\vartheta_c$:

$$\frac{d\phi}{dzdv} = \frac{l_d}{2\pi} \int_0^{f(z)} du \int_{(1+r)^{-1}}^{1} dt \cdot t^{-1} \exp\left\{-\frac{t}{2\bar{\Delta}_0^2} B(z, u, v)\right\}, \tag{23.18}$$

where $z = \varepsilon_\perp/U_0$, $\tilde{\Delta}_0 = \Delta_0/\vartheta_c$, $\tau = L/l_d\Delta_0^2$, L is the thickness of a crystal, $B(z,u,v) = (1-z)\vartheta(1-z)\text{sh}^2 u + (z-1)\vartheta(z-1)\text{ch}^2 u + v^2$, $\vartheta_c = (2U_0/\varepsilon)^{1/2}$, $f(z) = (1/2)\ln|1+\sqrt{z})/(1-\sqrt{z})|$, $l_d = 2\vartheta_c^2/\dot{\vartheta}_s^2 \simeq (\alpha\varepsilon U_0/\pi m^2)L_{\text{rad}}$, L_{rad} is the radiation length in the amorphous medium.

The calculation of the spectral intensity (23.15) of the radiation is considerably simplified by the use of a circular collimator. If the z axis points to the center of a circular collimator with angular aperture η, the region Ω is defined by the relation $n_x^2 + n_y^2 \le \eta^2$. Carrying out the integration over n_x and n_y in formula (23.15) for this case, we obtain the following expression for the spectral intensity of the radiation for fixed $\varepsilon_\perp$ and V_y

$$dI = \frac{16e^2\pi^2}{T^4}\gamma^4\xi d\xi \sum_{k=1}^{\infty} \frac{|v_k|^2}{1+2r_k}\Bigg\{\vartheta(\eta_1)\vartheta(\eta_1^2 - x_k)\vartheta(x_k)$$

$$+\vartheta(x_k - \eta_1^2)\vartheta(\eta_2^2 - x_k)\frac{1}{\pi}\left(y - \frac{V_x^2 - V_y^2}{V_x^2 + V_y^2} r_k \sin 2y\right)\Bigg\}, \quad (23.19)$$

where $\eta_{1,2} = \gamma(\eta \mp \theta)$, $\theta = (V_x^2 + V_y^2)^{1/2}$ is the angle between the time-average velocity of the particle and the z axis. For particles in the channel we have $V_x = 0$, and for particles above the barrier we have $V_x = d/T(\varepsilon_\perp)$; $x_k = k/\xi - 1$,

$$r_k = x_k/(1+x_k^2)\,, \quad \cos y = \frac{(x_k - \eta_1\eta_2)}{(\eta_s - \eta_1)\sqrt{x_k}}\,.$$

We shall give the analogous expression for a rectangular collimator where the z axis points toward its center and the angular dimensions are η_x and η_y. In this case the region Ω is given by the relations $\eta_x \ge n_x \ge -\eta_x$ and $\eta_y \ge n_y \ge -\eta_y$. Carrying out the integration in formula (23.15), we obtain

$$dI = \frac{4e^2\pi^2}{T^4}\gamma^4\xi d\xi \sum_{k=1}^{\infty} \frac{|v_k|^2}{1+2r_k}\psi_k(V_x, V_y)\,. \quad (23.20)$$

The quantity $\psi_k(V_x, V_y)$ is an even function of V_x and V_y and we give the expression for ψ_k for $V_x, V_y \ge 0$:

$$\psi_k(V_x, V_y) = \vartheta(V_y - \eta_y)[g(s_1) + g(s_2)\,\text{sign}\,(\eta_x - V_x)]$$

$$+\,\vartheta(\eta_y - V_y)[f(s_1) + f(s_2)\,\text{sign}\,(\eta_x - V_x)]\,. \quad (23.21)$$

Here

$$
\begin{aligned}
f(s) &= \varphi\left(\frac{t_1^2}{x_k}\right)\vartheta(x_k - t_1^2)\vartheta(t_1^2 + s^2 - x_k) \\
&\quad + \varphi\left(\frac{t_2^2}{x_k}\right)\vartheta(x_k - t_2^2)\vartheta(t_2^2 + s^2 - x_k) \\
&\quad - \varphi\left(1 - \frac{s^2}{x_k}\right)\vartheta(x_k - s^2)\left[\vartheta(t_1^2 + s^2 - x_k) + \vartheta(t_2^2 + s^2 - x_k)\right] \\
&\quad + \vartheta(x_k)[\vartheta(t_2^2 - x_k) + \vartheta(t_1^2 - x_k)]\,, \\
g(s) &= \varphi\left(\frac{t_1^2}{x_k}\right)\vartheta(x_k - t_1^2)\vartheta(t_1^2 + s^2 - x_k) \\
&\quad - \varphi\left(\frac{t_2^2}{x_k}\right)\vartheta(x_k - t_2^2)\vartheta(t_2^2 + s^2 - x_k) \\
&\quad - \varphi\left(1 - \frac{s^2}{x_k}\right)\vartheta(x_k - t_2^2 - s^2)\vartheta(t_1^2 + s^2 - x_k) \\
&\quad + \vartheta(t_1^2 - x_k)\vartheta(x_k - t_2^2)\,,
\end{aligned}
\tag{23.22}
$$

$$
\begin{aligned}
\varphi(y) &= \frac{2}{\pi}[\arcsin\sqrt{y} - 2r_k\sqrt{y(1-y)}]\,, \\
t_1^2 &= \gamma^2(\eta_y + V_y)^2, \quad t_2^2 = \gamma^2(\eta_y - V_y)^2\,, \\
s_1^2 &= \gamma^2(\eta_x + V_x)^2, \quad s_2^2 = \gamma^2(\eta_x - V_x)^2
\end{aligned}
\tag{23.23}
$$

The radiation produced by relativistic particles is concentrated around the direction of the average velocity in a cone of the apex angle $\sim 1/\gamma = m/\varepsilon$. Therefore, for a fixed angular aperture of the collimator, the radiation incident on the collimator comes from particles having restricted values of the parameters $\varepsilon_\perp$ and v. Then the right-hand side of (23.18) and the yield of radiation for $L(r) \to \infty$ increase as $\ln(L/l_d)$. For this L, the distribution of the particles depends less and less (with logarithmic accuracy) on the initial distribution and the spectral-density curves for different L are practically the same, with a similarity coefficient $\kappa = \ln(L_1/l_d)/\ln(L_2/l_d)$. This universality of behavior occurs to a still higher degree (to an accuracy $(l_d/L)^{1/2}$) in the case of slit collimator, when $\eta_y^2 > \Delta^2(1)$. Then the radiation intensity increases as $(L/l_d)^{1/2}$ and the coefficient of similarity is $\kappa \simeq (L_1/L_2)^{1/2}$. Thus, as an electron passes through a crystal the radiation spectrum is practically undeformed

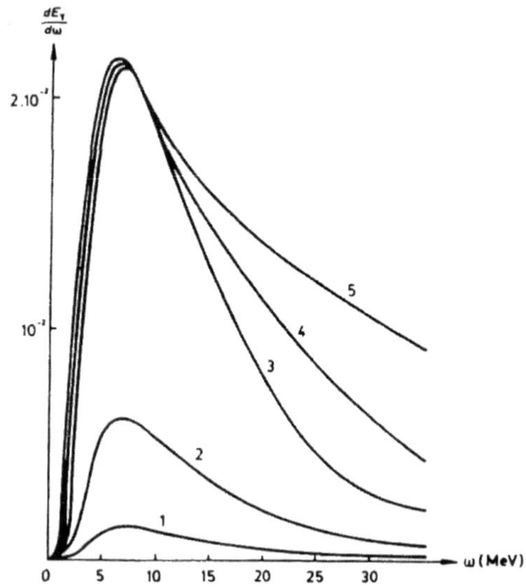

Fig. 23.8. Spectral distribution of the radiation intensity from electrons at the energy $\varepsilon =$ 1 GeV in diamond ((110) plane) for various values of crystal thickness and angular aperture of the circular collimator Δ.

(for a given collimator angle, of course). As the collimator angle η is increased, radiation from particles with larger and larger values of $\varepsilon_\perp$ is incident on the collimator. Since the frequency of the motion for large $\varepsilon_\perp$ increases as $\sqrt{\varepsilon}_\perp$, the contribution of the high-frequency radiation increases faster, for instance, than that from the frequencies in the region of the peak in the spectral density. Figure 23.8 illustrates what has been said above. In this figure the results obtained for the spectral distribution of radiation intensity are presented. The calculations have been made for the plane (110) in diamond at the energy $\varepsilon = 1$ GeV for various values of crystal thickness and the angular aperture of a circular collimator. The potential of the inverted-parabola type was used. DF was approximated by Eq. (23.18). The curve 1–3 are given for $\Delta = 1/2\gamma$ and $L/l_d = 1, 10, 880$, correspondingly. All three curves are almost similar, and it means that the contribution to the radiation is given by the particles with limited values of $\varepsilon_\perp$ and $v_\perp$. The curves 4 and 5 are given for $L/l_d = 880, \Delta =$ $1/\gamma$ and $\delta = 2/\gamma$. Their ordinates have been reduced by the factors 2.6 and 5.6, correspondingly because of comparison convenience. It is clear that with

Δ increasing not only the magnitude of the radiation intensity grows (in fact the growth is weaker than Δ^2), but there is a change of the spectral form discussed above.

24. NONDIPOLE RADIATION FROM ELECTRONS AND POSITRONS AT PLANAR CHANNELING

24.1. Radiation from Particles in Thin Crystals

With an arbitrary potential of planar channeling, for the radiation spectrum at a given values of $\varepsilon_\perp$ we have

$$\begin{aligned}\frac{dI}{d\omega} &= \frac{i\alpha\omega}{(2\pi\gamma)^2}\int_{-\infty}^{\infty}\frac{dt}{t-io} \\ &\times \int_{-\pi}^{\pi} d\tau\left[1+\frac{1}{2}a(\omega)(g(\tau-t)-g(\tau+t))^2\right] \\ &\times \exp\Bigg\{-i\xi\Bigg[2t(1+\langle g^2\rangle) \\ &+\int_{-t}^{t} d\varphi(g^2(\varphi+\tau)-\langle g^2\rangle) \\ &-\frac{1}{2t}\left(\int_{-t}^{t} d\varphi g(\varphi+\tau)\right)^2\Bigg]\Bigg\}, \qquad (24.1)\end{aligned}$$

using formula (5.31). Here ω is the photon frequency, $\xi = \omega\varepsilon/2\gamma^2\omega_0(\varepsilon-\omega)$, $\omega_0(\varepsilon_\perp)$ is the motion frequency; $g(\varphi) = \gamma[v_x(\varphi) - \langle v_x\rangle], \langle\cdots\rangle$ stands for an average value; $a(\omega) = 1 + \omega^2/2\varepsilon(\varepsilon-\omega)$; and the multipole parameter is $\rho = 2\langle g^2\rangle$. Formula (24.1) describes the radiation at any values of the parameter ρ, including the quantum region when the radiation recoil needs to be taken into account. In this section, the region of energy, in which the parameter $\rho = 2U_0\varepsilon/m^2 \sim 1$, will be considered when the description of radiation from particles in the channel can be made significantly simpler. The velocity $v_x(\varphi)$ (just as the function $g(\varphi)$) in Eq. (24.1) can be expanded in the Fourier series

$$\begin{aligned}v_x(\varphi) &= \sum_{l=1}^{\infty} v_{2l-1}\cdot\cos[(2l-1)\varphi] \\ v_{2l-1} &= \frac{1}{\pi}\int_0^{2\pi} d\varphi v_x(\varphi)\cos[(2l-1)\varphi]\,. \end{aligned} \qquad (24.2)$$

The values of v_{2l-1} for the appropriate potential are given in Sec. 9. The quantity $v_{(2l-1)}$ falls as the motion harmonic l grows, and the corresponding radiation frequency shifts to the right. Thus, the contribution of the higher motion harmonics l to the region of a maximum of the resultant spectrum ($\xi \sim 1$) proves to be suppressed for two reasons: (i) this maximum lies on the left slope of the spectral curve for a given l, (ii) even in its maximum the higher harmonic l contributes much less as compared with the first harmonic at $\xi \sim 1$. Correspondingly, at $\xi \sim 1$, for the purpose of describing the radiation from channeled particles one can restrict oneself to a few terms in (24.2). If we retain only the first term $v_1 \cos\varphi$, then substituting an appropriate $g(\varphi)$ into (24.1) we are led to (5.35), wherein the following substitution should be made: $\rho \to \rho_1 = \gamma^2 v_1^2$. We also present here an expression for the spectrum taking into account also a small addition contributed by the term $v_3 \cos 3\varphi$

$$\frac{dI}{d\omega} = \frac{i\alpha\omega}{2\pi\gamma^2}\int_{-\infty}^{\infty}\frac{dt}{t-io}\left\{J_0(z_0) + a(\omega)\rho_1 \sin^2 t(J_0(z_0)\right.$$

$$\left. - iJ_1(z_0)) + \rho_1\frac{v_3}{v_1}G(t)\right\}\exp(-iZ_1), \tag{24.3}$$

where

$$G(t) = 2a(\omega)\sin t \sin 3t[J_2(z_0) + iJ_1(z_0)]$$

$$+ \xi\left(\sin 2t - \frac{2\sin t \sin 3t}{3t}\right)\cdot[bJ_1(z_0) + i(b-1)J_1'(z_0)]$$

$$+ \xi\left(\sin 2t \cdot \cos 2t - \frac{2\sin t \cdot \sin 3t}{3t}\right)[ibJ_2(z_0)$$

$$- (b-1)J_2'(z_0)], \quad b = 1 + a(\omega)\rho_1 \sin^2 t,$$

$$z_0 = \xi\rho_1\left(\frac{\sin^2 t}{t} - \frac{\sin 2t}{2}\right), \quad z_1 = \xi\left[2t + \rho_1\left(t - \frac{\sin^2 t}{t}\right)\right] \tag{24.4}$$

To make the convergence of the integral in (24.3) better, one can use the subtraction procedure described in Sec. 5.

The presented approach calls for a knowledge only of the Fourier harmonics of velocity and hence, is applicable to the description of radiation from both electrons and positrons in any potential. We note that the dependence $T(\varepsilon_\perp)$ is taken into account exactly and the quantity ρ_1 is close to ρ.

As well as Eq. (24.1) we shall use another form of spectral distribution. If we base ourselves on the formula (5.24), describing also the polarization properties of the radiation , then we obtain for the radiation spectrum an expression in the form of a sum over harmonics

$$\frac{dI}{d\omega} = \frac{d\omega}{(2\pi\gamma)^2} \sum_{n=1}^{\infty} \vartheta\left(n - \xi\left(1 + \frac{\rho}{2}\right)\right) \int_{-\pi}^{\pi} dt_1 dt_2 J_0\left(2 \int_{t_2}^{t_1} d\varphi g(\varphi)\right.$$

$$\times \left.\sqrt{\xi(n - \xi(1 + \rho/2)}\right) \left[1 + \frac{a(\omega)}{2}(g(t_2) - g(t_1))^2\right]$$

$$\times \exp\left\{i\left(n - \frac{\xi\rho}{2}\right)(t_1 - t_2) + i\xi \int_{t_2}^{t_1} d\varphi g^2(\varphi)\right\}, \tag{24.5}$$

where J_0 is the Bessel function and $\vartheta(x) = 1, x > 0; \vartheta(x) = 0, x < 0$. For the rest of the notation see Eq. (24.1). Formulae (24.1) and (24.5) are completely equivalent and describe the radiation from both electrons and positrons. To derive explicit expressions, it is necessary to find $v_x(\varphi)$, i.e. to solve the problem of a one-dimensional motion. This is possible in any potential.

We note that the formula (24.1), as (5.31), is straightforwardly derived from the results of Baier, Katkov, Milstein and Strakhovenko, 1975 which presents a quantum theory of the interaction between charged particles and a field of a plane electromagnetic wave.

In the case of relativistic transverse motion ($\rho_0 \geq 1$) the frequency of the emitted photon is defined by Eq. (5.7), then one or a few first harmonics are emitted and the spectrum is concentrated in the range of frequencies $\omega \sim \omega_c$

$$\omega_c = \frac{2\gamma^2\omega_0}{1 + \frac{\rho}{2}} \tag{24.6}$$

and $\omega_c/\varepsilon \sim \lambda_c/d \ll 1$. Hence, the radiation can be described in terms of classical electrodynamics; below we shall neglect the quantum corrections $\sim \omega/\varepsilon$.

The parameter ρ depends substantially on the integral of motion $\varepsilon_\perp$ For example, in the oscillator potential for positrons

$$\frac{\rho}{\rho_0} = z + \vartheta(z - 1)\left[\frac{\sqrt{z-1}}{\arcsin(\frac{1}{\sqrt{z}})} - \frac{2}{(\arcsin(\frac{1}{\sqrt{z}}))^2}\right], \tag{24.7}$$

where $z = \varepsilon_\perp/U_0$. For particles in the channel ($\varepsilon_\perp < U_0$, label "c")$\rho \propto z, \rho$ grows with the increase of the transverse energy of the particle and reaches the maximum, $\rho = \rho_0$, at $\varepsilon_\perp = U_0$. At $z > 1$, when particles move above the barrier ($\varepsilon_\perp > U_0$, label "nc") and acquire a nonzero average velocity projection onto a direction normal to the corresponding plane, the quantity $\rho(z)$ decreases very fast, for example,

$$\rho_{nc}(z = 1.01) \simeq \rho_0/6.5 \tag{24.8}$$

and at $z \gg 1$

$$\rho_{nc}(z) \simeq 2\rho_0/45z\,. \tag{24.9}$$

Such behavior of $\rho(z)$ enables one to calculate, at $\rho \sim 1$, the emission characteristics of the above-barrier particles in a dipole approximation.

The frequency of radiation ω at a given value of the transverse positron energy $\varepsilon_\perp$ depends on ω_0 and ρ (24.6). In the dipole approximation ($\rho \ll 1$) the difference in emission frequencies for various $\varepsilon_\perp$ is determined only by the $\varepsilon_\perp$-dependence of the motion frequency ω_0. For the oscillator potential there is no such dependence, and the spectral characteristic of the radiation from positrons with different values of the transverse energy differ by their intensity only. For this reason, if $\rho \ll 1$, the difference of the potential $U(x)$ from the oscillator one is essential even when the spread of frequencies ω_0 for the particles moving with various $\varepsilon_\perp$ is not too large (see Sec. 22). In the nondipole case ($\rho \geq 1$), the more important factor determining the difference of the radiation frequencies for various $\varepsilon_\perp$ is the dependence of ω on $\rho(\varepsilon_\perp)$ and, hence, in this case the oscillator potential is suited, which will be used below. For electrons it corresponds to an "inverse parabola" potential (9.47). The total radiation intensity from electrons in the potential under consideration is

$$\begin{aligned} I_e(z) &= I_e\left(1 - z + \frac{\sqrt{z}}{f(z)}\right) \\ \frac{\rho_e(z)}{\rho_0} &= z + \frac{\sqrt{z}}{f(z)} - 1 - \frac{2\vartheta(z-1)}{f^2(z)}\,, \end{aligned} \tag{24.10}$$

where $I_e = 16e^2U_0^2\gamma^2/3m^2d^2$ The quantity $I_e(z)$ decreases from $2I_e$ at $z = 0$ to 0 at $z = 1$, and then increases, at $z > 1$, approaching $I_{as} = 2/3I_e$; $f(z) = 1/2\ \ln|(1+\sqrt{z})/(1-\sqrt{z})|$. We are reminded that with an increase of z from 0 to 1 the period of an electron motion in the channel increases and above the barrier, it decreases monotonously, and at $z \gg 1, T_e \propto 1/\sqrt{z}$; in the vicinity of the point $z = 1$ the period $T_e \propto \ln|1-z|$. The parameter $\rho(z)$ increases in the channel with growing z, attains a maximum, $0.43\rho_0$ at $z \simeq 0.85$, and then reduces to 0. Above the barrier, $\rho(z)$ grows from 0, takes the value $\rho_{\max} =$

$0.13\rho_0$ at $z \simeq 1.001$, then decreases, and $\rho/\rho_0 \simeq 2/45z$ at $z \gg 1$. This result coincides with that for the positron case because $\rho \simeq \gamma^2(\langle U^2\rangle - \langle U\rangle^2)/\varepsilon\varepsilon_\perp$ high above the barrier and is independent of the sign of U. In general, at $z \gg 1$ the radiation does not depend on the sign of the charge, since in this limit the result coincides with that of Born approximation. Since for electrons at $z \to 1$ the period $T(z)$ tends to infinity in any adequate potential, the non-harmonicity of motion will be largest at $z \to 1$. Thus, in the "inverse parabola" potential, at $z = (0.3;\ 0.6$ and $0.9) v_3/v_1$ is equal, correspondingly, to (0.13, 0.15 and 0.22) and $v_5/v_1 =$ (0.045, 0.056 and 0.089). However, because at $z \to 1$ the intensity (24.10) $I_e(z) \to 0$, the radiation from particles which underwent a significant non-harmonic motion, may be neglected for the finite population at $z \to 1$. Even in the case when the population in the channel is $\propto f(z)$ (thin crystal, the particles are uniformly distributed over the angle of incidence $\vartheta_0 \le \vartheta_c$; very thick crystal) and $I_e(z)f(z) \to I_e$ at $z \to 1$, the fraction of particles in the vicinity of $z = 1$ is small and corresponding radiation is very soft.

In the parabolic ("inverse parabola" for electrons) potential we have from (24.5) for the radiation spectrum at a given energy $\varepsilon_\perp$

$$\begin{aligned}\frac{dI}{d\omega} &= -\frac{\alpha}{\pi^2 d}\vartheta_c\xi_0\left(1-\frac{\omega}{\varepsilon}\right)\sum_{n=1}^{\infty}\vartheta\left(\frac{n}{S}-\xi_0\left(1+\frac{\rho}{2}\right)\right)\\ &\times\int_{-\pi}^{\pi}dt_1dt_2\,J_0\left[2\sqrt{\rho_0\xi_0\left(\frac{n}{S}-\xi_0\left(1+\frac{\rho}{2}\right)\right)}(q_1-q_2)\right]\\ &\times\left[1+\frac{1}{2}\rho_0 a(\omega)(\psi_1-\psi_2)^2\right]\cdot\cos\left[n(t_1-t_2)+\frac{1}{2}\xi_0\rho_0(r_1-r_2)\right],\end{aligned} \tag{24.11}$$

where $\xi_0 = (\omega/2\gamma^2\omega_{os})(1-\omega/\varepsilon)^{-1}, \omega_{os}$ is the $\varepsilon_\perp$-independent frequency of motion in the parabolic potential: $\omega_{os} = 2\vartheta_c/d$, d is the distance between planes forming the channel, $S = T/T_{os}$, T is the period of motion, $T_{os} = 2\pi/\omega_{os}$; index (1, 2) indicates the dependence of the function on t_1 or t_2. The explicit expressions for the functions $q(t), \psi(t)$, and $r(t)$ are given in Table 24.1 for $t > 0$; at $t < 0$ we have $\psi(-t) = \psi(t), q(-t) = -q(t)$, and $r(-t) = -r(t)$; the table also presents $S = S(z)$ and $\rho(z)$.

Let us consider the radiation in thin crystals. The spectra $dI/d\omega$, averaged with initial distribution (9.24) over $\varepsilon_\perp$, are presented in Figs. 24.1 and 24.2, where the frequency is given in ξ_0 units. The case is considered when the initial beam is uniformly distributed over the angle of incidence in a given range.

Table 24.1. The functions involved in (24.11).

Function	Particles and type of motion			
	e_c^+	e_{nc}^+	e_c^-	e_{nc}^-
$\Psi(t)$	$\sqrt{\mathbf{z}}\cos St$		$\sqrt{\mathbf{z}}\cosh Sh - \sinh St$	
$q(t)$	$\sqrt{\mathbf{z}}\sin t$	$\sqrt{\mathbf{z}}\sin St - \frac{t}{\pi}$	$1 - \cosh St + \sqrt{\mathbf{z}}\sinh St$	$1 - \cosh St + \sqrt{\mathbf{z}}\sinh St$ $-\frac{t}{\pi}$
$r(t)$	$\frac{1}{2}\mathbf{z}\sin 2t$	$\frac{1}{2}\sin 2St$ $-4q(t)/\pi S$ $-t\sqrt{z-1}/\pi$	$\sqrt{z}(1-\cosh 2St)$ $-2t\sqrt{\mathbf{z}}/\pi$ $\frac{1}{2}(1+\mathbf{z})\sinh 2St$	$\sqrt{\mathbf{z}}(1-\cosh 2St)$ $\frac{1}{2}(1+\mathbf{z})\sinh 2St$ $-\frac{t}{\pi}\sqrt{\mathbf{z}} - \frac{4q(t)}{\pi S}$
$S(\mathbf{z})$	1	A/π	$2f/\pi$	f/π
$\frac{\rho^{(z)}}{\rho_0}$	$\mathbf{z}$	$\mathbf{z} + \frac{\sqrt{\mathbf{z}-1}}{A} - \frac{2}{A^2}$	$\mathbf{z} + \frac{\sqrt{\mathbf{z}}}{f} - 1$	$\mathbf{z} + \frac{\sqrt{\mathbf{z}}}{f} - 1 - \frac{2}{f^2}$

In the table the following notations are used: $A = \arcsin(1/\sqrt{\mathbf{z}})$, $f = \frac{1}{2}\ln|(1+\sqrt{\mathbf{z}})/(1-\sqrt{\mathbf{z}})|$.

Figure 24.1 demonstrates the spectra in Si (110) at $\varepsilon = 10$ GeV, ($\xi_0 = 1$ corresponds to $\omega = 110$ MeV) for the interval of incident angles (0 to 20) μrad (Fig. 24.1)(a), (0 to 50) μrad (b) and (40 to 60) μrad (c); the Lindhard angle is $\vartheta_c = 71\,\mu$rad. Fig. 24.2 shows the spectra in diamond (110), $\varepsilon = 4.3$ GeV ($\xi_0 = 1$ corresponds to $\omega = 50$ GeV, $\vartheta_c = 113\,\mu$rad, $\rho_0 = 0.9$) for the intervals of angles (-90 to 90), (0 to 180) and (90 to 270) μrad. In Fig. 24.1, for curve 1, $dI^{nc}/d\omega$ has been calculated by means of the dipole formulae, and $dI^c/d\omega$ has been calculated according to Eq. (24.3) (with the term $\propto v_3$ omitted). Curve 2 is a result of calculations by means of exact formula (24.11). It is seen that taking into account only one harmonic of motion (at a given $\varepsilon_\perp$) is a good approximation. In addition, the discrepancy between curves 1 and 2 is more noticeable on their right slopes. For comparison, we also present the radiation spectrum from positrons (curve 3) obtained under the same conditions. Just as in the dipole case, the spectral curve for electrons is wider and has a lower height in its maximum (cf. Fig. 22.1 – 22.6) when compared with that for positrons, although the difference in height in the nondipole region becomes not so significant. It is seen in Fig. 24.1(b) that the angular broadening of the incident beam reduces the height of the peak, since the number of particles in the channel decreases and these particles give the main contribution in the maximum of spectral curves. The fraction of particles above

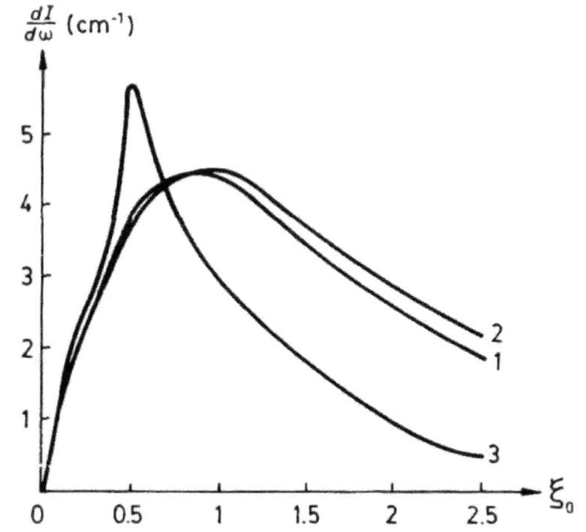

(a)

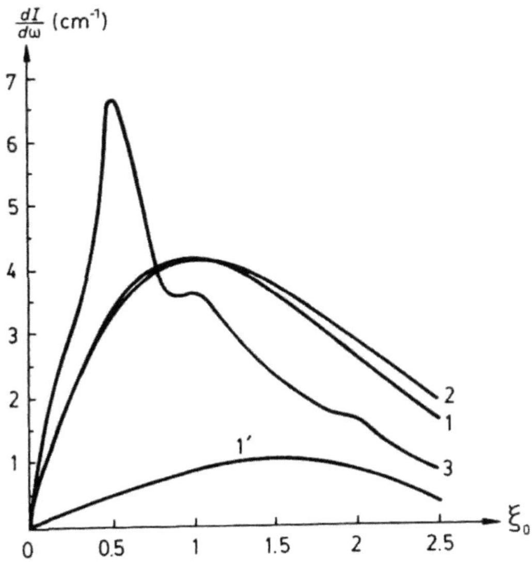

(b)

Fig. 24.1. Spectral intensity of radiation from electrons with the energy $\varepsilon = 10$ GeV upon their motion along the (110) plane of a Si crystal when the initial beam is uniformly distributed over the angle of incidence; 1′–contribution of above-barrier and 1″–channeled electrons.

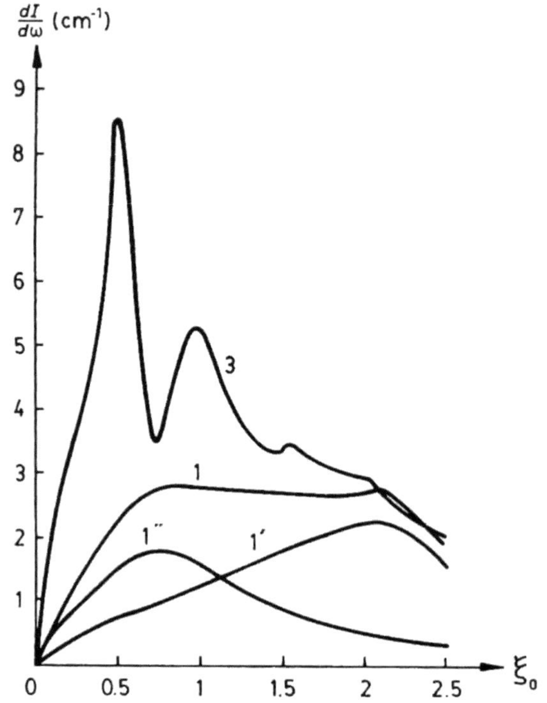

(c)

Fig. 24.1 (*Continued*)

the barrier, which radiation spectrum is presented by curve 1′, grows and it is one of the reasons why the maximum of the summary curve 1 shifts somewhat to the right. It is seen in Fig. 24.1(c) that the rotation of the narrow beam by the angle $\sim \vartheta_c$ leads to sweeping changes in the spectrum. The channeled (1″) and above barrier (1′) particles emit different frequencies and their contributions are approximately equal in magnitude; as a consequence, the summary spectrum (1) has a wide plateau. It is seen in Fig. 24.2 that when rotating a wide beam by an angle of about ϑ_c a maximum (keeping its position) from the channeled particles is still clearly seen in the radiation spectrum.

Let us turn now to a qualitative structure of the spectral intensity distribution from positrons at $\rho_0 \sim 1$. In this case the largest portion of the radiation intensity comes from the first harmonic. Positrons radiate most effectively at $\varepsilon_\perp \simeq U_0$ so that, according to (24.6), the radiation (after averaging over

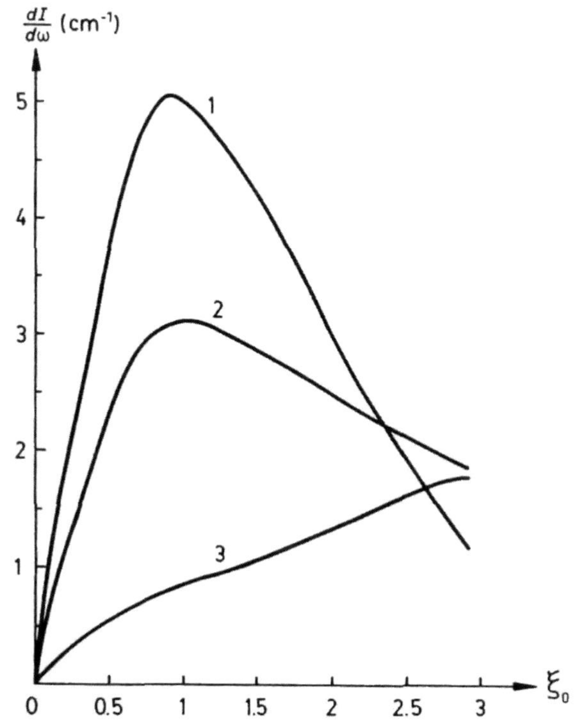

Fig. 24.2. Spectral intensity of radiation from electrons with the energy $\varepsilon = 4.3$ GeV upon their motion along the (110) plane of a diamond crystal when the initial beam is uniformly distributed over the angle of incidence in the (1) -90 to 90, (2) 0 to 180, (3) 90 to 270 μrad range.

the transverse energy distribution) attains a maximum at $\omega = \omega_{\max}^{(1)}$, where

$$\begin{aligned} \omega_{\max}^{(1)} &= 2\gamma^2\omega_0\xi_0 \,, \\ \xi_0 &= (1+\rho_0/2)^{-1} \end{aligned} \tag{24.12}$$

The minimum of the spectral density is on the boundary of the first harmonic:

$$\begin{aligned} \omega_{\min} &= 2\gamma^2\omega_0\xi_1 \,, \\ \xi_1 &= (1+\rho_1/2)^{-1} \,, \end{aligned} \tag{24.13}$$

where ρ_1 is a value of ρ at as low as possible value of $\varepsilon_\perp$ For example, for the particles entering a crystal at an angle $\vartheta = \vartheta_0 < \vartheta_c = \sqrt{2U_0/\varepsilon}$ to the crystal plane $z_1 = \vartheta_0^2/\vartheta_c^2$, $\rho_1 = \rho(z_1)$ (see Eq. (24.7)). The appearance of the second maximum in the spectral distribution is due to the contribution from the second harmonic of radiation, $\omega_{\rm max}^{(2)} = 2\omega_{\rm max}^{(1)}$. Such a picture is clearly observed if $2\xi_0 > \xi_1$. If the inverse inequality holds, the minimum of the average intensity lies (in terms of ξ) between the point ξ_0 and $2\xi_0$ and is to the left of the point ξ_1. In the particular case $2\xi_0 \simeq \xi_1$, these structures can be washed out. Such a situation occurs, for example, at the channeling of positrons with $\varepsilon = 10$ GeV and $\vartheta_0 = 0$ in the plane (110) of the diamond and silicon single crystals; in this case, $\rho_1 = 0$ and $\rho_0/2 \simeq 1$ (see Fig. 24.1(a), curve 3). However, in the case, when $\vartheta_{\rm min} = 40$ μrad (curve 3 in Fig. 24.1(c)), $2\omega_{\rm max}^{(1)} > \omega_{\rm min}^{(1)}$ and $\rho_1 = (\vartheta_{\rm min}/\vartheta_c)^2 \simeq 0.61$, so that one can show right away the point of the spectrum minimum corresponding to the frequency boundary of the first harmonic radiation, $\omega_{\rm min}^{(1)} \simeq 84$ MeV ($\xi \simeq 0.76$).

For a given angle of incidence of the particles into a crystal (ϑ), after averaging over the point of entry, we obtain

$$\begin{aligned}\langle I_c(\vartheta)\rangle &= A\int_{b^2}^{1} \frac{dz\cdot z}{2\sqrt{z-b^2}} \\ &= \frac{A}{3}\sqrt{1-b^2}(1+2b^2)\vartheta(1-b^2)\,, \end{aligned} \tag{24.14}$$

for the total radiation intensity of positrons trapped into the channel, where $A = 2e^2\rho_0^2/3d^2, b = \vartheta/\vartheta_c$. Let the particles of the incident beam be uniformly distributed over the angle ϑ in the interval $\vartheta_0 - \Delta \le \vartheta \le \vartheta_0 + \Delta$. Then the intensity (24.14) should be averaged over the angle ϑ.

$$\langle I_c(\tilde\vartheta_0, \tilde\Delta)\rangle_\vartheta = \frac{A}{6\tilde\Delta}\int_{\tilde\vartheta_0-\tilde\Delta}^{\tilde\vartheta_0+\tilde\Delta} dx\sqrt{1-x^2}(1+2x^2)\vartheta(1-x^2)\,, \tag{24.15}$$

where $\tilde\vartheta_0 = \vartheta_0/\vartheta_c$, $\tilde\Delta = \Delta/\vartheta_c$. The quantity $\langle I_c(\tilde\vartheta_0, \tilde\Delta)\rangle_\vartheta$ is an even functon of ϑ_0. In order to obtain its behavior at $\vartheta_0 = 0$, it is necessary to calculate the second derivative. For $\tilde\Delta < 1$ we find

$$\left.\frac{\partial^2\langle I_c(\tilde\vartheta_0, \tilde\Delta)\rangle_\vartheta}{\partial\tilde\vartheta_0^2}\right|_{\tilde\vartheta_0=0} = A\frac{1-2\tilde\Delta^2}{\sqrt{1-\tilde\Delta^2}}\,. \tag{24.16}$$

At $\tilde\Delta < 1/\sqrt{2}$ the radiation intensity of the channeled positrons has the minimum at $\vartheta_0 = 0$ and vanishes when $\tilde\vartheta_0 = 1 + \tilde\Delta$, i.e. it has the maximum between these points. In the case of a wide beam, $\tilde\Delta > 1/\sqrt{2}$, the radiation intensity decreases with increasing ϑ_0 and it has the maximum at $\vartheta_0 = 0$. Thus, the behavior of the radiation intensity as a function of the angle of incidence depends, to a great extent, on the angular spread in the beam.

If the mean angle of incidence of the particle $\vartheta_0 = 0$, the radiation intensity of positrons in the channel has the form

$$\langle I_c(0,\tilde{\Delta})\rangle_\vartheta = \frac{A}{4}\left\{\frac{\pi}{2\tilde{\Delta}}\vartheta(\tilde{\Delta}-1) + \vartheta(1-\tilde{\vartheta}) \times \left[\frac{\arcsin\tilde{\Delta}}{\tilde{\Delta}} + \frac{1}{3}\sqrt{1-\tilde{\Delta}^2}(1+2\tilde{\Delta}^2)\right]\right\} \tag{24.17}$$

This function has a minimum at $\tilde{\Delta} = 0$, reaches a maximum at $\tilde{\Delta} \simeq 0.77$, and decreases as $A\pi/8\tilde{\Delta}$ at $\tilde{\Delta} > 1$.

Since the calculation of the spectral intensity turns out to be a fairly tedious procedure, it seems useful to find a simple method for estimating the main maximum of this distribution when changing the magnitudes of ϑ and Δ. Since the maximum position determined by formula (24.12) is unchanged, one can expect that its height will be connected with the total intensity of particles trapped in the channel. The numerical comparison has shown that these quantities are indeed related by a proportional dependence. The latter makes it possible, starting with the quantity $d\langle I_c(\omega)\rangle/d\omega|_{\max}$ at fixed ϑ_0 and Δ, to obtain the value of $d\langle I_c(\omega)\rangle/d\omega|_{\max}$ for any values of these parameters (we denote them via ϑ_{01} and Δ_1) with the help of the following equation

$$\left.\frac{d\langle I_c(\omega,\vartheta_{01},\Delta_1)\rangle}{d\omega}\right|_{\max} = \left.\frac{d\langle I_c(\omega,\vartheta_0,\Delta)\rangle}{d\omega}\right|_{\max} \cdot \frac{\langle I_c(\vartheta_{01},\Delta_1)\rangle}{\langle I_c(\vartheta_0,\Delta)\rangle}\,. \tag{24.18}$$

The above relation proves to be helpful in the analysis since it allows us to compare the heights of the maximum of the spectral radiation distribution for various ϑ_0 and Δ by using the modified potentials. One of these modifications is, for example, a smoothing of the potential barrier on the channel's boundaries near the plane as a result of the lattice's zero and temperature vibrations. The analysis, confirming the validity of the relation (24.18), has been made by Baier, Katkov and Strakhovenko 1981d. In this paper, the detailed comparison of the obtained results with the experimental data of Miroshnichenko *et al.*, 1979 has been made. These data in a more complete form have been published in paper by Avakyan *et al.*, 1982. In this experiment, in particular, the spectral radiation distributions have been measured for positrons with $\varepsilon = 4, 6, 10$ GeV moving along the plane (110) of a diamond single crystal

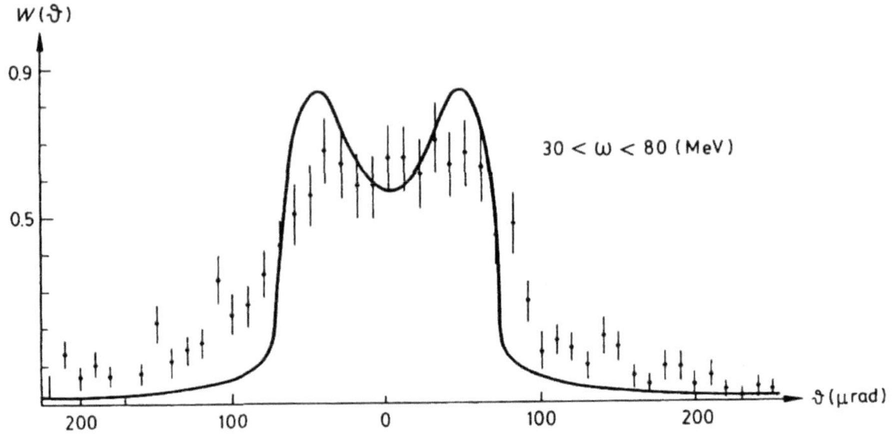

Fig. 24.3. Number of emitted photons (in arbitrary units) in the frequency interval $30 < \omega < 80$ MeV at planar channeling of positrons with the energy $\varepsilon = 10$ GeV in (110) Si plane versus the angle of incidence ϑ. The theoretical curve is given for $\Delta = 5$ μrad and the experimental results are taken from Filatova *et al.*, 1982.

of thickness 80 μm. The above considered qualitative picture of the spectrum has been clearly manifested in the experiment. The theoretical shape of the spectrum is consistent with the experiment, but the magnitude of the theoretical results is approximately three times more than the experimental ones. Perhaps the discrepancy will decrease if to calculate the potential in diamond with better accuracy, and consistently to take into account thermal vibrations and dechanneling. The last factor is very important for positrons at values $\varepsilon_\perp \simeq U_0$.

The experiment described in the paper by Filatova *et al.*, 1982 dealt with the emission from positrons of the energy 10 GeV during their passage through a silicon single crystal of thickness $L = 90$ μm along the plane (110). The depth of the potential for Si (110), (with allowance for thermal vibrations) $U_0 \simeq 25$ eV, then the Lindhard angle is $\vartheta_c = 71$ μrad, $\rho_0 = 1.9$. If one estimates l_d, according to formula (10.1), then $L/l_d \simeq 1/4$.

Figure 24.3 shows the theoretical graph and the experimental values (in arbitrary units) for the number of emitted photons in the interval $30 < \omega < 80$ MeV versus the incident angle of positrons with respect to the plane (110). Since the radiation of the first harmonic has a maximum at the frequency $\omega_{\rm max}^{(1)} \simeq 60$ MeV (see Eq. (24.12)), this interval of photon energies corresponds to the characteristic radiation energies of the first harmonic from positrons moving in the channel. In Fig. 24.3, theory and experiment are consistent satisfactorily with each other, and the experimental values of $W(\vartheta_0)$ do not, at least, decrease with increasing $|\vartheta_0|$ up to the value of 50 μrad. Further experiments (Bak *et al.*, 1985) carried out with better statistics and angular resolution clearly shows the existence of the dip in Fig. 24.3.

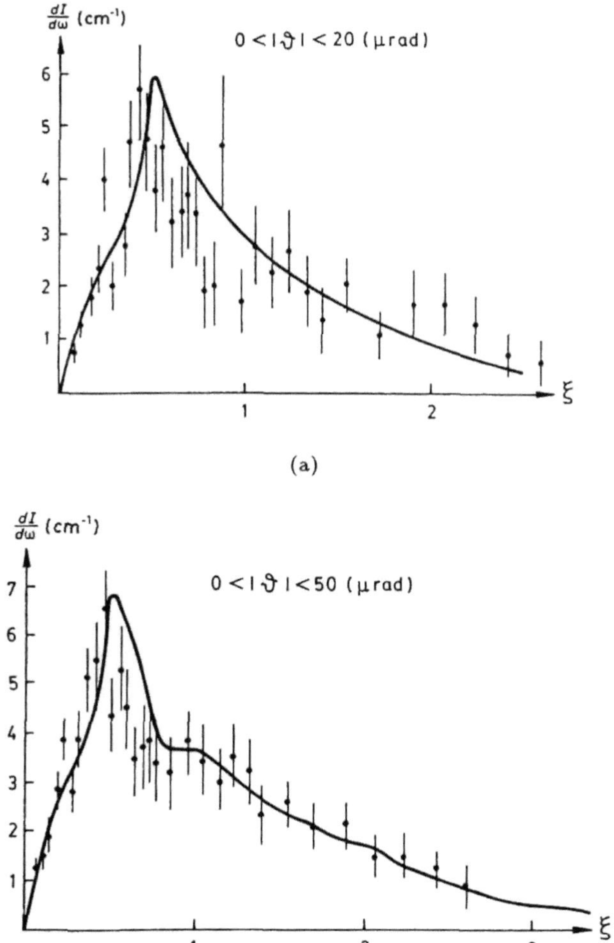

(b)

Fig. 24.4. Spectral intensity of radiation from positrons at planar channeling in Si (110) with the energy $\varepsilon = 10$ GeV for the incident angles interval: (a) $|\vartheta| < 20$ μrad, (b) $|\vartheta| < 50$ μrad.

In Fig. 24.4 our results are given by the solid curve and the experimental values are increased by the factor 1.2 in Fig. 24.4(a) and 1.7 in Fig. 24.4(b). It was made for the sake of convenient comparison. The shape of the theoretical curves and the positions of maxima and minima are in satisfactory agreement with the experiment. Note once more that for $\vartheta_0 = 0$, $2\omega^{(1)}_{\max} = \omega^{(1)}_{\min} = 2\gamma^2\omega_0$ in these graphs and a minimum position ceases to be connected to the lower boundary of the particles distribution over transverse energy.

For the magnitude of the spectral intensity, the experimental and theoretical results are in agreement within their accuracy for the interval of angles $|\vartheta_0| < 20\,\mu$rad (Fig. 24.6). The theoretical value of the spectral intensity in the interval of angles $|\vartheta_0| < 50\,\mu$rad is larger than that in the interval $|\vartheta_0| < 20\,\mu$rad and as we have shown in (1981) the change of the shape of the potential due to the thermal vibrations does not affect this tendency. Experimental data of the paper under discussion indicate the opposite tendency, that contradicts, in our opinion, the data of the same experiment shown in Fig. 24.3, where as it has already been mentioned, radiation probability at least does not decrease with increasing $|\vartheta_0|$ in the interval ϑ_0 under discussion. In the following works (see, for example, Bulgakov (*et al.*, 1983) the existence of minimum $\vartheta_0 = 0$, on the curve of the type presented in Fig. 24.3, has been determined clearly, and that is the experimental confirmation of the theoretical prediction, about the dependence of radiation characteristics on the incident angle of channeled positrons, obtained by Baier, Katkov, Strakhovenko, 1981d.

The simple rules have been indicated above to find the characteristic points (maxima and minima) of the spectral curve for positrons, which are based on an analysis of the z-dependence of the intensity and of the argument of the ϑ-function in (24.11). For electrons, these structures are washed out on account of a large width of the spectral curves which is due to the essential spread of the motion periods. The estimates become less definite as a result. The spectral curve has, as a rule, one maximum which is due to the first harmonic (term with $n = 1$ in (24.11)) of radiation from channeled particles (in Fig. 24.1 and 24.2 it is at $\xi_0 \sim 1$). It follows from (24.11) that at $n = 1$ the radiation frequencies lie within the interval $0 \leqslant \xi_0 \leqslant \xi_1$ where $\xi_1 = [S(1+\rho/2)]^{-1}$ The quantity $S(1+\rho/2)$ grows monotonously with increasing $z(0 < z < 1)$, and ξ_1 shifts to the left with increasing z. For estimations, it is convenient to introduce a "weighted intensity": $F(z) = I_e(z)dN/dz$. If $\max F = F(z_1)$, then as z increases from 0 to z_1 the boundary frequency shifts to the left and the peak becomes higher ($\sim F(z)$). At larger z both functions $F(z)$ and ξ_1 decrease. As a result, the maximum of the resultant distribution $dI_c/d\omega$ will lie to the left from the point $\xi_m^c(z_1) = [S(z_1)(1+\rho/2)]^{-1}$ If the function $F(z)$ has a sharp peak at z_1, then the maximum $dI^c/d\omega$ will lie near $\xi_m^c(z_1)$. If at $z > 1$, max $F(z_2)$, then, similarly, the maximum of $dI^{nc}/d\omega$ is to the left from $\xi_m^{nc}(z_2) = [S(z_2)(1+\rho/2)]^{-1}$.

For illustration, we consider the case when a narrow beam is incident at ϑ_0 near ϑ_c (Fig. 24.1(c)). Here the function $F(z)$ has a clear maximum both in the channel ($z_1 \simeq 0.7$) and above the barrier ($z \simeq 1.4$); the corresponding frequencies are: $\xi_m^c \simeq 0.94$ and $\xi_m^{nc} \simeq 2.4$. In Fig. 24.1(c) the maxima of the spectral curves correspond to $\xi_{\max}^c \simeq 0.8$ and $\xi_{\max}^{nc} \simeq 2.1$. We would like also to discuss an important case when $dN/dz \propto f(z)$. As shown in Sec. 10, such distribution is established in thick crystals in the channel and above the barrier at not too large $z = \varepsilon_\perp/U_0$. This case is realized also in thin crystals for eletrons moving in the channel at the uniform angular distribution of incident electrons in the interval $\vartheta_c(|\vartheta_0| \le \vartheta_c)$; whereas if the upper limit of the incident angles $\simeq \vartheta_c$, then the distribution of channeled electrons is not strongly different, in configuration, from $f(z)$. The function $F(z) = I(z)f(z)$ has a maximum at $z_1 \simeq 0.695$, and $\xi_m^c(z_1) \simeq 1.3/(1+\rho_0/5)$. For the (110) plane, at $\varepsilon = 10$ GeV in silicon $\xi_m^c \simeq 0.94$, and at $\varepsilon = 4.3$ GeV in diamond $\xi_m^c \simeq 1.1$ (cf. Fig. 24.1 and 24.2; curve 1).

24.2. Radiation from Electrons in Thick Crystals

We shall use the interpolating distribution (22.17). If the collimation of a photon beam is absent, then for potential (9.47) the averaged (over crystal thickness) L distribution over z can be obtained by making an elementary integration over ϑ in Eq. (22.18):

$$\left\langle \frac{dN}{dz} \right\rangle = \frac{1}{L}\int_0^L dl \frac{dN(l)}{dz}$$

$$= \sqrt{\frac{2}{\pi\lambda}} \cdot \frac{1}{1-\beta} \cdot \Bigg\{ f(z) \cdot \left[e^{-\frac{z}{2\lambda}} - \sqrt{\beta} e^{-\frac{z}{2\tilde{\Delta}_0^2}} \right]$$

$$- \frac{1}{\lambda} \int_{\sqrt{z}}^{\sqrt{\frac{z}{\beta}}} du\, e^{-\frac{u^2}{2\lambda}} + \frac{1}{\sqrt{\lambda}} \int_0^{f(z)} du\, g_1(ug_2 - g_1)$$

$$\times \left[\frac{1}{\sqrt{\lambda}} e^{\frac{-g_2^2}{2\lambda}} - \frac{1}{\tilde{\Delta}_0} e^{-\frac{g_2^2}{2\tilde{\Delta}_0^2}} \right] \Bigg\}, \qquad (24.19)$$

where

$$\lambda = \frac{L}{l_d} + \tilde{\Delta}_0^2,\ \tilde{\Delta}_0 = \frac{\Delta_0}{\vartheta_c},\ \beta = \frac{\tilde{\Delta}_0^2}{\lambda},$$

$$g_1(u) = \sqrt{|1-z|}\{\vartheta(1-z)\cosh + \vartheta(z-1)\sinh\}, \tag{24.20}$$

$$g_2(u) = \frac{dg_1}{du}$$

From (24.19), there is no difficulty in deriving an asymptotic distribution at $L \gg l_d$. In this limit, $\beta \to 0$ and $\lambda \to \infty$, and we find, within an accuracy up to terms $\sim 1/\sqrt{\lambda}$

$$\left\langle \frac{dN}{dz} \right\rangle = \sqrt{\frac{2l_d}{\pi L}} \cdot f(z)e^{-\frac{z}{2\lambda}} \tag{24.21}$$

so that for $z \ll \lambda$ we have $\langle dN/dz \rangle \propto f(z)$.

In the foregoing, we have estimated the position of the maximum of the spectral curve at $L \gg l_d$. Let us now show that these curves prove to be similar at different (large) thicknesses. Indeed, the resultant spectrum is given by the integral

$$\int_0^\infty \left\langle \frac{dN}{dz} \right\rangle \frac{dI}{d\omega} dz\,. \tag{24.22}$$

Consider now the case of large L when it is possible to use (24.21). We split the integration region into two parts: 0 to z_0 and z_0 to ∞; with $1 \ll z_0 \ll \lambda$. After that, in the first part, one can substitute $\exp(-z/2\lambda) \to 1$ and we obtain a contribution, including L only as the factor $L^{-1/2}$ It is the region ($z \simeq 1$) that contributes to the maximum of the spectrum ($\xi_0 \simeq 1$). In the second part ($z \gg 1$) the particles are in the high-lying above-barrier states and the dipole formulae may be used. Making use of formula (7) in the paper by Baier, Katkov, Strakhovenko, 1979b, we have roughly, at $z \gg 1 : dI/d\omega \propto \xi_0/z$ (photons with a frequency up to $\xi_0 \sim 2\sqrt{z}$ are emitted). Although the contribution of these particles breaks the similarity, it is strongly suppressed at $\xi_0 \sim 1$. The region up to $z \sim \lambda$ contributes to the integral $\int_{z_0}^\infty dz \cdots$, so that the frequency boundary of the spectrum corresponds to $\xi_0 \sim 2\sqrt{\lambda}$. Thus at $\xi_0 \sim 1$ the shape of the spectrum proves to be universal, and the difference in height is determined by the scaling factor $\sim L^{-1/2}$ (upon collimation of the photon beam to an angle $\leq \vartheta_c$, a factor $\sim (1/L)\ln(L/l_d)$ arises). The accuracy of this expression is not worse than $\lambda^{-1/2}$ (the order of the terms omitted in going from (24.19) to (24.21)). The similarity of electron radiation spectra in thick crystals has been found in Sec. 23 for other situations: the dipole radiation at the collimation condition. Actually, the similarity is caused by (i) at large thicknesses, the asymptotic distribution of (24.21) type is established; (ii) in both situations under consideration, the contribution to the spectrum is given by the particles with limited values of $\varepsilon_\perp$ In one of the cases this behavior is provided by the collimation of radiation, in the other it does so by the range of frequencies under consideration (in the region of maximum).

For illustration (Fig. 24.5) let us consider the radiation from electrons upon channeling in the (110) plane of a diamond single crystal at $\varepsilon = 4.5$ GeV; then $\vartheta_c \simeq 110$ μrad, $\rho_0 \simeq 0.95$ and $\omega = 54$ MeV at $\xi_0 = 1$; this corresponds

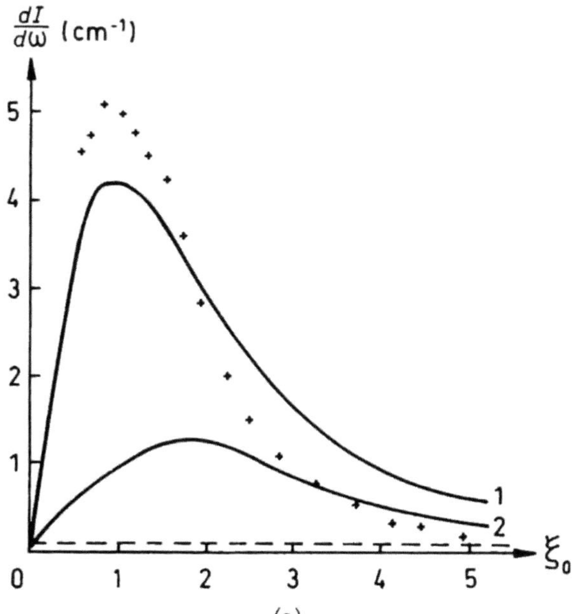

(a)

Fig. 24.5. Spectral intensity of radiation, per unit thickness, from electrons with $\varepsilon =$ 4.5 GeV upon the motion along the (110) plane in diamond. The dashed line shows the bremsstrahlung contribution; (2) is the contribution of the above-barrier electrons, and (1) is the resulting spectrum; the experimental data are from Avakyan *et al.*, 1983 (a) for thickness $L_1 = 0.1$ mm, (b) $L_2 = 1$ mm, (c) $L_3 = 1.7$ mm.

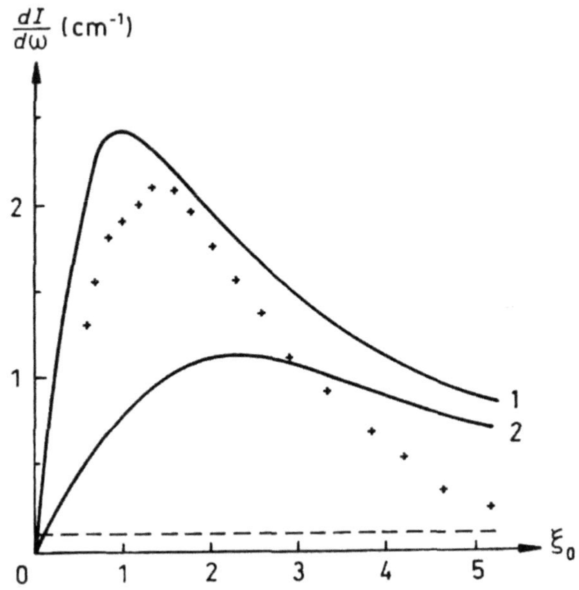

(b)

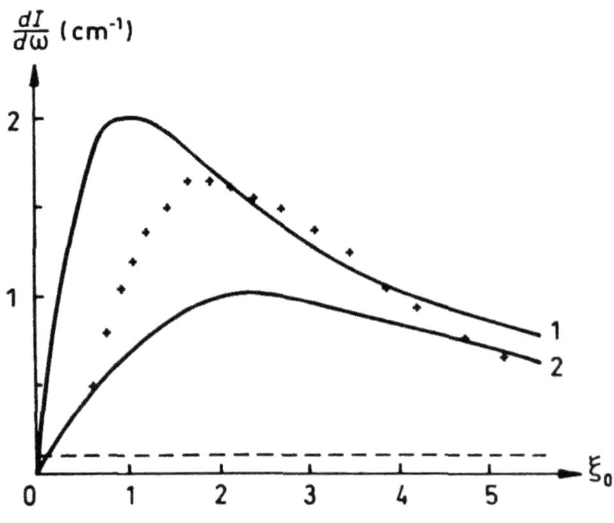

(c)

Fig. 24.5 (*Continued*)

to the experimental conditions of the work by Avakyan *et al.*, 1983. The authors of the cited publication estimate the angular width of the incident beam as $\Delta_0 \simeq 60\ \mu$rad. The theoretical curve 1 in Fig. 24.5 is that of the contribution from radiation in a continuous potential of the plane ($\langle dN/dz \rangle$ and is taken in the form (24.19)) and the bremsstrahlung (the latter contribution is depicted as the dashed line; as seen; it is negligibly small). The crystal thicknesses L_2 and L_3 are large enough; for them (see (24.20)) λ is: $\lambda_2 = 7.7, \lambda_3 = 13$. The expected accuracy of the similarity of the spectral curves is about $\lambda^{-1/2}$ (30 to 40%) and the asymptotic scale factor is $(L_2/L_3)^{-1/2} \simeq 1.3$. For $\xi_0 \leq 2$ neglecting bremsstrahlung the calculated curves are similar within an accuracy better than 5% and the scaling factor is 1.2; its difference from $(L_2/L_3)^{-1/2}$ means that the asymptotic regime does not yet occur. Nevertheless, the accuracy of similarity, at $\xi_0 \leq 2$, has proved to be much better than the estimate with respect to $\lambda^{-1/2}$ Curve 2 in Fig. 24.5 is the contribution given by the above-barrier particles, the maximum on it shifts to the right by increasing the thickness, whereas the point of peak for the contribution of particles in the channel is constant ($\xi_0 \simeq 0.8$). Practically the same is true for the point of maximum of the resultant spectrum (curve 1).

In Fig. 24.5, the experimental data are also presented. For the thickness $L_1 = 0.1$ mm the theory and the experiment are in the satisfactory agreement for $\xi_0 \leq 2$. However the maximum of spectrum moves to the right for the thicker crystals. The probable cause of this shift is the multiple scattering and the particularity of the spectrometer registering two and more photons, emitted at the same time, as one with summary energy, in the opinion of the authors of the experiment under discussion. Since we have taken into account the multiple scattering, the reason for the difference from the theory can be really connected with detector particularities. The probability of photon emission with the frequency $\xi_0 \leq 2$ can be estimated using the spectral curves in Fig. 24.5. We have $W \simeq 0.08, 0.46, 0.65$ for thickness L_1, L_2 and L_3 correspondingly, so that the probability of emission of two and more photons by electrons is not small. It is interesting to estimate quantitatively the accuracy of the dipole approximation by calculation in this approximation and by means of an exact formulae. Consider the radiation in the (100) plane in Si at $\varepsilon = 1.2$ GeV ($\vartheta_c = 150\ \mu$rad, $\rho_0 = 0.13$; $\xi_0 = 1$ corresponds to $\omega = 4.8$ MeV), $L = 0.24$ mm, $\Delta_0 = 100\ \mu$rad, that is in accord with the experimental condition (Ganenko *et al.*, 1983).

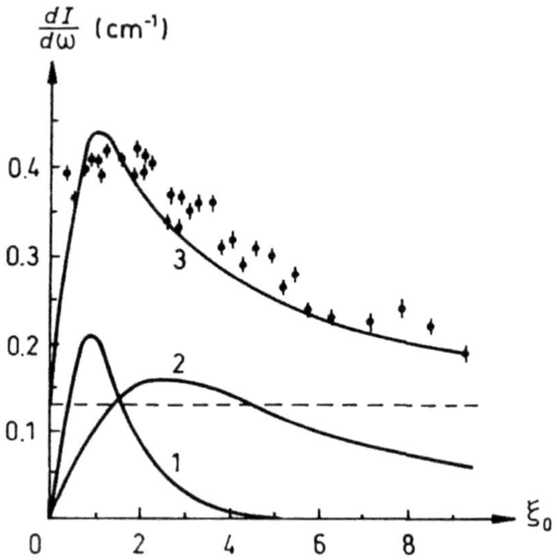

Fig. 24.6. Spectral intensity of radiation, per unit thickness, from electrons with the energy $\varepsilon = 1.2$ GeV moving along the (100) plane in a Si crystal of thickness $L = 0.24$ mm. The bremsstrahlung contribution is shown by the dashed line. (1) is the contribution from the channeled electrons, (2) is that from the above-barrier ones, and (3) is the resultant spectrum; the experimental data are from Ganenko *et al.*, 1983.

The calculated results with the use of (24.19) are given in Fig. 24.6. In the dipole approximation, the radiation has been calculated using the formulae from Sec. 23. Comparison with curves 1 and 2 in Fig. 24.6 has shown that the dipole curves lie 5 to 10% higher ($\sim \rho_0$). We would like to attract the reader's attention to the fact that the shape of the spectrum in Fig. 24.6 is mainly determined, in the region of maximum, by the contribution from the particles in the channel. The above-barrier particles give a dominant contribution only on the right slope (at $\xi_0 > 2$).

In the work by Aganiants *et al.*, 1983 the radiation from electrons with energy $\varepsilon = 4.3$ GeV has been measured in diamond. The crystal thickness was $L_4 = 0.1$ mm ((011) plane) and $L_5 = 0.47$ mm ((011) and (001) planes). In the case of the (011) plane the Lindhard angle $\vartheta_c = 113$ μrad and $\rho_0 = 0.91$; $\xi_0 = 1$ corresponds to $\omega = 50$ MeV. For the plane (001) the corresponding values are $\vartheta_c = 79$ μrad, $\rho_0 = 0.44$ and $\omega = 49$ MeV. The divergence of the incident beam was $\Delta_0 \simeq 100$ μrad and the beam of the emitted photons was collimated

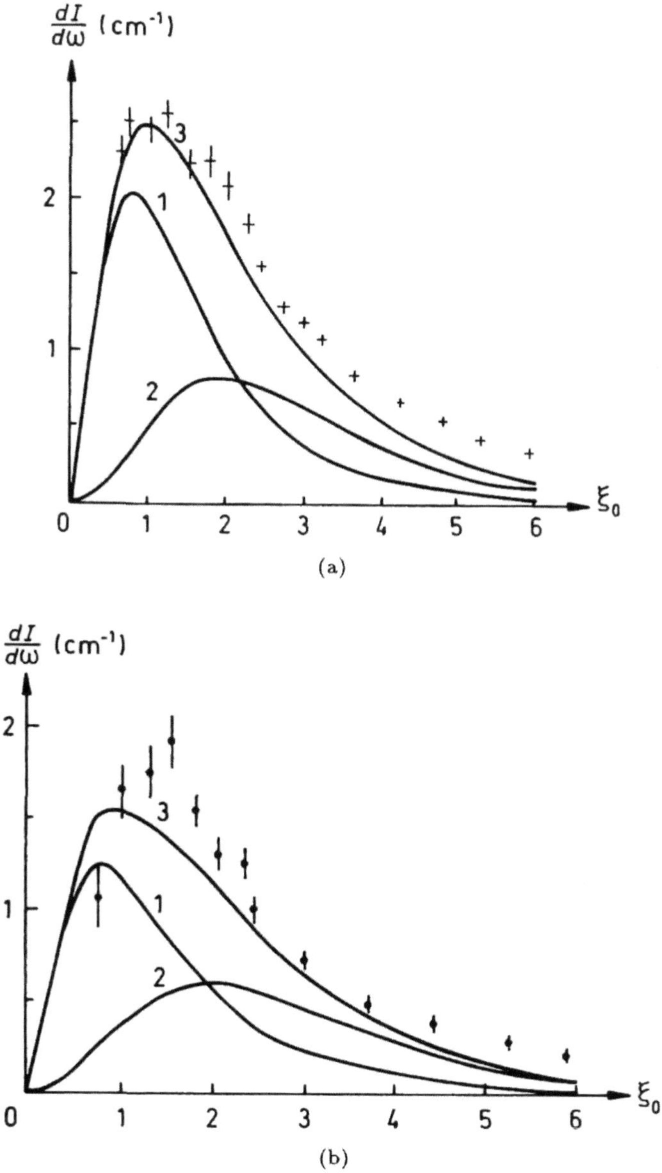

Fig. 24.7 Spectral intensity of radiation, per unit thickness, from electrons with the energy $\varepsilon = 4.3$ GeV moving along the (011) plane in diamond crystal of thickness (a) $L_4 = 0.1$ mm, (b) $L_5 = 0.47$ mm and of thickness L_5, but for (001) plane (c).

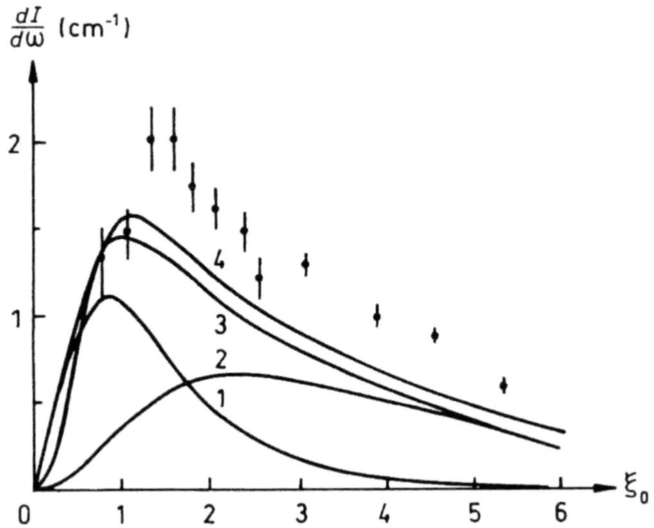

(c)

Fig. 24.7 (*Continued*)

to a solid angle 10^{-7} ster. The results in the above-mentioned work are presented in terms of dI/dI^{br} We give them in Fig. 24.7 in an absolute normalization using ($\omega \ll \varepsilon$) the value of $dI^{br}/d\omega \simeq 4k/3L_{rad}$, the coefficient k takes into account the influence of the collimator for a disoriented crystal. Within the frame of the accepted multiple-scattering model it is easy to find that $k \simeq 0.7$ at $L = 0.1$ mm and $k \simeq 0.45$ at $L = 0.47$ mm.

In the works by Atkinson *et al.*, 1982; Bak *et al.*, 1982, the authors have analyzed the radiation from channeling electrons in the (110) plane of a silicon crystal at the energy 10 GeV and 7 GeV, respectively. The crystal thickness was 0.135 mm and 0.1 mm; the Lindhard angle was 71 μrad and 95 μrad; the nondipole parameter was 1.9 and 1.33; and the characteristic frequency was 110 MeV and 65 MeV. Figure 24.8 gives the results of our calculations and experimental data for these two cases which we have represented in absolute units.

Thus, we can see that the developed theory of radiation from electrons at planar channeling describes quite well the experiments for various media, planes, and thickness in a broad interval of energies. It follows from this, in particular, that the used simple dechanneling model is quite adequate.

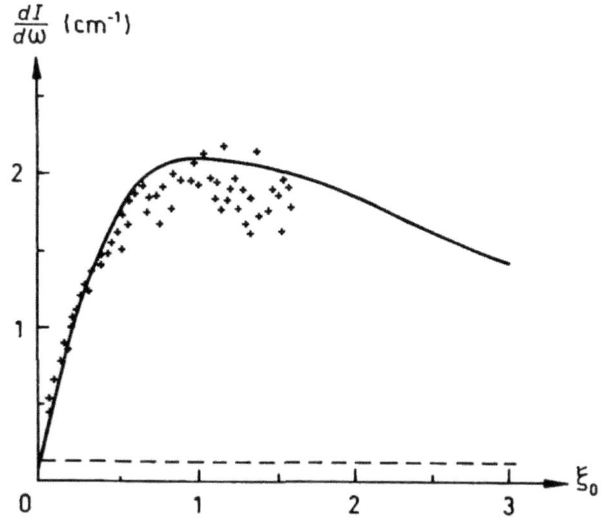

(a)

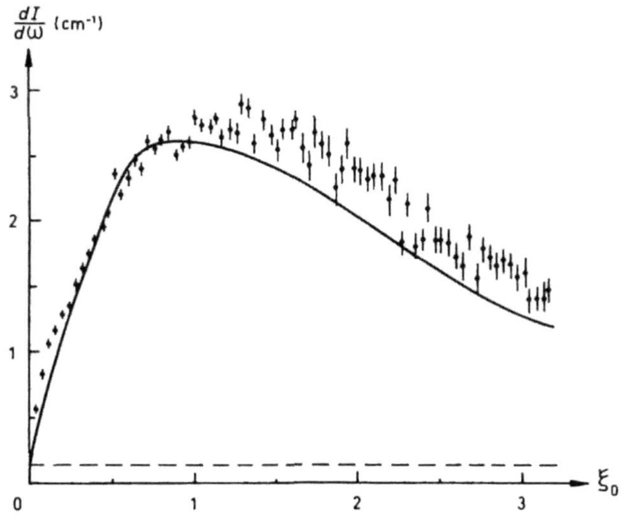

(b)

Fig. 24.8. Spectral intensity of radiation, per unit thickness, from electrons moving along the (110) plane in silicon crystal. (a) For $\varepsilon = 10$ GeV and $L = 0.135$ mm, the experimental data from Atkinson *et al.*, 1982; (b) for $\varepsilon = 7$ GeV and $L = 0.1$ mm, the experimental data from Bak *et al.*, 1982. The dashed line shows the bremsstrahlung contribution, the solid curve is the summary spectrum.

25. QUANTATIVE PICTURE OF RADIATION AT CHANNELING OF LOW ENERGY PARTICLES

25.1. Description of Particle Motion in Quantum Region

With decrease of the energy of the particles ε the wavelength of the transverse motion ($\lambda \sim 2\pi/p_\perp \sim 2\pi/\sqrt{2U_0\varepsilon}$) increases and at rather low energies becomes equal to the channel dimensions d. At these energies the quantum numbers of a transverse motion are not large and quantum effects in this motion become essential. With further decrease of energy the probability of particle penetration through the potential barrier becomes large and instead of an energy level in the potential well an energy zone appears.

Since the case of a motion in the isolated potential (channeling) is of interest, when the probability of penetration through the potential barrier and hence the width of the energy zone are small, it is necessary to have several levels in the potential well. Thus taking into account the estimation for the number of bound states $n \sim \sqrt{\gamma}$ (9.55), we shall consider relativistic particles ($\gamma \gg 1$), with the energy at not more than a few tens of MeV.

To quantize the transverse motion, the coordinate and the momentum of a particle should be substituted by the operators (in the coordinate representation $p_x \to -i\partial/\partial x$). The corresponding equation for eigenfunctions and eigenvalues has the form of the nonrelativistic Schrödinger equation taking into account the substitution of the particle mass by its energy $\varepsilon \simeq (p_z^2 + p_y^2 + m^2)^{1/2}$ (see Sec. 9.3)

$$\hat{H}\Psi_n(x) = -\frac{1}{2\varepsilon}\Psi_n'' + U(x)\Psi_n = \varepsilon_n\Psi_n\,. \tag{25.1}$$

The features of this equation and its solutions are well known (see, for example, Landau and Lifshits, 1974). The longitudinal motion of a particle is described by the plane wave, corresponding to the state with definite momentum.

In the quasiclassical approximation the main features of the quantization of transverse motion appear. The eigenfunctions of Eq. (25.1) have the form

$$\Psi_n(x) = \sqrt{\frac{2\omega_0\varepsilon}{\pi p_n(x)}}\sin\left\{\int_{-x_c}^{x} p_n(x)dx + \frac{\pi}{4}\right\}\,. \tag{25.2}$$

Here $p_n(x) = [2\varepsilon(\varepsilon_n - U(x))]^{1/2}$ is the classical momentum at a given energy ε_n; ω_0 is the motion frequency for this energy. The eigenvalues ε_n are obtained from the equation

$$n + \frac{1}{2} = \frac{1}{2\pi} \oint p(x)dx\,. \qquad (25.3)$$

For example, in the case of the motion in potential (9.33) (without allowance for the thermal vibrations) Eq. (25.3) reads

$$n_+ + \frac{1}{2} = \frac{4d}{\pi \lambda_c} \sqrt{\frac{V}{m}\gamma} \cdot \frac{1}{\delta} \sqrt{1 + z \sinh^2 \frac{\delta}{2}} (K(k(z)) - E(k(z)))\,, \qquad (25.4)$$

where

$$z = \frac{\varepsilon_\perp}{U_0} = \frac{\varepsilon_\perp}{2V \sinh^2 \frac{\delta}{2}}, k = \left(1 + \frac{1}{z \sinh^2 \frac{\delta}{2}}\right)^{-\frac{1}{2}},$$

K and E are the complete elliptic integrals, δ and V are the parameters of the potential (see Table 9.2), U_0 is the depth of a potential well. One can estimate the total number of levels by setting in Eq. (25.4) $z = 1$

$$N_+ \simeq \frac{4d}{\pi \lambda_c} \sqrt{\frac{V}{m}\gamma} \cdot \frac{1}{\delta} \cosh \frac{\delta}{2} \left(K \left(\tanh \frac{\delta}{2} \right) - E \left(\tanh \frac{\delta}{2} \right) \right) \qquad (25.5)$$

For electrons, the corresponding calculation of the number of levels in the potential (9.42) gives

$$n_- + \frac{1}{2} = \frac{4d}{\pi \lambda_c} \sqrt{\gamma \frac{V}{m}} \frac{1}{\delta} \left[\sqrt{z} \cosh \frac{\delta}{2} \right.$$
$$\left. - \sqrt{1 + (1 - z) \sinh^2 \frac{\delta}{2}}\, E\left(\arccos(\sqrt{1 - z}), \frac{1}{\sqrt{1 + (1 - z) \sinh^2 \frac{\delta}{2}}} \right) \right], \qquad (25.6)$$

where $E(\varphi, k)$ is the incomplete elliptic integral of the second kind. The estimation for the total number of levels for electrons has the form ($z = 1$, $E(\pi/2, 1) = E(1) = 1$)

$$N_- \simeq \frac{4d}{\pi \lambda_c} \sqrt{\gamma \frac{V}{m}} \frac{1}{\delta} \left(\cosh \frac{\delta}{2} - 1 \right) \qquad (25.7)$$

In Table 25.1 the quasiclassical results for level numbers of electrons and positrons obtained in the potential (9.14) are given for (110) plane of various

Table 25.1. Number of levels for positrons (+) and electrons (−).

	C	Si	Ge	Fe	W
$N_+/\sqrt{\gamma}$	0.76	1.2	1.65	2.25	3.5
$N_-/\sqrt{\gamma}$	0.48	0.70	0.94	1.2	1.7

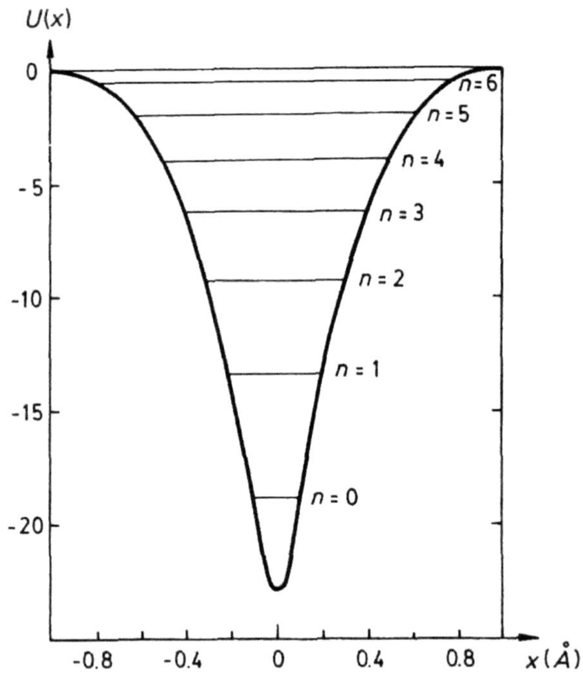

Fig. 25.1. Scheme of the energy levels of electrons with $\varepsilon = 54.4$ MeV in Si (110).

media with allowance for the thermal vibrations. In Fig. 25.1 the scheme of the electron levels is shown at energy 54.4 MeV in the planar potential of Si (110) at room temperature.

The solutions of Eq. (25.1) are required for the calculation of radiation characteristics. In connection with this we note the following. According to what has been said, in order for the energy levels to be narrow, it is necessary to have rather large phase volume in an isolated potential well, i.e. the existence of several levels in it. Therefore a sufficiently clear picture of the transition between the levels at radiation can be revealed only in those experiments where the number of levels is more or equal to four. In this case, the rather good approximation for the description of the transverse motion is the quasiclassical one, but the accuracy may not be completely sufficient for electron transitions to the lowest level. The radiation from positrons in the lower levels is somewhat suppressed while the potential for these levels is almost of the oscillator type. This leads to the overlapping of the characteristic spectral lines of radiation at transitions between these levels even if there are rather small widths. As the result it is impossible to select the contribution of the transitions to the lowest level. In addition, the use of the quasiclassical approximation for the oscillator potential gives a rather good accuracy even for the ground state.

25.2. Radiation at Low Energies

In the energy region under consideration the radiation, of course, is soft and dipole ($\omega \ll \varepsilon$ and $\rho \ll 1$), since it acquired these properties still in the region of middle energies, and with further decrease of energy the photon frequency and the dipole parameter continue to decrease ($\omega \propto \gamma^{3/2}$ and $\rho \propto \gamma$). The radiation characteristics at low energies can be obtained by using the equations of the classical theory and the correspondence principle. For this, it should be taken into account that for the large quantum numbers of transverse particle motion the dipole matrix element has the form

$$x_{fi} \simeq \frac{1}{T} \oint dt x(t) e^{im\omega_0 t} \equiv \frac{x_m}{T}, \tag{25.8}$$

where

$$m = i - f,\ m\omega_0 = \varepsilon_i - \varepsilon_f = \Delta\varepsilon_\perp$$

Equation (25.8) can be obtained by calculating the matrix element with quasiclassical wave functions (25.2) to make the expansion of phases in the terms where these phases are compensated

$$\int [p_i(x) - p_f(x)]dx \simeq \int \frac{\partial p_\perp}{\partial \varepsilon_\perp} \Delta\varepsilon_\perp dx\,;$$

$$\Delta\varepsilon_\perp \simeq \frac{\partial \varepsilon_i}{\partial i}(i - f) \simeq \omega_0(i - f)\,,$$

$$\frac{\partial p_\perp}{\partial \varepsilon_\perp} dx = \frac{dx}{v_\perp(x)} = dt\,. \tag{25.9}$$

The expression for $\partial\varepsilon_i/\partial i$ in Eq. (25.9) is obtained by the differentiation of formula (25.3). It is helpful also to take into account the following relation

$$|(\varepsilon_i - \varepsilon_f)x_{fi}| = |v_{fi}| = \left|\frac{w_{fi}}{\varepsilon_i - \varepsilon_f}\right| \tag{25.10}$$

which can be derived from matrix elements of the commutators $[\mathcal{H}, x], [\mathcal{H}, v]$.

Using Eq. (1.127) for the spectral intensity of radiation, formulae (25.8) and (25.10), and the correspondence principle, we obtain in the quantum region of the transverse motion the following expression

$$dW = \alpha\frac{d^3k}{2\pi\omega}\sum_f |\mathbf{e}^*(\boldsymbol{\rho}_{fi}\omega_{if} + \mathbf{V}(\mathbf{k}\boldsymbol{\rho}_{fi}))|^2\delta(\omega(1 - \mathbf{nV}) - \omega_{if}) \tag{25.11}$$

for the probability of radiation transition per unit time, where $\omega_{if} = \varepsilon_i - \varepsilon_f$, $\mathbf{V}$ is the mean value of the particle velocity, $\boldsymbol{\rho}_{fi}$ is the matrix element of the transverse coordinate

$$\boldsymbol{\rho}_{fi} = \int \Psi_f^*(\boldsymbol{\rho})\boldsymbol{\rho}\Psi_i(\boldsymbol{\rho})d^2\rho\,. \tag{25.12}$$

Analogously one can extend on the region of low energies all the results obtained in Sec. 22.

25.3. Widths and Positions of Spectral Lines

In the experiments with low energy particles (see, for example, Algard *et al.*, 1980; Swent *et al.*, 1979; Park *et al.*, 1983) the rather large width of the spectral lines has been observed which is the sum of many partial widths. Some of them depend on experimental conditions, such as the crystal thickness, the divergence of the incident beam, the collimation angle of the photon beam. For example, the finiteness of the flight time, through a crystal, contributes to the partial width $\Gamma_L \simeq 4\pi\gamma^2/L$, the collimation of radiation gives $\Gamma_{\text{col}} \simeq \gamma^2\vartheta_{\text{col}}^2\omega$. A multiple scattering angle ϑ_y at the planar channeling depends on the crystal thickness L, thus $\Gamma_y \simeq \gamma^2\vartheta_y^2\omega \simeq \gamma^2\dot{\vartheta}_s^2 L\omega/2$. In addition, the width of the transverse motion levels depending directly on the atomic scattering, should be taken into account.

Note that the width of discrete levels of the transverse motion is rather small and independent of the experimental conditions. Therefore the contributions to the width connected with these conditions should be relatively small to resolve clearly the spectral lines, what put definite restrictions on experimental conditions. In particular, the crystal thickness and the collimation angle should be properly chosen. For example, in the experiment (Gouanere *et al.*, 1982) satisfying these conditions, the crystal thickness was 20 μm, and the collimation angle was $\sim 10^{-3}$ rad, what for electron energy 54 MeV gives $\Gamma_L \sim \Gamma_{\text{col}} \sim 0.01\ \omega\ (\vartheta_{\text{col}}^2 \gamma^2 \sim 10^{-2})$. A few percent of the photon frequency are about Γ_y. Generally, in all experiments where the isolated peaks have been observed on the spectral curves, the collimation of the photon beam was hard: $\vartheta_{\text{col}}^2 \gamma^2 \ll 1$. At this collimation and for the relatively small number of levels ($n < 7$) the classical theory gives only the envelope of the spectral curve. At the absence of collimation, when the spectral peaks are spread, the classical theory can be used with satisfactory accuracy. In the case $n \geq 10$ the spectral lines are unsolvable practically even for the hard collimation of photons and the results of the classical radiation theory may be used (see Sec. 5). The classical limit is formerly for positrons because of the specific form of the potential well.

Although the probability of the photon emmision at a given angle depends substantially on the partial widths under discussion, the positions of maxima on the spectral curves are defined by the frequencies $\omega_m = 2\gamma^2\omega_{if}$ with good accuracy. Even the frequencies ω_{if}, calculated by means of quasiclassical Eq. (25.3) are in a rather good agreement with the experiment, what can be verified by comparing the position of peaks in the experiment by Park *et al.*, 1983) with the scheme of the levels shown in Fig. 25.1.

The more accurate results (especially for the transition frequencies) are given by using the solutions of Eq. (25.1). This calculation has been made, for example, in the works by Andersen *et al.*, 1983]. However, it should be noted that there are no analytical solutions of Eq. (25.1) in a realistic potential of the crystal planes, and all calculations are numerical. Because of this the quantum approach becomes complicated and cumbersome while the quasiclassical description is very simple although it gives the worst accuracy ($\sim 10\%$). All the more it is important for the description of experiments, when along with transition frequencies, it is necessary to carry out the calculation of the spectral line widths (where, as it has been noted, the scattering processes dominate). This problem is complicated for the consistent quantum analysis but it is simplified in the quasiclassical approximation.

Chapter 6

SOME ADDITIONAL PROBLEMS OF THE THEORY OF ELECTROMAGNETIC PROCESSES IN ALIGNED SINGLE CRYSTALS

26. INDUCED RADIATION OF ELECTRON AND POSITRON BEAMS TRAVERSING ORIENTED SINGLE CRYSTALS

26.1. Free-electron Lasers

As is known, a generation of the coherent radiation using beams of relativistic particles is realized in devices called free-electron lasers (FELs).* The FEL action is described in terms of the classical theory and for a wide range of parameters one can apply the so-called single-particle theory where the interaction between the charged particles in the beam is neglected (Baier and Milstein, 1977; Colson, 1977; see Appendix E). Because of the interaction of the electron beam moving in a periodic magnetic structure† (undulator) with the radiation field, the beam particles are spatially bunched. The important feature of the process is the phase stability which in turn ensures the bunching stability, thereby creating the conditions for the generation of coherent beam

*The action of the FELs has been considered in many papers (see, for example, the special issue by Granatstein and Roberson, 1985) in which there is a detailed bibliography.
†For FEL action, a periodicity in the beam motion is really needed.

radiation. In the region of low gains G ($G \ll 1$) the analysis is considerably simplified since the phase shift of the radiation field may be neglected (Baier and Milstein, 1977; Colson, 1977), while for high gains ($G \geq 1$) this shift is essential (Baier and Milstein, 1980a, b). Sometimes the region $G \ll 1$ is described within the quantum theory.

So far a lot of FELs have been created, which operate in the range from centimeter waves to visible light (see, e.g. the special issue mentioned in the footnote). The wavelength of the emitted radiation in an FEL λ is given by the standard undulator condition (the radiation in a forward direction, the first harmonics), (1.118) and (1.180)

$$\lambda = (\lambda_0/2\gamma^2)\,(1 + \rho/2)\,, \tag{26.1}$$

where λ_0 is the lattice wavelength. since the minimum wavelength in a magnetic undulator is of the order of 1 cm, a transition to the short-wavelength region is possible only due to an increase of the particle energy. The difficulty in realization into practice of this way is connected with the following dependence of the gain $G \propto \gamma^{-3}$ (see the section below).

Of great interest is a search for the objects where the particles are in a periodic motion with small wavelengths. One of such possibilities takes place when charged particles are channeled in single crystal (see, Sec. 9). The wavelength of a particle oscillations at channeling may be estimated as follows (cf. (9.37), (9.46), (9.49), (9.53))

$$\lambda_0 \approx 2\pi\gamma d/\sqrt{\rho}\,, \quad \rho = 2U_0\varepsilon/m^2\,, \tag{26.2}$$

where d is the characteristic size of the channel and U_0 is the depth of the potential well of the channel. For instance, at channeling in the (110) plane in Si ($U_0 = 25$ eV, $d \approx 0.1$ nm) we have at energy $\varepsilon = 100$ MeV that $\lambda_0 \sim 10^4 d \sim 1\,\mu$m. For this case the photon energy for the wavelength (26.1) will be $\omega \approx 100$ KeV. So, the problem under consideration is the induced radiation in the x-ray and gamma range.

26.2. Electromagnetic Wave Amplification by an Electron Beam in Single Crystal Basic Equations

We analyze the interaction between the linearly polarized plane electromagnetic wave propagating along the single crystal's planes and a beam of ultrarelativistic particles moving in the same planar channels, with the specific features of particle motion in single crystals taken into account (Baier and Milstein, 1986). In the axial case the problem becomes much more complicated technically, but as we will see from the results obtained, this case does not seem to be of great interest.

Let the particle beam is falling on the crystal in such a way that the momenta of the electrons lie near the crystallographic planes and interacts with the linearly polarized plane electromagnetic wave propagating along the same planes; its electric field is perpendicular to these planes. Our set of coordinates is: the x^1-axis is perpendicular to the planes forming a channel, and x^3-axis is chosen parallel to these planes such that the velocity along the x^2-axis is $v^2 = 0$. It is convenient to use the equation of motion in covariant form ($du^\mu/ds = (e/m)F^{\mu\nu}u_\nu$) and then for our problem we obtain

$$\begin{aligned} &d(u^0 + V)/ds = \Omega u^1 \sin\varphi, \\ &du^1/ds = -dV/dx u^0 + \Omega(u^0 - u^3) \sin\varphi, \\ &du^3/ds = \Omega u^1 \sin\varphi, \\ &d\Omega/ds = -\omega_p^2 \overline{u^1 \sin\varphi}, \end{aligned} \tag{26.3}$$

where u^μ is the component of the four-velocity, $V(x) = eU(x)/m, U(x^1)$ is the inter-planar potential, $\Omega = eE_w/m, E_w$ is the strength of the electric wave field given by the vector-potential

$$A_w^1 = (E_w/\nu)\cos\varphi, \varphi = \nu(t - x^3) + \psi, \omega_p^2 = 4\pi n e^2/m,$$

here n is the particle density in the beam, and $m(e)$ is the mass (charge) of an electron. The last Eq. (26.3) follows from the energy conservation law and the averaging sign means the averaging over the wave phase

$$\bar{a} = (1/2\pi)\int_0^{2\pi} d\psi a(\psi).$$

The set of equations (26.3) is valid only in the case when the gain is $G \ll 1$ and may be solved here in an adiabatic approximation, i.e. the first three equations in (26.3) can be solved assuming that $\Omega = \text{const}$.

The integral of motion follows from the first and the second equation in (26.3)

$$u^0 - u^3 + V(x^1) = \text{const}. \tag{26.4}$$

We shall consider the particle motion in a channel, i.e. $(u^1)^2 \leq 2V_{\max}\gamma$ where $V_{\max} = eU_0/m$, under the assumption that the transverse momentum is small:$|u^1| \ll |u^3|$. The problem of particle interaction with the wave field will be solved in an approximation linear with respect to the field strength of the wave. The inter-planar potential is convenient to represent as follows

$$V(x^1) = V_0 f(y), \quad y = x^1/d, \tag{26.5}$$

where $2d$ is the inter-planar distance, $V_0/2 = V_{\max} = eU_0/m$, and U_0 is the depth of the potential well. With the wave being absent, we write the solution of Eqs. (26.3)

$$x^1(s) = X(s), \quad u^3(s) = \gamma, \quad \varphi(s) = \Phi(s).$$

With the wave switched on, perturbation of this motion occurs:

$$x^1(s) = X(s) + \tilde{\xi}, \quad u^3 = \gamma + \tilde{\eta}, \quad \varphi = \Phi + \tilde{\chi}. \tag{26.6}$$

All the increments, $\tilde{\xi}, \tilde{\eta}$ and $\tilde{\chi}$ are proportional to Ω. The initial conditions being the following

$$\begin{aligned} &\tilde{\xi}(0) = d\tilde{\xi}(0)/ds = \tilde{\eta}(0) = \tilde{\chi}(0) = 0, \\ &X(0) = X_0, \quad dX(0)/ds = u^1(0), \quad \Phi(0) = \psi. \end{aligned} \tag{26.7}$$

We introduce the quantities, the using of which enables the expressions to be simplified considerably

$$\Omega_0^2 = \gamma V_0/d^2, \quad \rho = (\Omega_0 d)^2 = \gamma V_0, \quad \delta = \nu/(2\Omega_0\gamma). \tag{26.8}$$

It is convenient to change over to the dimensionless variables

$$\tau = \Omega_0 s, \quad \tilde{\xi} = \xi\Omega d^2/\gamma, \quad \tilde{\eta} = \Omega\eta d, \quad \tilde{\chi} = \chi 2\rho\Omega d\delta/\gamma. \tag{26.9}$$

Substituting Eqs. (26.5), (26.6), (26.8), (26.9) into the set (26.3), performing appropriate expansions and collecting the zero- and first-order terms with respect to Ω, we have, with Eq. (26.4) taken into account, the final set of equations for dimensionless variables:

$$d^2y/d\tau^2 + df(y)/dy = 0 \tag{26.10a}$$

$$d\Phi/d\tau = \delta(1 + \rho(dy/d\tau)^2) \tag{26.10b}$$

$$d\eta/d\tau = dy/d\tau \sin\Phi \tag{26.10c}$$

$$d\chi/d\tau = -\xi\, df(y)/dy \tag{26.10d}$$

$$d^2\xi/d\tau^2 = -\xi\, d^2 f(y)/dy^2 - \eta\, df(y)/dy + d\Phi/d\tau \sin\Phi/(2\rho\delta) \tag{26.10e}$$

$$dG/d\tau = -(2\omega_p^2 d^2/\gamma)(\overline{d\xi/d\tau \sin\Phi + (dy/d\tau)^2 2\rho\chi\delta \cos\Phi}), \tag{26.10f}$$

here we introduced the gain

$$G \equiv (\Omega^2(\tau) - \Omega^2(0))/\Omega^2(0) \approx 2(\Omega(\tau) - \Omega(0))/\Omega(0). \tag{26.11}$$

It is remarkable that despite the apparent complexity, the set (26.10) is solvable in quadrature for any inter-planar potential $U(x)$. Let us start to solve the set (26.10). The equation (26.10a) is solved in quadrature and we denote its solution as $y(\tau)$. This solution is a periodic function of period T. From Eq. (26.10b) we have

$$\begin{aligned}\Phi(\tau) &= \psi + \delta \int_0^\tau (1 + \rho(dy/d\tau)^2)d\tau \\ &= \psi + \tau\delta(1 + \rho\langle(dy/d\tau)^2\rangle) + \delta \int_0^\tau \rho((dy/d\tau)^2 \\ &\quad - \langle(dy/d\tau)^2\rangle)d\tau ,\end{aligned} \tag{26.12}$$

where $\langle(dy/d\tau)^2\rangle$ is a squared velocity averaged over the motion period. Equation (26.10c) gives

$$\eta = \int_0^\tau dy(\tau')/d\tau' \sin\Phi(\tau')d\tau' \tag{26.13}$$

Equation (26.10e) is the most complicated one. To solve it, we use an artificial trick: we differentiate Eq. (26.10a) over τ: $d^3y/d\tau^3 = -d^2 f(y)/dy^2 dy/d\tau$, multiply both parts of Eq. (26.10e) by $dy/d\tau$ and express $d^2 f(y)/dy^2 dy/d\tau$ via $d^3y/d\tau^3$ Using Eqs. (26.10b) and (26.10c) and taking integral over τ of both parts, we have

$$dy/d\tau d\xi/d\tau - d^2y/d\tau^2\xi = (\eta/2)(1/\rho + (dy/d\tau)^2). \tag{26.14}$$

It is seen from Eq. (26.14) that the expression for $\xi(\tau)$ may be found in quadrature. It is useful to introduce a periodic function $Z(\tau)$

$$Z(\tau) = dy(\tau)/d\tau \int_0^{\tau^*} 1/(dy(\tau')/d\tau')^2 d\tau' \tag{26.15}$$

Here $\tau^* = \tau - kT/2$ where k is the number of turning points on the trajectory of a particle in its transverse motion for the time τ. Using this function the solution of Eq. (26.14) may be presented in the form, taking Eq. (26.13) into account:

$$\xi(\tau) = (1/2\rho) \int_0^{\tau} ds \sin \Phi(s) W(s, \tau) \tag{26.16}$$

where

$$W(s, \tau) = Z(\tau) dy(s)/ds - Z(s) dy(\tau)/d\tau$$
$$+ \rho(\tau - s) dy(s)/ds dy(\tau)/d\tau\,. \tag{26.17}$$

The solution of Eq. (26.10d) is

$$\chi(\tau) = \int_0^{\tau} ds \xi(s) d^2y/(ds)^2\,. \tag{26.18}$$

Substituting our solutions of Eqs. (26.10a)–(26.10e) into Eq. (26.10f) and averaging over ψ we have for the gain

$$dG/d\tau = -G_0 \left[\int_0^{\tau} \partial W(s, \tau)/\partial\tau \cos(\Phi(\tau) - \Phi(s)) ds \right.$$
$$+ 2\delta\rho dy(\tau)/d\tau \int_0^{\tau} ds \int_0^{s} ds' d^2y(s)/(ds)^2$$
$$\left. \times W(s', s) \sin(\Phi(s') - \Phi(s)) \right], \tag{26.19}$$

where

$$G_0 = \omega_p^2 d^2/(2\gamma\rho)\,. \tag{26.20}$$

The expression (26.19) solves, in general form, the problem of amplification of a plane electromagnetic wave with linear polarization as a result of its interaction with charged particles moving in the single crystal's planar channels.

26.3. The Gain G for a Nonrelativistic Transverse Motion

In the region of comparatively low energies (tens and hundreds of MeV) the parameter $\rho = \gamma V_0 \ll 1$ (in the (110) plane, at room temperature we have $V_0 \approx 10^{-4}$ for silicon and diamond, $V_0 \approx 2 \cdot 10^{-4}$ for germanium and $V_0 \approx 5 \cdot 10^{-4}$ for tungsten). Then the transverse motion of the particles in the channel is nonrelativistic: $(u^1)^2 < \rho \ll 1$. As we shall see below, it is the energy range which is of special interest from the point of view of the radiation generation. By virtue of the said, in what follows we confine ourselves to an analysis of the case $\rho \ll 1$. But the "time" τ determined by the number of particle oscillations in the channel) is assumed to be large so that $\rho\tau \geq 1$.

In the accepted approximation, in Eq. (26.19) one can put, according to Eq. (26.12)

$$\Phi(\tau) - \Phi(s) = (\tau - s)\delta \,. \tag{26.21}$$

Since we analyze the situation in which the particle oscillates many times in the channel, the integral in (26.19) is convenient to reduce to an integral over period and summation over the number of oscullations, making allowances for the fact that the functions $dy(\tau)/d\tau$ and $Z(\tau)$, entering into $W(s,\tau)$ are periodic. There is no difficulty in performing this summation using elementary formula

$$\sum_{m=0}^{n-1} \cos(\varphi + \alpha m) = \frac{\sin(n\alpha/2)}{\sin\alpha/2} \cos\left[\varphi + \alpha(n-1)/2\right] . \tag{26.22}$$

The integrals entering into Eq. (26.19) are "large" only if resonance condition

$$\delta/2 = k\pi/T + \zeta/\tau \,, \quad k = 1, 2, 3, \ldots \,, \quad \zeta/\tau \ll 1 \tag{26.23}$$

is satisfied. After simple calculations using Eqs. (26.21)–(26.23), for the gain during the "time" τ we obtain

$$G(\tau) = -G_0\tau^2\delta[\mathrm{Im}(Z_k^*(dy/d\tau)_k) - (\rho\tau/2)|(dy/d\tau)_k|^2 d/d\zeta]R(\zeta) \tag{26.24}$$

where $R(\zeta) = \sin^2\zeta/\zeta^2$ determines the shape of the spectral line and Z_k and $(dy/d\tau)_k$ are the Fourier harmonics

$$\begin{aligned} Z_k &= \frac{1}{T}\int_0^T ds Z(s)\exp(-i2\pi ks/T) \\ (dy/d\tau)_k &= \frac{1}{T}\int_0^T dsdy(s)/ds \exp(-i2\pi ks/T) \,. \end{aligned} \tag{26.25}$$

The quantity G_0 is given by the Eq. (26.20).

According to formulae (26.8) and (26.23), for the resonance frequency we have

$$\nu_{\rm res} = 2\gamma\Omega_0\, 2k\pi/T\,. \tag{26.26}$$

In addition, in formula (26.24) an averaging over the coordinates of the point of incidence of the particle ($y(0) = x^1(0)/d$) should be performed. Changing $y(0)$ we have different values of the transverse energy (see Eq. (26.10a)

$$\varepsilon_\perp = (dy/d\tau)^2/2 + f(y) = (dy(0)/d\tau)^2/2 + f(y(0)) \tag{26.27}$$

and, generally speaking, different periods of motion T.

For averaging it is necessary to take an integral over $y(0)$ in the domain where $\varepsilon_\perp < 1/2$; this corresponds to the condition of presence of the particle in the channels (see Eq. (26.5))

$$\bar{g} = (1/2)\int_{-a}^{a} dy(0) g(y(0))\,, \quad (dy(0)/d\tau)^2/2 + f(a) = 1/2\,. \tag{26.28}$$

In the particular case of an oscillator potential $f(y) = y^2/2$, we find $Z(\tau) = \sin\tau/(dy(0)/d\tau)$. In this case there is one harmonic $k = 1, T = 2\pi$ and for resonance frequency (26.26) we obtain $\nu_{\rm res} = 2\Omega_0\gamma$. After the simple calculation we find the gain for the oscillatior potential after the averaging over the coordinates of the point of incidence

$$G^{os} = -G_0\tau^2/4\Big[1 - (\rho\tau/6)$$

$$\times\,(1 - (dy(0)/d\tau)^2)^{1/2}[1 + 2(dy(0)/d\tau)^2/3]\frac{d}{d\zeta}\Big]R(\zeta)\,. \tag{26.29}$$

Note that only in the case of the oscillator potential that the averaging over the coordinates in the point of incidence does not affect the spectral curve.

26.4. General Analysis of The Stimulated Radiation in Oriented Single Crystals

We proceed now to the discussion of the results obtained. The first term in the expression for the gain (26.24) always describes the absorption. It is easy to see that this expression is a radiation absorption coefficient of the system of relativistic oscillators. The structure of the second term is completely similar to the gain G for an FEL and can be compared with this G if one takes into account the similarity of particle trajectories (Baier and Milstein, 1977, Appendix E), namely: $(\nu_0)_{\rm FEL} \to \Omega_0/\gamma, (\Omega_0/\nu_0)_{\rm FEL} \to \Omega_0 d = \rho^{1/2}$ Note that under typical FEL conditions $(\Omega_0/\nu_0)_{\rm FEL} \sim 1$, the second term in the expression for G dominates and the first term is usually omitted. However, in the energy range under discussion the first term may be roughly equal to the second and it must be taken into account. It is seen from the expressions (26.24) and (26.28) that the amplification of the wave is possible actually only if $\rho\tau \geq 1$, otherwise the amplification can occur only in a very narrow band of the spectral curve $|\pi - \zeta| \ll 1$ where an additional suppression of G takes place and where it is readily smeared by the spread over the periods of motion. The wave will be amplified only if the condition $dR(\zeta)/d\zeta > 0$ is satisfied ($\zeta < 0$) and the amplification must exceed the absorption.

There are a number of conditions which limit considerably the possibilities of selecting the parameters for this problem. Let us dwell upon these conditions.

We start with an analysis of the dechanneling effects. Within the energy range under analysis the basic effect is a multiple scattering of the particles. Under the action of this mechanism the particle acquires an additional transverse momentum. And, as a result, they leave the channel after passing the length l_d. If the scattering is assumed to occur in the same way as in amorphous matter, then the dechanneling length is given by Eq. (10.1). This estimation is, strictly speaking, overestimated for electrons and underestimated for positrons ($l_d^{e-} < l_d^{e+}$). If allowance is made for the fact that during the interaction with the radiation the basic role is played by the particles whose oscillation amplitude is comparable with the size of the channel, then the above estimation seems to be rather satisfactory. The partameter $\rho\tau_d$, corresponding to l_d, is

$$\rho\tau_d \approx (\alpha/2\pi)(V_0^{5/2}\gamma^{3/2}/d)L_{\rm rad}\,. \tag{26.30}$$

It is obvious that it should be $\tau < \tau_d$. To reach the region where $\rho\tau_d \sim 1$, it is necessary that the energy be fairly high. The estimates show that for the usually used crystals (diamond, silicon, germanium, the (110) plane) we have $\rho\tau_d \sim 1$ at $\varepsilon = 200$ MeV. For this energy, for the (110) plane $\rho \approx 0.04$ in diamond and Si and $\rho \approx 0.08$ in Ge. Note that at the indicated energy one can use the classical theory with a large margin (cf. Sec. 25).

Another effect limiting τ is the dependence of the period T on the transverse energy. Note once more that the quantity ζ entering into expression for the spectral curve in Eq. (26.24) is $\zeta = \tau(\delta/2 - k\pi/T)$ (see Eq. (26.23)). In order that the amplification effect occurs, the increment of the quantity ζ must be $\Delta\zeta \ll \pi$, at averaging over the point of incidence $y(0)$; the latter can be represented as the averaging over period T. This imposes the limitation on a possible spread with respect to the periods of the motion

$$\Delta T/T \ll T/\tau \sim 2\pi/\tau \sim 1/N \tag{26.31}$$

(N is the number of oscillations). We discuss the region $\rho\tau \geq 1, \rho \ll 1$ and $\tau \gg 1$. The effect is possible only for potentials close to the oscillator ones; in addition, as follows from (26.31), the parameter τ cannot be very large. The inequality (26.31) excludes, in practice, the utilization of the planar channeling of electrons because $\Delta T/T \sim 1$ for them. Since the criterion (26.31) follows, in fact, from the coherency condition, it should be satisfied also for the induced radiation at axial channeling. This excludes actually the use of the axial channeling of electrons and positrons since for this channeling $\Delta T \sim T$. For the planar channeling of positrons the potentials are close to the oscillator ones. Unfortunately, even for this case, under the most favorable conditions $\Delta T/T \approx 0.05$, then $\tau \leq 100$ (for the potential (9.33) the period of motion is given by Eq. (9.37), one has to select crystals and planes, where the parameter δ is a minimum in the last formula).

We should like to dwell upon one more restriction due to the energy spread in the incident beam and to the energy losses during its passage through a single crystal. In the energy range under consideration the radiation losses dominate. Then, substituting the quantities entering in ζ (26.23), we find

$$\Delta\gamma/\gamma \ll \pi/\tau \tag{26.32}$$

whence, we have

$$\tau \ll (\pi L_{\text{rad}} V_0^{1/2}/(d\gamma^{1/2}))^{1/2} \tag{26.33}$$

For silicon, Eq. (26.33) gives $\tau \ll 10^3$, i.e. the restriction is weaker than as discussed above. It is seen from Eq. (26.32) that the requirement for the monochromaticity of the incident beam is rather weak. Note that the width of the spectral curve follows also from the analysis of the expression for ζ (26.23) (see Appendix E) and we have $\Delta\nu/\nu \sim 1/\tau$.

The analysis above enables one to estimate the magnitude of the amplification effects. After the substitution of $n = 10^{14}\,\text{cm}^{-3}$, $\tau = 100$, $d = 10^{-8}\text{cm}$

and $\gamma = 400$ into Eq. (26.28) we find that $G \approx 10^{-12}$ Thus, even at such a high particle density of the electron beam the wave gain turns out to be negligibly low. The extremely high requirements for the particle density in the beam are, to a known degree, evident and are due to a microscopically small amplitude of the particle oscillations in the channel. The rigid limitations on the magnitude of τ are much less evident; meanwhile this does not allow one to compensate the smallness of the quantity G_0 by a high τ (we would have $G \sim 1$ at $\tau \sim 10^6$).

The conclusion we have drawn from the above analysis is pessimistic since there is a large gap between the densities of the particle beams with an energy of hundreds MeV, generated by modern accelerators, and the densities needed for the FEL action on electrons and positrons moving in the channels of a single crystal. The possibility of increasing the gain at the expense of a growth of the number of oscillations in the channel (τ), has proved to be rigidly restricted by the crystalline effects.

In conclusion we note that the stimulated radiation of particles moving in the single crystal's channels is discussed in a series of papers, Beloshitsky and Kumakhov, [1978]; Visotsky and Kuzmin, [1983]; Ohtsuki, [1984] who used for the consideration of the gain an expression derived in an analysis of entirely different physical situations: propagation of a signal in a medium with an inverse population. Naturally, in this analysis no allowance has been made for the dynamics of an interaction of the particle beam with the electromagnetic wave and, therefore, features such as the shape of a spectral line and functional dependence on an interaction time have not been reproduced. In a paper by Ternov, Khalilov and Khalomai, [1985] this dynamics has already been taken into consideration, but in this paper the induced radiation has been considered at axial channeling under the assumption that the particles move along the circular trajectories of a given radius. Their weight in the set of trajectories has, however, the set of measure zero so that this analysis is not suitable for the analysis of a realistic situation.

27. HIGH ORDER EFFECTS IN CRYSTALS

In Sec. 6, we have considered creation of neutrino-antineutrino pairs and production of charged particles pairs by an electron moving in an external field, and photon splitting in a strong electromagnetic field. In this section on the basis of results obtained there, the analysis of these processes in aligned crystals have been carried out.

27.1. Photon Splitting in Aligned Crystals

As is shown in Secs. 11 and 12 at the incident angle of a photon $\vartheta_0 \ll V_0/m$ the process of pair creation in the axis (or plane) field can be described in the constant field (on formation length) approximation. To obtain the probability for a single crystal, the following integration should be made (for definiteness we shall consider the axial case)

$$W_{ax} = \frac{1}{S}\int d^2\rho W(E(\rho)), \tag{27.1}$$

where $W(E)$ is the probability of a process in the field E, S is the area of the transverse cross section per axis.

Since the photon splitting occurs via virtual electron-positron pair creation (see Fig. 6.7), then in the quasiclassical region under discussion ($E/E_0 \ll 1$) the dependence of the photon splitting probability on the angle of incidence in a crystal is the same as for pair creation. Thus for $\vartheta_0 \ll V_0/m$ the process of photon splitting can be considered in the constant field approximation, i.e. the probabilities obtained in Sec. 6.4 should be substituted in Eq. (27.1), and for $\vartheta_0 \gg V_0/m$ the process takes place that can be called coherent photon splitting.

In a media the photon splitting process occurs on the background of incoherent splitting of photons on the isolated nuclei, whose cross section for the unpolarized photons is given by the following equation (see [Baier *et al.* [1974]):

$$\begin{aligned}\sigma_{Nuc}^{\gamma\to\gamma\gamma} &= \frac{Z^2\alpha}{\pi}\ln(183Z^{-\frac{1}{3}})\cdot 25\frac{\alpha^4}{\pi^2m^2} \\ &= \frac{Z^2\alpha}{\pi}\ln(183Z^{-\frac{1}{3}})\cdot 10^{-29}\text{cm}^2 \end{aligned}\tag{27.2}$$

One can see that $\sigma_{Nuc}^{\gamma\to\gamma\gamma}$ (i) does not depend on energy (in the case of complete screening) (ii) is amplified numerically (factor 25), that is connected with the features of the amplitudes of light-light scattering.

Let us consider the incident angles region $\vartheta_0 \ll V_0/m$, when the results of Sec. 6.4 may be used. In the region $\kappa \ll 1$ the amplitudes of photon splitting are given by Eq. (6.85), and the corresponding probability is much less than that given by Eq. (27.2), and thus the crystal effects in this region are small. Figures 6.8–6.10 show that the probability of photon splitting in a field increases strongly at $\kappa \gg 1$ and reaches the maximum values in the vicinity of $\kappa \sim 10^2$ Just this region is of the main interest.

It is obvious that the process of photon splitting will occur at the condition of the large absorption due to the photoproduction in the field of the axis and on the isolated nuclei. Therefore to observe this process it is reasonable to use the crystals with thickness less than $L_{\rm ph} = 1/W_e^F$ (see Eqs. (12.16) and (12.19)). For the description of the process it is necessary to use the equations obtained for the absorption conditions. Its features are discussed after Eq. (6.78). Note that in the case when $\gamma \sim \beta$ (which take place at $\kappa \geq 1$) even if $\gamma L \ll 1$ (the absorption is small) we have $\rho(L) = L$, the ϑ-function is absent and all transitions, allowed by selection rules (6.76), are possible. The maximum effect is attained in the region, where the probability $W^{\gamma\to\gamma\gamma}$ is about its maximum value (it is seen from Fig. 6.10 that the transition $I \to IIII$ dominates here).

We have calculated the photon splitting probability under the assumption that $\gamma L \ll 1$. The form of the potential (9.13) has been used. The results of the calculations are presented in Figs. 27.1 and 27.2, which give the

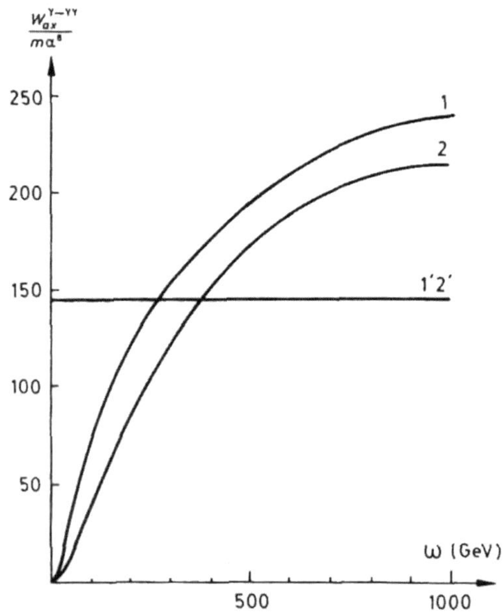

Fig. 27.1. The probability of a photon splitting for the axis $\langle 111 \rangle$ as a function of the incident photon energy in tungsten at $T = 77$ K (1), 293 K (2).

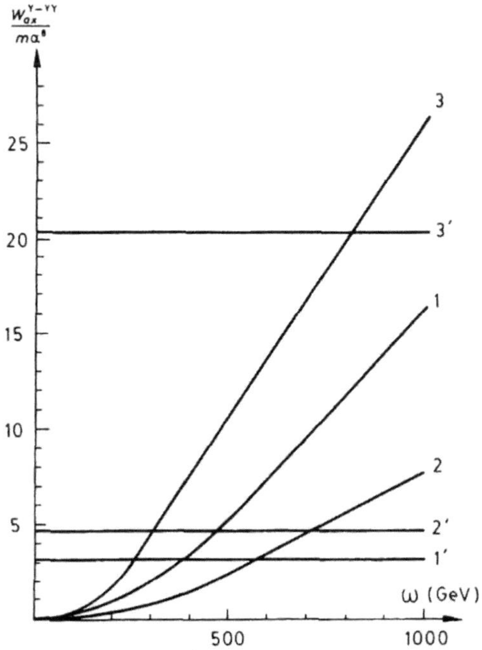

Fig. 27.2. The probability of a photon splitting for the axis ⟨111⟩, $T = 293$ K for diamond (1), silicon (2) and germanium (3). The lines $1', 2', 3'$ give the probability of a photon splitting in the fields of isolated nuclei.

quantity $W_{ax}^{\gamma\to\gamma\gamma}$ in the units $m\alpha^8 (m\alpha^8 = 2.1 \times 10^{-7}\text{cm}^{-1})$. Figure 27.1 illustrates the dependence of the splitting probability in tungsten for the photon incident along ⟨111⟩ axis at two temperatures: $T = 77$ K (curve 1) and $T = 293$ K (curve 2). The lines $1'$ and $2'$ are, respectively, the probabilities of the photon splitting on the isolated nuclei.

Figure 27.2 shows the same dependencies but for other crystals: diamond (1), silicon (2) and germanium (3). It is seen in these figures that $W_{ax}^{\gamma\to\gamma\gamma}$ is larger than $W_{Nucl}^{\gamma\to\gamma\gamma}$ only for $\kappa \gg 1$, and the maximum gain is achieved in diamond in just the same way as for the pair photoproduction probabilities (see Eq. (12.20)). In this region we have the following estimation

$$\frac{W_{ax}^{\gamma\to\gamma\gamma}}{W_{ax}^{\gamma\to e^+e^-}} \sim 0.3 \left(\frac{\alpha}{\pi}\right)^2 \tag{27.3}$$

27.2. Creation of Neutrino–antineutrino Pairs in Crystals

In Sec. 6.1 the process of the $\nu\bar{\nu}$ pair creation has been considered for electrons moving in an external electromagnetic field. Since the formation of the $\nu\bar{\nu}$ pair has the same structure (see Eq. (6.1)) as the photon formation with the mass equal to the invariant mass of a pair, the results of the constant field approach remain valid for the electron incident angles $\vartheta_0 \ll V_0/m$ (see Sec. 11.1). Then in the case of the uniform distribution of electrons (wide angles of the incident beam, thick crystals), the probability of $\nu\bar{\nu}$ pair creation is given by Eq. (27.1) where the probabilities (6.12), (6.15) and (6.17) should be substituted. In thin crystals and for narrow beams ($\Delta\vartheta_0 \ll \vartheta_0$) there is the maximum redistribution of the particles flux causing the appreciable enhancement of the effect for electrons. Then in the case of axial channeling and for the axially-symmetric potential, the probability of $\nu\bar{\nu}$ pair creation in a crystal reads (cf. Eq. (17.7))

$$W_{ax}^{\nu\bar{\nu}} = \frac{1}{x_0}\int_0^{x_0} dx \ln\left(\frac{x_0}{x}\right) W_{\nu\bar{\nu}}(\chi(x)), \tag{27.4}$$

where the probabilities (6.12), (6.15) and (6.17) should be substituted. The factor $\ln(x_0/x)$ in the above equation describes the flux redistribution. However the process probability is very small (see Eq. (6.19)) and thick crystals should be used. Therefore we shall consider the uniform beam of electrons. The general expression of the probability can be obtained if in Eq. (27.1) one substitutes Eq. (6.12).

For $\chi_s \ll 1$, substituting (6.15) into (27.1), we obtain the following expression for the probability of the $\nu\bar{\nu}$ pair creation in a crystal

$$W_{ax}^{\nu\bar{\nu}} = \frac{35G^2m^6\chi_s^5(7g^2 + 54g_A^2)}{2^7\sqrt{3}(6\pi)^2\varepsilon x_0(\eta + \sqrt{1+\eta})^9} \times \frac{1}{\eta(1+\eta)}\left(9 + \frac{2\eta + 1}{\sqrt{\eta(1+\eta)}}\right), \tag{27.5}$$

where we have used the same notation as in Eqs. (6.15), (17.9) and (17.10); the terms of relative smallness $1/x_0$ are omitted.

For $x_s \gg 1$, substituting (6.17) into (27.1), we have for the probability of the process with the logarithmic accuracy ($\ln\chi_s \gg 1$)

$$W_{ax}^{\nu\bar{\nu}} = \frac{4G^2g^2m^6\chi_s^2}{(6\pi)^3\varepsilon x_0}\ln(2\chi_s)\left((1+2\eta)\ln\frac{1+\eta}{\eta} - 2\right), \tag{27.6}$$

where the terms $\sim 1/x_0$ are also omitted. As in the case of external fields, these probabilities should be compared with the radiation emission probability for the same conditions (cf. Eq. (6.19)). We obtain by using Eqs. (27.5), (27.6) and (17.20), (17.21) the following equations:

For $\chi_s \ll 1$

$$\frac{W_{ax}^{\nu\bar{\nu}}}{W_{ax}^{\gamma}} = \frac{G^2 m^4}{\alpha} \chi_s^4 \frac{35(7g^2 + 54g_A^2)}{2^7\sqrt{3}(6\pi)^2} \varphi_1(\eta) \simeq 10^{-23}\, \chi_s^4 \varphi_1(\eta)\,,$$

$$\varphi_1(\eta) = \frac{1}{(\sqrt{\eta} + \sqrt{1+\eta})^9} \frac{1}{(\sqrt{1+\eta} - \sqrt{\eta})\eta(1+\eta)} \left(9 + \frac{2\eta + 1}{\sqrt{\eta(\eta+1)}}\right) \tag{27.7}$$

For $\chi_s \gg 1$

$$\frac{W_{ax}^{\nu\bar{\nu}}}{W_{ax}^{\gamma}} = \frac{G^2 m^4}{\alpha} \chi_s^{\frac{4}{3}} \frac{3}{(2\pi)^3 \cdot 7 \cdot 6^{\frac{2}{3}} \Gamma(\frac{2}{3})} \varphi_2(\eta) \simeq 5 \cdot 10^{-25} \chi_s^{\frac{4}{3}} \varphi_2(\eta)\,,$$

$$\varphi_2(\eta) = (1 + 2\eta) \ln\frac{1+\eta}{\eta} - 2\,.$$

The magnitudes of the functions $\varphi_1(\eta)$ and $\varphi_2(\eta)$ in these equations are (for example, $\langle 111 \rangle$ axis, $T = 293$ K)

$$\varphi_1 = 172, \quad \varphi_2 = 1.9 \,\text{at}\, \eta = 0.025 \,(\text{diamond});$$

$$\varphi_1 = 3.4, \quad \varphi_2 = 0.65 \,\text{at}\, \eta = 0.15 \,(\text{silicon})\,.$$

One can see that for all reasonable values of energy the probability of $\nu\bar{\nu}$ pair creation is negligibly small. For example, even at the energy 100 TeV we have in tungsten ($\langle 111 \rangle$ axis, $T = 293$ K) $W_{ax}^{\nu\nu}/W_{ax}^{\gamma} \sim 10^{-20}$. Since an electron loses the energy on the length L_{ch}, it is clear that crystals with larger thickness must not be used.

27.3. Electroproduction of Charged Particle Pairs in Crystals ($e^{\pm} \to e^{\pm}\, e^-\, e^+$)

This process (direct electroproduction) occurs by means of a virtual photon (v), and its properties (spectra, angular distributions) are similar to those of the successive process $e^{\pm} \to e^{\pm}\gamma, \gamma \to e^+e^-$ which was taken into account in the discussion of the cascade. The probability of the successive process is proportional to the squared crystal thickness $W_r \propto L^2$, while the probability of the direct electroproduction W_v depends linearly on length L. The ratio of the latter probability to W_r can be estimated as $W_v/W_r \sim \ell_c/L$, where ℓ_c is the formation length. It is clear that for $L \gg \ell_c$ the direct process contribution is relativelly small. Therefore the experimental observation of the direct process becomes possible only for thickness L of the order of a few formation lengths. For these conditions both the processes occur practically in a constant field, since while traversing a crystal, a particle shifts a little in the transverse direction. The pair creation in external fields has been considered in Sec. 6. For the calculation of probability in crystals one can use the results of this section, averaging the obtained probabilities over the fields in a crystal.

27.4. Production of π^0 Meson by a Photon

Production of a π^0 by a photon in a Coulomb field (the Primakoff effect) is used for the determination of the lifetime τ_π of this meson. In a crystal (cf. Kimball and Cue, 1986) a similar coherent effect is possible.

If we neglect that τ_π is finite (the corresponding decay width of the π^0 is $\Gamma \simeq 7.6$ eV), then in a two-photon process the conservation law $Q + K = p_\pi$ must be satisfied, or

$$2\omega|q_\parallel| = \mu^2 \,, \tag{27.8}$$

where μ is the mass of a π^0 meson. We recall that $q_\parallel = \mathbf{q}\boldsymbol{\nu} = q_n \cos\vartheta_0, \boldsymbol{\nu} = \mathbf{k}/\omega$, and $q_n = (2\pi/d)n_z$, where n_z is an arbitrary integer. We shall assume that $\vartheta_0 \ll 1$, then $q_\parallel \simeq q_n + \mathbf{q}_\perp\boldsymbol{\nu}$ and $\mathbf{q}_\perp\boldsymbol{\nu} \sim |\mathbf{q}|\vartheta_0 \ll |q_n|$ for $n_z \neq 0$. Since in a crystal values of $|q_\parallel|$ are limited on the high side by the existence of the factor $\exp(-\mathbf{q}^2 u_1^2)$ (see Eq. (9.7)), we obtain from (27.8) an estimate for the threshold frequency of the photon $\omega_{th} \simeq \mu^2 u_1/2$. For example, in diamond at room temperature $\omega_{th} \simeq 200$ GeV. We note that for $n_z = 0$ this threshold lies inaccessibly high, i.e., in contrast to the coherent processes discussed in Sec. 4, the term with $n_z = 0$ has no effect. We can analyze the effect in terms of equivalent photons, writing its probability in the following way:

$$\begin{aligned} W^{\text{coh}}_{\gamma\to\pi^0} &= \sum_q \frac{|G(\mathbf{q})|^2\mathbf{q}^2}{4\pi\alpha|q_\parallel|}\sigma_{\gamma\gamma}(|q_\parallel|,\omega)\,, \\ W^{nc}_{\gamma\to\pi^0} &= \frac{Z^2\alpha N}{2\pi^2}\int \frac{d^2q_\perp \mathbf{q}_\perp^2}{(\mathbf{q}^2+a^{-2})^2}\frac{dq_\parallel}{|q_\parallel|}\sigma_{\gamma\gamma}(|q_\parallel|,\omega)\,. \end{aligned} \tag{27.9}$$

In this equation, we made use of the expression for the flux (11.1) and gave also the probability of a similar incoherent process, where N is the average density of atoms and a is the corresponding screening radius. The cross section for the production of a π^0 meson by two photons $\sigma_{\gamma\gamma}(|q_\parallel|, \omega)$ has the form of a Breit–Wigner resonance distribution

$$\sigma_{\gamma\gamma}(|q_\parallel|, \omega) = \frac{8\pi\Gamma^2}{(2\omega|q_\parallel| - \mu^2)^2 + \Gamma^2\mu^2} \,. \tag{27.10}$$

The coherent Primakoff effect, as a result of the discreteness of the equivalent photon spectrum (of the values of $q_\parallel$), occurs only for certain frequencies $\omega_n = \mu^2/2|q_\parallel|$. If the angular and energy spreads in the incident beam of photons satisfy the condition $\Delta\vartheta_0, \Delta\omega/\omega \le \Gamma/\mu$, then the probability $W^{\text{coh}}_{\gamma\to\pi}$ will not depend on the width Γ and will significantly exceed the value of $W^{nc}_{\gamma\to\pi}$. However, since $\Gamma/\mu < 6 \cdot 10^{-8}$, the inverse equality is actually realized; $\Delta\vartheta_0$, $\Delta\omega/\omega \gg \Gamma/\mu$, and usually $\Delta\omega/\omega \gg \Delta\vartheta_0$ and $\Delta\omega \ll \omega_n - \omega_{n\pm1}$. Under these assumptions and for a uniform distribution, averaging (27.9) over frequency in an interval $\Delta\omega$ near ω_n, we find

$$\langle W^{\text{coh}}_{\gamma\to\pi^0}\rangle \simeq \frac{2\pi\Gamma}{\alpha\mu^3}\frac{\omega_n}{\Delta\omega}\frac{1}{|q_n|}\sum_{\mathbf{q}_\perp}|G(\mathbf{q}_\perp, q_\parallel = q_n)|^2\mathbf{q}_\perp^2 \,,$$

$$\langle W^{nc}_{\gamma\to\pi^0}\rangle \simeq \frac{8\pi Z^2\alpha\Gamma}{\mu^3}N\int_{(q_\perp^{\min})^2}^{(q^{\max})^2}\frac{x\,dx}{(x + q_n^2 + a^{-2})^2} \tag{27.11}$$

For a threshold energy $\omega_n \simeq \omega_{th}$, setting in (27.11) $|q_n| \simeq u_1^{-1}$ and replacing the summation by an integral, we obtain a rough estimate for the ratio of the coherent and incoherent effects:

$$\langle W^{\text{coh}}_{\gamma\to\pi^0}\rangle/\langle W^{nc}_{\gamma\to\pi^0}\rangle \sim (\omega_n/\Delta\omega)(u_1/l), \quad \omega \simeq \omega_{th} \,, \tag{27.12}$$

where l is the lattice constant. It can can be seen that the enhancement factor which arises as a result of the discreetness of the spectrum $(\omega_n/\Delta\omega)$ turns out to be compensated to a significant degree, e.g. in diamond $u_1/l \sim 10^{-2}$ Nevertheless for a photon beam with $\Delta\omega/\omega \le 10^{-2}$ observation of the coherent effect is already possible at the present time.

28. DETECTOR OF PHOTONS AND ELECTRONS OF HIGH ENERGIES USING ALIGNED SINGLE CRYSTALS

It has been shown above that at high particle energies the probability of electron-positron pairs production by photons and radiation from charged particles for certain directions of incidence exceeds considerably those in the case of amorphous medium. The maximum gain $r^{\max}$ is given by Eq. (12.20) in the case of pair creation and by Eq. (17.19) for the radiation. For example, (see Table 12.1) for the orientation along $\langle 110 \rangle$ axis in crystals of silicon and germanium, which is of the main practical interest, $r^{\max}$ is 78 and 29, accordingly. The dependence of the pair creation probability W_e on the photon frequency ω and the inverse value of the characteristic length L_{ch}^{-1} of particle energy ε losses in these crystals are shown in Fig. 28.1. Since the characteristic angles of pair creation ($\sim m/\omega$) and radiation emission ($\sim m/\varepsilon$) are significantly smaller than the angle V_0/m, which characterises the orientational dependence of these processes, ($\rho_c \sim V_0\omega/m^2 \sim V_0\varepsilon/m^2 \gg 1$) then particles incident along the axis cause the specific electron-photon shower (see Sec. 20), developing on the lengths tens times shorter than in the according amorphous medium.

In the work by Baier, Katkov, Strakhovenko, [1986f] we have suggested to use the mentioned physical phenomena for the creation of a comparatively

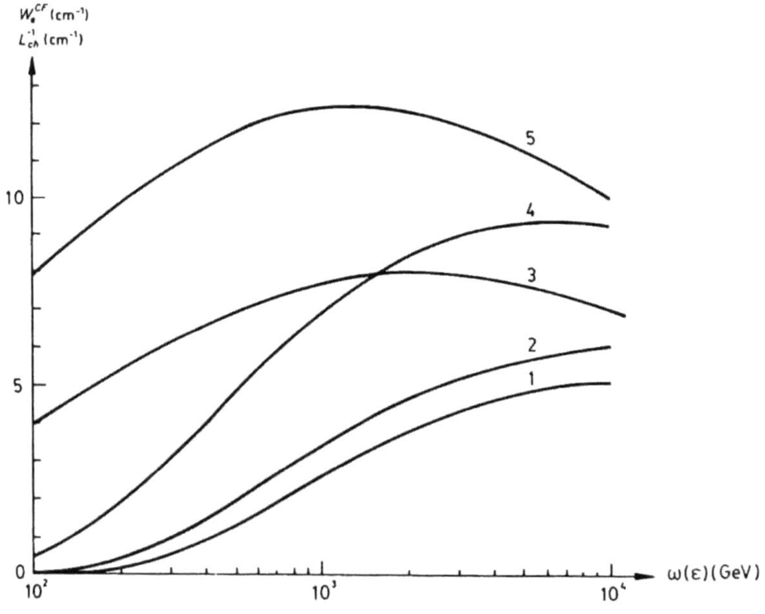

Fig. 28.1. Total pair creation probability versus the initial energy for Si $\langle 111 \rangle$ (curve 1), Si $\langle 110 \rangle$ (curve 2), and Ge $\langle 110 \rangle$ (curve 4), as well as the inverse characteristic length of the energy losses for Si $\langle 110 \rangle$ (curve 3) and Ge $\langle 110 \rangle$ (curve 5). All results are given for room temperature.

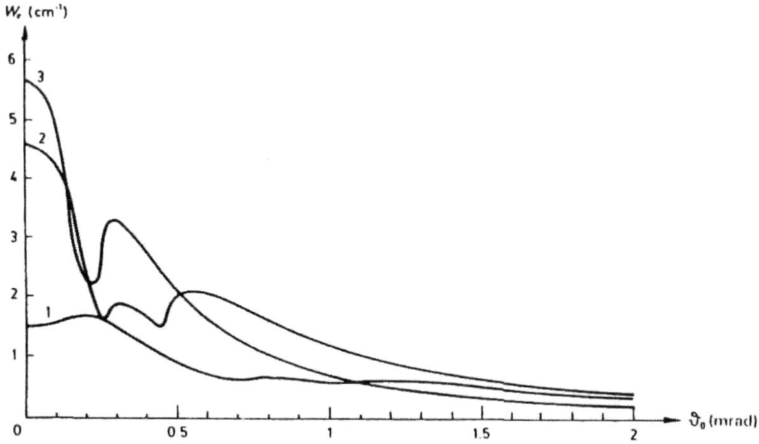

(a)

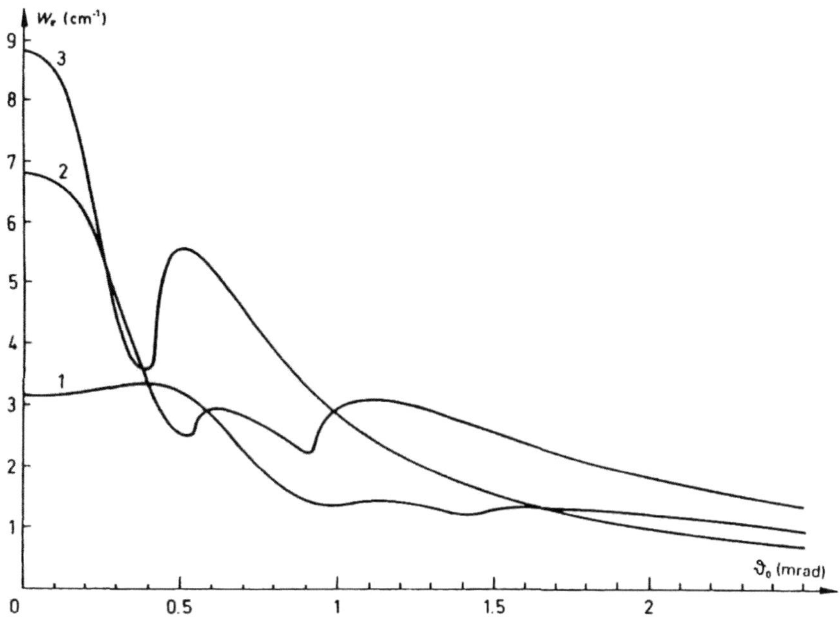

(b)

Fig. 28.2. Pair creation probability versus the angle of incidence with respect to the ⟨110⟩ axis. (a) in Si, $T = 293$ K for $\omega = 0.4$ TeV (curve 1), 2 TeV (curve 2) and 5 TeV (curve 3); (b) in Ge, $T = 280$ K for $\omega = 0.3$ TeV (curve 1), 1 TeV (curve 2) and 3 TeV (curve 3).

simple detector of high energies photons and electrons with high angular resolution. The energy of an incident photon should be larger than the value ω_t, at which the probability W_e is equal to the magnitude of W_{BH}. For definiteness one can use the value ω_{10}, whose magnitudes together with the values of the maximum gain $r^{\max}$ are given in Table 12.1; $W_e(\omega_{10}) = 10W_{BH}$. We remind that the maximum decrease of the characteristic length of particle energy losses with respect to the amorphous value L_{rad} is equal to $r^{\max}$ approximately.

The angular resolution of the suggested detector is determined by the orientation dependence of the radiation emission and pair creation processes considered above. Figures 28.2(a) and 28.2(b) illustrate the orientation dependence of the pair creation probability $W_e(\vartheta_0)$ relative to the $\langle 110 \rangle$ axes in Si and Ge respectively, at various photon energies. To describe the behavior of the function $W_e(\vartheta_0)$, we introduce the angles ϑ_0^{BH} and $\vartheta_0^{(1/2)}$, given in Table 28.1. By definition, (1) $W_e(\vartheta_0^{BH}) = W_{BH}$, i.e. the angle ϑ_0^{BH} may be regarded as the boundary at which the coherent effects manifest themselves. For Si and Ge the approximate formula $\vartheta_0^{BH} \simeq 30\, V_0/\sqrt{\omega\, W_{BH}}$ holds (here V_0 is taken in eV, ω in MeV, and W_{BH} in cm^{-1}) and (2)$W_e(\vartheta_0^{(1/2)}) = \frac{1}{2}\, W_e(\vartheta_0 = 0)$, i.e. the angle $\vartheta_0^{(1/2)}$ characterizes the angular resolution. As is seen from Figs. 28.2(a) and 28.2(b) near the chosen threshold $\omega \geq \omega_{10}$ the pair creation probability in the axis field ($\vartheta_0 < V_0/m$) is close to the coherent creation probability ($\vartheta_0 > V_0/m$) so that there is a wide plateau in the orientation dependence and, correspondingly, the angle $\vartheta_0^{(1/2)}$ is fairly large. As the energy increases, at small angles the peak in the function $W_e(\vartheta_0)$ appears. This peak grows while the photon frequency is less than ω_m and, as seen in Fig. 28.2(a), (b) and Table 28.1, its width decreases.

The detector would be a matrix of aligned single crystals of small thickness L. For definiteness, we take Si with $L = 1$ cm $= \frac{1}{13} L_{\text{ph}}^{BH}$ ($L_{\text{ph}}^{BH} = 1/W_{BH}$). When a photon moves at a large angle to the axis the pair creation

Table 28.1. The angles (μrad) characterizing orientational dependence of pair creation by a photon.

Si $\langle 110 \rangle$			Ge $\langle 110 \rangle$		
ω (TeV)	ϑ_0^{BH}	$\vartheta_0^{(1/2)}$	ω (TeV)	ϑ_0^{BH}	$\vartheta_0^{(1/2)}$
0.4	11	1.2	0.3	10	2
2	6	0.6	1	6	1.2
5	4	0.4	3	4	0.7

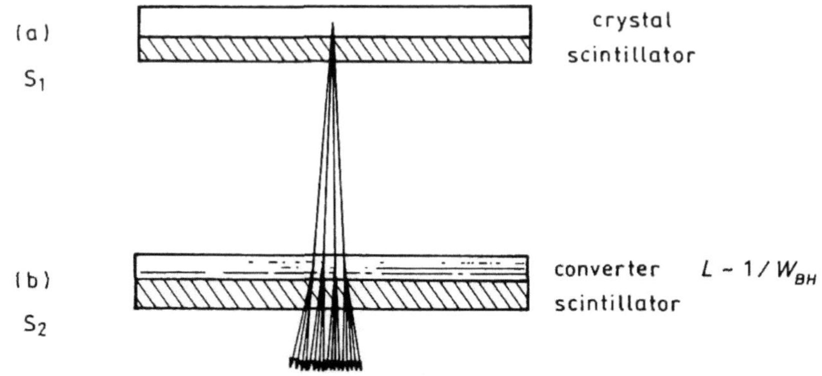

Fig. 28.3. Possible scheme of a detector using an oriented single crystal. An incident photon or electron hit the single crystal plate (a) whose thickness should be larger than L_{ch}. Charged particles, emitting some photons, are recorded by the scintillation counter S_1, at some distance from which the converter (c) is placed with thickness $L_c \sim 1/W_e^{BH}$, in which the photons are converted into e^+e^- pairs. All the charged particles are recorded by the second scintillation counter S_2.

probability is, correspondingly, low whereas when a photon is incident inside the characteristic angle discussed above, the pair creation probability is of order of unity, beginning with $\omega \geq 250$ GeV. The angular resolution of the detector is energy-dependent and is characterized by the angle $\vartheta_0^{(1/2)}$ (Table 28.1). The particles of the produced pair are detected by a detector of small thickness (in radiation lengths), for example, by a thin scintillator. In addition, there is the possibility of detecting the signal directly from the single crystal. Already at $\omega = \omega_{10}$ the particles of the produced pair radiate intensively. Behind the thin scintillator which is fixing the fact of pair (or pairs) creation in the single crystal, a standard (for example, lead) converter of $L_{\rm ph}^{BH}$ thickness is placed. In this converter the emitted photons are transformed into the pairs, and a strong signal caused by hundreds of particles will be registered in the second scintillator next to the converter (see Fig. 28.3).

Using the results of the cascade theory obtained in Sec. 20, it is possible to measure the energy of an initial photon based on the properties of this signal. For example, as it follows from Fig. 20.2, the number of photons flying out the crystal with the frequency $\omega_f > 100$ MeV is $N_\gamma = 40$, 90 and 160 at the energy of the incident photon $\omega_i = 0.4$, 1 and 2 TeV, correspondingly. When energy increases the number of charged particles produced in the crystal itself becomes large enough to omit the additional converter and the scheme of registration can be simplified.

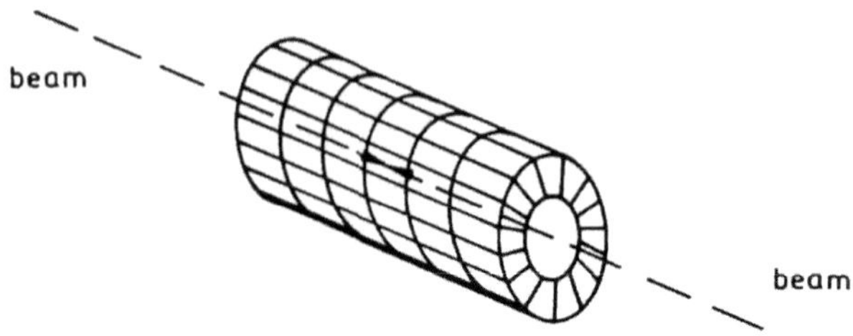

Fig. 28.4. Insertion in the universal detector for colliding beam experiments, using single crystal plates, gives the possibility of recording electrons, positrons and photons, with energy larger than 10 GeV (using tungsten crystals), and to differentiate them from hadrons.

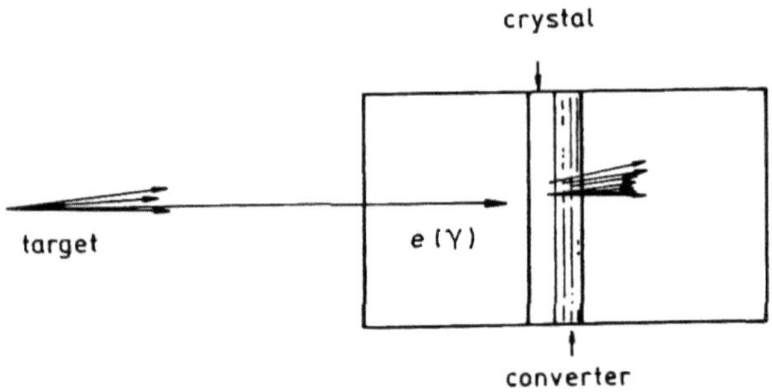

Fig. 28.5. Insertion of an oriented single-crystal plate in the detector with fixed target gives the possibility of distinguishing high-energy electrons, positrons and photons in the flux of secondary particles.

In high energy physics this device can be used as a thin directed converter. As an illustration, let us consider a tungsten crystal ($\langle 111 \rangle$ axis) of thickness $L = L_{\rm ph}^{BH}/5 \simeq 0.9$ mm. Beginning with the photon energy $\omega \simeq (50) \div 80)$ GeV (depending on the temperature) the pair creation probability in it will be of the order of unity, at a relatively wide acceptance angle $\vartheta_0^{(1/2)}$ (see Fig. 14.2). Similarly, electrons incident on this converter near the axis will radiate intensively (the characteristic angles for radiation emission and pair creation are of the same order) and this can be used for their detection and for the measurement of their energy. Figure 28.4 presents the sketch of the single crystal putting in any universal detector for the colliding beam experiments.

This construction with single crystals (see Fig. 28.5) may be useful when the particles flying out move in a small solid angle (experiment with a fixed target, HERA-experiments and so on). The main advantages of a single crystal as the particle detector are side by side with a small quantity of matter and high angular resolution (1) the registration of electrons and photons and their separation from hardrons, which do not cause a signal in the system; (2) the measurement of $e^{\pm}$ and γ energies.

Another interesting region of application of the detector is γ-astronomy. Let us discuss the problem of the detection of superhard photons from a point source, for example, Cygnus X3. The data on its emission of photons with an energy of $1 \div 10^3$ TeV have been obtained by observing the Cherenkov radiation in the atmosphere, wide atmospheric showers, as well as the muons in deep underground detectors. The experimental data obtained are quite different. It is recognized that the averaged flux of photons with an energy higher than ω behaves as $1/\omega$ and the flux is $I\,(\omega > 250\text{Gev}) = 10^{-9} \div 10^{-10}$ photons/cm^{2}s (see, for example, Stepanyan [1983]). With such a flux, a Si detector (the $\langle 110 \rangle$ axis) of $1\,\text{m}^2$ area will register $300 \div 30$ events with $\omega > 250$ GeV per year. The detector should be oriented to the source with an accuracy of about 10^{-3} The events of interest can be separated using the system described above.

Appendix A

NOTATIONS, DESCRIPTION OF FREE PARTICLE MOTION

A.1. Notations

The "natural" system of units $\hbar = c = 1$ is used in this book. Then dimensions of physical quantities are expressed through some powers of units of length. In particular, an energy, a momentum and a mass have the dimension of inverse length. Going over to the CGS-system, one should bear in mind, that the length is expressed in units of the Compton wave length $\hbar/mc$, the time in $\hbar/mc^2$ and the velocity in v/c. To emphasize the transition from quantum to classical quantities, the Planck constant $\hbar$ is reinserted when suitable and sometimes the full dimension is given in parentheses. The Gaussian system of units $e^2 = (e^2/\hbar c) = 1/137.036$ is used.

Four-vector and tensor indices are denoted by Greek letters $\mu, \nu, \dots$, running values 0, 1, 2, 3. The contravariant four-vector is represented in the form $a^\mu \equiv (a^0, \mathbf{a})$. Examples

$$
\begin{aligned}
x^\mu &\equiv (x^0, \mathbf{x}) \equiv (x^0, x^1, x^2, x^3) \equiv (t, x, y, z)\,; \\
p^\mu &\equiv (\varepsilon, \mathbf{p}) \equiv (p^0, p^1, p^2, p^3) \equiv (\varepsilon, p_x, p_y, p_z)\,.
\end{aligned}
\tag{A.1}
$$

Three-vector and tensor indices are denoted by Latin letters, the scalar product of two 3-vectors is $\mathbf{ab}$, and the vector product has the form $[\mathbf{ab}]$. The pseudo-Euclidean metric is assumed when the metric tensor $g_{\mu\nu}$ has the components $g_{00} = 1$, $g_{11} = g_{22} = g_{33} = -1$ and $g_{\mu\nu} = 0$ for $\mu \neq \nu$. Covariant and contravariant vectors are connected by the relation $a_\mu = g_{\mu\nu}a^\nu$ (summation over repeated indices is assumed). Examples

$$x_\mu = (x_0, x_1, x_2, x_3) \equiv (t, -\mathbf{x}), \quad \frac{\partial}{\partial x^\mu} \equiv \partial_\mu \equiv (\partial_t, \boldsymbol{\nabla}). \tag{A.2}$$

The product of two four-vectors is

$$\begin{aligned} ab = a^\mu b_\mu &= g_{\mu\nu}a^\mu b^\nu = a^0 b_0 + a^i b_i \\ &= a^0 b^0 - a^i b^i = a^0 b^0 - \mathbf{ab}. \end{aligned} \tag{A.3}$$

Examples

$$\begin{aligned} px &= p^\mu x_\mu = \varepsilon t - \mathbf{px}, \\ \gamma^\mu \partial_\mu &\equiv \gamma^\mu \frac{\partial}{\partial x^\mu} = \gamma^0 \frac{\partial}{\partial t} + \boldsymbol{\gamma}\boldsymbol{\nabla}, \\ \partial_\mu \partial^\mu &\equiv \Box = \partial_t^2 - \boldsymbol{\nabla}^2 \end{aligned} \tag{A.4}$$

The momentum operator in coordinate representation has the form

$$p^\mu \equiv i\partial^\mu = i\frac{\partial}{\partial x_\mu} = i\left(\frac{\partial}{\partial t}, -\frac{\partial}{\partial \mathbf{x}}\right). \tag{A.5}$$

A.2. Electrons and Positrons

The Dirac equation in covariant notation reads

$$\begin{aligned} (i\gamma^\mu \partial_\mu - m)\psi(x) &= 0 \\ i\partial_\mu \bar{\psi}(x)\gamma^\mu + m\bar{\psi}(x) &= 0, \end{aligned} \tag{A.6}$$

where $\bar{\psi}(x) = \psi^+(x)\gamma^0$ The matrices γ^μ satisfy the anti-commutation relation

$$\{\gamma^\mu, \gamma^\nu\} \equiv \gamma^\mu\gamma^\nu + \gamma^\nu\gamma^\mu = 2g^{\mu\nu}. \tag{A.7}$$

The standard representation for γ-matrices can be conveniently used

$$\gamma^\mu \equiv (\gamma^0, \gamma^k), \quad \gamma^0 = \begin{pmatrix} I & 0 \\ 0 & -I \end{pmatrix}, \quad \gamma^k = \begin{pmatrix} 0 & \sigma^k \\ -\sigma^k & 0 \end{pmatrix}, \tag{A.8}$$

where σ^k are the Pauli matrices, I is the unit matrix of rank two. The solution to the Dirac equation in the form of plane waves is

$$\psi(x) = \frac{1}{(2\pi)^{\frac{3}{2}}}\sqrt{\frac{m}{\varepsilon}}[u(p)e^{-ipx} + v(p)e^{ipx}], \tag{A.9}$$

where $\varepsilon \equiv \varepsilon_p = (\mathbf{p}^2 + m^2)^{1/2}$. The bi-spinors $u(p)$ and $v(p)$ satisfy the equation

$$\begin{aligned} (\hat{p} - m)u(p) &= \bar{u}(p)(\hat{p} - m) = 0\,, \\ (\hat{p} + m)v(p) &= \bar{v}(p)(\hat{p} + m) = 0\,, \end{aligned} \tag{A.10}$$

where $\hat{p} \equiv \gamma^\mu p_\mu = \gamma^0 p^0 - \gamma^k p^k, p^0 = \varepsilon_p$. The normalization and orthogonality conditions for $u(p)$ and $v(p)$ are

$$\begin{aligned} &\bar{v}^s(p) u^r(p) = 0\,, \\ &\bar{u}^r(p) u^s(p) = \bar{v}^r(p) v^s(p) = \delta^{rs}\,, \\ &r, s = 1, 2\,. \end{aligned} \tag{A.11}$$

The explicit form of bi-spinors is

$$u^r(p) = \frac{\hat{p} + m}{\sqrt{2m\,(\varepsilon + m)}} \begin{pmatrix} \varphi^r \\ 0 \end{pmatrix}, \quad v^r(p) = \frac{m - \hat{p}}{\sqrt{2m(\varepsilon + m)}} \begin{pmatrix} 0 \\ \varphi^r \end{pmatrix}$$

or

$$u^r(p) = \sqrt{\frac{\varepsilon + m}{2m}} \begin{pmatrix} \varphi^r \\ \frac{\sigma \mathbf{p}}{\varepsilon + m} \varphi^r \end{pmatrix}, \quad v^r(p) = \sqrt{\frac{\varepsilon + m}{2m}} \begin{pmatrix} \frac{\sigma \mathbf{p}}{\varepsilon + m} \varphi^r \\ \varphi^r \end{pmatrix}, \tag{A.12}$$

where two-component spinors φ^r describe the polarization states. Let the quantization axis be the unit vector $\boldsymbol{\zeta}(\vartheta, \varphi)$, then the spinors with a definite projection (along or opposite with respect to this axis) have the form

$$\begin{aligned} \varphi_\uparrow(\boldsymbol{\zeta}) \equiv \varphi_1(\boldsymbol{\zeta}) &= \begin{pmatrix} \cos\frac{\vartheta}{2} e^{-i\varphi/2} \\ \sin\frac{\vartheta}{2} e^{i\varphi/2} \end{pmatrix}, \\ \varphi_\downarrow(\boldsymbol{\zeta}) \equiv \varphi_2(\boldsymbol{\zeta}) &= \begin{pmatrix} -\sin\frac{\vartheta}{2} e^{-i\varphi/2} \\ \cos\frac{\vartheta}{2} e^{i\varphi/2} \end{pmatrix}, \end{aligned} \tag{A.13}$$

and we have

$$\begin{aligned} \varphi_{1,2}^+(\boldsymbol{\zeta}) \boldsymbol{\sigma} \varphi_{1,2}(\boldsymbol{\zeta}) &= \pm \boldsymbol{\zeta}\,, \\ (\boldsymbol{\sigma\zeta}) \varphi_{1,2}(\boldsymbol{\zeta}) &= \pm \varphi_{1,2}(\boldsymbol{\zeta})\,, \end{aligned} \tag{A.14}$$

i.e. the quantity $\boldsymbol{\sigma\zeta}$ is the polarization operator in the particle rest frame. The same operator in 4-component notation is $\boldsymbol{\zeta}\mathbf{O}_0 = \gamma^0(\boldsymbol{\zeta}\ \boldsymbol{\Sigma})$, where $\boldsymbol{\Sigma} =$

$\begin{pmatrix} \sigma & 0 \\ 0 & \sigma \end{pmatrix}$. In an arbitrary frame of reference we can use any of the following polarization operators

$$\mathbf{O}\boldsymbol{\zeta} = S^{\mu} s_{\mu} = R^{\mu\nu} t_{\mu\nu} ,$$

where

$$S^{\mu} = -\gamma^5 \gamma^{\mu} , \quad R^{\mu\nu} = \frac{i}{2} [\sigma^{\mu} , \sigma^{\nu}] ,$$

$$\mathbf{O} = \gamma^0 \boldsymbol{\Sigma} + \mathbf{p}/m \left[\gamma^0 \frac{\boldsymbol{\Sigma}\mathbf{p}}{\varepsilon + m} - \gamma^5 \gamma^0 \right] .$$

Here $\gamma^5 = -i\gamma^0\gamma^1\gamma^2\gamma^3 = \begin{pmatrix} 0 & -I \\ -I & 0 \end{pmatrix}$, s^{μ} is the four-vector of polarization, satisfying the condition $sp = 0$. In the rest frame $s \to (0, \boldsymbol{\zeta})$ and in the arbitrary frame of reference, the components of this vector are

$$\mathbf{s} = \boldsymbol{\zeta} + \frac{\mathbf{p}(\boldsymbol{\zeta}\mathbf{p})}{m\,(\varepsilon + m)} , \quad s^0 = \frac{(\boldsymbol{\zeta}\mathbf{p})}{m} . \tag{A.15}$$

The polarization density matrix is defined for electrons and positrons by the following relations

$$\rho^{(+)}_{\mu\nu} = \overline{u_{\mu}(p)\bar{u}_{\nu}(p)} , \quad \rho^{(-)}_{\mu\nu} = -\overline{v_{\mu}(p)\bar{v}_{\nu}(p)} .$$

For the explicit form of this matrix we have

$$\rho^{(\pm)}_s(p) = \frac{m \pm \hat{p}}{2m} \cdot \left(\frac{1 - \gamma^5 \hat{s}}{2} \right) ,$$

and $\boldsymbol{\zeta}$ is a unit vector $|\boldsymbol{\zeta}| = 1$ for pure states, while for mixed states (partially polarized) we have $0 \le |\boldsymbol{\zeta}| \le 1$.

A matrix element of the general form can be rewritten in the two-component notation

$$M_{fi} = \bar{u}_{s_2}(p_2)\, O u_{s_1}(p_1) = \varphi^{+}(\zeta_2) \mathcal{G} \varphi(\zeta_1) ,$$

where $\varphi(\zeta_2)$ and $\varphi(\zeta_1)$ are the spinors (A.14) which describe the polarization of final and initial particles, $\mathcal{G}$ is a 2×2 matrix which depends on the vectors of the problem. According to Eq. (A.14), the projection operator onto the state with the projection +1 with respect to the $\boldsymbol{\zeta}$-axis is

$$\varphi(\boldsymbol{\zeta})\varphi^{+}(\boldsymbol{\zeta}) = \frac{(1 + \boldsymbol{\sigma}\boldsymbol{\zeta})}{2} .$$

To describe the mixed states, the two-component matrix is introduced

$$\rho(\zeta) = \overline{\varphi(\zeta)\varphi^+(\varphi)} = \frac{(1+\sigma\zeta)}{2}, \tag{A.16}$$

where $0 \leq |\zeta| \leq 1$. Any 2×2 matrix can be decomposed in the sum of the unit matrix and Pauli matrices: $\mathcal{G} = A + i\sigma\mathbf{B}$. Then we have

$$\begin{aligned} |M_{fi}|^2 &= \varphi^+(\zeta_2)\mathcal{G}\varphi(\zeta_1)\,\varphi^+(\zeta_1)\,\mathcal{G}^+\varphi(\zeta_2) \\ &= \left(\frac{1}{4}\right) Sp[(1+\sigma\zeta_1)(A^* - i\sigma\mathbf{B}^*)(1+\sigma\zeta_2)(A+i\sigma\mathbf{B})]\,. \end{aligned} \tag{A.17}$$

Taking trace in Eq. (A.17), we can obtain the explicit form of the quantity $|M_{fi}|^2$. The two-component form of the density matrix of the final state reads

$$\rho_f = \frac{\mathcal{G}\rho(\zeta_1)\mathcal{G}^+}{[Sp(\mathcal{G}\rho(\zeta_1)\mathcal{G}^+)]},$$

and taking into account that $\zeta = Sp(\rho\sigma)$, we find for the spin in the final state

$$\zeta_f = \frac{Sp[\mathcal{G}\rho(\zeta_1)\mathcal{G}^+\sigma]}{Sp[\mathcal{G}\rho(\zeta_1)\mathcal{G}^+]}\,. \tag{A.18}$$

This polarization can be found also from Eq. (A.17), if we represent it in the form

$$|M_{fi}|^2 = a + \mathbf{b}\zeta_2\,. \tag{A.19}$$

On the other hand, we have

$$\begin{aligned} |M_{fi}|^2 &\sim Sp[\rho\,(\zeta_2)\rho(\zeta_f)] \\ &= \frac{1}{2}\,Sp[(1+\sigma\zeta_2)(1+\sigma\zeta_f)] = 1 + \zeta_2\zeta_f \end{aligned}$$

Comparing this expression with Eq. (A.19), we obtain

$$\zeta_f = \frac{\mathbf{b}}{a} \tag{A.20}$$

The same expression can be obtained from Eq. (A.18).

A.3. Electromagnetic Field

In the Lorentz gauge we have

$$\partial_\nu A^\nu(x) = 0\,, \tag{A.21}$$

where the 4-vector potential of an electromagnetic field $A^\nu(x) = (\varphi, \mathbf{A})$ satisfies the inhomogeneous wave equation

$$\partial_\nu \partial^\nu A^\mu(x) = 4\pi j^\mu(x)\,. \tag{A.22}$$

At the description of electromagnetic wave (electromagnetic field in vacuum) we can always choose the gauge when $\varphi = 0$. Then the Lorentz condition (A.21) goes over into the non-covariant condition

$$\operatorname{div}\mathbf{A} = 0\,, \tag{A.23}$$

usually called the Coulomb (or transverse) gauge.

The retarded potential induced by a point-like charge moving along the trajectory $\mathbf{r}_0(t)$ with the velocity $\mathbf{v}(t)$ at large distances from the particle (retaining only the main terms of an expansion in r^{-1}) can be represented as the Fourier transform

$$\mathbf{A}_{ret}(\omega, r) = (e/r)e^{i\omega r}\int \mathbf{v}(t)\, e^{i(\omega t - \mathbf{k}\mathbf{r}_0(t))}dt\,, \tag{A.24}$$

where ω is the frequency and $\mathbf{k} = \mathbf{n}\omega$ is the wave vector. The magnetic $\mathbf{H}$ and electric $\mathbf{E}$ fields are obtained by means of the formula

$$\mathbf{H} = [\dot{\mathbf{A}}\,\mathbf{n}]\,, \quad \mathbf{E} = [[\dot{\mathbf{A}}\,\mathbf{n}]\mathbf{n}]\,, \quad \dot{\mathbf{A}} = \frac{\partial \mathbf{A}}{\partial t} \tag{A.25}$$

The explicit form of these fields is

$$\begin{aligned} \mathbf{H}(\omega) &= \frac{iew}{r}e^{i\omega r}\int \exp[i(\omega t - \mathbf{k}\mathbf{r}_0)][\mathbf{n}d\mathbf{r}_0]\,, \\ \mathbf{E}(\omega) &= \frac{ie\omega}{r}e^{i\omega r}\int (\mathbf{v} - \mathbf{n})\exp[i(\omega t - \mathbf{k}\mathbf{r}_0)]dt\,. \end{aligned} \tag{A.26}$$

Since the electromagnetic field is described by the vectors $\mathbf{E}$ and $\mathbf{H}$ which are perpendicular to the direction of propagation $\mathbf{n}$, the properties of radiation are determined side by side with its spectral composition by the direction of these vectors. For definiteness, the vector $\mathbf{E}$ is usually chosen to characterize this important property of electromagnetic waves called the polarization. To describe quantitatively the polarization of radiation, we can project the Fourier component of an electric field onto two mutually orthogonal and, generally speaking, complex vectors $\mathbf{e}_{(\lambda)}$ $(\lambda = 1, 2)$ of unit length, lying in the plane which is perpendicular to the direction of the wave propagation $\mathbf{n}$

$$\mathbf{e} = e_1\mathbf{e}_{(1)} + e_2\mathbf{e}_{(2)} = \mathbf{e}_{(1)} \cos \mathbf{d} + \mathbf{e}_{(2)} \sin \alpha e^{i\beta} \,. \tag{A.27}$$

The description of polarization properties is the same within the classical and quantum mechanical theory. In the classical theory the quantities e_1 and e_2 are proportional to the magnitudes of the components of the wave electric field (a square root from the energy flux density is usually removed as a common factor). In the quantum theory the quantities $|e_1|^2$ and $|e_2|^2$ are the probabilities of the photon being polarized along $\mathbf{e}_{(1)}$ and $\mathbf{e}_{(2)}$, correspondingly. At $\beta = 0$, the wave is linearly polarized at the angle α to $\mathbf{e}_{(1)}$. At $\beta = \pm\pi/2, \alpha = \pi/4$, the wave is right (R) or left (L) circularly polarized. For a particle moving with the velocity of light, there is the axial symmetry, when only chirality, the projection of the total angular momentum onto the direction of propagation, is conserved. The circularly polarized states are the states with definite values of chirality.

Physical quantities contain the polarization of a wave (of a photon) in the combination $\rho_{ik} = e_i(k, \lambda)e_k^*(k, \lambda)$. Using the unit matrix I $(I_{ik} = \delta_{ik})$ and the Pauli matrices σ_k

$$\sigma_1 = \begin{pmatrix} 0 & 1 \\ 1 & 0 \end{pmatrix}; \quad \sigma_2 = \begin{pmatrix} 0 & -i \\ i & 0 \end{pmatrix}; \quad \sigma_3 = \begin{pmatrix} 1 & 0 \\ 0 & -1 \end{pmatrix}, \tag{A.28}$$

we can represent the density matrix in the form

$$\rho_{ik} = \frac{E_i(\omega)E_k^*(\omega)}{(\mathbf{E}\,\mathbf{E}^*)} = \frac{1}{2}\left[\delta_{ik} + \sum_n (\sigma_n)_{ik}\xi_n\right] = \frac{1}{2}(1 + \boldsymbol{\sigma\xi})_{ik} \,. \tag{A.29}$$

As ρ_{ik} is the Hermitian matrix, the parameters ξ_n are real. For a pure state, by definition, $\det \rho = 0$, and we have from Eq. (A.29)

$$\det \rho = \frac{1}{4} \det \begin{pmatrix} 1 + \xi_3 & \xi_1 - i\xi_2 \\ \xi_1 + i\xi_2 & 1 - \xi_3 \end{pmatrix} = \frac{1}{4}(1 - \boldsymbol{\xi}^2) = 0 \,. \tag{A.30}$$

So, the polarization properties of a monochromatic plane wave which, by definition, is completely polarized, can be described by three real parameters ξ_n called the Stokes parameters and satisfying the condition $\boldsymbol{\xi}^2 = \Sigma_{n=1}^3 \xi_n^2 = 1$.

Let us elucidate the physical meaning of these parameters. Let the wave propagate along the x_3-axis. The wave is circularly polarized (see Eq. (A.27)) if $\mathbf{e}_\pm = (\mathbf{e}_{(1)} \pm i\mathbf{e}_{(2)}) \cdot 2^{-1/2}$ where the signs $+, (-)$ correspond to the right (left) polarization. In this case Eq. (A.29) takes the form

$$\rho = \frac{1}{2}\begin{pmatrix} 1 & \mp i \\ \pm i & 1 \end{pmatrix} = \frac{1}{2}(I \pm \sigma_2)\,, \tag{A.31}$$

i.e. at circular (right and left) polarization, the Stokes parameters are $\xi_1 = \xi_3 = 0,\ \xi_2 = \pm 1$.

At linear polarizations along x_1-axis (along $\mathbf{e}_{(1)}$) or along x_2-axis (along $\mathbf{e}_{(2)}$), we have from Eq. (A.29)

$$\rho = \begin{pmatrix} 1 & 0 \\ 0 & 0 \end{pmatrix} = \frac{1}{2}(I + \sigma_3)\,, \quad \text{or} \quad \rho = \begin{pmatrix} 0 & 0 \\ 0 & 1 \end{pmatrix} = \frac{1}{2}(I - \sigma_3)\,. \tag{A.32}$$

In this case $\xi_1 = \xi_2 = 0,\ \xi_3 = \pm 1$ (the signs $+$ or $-$ correspond to the polarization along x_1- or x_2-axes).

Consider, finally, the linear polarization at the angle $\pm\pi/4$ to $\mathbf{e}_{(1)}$, when we find from Eq. (A.29)

$$\rho = \frac{1}{2}\begin{pmatrix} 1 & \pm 1 \\ \pm 1 & 1 \end{pmatrix} = \frac{1}{2}(I \pm \sigma_1)\,, \tag{A.33}$$

i.e. the Stokes parameters of the wave linearly polarized at the angle $\pm\pi/4$ to x_1-axis are $\xi_2 = \xi_3 = 0,\ \xi_1 = \pm 1$. For linear polarization at the angle φ to the x_1-axis, we have

$$\rho = \frac{1}{2}\begin{pmatrix} 1 + \cos 2\varphi & \sin 2\varphi \\ \sin 2\varphi & 1 - \cos 2\varphi \end{pmatrix} = \frac{1}{2}(I + \sigma_3 \cos 2\varphi + \sigma_1 \sin 2\varphi) \tag{A.34}$$

or $\xi_1 = \sin 2\varphi\,,\ \xi_3 = \cos 2\varphi$ and $\xi_1^2 + \xi_3^2 = 1$, i.e., at rotation in the plane which is perpendicular to the direction of propagation $\mathbf{n}$, the parameters ξ_1 and ξ_3 change, while the sum $\xi_1^2 + \xi_3^2$ remains constant. The circular polarization does not change at such rotations. So, the quantities $\xi_1^2 + \xi_3^2$ and ξ_2 characterize a degree of linear and circular polarization and do not change not only at indicated rotations but at any Lorentz transformations.

For non-monochromatic waves, the polarization properties are described by an average value of the tensor $\rho_{ik} = \overline{e_i e_k^*}$ (at the quantum description ρ_{ik} is a density matrix) whereas Eq. (A.29) holds. However, the matrix ρ_{ik} cannot be reduced now to a product of vector components. Then, in general, the condition $\det \rho = 0$ and its consequence, Eq. (A.30), are violated. For unpolarized waves, all the directions in the plane perpendicular to **n** are equivalent, which yields

$$\rho_{ik} = \frac{1}{2}\delta_{ik}, \quad \det\rho = \frac{1}{4}, \quad \xi_n = 0. \tag{A.35}$$

In the general case we have

$$0 \le \det\rho \le \frac{1}{4}, \tag{A.36}$$

and the value *det* $\rho = 0$ corresponds to the completely polarized wave, while det $\rho = 1/4$ corresponds to the unpolarized one. For arbitrary polarization, the matrix ρ (see Eq. (A.29)) can be rewritten as follows

$$\rho = \frac{1}{2}\left((I - |\boldsymbol{\xi}|) + |\boldsymbol{\xi}|\left(1 + \frac{\boldsymbol{\xi}\boldsymbol{\sigma}}{|\boldsymbol{\xi}|}\right)\right), \tag{A.37}$$

where the first term corresponds to the unpolarized part of the wave and the second term corresponds to the completely polarized part. Evidently, the polarization degree is given by the formula

$$|\boldsymbol{\xi}| = \sqrt{\boldsymbol{\xi}^2} = \sqrt{\xi_1^2 + \xi_2^2 + \xi_3^2}. \tag{A.38}$$

If we know the polarization matrix ρ, then the "vector" characterizing the polarization is

$$\boldsymbol{\xi} = Sp\,(\rho\boldsymbol{\sigma}). \tag{A.39}$$

It can be easily proved, if we take into account that $Sp\,\boldsymbol{\sigma} = 0$ and $Sp\,\sigma_i\,\sigma_j = 2\delta_{ij}$.

If the radiation intensity matrix dI_{ik} is known, we have

$$\rho = \frac{1}{dI}\begin{pmatrix} dI_{11} & dI_{12} \\ dI_{21} & dI_{22} \end{pmatrix} = \frac{1}{2}\left(I + \sum_n \xi_n\sigma_n\right), \tag{A.40}$$

where dI is the differential radiation intensity summed up over polarizations. Substituting (A.40) into (A.39), we obtain

$$\xi_1 = \frac{dI_{12} + dI_{21}}{dI}, \quad \xi_2 = \frac{i\,(dI_{12} - dI_{21})}{dI},$$
$$\xi_3 = \frac{dI_{11} - dI_{22}}{dI}, \quad dI = dI_{11} + dI_{22}. \tag{A.41}$$

Note yet that for a monochromatic wave

$$\xi_2 = i[\mathbf{e}\mathbf{e}^*]\mathbf{n}\,, \tag{A.42}$$

and at spatial inversion $\xi_2 \to -\xi_2$ while at the inversion of time $\xi_2 \to \xi_2$. A summation over photon polarizations can be carried out by two equivalent ways

$$\sum_{\lambda} e_{\mu}(\lambda)e_{\nu}^*(\lambda) = -g_{\mu\nu} \tag{A.43}$$

in arbitrary gauge, and

$$\sum_{\lambda} e_i(\lambda)e_k^*(\lambda) = \delta_{ik} - n_i n_k \tag{A.44}$$

in the Coulomb gauge.

Appendix B

PARTICLES IN AN EXTERNAL ELECTROMAGNETIC FIELD

B.1. General Relations

The Lagrangian of a classical charged particle in the external field $A^\mu(\varphi, \mathbf{A})$ has the form

$$L = -m\sqrt{1 - \mathbf{v}^2} + e\mathbf{A}\mathbf{v} - e\varphi\,. \tag{B.1}$$

The canonical momentum $\mathbf{p}$ is

$$\mathbf{p} = \frac{\partial L}{\partial \mathbf{v}} = \frac{m\mathbf{v}}{\sqrt{1 - \mathbf{v}^2}} + e\mathbf{A} = \boldsymbol{\mathcal{P}} + e\mathbf{A}\,. \tag{B.2}$$

Here $\boldsymbol{\mathcal{P}}$ is the usual (kinematic) momentum. The Hamiltonian of a charged particle in the field A^μ has the form

$$\mathcal{H} = \mathbf{v}\frac{\partial L}{\partial \mathbf{v}} - L = \sqrt{(\mathbf{p} - e\mathbf{A})^2 + m^2} + e\varphi\,. \tag{B.3}$$

In a quantum theory the canonical momentum becomes the operator: $\mathbf{p} \to -i\hbar\boldsymbol{\nabla}(p_\mu \to i\hbar\partial_\mu)$. In the presence of an external field, we have

$$\mathcal{P}_\mu = p_\mu - eA_\mu = i\hbar\partial_\mu - eA_\mu\,. \tag{B.4}$$

In particular, for the Dirac equation in an external field, we have (cf. Eq. (A.6))

$$(\gamma^\mu \mathcal{P}_\mu - m)\psi = [\gamma^\mu(i\hbar\partial_\mu - eA_\mu) - m]\psi = 0\,. \tag{B.5}$$

If we apply the operator $(\gamma^\mu \mathcal{P}_\mu + m)$ to Eq. (B.5), we obtain the "squared" Dirac equation

$$\left[\mathcal{P}_\mu \mathcal{P}^\mu - m^2 - e\hbar\sigma^{\mu\nu}\frac{F_{\mu\nu}}{2}\right]\psi = 0\,, \tag{B.6}$$

where $F_{\mu\nu}$ is the tensor of an electromagnetic field, $\sigma^{\mu\nu} = i[\gamma^\mu, \gamma^\nu]/2$. The Klein–Gordon equation in an electromagnetic field has the form

$$(\mathcal{P}_\mu \mathcal{P}^\mu - m^2)\varphi(x) = 0\,. \tag{B.7}$$

The additional term $e\hbar\sigma^{\mu\nu}F_{\mu\nu}/2$ in (B.6) is connected with the electron spin.

B.2. Analysis of Quantum Effects at Particle Motion in a Magnetic Field

When charged particles are moving in a magnetic field, the basic features of such motion are independent of the particle spin. The addition $e\hbar H$ connected with the electron spin in the squared energy (see Eq. (B.6)) is relatively small at high energies (the relative magnitude of this addition is $\hbar\omega_0/\varepsilon$, where $\omega_0 = eH/\varepsilon$ is the Larmour frequency) and can be neglected. For the sake of simplicity, we consider the quantum features of the motion in a magnetic field for a spin zero particle, whose Hamiltonian is obtained by going over to the operators in a classical expression

$$\mathcal{H} = \sqrt{\mathcal{P}^2 + m^2}\,, \tag{B.8}$$

where $\mathcal{P} = -i\hbar\nabla - e\mathbf{A}$. The velocity operator has the form

$$\mathbf{v} = \dot{\mathbf{r}} = \frac{1}{i\hbar}[\mathbf{r}, \mathcal{H}]\,. \tag{B.9}$$

Using the identity

$$[\mathbf{r}, \mathcal{H}^2] = \{[\mathbf{r}, \mathcal{H}], \mathcal{H}\}\,, \tag{B.10}$$

we obtain the operator equation for $\mathbf{v}$

$$\mathbf{v} + \frac{1}{2}[\mathcal{H}, \mathbf{v}]\,\mathcal{H}^{-1} = \mathcal{P}\,\mathcal{H}^{-1} \tag{B.11}$$

As the commutator in Eq. (B.11) is proportional to the Planck constant, it can be conveniently iterated in powers of $\hbar$. The solution to Eq. (B.11) reads

$$\mathbf{v} = \mathcal{P}\,\mathcal{H}^{-1} + \sum_{n=1}^{\infty}\frac{(-1)^n}{2^n}\underbrace{[\mathcal{H}, [\mathcal{H}, \ldots [\mathcal{H}}_{n}, \mathcal{P}]\ldots]\,\mathcal{H}^{-n-1}\,. \tag{B.12}$$

Components of the operator $\mathcal{P}_n$ satisfy the commutation relations

$$[\mathcal{P}_m, \mathcal{P}_n] = ie\hbar\varepsilon_{mnk}H_k \tag{B.13}$$

and

$$[\mathcal{P}^2, \mathcal{P}_n] = \sum_m [\mathcal{P}_m^2, \mathcal{P}_n] = 2ie\hbar\, \varepsilon_{mnk} H_k \mathcal{P}_m\,, \tag{B.14}$$

where $\mathbf{H}$ is the strength of the magnetic field; curl $\mathbf{H} = 0$.

Starting from the identity (B.10), we can obtain the equation

$$2\mathbf{b} + [\mathcal{H}, \mathbf{b}]\, \mathcal{H}^{-1} = \mathbf{c}\mathcal{H}^{-1} \tag{B.15}$$

This equation can be solved in the same way as Eq. (B.11)

$$\mathbf{b} \equiv [\mathcal{H}, \mathcal{P}] = \frac{1}{2}\mathbf{c}\mathcal{H}^{-1} + \sum_{n=1}^{\infty} \frac{(-1)^n}{2^{n+1}} \underbrace{[\mathcal{H}, [\mathcal{H}, \ldots [\mathcal{H}}_{n}, \mathbf{c}] \ldots]\, \mathcal{H}^{-n-1} \tag{B.16}$$

The commutator $\mathbf{c}$ is calculated according to Eq. (B.14)

$$\mathbf{c} = [\mathcal{H}^2, \mathcal{P}] = 2ie\hbar[\mathbf{HP}]\,. \tag{B.17}$$

Substituting this expression into Eq. (B.16), we obtain an equation containing the commutator $\mathbf{b}$ only. This equation can be solved by successive iterations. The corresponding solutions accurate within terms $\sim \hbar^3$ has the form

$$\begin{aligned}\mathbf{b} \equiv [\mathcal{H}, \mathcal{P}] &= ie\hbar[\mathbf{H}\mathcal{P}]H^{-1} - \frac{1}{2}(ie\hbar)^2(\mathbf{H}(\mathbf{H}\mathcal{P}) \\ &\quad - \mathbf{H}^2\mathcal{P})\mathcal{H}^{-3} - \frac{1}{2}(ie\hbar)^3\mathbf{H}^2[\mathbf{H}\mathcal{P}]\mathcal{H}^{-5} + \cdots \\ &\quad + e\frac{(i\hbar)^2}{4}\left[\left\{\frac{\partial \mathbf{H}}{\partial x_m}, \mathcal{P}_m\right\}\mathcal{P}\right]\mathcal{H}^{-3} + \cdots\,.\end{aligned} \tag{B.18}$$

This expression contains the terms of two different types, first of which depends on the strength $\mathbf{H}$ of the magnetic field, and, the second, depends on gradients of this field. The expansion is carried out for both types of terms in powers of

$$e\hbar\langle H\mathcal{H}^{-2}\rangle = \frac{H}{\gamma^2 H_0} = \frac{\hbar\omega_0}{\varepsilon}\,, \tag{B.19}$$

where

$$H_0 = \frac{m^2}{e\hbar} \left(= \frac{m^2 c^3}{e\hbar} \right) = 4.41 \cdot 10^{13}\, Oe \tag{B.20}$$

is the critical magnetic field (for an electron). Similarly, the critical electric field for an electron is

$$E_0 = \frac{m^2}{e\hbar} \left(= \frac{m^2 c^3}{e\hbar} \right) = 1.32 \cdot 10^{16} V/cm\,. \tag{B.21}$$

Let us now find the commutator of the velocity components. From Eq. (B.12), we have

$$\begin{aligned}[v_m, v_n] = \{[\mathcal{P}_m, \mathcal{P}_n] - \mathcal{P}_m \mathcal{H}^{-1}[\mathcal{H}, \mathcal{P}_n] \\ + \mathcal{P}_n \mathcal{H}^{-1}[\mathcal{H}, \mathcal{P}_m]\}\mathcal{H}^{-2}\end{aligned} \tag{B.22}$$

Making use of Eqs. (B.13) and (B.18), we obtain in the lowest order in $\hbar$

$$[\mathbf{vv}] = ie\hbar[\mathbf{H}(1 - \mathbf{v}^2) + \mathbf{v}(\mathbf{vH})]\mathcal{H}^{-2}\,, \tag{B.23}$$

so that the uncertainty relation for the velocity components reads

$$\Delta v_i \Delta v_k \sim e\hbar H/\varepsilon^2 = \frac{H}{H_0 \gamma^2} = \frac{\hbar\omega_0}{\varepsilon} \tag{B.24}$$

Within the same accuracy, we have

$$[r_m, v_n] = i\hbar(\delta_{mn} - v_m v_n)\mathcal{H}^{-1} \tag{B.25}$$

It is seen from the above consideration that the magnitude of noncommutativity of dynamic particle variables is of the order of $H/(H_0\gamma^2) = \hbar\omega_0/\varepsilon$ (series in powers of $\hbar\omega_0/\varepsilon$ in a general case), which implies a decrease of the magnitude of quantum effects with increasing energy. So, the higher the energy, the more classical is the particle motion. This important property is systematically used in radiation theory.

B.3. Invariant Characteristics of Particle Motion in an External Field

As it was shown above, a motion of high energy particles in external fields is quasiclassical, and therefore the particle states in the field may be characterized by 4-momentum p_μ.

After summation over particle spin states, invariant characteristics of processes in a constant field may depend only on invariant combinations of 4-momentum p_μ and the external field tensor (multiplied by the particle charge e, since we consider the electromagnetic interaction) $eF_{\mu\nu}$. We can construct three dimensionless invariants

$$\chi^2 = -\frac{(e\hbar)^2(F^{\mu\nu}p_\nu)^2}{m^6} = (mH_0)^{-2}[([\mathbf{pH}] + p_0\mathbf{E})^2 - (\mathbf{pE})^2]\,,$$
$$f^2 = \frac{1}{2}\left(\frac{e\hbar F^{\mu\nu}}{m^2}\right)^2 = \frac{\mathbf{H}^2 - \mathbf{E}^2}{H_0^2}\,, \qquad \text{(B.26)}$$
$$g^2 = -\frac{e^2\hbar^2}{8m^2}F^{\lambda\mu}\varepsilon_{\lambda\mu\alpha\beta}F^{\alpha\beta} = \frac{\mathbf{EH}}{H_0^2}\,,$$

where H_0 is defined by Eq. (B.20). For the crossed external field, we have $f = g = 0$, which is the definition of such field; for the magnetic field we have $f = H/H_0, g = 0$. If the momentum of the ultra-relativistic particle ($p_0 \gg m$) forms the angle $\vartheta_0 \gg m/p_0$ with respect to the fields' directions $\mathbf{H}$ and $\mathbf{E}$, then the parameter χ is the largest: $\chi^2 \gg f^2, \chi^2 \gg g^2$ If, in addition, the strengths of the fields $\mathbf{H}$ and $\mathbf{E}$ are much less than the critical one ($H_0 = E_0 = m^2c^3/e\hbar$), we have $|f| \ll 1$ and $|g| \ll 1$. Then some invariant characteristic of a process can be expanded as follows:

$$W(\chi, f, g) = W_0(\chi, 0, 0) + fW_1(\chi, 0, 0) + gW_2(\chi, 0, 0) + \cdots\,, \qquad \text{(B.27)}$$

where $W_0(\chi)$ describes the process in a crossed field. In this sense, a characteristic of the process calculated in a crossed field describes this process in any constant field. It is connected with the fact that any constant field in the rest frame of the ultra-relativistic particle turns out to be (for $\vartheta_0 \gg 1/\gamma$) very similar to a crossed field. If there is a dependence on some other parameters, e.g., if the spin phenomena are taken into account, the statement about the universality of a crossed field description becomes invalid.

In inhomogeneous (in four dimension) external fields, additional invariants can be constructed, containing terms of the form $\partial F^{\alpha\beta}/\partial x^\mu$ (the terms of the type $\partial_\alpha F^{\alpha\beta}$ and $\varepsilon_{\alpha\beta\gamma\delta}\partial/\partial x_\beta F^{\gamma\delta}$ vanish on account of the Maxwell equation). However, if the quantities $F^{\mu\nu}$ and p^μ slightly change during the process formation time (length), the process characteristics will, as before, depend on invariants (B.26), where the local values of $F^{\mu\nu}$ and p^μ must be substituted. Then the above considerations hold for inhomogeneous external fields as well.

Let us give several forms of writing for the parameter χ in a magnetic field

$$\chi = \frac{H}{H_0}\frac{p_\perp}{mc} = \frac{\hbar}{mc^2}\frac{|\dot{\mathbf{v}}_\perp|}{c}\gamma^2 = \frac{\hbar\omega_0}{\varepsilon}\gamma^3\frac{|\mathbf{v}_\perp|}{c} \tag{B.28}$$

In consideration of the processes induced by a photon with 4-momentum k^μ in an external field, we introduce the invariant parameter, which is similar to the parameter χ

$$\begin{aligned}\kappa^2 &= -\frac{e^2\,\hbar^4}{m^6}\,(F_{\mu\nu}k^\nu)^2 \\ &= \left(\frac{\hbar}{m\,H_0}\right)^2 \left[([\mathbf{kH}] + k_0\mathbf{E})^2 - (\mathbf{kE})^2\right],\end{aligned} \tag{B.29}$$

in particular, we have for the transverse electric field

$$\kappa = \frac{E}{E_0}\frac{\hbar\omega}{m}, \tag{B.30}$$

where $\hbar\omega = \hbar k_0$ is the photon energy.

Appendix C

MASS OPERATOR, EMISSION OF PHOTONS AND CORRECTIONS TO THE MASS OF A CHARGED PARTICLE IN A PLANE WAVE FIELD

In this Appendix we will consider, in particular, the process of a photon emission by a charged particle in the field of an electromagnetic plane wave. In this connection, a comparison of results obtained for radiation at quasi-periodic motion (e.g. in undulators and crystals) with those in the wave field, will be of main interest. The standard representation for the photon emission probability containing the explicit summation over numbers of quanta absorbed from the wave (see for example, Ritus (1979) Nikishov (1979)) is not suitable for the above mentioned purposes. We will start therefore, according to Baier *et al.* (1975), with the calculation of the mass operator for 0 and 1/2 spin particles in the wave field taken exactly into account. If we know the mass operator, we can obtain not only a total probability and a radiation spectrum, but also corrections to the particle mass in the wave field.

As it was noted in Sec. 5, the emission of photons at quasi-periodic motion occurs in the same way as in the corresponding electromagnetic wave field. The correspondence rule between the quantities $\mathbf{p}(t)$ and $\mathbf{A}(\varphi)$ formulated there can be easily understood: within a quasiclasscial description of the radiation process, it is determined by the particle trajectory (by the dependence $p(t)$) while the indicated form of the vector potential provides the same trajectory in the wave field. Let us prove this statement for the case of a plane electromagnetic wave of a general shape described by the vector-potential A_μ

$$A_\mu(\varphi) = a_{1\mu}\Psi_1(\varphi) + a_{2\mu}\Psi_2(\varphi)\,, \tag{C.1}$$

where $\varphi = \kappa x = \kappa^0 x^0 - \boldsymbol{\kappa}\mathbf{x}$, while

$$\kappa^2 = \kappa a_1 = \kappa a_2 = a_1 a_2 = 0\,. \tag{C.2}$$

In the case of linear polarization $a_{2\mu} = 0$. The wave field strength is

$$F^{\mu\nu} = (a_1^\nu \kappa^\mu - a_1^\mu \kappa^\nu)\Psi_1'(\varphi) + (a_2^\nu \kappa^\mu - a_2^\mu \kappa^\nu)\Psi_2'(\varphi)\,, \tag{C.3}$$

where $\Psi'_{1,2}(\varphi) \equiv d\Psi_{1,2}/d\varphi$. It sometimes turns out to be convenient to introduce a "special" frame of reference where $A_0 = 0$ and the wave vector $\boldsymbol{\kappa}$ is directed along the x_3-axis, i.e. $\kappa^0 = |\boldsymbol{\kappa}| = \kappa^3$, while the field strength vectors of the wave lie in the (x_1, x_2)-plane. The equation of a charged particle motion

$$\frac{dp^\mu}{ds} = \frac{e}{m}\, F^{\mu\nu} p_\nu\,, \tag{C.4}$$

where the quantity $F_{\mu\nu}$ from Eq. (C.3) must be substituted, are easily solved in the "special" frame

$$\begin{aligned} p^i &= \langle p^i \rangle - eA^i, \quad i = 1,2\,; \\ p^3 &= \frac{m^2 + p_1^2 + p_2^2 - c_0^2}{2c_0} \\ c_0 &= p^0 - p^3 = \text{const}\,. \end{aligned} \tag{C.5}$$

The solution (C.5) implies the correspondence rule between $\mathbf{p}(t)$ and $\mathbf{A}(\varphi)$ indicated in Sec. 5.

We proceed now to the calculation of the mass operator. For a spin-zero (scalar) particle it can be represented in the form

$$M^{(0)} = -\frac{ie^2}{(2\pi)^4}\int \frac{d^4k}{k^2 + io}\,(2\mathcal{P} - k)^\mu \left[(\mathcal{P} - k)^2 - m^2 + io\right]^{-1} (2\mathcal{P} - k)_\mu\,, \tag{C.6}$$

where $\mathcal{P}_\mu = i\partial_\mu - eA_\mu(\varphi)$. Equation (C.6) has the same form as the mass operator for a scalar particle in the absence of an external field, however, the presence in (C.6) of non-commuting operators is essentially a new circumstance.

Therefore, before taking the integral over k, we must carry out the proper transformation of the integrand. Omitting the details of this non-trivial calculation, which can be found in the original paper, we give at once an expression for the conventionally renormalized mass operator of the scalar particle for the linear polarization of the wave

$$M_R^{(0)} = \frac{\alpha m^2}{2\pi}\int_0^\infty \frac{ds}{s}\int_0^1 du\,\Bigg\{\left[1+\frac{\mathcal{P}^2}{m^2}+\xi^2\Delta^2(\eta)+\frac{(\kappa\mathcal{P})}{m^2}\frac{d\beta}{d\varphi}\right] \times e^{-ism^2u}e^{i\beta}e^{i\eta\mathcal{P}^2} - e^{-ism^2u^2} \times\left[2+\left(\frac{\mathcal{P}^2}{m^2}-1\right)(1+2isu(1-u))\right]\Bigg\}, \tag{C.7}$$

where $\eta = u(1-u)s$, and

$$\Delta(\eta) = \Psi(\varphi - 2\eta(\kappa\mathcal{P})) - \Psi(\varphi),$$
$$\beta = -\xi^2 m^2 su^2\left[\int_0^1 dy\,\Delta^2(\eta y) - \left(\int_0^1 dy\,\Delta(\eta y)\right)^2\right]. \tag{C.8}$$

Here Ψ is the function in the definition of $A_\mu(\varphi)$ in Eq. (C.1) and

$$\xi^2 = -\frac{e^2a^2}{m^2} \tag{C.9}$$

is the parameter characterizing the intensity of the plane wave. A pure classical meaning can be given to this parameter: the work done by the field over a wavelength $\lambdabar = \kappa_0^{-1}$ is ξm. If we introduce an "interaction volume" with transverse dimensions $\lambdabar_c = 1/m$ and longitudinal dimensions $\lambdabar$, then $\xi^2 = \alpha N_{\text{ph}}$, where $\alpha = 1/137$ and N_{ph} is the number of photons in the indicated interaction volume. Note, that, besides the considered diagram, for a scalar particle there is one more (the so called contact) diagram, where a photon is emitted and absorbed in the same point. The contribution of this diagram vanishes when performing the renormalization procedure.

The average value of $M_R^{(0)}$ on the mass shell $\mathcal{P}^2\Phi = m^2\Phi$ is of considerable interest for physical applications. As $M_R^{(0)}$ depends only on $\mathcal{P}^2$ and $(\kappa\mathcal{P})$, it follows directly from (C.7) that

$$\langle M_R^{(0)} \rangle \equiv \int d^4x \Phi^+(x) M_R^{(0)} \Phi(x) = \frac{\alpha m^2}{\pi} \cdot \int \frac{d^4x}{V} \int_0^\infty \frac{ds}{s}$$

$$\times \int_0^1 du e^{-ism^2u^2} \left\{ \left[1 + \frac{1}{2}\xi^2 \Delta^2(\eta) + \frac{\lambda}{4}\frac{d\beta}{d\varphi} \right] e^{i\beta} - 1 \right\}, \tag{C.10}$$

where Φ is the solution to the Klein–Gordon equation in the field of the wave, V is the normalization 4-volume (we have used only the properties $\mathcal{P}^2\Phi = m^2\Phi$ and $2(\kappa\mathcal{P})\Phi = m^2\lambda\Phi$).

Above we have found the mass operator in the field of a linearly polarized wave. In the general case of an elliptically polarized wave the calculation is similar to the case of linear polarization, and terms containing the amplitudes of the wave a_1 and a_2 occur additively. As a result it turns out that we must make the substitution $\beta \to \beta_1 + \beta_2$

$$\beta_k = -\xi_k^2 sm^2u^2 \left[\int_0^1 dy\, \Delta_k^2(\eta y) - \left(\int_0^1 dy\, \Delta_k(\eta y) \right)^2 \right], \tag{C.11}$$

where

$$\Delta_k(\eta y) = \Psi_k(\varphi - 2\eta y(\kappa\tau)) - \Psi_k(\varphi), \quad \xi_k^2 = -\frac{e^2a_k^2}{m^2} \tag{C.12}$$

and, in addition, evident substitutions

$$\xi^2\Delta^2(\eta) \to \xi_1^2\Delta_1^2(\eta) + \xi_2^2\Delta_2^2(\eta), \quad \frac{d\beta}{d\varphi} \to \frac{d}{d\varphi}(\beta_1 + \beta_2) \tag{C.13}$$

should be made. As a result an explicit form for the mass operator of a scalar particle in the field of an elliptically polarized wave follows directly from Eq. (C.7) after making the substitutions (C.11) – (C.13).

To complete the integration over the variable x, we bear in mind that the integrand depends only on the variable φ (in the special system $\varphi = \kappa^0(x^0 - x^3)$). For a monochromatic plane wave

$$\Psi_1(\varphi) = \cos\varphi, \quad \Psi_2(\varphi) = \sin\varphi, \tag{C.14}$$

we have a periodic function of φ, so that we can restrict the integration over a single period. Using this, we have

$$\int \frac{d^4x}{V}[\cdots] = \frac{1}{2\kappa^0 L} \int_{-\kappa_0 L}^{\kappa^0 L} d\varphi[\cdots] = \frac{1}{2\pi} \int_{-\pi}^{\pi} d\varphi[\cdots]. \tag{C.15}$$

An important advantage of this approach is the possibility of obtaining the results for the general case of a plane wave of arbitrary configuaration (see Eqs. (C.7) and (C.10)) and only in the last stage of the calculation do we need to use the actual form of the wave (see (C.14)). After simple transformations we easily check that terms depending on φ are present only in the combination $(\xi_1^2 - \xi_2^2)$ cos $[2(\varphi - \eta(\kappa P)/m^2)]$. After separating this combination, the integration over φ is performed easily. For a monochromatic plane wave with elliptical polarization we have for the averaged mass operator of a scalar particle

$$\langle M_R^{(0)}\rangle_{el} = \frac{\alpha m^2}{\pi}\int_0^\infty \frac{dt}{t}\int_0^\infty \frac{dv}{(1+v)^2}\{e^{-iZ_1}[J_0(Z_0)(1+(\xi_1^2+\xi_2^2)\sin^2 t) - iJ_1(Z_0)(\xi_1^2-\xi_2^2)\sin^2 t] - \exp(-2ivt/\lambda)\}, \tag{C.16}$$

where $J_0(Z_0)$ and $J_1(Z_0)$ are Bessel functions,

$$Z_0 = \frac{vt}{\lambda}(\xi_1^2-\xi_2^2)\left[\left(\frac{\sin t}{t}\right)^2 - \frac{\sin 2t}{2t}\right],$$

$$Z_1 = \frac{2vt}{\lambda}\left[1+\frac{\xi_1^2+\xi_2^2}{2}\left(1-\frac{\sin^2 t}{t^2}\right)\right], \tag{C.17}$$

$$\lambda = \frac{2(\kappa P)}{m^2},$$

in Eq. (C.16) we changed to the new variables $v = u/(1-u)$ and $t = \lambda svm^2/2(l+v)^2$ The averaged mass operator (C.16) depends on the invariant characteristics of the wave intensity ξ_1^2 and ξ_2^2, and on the invariant integral of motion of the particle in the wave field, $\lambda = 2(\kappa P)/m^2$. If we put $\xi_2^2 = 0$ in (C.16), we get the case of linear polarization. In the particular case of a circularly polarized wave, $\xi_1^2 = \xi_2^2 \equiv \xi^2$, the expression (C.16) is simplified considerably

$$\langle M_R^{(0)}\rangle_{cz} = \frac{\alpha}{\pi}m^2\int_0^\infty \frac{dt}{t}\int_0^\infty \frac{dv}{(1+v)^2}\left\{(1+2\xi^2\sin^2 t)e^{-iZ_1^{cr}} - \exp\left(-\frac{2ivt}{\lambda}\right)\right\}, \tag{C.18}$$

where

$$Z_1^{cr} = \frac{2vt}{\lambda}\left[1+\xi^2\left(1-\frac{\sin^2 t}{t^2}\right)\right].$$

The imaginary part of $\langle M_R^{(0)} \rangle$ is connected with the total probability $W^{(0)}$ for the photon emission by a scalar charge particle in the wave field

$$W^{(0)} = -\frac{1}{\varepsilon} \, \mathrm{Im} \, \langle M_R^{(0)} \rangle \,, \tag{C.19}$$

where ε is the zeroth component of the average kinetic momentum (quasi-momentum) or the particle in the wave field. Substituting (C.16) or (C.18) into Eq. (C.19), we can obtain an explicit form for the total emission probability. For circularly polarized wave it is represented by a double integral containing only elementary functions

$$W_{cr}^{(0)} = \frac{\alpha}{\pi} \cdot \frac{m^2}{\varepsilon} \int_0^\infty \frac{dt}{t} \int_0^\infty \frac{dv}{(1+v)^2} \, \{(1 + 2\xi^2 \sin^2 t) \sin Z_1^{cr}$$
$$- \sin(2vt/\lambda)\} \,. \tag{C.20}$$

Expressions (C.19), (C.16) and (C.20) give the representation for the total emission probability which is convenient, in particular, for an analysis of the situation when $\xi^2 \sim 1$. If we take into account that the variable v has the meaning of a ratio $\omega/(\omega - \varepsilon)$ (ω is the frequency of the emitted photon), and throughout the calculation the integration by parts with respect to this variable has not been carried out, then Eqs. (C.19) and (C.16) give the spectrum of emitted photons, if we omit (e.g., in Eq. (C.20)) the integration over v. We can directly check now that the results obtained for the radiation spectrum in the wave coincide with those at quasi-periodic motions. So, in the case of an elliptic polarization, the obtained result should be compared with Eq. (5.33), where, as usual for scalar particles, the term $\propto u^2$ should be omitted $(1+u^2/2(1+u) \to 1)$ as well as the spin term $\propto (\boldsymbol{\zeta}\, \mathbf{v})$.

As far as the strength of interaction is characterized by the parameter ξ^2, the perturbation theory is applicable at $\xi_{1,2}^2 \ll 1$ and we can expand in powers of these small parameters. The first term of such expansion of the probability $W^{(0)}(\propto \xi_{1,2}^2)$ after dividing by a flux $J = \lambda m^4(\xi_1^2 + \xi_2^2)/(16\pi\alpha\varepsilon)$ gives the total cross section for the Compton scattering on a scalar particle

$$\sigma^{(0)} = \frac{4\pi\alpha^2}{m^2\lambda} \left\{ \frac{(2+\lambda)^2}{\lambda(1+\lambda)} - \left(\frac{2}{\lambda} + \frac{4}{\lambda^2} \right) \ln(1+\lambda) \right\} \,, \tag{C.21}$$

where $\lambda = 2(\kappa p)/m^2$. This result is obviously valid for any polarization of the wave.

The opposite case $\xi^2 \gg 1$ corresponds, as it is well known, to the constant field limit. After elementary calculation, we find from Eq. (C.10)

$$\langle M_R^{(0)} \rangle_c = \frac{\alpha}{\pi} m^2 \int_0^\infty \frac{dt}{t} \int_0^1 du \left\{ (1 + 2t^2(1-u)^2) \right.$$

$$\left. \times \exp\left[-\frac{iu}{\chi}\left(t + \frac{t^3\,(1-u)^2}{3} \right) \right] - \exp\left(-\frac{iut}{\chi} \right) \right\} \tag{C.22}$$

$$= \frac{\alpha m^2}{3\pi} \int_0^\infty \frac{dv}{(1+v)^2} (5 + 2v) \left[L_{\frac{2}{3}}\left(\frac{2v}{3\chi} \right) - \frac{i}{\sqrt{3}}\, K_{\frac{2}{3}}\left(\frac{2v}{3\chi} \right) \right],$$

where $K_{2/3}$ is the Macdonald function and $L_{2/3}$ is defined in Eq. (15.18),

$$\chi^2 = \frac{e^2 \langle \mathcal{P}\, F^2 \mathcal{P} \rangle}{m^6} = \frac{\xi^2 (\kappa\, p)^2}{m^4} \tag{C.23}$$

Equation (C.22) coincides with that obtained earlier directly within the constant field approximation (see for example, Baier, Katkov and Fadin (1973), p. 188).

The mass operator for a spin-1/2 particle in an external field can be represented in the form

$$M^{(\frac{1}{2})} = -\frac{ie^2}{(2\pi)^4} \int \frac{d^4 k}{k^2 + i0}\, \gamma^\mu\, \frac{\hat{\mathcal{P}} - \hat{k} + m}{(\hat{\mathcal{P}} - \hat{k})^2 - m^2 + i0} \gamma_\mu\,, \tag{C.24}$$

where $\hat{\mathcal{P}} = \mathcal{P}_\mu \gamma^\mu$, γ^μ are the Dirac matrices. In the spinor particle case the problem is more complicated because of the appearance of additional non-commuting quantities, γ-matrices, besides non-commuting components of the momentum $\mathcal{P}_\mu$. The detailed calculation was performed in the paper by Baer *et al.* (1975). Here we give at once an expression for the mass operator in the field of a linearly polarized wave

$$M_l^{(\frac{1}{2})} = \frac{\alpha}{2\pi} \int_0^\infty \frac{ds}{s} \int_0^1 du\, e^{-ism^2 u} \{ 2(m - \hat{\mathcal{P}})(1 - \hat{a}\hat{\kappa} H)$$

$$+ \hat{\mathcal{P}}(1+u) + (\gamma f \mathcal{P}) E - iuH\gamma^5 (\gamma f^* \mathcal{P}) + \xi^2 \hat{\kappa} G \} e^{i\beta} e^{i\eta \hat{\mathcal{P}}^2}\,, \tag{C.25}$$

where the notation is introduced

$$H = \frac{e\Delta(\eta)}{2(\kappa\mathcal{P})}, \quad E = \frac{e}{(\kappa\mathcal{P})}\left[u\int_0^1 dy\,\Delta(\eta y) - \frac{1}{2}\left(1+\frac{u}{2}\right)\Delta(\eta)\right],$$

$$G = \frac{m^2}{2(\kappa\mathcal{P})}\left[\Delta^2(\eta) + \frac{u}{1-u}\left[u\Delta(\eta)\int_0^1 dy\,\Delta(\eta y) \right.\right.$$

$$\left.\left. + \int_0^1 dy\,\Delta^2(\eta y) - 2u\left(\int_0^1 dy\,\Delta(\eta y)\right)^2\right]\right], \tag{C.26}$$

$$f_{1,2}^{\mu\nu} = \kappa^\mu a_{1,2}^\nu - \kappa^\nu a_{1,2}^\mu, \quad f_{\alpha\beta}^* = \frac{1}{2}\varepsilon_{\alpha\beta\gamma\delta}f^{\gamma\delta}.$$

A transition to the case of the elliptically polarized wave is determined on the whole by the substitutions (C.11) – (C.13)

$$M_{el}^{(\frac{1}{2})} = \frac{\alpha}{2\pi}\int_0^\infty \frac{ds}{s}\int_0^1 du\, e^{ism^2u}\{2(m-\hat{\mathcal{P}})(1-\hat{a}_1\hat{\kappa}H_1 - \hat{a}_2\hat{\kappa}H_2)$$

$$+ \hat{\mathcal{P}}(1+u) + (\gamma f_1\mathcal{P})E_1 + (\gamma f_2\mathcal{P})E_2 - i\gamma^5(\gamma f_1^*\mathcal{P})uH_1$$

$$- i\gamma^5(\gamma f_2^*\mathcal{P})uH_2 + \hat{\kappa}(\xi_1^2G_1 + \xi_2^2G_2) - i\gamma^5\,\Lambda\cdot N\}e^{i(\beta_1+\beta_2)}e^{i\eta\hat{\mathcal{P}}^2}, \tag{C.27}$$

where the quantities $E_{1,2}$, $G_{1,2}$ and $H_{1,2}$ are given by Eq. (C.26) with the substitution of Δ_1 and Δ_2, respectively,

$$\Lambda = \varepsilon_{\alpha\beta\gamma\delta}\gamma^\alpha a_1^\beta a_2^\gamma\,\kappa^\delta$$

$$N = \frac{e^2u}{2(\kappa\mathcal{P})}\left(1+\frac{1}{1-u}\right)\left[\Delta_2(\eta)\int_0^1 \Delta_1(\eta y) - \Delta_1(\eta)\int_0^1 dy\,\Delta_2(\eta y)\right] \tag{C.28}$$

We draw attention to the appearance of a term which contains $\gamma^5\Lambda$ and which does not occur in (C.25).

The expressions (C.25) and (C.27) found for the mass operator must be renormalized in the standard way

$$M_R^{(\frac{1}{2})} = M^{(\frac{1}{2})} - M^{(\frac{1}{2})}(\hat{\mathcal{P}} = m,\, A_\mu = 0)$$

$$- (\hat{\mathcal{P}} - m)\frac{\partial M^{(\frac{1}{2})}}{\partial\hat{\mathcal{P}}}(\hat{\mathcal{P}} = m,\, A_\mu = 0). \tag{C.29}$$

As a result we obtain the renormalized mass operator of an electron in the field of a plane wave:

$$M_R^{(\frac{1}{2})} = M^{(\frac{1}{2})} - \frac{\alpha}{2\pi}\int_0^\infty \frac{ds}{s}\int_0^1 du\, e^{-ism^2u^2}\{m(1+u) \\ - (\hat{\mathcal{P}} - m)(1-u)[1 - 2ism^2 u(1+u)]\}, \tag{C.30}$$

where we must use (C.27) (or (C.25) in the case of a linearly polarized wave) for $M^{(1/2)}$.

The average of the mass operator on the mass shell is of major interest for physical applications. Let us introduce the operator

$$\pi^\mu = \pi_0^\mu + i\kappa^\mu \frac{e\hat{\kappa}\hat{\mathrm{A}}'}{2(\kappa i\partial)}, \\ \pi_0^\mu = i\partial^\mu - \kappa^\mu\left(e\frac{(\mathrm{A}i\partial)}{(\kappa i\partial)} - \frac{e^2\mathrm{A}^2}{2(\kappa i\partial)}\right), \tag{C.31}$$

which satisfies the following relations

$$[\pi^\mu, \pi^\nu] = [\hat{\mathcal{P}}^2, \pi^\mu] = 0, \quad \pi^2 = \hat{\mathcal{P}}^2 \tag{C.32}$$

In accordance with (C.32) we choose states which satisfy the squared Dirac equation $(\hat{\mathcal{P}}^2 - m^2)\,\Phi = 0$, and also the equation

$$\pi^\mu\Phi = p^\mu\Phi, \tag{C.33}$$

where p^μ is a constant 4-vector ($p^2 = m^2$), i.e. there are three independent vector components ($p^1, p^2, p^3 - p^0$ in the special system). We can write the solution of the Dirac equation in the form

$$\Psi = (\hat{\mathcal{P}} + m)\Phi. \tag{C.34}$$

Let us perform a unitary transformation of Ψ

$$\Psi_u = U\Psi, \quad U = \exp\left\{-e\frac{\hat{\kappa}\hat{\mathrm{A}}}{2(\kappa\mathcal{P})}\right\} \tag{C.35}$$

changing the Dirac equation $(\hat{\mathcal{P}} - m)\,\Psi = 0$ into the equation

$$(\hat{\pi}_0 - m)\Psi_u = 0, \quad \pi_0^\mu\Psi_u = p^\mu\Psi_u. \tag{C.36}$$

With regard to Eqs. (C.32) and (C.36) spin states can be described now as in the case of free particles. We introduce then the spin 4-vector $S^\mu(S^2 = -1, pS = 0$, in operator form $p^\mu \to \pi_0^\mu)$ and use the operator $(-\gamma^5 \hat{S})$ to classify the spin states. Performing on this operator the transformation which is the inverse of (C.35), we get for the Dirac equation spin operator

$$R = -\gamma^5 \left\{ \hat{S} + \frac{e}{(\kappa\mathcal{P})} \left[(\mathrm{A}S)\hat{\kappa} - (\kappa S)\hat{\mathrm{A}} \right] - \frac{e^2}{2(\kappa\mathcal{P})^2} (\kappa S)\hat{\kappa}\hat{\mathrm{A}} \right\} \tag{C.37}$$

with eigenstates

$$R\,\Psi_p^{\pm S} = \pm\,\Psi_p^{\pm S}\,, \tag{C.38}$$

where p^μ is the constant 4-vector (C.33).

We note that all relations become completely obvious if we use the well-known form of the wavefunction of an electron in the field of a wave (see for example, Berestetskii, Lifshitz and Pitaevskii (1982)). One checks easily that the following relations hold for the states (C.38)

$$\begin{gathered} \bar{\Psi}_p^S \hat{\kappa} \Psi_p^S = \frac{(\kappa\,p)}{m^2}\,, \quad \bar{\Psi}_p^S (\gamma f_{1,2}\mathcal{P})\,\Psi_p^S = 0 \\ \bar{\Psi}_p^S \gamma^5 (\gamma f_{1,2}^* \mathcal{P})\,\Psi_p^S = (S f_{1,2}^* \mathcal{P})\,, \quad \bar{\Psi}_p^S \gamma^5 \Lambda \Psi_p^S = (S a_1 a_2 \kappa)\,, \end{gathered} \tag{C.39}$$

where $(a_1 a_2 a_3 a_4) = \varepsilon^{\alpha\beta\gamma\delta} a_{1\alpha} a_{2\beta} a_{3\gamma} a_{4\delta}$; in deriving the last equation one can start from $(\gamma f P) = [\hat{P}, \sigma f]/2$. Substituting these relations into (C.30), we have for the average (on the mass shell) of the renormalized mass operator in the field of an elliptically polarized plane wave

$$\begin{aligned} \langle M_R^{(\frac{1}{2})} \rangle_{el} &= \int d^4x \bar{\Psi}_p^S (M_R^{(\frac{1}{2})})_{el} \Psi_p^S \\ &= \frac{\alpha}{2\pi} \int \frac{d^4x}{V} \int_0^\infty \frac{ds}{s} \int_0^1 du\, e^{-ism^2u^2} \Big\{ [m(1+u) \\ &\quad + \frac{(\kappa p)}{m^2} (\xi_1^2 G_1 + \xi_2^2 G_2) - iu(H_1(S f_1^* p) + H_2(S f_2^* P)) \\ &\quad - i(S a_1 a_2 \kappa) N]\, e^{i(\beta_1+\beta_2)} - m(1+u) \Big\}\,, \end{aligned} \tag{C.40}$$

where we have used the same notation as in (C.27). The last two terms in the square bracket depend on the electron spin; they vanish for unpolarized electrons.

As in the scalar particle case, we need to use the actual form of the wave only at the last stage of the calculation when integrating over φ. For the monochromatic plane wave (C.14), we find the final form for the average of the renormalized mass operator of an electron

$$\langle M_R^{(\frac{1}{2})}\rangle_{el} = \frac{\alpha m}{2\pi}\int_0^\infty \frac{dt}{t}\int_0^\infty \frac{dv}{(1+v)^2}\left\{e^{-iZ_1}\left[\left(1+\frac{v}{1+v}\right.\right.\right.$$
$$\left. + g_1\frac{\xi_1^2+\xi_2^2}{2} + i\frac{e^2}{m^2}\frac{(Sa_1a_2\,\kappa)}{m}g_3\right)J_0(Z_0)$$
$$\left.\left. + ig_2\frac{(\xi_1^2-\xi_2^2)}{2}J_1(Z_0)\right] - \left(1+\frac{v}{1+v}\right)e^{-2ivt/\lambda}\right\} \tag{C.41}$$

Here

$$g_1 = 2\sin^2 t + v\left(1-\frac{\sin 2t}{2t}+\frac{v}{1+v}g\right),$$
$$g_2 = -2\sin^2 t + v\left[\frac{1}{2}\left(\cos 2t - \frac{\sin 2t}{2t}\right)+\frac{v}{1+v}g\right],$$
$$g = -\cos^2 t + \frac{\sin 2t}{t} - \frac{\sin^2 t}{t^2}, \tag{C.42}$$
$$g_3 = \frac{2v(2+v)}{\lambda(1+v)}\quad \sin t\left(\frac{\sin t}{t}-\cos t\right),\quad \lambda = \frac{2(\kappa p)}{m^2},$$

the variables and the remaining notation is the same as in (C.16). The average of the mass operator of an electron in the plane wave field depends not only on the characteristics ξ_1^2 and ξ_2^2 of the wave intensity and the invariant integral of motion λ, but also on the spin correlation term

$$\frac{e^2}{m^2}(Sa_1a_2\kappa) = -\frac{1}{2}(\xi_1^2+\xi_2^2)\cdot\tilde{\xi}_2(\kappa S), \tag{C.43}$$

where $\tilde{\xi}_2$ is the Stokes parameter characterizing the degree of circular polarization of the plane wave. Putting in Eq. (C.41) $\xi_2 = 0$, we have the case of linear polarization and when $\xi_1^2 = \xi_2^2 \equiv \xi^2$ the case of circular polarization. In the latter case, this expression simplifies considerably and depends only on elementary functions

$$\langle M_R^{(\frac{1}{2})} \rangle_{cr} = \frac{\alpha}{2\pi} m \int_0^\infty \frac{dt}{t} \int_0^\infty \frac{dv}{(1+v)^2} \left\{ \left(e^{-iZ_1^{cr}} - e^{-2ivt/\lambda}\right) \right.$$

$$\left. \times \left[1 - \frac{\lambda r}{4} \left(\frac{3}{(1+v)^2} - 1 \right) \right] + \xi^2 \sin^2 t \left(2 + \frac{v^2}{1+v} \right) e^{-iZ_1^{cr}} \right\},$$

$$r = -\tilde{\xi}_2 \frac{m(\kappa S)}{(\kappa p)}, \qquad \text{(C.44)}$$

where Z_1^{cr} is defined in (C.18) and the integral has been transformed, using the integration by parts.

Generally speaking, to obtain the radiation shift of energy levels, we should solve the modified Dirac–Schwinger equation $(\hat{\mathcal{P}} - m - M)\Psi = 0$, where M is the renormalized mass operator (C.30). In other words, the mass operator having in the chosen basis non-diagonal matrix elements owing to the presence of spin operators $\gamma^5(\gamma f_k^* \mathcal{P})$ and $\gamma^5 \hat{\kappa}$ should be diagonalized. This procedure was carried out by Baier and Milstein (1976). As a result it turns out that the values of radiation energy shift are given (up to the factor m/ε) by Eqs. (C.41)–(C.44), where the operator (κS) takes two values: $(\kappa S) = \pm(\kappa p)/m$. This result should be taken into account in the following.

The total probability of the photon emission in the field of a wave for a spin $-1/2$ particle is

$$W^{(\frac{1}{2})} = -\frac{2m}{\varepsilon} \operatorname{Im} \langle M_R^{(\frac{1}{2})} \rangle, \qquad \text{(C.45)}$$

where ε is the zeroth component of the average kinetic momentum (quasi-momentum) of an electron in the field of the wave. An explicit form for the probability $W^{(1/2)}$ is obtained by substitution of Eqs. (C.41) or (C.44) into (C.45). For a circularly polarized wave, the total probability is represented by a double integral containing elementary functions.

Opposite to the case of scalar particles, integration by parts over v was carried out in deriving Eqs. (C.41) and (C.44), so that omitting this integration does not give the true radiation spectrum. However, we can find this spectrum if we put the magnitude of an external field in Eq. (5.68) equal to zero. In particular, for circularly polarized wave we obtain the following radiation spectrum

$$dW_\gamma/dv = \frac{i\alpha m^2}{2\pi\varepsilon(1+v)^2}\int_{-\infty}^{+\infty}\frac{d\tau}{\tau - i0}\,e^{-iZ_1^{cr}}\left\{1+\xi^2\left[\left(2+\frac{v^2}{1+v}\right)\sin^2\tau\right.\right.$$

$$\left.\left.+\frac{irv(2+v)}{(1+v)}\left(\frac{\sin^2\tau}{\tau}-\frac{\sin 2\tau}{2}\right)\right]\right\}, \tag{C.46}$$

where $v = \omega/(\varepsilon-\omega)$ and the quantities r and Z_1^{cr} are defined in Eqs. (C.44) and (C.18).

We now proceed to consider the asymptotic expansions of the mass operator. In the case of low intensities, $\xi_{1,2}^2 \ll 1$, when perturbation theory is applicable, we can expand the mass operator (for the sake of simplicity we restrict ourselves to the case of a circularly polarized wave) in powers of ξ^2

$$\langle M_R^{(\frac{1}{2})}(\xi^2 \ll 1)\rangle_{cr} = \frac{\alpha m\xi^2}{2\pi}\left\{\left(\frac{1}{\lambda}+\frac{r}{4}\right)[S_2(1+\lambda)-S_2(1-\lambda)]\right.$$

$$+\left(\frac{1}{4}-\frac{r}{2\lambda}-\frac{2}{\lambda^2}\right)\left[\frac{\pi^2}{3}-S_2(1+\lambda)-S_2(1-\lambda)\right]$$

$$+\frac{3\lambda^2-\lambda r-4}{4(1-\lambda^2)}+\frac{\lambda^4-2\lambda^3 r-3\lambda^2}{4(1-\lambda^2)^2}\cdot\ln\lambda$$

$$\left.-i\pi\left[\frac{1}{8}+\frac{r}{2}+\frac{2}{\lambda}+\frac{\lambda^2 r-1}{8(1+\lambda)^2}\right]\right\}; \tag{C.47}$$

here

$$S_2(x) = -\int_0^1\frac{dt}{t}\ln(1-tx) \tag{C.48}$$

is the Spence function. As it was already noted, the imaginary part of $\langle M_R^{(1/2)}(\xi^2 \ll 1)\rangle_{cr}$ is connected with the total cross section of the Compton scattering of a photon with circular polarization by an electron

$$\sigma_{com}^{(\frac{1}{2})} = -\frac{16\pi\alpha}{m^2\lambda\xi^2}\,\mathrm{Im}\,\frac{\langle M_R^{(\frac{1}{2})}(\xi^2 \ll 1)\rangle_{cr}}{m}. \tag{C.49}$$

Bearing in mind that, for $\lambda > 0$, $\mathrm{Im}\, S_2(1+\lambda) = \pi(1+\lambda)$, we get from (C.47)

$$\sigma_{com}^{(\frac{1}{2})} = \frac{2\pi\alpha^2}{m^2\lambda}\left\{\left[1-r-\frac{2(r+2)}{\lambda}-\frac{8}{\lambda^2}\right]\ln(1+\lambda)\right.$$

$$\left.+\frac{1}{2}(1+4r)+\frac{8}{\lambda}+\frac{\lambda^2 r-1}{2(1+\lambda)^2}\right\}. \tag{C.50}$$

The same cross section can be obtained from the expressions given, e.g., in Berestetskii, Lifshitz and Pitaevskii (1982) by elementary integration. It is interesting that for $\lambda \gg 1$ the logarithmic term in (C.50) occurs with a factor $(1-r)$ and vanishes at $r=1$. This fact can easily be understood. To do this we note that the value $|r|=1$ is possible only if the vectors $\mathbf{n}=\boldsymbol{\kappa}/\kappa_0$ and $\boldsymbol{\zeta}$ (the direction of the spin quantization axis) are collinear (e.g. in the rest frame of the electron $r=\tilde{\xi}_2(\boldsymbol{\zeta}\,\mathbf{n})$). A contribution which is logarithmic at $\lambda \gg 1$ comes from the diagram where first the final photon is emitted and afterwards the initial one is absorbed, and is determined by the kinematic situation in which the final photon is scattered backward (this is most simply verified by analyzing the situation in the c.m.s.). If the vectors $\boldsymbol{\zeta}$ and $\mathbf{n}$ are collinear, the process is only possible for a single value of the chirality of the initial photon.

The real part of Eq. (C.47) gives the correction to the mass in the lowest order in ξ^2; its asymptotic expansions have the form

$$\mathrm{Re}\langle M_R^{(\frac{1}{2})}(\xi^2 \ll 1)\rangle_{cr} = \begin{cases} \frac{\alpha m\xi^2}{6\pi}\lambda^2\left(\ln\frac{1}{\lambda^2}-\frac{11}{24}\right), & \lambda \ll 1 \\ \frac{\alpha m\xi^2}{8\pi}\left[\ln^2\lambda+\ln\lambda+\frac{\pi^2}{2}\left(r+\frac{1}{3}\right)-3\right], & \lambda \gg 1\,. \end{cases} \tag{C.51}$$

It can be seen from (C.51) that for small λ this quantity increases with increasing λ from zero, while for large λ it is proportional to $\ln^2\lambda$ and depends weakly on r for any λ.

The behavior of the average mass operator for $\lambda \gg 1$ and arbitrary fixed value of the parameter ξ^2 is also of interest. The corresponding asymptotic expression has the form

$$\begin{aligned}\langle M_R^{(\frac{1}{2})}\rangle_{cr} \simeq \frac{\alpha m\xi^2}{8\pi}&\left\{\ln^2\tilde{\lambda}+\ln\tilde{\lambda}+\frac{\pi^2}{2}\left(r+\frac{1}{3}\right)+1-\frac{4}{\xi^2}\ln\left(1+\xi^2\right)\right.\\ &-i\pi\left[(1-r)\ln\tilde{\lambda}+\frac{1}{2}(1+5r)\right]-4\int_0^\infty dt\left[\frac{\sin^2 t}{t}\right.\\ &\left.\left.+ir\left(\frac{\sin^2 t}{t^2}-\frac{\sin 2t}{2t}\right)\right]\ln\left(1-\frac{\sin^2 t}{t^2}\frac{\xi^2}{1+\xi^2}\right)\right\},\end{aligned} \tag{C.52}$$

where $\tilde{\lambda}=\lambda/(1+\xi^2)=2(\kappa p)/m^2(1+\xi^2)$. We see that taking the wave intensity exactly into account in the logarithmic terms (Eq. (C.52)), as compared to perturbation theory (Eqs. (C.47), (C.51)), is provided for $\lambda \gg 1$ by the substitution $\lambda \to \tilde{\lambda}$ or $m^2 \to m_*^2 = m^2(1+\xi^2)$, where m_* is the effective mass of the electron in the field of a wave (the square of the quasi-momentum equals m_*^2). There is no such simple connection for the non-logarithmic terms given by a complicated function of ξ^2 in (C.52). We draw an analogy of this property of the asymptotic expression (C.52) with the approximation used in Sec. 18 at deriving the modified coherent theory, which essentially takes into account the distinction between m^2 and m_*^2.

Appendix D

PHOTON SCATTERING AND E^+E^- PAIR PRODUCTION IN THE FIELD OF A PLANE ELECTROMAGNETIC WAVE

There is a direct analogy of the pair production process in the field of a plane electromagnetic wave with that in other electromagnetic structures, e.g., in single crystals. As far as such correspondence is established for the photon emission, it follows for the photo-production owing to the close connection of the indicated processes. From the point of view of a quasiclassical description, a set of classical trajectories of final particles appears in the problem, which is the same (see Appendix C) in the corresponding wave. At last, in the reference frame moving sufficiently fast along the initial photon momentum any periodic external field may be presented as a flux of incident equivalent photons. We shall use the indicated analogy to analyze some aspects of the photo-production process in crystals in a more simple example of the monochromatic plane wave.

Let us find first the amplitude of a photon scattering in the field (C.1) of a plane wave. This amplitude allows one not only to obtain the probability of pair production in the form enabling the direct comparison with the results of a general quasiclassical consideration, but also to find the photon mass appearing in an external field. In deriving the photon scattering amplitude we refer to a paper by Baier, Milstein and Strakhovenko, (1975), omitting some details of the calculation.

The amplitude for the scattering of photon of momentum k_1 by the field of a plane wave (k_1+ wave $\rightarrow k_2$), taking into account the polarization of the vacuum by spinor particles, has the form

$$T = e_{1\mu}(k_1)\, e^*_{2\nu}(k_2)\, T^{\mu\nu}(k_1, k_2)$$

$$T^{\mu\nu}(k_1, k_2) = -\frac{e^2}{(2\pi)^4}\int d^4x\, Sp\,\langle x|\frac{1}{\hat{\mathcal{P}} - m}\gamma^{\mu}\, e^{-ik_1X} \times \frac{1}{\hat{\mathcal{P}} - m}\gamma^{\nu}\, e^{ik_2X}|x\rangle\,, \tag{D.1}$$

where $|x\rangle$ is the eigenvector of the operator for the x coordinate. Since a change in k_1 is determined by the interaction with the wave, then we always have $k_2 = k_1 + c\kappa$, where c is a scalar. Taking into account this and the relation (C.2), we have

$$\kappa k_1 = \kappa k_2 = \kappa k\,, \quad a_1 k_1 = a_1 k_2 = a_1 k\,, \quad a_2 k_1 = a_2 k_2 = a_2 k\,, \tag{D.2}$$

i.e. when the scalar products for the vectors k_1 and k_2 are the same we shall use the notation k.

Let us construct the vectors

$$\Lambda_1^{\mu} = \frac{(kf_1)^{\mu}}{(\kappa k)\sqrt{-a_1^2}}\,, \quad \Lambda_2^{\mu} = \frac{(kf_2)^{\mu}}{(\kappa k)\sqrt{-a_2^2}}\,, \quad \Lambda_3^{\mu} = \frac{\kappa^{\mu}k_1^2 - (\kappa k)k_1^{\mu}}{(\kappa k)\sqrt{k_1^2}}\,, \quad \Lambda_4^{\mu} = \frac{\kappa^{\mu}k_2^2 - (\kappa k)k_2^{\mu}}{(\kappa k)\sqrt{k_2^2}}\,, \tag{D.3}$$

where

$$f_{1,2}^{\mu\nu} = \kappa^{\mu}a_{1,2}^{\nu} - \kappa^{\nu}a_{1,2}^{\mu}\,, \quad (kf_{1,2})^{\mu} = k_{\nu}f_{1,2}^{\nu\mu}\,, \quad \Lambda_1^2 = \Lambda_2^2 = \Lambda_3^2 = \Lambda_4^2 = -1\,. \tag{D.4}$$

The sets $k_1^{\mu}(k_1^2)^{-1/2}$, $\Lambda_1^{\mu}, \Lambda_2^{\mu}, \Lambda_3^{\mu}$ and $k_2^{\mu}(k_2^2)^{-1/2}$, Λ_1^{μ}, Λ_2^{μ}, Λ_4^{μ} represent orthonormalized sets (completeness relations hold, e.g. $g^{\mu\nu} = k_1^{\mu}k_1^{\nu}/k_1^2 - \Lambda_1^{\mu}\Lambda_1^{\nu} - \Lambda_2^{\mu}\Lambda_2^{\nu} - \Lambda_3^{\mu}\Lambda_3^{\nu}$) in terms of which any vector of the problem can be expanded.

The amplitude $T^{\mu\nu}(k_1, k_2)$ (D.1) is gauge-invariant (strictly speaking, after renormalization has been carried out). Then by virtue of the foregoing, it can expanded in terms of the vectors (D.3)

$$T^{\mu\nu}(k_1, k_2) = c_1 \Lambda_1^\mu \Lambda_2^\nu + c_2 \Lambda_2^\mu \Lambda_1^\nu + c_3 \Lambda_1^\mu \Lambda_1^\nu + c_4 \Lambda_2^\mu \Lambda_2^\nu + c_5 \Lambda_3^\mu \Lambda_4^\nu \tag{D.5}$$

The coefficients in front of the other possible combinations constructed from the vectors Λ_n^μ vanish owing to the Furry theorem. The further problem consists of evaluating the coefficients $c_1 - c_5$. We make use of the fact that in the expression $\int d^4x\, Sp\, \langle x| \cdots |x\rangle$ one can cyclically permute the operators within the brackets and of the completeness theorem

$$\langle x|R(p)|x\rangle = \int d^4p\, R(p)\,. \tag{D.6}$$

After finding an explicit form of the coefficients $c_1 - c_5$ in Eq. (D.5), the obtained expression for the tensor $T^{\mu\nu}(k_1, k_2)$ must be renormalized. In order to do this we represent it in the form

$$T^{\mu\nu}(k_1, k_2) = [T^{\mu\nu}(k_1, k_2) - T^{\mu\nu}_{F=0}(k_1, k_2)] + T^{\mu\nu}_{F=0}(k_1, k_2)\,. \tag{D.7}$$

The first term vanishes when the field of the wave $F = 0$, while the second term (independent of the field and therefore with $k_1 = k_2$) must be renormalized in the standard way. After subtracting the term $T^{\mu\nu}_{F=0}$, we obtain the following expression for the coefficients $c_1 - c_5$ in (D.5)

$$c_n = -\frac{i\alpha}{2\pi} \int_{-1}^{1} dv \int_0^\infty \frac{d\tau}{\tau} \int d^4x b_n\, e^{i(k_2-k_1\,,x)} \times \exp\left\{-im^2\tau \left[1 - \frac{k_1^2\,(1-v^2)}{4m^2}\right]\right\}, \tag{D.8}$$

where

$$b_1 = 2\xi_1\xi_2 m^2 \left[D_1 \int_0^1 dy\, \Delta_2(\mu y) - \frac{\Delta_2(\mu)}{1-v^2} \int_0^1 dy\, \Delta_1(\mu y)\right] e^{i\tau\beta}$$

$$b_2 = 2\xi_1\xi_2 m^2 \left[D_2 \int_0^1 dy\, \Delta_1(\mu y) - \frac{\Delta_1(\mu)}{1-v^2} \int_0^1 dy\, \Delta_2(\mu y)\right] e^{i\tau\beta}\,,$$

$$b_3 = 2m^2 \left[\xi_1^2 D_1 \int_0^1 dy\, \Delta_1(\mu y) + \frac{\xi_2^2 \Delta_2(\mu)}{1-v^2} \int_0^1 dy\, \Delta_2(\mu y)\right] e^{i\tau\beta}$$

$$- \left(\frac{i}{\tau} + \frac{k_2^2}{2}\right)(e^{i\tau\beta} - 1),$$

$$b_4 = 2m^2 \left[\xi_2^2 D_2 \int_0^1 dy\, \Delta_2(\mu y) + \frac{\xi_1^2 \Delta_1(\mu)}{1-v^2} \int_0^1 dy\, \Delta_1(\mu y)\right] e^{i\tau\beta} \tag{D.9}$$

$$- \left(\frac{i}{\tau} + \frac{k_2^2}{2}\right)(e^{i\tau\beta} - 1),$$

$$b_5 = -\sqrt{k_1^2 k_2^2}\,(1-v^2)(e^{i\tau\beta} - 1)/2,$$

$$\beta = \beta_1 + \beta_2, \quad \xi_{1,2}^2 = -\frac{e^2 a_{1,2}^2}{m^2}, \quad \mu = \tau(1-v^2)/4,$$

here

$$\beta_{1,2} = e^2\, a_{1,2}^2 \left[\int_0^1 dy\, \Delta_{1,2}^2(\mu y) - \left(\int_0^1 dy\, \Delta_{1,2}(\mu y)\right)^2\right],$$

$$\Delta_{1,2}(\mu y) = \Psi_{1,2}(\varphi - 2(\kappa k)\mu y) - \Psi_{1,2}(\varphi), \tag{D.10}$$

$$D_{1,2} = \int_0^1 dy\, \Delta_{1,2}(\mu y) + \frac{v^2}{1-v^2} \cdot \Delta_{1,2}(\mu).$$

Since the coefficients b_n are the functions of $\varphi = \kappa x$, then the interaction of the photon described by the tensor $T^{\mu\nu}(k_1, k_2)$ is in the general case inelastic ($k_1 \neq k_2$), i.e. the plane wave of the form (C.1) is an optically active "medium" for an external photon.

In the case of an elliptically polarized monochromatic wave when

$$\Psi_1 = \cos\varphi, \quad \Psi_2 = \sin\psi, \tag{D.11}$$

the integrals appearing in b_n (D.9) can be easily evaluated. As a result we obtain for the coefficients c_n in the expression for $T_{\mu\nu}(k_1, k_2) - T^{\mu\nu}_{F=0}(k_1, k_2)$

(see Eq. (D.5))

$$\mathbf{c}_n = -i(2\pi)^4 \frac{\alpha}{\pi} m^2 \int_{-1}^{1} dv \int_0^\infty \frac{d\tau}{\tau} \cdot \exp\left\{ -\frac{2i\tau}{|\lambda|(1-v^2)} \times \left[1 - \frac{(k_1 k_2)(1-v^2)}{4m^2} + A(\xi_1^2 + \xi_2^2) \right] \right\} \left[\delta(k_1 - k_2) d_n + \sum_{l=-\infty}^{\infty} \delta(k_1 - k_2 - 2l\kappa) g_n^l \right], \tag{D.12}$$

where

$$\begin{aligned}
d_1 &= -d_2 = 2\tau\xi_1\xi_2 \frac{1+v^2}{1-v^2} A_0 J_0(z)\, sign\,(\lambda), \\
d_3 &= \left[A_1\xi_1^2 - \frac{\xi_1^2 - \xi_2^2}{1-v^2} \sin^2\tau \right] \quad [J_0(z) - iJ_0'(z)] \\
&\quad + \xi_1^2 \frac{1+v^2}{1-v^2} J_0(z)\sin^2\tau - \frac{1}{4}\left[\frac{(k_1 k_2)}{m^2} + \frac{i|\lambda|(1-v^2)}{\tau} \right] \quad [J_0(z) - e^{iy}], \\
d_4 &= d_3(\xi_1^2 \leftrightarrow \xi_2^2), \\
d_5 &= -\frac{(k_1 k_2)(1-v^2)}{4m^2}[J_0(z) - e^{iy}], \\
g_1^l &= \xi_1\xi_2 \left[2A_0\tau \frac{1+v^2}{1-v^2} sign\,(\lambda) - A_1 \frac{l}{z} \right] i^l J_l(z), \\
g_2^l &= g_1^l(A_0 \to -A_0, z \to z), \\
g_3^l &= \left[A_1\xi_1^2 + \frac{\xi_1^2 v^2 + \xi_2^2}{1-v^2}\sin^2\tau \right] i^l J_l(z) + \left[A_1\xi_1^2 - \frac{\xi_1^2 - \xi_2^2}{1-v^2} \cdot \sin^2\tau \right] i^{l-1} J_l'(z) \\
&\quad - \frac{1}{4}\left(\frac{(k_1 k_2)}{m^2} + \frac{i|\lambda|(1-v^2)}{\tau} \right) i^l J_l(z), \\
g_4^l &= g_3^l(\xi_1^2 \leftrightarrow \xi_2^2)(-1)^l, \\
g_5^l &= -\frac{\sqrt{k_1^2 k_2^2}\,(1-v^2)}{4m^2} i^l J_l(z).
\end{aligned} \tag{D.13}$$

Here $J_l(z)$ are the Bessel functions, the following notation has been used

$$A = \frac{1}{2}\left(1 - \frac{\sin^2\tau}{\tau^2}\right), \quad A_0 = \frac{1}{2}\left(\frac{\sin^2\tau}{\tau^2} - \frac{\sin 2\tau}{2\tau}\right),$$

$$A_1 = A + 2A_0, \; z = \frac{2A_0\tau(\xi_1^2 - \xi_2^2)}{|\lambda|\,(1-v^2)}, \; y = \frac{2A\tau(\xi_1^2 + \xi_2^2)}{|\lambda|\,(1-v^2)}, \qquad \text{(D.14)}$$

$$\lambda = \frac{(\kappa\, k)}{2m^2}$$

and the replacement $\tau \to 2\tau[(1-v^2)|\lambda|m^2]^{-1}$ has been carried out. The expressions for c_n are sums of the two terms. One of them (with coefficients d_n) describes the elastic scattering of a photon by the field of the wave, the other (containing g_n) describes inelastic scattering accompanied by emission (absorption) of the "photons" of the wave. In the case of a circularly polarized monochromatic wave $\xi_1^2 = \xi_2^2 \equiv \xi^2$ the coefficients appearing in (D.12) are essentially simplified. In this case the tensor $T^{\mu\nu}(k_1, k_2)$ (D.5) has the form (we subtract the tensor $T^{\mu\nu}$ for $F = 0$)

$$\frac{T^{\mu\nu}(k_1,k_2) - T^{\mu\nu}_{F=0}}{i(2\pi)^4} = \sqcap^{\mu\nu}_{(0)} \cdot \delta(k_1 - k_2)$$

$$+ \sqcap^{\mu\nu}_{(-)}\delta(k_1 - k_2 - 2\kappa) + \sqcap^{\mu\nu}_{(+)}\delta(k_1 - k_2 + 2\kappa),$$

$$\sqcap^{\mu\nu}_{(0)} = \alpha_1(\Lambda_1^\mu\Lambda_2^\nu - \Lambda_2^\mu\Lambda_1^\nu) + \alpha_3(\Lambda_1^\mu\Lambda_1^\nu + \Lambda_2^\mu\Lambda_2^\nu) \qquad \text{(D.15)}$$

$$+ \alpha_5\Lambda_3^\mu\Lambda_3^\nu,$$

$$\sqcap^{\mu\nu}_{(\pm)} = \alpha_0\,(\Lambda_1^\mu \pm i\Lambda_2^\mu)(\Lambda_1^\nu \pm i\Lambda_2^\nu),$$

where (see also (D.14))

$$\alpha_n = -\frac{\alpha m^2}{2\pi}\int_{-1}^{1} dv \int_0^\infty \frac{d\tau}{\tau}\exp\left\{-\frac{2i\tau}{|\lambda|\,(1-v^2)}\left[1 + \frac{(k_1k_2)(1-v^2)}{4m^2}\right.\right.$$

$$\left.\left. + 2A\xi^2\right]\right\}\omega_n, \quad \omega_0 = A_1\xi^2,$$

$$\text{(D.16)}$$

$$\omega_1 = 4A_0\,\xi^2\,\tau\frac{1+v^2}{1-v^2}sign\,(\lambda), \quad \omega_3 = 2\xi^2\frac{1+v^2}{1-v^2}\sin^2\tau$$

$$-\left[1 + \frac{k_1^2(1+v^2)}{4m^2}\right](1 - e^{iy}), \quad \omega_5 = -\frac{k_1^2(1-v^2)}{2m^2}(1 - e^{iy}).$$

Thus, from a circularly polarized wave only two photons can be emitted (absorbed), and this is determined by the fact that the "photons" of the wave have in this case a definite chirality and transitions are possible both without a change in the chirality of the incident photon ($l = 0$) and with a reversal of chirality ($l = \pm 1$). This can be seen from (D.15), since in the inelastic terms, characteristic combinations of chiral eigenvectors have been formed.

We introduce the "diagonal" polarization operator $\Pi_e^{\mu\nu}(k_1)$ related to the amplitude of the elastic scattering of a photon in the following manner

$$T_e^{\mu\nu}(k_1, k_2) = i(2\pi)^4 \delta(k_1 - k_2) \sqcap_e^{\mu\nu}(k_1) . \tag{D.17}$$

Its imaginary part determines the probability of the production of an electron-positron pair by a photon in the field of a wave

$$W = \frac{e_\mu e_\nu^*}{k_1^0} \text{Im } \sqcap_e^{\mu\nu}(k_1) , \tag{D.18}$$

where k_1^0 is the photon energy. Taking into account Eqs. (D.16), (D.5), (D.12), (D.15), we obtain from Eq. (D.18) the probability of production of a pair by an unpolarized real ($k_1^2 = 0$) photon in the field of a circularly polarized wave

$$\begin{aligned} W_{cr} = & -\frac{g^{\mu\nu}\text{Im}\sqcap_{cr}^{\mu\nu}}{2k_1^0} = \frac{\alpha m^2}{2\pi k_1^0} \text{Im} \int_1^\infty \frac{du}{u\sqrt{u(u-1)}} \\ & \times \int_0^\infty \frac{d\tau}{\tau} \{[1 - 2\xi^2(2u-1)\sin^2\tau]\, e^{-iZ_1^{cr}} - \exp[-2iu\tau/|\lambda|]\} \\ = & -\frac{\alpha m^2}{2k_1^0} \left\{ \text{Im} \int_0^\infty \frac{d\tau}{\tau}\, e^{-i\eta} [\eta(1 + 2\xi^2 \sin^2\tau)(H_0^{(2)}(\eta) \right. \\ & \left. + iH_1^{(2)}(\eta)) - 2i\,\xi^2 \sin^2\tau H_0^{(2)}(\eta)] + 1 \right\} , \end{aligned} \tag{D.19}$$

Here $Z_1^{cr} = 2u\eta, \eta = \tau[1 + \xi^2(1 - \sin^2\tau/\tau^2)]/|\lambda|$, $H_{0,1}^{(2)}$ are Hankel functions. In a similar manner one obtains the probability of pair production by polarized photons. Formula (D.19) gives the representation for the probability of pair production by a photon in the field of a plane wave analogous to Eq. (C.46) for the probability of a photon emission by an electron in this field.

The calculation of the contribution of 0-spin particles to the polarization of the vacuum is completely analogous to that carried out above, while the calculations themselves are considerably simpler. The result can be represented in the form (D.5), (D.8), where the coefficients b_n are given by

$$
\begin{aligned}
b_1^{(0)} &= -\xi_1\xi_2 m^2 \int_0^1 dy\,\Delta_2(\mu y)\left[\int_0^1 dy\,\Delta_1(\mu y) - \Delta_1(\mu)\right] e^{i\tau\beta}\,,\\
b_2^{(0)} &= b_1^{(0)}\,(\xi_1 \leftrightarrow \xi_2\,, \Delta_1 \leftrightarrow \Delta_2)\,,\\
b_3^{(0)} &= -\xi_1^2 m^2 \int_0^1 dy\,\Delta_1(\mu y)\left[\int_0^1 dy\,\Delta_1(\mu y)\right.\\
&\quad\left. - \Delta_1(\mu)\right] e^{i\tau\beta} + \frac{i}{2\tau}\,(e^{i\tau\beta} - 1)\,,\\
b_4^{(0)} &= b_3^{(0)}\,(\xi_1 \to \xi_2\,, \Delta_1 \to \Delta_2)\\
b_5^{(0)} &= -\sqrt{k_1^2 k_2^2}\cdot v^2(e^{i\tau\beta} - 1)/4\,.
\end{aligned}
\tag{D.20}
$$

For an external photon the field of a wave can be regarded as a material "medium". The propagation of a photon in this medium is described by solutions of the Maxwell equations, which with regard for the interaction with the field of a plane wave can be written in the momentum representation in the form

$$k_1^2 e^{\mu}(k_1) = -\frac{i}{(2\pi)^4}\int d^4k_2 T^{\mu\nu}(k_1, k_2)\, e_{\nu}(k_2)\,, \tag{D.21}$$

where $e_\mu(k)$ is the photon polarization vector. Let us investigate this equation in the case of a circularly polarized wave using Eq. (D.15). The polarization vector in Eqs. (D.21) and (D.15) can be expanded in terms of the orthonormal set $k_1^\mu/(k_1^2)^{-1/2}, \Lambda_1^\mu, \Lambda_2^\mu, \Lambda_3^\mu$. Correspondingly, we obtain four solutions; the longitudinal one and three transverse solutions. The longitudinal solution can be written in the form

$$e_\mu^l(k) = f(k)\delta(k^2)\, k_\mu\,, \tag{D.22}$$

where $f(k)$ is an arbitrary function. One of the transverse solutions is

$$e_\mu^{(1)}(k) = f^{(1)}(k)\delta(k^2 + \alpha_5)\,\Lambda_{3\mu} \tag{D.23}$$

since $(k^2 + \alpha_5)\, e_\mu^{(1)} = 0$. We seek the other two solutions in the form

$$e^\mu = f^{(2)}\Lambda_{(+)}^\mu + f^{(3)}\Lambda_{(-)}^\mu,\ \Lambda_{(\pm)}^\mu = \Lambda_1^\mu \pm i\Lambda_2^\mu\,. \tag{D.24}$$

Substituting this expression into Eq. (D.21), we find that the functions $f^{(3)}$ and $f^{(2)}$ satisfy the system of equations

$$\begin{aligned} &(k^2 + \alpha_3 + i\alpha_1)\, f^{(2)} + 2\alpha_0\, f^{(3)}_{(+)} = 0 \\ &(k^2 + \alpha_3 - i\alpha_1)\, f^{(3)} + 2\alpha_0\, f^{(2)}_{(-)} = 0\,, \end{aligned} \tag{D.25}$$

where the functions having the subscript $\pm$ depend on $k \pm 2\kappa$, and the other functions depend on k. This system has solutions if the determinant vanishes

$$\begin{vmatrix} k^2 + \alpha_3 + i\alpha_1 & 2\alpha_0 \\ 2\,\alpha_0^{(+)} & (k + 2\,\kappa)^2 + \alpha_3^{(+)} - i\alpha_1^{(+)} \end{vmatrix} = 0 \tag{D.26}$$

and we obtain the dispersion equation for these two solutions. As a result a photon acquires a mass ($k^2 \neq 0$). In the region $|\lambda| \ll 1$ all the effects are very weak. The index of refraction in this limit depends neither on the photon energy nor on the frequency of the wave. The effects will become noticeable close to the threshold of the two-photon reaction when $\lambda \geq 1 + \xi^2$

We proceed now to trace the basic features of the process of pair production by a photon in an aligned single crystal by considering the example of the production of an electron-positron pair in the field of a monochromatic circularly polarized plane wave for which the analysis is substantially simplified. The probability of the process in the wave field (see Eq. (D.19)) depends on the intensity of this wave ξ^2 and on the parameter $\lambda = (kq)/2m^2$ (k and q are the four-momenta of the photon and the wave, or of the equivalent photon in the crystal case). To obtain the dependence on ω and on the angle of incidence (on the ρ parameter), one should use the correspondence $\xi^2 \to \rho\,, \lambda \to \rho^{-1/2}\,\kappa/2$. The indicated probability then takes the form

$$W = \frac{\alpha m^2 \kappa}{2\pi\omega}\, F(\kappa\,, \rho)\,, \tag{D.27}$$

where

$$\begin{aligned} F(\kappa\,, \rho) = \frac{i}{\kappa} \int_0^1 dv \int_{-\infty}^{\infty} \frac{dt}{t} \Big\{ &\exp\left[i\mu t + i\mu\rho(t - \sin^2 t/t)\right] \\ &\times \left(1 - 2\rho \frac{1 + v^2}{1 - v^2} \sin^2 t\right) - e^{i\mu t} \Big\}, \quad \mu = \frac{4\sqrt{\rho}}{(1 - v^2)\,\kappa} \end{aligned} \tag{D.28}$$

When $\kappa \ll 1$, the function $F(\kappa, \rho)$ is exponentially small in the region $\rho \geq 1$. Only for $\rho \leq \kappa^2$ the pair production probability is not vanishing. This enables one to expand the exponent in Eq. (D.28) in powers of $\rho \ll 1$, which corresponds to the transition to the Born approximation. To first order in ρ (analogy of the coherent theory in a crystal) we have

$$F(\kappa,\rho) = \frac{\rho}{\kappa}\int_0^1 dv \int_{-\infty}^{\infty}\frac{dt}{t}\, e^{i\mu t}\left(\mu\,\frac{\sin^2 t}{t} - 2i\frac{1+v^2}{1-v^2}\sin^2 t\right)$$

$$= \frac{\pi\rho}{\kappa}\left\{\left(1+\mu_0-\frac{1}{2}\,\mu_0^2\right)\ln\frac{1+\sqrt{1-\mu_0}}{1-\sqrt{1-\mu_0}}\right.$$

$$\left.-\,(1+\mu_0)\sqrt{1-\mu_0}\right\}\vartheta(1-\mu_0)\,, \qquad \text{(D.29)}$$

where $\mu_0 = 2(\rho)^{1/2}/\kappa$.

As the photon energy grows, the parameter κ becomes comparable with unity. For such values of κ, the region $\rho \ll 1$ is described, as previously, by Eq. (D.29), and in the region $\rho \geq 1$ the probability of the process is no longer small. When $\rho \gg 1$ the constant field approximation becomes applicable; for such ρ the contribution to the integral (D.28) comes from small values of $t \sim (\rho)^{-1/2}$ Carrying out the appropriate expansion, we obtain after integration

$$F(\kappa,\rho) = F_1(\kappa) + \frac{1}{\rho}\,F_2(\kappa)\,, \qquad \text{(D.30)}$$

where

$$F_1(\kappa) = \frac{1}{4\sqrt{3}}\int_0^1 dv(9-v^2)\,K_{2/3}(s)\cdot s\,,$$

$$F_2(\kappa) = \frac{1}{60\sqrt{3}}\int_0^1 dv\left\{2(1+5v^2)\,s\cdot K_{2/3}(s)\right.$$

$$\left.+\left[3s^2(3+v^2)-\frac{28}{3}\,(1+v^2)\right]K_{\frac{1}{3}}(s)\right\}, \qquad \text{(D.31)}$$

$$s = \frac{8}{3\kappa(1-v^2)}\,.$$

Formula (D.30) contains the correction $\propto 1/\rho$ to the constant field limit. With increasing κ, the function $F_1(\kappa)$ grows, achieves a maximum at $\kappa \simeq 11.5$ and then decreases. As for the function $F_2(\kappa)$, it decreases monotonically with increasing κ, changes sign at the point $\kappa = \kappa_0 \simeq 4.3$ and is negative for $\kappa > \kappa_0$.

With a further increase of the photon energy when the condition $\kappa \gg 1$ is fulfilled, for large ρ one can use, as before, the expressions (D.30) and (D.31) for which the following asymptotic expansion holds in this region

$$
\begin{aligned}
F_1(\kappa) &= \frac{15\sqrt{3}}{7(3\kappa)^{\frac{1}{3}}}\frac{\Gamma^3\left(\frac{2}{3}\right)}{\Gamma\left(\frac{1}{3}\right)} - \frac{4\pi}{3\kappa} \simeq 2.39\,\kappa^{-\frac{1}{3}} - 4.19/\kappa\,, \\
F_2(\kappa) &= -\,\frac{98\,\Gamma^3\left(\frac{1}{3}\right)(3\,\kappa)^{\frac{1}{3}}}{9900\,\sqrt{3}\,\Gamma\left(\frac{2}{3}\right)} + \frac{46\,\Gamma^3\left(\frac{2}{3}\right)(3\,\kappa)^{-\frac{1}{3}}}{35\,\sqrt{3}\,\Gamma\left(\frac{1}{3}\right)} \\
&\quad - \frac{\pi}{5\kappa} \simeq -0.12\kappa^{\frac{1}{3}} + 0.49\kappa^{-\frac{1}{3}} - 0.63/\kappa\,.
\end{aligned}
\tag{D.32}
$$

The ratio of the leading terms in (D.32) is $F_2/F_1 \propto \kappa^{2/3}$ and it is clear that the approximation (D.30) is valid for $\rho \geq \kappa^{2/3}$; what can be directly established from the basic expression (D.28).

When κ is large, in the region $\rho \leq 1$, one has $\mu\rho \ll 1$ and, by virtue of the fact that the function $\sin^2 t/t < 1$, it is possible to expand the term $\exp\left(-i\,\mu\rho\,\sin^2 t/t\right)$ in Eq. (D.28). After integration, we find

$$
\begin{aligned}
F(\kappa\,,\rho) &= \frac{\pi\rho}{\kappa}\left[\left(1 + \frac{\mu_1}{1+\rho} - \frac{\mu_1^2}{2(1+\rho)}\right)\ln\frac{1+\sqrt{1-\mu_1}}{1-\sqrt{1-\mu_1}}\right. \\
&\quad \left. - \left(1 + \frac{\mu_1}{1+\rho}\right)\sqrt{1-\mu_1}\right]\vartheta(1-\mu_1)\,, \\
\mu_1 &= \frac{2\sqrt{\rho}(1+\rho)}{\kappa}\,.
\end{aligned}
\tag{D.33}
$$

The expression (D.33) is not more complicated than formula (D.29) for the first Born approximation, but differs from it by having a wider region of applicability with respect to ρ. This is due to the fact that when deriving Eq. (D.33) we have kept the term $i\,\mu\rho t$ in the exponent in (D.28). The estimation made of the contribution from the narrow region with respect to v to the integral in Eq. (D.28) ($\Delta v \sim \rho^{3/2}/\kappa \ll 1$) and from the terms containing the higher powers of ρ in the pre-exponent has shown that the relative accuracy of formula (D.33) is of the order of $\rho^{3/2}/\kappa$.

The analogy noted above between the processes of pair production by a photon in the crystal and in the wave field allows one to verify quantitatively, owing to a relative simplicity of the expression (D.28), the description of the orientation dependence (the dependence on ρ in the model) by approximate formulae of the type (D.30) and (D.33). Comparison of the results, obtained by means of these formulae, with the calculation according to formula (D.28) permits one to estimate the precision of the approximate formulae in the region of their applicability and to show that the interpolation procedure in the intermediate region is not arbitrary, as Fig. D.1 demonstrates.

The interpolation curve should start near the point at which $\mu_1(\rho) = 1$ and must join the curve 1 on the right-hand slope of the latter, not far from its maximum. It is seen from Fig. D.1 that the interpolation region narrows with increasing κ, according to the general analysis.

The structure, observed in the exact calculation, in the vicinity of the point $\mu_1(\rho) = 1$ is associated with the singularity of the derivative $dF/d\rho$ at this point. The character of this singularity can be found from Eq. (D.33) or directly from Eq. (D.28). Note that for the process in the crystal the indicated structure will smear out because of the "multifrequency" nature.

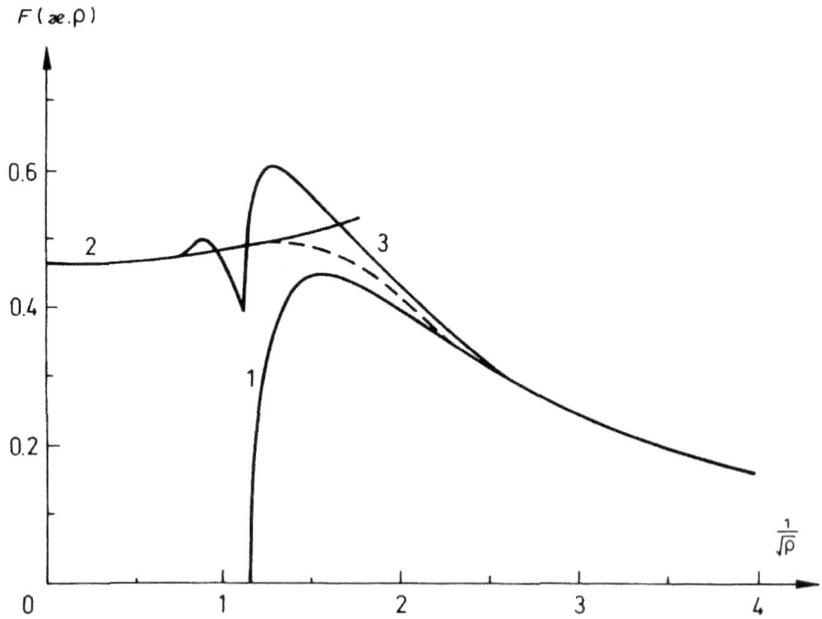

(a)

Fig. D.1. Dependence of the function $F(\kappa, \rho)$ on $(\rho)^{-1/2}$ which corresponds to the dependence on the incidence angle ϑ_0 in the crystal. Curves 1, 2 and 3 are the calculations according to formulae (D.33), (D.30) and (D.28), respectively; (a) for $\kappa = 3$; (b) for $\kappa = 27$.

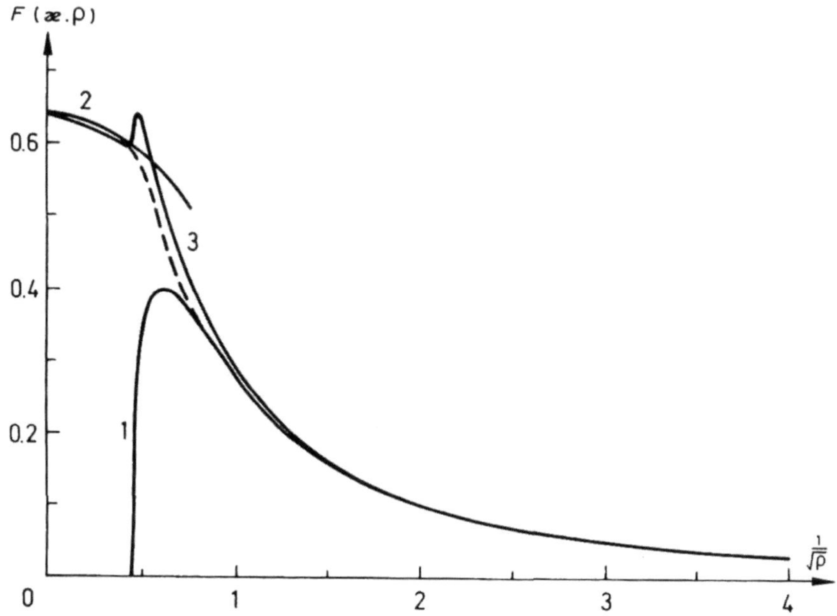

(b)

Fig. D.1. (*Continued*)

At intermediate values of parameters κ and ρ the total probability of pair creation (D.27) was calculated numerically (see Baier, Yokoya (1994)). Results are plotted in Fig. D.2, where the 3D graph of the total probability for the fixed magnitude of an electric field of the wave gives the complete picture of its behavior as a function of both parameters. It is seen that the total probability of pair creation depends on both parameters essentially non-monotonically. For fixed $\kappa = 2\lambda\sqrt{\rho}$ there exists a number of maxima which become less and less pronounced with increasing ρ. Such dependence on ρ is associated with the fact that the process of pair creation has thresholds at

$$\mu_n = \frac{2\sqrt{\rho}\,(1+\rho)}{\kappa n = 1} \tag{D.34}$$

(cf. μ_1 in Eq. (D.33)) where n is the number of photons absorbed from the wave. In other words, with increasing ρ the minimal number of photons which should be absorbed from the wave to start the process of pair creation is also increasing. For $\kappa = 1$ and $n = 1$ the solution of Eq. (D.34) is $\sqrt{\rho_1} = 0.424$,

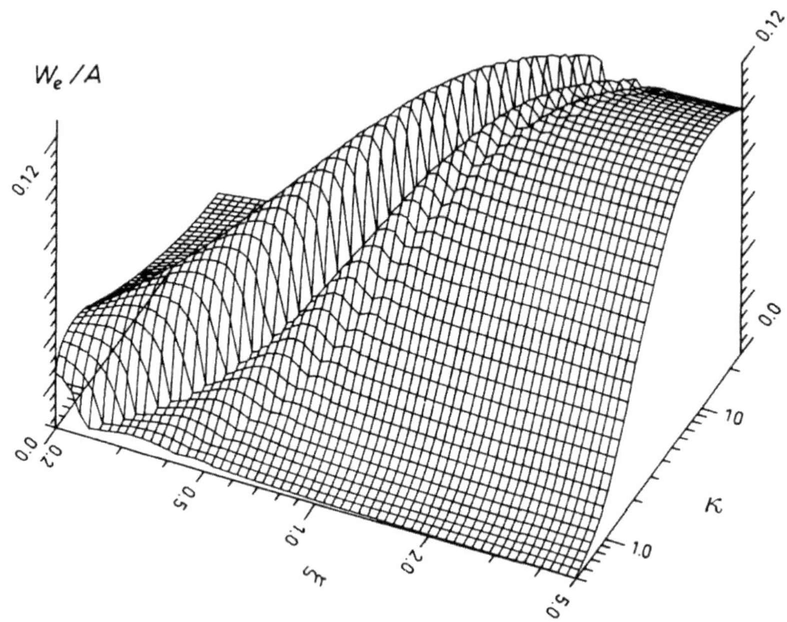

Fig. D.2. The total probability of pair creation in the field of circularly polarized waves as a function of the parameters $\sqrt{\rho} = \xi$ and κ in units of $A = \alpha mF/F_0$.

and for $n = 2\sqrt{\rho_2} = 0.682$. If $\rho > \rho_1$ two and more photons are absorbed from the wave in this process. The threshold cusp is situated just at $\rho = \rho_1$ and the next cusp is situated at the next threshold $\rho = \rho_2$. For relatively low ρ-values the probability of one photon absorption is dominated. It attains the maximum and then decreases because of the reduction of the phase space. When κ increases from $\kappa = 1$, the height of the maximum increases as well. The absolute maximum $W_e = 0.1171 \frac{\alpha mF}{F_0}$ is situated at $\xi = \sqrt{\rho} = 1.36$ and $\kappa = 9.3$. After it, the height of the maximum starts to decrease with κ increasing.

Appendix E

THEORY OF FREE-ELECTRON LASER

In Secs. 1.8–1.11, an incoherent (spontaneous) radiation at quasiperiodic motion of a relativistic particle has been investigated in detail. One of the examples is the radiation in the periodic magnetic field (undulator). Side by side with the spontaneous radiation a beam of relativistic electrons when traversing periodic magnetic field can radiate coherently. Such sources of the coherent radiation are called free-electron lasers (FEL). In the last few years, FELs became one of the most promising sources of coherent radiation in a very broad range from the sub-millimeter wave down to the ultraviolet light (see e.g. the special issue, Granatshtein and Roberson (1985), in which there is a detailed bibliography).

FELs have a series of unique features: continuous tunability in a broad range of frequencies with the possibility of generating radiation at any given frequency within this range; prospects of operating in a region of short wavelength (down to ultraviolet light if one uses magnetic undulators) and also the possibility of high power operation.

In typical FEL conditions, the electrons move in a periodic magnetic field (magnetic lattice) of two types: (1) with a helicoidal magnetic field, then the waves with the circular polarization are radiated (see Sec. 1.10(b)) for spontaneous radiation); (2) with planar magnetic field, then the waves with linear polarization are radiated (see Sec. 1.10(a)).

We consider first a motion in the helicoidal magnetic field which we write in the form (1.162)

$$\begin{aligned} \underset{e}{F}^{13} &= -\underset{e}{F}^{31} = H_y = H_0 \sin \nu_0 z \,, \\ \underset{e}{F}^{23} &= -\underset{e}{F}^{32} = H_x = H_0 \cos \nu_0 z \,, \end{aligned} \tag{E.1}$$

where $\nu_0 = 2\pi/\lambda_0$, λ_0 is the period of the lattice. In this lattice (see (1.164)) an electron is moving along a helical line with the constant velocity along the lattice axis. We consider the motion when the transverse momentum is small compared with the longitudinal momentum of the electron ($p_{\|} \gg p_{\perp} \sim m$), then in the dipole approximation ($\rho \ll 1$, see (1.180)) the spectrum of the incoherent (spontaneous) radiation is a narrow band (the first harmonic), the broadening of which is due to fact that the ultrarelativistic particles radiate in an angle $\sim 1/\gamma$, but not precisely ahead, and the radiated frequency is connected with the radiation angle (see (1.118) and (1.120)). The additional broadening of the spectrum arises when an undulator has a finite number of elements (see Fig. 1.10). So, the radiation ahead is monochromatic for infinite undulator and the first harmonic is radiated, but for finite undulator there will be broadening. Outside of the region of the dipole approximation ($\rho \sim 1, p_{\perp} \sim m$) a few first harmonics (see (1.180), (1.182)) will be radiated and the frequency of the radiation ahead is

$$\omega = 2\gamma^2 \omega_0 n/(1+\rho/2)\,, \quad n = 1, 2, 3 \ldots \tag{E.2}$$

This equality is in a certain sense a kinematic one and valid both for spontaneous and stimulated radiation.

E.1. The Case of Small Gains[a]

Consider now the FEL action. Let an electron beam move through the helicoidal magnetic lattice of period λ_0 and interact with a circularly polarized electromagnetic wave propagating along the lattice axis in the direction of the beam motion. The vector potential of the wave we represent is as follows

$$A_\mu(\varphi) = e_{1\mu}\, a\, \psi_1(\varphi) + e_{2\mu}\, a\, \psi_2(\varphi)\,, \tag{E.3}$$

where $\psi_1(\varphi) = \cos\varphi\,; \psi_2(\varphi) = \sin\varphi\,; \varphi = \nu\kappa x + \psi = \nu(t-z) + \psi$; The definition of the 4-vector κ is: $\kappa^\mu = (1;0,0,1)\,, \kappa^2 = 0\,, e_{1\,,2}\kappa = 0$. The field strength of

[a] See Baier and Milstein (1977).

the wave has the form

$$F_w^{\mu\nu} = H_w\left[(e_1^\nu\,\kappa^\mu - e_1^\mu\,\kappa^\nu)\,(d\psi_1/d\varphi) + (e_2^\nu\,\kappa^\mu - e_2^\mu\,\kappa^\nu)\,(d\psi_2/d\varphi)\right], \qquad \text{(E.4)}$$

where $H_w = a\nu$. The covariant equations of motion $du^\mu/d\tau = (e/m)\,F^{\mu\nu}\,u_\nu$ in the superposition of the fields (E.1) and (E.4), $F^{\mu\nu} = F_e^{\mu\nu} + F_w^{\mu\nu}$ have the following form:

$$\begin{aligned}
du^0/d\tau &= \Omega(u^1\,\sin\,\varphi - u^2\,\cos\,\varphi)\,,\\
du^1/d\tau &= \Omega(u^0 - u^3)\,\sin\,\varphi + \Omega_0 u^3\,\sin\,\nu_0 z\,,\\
du^2/d\tau &= -\Omega\,\cos\,\varphi(u^0 - u^3) - \Omega_0 u^3\,\cos\,\nu_0 z\,,\\
du^3/d\tau &= du^0/d\tau - \Omega_0(u^1\,\sin\,\nu_0 z - u^2\,\cos\,\nu_0 z)\,,\\
d\Omega/ds &= -\omega_p^2\,\overline{(u^1\,\sin\,\varphi - u^2\,\cos\,\varphi)}/2\,,
\end{aligned} \qquad \text{(E.5)}$$

where $\Omega = eH_w/m$, $\Omega_0 = eH_0/m$, $\omega_p^2 = 4\pi e^2 N_e/(mV), \tau$ is the proper times, V is the volume of the electromagnetic radiation bunch, N_e is the number of electrons within this bunch. The bar in the last of Eqs. (E.5) denotes averaging over all the electrons in the interaction region. This equation describes the change of the wave field because of the interaction with all the electrons and follows from the law of conservation of energy

$$md/d\tau \sum u^0 = -d/d\tau(VH_w^2/4\pi)\,. \qquad \text{(E.6)}$$

Assuming that Ω is a smooth function of time, we integrate the second and third equations of the set (E.5)

$$\begin{aligned}
u^1 &= -\left[\frac{\Omega}{\nu}\cos\varphi + \frac{\Omega_0}{\nu_0}\cos\nu_0 z\right] + a^1\,,\\
u^2 &= -\left[\frac{\Omega}{\nu}\sin\varphi + \frac{\Omega_0}{\nu_0}\sin\nu_0 z\right] + a^2
\end{aligned} \qquad \text{(E.7)}$$

As will be seen below, near the resonance the wave frequency $\nu \sim 2\gamma^2\,\nu_0$ (see (E.2)) and of interest is the situation when $\Omega/\nu \ll \Omega_0/\nu_0$. In order that a transverse displacement of the particle in the lattice does not exceed the oscillation amplitude it is necessary that $a^{1,2} \ll (\Omega_0/\nu_0)(1/2\pi\,N)$ (here N is the number of lattice elements), then $u^1\,(0) = -\Omega_0/\nu_0\,, u^2\,(0) = 0$.

Substituting the solutions (E.7) into Eq. (E.5) we get the set of equations[b]

$$\frac{d^2\Phi}{d\tau^2} = 2\Omega\Omega_0 \sin\Phi$$
$$\frac{d\Omega}{d\tau} = \left(\frac{\Omega_0\omega_p^2}{2\nu_0}\right)\overline{\sin\Phi}\,, \tag{E.8}$$

where $\Phi = \varphi - \nu_0 z$. The terms $\sim 1/\gamma^2$ are discarded in Eq. (E.8). (E.9)

We carry out in Eq. (E.8) the substitutions: $\Omega = \Omega_w y, \tau(\Omega_0\Omega_w)^{1/2} = s$, note that $y(0) = 1$. Then these equations take the form

$$\frac{d^2\Phi}{ds^2} = 2y\sin\Phi, \quad \frac{dy}{ds} = \beta\overline{\sin\Phi}\,, \tag{E.10}$$

where $\beta = (\omega_p^2/2\nu_0\,\Omega_w)(\Omega_0/\Omega_w)^{1/2}$.

The initial condition for $\Phi' \equiv d\Phi/ds$ (See Eq. E.9)) is

$$\kappa = \Phi'(0) = \left(\frac{\nu_0\gamma}{(\Omega_0\Omega_w)^{1/2}}\right)(\delta - 1)\,, \tag{E.11}$$

here $\delta = \frac{\nu}{2\nu_0\gamma^2}\,(1 + \Omega_0^2/\nu_0^2)$. In the center of the resonance line $\delta = 1$, i.e.

$$\lambda = \frac{\lambda_0}{2_{\gamma^2}}\left(1 + \frac{\Omega_0^2}{\nu_0^2}\right). \tag{E.12}$$

When passing through the lattice, the time increment is equal to

$$s_N = \left(\frac{2\pi N}{\gamma\nu_0}\right)(\Omega_0\Omega_w)^{1/2} \tag{E.13}$$

If $\kappa \gg 1\,, s_N/\kappa \ll 1$, which is valid when the wave field is weak, the solution of the set (E.10) may be found in an analytical form by using the perturbation expansion in inverse powers of κ. In this case

$$y(s) = 1 + \beta s^3\, R\left(-\frac{\kappa s}{2}\right), \tag{E.14}$$

where

$$R(z) = \frac{1}{2}\frac{\sin z}{z^2}\left(\frac{\sin z}{z} - \cos z\right) = -\frac{1}{4}\frac{d}{dz}\left(\frac{\sin^2 z}{z^2}\right). \tag{E.15}$$

---

[b]Note that expression $u^0(1 + \nu_0/\nu) - u^3 = C$ is the integral of the motion in the problem under consideration.

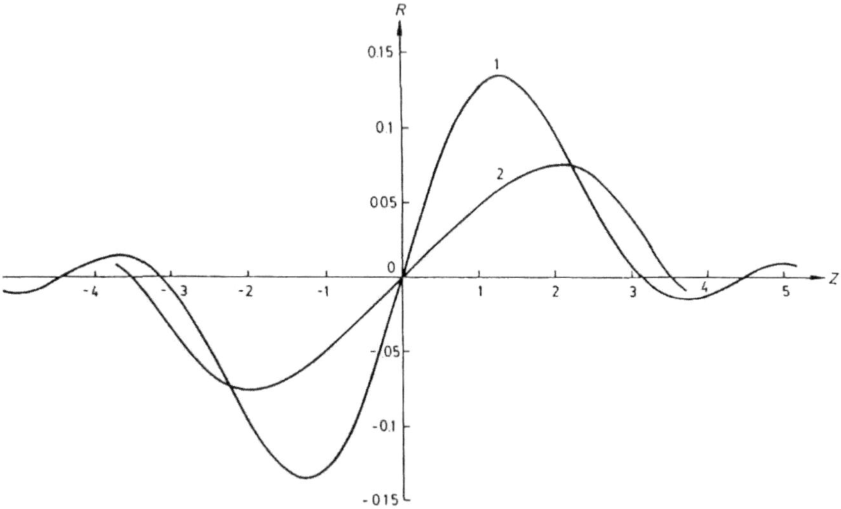

Fig. E.1. Function $R(z)$, $z = -\kappa x/2$. (1) Theoretical curves for weak wave field (Eq. (E.15)); (2) The curve for strong wave fields ($\beta = 0.11$, $H_w/H_0 = 7.10^{-2}$; $s = 2.25$).

In deriving Eq. (E.14) the electrons are assumed to be uniformly distributed over the initial phase at $s = 0$. The function $R(z)$ (see Fig. E.1) describes the dependence of the gain on the distance from the resonance (the form of the spectral curve)[c] while the maximum value $R_{\max} = 0.135$ being at $z = 1.3$. Taking into account Eqs. (E.11) and (E.13), we have that for maximum R the function $1 - \delta = 1.3/(\pi N) \ll 1$. At a single passing (amplification mode) the resonance linewidth is determined by the function R (Fig. E.1) and the $1/e$ half-width is $\Delta\nu/\nu = 2.2/(\pi N)$. In the generation mode an additional narrowing of the line takes place due to multiple passing.

The wave is naturally amplified with the same polarization as the initial one. Moreover, this polarization should correspond to the lattice symmetry. From Eq. (E.14) we have for the gain

$$G \equiv y^2 - 1 = 2\eta\left(1 + \frac{\eta}{2}\right), \tag{E.16}$$

[c]The function $\sin^2 z/z^2$ gives the form (and the width) of the spectral line of the spontaneous (incoherent) radiation as one can see from Eqs. (1.148), (1.152) and (1.166). It follows from Eq. (E.15) that amplification (attenuation) of the electromagnetic wave is connected with "coupling strength" with the radiation field at different frequencies on the slope of the spectral line.

where

$$2\eta = \frac{R(z)}{2\pi}\frac{\omega_p^2\Omega_0^2 N^3\lambda_0^4}{\gamma^3}, \qquad z = \pi N(1-\delta).$$

The formulae presented above may be applied directly to the description of the wave amplification. In the case of low amplification, when $y - 1 \ll 1$, in Eq. (E.16) it is possible to confine ourselves to the term linear in η. In this case formula (E.16) presents the dependence of G on ω_p^2, N, λ_0 and γ, as well as the form of the resonance curve.

If $G \ll 1$, the generation mode may be implemented only at a multiple interaction of the radiation with electrons, when the bunch of the electromagnetic radiation moving in wave cavity successively interacts with the electron bunches. In the beginning of the generation, the wave field is very small, i.e. $\kappa \gg 1$ (see Eq. (E.11)), then Eqs. (E.14) – (E.16) are valid. The generation is possible if the gain $G > \kappa_{ef'}$ where κ_{ef} is the total loss coefficient (transmission of mirrors, diffractional losses, etc.). For confocal cavity where the transverse cross section of the radiation bunch is $L\lambda/4$ (L is the distance between mirrors) this inequality may be written as

$$\left(\frac{\Omega_0^2}{\nu_0^2}\right)\left(1+\frac{\Omega_0^2}{\nu_0^2}\right)^{-1} 16R(z)\frac{J}{e}\frac{r_0}{L}\,\frac{(N\lambda_0)^3\nu_0^2}{\gamma c^3} > \kappa_{ef}\,, \tag{E.17}$$

where $r_0 = e^2/mc^2$ is the classical electron radius, J is the electron current. This expression determines, particularly, a threshold magnitude of the current.

As the wave strength is increasing, the inequalities determining the applicability of the Eq. (E.14) ceases to be fulfilled. In this case the set (E.10) has been solved numerically for the parameters close to those in the pioneer experiment of Deacon *et al.* (1977). This solution shows that the gain G for which a weak field (Eq. (E.16)) is independent of the field strength H_w, begins to fall down with increasing H_w. In Fig. E.2 the dependence of the gain on H_w/H_0 is shown. For strong fields the spectral curve is also distorted (see Fig. E.1). The decrease of the gain finally stops the wave-field increase. After this, the generation becomes stationary.

All the obtained results concern the electron bunches with uniform initial phase distribution. In the course of the electrons interaction with the electromagnetic wave and the magnetic lattice the phase distribution transforms in such a way that at an electron energy higher than the resonance one (see

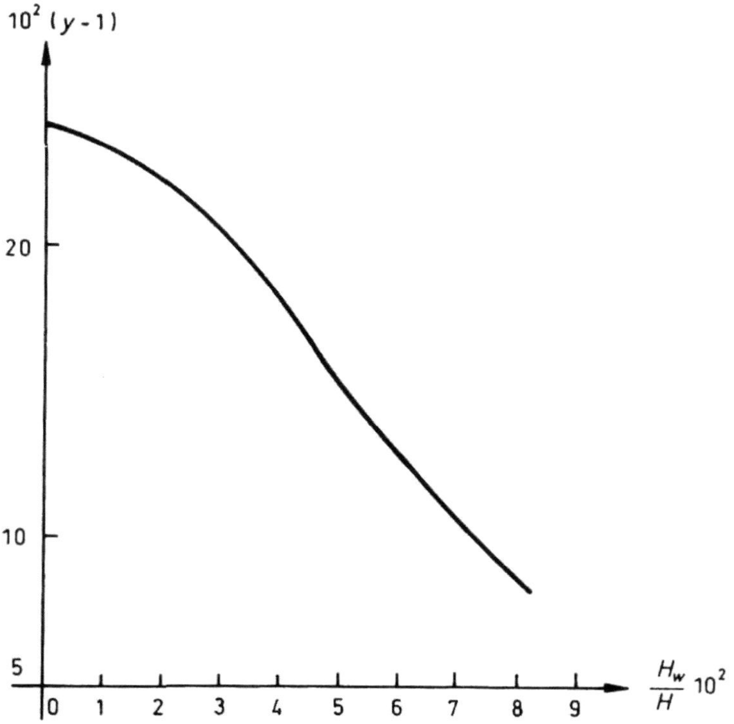

Fig. E.2. Dependence of the gain on the field strength of the wave ($\beta s^3 = 1.78$).

(E.12)) the radiation of the electron bunch exceeds the absorption, and at an electron energy lower than the resonance one the absorption dominates. This means that in weak-field domains, a specific phase stability arises. In the strong-signal domain, a phase oscillation appear which disturbs this phase stability and leads to a significant decrease of the resonance energy transfer from particle to wave. The most efficient transfer of the electron energy to the wave field would take place in the case when at the beginning all the electrons were phased in an appropriate manner, and the increment of the phase when the electron bunch passes through the lattice was $\delta\Phi \ll 1$. In such a situation the gain is

$$G = \frac{\omega_p^2}{\nu_0^2}\frac{\Omega_0}{\Omega}\frac{2\pi N}{\gamma}\sin\psi\,, \tag{E.18}$$

so that at $0 < \psi < \pi$ the resonance loss of the electron energy occurs, and at $-\pi < \psi < 0$ we have the opposite situation. This result may be considered as an upper limit for the resonant energy transfer from the particles in the wave and inversely.

Above the FEL action was considered for the undulator with the helicoidal field and correspondingly with the circular polarization of the radiation. The planar undulator (where the radiation with the linear polarization is amplified or generated) may be considered in the same way. In this case, all the basic behavior is the same including a dependence G on the parameters (see (E.16)). Note, that in Sec. 26 the planar problem was solved.

E.2. The Case of Large Gains[d] $G \geq 1$

Suppose that the electron beam is moving through a helicoidal magnetic lattice (E.1). We consider the interaction of the electron beam with a circularly polarized electromagnetic wave propagating along the axis of the lattice in the direction of the motion of the beam. We represent the vector potential of the wave in the form (E.3), where

$$a = a\,(t), \varphi = \nu\kappa x + \chi(t) = \nu(t - z) + \chi(t)\,, \tag{E.19}$$

here $a\,(t)$ and $\chi\,(t)$ are the slowly varying amplitude and phase of the wave, respectively. At small gains $G \ll 1$, we can neglect the difference between the phase velocity of the wave and the velocity of the light and associated with it, phase increment $\chi(t)$. For gains $G \geq 1$, the phase increment becomes significant.

The transverse part $\mathbf{P}_\perp(P^1, P^2)$ of the canonical momentum is conserved in the superposition of fields (E.1), (E.3) and (E.19), so that we can set $\mathbf{P}_\perp = 0$. Then expressing the kinetic momentum $\mathbf{p}_\perp$ (and current) in terms of the vector potential, inserting this current into the Maxwell equation for the transverse component of the vector potential and carrying out the appropriate averaging over the phase, we arrive at the set of equations

$$\begin{aligned} \Omega' &= \frac{\Omega_0\omega_p^2}{2\nu_0}\,\overline{\sin(\Phi + \chi)}\,, \\ \Omega\,\chi' &= \frac{\Omega_0\omega_p^2}{2\nu_0}\,\overline{\cos(\Phi + \chi)} + \frac{\omega_p^2\Omega}{(2\,\nu)}\,, \end{aligned} \tag{E.20}$$

[d]This section is based on papers by Baier and Milstein (1980, a, b).

where we differentiate over the proper time, $\Phi = \nu(t - z) - \nu_0 z + \Phi_0$ (cf. (E.9)). We use the same notations as in Eqs. (E.4) and (E.5). Remember also that $n = N_e/V$ is the electron density in the interaction volume and the bar denotes the averaging over the initial phases. The term $\omega_p^2 \Omega/2\nu$ in the second equation (E.20) allows for the constant component of the dielectric permittivity of the electron beam. To obtain the complete set of equations, we need the equation for phase Φ. We find it using the electron equations of motion (see Eqs. (E.5) – (E.8))

$$\Phi'' = 2\Omega\Omega_0 \sin(\Phi + \chi). \tag{E.21}$$

The set of equations (E.20) and (E.21) is complete and describes both the amplification and generation for a small and a large gain.

To find the universal characteristics of the electromagnetic wave evolution in the considered system, the change to a new dimensionless variables is convenient

$$s = \Omega_0 \tau a^{1/3}, \quad y = \frac{\Omega}{\Omega_0} a^{-2/3}, \tag{E.22}$$

where $a = \omega_p^2/(2\nu_0\Omega_0)$. The set (E.20) and (E.21) takes the following form in these variables:

$$\begin{aligned} \frac{dy}{ds} &= \overline{\sin\,(\Phi + \chi)}\,. \\ y\frac{d\chi}{ds} &= \overline{\cos\,(\Phi + \chi)} + y\delta \\ \frac{d^2\Phi}{ds^2} &= 2y\,\sin(\Phi + \chi)\,, \end{aligned} \tag{E.23}$$

where $\delta = \left(\frac{\nu_\Omega}{\nu}\right) a^{2/3}$

The region $G \ll 1$ corresponds to $s \ll 1$. In this case the phase increment χ may be neglected, and from the set (E.23) we obtain the set (E.10).

Description of the operation regime when $y \ll 1$ (linear regime) can be made in an analytical form. Let $d\Phi_0/ds = \kappa$ (see (E.20)), then

$$\kappa = \gamma \frac{\nu_0}{\Omega_0 a^{\frac{1}{3}}} \left[\frac{\nu}{2\nu_0\,\gamma^2} \left(1 + \frac{\Omega_0^2}{\nu_0^2} \right) - 1 \right] \tag{E.24}$$

We put

$$\Phi = \Phi_0 + \kappa s + \Phi_1$$

At $y \ll 1$ the value $\Phi_1 \ll 1$. From the last equation of the set (E.23) we have

$$\Phi_1 = 2 \int_0^s y(s')(s - s') \sin[\Phi_0 + \kappa s' + \chi(s')] ds' . \tag{E.25}$$

After substituting this expression into (E.25) and then the result into the first two equations of (E.23), and averaging over Φ_0, we get

$$\begin{aligned} \frac{dy}{ds} &= \int_0^s y(s')(s - s') \sin[\kappa(s - s') + \chi(s') - \chi(s)] ds' \\ y\frac{d\chi}{ds} &= \int_0^s y(s')(s - s') \sin[\kappa(s - s') + \chi(s) - \chi(s')] ds' + y\delta . \end{aligned} \tag{E.26}$$

We present the solution of the set (E.26) in the form $z = y \exp(i(\chi + \kappa s))$. Then for the function $z(s)$ the following equation may be obtained from the set (E.26)

$$\frac{dz}{ds} - iz(\kappa + \delta) + i \int_0^s (s - s') z(s') ds' = 0 . \tag{E.27}$$

One should bear in mind that the process here is considered in a single-particle approximation, when the electrons interact with each other through an electromagnetic wave, and electrostatic interaction between the electrons in a bunch is not taken into account. Since $G \geq 1$ is attained at high density of electrons n, the type of the operation determined by the plasma effects may be also realized (e.g. three-wave instability). The accepted approximation is valid if a wavelenth of the Langmuir oscillations is long as compared to a characteristic length of the electromagnetic wave growth l_g (l_g is estimated below, Eq. (E.38)), or

$$\frac{\omega_p l_g}{c\gamma^{\frac{3}{2}}} \ll 1 . \tag{E.28}$$

Under this condition the allowance for the parameter δ leads to an insignificant shift of the resonance curve. In what follows it will be omitted. A subsequent differentiation of Eq. (E.27) yields

$$\frac{d^3 z}{ds^3} - i\kappa \frac{d^2 z}{ds^2} + iz = 0 . \tag{E.29}$$

In the considered case, the initial conditions are

$$z(0) = y(0), \quad dz(0)/ds = i\kappa y(0), \quad \frac{d^2 z(0)}{ds^2} = -\kappa^2 y(0) .$$

We put $z(s) = z(0)\ \exp\,(-i\lambda s)$ and obtain the characteristic equation

$$\lambda^3 + \kappa\lambda^2 + 1 = 0 \tag{E.30}$$

similar to the known equation for the travelling wave tube. The solution of Eq. (E29) for the given initial conditions is

$$z(s) = z(0)[B_1\ \exp\,(-i\lambda_1 s) + B_2\ \exp\,(-i\lambda_2 s) + B_3\ \exp\,(-i\lambda_3 s)]\,. \tag{E.31}$$

Here $\lambda_1, \lambda_2, \lambda_3$ are the roots of the characteristic equation,

$$B_1 = \frac{\kappa^2 + \kappa(\lambda_2 + \lambda_3) + \lambda_2\,\lambda_3}{\lambda_1^2 - \lambda_1(\lambda_2 + \lambda_3) + \lambda_2\,\lambda_3}$$

and B_2 and B_3 are obtained from B_1 by cyclic permutation of the indexes. The signal amplification on a "length" s is characterized by $|z/y(0)|^2 = 1 + G$ the explicit representation for which follows from (E.31). It has a simple analytical form when $\kappa \ll 1$ and $\kappa \gg 1$

$$\left|\frac{z}{y(0)}\right|^2 = \begin{cases} f + \kappa f_1 & |\kappa| \ll 1 \\ 1 + f_2 & |\kappa| \gg 1 \end{cases}, \tag{E.32}$$

where

$$\begin{aligned} f &= \frac{\left(1 + 4\ \cosh^2\left(\frac{\sqrt{3}\,s}{2}\right) + 4\ \cosh\left(\frac{\sqrt{3}\,s}{2}\right)\ \cos\left(\frac{3s}{2}\right)\right)}{9}\,, \\ f_1 &= \frac{4}{27}\left[-\cosh(\sqrt{3}\,s) + \cosh\left(\frac{\sqrt{3}\,s}{2}\right)\ \cos\left(\frac{3s}{2}\right)\right. \\ &\quad \left. + \sqrt{3}\sinh\left(\frac{\sqrt{3}\,s}{2}\right)\ \sin\left(\frac{3s}{2}\right)\right], \\ f_2 &= \frac{4}{\kappa^3}\left[\frac{1}{4}\ \sinh^2\left(\frac{s}{\kappa^{\frac{1}{2}}}\right) + \frac{\kappa^{\frac{3}{2}}}{2}\ \sinh\left(\frac{s}{\kappa^{\frac{1}{2}}}\right)\ \sin\,\kappa s\right. \\ &\quad \left. + \cosh\left(\frac{s}{\kappa^{\frac{1}{2}}}\right)\ \cos\,\kappa s - 1\right]. \end{aligned} \tag{E.33}$$

Note that at $s \ll 1 : f = 1 + s^6/40,\ f_1 = s^4/6$; and at $s \gg 1$:

$$f = \frac{1}{9}\,\exp(\sqrt{3}\,s)\,,\ f_1 = -\frac{2}{27}\ \exp(\sqrt{3}\,s)\,.$$

In the case, when $s \ll 1\,(G \ll 1)$ it is quite easy to get a general expression for the gain G (see Eq. (E.16))

$$\left|\frac{z}{y(0)}\right|^2 = 1 + G = 1 + 2s^3\,R\left(-\frac{\kappa s}{2}\right), R\,(z) = -\frac{1}{4}\,\frac{d}{dz}\left(\frac{\sin^2 z}{z^2}\right). \qquad \text{(E.34)}$$

Note, that under conditions of the above mentioned experiment (Deacon *et al.* (1977)), $s = 0.48$. In the case, when $s \gg 1$ the gain G may reach very large values

$$\left|\frac{z}{y(0)}\right|^2 = 1 + G = \frac{1}{9}\,\exp\left(\sqrt{3}\,s\right).$$

The linear regime takes place until the phase increment Φ_1 becomes of the order of unity. In the case of a single passage (super-radiant oscillator) for the outgoing signal $y \sim 1$, independently upon $y(0)$. Taking into account (E.22), we have the estimate for the maximum strength of the outgoing electromagnetic field

$$H = H_0 a^{\frac{2}{3}}. \qquad \text{(E.35)}$$

Due to the build-up of the phase oscillations, the growth of the wave amplitude is limited. The efficiency $\eta = \Delta\,\gamma/\gamma$ can be estimated using (E.35)

$$\eta \approx \frac{\Omega_0}{2\gamma\nu_0} a^{\frac{1}{3}}. \qquad \text{(E.36)}$$

The spectral distribution of the emitted radiation is determined, for fixed length s, by the dependence of $|z/y\,(0)|^2$ on κ (see (E.24)). In a linear regime this dependence is expressed by (E.31), and in the cases $\kappa \ll 1$ and $\kappa \gg 1$ we have obtained the explicit expressions (E.32) and (E.33). In general, however, the set (E.23) can be solved only numerically. The results of calculations for $y\,(0) = 0.01$ are shown in Fig. E.3, which depicts the evolution of the spectral curves for various values of s ($s = 5;\ 6;\ 7.5$). Note, that in the region $s \ll 1$, which was investigated in the previous subsection (see Fig. E.1), the maximum of the spectral curve lies at $\kappa < 0$ (with $|\kappa| s/2 \sim 1$), while the gain vanishes for $\kappa = 0$ (compare, in this connection, the asymptotic expressions for the functions f and f_1 (E.33) at $s \ll 1$). It can be seen from Fig. E.3 (and also from expressions (E.32) and (E.33)) that at $s \geq 1$ the spectral curve is deformed

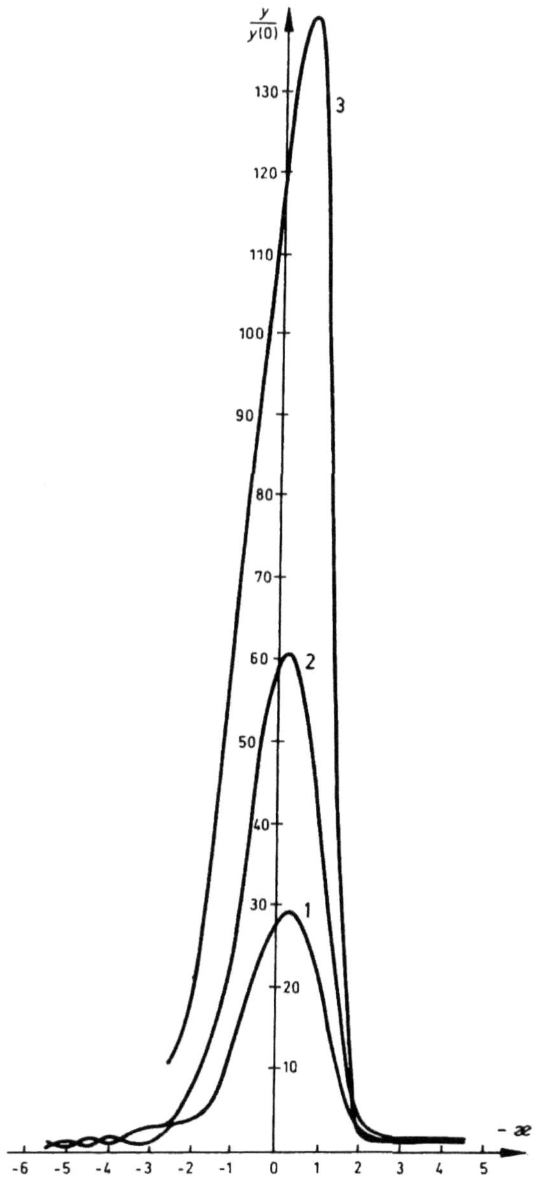

Fig. E.3. Dependence of emitted field on κ for different s: (1) for $s = 5$; (2) for $s = 6$; (3) for $s = 7.5$.

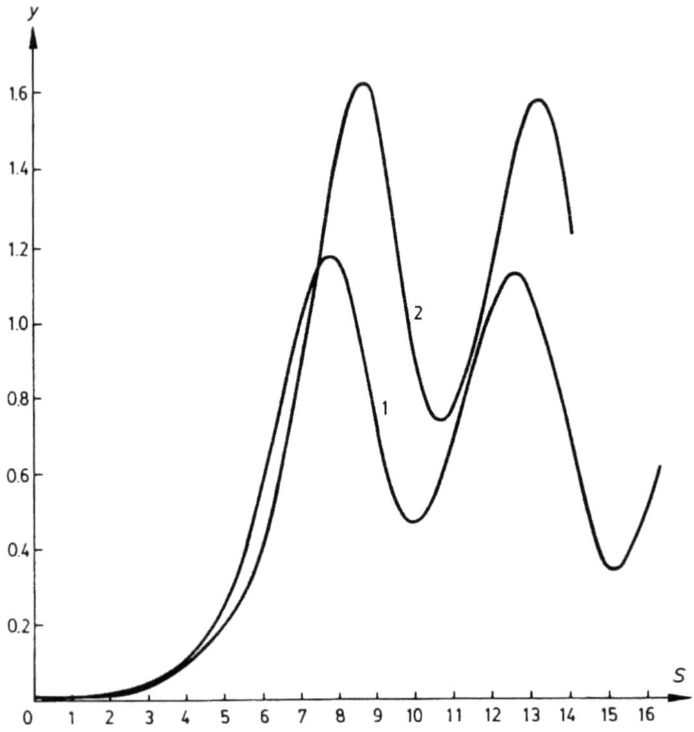

Fig. E.4. Dependence of emitted field on the length s for fixed values of κ : $\kappa = 0$ (1) and $\kappa = -1$ (2); $y\ (0) = 0.01$.

considerably. Now its maximum lying, as before, at $\kappa < 0$, however, the curve is converted from an antisymmetric one into an almost symmetric curve, and the point $\kappa = 0$ lies not far from the maximum. The width of the resonance curve is determined by the condition $|\kappa| \sim 1$ or

$$\frac{\Delta\,\nu}{\nu} \approx \frac{2a^{\frac{1}{3}}\,\Omega_0}{\gamma\nu_0} \sim \eta\,. \tag{E.37}$$

Figure E.4 shows the dependence of the emitted wave (for $y(0) = 0.01$) on the "length" s for the fixed values $\kappa = 0$ and $\kappa = -1$ (i.e., near the maximum of the spectral curve). It can be seen that for $s > 1$ the strength of the emitted wave rapidly (exponentially) grows to values $y \sim 1$, when the phase oscillations develop, leading to the oscillation of the strength. Since

$|z/y(0)| \sim (1/3)\ \exp(\sqrt{3}\, s/2)$, the characteristic growth length is $s = 2/\sqrt{3}$ or (see (E.22))

$$l_g = \frac{2}{\sqrt{3}} \frac{c\gamma}{\Omega_0 a^{\frac{1}{3}}}, \tag{E.38}$$

while the length over which $y \sim 1$ is reached is $L = l_g \ln(3/y(0))$. The curve 3 in Fig. E.3 gives the shape of the emitted wave at $s = 7.5$, when $\kappa = -0.8$, for this value of s the gain is maximum. Note, that values $s = 5$ and $s = 6$ are lying in the region of the linear regime (see Eqs. (E.31) – (E.33)), while at $s = 7.5$ the regime is essentially nonlinear.

If the length of a magnetic lattice $L_e \sim L$, the maximum value of the outgoing electromagnetic wave is already attained upon a single passage, as it has been discussed above. If $L_e \ll L$, it is fairly reasonable to use a multiple amplification of the wave, for instance, using the quasi-optic type cavity.

For illustration let us consider the following example. Let the current $J = 3$ mA with cross section $S = 0.2\ \text{cm}^2$, bunch length $l_b = 0.25$ cm, and the energy $\varepsilon = 100$ MeV ($\gamma = 200$) pass through the magnetic lattice with $H_0 = 2.4$ kg and $\lambda_0 = 3.2$ cm. In these conditions the growth length is $l_g = 76$ cm, and the criterion (E.28) is fulfilled. Let an incident wave appear due to incoherent radiation when the bunch passes through the initial part of the lattice. Then the length on which the strength of the emitted wave attains its maximum is 10.5 m. Radiated wave length is $\lambda = 6.10^{-5}$ cm, radiated power is $W \sim 10^3$ W, and the efficiency is $\sim 4.10^{-3}$ (see Eq. (E.36)). In accordance with (E.37), the relative line width $\Delta\nu/\nu$ is of the same order of magntiude.

The foregoing study has been carried out for a given energy γ and angle of incidence of the electron into the lattice ϑ. Using formula (E.24) and Fig. E.3, we have for the permissible spreads in energy and angle:

$$\frac{\delta\gamma}{\gamma} \leq \frac{c}{\nu_0\, l_g} \sim \eta, \quad \delta\vartheta \leq \frac{c}{\Omega_0\, \gamma\, l_g}\left(1 + \frac{\Omega_0^2}{\nu_0^2}\right). \tag{E.39}$$

Only those electrons in the beam that satisfy these conditions participate in the generation of the electromagnetic wave.

REFERENCES

1. Avakyan, A. L., Zhevago, N. K. and Yan, Shi, (1982) "Radiation from electrons and positrons at axial quasi channeling", *Zh. Eksp. Teor. Fiz.* **82**, 573; also in *Soviet Physics JETP* **55**, 341.
2. Avakyan, R. O., Miroshnichenko, I. I., Murrey, D. and Figut, T. (1982) "Radiation from ultra-relativistic positrons moving in a single crystal near axes and planes", *Zh. Eksp.Teor. Fiz.* **82**, 1825.
3. Avakyan, R. O., Avakyan, E. O., Avetisyan A. E. et al. (1983) "Radiation from high-energy electrons moving near crystalographic axes and planes in diamond crystal", *Second All-Union Conference on Radiation of Relativistic Particles* in *Crystals. Cheget.*
4. Aganyanz, A. O., Vartanov, Yu. A., Vartapetyan, G. A. et al. (1983) "Spectral-angular distribution of radiation from electrons with energy 4.3 GeV for small angles of incidence in diamond crystals", Erevan, 1983, preprint Erevan Physics Institute No. 666(56).
5. Adeishwilli, D. I., Bochek, G. L., Vitjko, V. I. et al. (1984) "Ionization and radiation energy losses of relativistic electrons in silicon single crystals", *Yad. Fiz.* **40**, 318.
6. Adler, S. L., Bahcall, J. N., Callan, C. G. and Rosenbluth, M. N. (1970) "Photon splitting in a strong magnetic field", *Phys. Rev. Lett.* **25**, 1061.
7. Adler, S. L. (1971) "Photon splitting and photon dispersion in a strong magnetic field", *Ann. Phys.* **67**, 599.
8. Alguard, M. J., Swent, R. L., Pantell, R. H. et al. (1980) "Radiation from channeled leptons", *Nucl. Instr. Meth.* **170**, 7.
9. Alferov, D. F., Bashmakov, Yu. A. and Bessonov, E. G. (1972) "On a radiation of relativistic particles in undulators", *Zh. Tech. Fiz.* **42**, 1921.
10. Alferov, D. F., Bashmakov, Yu. A. and Bessonov, E. G., (1975) "Undulator radiation", *Proc. (Tr.)* Lebedev Phys. Inst. **80**, 100.

11. Andersen J. U., Andersen S. K. and Augustyniak W. (1977) "Channeling of electrons and positrons", *Mat. Fys. Medd. Dan. Seals.* **39**, (10), 3.
12. Andersen, J. U., Bonderup, E., Laegsgaard, E. et al. (1982) "Axial channeling radiation from MeV electrons", *Nucl. Instr. Meth.* **194**, 209.
13. Animalu A. (1977) *Quantum Theory of Crystalline Solids*, Prentice-Hall, New Jersey.
14. Arzimovich, L. A. and Pomeranchuk, I. Ya. (1946) "Radiation of fast electrons in a magnetic field", *Zh. Eksp. Teor. Fiz.* **16**, 379.
15. Atkinson, M., Bak, J. F., Bussey, P. J. et al. (1982) "Radiation from planar channeled 5-55 GeV/c positrons and electrons", *Phys. Lett.* **110B**, 162.
16. Achiezer, A. I. and Berestetskii, V. B. (1963) *Quantum Electrodynamics*, 2nd edn., J. Wiley, New York.
17. Achiezer, A. I., Boldyshev, V. F. and Shulga, N. F. (1979) "Theory of elastic scattering and bremsstrahlung of fast charged particles in crystals", *Fiz. Elem. Chastits i Yadra* **10**, 51; see also *Soviet Phys. J. Part. Nucl.* **10**, 19.
18. Achiezer, A. I. and Shulga, N. F. (1982) "Radiation from relativistic particles in single crystals", *Uspekhi Fiz. Nauk* **137**, 561; see also *Soviet Phys. Uspekhi* **25**, 541.
19. Achiezer, A. I. and Shulga N. F. (1983) "To the theory of electromagnetic showers in crystals", *Zh. Eksp. Teor. Fiz.* **85**, 94; see also *Soviet Phys. JETP* **58**, 55.
20. Bhabha, H. and Heitler W. (1937) "The passage of fast e^- and the theory of cosmic showers", *Proc. Roy. Soc.* **159A**, 432.
21. Bazylev, V. A. and Zhevago, N. K. (1987) *Radiation from Fast Particles in Solids and External Fields* "Nauka", Moscow, (in Russian).
22. Bazylev V. A., Golovisnin, V. V. and Demura, A. V. (1987) "Bremsstrahlung of relativistic electrons and positrons moving at a small angle to axes and planes of crystals", *Zh. Eksp. Teor. Fiz.* **92**, 1946.
23. Baier, V. N. and Katkov, V. M. (1966) "Neutrino pair production associated with the motion of an electron in a magnetic field", *Dokl. Akad. Nauk SSSR* **171**, 313; see also *Soviet Phys. -Doklady* **11**, 947 (1967).
24. Baier, V. N. and Katkov, V. M. (1967a) "Radiative polarization of electrons in a magnetic field", *Zh. Eksp. Teor. Fiz.* **52**, 1422; see also *Soviet Phys. JETP* **25**, 944.
25. Baier, V. N. and Katkov, V. M. (1967b) "Quantum effects in a magnetic bremsstrahlung", *Phys. Lett.* **25A**, 492.
26. Baier, V. N. and Katkov V. M. (1967c) "Processes involved in the motion of high energy particles in a magnetic field", *Zh. Eksp. Teor. Fiz.* **53**, 1478; see also *Soviet Phys. JETP* **26**, 854 (1968).
27. Baier, V. N. and Katkov, V. M. (1968) "Quasiclassical theory of bremsstrahlung by relativistic particles", *Zh. Eksp. Teor. Fiz.* **55**, 1542; see also *Soviet Phys. JETP* **28**, 807 (1969).
28. Baier, V. N. (1971) "Radiative polarization of electrons in storage rings", *Uspekhi Fiz. Nauk* **105**, 441; see also *Soviet Phys. Uspekhi* **14**, 695 (1972).

29. Baier, V. N., Katkov, V. M. and Strakhovenko, V. M. (1971a) "Radiative effects in an external electromagnetic field", *Dokl. Akad. Nauk SSSR* **197**, 66; see also *Soviet Phys.-Doklady* **16**, 230.
30. Baier, V. N., Katkov, V. M. and Strakhovenko, V. M. (1971b) "Higher order effects in external field: pair production by a particle", *Yad. Fiz.* **14**, 1020; see also *Soviet Phys. Nucl. Phys* **14**, 572 (1972).
31. Baier, V N. and Katkov, V. M. (1972) "Production of bremsstrahlung during collision of high-energy particles in a magnetic field", *Dokl. Akad. Nauk SSSR* **207**, 68; see also *Soviet Phys.-Doklady* **17**, 1068.
32. Baier, V. N., Katkov, V. M. and Strakhovenko, V. M. (1972) "Radiation emitted by relativistic particles in periodic structures", *Zh. Eksp. Teor. Fiz.* **63**, 2121; see also *Soviet Phys. JETP* **36**, 1120 (1973).
33. Baier, V. N., Katkov, V. M. and Fadin, V. S. (1973) *Radiation from Relativistic Electrons*, Atomizdat. Moscow, (in Russian).
34. Baier, V. N., Fadin, V. S., Katkov, V. M and Kuraev E. A. (1974) "Photon splitting into two photons in a Coulomb field", *Phys. Lett.* **49B**, 385.
35. Baier, V. N., Katkov, V. M. and Strakhovenko, V. M. (1974) "Operator approach to quantum electrodynamics in an external field: The mass operator", *Zh. Eksp. Teor. Fiz.* **67**, 453; see also *Soviet Phys. JETP* **40**, 225 (1975).
36. Baier, V. N., Katkov, V. M. and Strakhovenko, V. M. (1975) "Operator approach to quantum electrodynamics in an external field. Electron loops", *Zh. Eksp. Teor. Fiz.* **68**, 405; see also *Soviet Phys. JETP* **41**, 198.
37. Baier, V. N., Katkov, V. M., Milstein, A. I. and Strakhovenko V. M. (1975) "To the theory of quantum processes in a field of intense electromagnetic wave", *Zh. Eksp. Teor. Fiz.* **69**, 783; see also *Soviet Phys. JETP* **42**, 400.
38. Baier, V. N., Milstein A.I. and Strakhovenko V. M. (1975) "Interaction between a photon and an intense electromagnetic wave", *Zh. Eksp. Teor. Fiz.* **69**, 1893; see also *Soviet Phys. JETP* **42**, 961 (1976).
39. Baier, V. N. and Katkov, V. M. (1976) "Corrections to the eikonal approximation and the bremsstrahlung of electrons at high energies", *Dokl. Akad. Nauk SSSR* **227**, 325; see also *Soviet Phys.-Doklady* **21**, 150.
40. Baier V. N., Katkov, V. M and Strakhovenko, V. M. (1976) "Anomalous magnetic moment of the electron in a magnetic field", *Lett. al Nuovo Cimento* **15**, 149.
41. Baier V. N. and Milstein, A. I. (1976) "Radiation effects in the field of an electromagnetic wave", *Dokl. Akad. Nauk SSSR* **231**, 1100; see also *Soviet Phys.-Doklady* **21**, 734.
42. Baier, V. N. and Milstein, A. L. (1977) "To the theory of free electron laser", Preprint INP (Novosibirsk) 77-66 (1977); see also *Phys. Lett.* **65A**, 319 (1978).
43. Baier, V. N., Katkov, V. M. and Strakhovenko, V. M. (1979a) "On radiation of relativistic positrons at channeling", *Phys. Lett.* **73A**, 414.
44. Baier, V. N., Kathov, V. M. and Strakhovenko, V. M. (1979b) "Electron radiation during a planar channeling", *Dokl. Akad. Nauk SSSR* **246**, 1347; see also *Soviet Phys.-Doklady* **24**, 469.

45. Baier, V. N. and Milstein, A. I. (1980a) "Operation of the free electron laser at large gains", *Dokl. Akad. Nauk SSSR* **250**, 1347; see also *Soviet Phys.-Doklady* **25**, 112.
46. Baier, V. N. and Milstein, A. I. (1980b) "Nonlinear wave interaction in a free electron laser", *Phys. Lett.* **79A**, 77.
47. Baier, V. N., Katkov, V. M. and Strakhovenko, V. M. (1980) *Radiation from Relativistic Particles at Planar Channeling* Preprint INP (Novosibirsk) 80–03, (in Russian).
48. Baier, V. N., Katkov, V. M. and Strakhovenko, V. M. (1981a) "Radiation of relativistic particles moving quasiperiodically", *Zh. Eksp. Teor. Fiz.* **80**, 1348; see also *Soviet Phys. JETP* **53**, 688.
49. Baier V. N. Katkov, V. M. and Strakhovenko, V. M. (1981b) "Radiation of relativistic particles in planar channeling", *Yad. Fiz.* **33**, 1284; see also *Soviet Phys. J. Nucl. Phys.* **33**, 686.
50. Baier, V. N., Katkov, V. M. and Strakhovenko, V. M. (1981c) "Influence of macroscopic factors on the bremsstrahlung in the collision of an electron and positron", *Dokl. Akad. Nauk SSSR* **260**, 861; see also *Soviet Phys.-Doklady* **26**, 961.
51. Baier, V. N., Katkov, V. M. and Strakhovenko, V. M. (1981d) "Nondipole radiation in planar channeling of positrons. Comparison of theory with experiment", Preprint INP (Novosibirsk) 81–139, *Dokl. Akad. Nauk. SSSR* **266**, 605; see also *Soviet Phys.-Doklady* **27**, 738 (1982).
52. Baier, V. N., Fadin, V. S., Khoze, V. A., Kuraev, E. A. (1981) "Inelastic processes in high energy quantum electrodynamics", *Phys. Reports* **78**, 293.
53. Baier, V. N., Katkov, V. M. and Strakhovenko, V. M. (1982) "Role of geometrical factors in bremsstrahlung in colliding e^-e^+ beams", *Yad. Fiz.* **36**, 163; see also *Soviet J. Nucl. Phys.* **36** 95.
54. Baier, V. N., Katkov V. M. and Strakhovenko, V. M. (1983a) "Radiation of ultra-relatistic particles moving near crystalline axes", *Phys. Lett.* **95A**, 403.
55. Baier, V. N., Katkov, V. M. and Strakhovenko, V. M. (1983b) "Radiation at planar channeling of relativistic electrons in thick crystals", *Phys. Stat. Sol. (b)* **118**, 499.
56. Baier, V. N., Katkov, V. M. and Strakhovenko, V. M. (1983c) "Radiation spectrum in planar channeling of electrons with allowance for dechanneling and collimation," *Dokl. Akad. Nauk SSSR* **268**, 1365; see also *Soviet Phys.-Doklady* **28**, 167.
57. Baier, V. N., Katkov, V. M. and Strakhovenko, V. M. (1983d) "Radiation in motion of high energy particles near crystalline axes in thick crystals", Preprint INP (Novosibirsk) 83–70, (in Russian).
58. Baier, V. N., Katkov, V. M. and Strakhovenko, V. M. (1984a) "Radiation intensity from ultra-relativistic particles during axial channeling and its orientation dependence", *Nucl. Instrum. Methods* **B4**, 346.

59. Baier, V. N., Katkov, V. M. and Strakhovenko, V. M., (1984b) "Intensity of radiation in axial channeling of high-energy particles in thick crystals," *Dokl. Akad. Nauk SSSR* **275**, 1369; see also *Sov. Phys. Dokl.* **29**, 306.
60. Baier, V. N., Katkov, V. M. and Strakhovenko, V. M. (1984c) "Radiation yield of high-energy electrons in thick crystals", Preprint INP 84-11, Novosibirsk, *Phys. Stat. Solidi (b)* **133**, 583 (1986).
61. Baier, V. N., Katkov, V. M. and Strakhovenko, V. M. (1984d) "Nondipole Radiation of electrons at planar channeling", Preprint INP 84-116, Novosibirsk, *Phys. Stat. Solidi (b)* **133**, 211 (1986).
62. Baier, V. N., Katkov, V. M. and Strakhovenko, V. M. (1984e) "Mechanism of electron–positron pair production by high-energy photons in a single crystal", *Phys. Lett.* **104A**, 231.
63. Baier, V. N., Katkov, V. M. and Strakhovenko, V. M. (1985a) "Production of a pair of particles by a photon entering a single crystal along an axis or a plane", *Dokl. Acad. Nauk SSSR* **282**, 851; see also *Sov. Phys. Dokl.* **30**, 474 (1986).
64. Baier, V. N., Katkov, V. M. and Strakhovenko, V. M. (1985b) "Electron–positron pair creation by photons in an aligned single crystal: comparison of theory with experiment", *Phys. Lett.* **109A**, 179.
65. Baier, V. N., Katkov, V. M. and Strakhovenko, V. M. (1986a) "Production of electron–positron pairs by high-energy photons in oriented crystals", *Zh. Eksp. Theor. Fiz.* **90**, 801; see also *Sov. Phys. JETP* **63**, 467.
66. Baier, V. N., Katkov, V. M. and Strakhovenko, V. M. (1986b) "Theory of pair creation in aligned single crystals", *Nucl. Instrum. Methods* **B16**, 5.
67. Baier, V. N., Katkov, V. M. and Strakhovenko, V. M. (1986v) "On the theory of pair production in oriented single crystals", *Dokl. Akad. Nauk SSSR* **289**, 75; see also *Sov. Phys. Dokl.* **31**, 552 (1987).
68. Baier, V. N., Katkov, V. M. and Strakhovenko, V. M. (1986d) "Quantum effects in radiation emitted by ultra-high energy electrons in aligned crystals", *Phys. Lett.* **114A**, 511.
69. Baier, V. N., Katkov, V. M. and Strakhovenko, V. M. (1986e) "On the radiation theory of high energy electrons in aligned single crystals", *Phys. Lett.* **117A**, 251.
70. Baier, V. N., Katkov, V. M. and Strakhovenko, V. M. (1986f) "A detector of superhard photons using aligned single crystals", *Nucl. Instrum. Methods* **A250**, 514.
71. Baier, V. N. and Milstein, A. I. (1986) "Electromagnetic-wave amplification by an electron beam in aligned single crystals", *Nucl. Instrum. Methods* **B17**, 25.
72. Baier, V. N., Milstein, A. I. and Shaisultanov, R. Zh. (1986) "Photon splitting in a strong electromagnetic field", *Zh. Eksp. Teor. Fiz.* **90**, 1141; see also *Sov. Phys. JETP* **63**, 665.
73. Baier, V. N., Milstein, A. I. and Shaisultanov, R. Zh. (1987) "Photon splitting in a strong electromagnetic field and the possibility of its observation in single crystals", *Phys. Lett.* **120A**, 255.

74. Baier, V. N., Katkov, V. M. and Strakhovenko, V. M. (1987a) "Emission of radiation by high-energy particles in oriented single crystals", *Zh. Eksp. Teor. Fiz.* **92**, 1228; see also *Sov. Phys. JETP* **65**, 686.
75. Baier, V. N., Katkov, V. M. and Strakhovenko, V. M. (1987b) "Cascade processes in the field of the axes of aligned single crystals", *Nucl. Instrum. Methods* **B27**, 360.
76. Baier, V. N., Katkov, V. M. and Strakhovenko, V. M. (1988a) "Radiation from relativistic particles colliding in a medium in the presence of an external field", *Zh. Eksp. Teor. Fiz.* **94**, 125; see also *Sov. Phys. JETP* **67**, 70.
77. Baier, V. N., Katkov, V. M. and Strakhovenko, V. M. (1988b) "Electron–positron pair creation by high-energy photons incident near crystalline plane", *Nucl. Instrum. Methods* **B35**, 21.
78. Baier, V. N. Katkov, V. M. and Strakhovenko, V. M. (1988c) "Incoherent radiation and pair creation in crystals", *Phys. Stat. Solidi (b)* **149**, 403.
79. Baier, V. N., Katkov, V. M. and Strakhovenko, V. M. (1988d) "Theory and experiments for pair production in crystals", *Nucl. Instrum. Methods* **B34**, 521.
80. Baier, V. N., Katkov, V. M. and Strakhovenko, V. M. (1988e) "Radiation in oriented single crystals: status of theory and experiment", *Phys. Lett.* **132A**, 211.
81. Baier, V. N., Baskov, V. A., Ganenko, V. B. et al. (1989) "Radiation from electrons at energy 28 GeV in a thick crystal of tungsten", *Pisma Zh. Eksp. Teor. Fiz.* **49**, 533.
82. Baier, V. N., Katkov, V. M. and Strakhovenko, V. M. (1989a) "Interaction of high-energy electrons and photons with crystals", *Usp. Fiz. Nauk* **159**, 455; see also *Sov. Phys. Usp.* **32**, 972.
83. Baier, V. N., Katkov, V. M. and Strakhovenko, V. M. (1989b) "Pair creation by a photon in oriented single crystal", *Nucl. Sci. Appl.* **3**, 245.
84. Baier, V. N., Katkov, V. M. and Strakhovenko, V. M. (1991a) "Electroproduction of e^+e^- pair in an external field", *Yad. Fiz.* **53**, 1021.
85. Baier, V. N., Katkov, V. M. and Strakhovenko, V. M. (1991b) "Quasiclassical theory of radiation and pair creation in crystals at high energy", *Rad. Eff.* **122–123**, 527.
86. Baier, V. N., Katkov, V. M. and Strakhovenko, V. M. (1991c) "Semiclassical theory of electromagnetic processes in a plane wave and a constant field", *Zh. Eksp. Teor. Fiz.* **100**, 1713; see also *Sov. Phys. JETP* **73**, 945.
87. Baier, V. N., Katkov, V. M. and Strakhovenko, V. M. (1991d) "A simple model of kinetic phenomena for the description of radiation", *Nucl. Instrum. Methods* **B62**, 213.
88. Baier, V. N., Katkov, V. M. and Strakhovenko, V. M. (1992a) "Quantum effects in channeling radiation", *Phys. Lett.* **162A**, 429.
89. Baier, V. N., Katkov, V. M. and Strakhovenko, V. M. (1992b) "Hard photon emission from high energy electrons and positrons in single crystals", *Nucl. Instrum. Methods* **B69**, 258.

90. Baier, V. N., Baskov, V. A., Ganenko, V. B. et al. (1992c) "Radiation from electrons in tungsten crystals", *Zh. Eksp. Teor. Fiz.* **101**, 1351.
91. Baier, V. N., Katkov, V. M. and Strakhovenko, V. M. (1993) "Integral characteristics of radiation in QED in external field", *Z. Phys.* **60**, 671.
92. Baier, V. N., and Yokoya, K. (1994) "Interaction of high energy electrons and photons with intense electromagnetic wave", *Particle Accelerators* **44**, 77.
93. Bak, J., Komarov, F., Meyer, Ph. et al. (1982) "Measurement of channeling radiation and its polarization and x-ray excitation, together with deviations from Landay distribution", Preprint /CERN/PSCC/82-94, Geneva.
94. Bak, J., Ellison, J. A., Marsh, B. et al. (1985b) "Channeling radiation from 2–55 Gev/c electrons and positrons (1). Planar case", *Nucl. Phys.* **B254**, 491.
95. Bak, J. F., Barberis, D., Brodbeck. T. J. et al. (1988a) "e^+e^- pair creation by 40–150 GeV photons incident near the $\langle 110 \rangle$ axis in a germanium crystal", *Phys. Lett.* **202B**, 615.
96. Bak, J. F., Ellison, J. A., Marsh, B. et al. (1988b) "Channeling radiation from 2 to 20 Gev/c electrons and positrons (II). Axial case", *Nucl. Phys.* **B302**, 525.
97. Baryshevskii, V. G. (1982) *Channeling, Radiation and Reaction in Crystals at High Energy*, Byelorussian State University, Minsk, (in Russian).
98. Baryshevskii, V. G. and Tichomirov, V. V. (1982) "Birefringence of high-energy quanta in crystals", *Yad. Fiz.* **36**, 697.
99. Baryshevskii, V. G. and Tichomirov, V. V. (1983) "Creation of the transverse polarized electrons and positrons of high energy in crystals", *Zh. Eksp. Teor. Fiz.* **85**, 232.
100. Baryshevskii, V. G. and Tichomirov, V. V. (1985) "Pair production in a slowly varying electromagnetic field and the pair production process", *Phys. Lett.* **113A**, 335.
101. Baryshevskii, V. G. and Tichomirov, V. V. (1986) "Role of the incoherent scattering in radiation processes at small angles of particles incidence on crystal axes and planes", *Zh. Eksp. Teor. Fiz.* **90**, 1908.
102. Bateman, H. and Erdelyi, A. (1953) *Higher Transcendental Functions*, v.II, McGraw-Hill, New York.
103. Belkacem, A., Bologna, G., Chevallier, M. et al. (1984) "Observation of enhanced pair creation for 50–110 GeV photons in an aligned Ge crystal", *Phys. Rev. Lett.* **53**, 2371 and Erratum **54**, 852 (1985).
104. Belkacem, A., Cue, N. and Kimball, J. C. (1985) "Theory of crystal-assisted radiation and pair creation for imperfect alignment", *Phys. Lett.* **111A**, 86.
105. Belkacem, A., Bologna, G., Chevallier, M. et al. (1986a) "Measurement of axial effects on the pair creation by 30-150 GeV photons and on the radiation emitted by 150 GeV electrons and positrons in Ge crystal", *Nucl. Instrum. Methods* **B13**, 9.
106. Belkacem, A., Bologna, G., Chevallier, M. et al. (1986b) "New channeling effects in the radiative emission of 150 GeV electrons in a thin germanium crystal", *Phys. Lett.* **117B**, 211.

107. Belkacem, A., Bologna, G., Chevallier, M. et al. (1987) "Study of e^+e^- pair creation by 20–150 GeV photons incident on a germenium crystal in alignment conditions", *Phys. Rev. Lett.* **58**, 1196.
108. Beloshytskii, V. V. and Kumachov, M. A. (1973) "Multiple scattering of channeled ions in a crystal", *Dokl. Akad. Nauk SSSR* **212**, 846.
109. Beloshytskii, V. V. and Kumachov, M. A. (1978) "Quantum theory of spontaneous and stimulated radiation from channeled electrons and positrons", *Zh. Eksp. Teor. Fiz.* **74**, 1244.
110. Beloshytskii, V. V. and Kumachov, M. A. (1982) "Multiple scattering of electrons at axial channeling", *Zh. Eksp. Teor. Fiz.* **82**, 462.
111. Bialynicka-Birula, Z. and Bialynicka-Birula, I., (1970) "Non-linear effects in quantum electrodynamics. Photon propagation and photon splitting in an external field", *Phys. Rev.* **D2**, 2341.
112. Berestetskii, V. B., Lifshitz, E. M. and Pitaevskii, (1982) *Quantum Electrodynamics*, 2nd edn. Pergamon Press, Oxford.
113. Bogolubov, N. N. and Shirkov, D. V. (1959) *Introduction to the Theory of Quantized Fields*, Interscience, New York.
114. Borisov A. V. and Zhukovskii V. Ch. (1975) "Photoproduction of electron–positron pairs on nuclei at presence of a constant external field", *Yad. Fiz.* **21**, 579.
115. Bochek G. L., Vitko, V. I., Ganenko, V. B. et al. (1983) "Investigation of radiation energy losses of relativistic electrons in single crystal of germanium", *Problem of Atomic Science and Technology* **1**, 83.
116. Bulgakov, N. K., Vodopyanov, A. S., Golovatyuk, V. M. et al. (1983)"Radiation from electrons at energy 10 GeV in single crystal of silicon during axial channeling process", Preprint 1-83-640, Dubna.
117. Bulgakov, N. K., Vodopyanov, A. S., Voitkovich, I. et al. (1986)"Experimental investigation of gamma radiation emitted during process of axial and planar channeling of electrons in silicon", *Zh. Eksp. Teor. Fiz.* **90**, 1527.
118. Cabibo, N., Da Prato, G., De Franceschi, G. and Mosco, U. (1962a) "New method for producing and analyzing linearly polarized gamma-ray beam", *Phys. Rev. Lett.* **9**, 270.
119. Cabibo, N., Da Prato, G., De Franceschi, G. and Mosco, U. (1962b) "Circular polarization of high-energy X-ray by birefringence in crystals", *Phys. Rev. Lett.* **9**, 435.
120. Carlson, J. F. and Oppenheimer, J. R. (1937) "On multiplicative showers", *Phys. Rev.* **51**, 220.
121. Colson, W. B. (1977) "One-body electron dynamics in a free electron laser", *Phys. Lett.* **64A**, 190.
122. Cox, H. L. and Bonham, R. A. (1967) "Elastic electron scattering amplitudes for neutral atoms calculated using the partial wave methods at 10, 40, 70, 100 KeVc for $Z = 1 + 54$", *J. Chem. Phys.* **47**, 2599.
123. Diambrini Palazzi, G. (1968) "High-energy bremsstrahlung and electron pair production in thin crystals", *Rev. Mod. Phys.* **40**, 611.

124. Deacon, D. A., Elias, L. R., Madey, J. M. et al. (1977) "First operation of a free electron laser", *Phys. Rev. Lett.* **38**, 892.
125. Doyle, P. A. and Turner, P. S. (1968) "Relativistic Hartree–Fock X-ray and electron scattering factor;", *Acta Crystallog.* **24A**, 390.
126. Ferretti, B. (1950) "On "bremsstrahlung" in crystals", *Nuovo Cimento* **7**, 118.
127. Filatova, N. A., Golovatyuk, V. M., Isakov, A. N. et al. (1982)"Radiation from the channeling of 10 GeV positrons by silicon single crystals, *Phys. Rev. Lett.* **48**, 488.
128. Ganenko, V. B., Nasonov, N. N., Pegushin, E. V. et al. (1983) "On radiation from relativistic electrons moving along crystal axis or plane", *Ukr. Fiz. Zh.* **28**, 302.
129. Ganenko, V. B., Zhebrovskii, Yu. V., Kolesnikov, L. Ya. et al. (1987) "Orientation dependencies of hard bremsstrahlung yields from electrons moving near axes of single crystals", *Theses of 17th National Meeting Charged Particle-Crystal Interaction Physics*, p.85, Moscow State University.
130. Gemmel, D. S. (1974) "Channeling and related effects in the motion of charged particles through crystal", *Rev. Mod. Phys.* **46**, 129.
131. Ginzburg, V. L. (1947) "On radiation of micro-radio-waves and their absorption in air", *Izvestiya Akad. Nauk SSSR, Ser. Fiz.* **II**, 165.
132. Gouanere, M., Sillu, D., Spighel, M. et al. (1982) "Sharp-line and broad-continuum radiation from electrons channeled in diamond", *Nucl. Instrum. Methods* **194**, 225.
133. Granatshtein, V. L. and Roberson, C. W., eds (1985) *Third Special Issue on Free-Electron Lasers. IEEE J. Quantum Electronics* **QE-21**, nr.7.
134. Gradshteyn, I. S. and Ryzhik, I. M. (1965) *Tables of Integrals, Series, and Products,* 4th edn. Academic Press, New York.
135. Ivanenko, D. D. and Sokolov, A. A. (1965) "On the theory of "radiant" electron", *Dokl. Akad. Nauk SSSR* **59**, 1551.
136. Izakson, C. and Zuber, J. (1980) *Quantum Field Theory*, McGraw-Hill.
137. Jackson, J. D. (1962) *Classical Electrodynamics*, Wiley, New York-London,.
138. Kalashnikov, N. P. (1981) *Coherent Interaction of Charged Particles in Single Crystals*, Atomizdat, Moscow, (in Russian).
139. Kimball, J. C., Cue, N., Roth, L. M. and Marsh, B. B. (1983) "New crystal assisted pair creation process", *Phys. Rev. Lett.* **50**, 950.
140. Kimball, J. C. and Cue, N. (1984) "Synchrotron radiation and channeling of ultrarelativistic particles", *Phys. Rev. Lett.* **52**, 1747.
141. Kimball, J. C. and Cue, N. (1985) "Quantum electrodynamics and channeling in crystals", *Phys. Rep.* **125**, 69.
142. Kimball, J. C., Cue, N. and Belkacem, A. (1986)" Crystal assisted quantum electrodynamics pair production and radiation", *Nucl. Instrum. Methods* **B13**, 1.
143. Kittel, C. (1956) *Introduction to Solid State Physics*, 2nd edn., John Wiley and Sons, London.

144. Klepikov, N. G. (1954) "Emission of photons and electron–positron pairs in a magnetic field", *Zh. Eksp. Teor. Fiz.* **26**, 19.
145. Kumakhov, M. A. (1976a) "On the theory of electromagnetic radiation of charged particles in a crystal", *Phys. Lett.* **57A**, 17.
146. Kumakhov, M. A. (1976b) "On the possibility of spontaneous emission of γ-quanta by relativistic channeled particles", *Dokl. Akad. Nauk SSSR* **230**, 1077.
147. Kumakhov, M. A. (1986) *Radiation from Channeled Particles in Crystals*, Energoatomizdat, Moscow, (in Russian).
148. Landau, L. D. and Lifshitz, E. M. (1975) *Classical Field Theory*, Pergamon, New York.
149. Landau, L. D. and Lifshitz, E. M. (1965) *Quantum Mechanics, Nonrelativistic Theory*, Pergamon Press, Oxford.
150. Landau, L. D. and Lifshitz, E. M. (1960) *Electrodynamics of Continuous Media.* Addison-Wesley, Reading, Massachusetts.
151. Landau, L. D. and Pomeranchuk, I. Ya (1953) "Limitations on applicability of bremsstrahlung and pair production theory at high energies. Electron-shower processes at ultra-high energies", *Dokl. Akad. Nauk SSSR* **92**, 535, 735.
152. Landau, L. D. and Rumer, Yu. B. (1937) "Cascade theory of electron showers", *Proc. R. Soc.* **166**, 213.
153. Lasukov, V. V. and Vorobiev, S. A. (1984) "Neutrino radiation by ultrarelativistic electrons at channeling in crystal", *Phys. Lett.* **106A**, 179.
154. Lindhard, J. (1965) "Influence of crystal lattice on motion of energetic charged particles", *Kong. Danske Vid. Selsk., Mat. Fys. Med.* **34**, 14.
155. Medenwaldt, R., Moller, S. P., Sorensen, A. H. et al. (1989a) "Hard photon emission from 150 GeV electrons incident on Si and Ge single crystals near axial directions", *Phys. Rev. Lett.* **63**, 2827.
156. Medenwaldt, R., Moller, S. P., Tang-Petersen, S. et al. (1989b) "Detailed investigation of shower formation in Ge- and W-crystals by 40–287 GeV/c electrons", *Phys. Lett.* **B227**, 483.
157. Medenwaldt, R., Moller, S. P., Jensen, B. N. et al. (1992) "Experimental investigations of hard photon emission from strong crystalline fields", *Phys. Lett.* **B281**, 153.
158. Merenkov, N. P. (1985) "On neutrino emission by an electron in the field of a plane electromagnetic field", *Yad. Fiz.* **42**, 1484.
159. Migdal, A. B. (1955) "Quantum kinetic equation for multiple scattering", *Dokl. Akad. Nauk. SSSR* **105**, 77.
160. Migdal, A. B. (1957) "Bremsstrahlung and pair production in condensed media at high energies", *Zh. Eksp. Teor. Fiz.* **32**, 633.
161. Miroshnichenko, I. I., Murray, J., Avakyan, R. O. and Fieguth T. Ch. (1979) "Experimental investigation of radiation from channeled positrons", *Zh. Eksp. Teor. Fiz., Pisma* **29**, 786.
162. Narozhnyi, N. B. and Nikishov, A. I. (1972) "Electron positron pair production by a Coulomb center in the presence of a constant field", *Zh. Eksp. Teor. Fiz.* **63**, 1135.

163. Nikishov, A. I. and Ritus, V. I. (1964a) "Quantum processes in the field of an electromagnetic wave and a constant field I", *Zh. Eksp. Teor. Fiz.* **46**, 776; see also *Sov. Phys. -JETP* **19**, 529.
164. Nikishov, A. I. and Ritus, V. I. (1964b) "Quantum processes in the field of an electromagnetic wave and a constant field II", *Zh. Eksp. Teor. Fiz.* **46**, 1768.
165. Nikishov, A. I. (1979) "Quantum electrodynamics of phenomena in intense fields", *Tr. Fiz. Inst. Akad. Nauk. SSSR* **111**, 152; see also *Proc. P. N. Lebedev Phys. Inst.* **111**.
166. Okun, L. B. (1982) *Leptons and Quarks*, North-Holland, Amsterdam.
167. Olsen, H. and Maximon, L. C. (1959) "Photon and electron polarization in high-energy bremsstrahlung and pair production with screening", *Phys. Rev.* **114**, 887.
168. Ohtsuki, Y. H. (1984) "X-ray laser by channeling radiation", *Nucl. Instrum. Methods* **B2**, 80.
169. Ohtsuki, Y. H. (1983) *Charged Beam Interaction With Solids*, Taylor and Francis, London and New York.
170. Panofsky, W. K. H. and Phillips M. (1955) *Classical Electricity and Magnetism*, Addison-Wesley, Reading, Massachusetts.
171. Papanyan, V. O. and Ritus, V. I. (1971) "Vacuum polarization and photon spitting in an intense field", *Zh. Eksp. Teor. Fiz.* **61**, 2231.
172. Papanyan, V. O. and Ritus, V. I. (1973) Three-photon interaction in an intense field and the scale invariance", *Zh. Eksp. Teor. Fiz.* **65**, 1756.
173. Park, H., Swent, R. L., Kephart, J. O. et al. (1983) "Positron and electron channeling radiation from germanium", *Phys. Lett.* **96A**, 45.
174. Ritus, V. I. (1969) "Influence of an electromagnetic field on particle decays", *Zh. Eksp. Teor. Fiz.* **56**, 986.
175. Ritus, V. I. (1972a) "Radiative corrections in quantum electrodynamics with intense field and their analytical properties", *Ann. Phys.* **69**, 555.
176. Ritus, V. I. (1972b) "Vacuum polarization correction to elastic electron and muon scattering in an intense field and pair electro- and muon-production", *Nucl. Phys.* **44B**, 236.
177. Ritus, V. I. (1979) "Quantum electrodynamics of phenomena in intense fields", *Tr. Fiz. Inst. Akad. Nauk. SSSR* **111**, 5; see also *Proc. P. N. Lebedev Phys. Inst.* **111**.
178. Rossi, B. (1952) *High Energy Particles*, Prentice-Hall, New York.
179. Schwinger, J. (1949) "On the classical radiation of accelerated electrons", *Phys. Rev.* **75**, 1912.
180. Schwinger, J. (1951) "On gauge invariance and vacuum polarization", *Phys. Rev.* **82**, 664.
181. Schwinger, J. (1954) "The quantum correction in the radiation by energetic accelerated electrons", *Proc. Nat. Acad. Sci.* **40**, 132.
182. Schwinger, J. (1973) "Classical radiation of accelerated electrons. II. A quantum view point", *Phys. Rev.* **D7**, 1696.

183. Sokolov, A. A., Klepikov, N. G. and Ternov, I. M. (1952) "On the theory of radiant electron", *Zh. Eksp. Teor. Fiz.* **23**, 632.
184. Sokolov, A. A. and Ternov, I. M. (1983) *Relativistic Electron* Nauka, Moscow, (in Russian).
185. Stepanyan, (1983) "Gamma-astronomy at ultra-high energies",*Itogi Nauki i Tekhniki, Ser. Astronomy* **24**, 205.
186. Swent, R. L., Pantell, R. H., Alguard, M. J. et al. (1979) "Observation of channeling radiation from relativistic electrons", *Phys. Rev. Lett.* **43**, 1723.
187. Ter-Mikaelyan, M. A. (1972) *High-Energy Electromagnetic Processes in Condensed Media*, Wiley-Interscience New York.
188. Ternov, I. M., Khalivov, V. P. and Kholomai, B. V. (1985) "Induced radiation at axial channeling", *Zh. Eksp. Teor. Phys.* **88**, 329.
189. Tikhomirov, V. V. (1983) "Radiation spectrum from channeled particles at high energies", Vestnik Byelorussian State University, Series **1**, 6.
190. Tsai, W. and Yildiz, A. (1973) "Motion of an electron in a homogeneous magnetic field—modified propagation function and synchrotron radiation", *Phys. Rev.* **D8**, 3446.
191. Tsai, W. (1973) "Magnetic bremsstrahlung and modified propagation function. Spin 0 charged particles in a homogeneous magnetic field", *Phys. Rev.* **D8**, 3460.
192. Uberall, H. (1956) "High-energy interference effect of bremsstrahlung and pair production in crystals", *Phys. Rev.* **103**, 1055.
193. Vorobiev, S. A. (1984) *Channeling of Electron Beams*, Energoatomizdat, Moscow, (in Russian).
194. Vorobiev, S. A., Kalinin, B. N., Kurkov, A. A., and Potylitsin, A. P. (1986) "Electromagnetic radiation from relativistic electrons in thick crystals", *Izvestiya Vuzov, Fiz.* **XXIX**, 32.
195. Vysotskii, V. I. and Kuzmin, R. N. (1983) "Induced short-wave radiation from charged particles in natural empty channels", *Zh. Tekh. Fiz.* **53**, 1254.

INDEX